Invariant Subspaces
of Matrices
with Applications

Invariant Subspaces of Matrices with Applications

I. GOHBERG

School of Mathematical Sciences
Tel Aviv University
Tel Aviv, Israel

P. LANCASTER

Department of Mathematics and Statistics
The University of Calgary
Calgary, Alberta, Canada

L. RODMAN

School of Mathematical Sciences
Tel Aviv University
Tel Aviv, Israel
and
Department of Mathematics
Arizona State University
Tempe, Arizona

A Wiley-Interscience Publication

JOHN WILEY & SONS

New York · Chichester · Brisbane · Toronto · Singapore

Library of Congress Cataloging in Publication Data:

Gohberg, I. (Israel), 1928-
  Invariant subspaces of matrices with applications.

  (Canadian Mathematical Society series of monographs
and advanced texts)
  Bibliography; p.
  1. Invariant subspaces.   2. Matrices. I. Lancaster,
Peter, 1929-   . II. Rodman, L. III. Title.
IV. Series.
QA322.G649 1986   515.7'246   86-11091
ISBN 0-471-84260-5

Printed in the United States of America

10 9 8 7 6 5 4 3 2 1

To our wives

Bella, Edna, and Ella

Preface

This is a book in advanced linear algebra in which invariant subspaces of matrices are the central notion and the main tool. To the authors' knowledge it is the first book written with such a theme. It contains a reasonably comprehensive treatment of geometrical, algebraic, topological and analytical properties of invariant subspaces. As well, an important part of the work consists of applications to matrix polynomials, rational matrix functions, linear systems, and matrix quadratic equations.

Parts of the book are written like a textbook and are easily accessible for undergraduate students. Gradually, the exposition changes to approach the style, and admit the content, of a monograph. Here, recent achievements and some unsolved problems are presented. A large portion of the content of the book has not appeared before in books. The fundamental character of the mathematics, its accessibility, and its importance in applications should make this a widely useful work for experts and students in mathematics, science, and engineering.

This is the third book written jointly by the authors. The first book is *Matrix Polynomials*, published by Academic Press in 1982, and the second is *Matrices and Indefinite Scalar Products*, published by Birkhäuser Verlag in 1983. These three books are connected and, to some extent, one led to another. Material that could not be included in one of the books became the starting point for the next. Moreover, invariant subspaces play an important role in the first two books and indicated to us the need for a systematic treatment of this subject.

The authors are pleased to acknowledge continuing support, throughout the development of this work, from the Natural Sciences and Engineering Research Council of Canada and from the Nathan and Lily Silver Chair in Mathematical Analysis and Operator Theory of Tel Aviv University. Constant support has also been provided by the staffs of the School of Mathematical Sciences, Tel Aviv University, and the Department of Mathematics and Statistics of the University of Calgary. We are especially grateful to Jacqueline Gorsky in Israel and Pat Dalgetty in Canada for skillful and speedy work on our typescripts. In addition, the second named author is pleased to acknowledge a Killam Resident Fellowship awarded for the Winter Term of 1985 to assist in the completion of the book, and the third named author gratefully acknowledges support from the Basic Research Fund administered by the Israel Academy for Science and Humanities, and also from the U.S. National Science Foundation.

I. GOHBERG
P. LANCASTER
L. RODMAN

Calgary, Alberta, Canada
July 1986

Contents

ix

Invariant Subspaces
of Matrices
with Applications

Introduction

Invariant subspaces are a central notion of linear algebra. However, in existing texts and expositions the notion is not easily or systematically followed. Perhaps because the whole structure is very rich, the treatment becomes fragmented as other related ideas and notions intervene. In particular, the notion of an invariant subspace as an entity is often lost in the discussion of eigenvalues, eigenvectors, generalized eigenvectors, and so on. The importance of invariant subspaces becomes clearer in the context of operator theory on spaces of infinite dimension. Here, it can be argued that the structure is poorer and this is one of the few available tools for the study of many classes of operators. Probably for this reason, the first books on invariant subspaces appeared in the framework of infinite-dimensional spaces. It seems to the authors that now there is a case for developing a treatment of linear algebra in which the central role of invariant subspace is systematically followed up.

The need for such a treatment has become more apparent in recent years because of developments in different fields of application and especially in linear systems theory, where concepts such as controllability, feedback, factorization, and realization of matrix functions are commonplace. In the treatment of such problems new concepts and theories have been developed that form complete new chapters in the body of linear algebra. As examples of new concepts of linear algebra developed to meet the needs of systems theory, we should mention invariant subspaces for nonsquare matrices and similarity of such matrices.

In this book the reader will find a treatment of certain aspects of linear algebra that meets the two objectives: to develop systematically the central role of invariant subspaces in the analysis of linear transformations and to include relevant recent developments of linear algebra stimulated by linear systems theory. The latter are not dealt with separately, but are integrated into the text in a way that is natural in the development of the mathematical structure.

1

The first part of the book, taken alone or together with selections from the other parts, can be used as a text for undergraduate courses in mathematics, having only a first course in linear algebra as prerequisite. At the same time, the book will be of interest to graduate students in science and engineering. We trust that experts will also find the exposition and new results interesting. The authors anticipate that the book will also serve as a valuable reference work for mathematicians, scientists, and engineers. A set of exercises is included in each chapter. In general, they are designed to provide illustrations and training rather than extensions of the theory.

The first part of the book is devoted mainly to geometric properties of invariant subspaces and their applications in three fields. The fields in question are matrix polynomials, rational matrix functions, and linear systems theory. They are each presented in self-contained form, and—rather than being exhaustive—the focus is on those problems in which invariant subspaces of square and nonsquare matrices play a central role. These problems include factorization and linear fractional decompostions for matrix functions; problems of realization for rational matrix functions; and the problem of describing connections, or cascades, of linear systems, pole assignment, output stabilization, and disturbance decoupling.

The second part is of a more algebraic character in which other properties of invariant subspaces are analyzed. It contains an analysis of the extent to which the invariant subspaces determine the parent matrix, invariant subspaces common to commuting matrices, and lattices of subspaces for a single matrix and for algebras of matrices.

The numerical computation of invariant subspaces is a difficult task as, in general, it makes sense to compute only those invariant subspaces that change very little after small changes in the transformation. Thus it is important to have appropriate notions of "stable" invariant subspaces. Such an analysis of the stability of invariant subspaces and their generalizations is the main subject of Part 3. This analysis leads to applications in some of the problem areas mentioned above.

The subject of Part 4 is analytic families of invariant subspaces and has many useful applications. Here, the analysis is influenced by the theory of complex vector bundles, although we do not make use of this theory. The study of the connections between local and global problems is one of the main problems studied in this part. Within reasonable bounds, Part 4 relies only on the theory developed in this book. The material presented here appears for the first time in a book on linear algebra and is thereby made accessible to a wider audience.

Part One

Fundamental Properties of Invariant Subspaces and Applications

Part 1 of this work comprises almost half of the entire book. It includes what can be described as a self-contained course in linear algebra with emphasis on invariant subspaces, together with substantial developments of applications to the theory of polynomial and rational matrix-valued functions, and to systems theory. These applications demand extensions of the standard material in linear algebra that are included in our treatment in a natural way. They also serve to breathe new life into an otherwise familiar body of knowledge. Thus there is a considerable amount of material here (including all of Chapters 3, 4, and 6) that cannot be found in other books on linear algebra.

Almost all of the material in this part can be understood by readers who have completed a beginning course in linear algebra, although there are places where basic ideas of calculus and complex analysis are required.

Chapter One

Invariant Subspaces: Definition, Examples, and First Properties

This chapter is mainly introductory. It contains the simplest properties of invariant subspaces of a linear transformation. Some basic tools (projectors, factor spaces, angular transformations, triangular forms) for the study of invariant subspaces are developed. We also study the behaviour of invariant subspaces of a transformation when the operations of similarity and taking adjoints are applied to the transformation. The lattice of invariant subspaces of a linear transformation—a notion that will be important in the sequel—is introduced. The presentation of the material here is elementary and does not even require use of the Jordan form.

1.1 DEFINITION AND EXAMPLES

Let $A: \mathbb{C}^n \to \mathbb{C}^n$ be a linear transformation. A subspace $\mathcal{M} \subset \mathbb{C}^n$ is called *invariant* for the transformation A, or A *invariant*, if $Ax \in \mathcal{M}$ for every vector $x \in \mathcal{M}$. In other words, $\mathcal{M}$ is invariant for A means that the image of $\mathcal{M}$ under A is contained in $\mathcal{M}$; $A\mathcal{M} \subset \mathcal{M}$. Trivial examples of invariant subspaces are $\{0\}$ and $\mathbb{C}^n$. Less trivial examples are the subspaces

$$\text{Ker } A = \{x \in \mathbb{C}^n \mid Ax = 0\}$$

and

$$\text{Im } A = \{Ax \mid x \in \mathbb{C}^n\}$$

Indeed, as $Ax = 0 \in \text{Ker } A$ for every $x \in \text{Ker } A$, the subspace $\text{Ker } A$ is A invariant. Also, for every $x \in \mathbb{C}^n$, the vector Ax belongs to $\text{Im } A$; in particular, $A(\text{Im } A) \subset \text{Im } A$, and $\text{Im } A$ is A invariant.

5

More generally, the subspaces

$$\text{Ker } A^m = \{x \in \mathbb{C}^n \mid A^m x = 0\}, \qquad m = 1, 2, \dots$$

and

$$\text{Im } A^m = \{A^m x \mid x \in \mathbb{C}^n\}, \qquad m = 1, 2, \dots$$

are A invariant. To verify this, let $x \in \text{Ker } A^m$, so $A^m x = 0$. Then $A^m(Ax) = A(A^m x) = 0$, that is, $Ax \in \text{Ker } A^m$. This means that $\text{Ker } A^m$ is A invariant. Further, let $x \in \text{Im } A^m$, so $x = A^m y$ for some $y \in \mathbb{C}^n$. Then $Ax = A(A^m y) = A^m(Ay)$, which implies that $Ax \in \text{Im } A^m$. So $\text{Im } A^m$ is A invariant as well.

When convenient, we shall often assume implicitly that a linear transformation from $\mathbb{C}^m$ into $\mathbb{C}^n$ is given by an $n \times m$ matrix with respect to the standard orthonormal bases $e_1 = \langle 1, 0, \dots, 0 \rangle$, $e_2 = \langle 0, 1, 0, \dots, 0 \rangle$, $e_n = \langle 0, 0, \dots, 0, 1 \rangle$ in $\mathbb{C}^n$, $e_1, \dots, e_m$ in $\mathbb{C}^m$.

The following three examples of transformations and their invariant subspaces are basic and are often used in the sequel.

EXAMPLE 1.1.1.   Let

$$A = \begin{bmatrix} \lambda_0 & 1 & \cdots & & 0 \\ 0 & \lambda_0 & & & \vdots \\ \vdots & & & & \vdots \\ & & & & 1 \\ 0 & \cdots & & 0 & \lambda_0 \end{bmatrix}, \qquad \lambda_0 \in \mathbb{C}$$

(the $n \times n$ *Jordan block* with λ_0 on the main diagonal). Every nonzero A-invariant subspace is of the form $\text{Span}\{e_1, \dots, e_k\}$, where e_i is the vector $\langle 0, \dots, 0, 1, 0, \dots, 0 \rangle$ with 1 in the ith place. Indeed, let $\mathcal{M}$ be a nonzero A-invariant subspace, and let

$$x = \sum_{i=1}^n \alpha_i e_i, \qquad \alpha_i \in \mathbb{C}$$

be a vector from $\mathcal{M}$ for which the index $k = \max\{m \mid 1 \le m \le n, \alpha_m \ne 0\}$ is maximal. Then clearly

$$\mathcal{M} \subset \text{Span}\{e_1, \dots, e_k\}$$

On the other hand, the vector $x = \sum_{i=1}^k \alpha_i e_i$, $\alpha_k \ne 0$ belongs to $\mathcal{M}$. Hence, since $\mathcal{M}$ is A invariant, the vectors

$$x_1 = Ax - \lambda_0 x = \sum_{i=2}^{k} \alpha_i e_{i-1}$$

$$x_2 = Ax_1 - \lambda_0 x_1 = \sum_{i=3}^{k} \alpha_i e_{i-2}$$

$$\vdots$$

$$x_{k-1} = Ax_{k-2} - \lambda_0 x_{k-2} = \alpha_k e_1$$

also belong to $\mathcal{M}$. Hence the vectors

$$e_1 = \frac{1}{\alpha_k} x_{k-1}$$

$$e_2 = \frac{1}{\alpha_k}(x_{k-2} - \alpha_{k-1} e_1)$$

$$\vdots$$

$$e_k = \frac{1}{\alpha_k}\left(x - \sum_{i=1}^{k-1} \alpha_i e_i\right)$$

belong to $\mathcal{M}$ as well. So

$$\text{Span}\{e_1, \ldots, e_k\} \subset \mathcal{M}$$

and the equality

$$\text{Span}\{e_1, \ldots, e_k\} = \mathcal{M}$$

follows. As for every $y = \sum_{i=1}^{k} \beta_i e_i \in \text{Span}\{e_1, \ldots, e_k\}$ we have

$$Ay = \lambda_0 y + \sum_{i=2}^{k} \beta_i e_{i-1} \in \text{Span}\{e_1, \ldots, e_k\}$$

The subspace $\text{Span}\{e_1, \ldots, e_k\}$ is indeed A invariant. The total number of A-invariant subspaces (including $\{0\}$ and $\mathbb{C}^n$) is thus $n + 1$.

In this example we have

$$\text{Ker } A = \begin{cases} \{0\} & \text{if} \quad \lambda_0 \neq 0 \\ \text{Span}\{e_1\} & \text{if} \quad \lambda_0 = 0 \end{cases}$$

and

$$\text{Im } A = \begin{cases} \mathbb{C}^n & \text{if} \quad \lambda_0 \neq 0 \\ \text{Span}\{e_1, \ldots, e_{n-1}\} & \text{if} \quad \lambda_0 = 0 \end{cases}$$

As expected, these subspaces are A invariant. $\square$

EXAMPLE 1.1.2. Let $A = \lambda_0 I$, where I is the $n \times n$ identity matrix. Clearly, every subspace in $\mathbb{C}^n$ is A invariant. Here the number of A-invariant subspaces is infinite (if $n > 1$).

Note that the set $\text{Inv}(A)$ of all A-invariant subspaces is uncountably infinite. Indeed, for linearly independent vectors $x, y \in \mathbb{C}^n$ the one-dimensional subspaces $\text{Span}\{x + \alpha y\}$, $\alpha \in \mathbb{R}$ are all different and belong to $\text{Inv}(A)$. So they form an uncountable set of A-invariant subspaces.

Conversely, if every one-dimensional subspace of $\mathbb{C}^n$ is A invariant for a linear transformation A, then $A = \lambda_0 I$ for some λ_0. Indeed, for every $x \neq 0$ the subspace $\text{Span}\{x\}$ is A invariant, so $Ax = \lambda(x)x$, where $\lambda(x)$ is a complex number that may, *a priori*, depend on x. Now if $\lambda(x_1) \neq \lambda(x_2)$ for linearly independent vectors x_1 and x_2, then $\text{Span}\{x_1 + x_2\}$ is not A invariant, because

$$A(x_1 + x_2) = \lambda(x_1)x_1 + \lambda(x_2)x_2 \notin \text{Span}\{x_1 + x_2\}$$

Hence we must have $\lambda_0 = \lambda(x)$ is independent of $x \neq 0$, so actually $A = \lambda_0 I$. $\square$

Later (see Proposition 2.5.4) we shall see that the set of all A-invariant subspaces of on $n \times n$ complex matrix A is never countably infinite; it is either finite or uncountably infinite.

EXAMPLE 1.1.3. Let

$$A = \begin{bmatrix} \lambda_1 & 0 & \cdots & 0 \\ \vdots & \lambda_2 & & \vdots \\ & & \ddots & \\ 0 & \cdots & & \lambda_n \end{bmatrix} \qquad (n \geq 2)$$

where the complex numbers $\lambda_1, \ldots, \lambda_n$ are distinct. For any indices $1 \leq i_1 < \cdots < i_k \leq n$ the subspace $\text{Span}\{e_{i_1}, \ldots, e_{i_k}\}$ is A invariant. Indeed, for

$$x = \sum_{j=1}^{k} \alpha_j e_{i_j} \in \text{Span}\{e_{i_1}, \ldots, e_{i_k}\}$$

we have

$$Ax = \sum_{j=1}^{k} \alpha_j \lambda_{i_j} e_{i_j} \in \text{Span}\{e_{i_1}, \ldots, e_{i_k}\}$$

It turns out that these are all the invariant subspaces for A. The proof of this fact for a general n is given later in a more general framework. So the total number of A-invariant subspaces is

$$\sum_{k=0}^{n} \binom{n}{k} = 2^n$$

Here we shall check only that the 2×2 matrix

$$A = \begin{bmatrix} \lambda_1 & 0 \\ 0 & \lambda_2 \end{bmatrix}, \qquad \lambda_1 \neq \lambda_2$$

has exactly two nontrivial invariant subspaces, $\mathrm{Span}\{e_1\}$ and $\mathrm{Span}\{e_2\}$. Indeed, let $\mathcal{M}$ be any one-dimensional A-invariant subspace

$$\mathcal{M} = \mathrm{Span}\{x\}, \qquad x = \alpha_1 e_1 + \alpha_2 e_2 \neq 0$$

Then $Ax = \alpha_1 \lambda_1 e_1 + \alpha_2 \lambda_2 e_2$ should belong to $\mathcal{M}$ and thus is a scalar multiple of x_1:

$$\alpha_1 \lambda_1 e_1 + \alpha_2 \lambda_2 e_2 = \beta_{\alpha_1} e_1 + \beta_{\alpha_2} e_2$$

for some $\beta \in \mathbb{C}$. Comparing coefficients, we see that we obtain a contradiction $\lambda_1 = \lambda_2$ unless $\alpha_1 = 0$ or $\alpha_2 = 0$. In the former case $\mathcal{M} = \mathrm{Span}\{e_2\}$ and in the latter case $\mathcal{M} = \mathrm{Span}\{e_1\}$.

In this example we have $\mathrm{Ker}\, A = \mathrm{Span}\{e_{i_0}\}$ (when $\det A = 0$), where i_0 is the index for which $\lambda_{i_0} = 0$ (as we have assumed that the λ_i are distinct and $\det A = 0$, there is exactly one such index), and $\mathrm{Im}\, A = \mathrm{Span}\{e_i \mid i \neq i_0\}$. $\square$

The following observation is often useful in proving that a given subspace is A invariant: A subspace $\mathcal{M} = \mathrm{Span}\{x_1, \ldots, x_k\}$ is A invariant if and only if $Ax_i \in \mathcal{M}$ for $i = 1, \ldots, k$. The proof of this fact is an easy exercise.

For a given transformation $A: \mathbb{C}^n \to \mathbb{C}^n$ and a given vector $x \in \mathbb{C}^n$, consider the subspace

$$\mathcal{M} = \mathrm{Span}\{x, Ax, A^2x, \ldots\}$$

We now appeal to the Cayley–Hamilton theorem, which states that $\sum_{j=0}^n \alpha_j A^j = 0$, where the complex numbers $\alpha_0, \ldots, \alpha_n$ are the coefficients of the characteristic polynomial $\det(\lambda I - A)$ of A:

$$\det(\lambda I - A) = \sum_{j=0}^n \alpha_j \lambda^j$$

(By writing A as an $n \times n$ matrix in some basis in $\mathbb{C}^n$, we easily see from the definition of the determinant that $\det(\lambda I - A)$ is a polynomial of degree n with $\alpha_n = 1$.) Hence $A^k x$ with $k \geq n$ is a linear combination of $x, Ax, \ldots, A^{n-1}x$, so actually

$$\mathcal{M} = \mathrm{Span}\{x, Ax, A^2x, \ldots, A^{n-1}x\}$$

The preceding observation shows immediately that $\mathcal{M}$ is A invariant. Any

A-invariant subspace $\mathcal{L}$ that contains x also contains all the vectors $Ax, A^2x, \ldots$, and hence contains $\mathcal{M}$. It follows that $\mathcal{M}$ is the smallest A-invariant subspace that contains the vector x.

We conclude this section with another useful fact regarding invariant subspaces. Namely, a subspace $\mathcal{M} \subset \mathbb{C}^n$ is A invariant for a transformation $A: \mathbb{C}^n \to \mathbb{C}^n$ if and only if it is $(\alpha A + \beta I)$ invariant, where α, β are arbitrary complex numbers such that $\alpha \neq 0$. Indeed, assume that $\mathcal{M}$ is A invariant. Then for every $x \in \mathcal{M}$ we see that the vector

$$(\alpha A + \beta I)x = \alpha Ax + \beta x$$

belongs to $\mathcal{M}$. So $\mathcal{M}$ is $(\alpha A + \beta I)$ invariant. As

$$A = \frac{1}{\alpha}(\alpha A + \beta I) - \frac{\beta}{\alpha}I$$

the same reasoning shows that any $(\alpha A + \beta I)$ invariant subspace is also A invariant.

1.2   EIGENVALUES AND EIGENVECTORS

The most primitive nontrivial invariant subspaces are those with dimension equal to one. For a transformation $A: \mathbb{C}^n \to \mathbb{C}^n$ and some nonzero $x \in \mathbb{C}^n$, therefore, we consider an A-invariant subspace of the form $\mathcal{M} = \text{Span}\{x\}$. In this case there must be a $\lambda_0 \in \mathbb{C}$ such that $Ax = \lambda_0 x$. Since we then have $A(\alpha x) = \alpha(Ax) = \lambda_0(\alpha x)$ for any $\alpha \in \mathbb{C}$, the number λ_0 does not depend on the choice of the nonzero vector in $\mathcal{M}$. We call λ_0 an *eigenvalue* of A, and, when $Ax = \lambda_0 x$ with $0 \neq x \in \mathbb{C}^n$, we call x an *eigenvector* of A (corresponding to the eigenvalue λ_0). Observe that, since $(\lambda_0 I - A)x = 0$, the eigenvalues of A can also be characterized as the set of complex zeros of the *characteristic polynomial* of A; $\varphi_A(\lambda) \overset{\text{def}}{=} \det(\lambda I - A)$.

The set of all eigenvalues of A is called the *spectrum* of A and is denoted by $\sigma(A)$. We have seen that any one-dimensional A-invariant subspace is spanned by some eigenvector. Conversely, if x_0 is an eigenvector of A corresponding to some eigenvalue λ_0, then $\text{Span}\{x_0\}$ is A invariant. (In other words, A is the operator of multiplication by λ_0 when restricted to $\text{Span}\{x_0\}$.)

Let us have a closer look at the eigenvalues. As the characteristic polynomial $\varphi_A(\lambda) = \det(\lambda I - A)$ is a polynomial of degree n, by the fundamental theorem of algebra, $\varphi_A(\lambda)$ has n (in general, complex) zeros when counted with multiplicities. These zeros are exactly the eigenvalues of A. Since the characteristic polynomial and eigenvalues are independent of the choice of basis producing the matrix representation, they are properties of the underlying transformation. So a transformation $A: \mathbb{C}^n \to \mathbb{C}^n$ has exactly

n eigenvalues when counted with multiplicities, and, in any event, the number of *distinct* eigenvalues of A does not exceed n. Note that this is a property of transformations over the field of complex numbers (or, more generally, over an algebraically closed field). As we shall see later, a transformation from $\mathcal{R}^n$ into $\mathcal{R}^n$ does not always have (real) eigenvalues. Since at least one eigenvector corresponds to any eigenvalue λ_0 of A it follows that every linear transformation $A: \mathbb{C}^n \rightarrow \mathbb{C}^n$ has at least one one-dimensional invariant subspace. Example 1.1.1 shows that in certain cases a linear transformation has exactly one one-dimensional invariant subspace.

We pass now to the description of two-dimensional A-invariant subspaces in terms of eigenvalues and eigenvectors. So assume that $\mathcal{M}$ is a two-dimensional A-invariant subspace. Then, in a natural way, A determines a transformation from $\mathcal{M}$ into $\mathcal{M}$. We have seen above that for every transformation in a (complex) finite-dimensional vector space (which can be identified with $\mathbb{C}^m$ for some m) there is an eigenvalue and a corresponding eigenvector. So there exists an $x_0 \in \mathcal{M} \backslash \{0\}$ and a complex number λ_0 such that $Ax_0 = \lambda_0 x_0$. Now let x_1 be a vector in $\mathcal{M}$ for which $\{x_0, x_1\}$ is a linearly independent set; in other words, $\mathcal{M} = \text{Span}\{x_0, x_1\}$. Since $\mathcal{M}$ is A invariant it follows that

$$Ax_1 = \mu_0 x_0 + \mu_1 x_1$$

for some complex numbers μ_0 and μ_1. If $\mu_0 = 0$, then x_1 is an eigenvector of A corresponding to the eigenvalue μ_1. If $\mu_0 \neq 0$ and $\mu_1 \neq \lambda_0$, then the vector $y = -\mu_0 x_0 + (\lambda_0 - \mu_1) x_1$ is an eigenvector of A corresponding to μ_1 for which $\{x_0, y\}$ is a linearly independent set. Indeed

$$Ay = -\mu_0 Ax_0 + (\lambda_0 - \mu_1) Ax_1 = -\mu_0 \lambda_0 x_0 + (\lambda_0 - \mu_1)(\mu_0 x_0 + \mu_1 x_1)$$
$$= (\lambda_0 - \mu_1)\mu_1 x_1 - \mu_1 \mu_0 x_0 = \mu_1 y$$

Finally, if $\mu_0 \neq 0$ and $\mu_1 = \lambda_0$, then x_0 is the only eigenvector (up to multiplication by a nonzero complex number) of A in $\mathcal{M}$. To check this, assume that $\alpha_0 x_0 + \alpha_1 x_1$, $\alpha_1 \neq 0$, is an eigenvector of A corresponding to an eigenvalue ν_0. Then

$$A(\alpha_0 x_0 + \alpha_1 x_1) = \nu_0 \alpha_0 x_0 + \nu_0 \alpha_1 x_1 \qquad (1.2.1)$$

But the left-hand side of this equality is

$$\alpha_0 Ax_0 + \alpha_1 Ax_1 = \alpha_0 \lambda_0 x_0 + \alpha_1 (\mu_0 x_0 + \lambda_0 x_1)$$

and comparing this with equality (2.1), we obtain

$$\lambda_0 \alpha_1 = \nu_0 \alpha_1, \qquad \alpha_0 \lambda_0 + \alpha_1 \mu_0 = \nu_0 \alpha_0$$

which (with $\alpha_1 \neq 0$) implies $\lambda_0 = \nu_0$ and $\alpha_1 \mu_0 = 0$, a contradiction with the assumption $\mu_0 \neq 0$. However, note that the vectors $z = (1/\mu_0)x_1$ and x_0 form a linearly independent set and z has the property that $Az - \lambda_0 z = x_0$. Such a vector z will be called a *generalized eigenvector* of A corresponding to the eigenvector x_0.

In conclusion, the two-dimensional invariant subspace $\mathcal{M}$ is spanned by two eigenvectors if and only if either $\mu_0 = 0$ or $\mu_0 \neq 0$ and $\mu_1 \neq \lambda_0$. If $\mu_0 \neq 0$ and $\mu_1 = \lambda_0$, then $\mathcal{M}$ is spanned by an eigenvector and a corresponding generalized eigenvector.

A study of invariant subspaces of dimension greater than 2 along these lines becomes tedious. Nevertheless, it can be done and leads to the well-known Jordan normal form of a matrix (or transformation) (see Chapter 2).

Using eigenvectors, one can generally produce numerous invariant subspaces, as demonstrated by the following proposition.

Proposition 1.2.1

Let $\lambda_1, \ldots, \lambda_k$ be eigenvalues of A (not necessarily distinct), and let x_i be an eigenvector of A corresponding to λ_i, $i = 1, \ldots, k$. Then $\mathrm{Span}\{x_1, \ldots, x_k\}$ is an A-invariant subspace.

Proof. For any $x = \sum_{i=1}^{k} \alpha_i x_i \in \mathrm{Span}\{x_1, \ldots, x_k\}$, where $\alpha_i \in \mathbb{C}$, we have

$$Ax = \sum_{i=1}^{k} \alpha_i A x_i = \sum_{i=1}^{k} \alpha_i \lambda_i x_i$$

so indeed $\mathrm{Span}\{x_1, \ldots, x_k\}$ is A invariant. $\square$

For some transformations all invariant subspaces are spanned by eigenvectors as in Proposition 1.2.1, and for some transformations not all invariant subspaces are of this form. Indeed, in Example 1.1.1 only one of the n nonzero invariant subspaces is spanned by eigenvectors. On the other hand, in Example 1.1.2 every nonzero vector is an eigenvector corresponding to λ_0, so obviously every A-invariant subspace is spanned by eigenvectors.

1.3 JORDAN CHAINS

We have seen in the description of two-dimensional invariant subspaces that eigenvectors alone are not always sufficient for description of all invariant subspaces. This fact necessitates consideration of generalized eigenvectors as well. Let us make a general definition that will include this notion. Let λ_0 be an eigenvalue of a linear transformation $A: \mathbb{C}^n \to \mathbb{C}^n$. A chain of vectors

$x_0, x_1, \ldots, x_k$ is called a *Jordan chain* of A corresponding to λ_0 if $x_0 \neq 0$ and the following relations hold:

$$Ax_0 = \lambda_0 x_0$$

$$Ax_1 - \lambda_0 x_1 = x_0$$

$$Ax_2 - \lambda_0 x_2 = x_1 \tag{1.3.1}$$

$$\vdots$$

$$Ax_k - \lambda_0 x_k = x_{k-1}$$

The first equation (together with $x_0 \neq 0$) means that x_0 is an eigenvector of A corresponding to λ_0. The vectors $x_1, \ldots, x_k$ are called *generalized eigenvectors* of A corresponding to the eigenvalue λ_0 and the eigenvector x_0.

For example, let

$$A = \begin{bmatrix} \lambda_0 & 1 & \cdots & & 0 \\ 0 & \lambda_0 & & & \vdots \\ \vdots & & & & \\ & & & & 1 \\ 0 & & \cdots & 0 & \lambda_0 \end{bmatrix}, \qquad \lambda_0 \in \mathbb{C}$$

as in Example 1.1.1. Then e_1 is an eigenvector of A corresponding to λ_0, and $e_1, e_2, \ldots, e_n$ is a Jordan chain. This Jordan chain is by no means unique; for instance, $e_1, e_2 + \alpha e_1, \ldots, e_n + \alpha e_{n-1}$ is again a Jordan chain of A, where $\alpha \in \mathbb{C}$ is any number.

In Example 1.1.3 the matrix A does not have generalized eigenvectors at all; that is, every Jordan chain consists of an eigenvector only. Indeed, we have $A = \text{diag}[\lambda_1, \lambda_2, \ldots, \lambda_n]$, where $\lambda_1, \ldots, \lambda_n$ are distinct complex numbers; therefore

$$\det(\lambda I - A) = (\lambda - \lambda_1)(\lambda - \lambda_2) \cdots (\lambda - \lambda_n)$$

So $\lambda_1, \ldots, \lambda_n$ are exactly the eigenvalues of A. It is easily seen that any eigenvector of A corresponding to λ_{i_0} is of the form αe_{i_0} with a nonzero scalar α. Assuming that there is a Jordan chain $\alpha e_{i_0}, x$ of A corresponding to λ_{i_0}, equations (1.3.1) imply

$$Ax - \lambda_{i_0} x = \alpha e_{i_0} \tag{1.3.2}$$

Write $x = \sum_{i=1}^{n} \beta_i e_i$, then $Ax = \sum_{i=1}^{n} \lambda_i \beta_i e_i$, and equality (1.3.2) gives

$$\sum_{i=1}^{n} (\lambda_i - \lambda_{i_0}) \beta_i e_i = \alpha e_{i_0} \tag{1.3.3}$$

As $\lambda_i \neq \lambda_{i_0}$ for $i \neq i_0$, we find immediately that $\beta_i = 0$ for $i \neq i_0$. But then the left-hand side of equation (1.3.3) is zero, a contradiction with $\alpha \neq 0$. So there are no generalized eigenvectors for the transformation A.

Jordan chains allow us to construct more invariant subspaces.

Proposition 1.3.1

Let $x_0, \ldots, x_k$ be a Jordan chain of a transformation A. Then the subspace $\mathcal{M} = \text{Span}\{x_0, \ldots, x_k\}$ is A invariant.

Proof. We have

$$Ax_0 = \lambda_0 x_0 \in \mathcal{M}$$

where λ_0 is the eigenvalue of A to which $x_0, \ldots, x_k$ corresponds; and for $i = 1, \ldots, k$

$$Ax_i = \lambda_0 x_i + x_{i-1} \in \mathcal{M}$$

Hence the A invariance of $\mathcal{M}$ follows. $\square$

The following proposition shows how the Jordan chains behave under a linear change in the matrix A.

Proposition 1.3.2

Let $\alpha \neq 0$ and β be complex numbers. A chain of vectors $x_0, x_1, \ldots, x_k$ is a Jordan chain of A corresponding to the eigenvalue λ_0 if and only if the vectors

$$x_0, \frac{1}{\alpha} x_1, \ldots, \frac{1}{\alpha^k} x_k \tag{1.3.4}$$

form a Jordan chain of $\alpha A + \beta I$ corresponding to the eigenvalue $\alpha \lambda_0 + \beta$ of $\alpha A + \beta I$.

Proof. Assume that $x_0, \ldots, x_k$ is a Jordan chain of A corresponding to λ_0, that is, equalities (1.3.1) hold. Then we have

$$(\alpha A + \beta I) x_0 = \alpha A x_0 + \beta x_0 = \alpha \lambda_0 x_0 + \beta x_0 = (\alpha \lambda_0 + \beta) x_0$$

$$(\alpha A + \beta I) \frac{1}{\alpha} x_1 - (\alpha \lambda_0 + \beta) \frac{1}{\alpha} x_1 = A x_1 - \lambda_0 x_1 = x_0$$

and in general for $i = 1, \ldots, k$

$$(\alpha A + \beta I) \frac{1}{\alpha^i} x_i - (\alpha \lambda_0 + \beta) \frac{1}{\alpha^i} x_i = \frac{1}{\alpha^{i-1}} (A x_i - \lambda_0 x_i) = \frac{1}{\alpha^{i-1}} x_{i-1}$$

So by definition the vectors in equality (1.3.4) form a Jordan chain of $\alpha A + \beta I$ corresponding to $\alpha \lambda_0 + \beta$.

Conversely, assume that equality (1.3.4) is a Jordan chain of $\alpha A + \beta I$ corresponding to $\alpha \lambda_0 + \beta$. As

$$A = \frac{1}{\alpha}(A + \beta I) - \frac{\beta}{\alpha} I$$

the first part of the proof shows that the vectors

$$x_0, \; \alpha\left(\frac{1}{\alpha} x_1\right) = x_1, \ldots, \; \alpha^k\left(\frac{1}{\alpha^k} x^k\right) = x_k$$

form a Jordan chain of A corresponding to the eigenvalue $(1/\alpha)(\alpha \lambda_0 + \beta) - (\beta/\alpha) = \lambda_0$. $\square$

Two corollaries from Proposition 1.3.2 will be especially useful in the sequel.

Corollary 1.3.3

(a) *The vector x_0 is an eigenvector of A corresponding to λ_0 if and only if x_0 is an eigenvector of $\alpha A + \beta I$ (here $\alpha \neq 0$, β are complex numbers) corresponding to $\alpha \lambda_0 + \beta$; (b) the vectors $x_0, \ldots, x_k$ form a Jordan chain of A corresponding to λ_0 if and only if these vectors constitute a Jordan chain of $A + \beta I$ corresponding to $\lambda_0 + \beta$ for any complex number β.*

In many instances Corollary 1.3.3 allows us to reduce the consideration of eigenvalues and Jordan chains to cases when the eigenvalue is zero. Our first example of this device appears in the proof of the following proposition.

Proposition 1.3.4

The vectors in a Jordan chain $x_0, \ldots, x_k$ of A are linearly independent.

Proof. Assume the contrary, and let x_p be the first generalized eigenvector in the Jordan chain that is a linear combination of the preceding vectors:

$$x_p = \sum_{i=0}^{p-1} \alpha_i x_i ; \qquad \alpha_i \in \mathbb{C}$$

We can assume that the eigenvalue λ_0 of A to which the Jordan chain $x_0, \ldots, x_k$ corresponds is zero. (Otherwise, in view of Corollary 1.3.3b, we consider $A - \lambda_0 I$ in place of A.) So we have $Ax_p = x_{p-1}$. On the other hand, we have

$$Ax_p = \sum_{i=0}^{p-1} \alpha_i Ax_i = \sum_{i=1}^{p-1} \alpha_i x_{i-1}$$

Comparing both expressions, we see that x_{p-1} is a linear combination of the vectors $x_0, \ldots, x_{p-2}$. This contradicts the choice of x_p as the *first* vector in the Jordan chain that is a linear combination of the preceding vectors. $\square$

1.4   INVARIANT SUBSPACES AND BASIC OPERATIONS ON LINEAR TRANSFORMATIONS

In this section we first consider questions concerning invariant subspaces of sums, compositions, and inverses of linear transformations. We shall also develop the connection between invariant subspaces for a linear transformation and those of similar and adjoint transformations.

The basic result for the first three algebraic operations is given in the following proposition.

Proposition 1.4.1

Let $A, B: \mathbb{C}^n \to \mathbb{C}^n$ be transformations, and let $\mathcal{M} \subset \mathbb{C}^n$ be a subspace which is simultaneously A invariant and B invariant. Then $\mathcal{M}$ is also invariant for $\alpha A + \beta B$ (with any $\alpha, \beta \in \mathbb{C}$) and for AB. Further, if A is invertible, then $\mathcal{M}$ is also invariant for A^{-1}.

Proof.   For every $x \in \mathcal{M}$ we have

$$(\alpha A + \beta B)x = \alpha(Ax) + \beta(Bx) \in \mathcal{M}$$

and $(AB)x = A(Bx) \in \mathcal{M}$ because $Bx \in \mathcal{M}$.

Assume now that A is invertible, and let $x_1, \ldots, x_p$ be a basis in $\mathcal{M}$. Then the vectors $y_1 = Ax_1, \ldots, y_p = Ax_p$ are linearly independent (because A is invertible) and belong to $\mathcal{M}$ (because $\mathcal{M}$ is A invariant). So $y_1, \ldots, y_p$ is also a basis in $\mathcal{M}$. Now

$$A^{-1}\mathcal{M} = A^{-1} \operatorname{Span}\{y_1, \ldots, y_p\} = \operatorname{Span}\{x_1, \ldots, x_p\} = \mathcal{M} \quad \square$$

For any transformation A, we denote by $\operatorname{Inv}(A)$ the set of all A-invariant subspaces. Then Proposition 1.4.1 means, in short, that

$$\operatorname{Inv}(A) \cap \operatorname{Inv}(B) \subset \operatorname{Inv}(\alpha A + \beta B) \tag{1.4.1}$$

$$\operatorname{Inv}(A) \cap \operatorname{Inv}(B) \subset \operatorname{Inv}(AB) \tag{1.4.2}$$

$$\operatorname{Inv}(A) \subset \operatorname{Inv}(A^{-1}) \quad \text{(if } A \text{ is invertible)} \tag{1.4.3}$$

By applying equality (1.4.3) with A replaced by A^{-1}, we get $\text{Inv}(A^{-1}) \subset \text{Inv}(A)$, so actually equality holds in (1.4.3). It is very easy to produce examples when the equality fails in (1.4.1) or (1.4.2). For instance:

EXAMPLE 1.4.1. Let $A: \mathbb{C}^n \to \mathbb{C}^n$ be a transformation that is not of the form γI for some $\gamma \in \mathbb{C}$ (if $n \geq 2$, such transformations obviously exist). By Example 1.1.2, not all subspaces in $\mathbb{C}^n$ are A invariant. On the other hand, take $B = A$ and $\alpha + \beta = 0$ in (1.4.1). Then the right-hand side of (1.4.1) is the zero transformation for which every subspace in $\mathbb{C}^n$ is invariant. $\square$

To give an example where the inclusion in (1.4.2) is strict, put

$$A = B = \begin{bmatrix} 0 & 1 \\ 0 & 0 \end{bmatrix}$$

The following example of strict inclusion in (1.4.2) is also instructive.

EXAMPLE 1.4.2. Let

$$A = \begin{bmatrix} \lambda_0 & 1 \\ 0 & \lambda_0 \end{bmatrix}, \qquad B = \begin{bmatrix} \lambda_0 & 0 \\ 1 & \lambda_0 \end{bmatrix}, \qquad \lambda_0 \in \mathbb{C}$$

An easy analysis (using Example 1.1.1) shows that A and B have no nontrivial common invariant subspaces. Thus $\text{Inv}(A) \cap \text{Inv}(B) = \{\{0\}, \mathbb{C}^2\}$. On the other hand, $\text{Inv}(AB)$ must have an eigenvector that spans a nontrivial AB-invariant subspace. Again, the inclusion (1.4.2) is strict. $\square$

Consider now the notion of similarity. Recall that two transformations A and B on $\mathbb{C}^n$ are called *similar* if $A = S^{-1}BS$ for some invertible transformation S (called a *similarity transformation* between A and B). Evidently, similar transformations have the same characteristic polynomial and, consequently, the same eigenvalues. The next proposition reveals the close connection between invariant subspaces of similar transformations.

Proposition 1.4.2

Let transformations A and B be similar, with the similarity transformation $S: A = S^{-1}BS$. Then a subspace $\mathcal{M} \subset \mathbb{C}^n$ is A invariant if and only if the subspace

$$S\mathcal{M} = \{Sx \mid x \in \mathcal{M}\} \subset \mathbb{C}^n$$

is B invariant.

Proof. Let $\mathcal{M}$ be A invariant, and let $x \in S\mathcal{M}$, so that $x = Sy$ for some $y \in \mathcal{M}$. Then $Bx = BSy = SAy$, and since $Ay \in \mathcal{M}$, we find that $Bx \in S\mathcal{M}$. So $S\mathcal{M}$ is B invariant.

Conversely, assume that $S\mathcal{M}$ is B invariant. Then for $y \in \mathcal{M}$ we have $BSy \in S\mathcal{M}$ and thus

$$Ay = S^{-1}BSy \in S^{-1}(S\mathcal{M}) = \mathcal{M}$$

So $\mathcal{M}$ is A invariant. $\square$

Proposition 1.4.2 shows, in particular, that there is a natural correspondence between the sets of invariant subspaces of similar transformations. Let us check this correspondence more closely in some of the examples of invariant subspaces already introduced.

Proposition 1.4.3

Let A and B be similar, with the similarity transformation S. Then (a) $\operatorname{Im} B = S(\operatorname{Im} A)$; (b) $\operatorname{Ker} B = S(\operatorname{Ker} A)$; (c) if $x_0, x_1, \ldots, x_k$ is a Jordan chain of A corresponding to λ_0, then $Sx_0, Sx_1, \ldots, Sx_k$ is a Jordan chain of B corresponding to the same λ_0.

Proof. The proof is straightforward. Let us check (b). Take $x \in \operatorname{Ker} A$, so $Ax = 0$. Then $Ax = S^{-1}BSx = 0$, and as S is invertible, $BSx = 0$, that is, $Sx \in \operatorname{Ker} B$. Reversing the order of this argument, we see that if $Sx \in \operatorname{Ker} B$ for some $x \in \mathbb{C}^n$, then $x \in \operatorname{Ker} A$. The proofs of (a) and (c) proceed in a similar way. $\square$

Consider now the operation of taking adjoints. Let $A: \mathbb{C}^n \to \mathbb{C}^n$ be a transformation. Recall that the adjoint transformation $A^*: \mathbb{C}^n \to \mathbb{C}^n$ is defined by the relation

$$(Ax, y) = (x, A^*y), \quad \text{for all } x, y \in \mathbb{C}^n$$

where $(\cdot, \cdot)$ is the standard scalar product in $\mathbb{C}^n$:

$$(x, y) = \sum_{i=1}^{n} x_i \bar{y}_i, \quad x = \langle x_1, \ldots, x_n \rangle, \quad y = \langle y_1, \ldots, y_n \rangle$$

More generally, if $\mathcal{T}_1, \mathcal{T}_2$ are subspaces in $\mathbb{C}^n$ and $A: \mathcal{T}_1 \to \mathcal{T}_2$ is a linear transformation, its adjoint $A^*: \mathcal{T}_2 \to \mathcal{T}_1$ is defined by the relation

$$(Ax, y) = (x, A^*y) \quad \text{for all } x \in \mathcal{T}_1, y \in \mathcal{T}_2$$

It is not difficult to check that the adjoint transformation always exists and is unique. It is easily verified that for any linear transformations A and B on $\mathbb{C}^n$ and any $\alpha \in \mathbb{C}$

$$(A + B)^* = A^* + B^*, \qquad (\alpha A)^* = \bar{\alpha} A^*$$

$$(AB)^* = B^* A^*, \qquad (A^*)^* = A$$

If (in the standard basis $e_1, \ldots, e_n$)

$$A = \begin{bmatrix} a_{11} & a_{12} & \cdots & a_{1n} \\ \vdots & \vdots & & \vdots \\ a_{n1} & a_{n2} & \cdots & a_{nn} \end{bmatrix}$$

then the adjoint transformation is given by the formula

$$A^* = \begin{bmatrix} \bar{a}_{11} & \bar{a}_{21} & \cdots & \bar{a}_{n1} \\ \vdots & \vdots & & \vdots \\ \bar{a}_{1n} & \bar{a}_{2n} & \cdots & \bar{a}_{nn} \end{bmatrix}$$

The same formula also holds for the transformation A written as a matrix in any orthogonal basis in $\mathbb{C}^n$ as long as A^* is considered as a matrix in the same basis.

There is a simple and useful characterization of the invariant subspaces of the adjoint transformation A^* in terms of the invariant subspaces of A, as follows.

Proposition 1.4.4

Let $A: \mathbb{C}^n \to \mathbb{C}^n$ be a linear transformation. A subspace $\mathcal{M} \subset \mathbb{C}^n$ is A^* invariant if and only if its orthogonal complement $\mathcal{M}^\perp$ is A invariant.

Proof. Assume that $\mathcal{M}$ is A^* invariant, and let $x \in \mathcal{M}^\perp$. We must prove that $Ax \in \mathcal{M}^\perp$. Indeed, for every $y \in \mathcal{M}$ we have

$$(Ax, y) = (x, A^*y) = 0$$

because $A^*y \in \mathcal{M}$ and $x \in \mathcal{M}^\perp$. Conversely, assume that $\mathcal{M}^\perp$ is A invariant, and take $y \in \mathcal{M}$. Then for every $x \in \mathcal{M}^\perp$ we have

$$(A^*y, x) = (y, Ax) = 0$$

which means that $A^*y \in \mathcal{M}$. So $\mathcal{M}$ is A^* invariant. $\square$

Note the following equalities for the A-invariant subspaces Ker A and Im A and the A^*-invariant subspaces Ker A^* and Im A^*:

$$(\text{Ker } A)^\perp = \text{Im } A^*; \qquad \text{Ker } A^* = (\text{Im } A)^\perp \tag{1.4.4}$$

Indeed, let $x = A^*y$ and $z \in \text{Ker } A$. Then $(x, z) = (A^*y, z) = \overline{(z, A^*y)} = \overline{(Az, y)} = 0$; so $x \in (\text{Ker } A)^\perp /$ Hence we have proved that

$$\text{Im } A^* \subset (\text{Ker } A)^\perp \tag{1.4.5}$$

On the other hand, let x be orthogonal to $\text{Im } A^*$. Then for every $y \in \mathbb{C}^n$, we have $(Ax, y) = (x, A^*y) = 0$; so $Ax \perp \mathbb{C}^n$, and thus $Ax = 0$, or $x \in \text{Ker } A$. So $(\text{Im } A^*)^\perp \subset \text{Ker } A$. Taking orthogonal complements, we obtain $\text{Im } A^* \supset (\text{Ker } A)^\perp$. Combining with (1.4.5), we obtain the first equality in (1.4.4). The second equality follows from the first one applied to A^* instead of A [recall that $(A^*)^* = A$].

Later, we shall also need the following property:

$$\text{Im } A = \text{Im}(AA^*)$$

Here, the inclusion $\supset$ is clear. For the opposite inclusion, let $x \in \text{Im } A$. Then $x = Ay$ for some y. If z is the projection of y onto $\text{Ker } A$, then $y - z \in (\text{Ker } A)^\perp$ and also $x = A(y - z)$. Then (1.4.1) implies that $y - z \in \text{Im } A^*$ and so $x \in \text{Im}(AA^*)$, as required.

A transformation $A: \mathbb{C}^n \to \mathbb{C}^n$ is called *self-adjoint* if $A = A^*$. It is easily seen that A is self-adjoint if and only if it is represented by a hermitian matrix in some orthogonal basis (recall that a matrix $[a_{jk}]_{j,k=1}^n$ is called *hermitian* if $a_{jk} = \bar{a}_{kj}$, $j, k = 1, \ldots, n$). For this important class of transformations we have the following corollary of Proposition 1.4.4.

Corollary 1.4.5

If A is self-adjoint, then $\mathcal{M}^\perp$ is A invariant if and only if $\mathcal{M}$ is A invariant.

1.5 INVARIANT SUBSPACES AND PROJECTORS

A linear transformation defined by $P: \mathbb{C}^n \to \mathbb{C}^n$ is called a *projector* if $P^2 = P$. The important feature of projectors is that there exists a one-to-one correspondence between the set of all projectors and the set of all pairs of complementary subspaces in $\mathbb{C}^n$. This correspondence is described in Theorem 1.5.1.

Recall first that if $\mathcal{M}, \mathcal{L}$ are subspaces of $\mathbb{C}^n$, then $\mathcal{M} + \mathcal{L} = \{z \in \mathbb{C}^n \mid z = x + y, x \in \mathcal{M}, y \in \mathcal{L}\}$. This sum is said to be *direct* if $\mathcal{M} \cap \mathcal{L} = \{0\}$, in which case we write $\mathcal{M} \dotplus \mathcal{L}$ for the sum. The subspaces $\mathcal{M}, \mathcal{L}$ are *complementary* (*are direct complements of each other*) if $\mathcal{M} \cap \mathcal{L} = \{0\}$ and $\mathcal{M} \dotplus \mathcal{L} = \mathbb{C}^n$.

Nontrivial subspaces $\mathcal{M}, \mathcal{L}$ are *orthogonal* if for each $x \in \mathcal{M}$ and $y \in \mathcal{L}$ we have $(x, y) = 0$ and they are *orthogonal complements* if, in addition, they are complementary. In this case, we write $\mathcal{M} = \mathcal{L}^\perp$, $\mathcal{L} = \mathcal{M}^\perp$.

Theorem 1.5.1

Let P be a projector. Then $(\text{Im } P, \text{Ker } P)$ is a pair of complementary subspaces in $\mathbb{C}^n$. Conversely, for every pair $(\mathscr{L}_1, \mathscr{L}_2)$ of complementary subspaces in $\mathbb{C}^n$, there exists a unique projector P such that $\text{Im } P = \mathscr{L}_1$, $\text{Ker } P = \mathscr{L}_2$.

Proof. Let $x \in \mathbb{C}^n$. Then $x = (x - Px) + Px$. Clearly, $Px \in \text{Im } P$ and $x - Px \in \text{Ker } P$ (because $P^2 = P$). So $\text{Im } P + \text{Ker } P = \mathbb{C}^n$. Further, if $x \in \text{Im } P \cap \text{Ker } P$, then $x = Py$ for some $y \in \mathbb{C}^n$ and $Px = 0$. So

$$x = Py = P^2 y = P(Py) = Px = 0$$

and $\text{Im } P \cap \text{Ker } P = \{0\}$. Hence $\text{Im } P$ and $\text{Ker } P$ are indeed complementary subspaces.

Conversely, let $\mathscr{L}_1$ and $\mathscr{L}_2$ be a pair of complementary subspaces. Let P be the unique linear transformation in $\mathbb{C}^n$ such that $Px = x$ for $x \in \mathscr{L}_1$ and $Px = 0$ for $x \in \mathscr{L}_2$. Then clearly $P^2 = P$, $\mathscr{L}_1 \subset \text{Im } P$, and $\mathscr{L}_2 \subset \text{Ker } P$. But we already know from the first part of the proof that $\text{Im } P \dotplus \text{Ker } P = \mathbb{C}^n$. By dimensional considerations we have, consequently, $\mathscr{L}_1 = \text{Im } P$ and $\mathscr{L}_2 = \text{Ker } P$. So P is a projector with the desired properties. The uniqueness of P follows from the property that $Px = x$ for every $x \in \text{Im } P$ (which, in turn, is a consequence of the equality $P^2 = P$). $\square$

We say that P is the projector *on* $\mathscr{L}_1$ *along* $\mathscr{L}_2$ if $\text{Im } P = \mathscr{L}_1$, $\text{Ker } P = \mathscr{L}_2$. A projector P is called *orthogonal* if $\text{Ker } P = (\text{Im } P)^{\perp}$. Thus the corresponding complementary subspaces are mutually orthogonal. Orthogonal projectors are particularly important and can be characterized as follows.

Proposition 1.5.2

A projector P is orthogonal if and only if P is self-adjoint, that is, $P^* = P$.

Proof. Suppose that $P^* = P$, and let $x \in \text{Im } P$, $y \in \text{Ker } P$. Then $(x, y) = (Px, y) = (x, Py) = (x, 0) = 0$, that is, $\text{Ker } P$ is orthogonal to $\text{Im } P$. Since by Theorem 1.5.1 $\text{Ker } P$ and $\text{Im } P$ are complementary, it follows that in fact $\text{Ker } P = (\text{Im } P)^{\perp}$.

Conversely, let $\text{Ker } P = (\text{Im } P)^{\perp}$. To prove that $P^* = P$, we have to check the equality

$$(Px, y) = (x, Py) \quad \text{for all} \quad x, y \in \mathbb{C}^n \tag{1.5.1}$$

Because of the sesquilinearity of the function (Px, y) in the arguments $x, y \in \mathbb{C}^n$, and in view of Theorem 1.5.1, it is sufficient to prove equation (1.5.1) for the following four cases: (a) $x, y \in \text{Im } P$; (b) $x \in \text{Ker } P$, $y \in \text{Im } P$; (c) $x \in \text{Im } P$, $y \in \text{Ker } P$; (d) $x, y \in \text{Ker } P$. In case (d), equality (1.5.1) is trivial because both sides are 0. In case (a) we have

$$(Px, y) = (Px, Py) = (x, Py)$$

and (1.5.1) follows. In case (b), the left-hand side of equation (1.5.1) is zero (since $x \in \text{Ker } P$) and the right-hand side is also zero in view of the orthogonality $\text{Ker } P = (\text{Im } P)^{\perp}$. In the same way, one checks (1.5.1) in case (c).

So (1.5.1) holds, and $P^* = P$. $\square$

Note that if P is a projector, so is $I - P$. Indeed, $(I - P)^2 = I - 2P + P^2 = I - 2P + P = I - P$. Moreover, $\text{Ker } P = \text{Im}(I - P)$ and $\text{Im } P = \text{Ker}(I - P)$. It is natural to call the projectors P and $I - P$ *complementary projectors*.

We now give useful representations of a projector with respect to a decomposition of $\mathbb{C}^n$ into a sum of two complementary subspaces. Let $T: \mathbb{C}^n \to \mathbb{C}^n$ be a transformation and let $\mathcal{L}_1, \mathcal{L}_2$ be a pair of complementary subspaces in $\mathbb{C}^n$. Denote $m_i = \dim \mathcal{L}_i$ ($i = 1, 2$); then $m_1 + m_2 = n$. The transformation T may be written as a 2×2 block matrix with respect to the decomposition $\mathcal{L}_1 \dotplus \mathcal{L}_2 = \mathbb{C}^n$:

$$T = \begin{bmatrix} T_{11} & T_{12} \\ T_{21} & T_{22} \end{bmatrix} \tag{1.5.2}$$

Here T_{ij} ($i, j = 1, 2$) is an $m_i \times m_j$ matrix that represents in some basis the transformation $P_i T|_{\mathcal{L}_j}: \mathcal{L}_j \to \mathcal{L}_i$, where P_i is the projector on $\mathcal{L}_i$ along $\mathcal{L}_{3-i}$ (so $P_1 + P_2 = I$).

Suppose now that $T = P$ is a projector on $\mathcal{L}_1 = \text{Im } P$. Then representation (1.5.2) takes the form

$$P = \begin{bmatrix} I & X \\ 0 & 0 \end{bmatrix} \tag{1.5.3}$$

for some matrix X. In general, $X \neq 0$. One can easily check that $X = 0$ if and only if $\mathcal{L}_2 = \text{Ker } P$. Analogously, if $\mathcal{L}_1 = \text{Ker } P$, then (1.5.2) takes the form

$$P = \begin{bmatrix} 0 & Y \\ 0 & I \end{bmatrix} \tag{1.5.4}$$

and $Y = 0$ if and only if $\mathcal{L}_2 = \text{Im } P$. By the way, the direct multiplication $P \cdot P$, where P is given by (1.5.3) or (1.5.4), shows that P is indeed a projector: $P^2 = P$.

Consider now an invariant subspace $\mathcal{M}$ for a transformation $A: \mathbb{C}^n \to \mathbb{C}^n$. For any projector P with $\text{Im } P = \mathcal{M}$ we obtain

$$PAP = AP \tag{1.5.5}$$

Indeed, if $x \in \operatorname{Ker} P$, we obviously have

$$PAPx = APx$$

If $x \in \operatorname{Im} P = \mathcal{M}$, we see that Ax belongs to $\mathcal{M}$ as well and thus

$$PAPx = PAx = Ax = APx$$

once more. Since $\mathbb{C}^n = \operatorname{Ker} P \dotplus \operatorname{Im} P$, (1.5.5) follows. Conversely, if P is a projector for which (1.5.5) holds, then for every $x \in \operatorname{Im} P$ we have $PAx = Ax$; in other words, $\operatorname{Im} P$ is A invariant. So *a subspace $\mathcal{M}$ is A invariant if and only if it is the image of a projector P for which (1.5.5) holds.*

Let $\mathcal{M}$ be an A-invariant subspace and let P be a projector on $\mathcal{M}$ [so that (1.5.5) holds]. Denoting by $\mathcal{M}'$ the kernel of P, represent A as a 2×2 block matrix

$$\begin{bmatrix} A_{11} & A_{12} \\ A_{21} & A_{22} \end{bmatrix}$$

with respect to the direct sum decomposition $\mathbb{C}^n = \mathcal{M} \dotplus \mathcal{M}'$. Here A_{11} is a transformation $PAP|_{\mathcal{M}} : \mathcal{M} \to \mathcal{M}$, A_{12} is a linear transformation $PA(I-P)|_{\mathcal{M}'} : \mathcal{M}' \to \mathcal{M}$,

$$A_{21} = (I-P)AP|_{\mathcal{M}} : \mathcal{M} \to \mathcal{M}'$$

$$A_{22} = (I-P)A(I-P)|_{\mathcal{M}'} : \mathcal{M}' \to \mathcal{M}'$$

and all these transformations are written as matrices with respect to some chosen bases in $\mathcal{M}$ and $\mathcal{M}'$. As $\mathcal{M}$ is A invariant, equation (1.5.5) implies that $(I-P)AP = 0$, that is, $A_{21} = 0$. Hence

$$A = \begin{bmatrix} A_{11} & A_{12} \\ 0 & A_{22} \end{bmatrix} \tag{1.5.6}$$

Using this representation of the matrix A, we can deduce some important connections between the restriction $A|_{\mathcal{M}} = A_{11}$ and the matrix A itself.

Proposition 1.5.3

Let $x_0, \ldots, x_k$ be a Jordan chain of $A|_{\mathcal{M}}$ corresponding to the eigenvalue λ_0 of $A|_{\mathcal{M}}$. Then $x_0, \ldots, x_k$ is also a Jordan chain of A corresponding to λ_0. In particular, all eigenvalues of $A|_{\mathcal{M}}$ are also eigenvalues of A.

Proof. We have $x_0 \neq 0$; $x_i \in \mathcal{M}$ for $i = 0, \ldots, k$, and

$$A|_{\mathcal{M}} x_0 = \lambda_0 x_0$$

$$A|_{\mathcal{M}} x_i - \lambda_0 x_i = x_{i-1}, \qquad i = 1, \ldots, k$$

As $A|_{\mathcal{M}} = PAP|_{\mathcal{M}} = AP|_{\mathcal{M}}$, these relations can be rewritten as

$$APx_0 = \lambda_0 x_0, \qquad APx_i - \lambda_0 x_i = x_{i-1}, \qquad i = 1, \ldots, k$$

But $Px_i = x_i$, $i = 0, 1, \ldots, k$, and we obtain the relations defining $x_0, \ldots, x_k$ as a Jordan chain of A corresponding to λ_0. $\square$

The last statement in Proposition 1.5.3 can also be proved in the following way. Suppose that $\lambda_0 \in \sigma(A_{11})$, that is, $\text{Ker}(\lambda_0 I - A_{11}) \neq \{0\}$. The representation (1.5.6) implies that any nonzero vector from $\text{Ker}(\lambda_0 I - A_{11})$ belongs to $\text{Ker}(\lambda_0 I - A)$. Thus $\text{Ker}(\lambda_0 I - A) \neq \{0\}$, and $\lambda_0 \in \sigma(A)$.

In fact, a more general result holds.

Proposition 1.5.4

Let $\mathcal{M}$ be an A-invariant subspace with a direct complement $\mathcal{M}'$ in $\mathbb{C}^n$, and let

$$A = \begin{bmatrix} A_{11} & A_{12} \\ 0 & A_{22} \end{bmatrix}$$

be the representation of A with respect to the decomposition $\mathbb{C}^n = \mathcal{M} \dot{+} \mathcal{M}'$. Then

$$\sigma(A) = \sigma(A_{11}) \cup \sigma(A_{22})$$

Proof. This follows immediately from the fact that $\det(\lambda I - A) = \det(\lambda I - A_{11}) \det(\lambda I - A_{22})$. $\square$

As an example in which projectors and the subspaces $\text{Im } A$ and $\text{Ker } A$ of a transformation A all play important roles, let us describe here a construction of generalized inverses for A.

Given a transformation $A: \mathbb{C}^n \to \mathbb{C}^m$, the transformation $X: \mathbb{C}^m \to \mathbb{C}^n$ is called a *generalized inverse* of A if the following holds: for any $b \in \text{Im } A$ the linear system $Ax = b$ has a solution $x = Xb$, and for any $b \in \text{Im } X$ the linear system $Xx = b$ has a solution $x = Ab$. So this is a natural generalization of the notion of the inverse transformation.

Observe that X is a generalized inverse of A if and only if $AXA = A$ and $XAX = X$. Indeed, let X be a generalized inverse of A. Then $AXb = b$ for every $b \in \text{Im } A$, that is, for every b of the form $b = Ay$. So $AXAy = Ay$ for all $y \in \mathbb{C}^n$, and $AXA = A$. Similarly, one checks that $XAX = X$. Conversely, if $AXA = A$, then for every b of the form $b = Ay$ the vector $Xb = XAy$ is obviously a solution of the linear equation $Ax = b$.

The description of all generalized inverses of A, which implies, in particular, that a generalized inverse of A always exists, is given by the following theorem.

Theorem 1.5.5

Let $A: \mathbb{C}^n \to \mathbb{C}^m$ be a transformation, let $\mathbb{C}^n = \operatorname{Ker} A \dotplus \tilde{N}$, $\mathbb{C}^m = \operatorname{Im} A \dotplus \tilde{R}$ for some subspaces $\tilde{N}$ and $\tilde{R}$, and let P be the projector on $\operatorname{Im} A$ along $\tilde{R}$, Q the projector on $\tilde{N}$ along $\operatorname{Ker} A$. Then (a) the transformation $A_1 = A|_{\tilde{N}}$ is a one-to-one transformation of $\tilde{N}$ onto $\operatorname{Im} A$; (b) the transformation A^I defined on $\mathbb{C}^m$ by $A^I y = A_1^{-1}(Py)$, for all $y \in \mathbb{C}^m$, is a generalized inverse of A for which $AA^I = P$ and $A^I A = Q$; (c) all generalized inverses of A are determined as $\tilde{N}$, $\tilde{R}$ range over all complementary subspaces for $\operatorname{Ker} A$, $\operatorname{Im} A$, respectively.

The proof of Theorem 1.5.5 is straightforward.

It is easily seen that, in the hypothesis of the theorem, complementary subspaces $\tilde{R}$, $\tilde{N}$ are simply the range and null-space of the generalized inverse that they determine.

Corollary 1.5.6

In the statement of Theorem 1.5.5, we have

$$\operatorname{Im} A^I = \tilde{N} \quad \text{and} \quad \operatorname{Ker} A^I = \tilde{R}$$

$$\operatorname{Ker} A \dotplus \operatorname{Im} A^I = \mathbb{C}^n \quad \text{and} \quad \operatorname{Im} A \dotplus \operatorname{Ker} A^I = \mathbb{C}^m$$

1.6 ANGULAR TRANSFORMATIONS AND MATRIX QUADRATIC EQUATIONS

In this section we study angular transformations and their connections with matrix quadratic equations and invariant subspaces. The correspondence between the invariant subspaces of similar transformations described in Proposition 1.4.2 is useful here.

This discussion can be seen as the first step in the examination of solutions of matrix quadratic equations. In this program, we first need the notion of a subspace "angular with respect to a projector." In Chapter 13 we discuss the topological properties of such subspaces in preparation for the applications to quadratic equations to be made in Chapters 17 and 20.

Let π be a projector defined on $\mathbb{C}^n$. Transformations acting on $\mathbb{C}^n$ in this section are written in 2×2 block matrix form with respect to the decomposition $\mathbb{C}^n = \operatorname{Ker} \pi \dotplus \operatorname{Im} \pi$.

A subspace $\mathcal{N}$ of $\mathbb{C}^n$ is said to be *angular with respect to π* if $\mathcal{N} \dotplus \operatorname{Ker} \pi = \mathbb{C}^n$. That is, if and only if $\mathcal{N}$ and $\operatorname{Ker} \pi$ are complementary subspaces of $\mathbb{C}^n$. Thus $\operatorname{Im} \pi$ is angular with respect to π, but more generally, if R is any transformation from $\operatorname{Im} \pi$ into $\operatorname{Ker} \pi$, then the subspace

$$\mathcal{N}_R \stackrel{\text{def}}{=} \{x \mid x = Ry + y, \, y \in \operatorname{Im} \pi\} \tag{1.6.1}$$

is angular with respect to π. To see this, observe first that $\mathcal{N}_R$ is indeed a subspace; that is, if $x_1, x_2 \in \mathcal{N}_R$, then for some $y_1, y_2 \in \text{Im } \pi$

$$x_1 + x_2 = (Ry_1 + y_1) + (Ry_2 + y_2) = R(y_1 + y_2) + (y_1 + y_2) \in \mathcal{N}_R$$

and if $\alpha \in \mathbb{C}$

$$\alpha x_i = \alpha(Ry_1 + y_1) = R(\alpha y) + (\alpha y) \in \mathcal{N}_R$$

Then $\mathbb{C}^n = \mathcal{N}_R + \text{Ker } \pi$ because, for any $y \in \mathbb{C}^n$, if $y_1 = \pi y$, $y_2 = (I - \pi)y$, then

$$y = y_1 + y_2 = (Ry_1 + y_1) + (y_2 - Ry_1)$$

and $Ry_1 + y_1 \in \mathcal{N}_R$, $y_2 - Ry_1 \in \text{Ker } \pi$.

Finally, if $z \in \mathcal{N}_R \cap \text{Ker } \pi$, then $z = Ry + y$, where $y \in \text{Im } \pi$ and also $\pi z = 0$. Thus

$$0 = \pi Ry + \pi y = \pi Ry + y$$

Since R is into $\text{Ker } \pi$, $\pi R = 0$ and it follows that $y = 0$. Hence $z = 0$ and $\mathbb{C}^n = \mathcal{N}_R \dotplus \text{Ker } \pi$.

The angular subspaces generated in this way are, in fact, all possible angular subspaces.

Proposition 1.6.1

Let $\mathcal{N}$ be a subspace of $\mathbb{C}^n$. Then $\mathcal{N}$ is angular with respect to π if and only if $\mathcal{N} = \mathcal{N}_R$ for some transformation R: $\text{Im } \pi \rightarrow \text{Ker } \pi$ that is uniquely determined by $\mathcal{N}$.

Proof. If $\mathcal{N} = \mathcal{N}_R$, we have already checked that $\mathcal{N}$ is angular. To prove the converse, assume that $\mathcal{N}$ is angular with respect to π, and let Q be the projector of $\mathbb{C}^n$ onto $\mathcal{N}$ along $\text{Ker } \pi$. Put

$$Rx = (Q - \pi)x , \qquad x \in \text{Im } \pi \tag{1.6.2}$$

Then $\mathcal{N} = \mathcal{N}_R$. Indeed

$$\pi(Rx) = (\pi Q - \pi)x = (\pi - \pi)x = 0$$

that is, R: $\text{Im } \pi \rightarrow \text{Ker } \pi$, and we have to show that $\mathcal{N} = \mathcal{N}_R$.

If $x \in \mathcal{N}_R$, then for some $y = \pi y$,

$$x = Ry + y = (Q - \pi)y + \pi y = Qy \in \mathcal{N}$$

Thus $\mathcal{N}_R \subset \mathcal{N}$. Conversely, if $y \in \mathcal{N}$ then

$$y = Qy = Q\pi y = (R + \pi)\pi y = R(\pi y) + (\pi y) \in \mathcal{N}_R$$

thus $\mathcal{N} = \mathcal{N}_R$, as required.

To prove the uniqueness of R, we show that any defining transformation R in (1.6.1) must have the form (1.6.2). Thus let $\mathcal{N}$ be angular with respect to π, and let $R: \operatorname{Im} \pi \to \operatorname{Ker} \pi$ satisfy (1.6.1). Let $y \in \operatorname{Im} \pi$ and $x = Ry + y \in \mathcal{N}$. Then, since $I - Q$ is onto $\operatorname{Ker} \pi$ *along* $\mathcal{N}$

$$0 = (I - Q)x = (I - Q)Ry + (I - Q)\pi y$$

But $QR = 0$ and $Q\pi = 0$ so that $Ry = (Q - \pi)y$. $\quad\square$

The transformation R appearing in the preceding proposition is called the *angular transformation* for $\mathcal{N}$. Note that R can be defined as the restriction of a difference of projectors:

$$R = (Q - \pi)|_{\operatorname{Im} \pi}$$

Consider now a transformation $T: \mathbb{C}^n \to \mathbb{C}^n$. As before, let $\pi: \mathbb{C}^n \to \mathbb{C}^n$ be a projector so that we have $\mathbb{C}^n = \operatorname{Im} \pi \dotplus \operatorname{Ker} \pi$. Then T has a representation with respect to this decomposition:

$$T = \begin{bmatrix} T_{11} & T_{12} \\ T_{21} & T_{22} \end{bmatrix} \tag{1.6.3}$$

It is clear that $\operatorname{Im} \pi$ is invariant under T if and only if $T_{21} = 0$. Similarly, $\operatorname{Ker} \pi$ is T invariant if and only if $T_{12} = 0$. More generally, what is the condition that a subspace $\mathcal{N}$ that is angular with respect to π be T invariant?

Theorem 1.6.2

Let $\mathcal{N}$ be an angular subspace with respect to the projector π. Let T have the representation (1.6.3) with respect to the decomposition $\mathbb{C}^n = \operatorname{Im} \pi \dotplus \operatorname{Ker} \pi$. Then $\mathcal{N}$ is T invariant if and only if the angular transformation R for $\mathcal{N}$ satisfies the matrix quadratic equation.

$$RT_{12}R + RT_{11} - T_{22}R - T_{21} = 0 \tag{1.6.4}$$

Proof. If I_1, I_2 are the identity transformations on $\operatorname{Im} \pi$ and $\operatorname{Ker} \pi$, respectively, then since $R: \operatorname{Im} \pi \to \operatorname{Ker} \pi$ we can define the transformation

$$E = \begin{bmatrix} I_1 & 0 \\ R & I_2 \end{bmatrix}: \qquad \mathbb{C}^n \to \mathbb{C}^n$$

which is written as a 2×2 matrix with respect to the decomposition $\mathbb{C}^n = \text{Im } \pi \dotplus \text{Ker } \pi$. The transformation E is obviously invertible and

$$E^{-1} = \begin{bmatrix} I_1 & 0 \\ -R & I_2 \end{bmatrix}$$

For every $x \in \text{Im } \pi$ we have $Ex = x + Rx \in \mathcal{N}$. So E maps $\text{Im } \pi$ onto $\mathcal{N}$ and E^{-1} maps $\mathcal{N}$ back onto $\text{Im } \pi$. By Proposition 1.4.2, $\mathcal{N}$ is T invariant if and only if $\text{Im } \pi$ is $E^{-1}TE$ invariant. Now observe that

$$E^{-1}TE = \begin{bmatrix} T_{11} + T_{12}R & T_{12} \\ -RT_{12}R - RT_{11} + T_{22}R + T_{21} & T_{22} - T_{12}R \end{bmatrix} \quad (1.6.5)$$

so $\text{Im } \pi$ is $E^{-1}TE$ invariant if and only if (1.6.4) holds. $\quad\square$

Another important observation follows from the similarity (1.6.5).

Corollary 1.6.3

If $\mathcal{N}$ is T invariant, then

$$\sigma(T) = \sigma(T_{11} + T_{12}R) \cup \sigma(T_{22} - T_{12}R) \quad (1.6.6)$$

and

$$\sigma(T|_\mathcal{N}) = \sigma(T_{11} + T_{12}R) \quad (1.6.7)$$

Proof. We have

$$\sigma(T) = \sigma(E^{-1}TE) = \sigma\begin{bmatrix} T_{11} + T_{12}R & T_{12} \\ 0 & T_{22} - T_{12}R \end{bmatrix}$$

Now use Proposition 1.5.4 to obtain (1.6.6). Further, $\sigma(T|_\mathcal{N}) = \sigma(E^{-1}TE|_{\text{Im } \pi}) = \sigma(T_{11} + T_{12}R)$. $\quad\square$

1.7   TRANSFORMATIONS IN FACTOR SPACES

Let $\mathcal{N} \subset \mathbb{C}^n$ be a subspace. We say that two vectors $x, y \in \mathbb{C}^n$ are *comparable modulo* $\mathcal{N}$ if $x - y \in \mathcal{N}$, and denote this by $x \equiv y \pmod{\mathcal{N}}$. In particular, $x \equiv 0 \pmod{\mathcal{N}}$ if and only if $x \in \mathcal{N}$. This relation is easily seen to be reflexive, symmetrical, and transitive. That is

$$x \equiv x \pmod{\mathcal{N}} \quad \text{for all} \quad x \in \mathbb{C}^n$$

$$x \equiv y \pmod{\mathcal{N}} \Rightarrow y \equiv x \pmod{\mathcal{N}}$$

$$x \equiv y \pmod{\mathcal{N}} \quad \text{and} \quad y \equiv z \pmod{\mathcal{N}} \Rightarrow x \equiv z \pmod{\mathcal{N}}$$

Thus we have an equivalence relation on $\mathbb{C}^n$. It follows that $\mathbb{C}^n$ is decomposed into disjoint classes of vectors with the properties that in each class the vectors are comparable modulo $\mathcal{N}$, and in different classes the vectors are not comparable modulo $\mathcal{N}$. We denote by $[x]_{\mathcal{N}}$ the class of vectors that are comparable modulo $\mathcal{N}$ to a given vector $x \in \mathbb{C}^n$. The set of all such classes of vectors defined by comparability modulo $\mathcal{N}$ is denoted $\mathbb{C}^n/\mathcal{N}$.

Proposition 1.7.1

Let set $\mathbb{C}^n/\mathcal{N}$ be a vector space over $\mathbb{C}$ with the following operations of addition and multiplication by a complex number:

$$[x]_{\mathcal{N}} + [y]_{\mathcal{N}} = [x + y]_{\mathcal{N}}, (x, y \in \mathbb{C}^n)$$

$$\alpha[x]_{\mathcal{N}} = [\alpha x]_{\mathcal{N}}, (x \in \mathbb{C}^n, \alpha \in \mathbb{C})$$

Proof. We have to check first that these definitions do not depend on the choice of the representatives $x \in [x]_{\mathcal{N}}$ and $y \in [y]_{\mathcal{N}}$. If $x_1 \in [x]_{\mathcal{N}}$ and $y_1 \in [y]_{\mathcal{N}}$, then

$$(x_1 + y_1) - (x + y) = (x_1 - x) + (y_1 - y) \in \mathcal{N}$$

that is, $x_1 + y_1 \in [x + y]_{\mathcal{N}}$. So indeed the class $[x + y]_{\mathcal{N}}$ does not depend on the choice of x and y. Similarly, one checks that $[\alpha x]_{\mathcal{N}}$ does not depend on the choice of x in the class $[x]_{\mathcal{N}}$ (for fixed α).

It is a straightforward but tedious task to verify that $\mathbb{C}^n/\mathcal{N}$ satisfies the following defining properties of a vector space over $\mathbb{C}$: The sum is commutative and associative: (a) $x + y = y + x$, $(x + y) + z = x + (y + z)$ for every $x, y, z \in \mathbb{C}^n/\mathcal{N}$; (b) there is a zero element $0 \in \mathbb{C}^n/\mathcal{N}$, that is, an element 0 such that $x + 0 = x$ for all $x \in \mathbb{C}^n/\mathcal{N}$; (c) for every $x \in \mathbb{C}^n/\mathcal{N}$ there is an additive inverse element $y \in \mathbb{C}^n/\mathcal{N}$, that is, such that $x + y = 0$; (d) for every $\alpha, \beta \in \mathbb{C}$ and $x, y \in \mathbb{C}^n/\mathcal{N}$ the following equalities hold: $\alpha(x + y) = \alpha x + \alpha y$, $(\alpha + \beta)x = \alpha x + \beta x$, $(\alpha\beta)x = \alpha(\beta x)$, and $1x = x$ (here 1 is the complex number). We leave the verification of all these properties to the reader. $\square$

The vector space $\mathbb{C}^n/\mathcal{N}$ is isomorphic to any direct complement $\mathcal{N}'$ of $\mathcal{N}$ in $\mathbb{C}^n$. Indeed, let $a \in \mathbb{C}^n/\mathcal{N}$; then there exists a unique vector $y \in \mathcal{N}'$ such that $a = [y]_{\mathcal{N}}$ and in fact, $y = Px$, where P is the projector on $\mathcal{N}'$ along $\mathcal{N}$ and x is any vector in the class a. This is easily checked. We have $y - x = -(I - P)x \in \mathcal{N}$, so $y \in a$. If there were two different vectors y_1 and y_2 from $\mathcal{N}'$ such that $[y_1]_{\mathcal{N}} = [y_2]_{\mathcal{N}} = a$, then $y_1 = y_2 \in \mathcal{N}' \cap \mathcal{N}$ and $y_1 \neq y_2$, which contradicts the choice of $\mathcal{N}'$ as a direct complement to $\mathcal{N}$ in $\mathbb{C}^n$. So we have constructed a map $\varphi: \mathbb{C}^n \to \mathcal{N}'$ defined by $\varphi(a) = y$. This map is easily seen to be a homomorphism of vector spaces; that is

$$\varphi(a + b) = \varphi(a) + \varphi(b) ; \qquad \varphi(\alpha a) = \alpha\varphi(a)$$

for every $a, b \in \mathbb{C}^n/\mathcal{N}$ and every $\alpha \in \mathbb{C}$. Moreover, if $\varphi(a) = \varphi(b)$, then the vector $y = \varphi(a) = \varphi(b)$ belongs to both classes a and b of comparable vectors modulo $\mathcal{N}$, and thus $a = b$. So φ is one-to-one. Taking any $y \in \mathcal{N}'$, we see that $\varphi([y]_{\mathcal{N}}) = y$, so φ is onto. Summing up, φ is an isomorphism between the two vector spaces $\mathbb{C}^n/\mathcal{N}$ and $\mathcal{N}'$. In particular, dim $\mathbb{C}^n/\mathcal{N} = n - \dim \mathcal{N}$. Assume now that $\mathcal{N}$ is A invariant for some transformation $A: \mathbb{C}^n \to \mathbb{C}^n$. Then the induced transformation $\hat{A}: \mathbb{C}^n/\mathcal{N} \to \mathbb{C}^n/\mathcal{N}$ is defined by $\hat{A}[x]_{\mathcal{N}} = [Ax]_{\mathcal{N}}$ for any $x \in \mathbb{C}^n$. This definition does not depend on the choice of the vector x in its class $[x]_{\mathcal{N}}$. Indeed, if $[x_1]_{\mathcal{N}} = [x_2]_{\mathcal{N}}$, then

$$Ax_1 - Ax_2 = A(x_1 - x_2) \in \mathcal{N}$$

because $x_1 - x_2 \in \mathcal{N}$ and $\mathcal{N}$ is A invariant.

We now present some basic properties of the induced linear transformation $\hat{A}$.

Proposition 1.7.2

If $\mathcal{N}$ is invariant for both transformations $A: \mathbb{C}^n \to \mathbb{C}^n$ and $B: \mathbb{C}^n \to \mathbb{C}^n$, then

$$\widehat{(\alpha A + \beta B)} = \alpha \hat{A} + \beta \hat{B} \quad \text{for any} \quad \alpha, \beta \in \mathbb{C}$$

$$\widehat{(AB)} = \hat{A}\hat{B} \tag{1.7.1}$$

If, in addition, A is invertible, then

$$\widehat{(A^{-1})} = (\hat{A})^{-1} \tag{1.7.2}$$

Proof. By Proposition 1.4.1, $\mathcal{N}$ is invariant for $\alpha A + \beta B$, AB, and A^{-1} (if A is invertible). For any $x \in \mathbb{C}^n$ we have

$$\widehat{(\alpha A + \beta B)}[x]_{\mathcal{N}} = [(\alpha A + \beta B)x]_{\mathcal{N}} = \alpha[Ax]_{\mathcal{N}} + \beta[Bx]_{\mathcal{N}}$$

$$= \alpha \hat{A}[x]_{\mathcal{N}} + \beta \hat{B}[x]_{\mathcal{N}}$$

Further, by definition of the induced transformation we have

$$\widehat{(AB)}[x]_{\mathcal{N}} = [ABx]_{\mathcal{N}}$$

and

$$\hat{A}\hat{B}[x]_{\mathcal{N}} = \hat{A}[Bx]_{\mathcal{N}} = [ABx]_{\mathcal{N}}$$

for every $x \in \mathbb{C}^n$. Finally, (1.7.2) is a particular case of (1.7.1) (with $B = A^{-1}$), taking into account the fact that $\hat{I} = I$. $\quad\square$

It may happen that A is not invertible but $\hat{A}$ is invertible. For instance, let $A: \mathbb{C}^n \to \mathbb{C}^n$ be any transformation with the property that $\mathbb{C}^n = \text{Ker } A \dotplus \text{Im } A$. (There are many transformations with this property; those represented by a diagonal matrix in some basis for $\mathbb{C}^n$, for example.) Put $\mathcal{N} = \text{Ker } A$. Then for every vector $x \in \mathbb{C}^n$ that is not in $\mathcal{N}$ we have $\hat{A}[x]_{\mathcal{N}} = [Ax]_{\mathcal{N}} \neq 0$. Thus $\text{Ker } \hat{A} = \{0\}$ and $\hat{A}$ is invertible. The following proposition clarifies the situation.

Proposition 1.7.3

If λ_0 is an eigenvalue of A and $\text{Ker}(A - \lambda_0 I)$ is not contained in $\mathcal{N}$, then λ_0 is also an eigenvalue of $\hat{A}$. Conversely, every eigenvalue λ_0 of $\hat{A}$ is an eigenvalue of A and $\text{Ker}(A - \lambda_0 I)$ is not contained in $\mathcal{N}$.

The proof is immediate: if $Ax = \lambda_0 x$ with $x \notin \mathcal{N}$, then $\hat{A}[x]_{\mathcal{N}} = \lambda_0 [x]_{\mathcal{N}}$ with $[x]_{\mathcal{N}} \neq 0$, and conversely.

1.8 THE LATTICE OF INVARIANT SUBSPACES

We start with the notion of a lattice of subspaces in $\mathbb{C}^n$. A set S of subspaces in $\mathbb{C}^n$ is called a *lattice* if $\{0\}$ and $\mathbb{C}^n$ belong to S and S contains the intersection and sum of any two subspaces belonging to S. The following are examples of lattices of subspaces: (a) $S = \{\{0\}, \mathcal{M}, \mathcal{M}^\perp, \mathbb{C}^n\}$, where $\mathcal{M}$ is a fixed subspace in $\mathbb{C}^n$; (b) $S = \{\{0\}, \text{Span}\{e_1, \dots, e_k\}$ for $k = 1, \dots, n\}$; (c) S is the set of all subspaces in $\mathbb{C}^n$. For us, the following example of a lattice of subspaces will be the most important.

Proposition 1.8.1

The set $\text{Inv}(A)$ of all invariant subspaces for a fixed transformation $A: \mathbb{C}^n \to \mathbb{C}^n$ is a lattice.

Proof. Let $\mathcal{M}, \mathcal{N} \in \text{Inv}(A)$. If $x \in \mathcal{M} \cap \mathcal{N}$, then because of the A invariance of $\mathcal{M}$ and $\mathcal{N}$ we have $Ax \in \mathcal{M}$ and $Ax \in \mathcal{N}$, so $\mathcal{M} \cap \mathcal{N}$ is A invariant. Now let $x \in \mathcal{M} + \mathcal{N}$, so that $x = x_1 + x_2$, where $x_1 \in \mathcal{M}$, $x_2 \in \mathcal{N}$. Then $Ax = Ax_1 + Ax_2 \in \mathcal{M} + \mathcal{N}$, and $\mathcal{M} + \mathcal{N}$ is A invariant as well. Finally, both $\{0\}$ and $\mathbb{C}^n$ obviously belong to $\text{Inv}(A)$. $\square$

Actually, examples (b) and (c) are particular cases of Proposition 1.8.1: (b) is just the set of all A-invariant subspaces for

$$A = \begin{bmatrix} 0 & 1 & 0 & & 0 \\ 0 & 0 & 1 & & \vdots \\ \vdots & & & \ddots & \\ \vdots & & & \ddots & 1 \\ 0 & & \cdots & & 0 \end{bmatrix}$$

and example (c) is the set of all invariant subspaces for the zero matrix.

In contrast, if $n > 2$, the lattice of example (a) is never the lattice of invariant subspaces of a fixed transformation A. Indeed, assuming the contrary, the restriction $A|_{\mathcal{M}}$ has a one-dimensional invariant subspace (a subspace spanned by an eigenvector; here we consider $A|_{\mathcal{M}}$ as a transformation from $\mathcal{M}$ into $\mathcal{M}$). By Proposition 1.5.3, this subspace is also an invariant subspace of A. Hence necessarily $\dim \mathcal{M} = 1$, and for the same reason $\dim \mathcal{M}^{\perp} = 1$. Since $\mathbb{C}^n = \mathcal{M} \dotplus \mathcal{M}^{\perp}$ we obtain a contradiction when $n > 2$.

In terms of the lattices of invariant subspaces, Propositions 1.4.2 and 1.4.4 can be restated as follows. We define $[\text{Inv}(A)]^{\perp}$ to be the set of subspaces $\mathcal{M}^{\perp}$ for which $\mathcal{M} \in \text{Inv}(A)$.

Proposition 1.8.2

Given a transformation $A: \mathbb{C}^n \to \mathbb{C}^n$ and an invertible transformation $S: \mathbb{C}^n \to \mathbb{C}^n$, we have

$$S[\text{Inv}(A)] = \text{Inv}(SAS^{-1})$$

and

$$\text{Inv}(A^*) = [\text{Inv}(A)]^{\perp}$$

We know that if $\mathcal{M}_1$ and $\mathcal{M}_2$ are A invariant, then so are $\mathcal{M}_1 + \mathcal{M}_2$ and $\mathcal{M}_1 \cap \mathcal{M}_2$. It is of interest to find out how the spectra of the restrictions $A|_{\mathcal{M}_1 + \mathcal{M}_2}$ and $A|_{\mathcal{M}_1 \cap \mathcal{M}_2}$ are related to the spectra of $A|_{\mathcal{M}_1}$ and $A|_{\mathcal{M}_2}$.

Theorem 1.8.3

If $\mathcal{M}_1$ and $\mathcal{M}_2$ are A-invariant subspaces, then

$$\sigma(A|_{\mathcal{M}_1 + \mathcal{M}_2}) = \sigma(A|_{\mathcal{M}_1}) \cup \sigma(A|_{\mathcal{M}_2}) \tag{1.8.1}$$

and

$$\sigma(A|_{\mathcal{M}_1 \cap \mathcal{M}_2}) \subset \sigma(A|_{\mathcal{M}_1}) \cap \sigma(A|_{\mathcal{M}_2}) \tag{1.8.2}$$

Recall that $\sigma(B)$ stands for the set of eigenvalues of a transformation B.

Proof. Proposition 1.5.3 shows that the inclusion $\supset$ holds in (1.8.1). To prove the opposite inclusion, write

$$\mathcal{M}_1 + \mathcal{M}_2 = \mathcal{M}_1' \dotplus (\mathcal{M}_1 \cap \mathcal{M}_2) \dotplus \mathcal{M}_2' \tag{1.8.3}$$

where $\mathcal{M}_1'$ is a subspace in $\mathcal{M}_1$ such that $\mathcal{M}_1' \dotplus (\mathcal{M}_1 \cap \mathcal{M}_2) = \mathcal{M}_1$, and $\mathcal{M}_2' \subset \mathcal{M}_2$ satisfies $\mathcal{M}_2' \dotplus (\mathcal{M}_1 \cap \mathcal{M}_2) = \mathcal{M}_2$. Write $A|_{\mathcal{M}_1 + \mathcal{M}_2}$ as the 3×3 block matrix with respect to decomposition (1.8.3):

$$A = \begin{bmatrix} A_{11} & A_{12} & A_{13} \\ A_{21} & A_{22} & A_{23} \\ A_{31} & A_{32} & A_{33} \end{bmatrix} : \quad \mathcal{M}_1 \dotplus \mathcal{M}_2 \to \mathcal{M}_1 \dotplus \mathcal{M}_2$$

Here, $A_{ij} = P_i A P_j$, and P_1 (resp. P_3) is the projector on $\mathcal{M}_1'$ along $(\mathcal{M}_1 \cap \mathcal{M}_2) \dotplus \mathcal{M}_2'$ [resp. on $\mathcal{M}_2'$ along $(\mathcal{M}_1 \cap \mathcal{M}_2) \dotplus \mathcal{M}_1'$], and $P_2 = I - P_1 - P_3$. As we have seen above, the A invariance of $\mathcal{M}_1$ implies $A_{31} = A_{32} = 0$, and the A invariance of $\mathcal{M}_2$ implies $A_{12} = A_{13} = 0$. So

$$A = \begin{bmatrix} A_{11} & 0 & 0 \\ A_{21} & A_{22} & A_{23} \\ 0 & 0 & A_{33} \end{bmatrix} \tag{1.8.4}$$

We find that

$$\det(\lambda I - A|_{\mathcal{M}_1 + \mathcal{M}_2}) = \det(\lambda I - A_{11}) \det(\lambda I - A_{22}) \det(\lambda I - A_{33})$$

and hence that $\lambda \in \sigma(A|_{\mathcal{M}_1 + \mathcal{M}_2})$ implies that $\lambda \in \sigma(A|_{\mathcal{M}_1})$ or $\lambda \in \sigma(A|_{\mathcal{M}_2})$.

For the proof of (1.8.2) note that $\mathcal{M}_1 \cap \mathcal{M}_2 \subset \mathcal{M}_1$, and hence by Proposition 1.5.3, $\sigma(A|_{\mathcal{M}_1 \cap \mathcal{M}_2}) \subset \sigma(A|_{\mathcal{M}_1})$. Similarly, $\sigma(A|_{\mathcal{M}_1 \cap \mathcal{M}_2}) \subset \sigma(A|_{\mathcal{M}_2})$, and (1.8.2) follows. $\square$

The following example shows that the inclusion in (1.8.2) may be strict.

EXAMPLE 1.8.1. Let

$$A = \begin{bmatrix} 1 & 0 & 0 \\ 0 & 0 & 0 \\ 0 & 0 & 0 \end{bmatrix}, \quad \mathcal{M}_1 = \mathrm{Span}\{e_1, e_2\}, \quad \mathcal{M}_2 = \mathrm{Span}\{e_1, e_3\}$$

Then $\mathcal{M}_1$ and $\mathcal{M}_2$ are A invariant and $\sigma(A|_{\mathcal{M}_1 \cap \mathcal{M}_2}) = \{1\}$; $\sigma(A|_{\mathcal{M}_1}) = \sigma(A|_{\mathcal{M}_2}) = \{1, 0\}$. $\square$

A set S of subspaces in $\mathbb{C}^n$ is called a *chain* if $\{0\}$ and $\mathbb{C}^n$ belong to S and either $\mathcal{M} \subset \mathcal{N}$ or $\mathcal{N} \subset \mathcal{M}$ (with proper inclusions) for every pair of different subspaces $\mathcal{M}, \mathcal{N} \in S$. Obviously, a chain is also a lattice. Also, a chain of subspaces is always finite (actually, it cannot contain more than $n + 1$ subspaces), in contrast to lattices that may be infinite, as in example (c) above.

Let

$$\{0\} \subset \mathcal{M}_1 \subset \mathcal{M}_2 \subset \cdots \subset \mathcal{M}_{k-1} \subset \mathcal{M}_k = \mathbb{C}^n \tag{1.8.5}$$

be a chain of different subspaces. We choose a direct complement $\mathcal{L}_i$ to

$\mathcal{M}_{i-1}$ in the subspace $\mathcal{M}_i$ $(i = 1, \ldots, k)$. Then we obtain a decomposition of $\mathbb{C}^n$ into a direct sum

$$\mathcal{L}_1 \dotplus \mathcal{L}_2 \dotplus \cdots \dotplus \mathcal{L}_k = \mathbb{C}^n \tag{1.8.6}$$

This means that for every vector $x \in \mathbb{C}^n$ there exists unique vectors $x_1 \in \mathcal{L}_1, \ldots, x_k \in \mathcal{L}_k$ such that $x = x_1 + x_2 + \cdots + x_k$. Now let P_i be the projector on $\mathcal{L}_i$ along

$$\mathcal{L}_1 \dotplus \mathcal{L}_2 \dotplus \cdots \dotplus \mathcal{L}_{i-1} \dotplus \mathcal{L}_{i+1} \dotplus \cdots \dotplus \mathcal{L}_k$$

The projectors P_i are mutually disjoint; that is, $P_i P_j = P_j P_i = 0$ for $i \neq j$, and $P_1 + \cdots + P_k = I$.

Now any transformation $A: \mathbb{C}^n \to \mathbb{C}^n$ can be written as a $k \times k$ block matrix with respect to the decomposition (1.8.6):

$$A = \begin{bmatrix} A_{11} & A_{12} & \cdots & A_{1k} \\ \vdots & \vdots & & \vdots \\ A_{k1} & A_{k2} & \cdots & A_{kk} \end{bmatrix} \tag{1.8.7}$$

where each transformation $A_{ij} = P_i A P_j|_{\mathcal{L}_j}: \mathcal{L}_j \to \mathcal{L}_i$ is written as a matrix in some fixed bases in $\mathcal{L}_j$ and $\mathcal{L}_i$.

Choose a basis $x_1, \ldots, x_n$ in $\mathbb{C}^n$ in such a way that

$$\operatorname{Span}\{x_1, \ldots, x_{p_i}\} = \mathcal{M}_i, \qquad i = 1, \ldots, k$$

where $0 < p_1 < p_2 < \cdots < p_k = n$, and let

$$\mathcal{L}_i = \operatorname{Span}\{x_{p_{i-1}}, x_{p_{i-1}+1}, \ldots, x_{p_i}\}$$

Then one can characterize all matrices for which (1.8.5) is a chain of (not necessarily all) invariant subspaces in terms of the $k \times k$ block representation as follows.

Proposition 1.8.4

All subspaces from the chain (1.8.5) are invariant for a transformation A if and only if A has the following form in the chosen basis $x_1, \ldots, x_n$:

$$A = \begin{bmatrix} A_{11} & A_{12} & \cdots & A_{1,k} \\ 0 & A_{22} & \cdots & A_{2,k} \\ \vdots & \vdots & & \vdots \\ 0 & 0 & \cdots & A_{kk} \end{bmatrix} \tag{1.8.8}$$

where A_{ij} is a $(p_i - p_{i-1}) \times (p_j - p_{j-1})$ matrix, $1 \leq i \leq j \leq k$ (and we define $p_0 = 0$).

Proof. Assume that A has the form (1.8.8), which means that in terms of the projectors $P_1, \ldots, P_k$ defined above the equalities $P_i A P_j = 0$ for $i > j$ hold. For a fixed j, it follows that

$$(P_{j+1} + \cdots + P_k) A (P_1 + \cdots + P_j) = 0$$

As $Q_j \stackrel{\text{def}}{=} P_1 + \cdots + P_j$ is a projector on $\mathcal{M}_j$ and $P_{j+1} + \cdots + P_k = I - Q_j$, we obtain $(I - Q_j) A Q_j = 0$, which means that $\mathcal{M}_j = \operatorname{Im} Q_j$ is A invariant.

Conversely, if $\mathcal{M}_1, \mathcal{M}_2, \ldots, \mathcal{M}_k$ are all A invariant, then the equality $(I - Q_j) A Q_j = 0$ holds for $j = 1, \ldots, k$. So $P_i A P_j = 0$ for $i > j$, and A has the form (1.8.8). $\square$

A chain of subspaces

$$\{0\} \subset \mathcal{M}_0 \subset \mathcal{M}_1 \subset \mathcal{M}_2 \subset \cdots \subset \mathcal{M}_k = \mathbb{C}^n \tag{1.8.9}$$

is called *maximal* (or *complete*) if it cannot be extended to a larger chain, that is, any chain of subspaces

$$\{0\} \subset \mathcal{L}_0 \subset \mathcal{L}_1 \subset \mathcal{L}_2 \subset \cdots \subset \mathcal{L}_l = \mathbb{C}^n$$

with the property that every $\mathcal{M}_i$ is equal to some $\mathcal{L}_j$, coincides with the chain (1.8.9). It is easily seen that a chain (1.8.9) is maximal if and only if $\dim \mathcal{M}_i = i$, $i = 1, \ldots, n$.

Now if (1.8.9) is a maximal chain, we may choose a basis $x_1, \ldots, x_n$ in $\mathbb{C}^n$ in such a way that

$$\mathcal{M}_i = \operatorname{Span}\{x_1, \ldots, x_i\}, \qquad i = 1, \ldots, n$$

As a particular case of Proposition 1.8.4, we find that all the subspaces $\mathcal{M}_1, \ldots, \mathcal{M}_n$ are A invariant for a transformation A if and only if A has upper triangular form in the basis $x_1, \ldots, x_n$:

$$A = \begin{bmatrix} a_{11} & a_{12} & \cdots & a_{1n} \\ 0 & a_{22} & \cdots & a_{2n} \\ \vdots & \vdots & & \vdots \\ 0 & 0 & \cdots & a_{nn} \end{bmatrix}$$

We conclude this section with a useful result on chains of invariant subspaces for a transformation having a basis of eigenvectors in $\mathbb{C}^n$. It turns out that such chains can be chosen to be complementary to any chain of subspaces given in advance.

Theorem 1.8.5

Let $A: \mathbb{C}^n \to \mathbb{C}^n$ be a transformation having a basis in $\mathbb{C}^n$ formed by eigenvectors of A. Then for every chain of subspaces $\mathcal{N}_1 \subset \cdots \subset \mathcal{N}_p$ in $\mathbb{C}^n$

there exists a chain of A-invariant subspaces $\mathcal{M}_1 \supset \cdots \supset \mathcal{M}_p$ such that $\mathcal{M}_j$ is a direct complement to $\mathcal{N}_j$, $j = 1, \ldots, p$.

Proof. Let $x_1, x_2, \ldots, x_n$ be a basis in $\mathbb{C}^n$ consisting of eigenvectors of A. We show first of all that there exists a set of indices $K_1 \subset \{1, \ldots, n\}$ such that the subspace $\mathcal{M}_1 \overset{\text{def}}{=} \{x_i \mid i \in K_1\}$ is a direct complement to $\mathcal{N}_1$ in $\mathbb{C}^n$. Let i_1 be the first index such that x_{i_1} does not belong to $\mathcal{N}_1$. If $i_1 < i_2 < \cdots < i_s$ ($\leq n$) are already chosen, let i_{s+1} be the first index such that $x_{i_{s+1}}$ does not belong to $\text{Span}\{x_{i_1}, \ldots, x_{i_s}\} \dotplus \mathcal{N}_1$. This process will stop after t steps (say) when the equality $\text{Span}\{x_{i_1}, \ldots, x_{i_t}\} \dotplus \mathcal{N}_1 = \mathbb{C}^n$ is reached. Now one can put $K_1 = \{i_1, \ldots, i_t\}$ to ensure that $\text{Span}\{x_i \mid i \in K_1\}$ is a direct complement to $\mathcal{N}_1$.

By the same token, there is a set $K_2 \subset K_1$ such that $\mathcal{M}_2 \overset{\text{def}}{=} \text{Span}\{x_i \mid i \in K_2\}$ is a direct complement to $\mathcal{M}_1 \cap \mathcal{N}_2$ in $\mathcal{M}_1$. As $\mathcal{N}_2 = (\mathcal{M}_1 \cap \mathcal{N}_2) \dotplus \mathcal{N}_1$, clearly $\mathcal{M}_2$ is a direct complement to $\mathcal{N}_2$ in $\mathbb{C}^n$. Let $\mathcal{M}_3 = \text{Span}\{x_i \mid i \in K_3\}$, where $K_3 \subset K_2$ and $\mathcal{M}_3$ is a direct complement to $\mathcal{M}_2 \cap \mathcal{N}_3$ in $\mathcal{M}_2$, and so on. Clearly, all the subspaces $\mathcal{M}_j$ are A invariant. $\square$

In connection with Theorem 1.8.5 we emphasize that not every transformation has a basis of eigenvectors. Indeed, we have seen in Example 1.1.1 a transformation with only one eigenvector (up to multiplication by a nonzero complex number); obviously one cannot form a basis in $\mathbb{C}^n$ from the eigenvectors of A. Furthermore, the transformation A of Example 1.1.1 does not satisfy the conclusion of Theorem 1.8.5. We leave it to the reader to verify the following fact concerning this transformation A: for a chain $\mathcal{N}_1 \subset \cdots \subset \mathcal{N}_p$ of subspaces in $\mathbb{C}^n$ there is a chain $\mathcal{M}_1 \supset \cdots \supset \mathcal{M}_p$ of A-invariant subspaces such that $\mathcal{M}_i \dotplus \mathcal{N}_i = \mathbb{C}^n$, $i = 1, \ldots, p$ if and only if each $\mathcal{N}_j$ is spanned by the vectors of type $e_n + f_n$, $e_{n-1} + f_{n-1}, \ldots, e_{n-r_j+1} + f_{n-r_j+1}$, where $r_j = \dim \mathcal{N}_j$ and the vectors $f_n, f_{n-1}, \ldots, f_{n-r_j+1}$ belong to $\text{Span}\{e_1, e_2, \ldots, e_{n-r_j}\}$. (As usual, e_k stands for the kth unit coordinate vector in $\mathbb{C}^n$.)

The converse of Theorem 1.8.5 is also true: if for every chain of subspaces $\mathcal{N}_1 \subset \cdots \subset \mathcal{N}_p$ in $\mathbb{C}^n$ there exists a chain of A-invariant subspaces $\mathcal{M}_1 \supset \cdots \supset \mathcal{M}_p$ such that $\mathcal{M}_j$ is a direct complement to $\mathcal{N}_j$, $j = 1, \ldots, p$, then there exists a basis of eigenvectors of A. However, a stronger statement holds.

Proposition 1.8.6

Let $A: \mathbb{C}^n \to \mathbb{C}^n$ be a transformation. If each subspace $\mathcal{N} \subset \mathbb{C}^n$ has a complementary subspace that is A invariant, there is a basis in $\mathbb{C}^n$ consisting of eigenvectors of A.

Proof. Let $\mathcal{N}_0$ be the subspace spanned by all the eigenvectors of A. We have to prove that $\mathcal{N}_0 = \mathbb{C}^n$. Assume the contrary: $\mathcal{N}_0 \neq \mathbb{C}^n$. The hypothesis of the theorem implies that there is an A-invariant subspace $\mathcal{M}_0$ that is a direct complement to $\mathcal{N}_0$. Clearly, $\mathcal{M}_0 \neq \{0\}$. Hence there exists an eigenvector x_0 of A in $\mathcal{M}_0$: $Ax_0 = \lambda_0 x_0$, $x_0 \neq 0$. Since $x_0 \not\in \mathcal{N}_0$, we contradict the definition of $\mathcal{N}_0$. $\quad\square$

1.9   TRIANGULAR MATRICES AND COMPLETE CHAINS OF INVARIANT SUBSPACES

The main result of this section is the following theorem on unitary triangularization of a transformation. It has important implications for the study of invariant subspaces.

Recall that a transformation $U: \mathbb{C}^n \to \mathbb{C}^n$ is called *unitary* if it is invertible and $U^{-1} = U^*$ or, equivalently, if $(Ux, Uv) = (x, y)$ for all $x, y \in \mathbb{C}^n$. Note that the seemingly weaker condition $\|Ux\| = \|x\|$ for all $x \in \mathbb{C}^n$ is also sufficient to ensure that U is unitary. Note also that the product of two unitary transformations is unitary again, and so is the inverse of a unitary transformation.

It will be convenient to write linear transformations from $\mathbb{C}^n$ into $\mathbb{C}^n$ as $n \times n$ matrices with respect to the standard orthonormal basis $e_1, \ldots, e_n$ in $\mathbb{C}^n$. We shall use the fact that a matrix is unitary if and only if its columns form an orthonormal basis in $\mathbb{C}^n$.

Theorem 1.9.1

For any $n \times n$ matrix A there exists a unitary matrix U such that

$$T = U^* A U = [t_{ij}]_{i,j=1}^n \tag{1.9.1}$$

is an upper triangular matrix, that is, $t_{ij} = 0$ for $i > j$, and the diagonal elements $t_{11}, \ldots, t_{nn}$ are just the eigenvalues of A.

Proof. Let λ_1 be an eigenvalue of A with an eigenvector x_1 and assume that $\|x_1\| = 1$. Let $x_2, \ldots, x_n$ be vectors in $\mathbb{C}^n$ that, together with x_1, form an orthonormal basis for $\mathbb{C}^n$. Then the matrix

$$U_1 = [x_1 \cdots x_n]$$

is unitary. Write U_1 in a block matrix form $U_1 = [x_1 V]$, where $V = [x_2 \cdots x_n]$ is an $n \times (n-1)$ matrix. Then because of the orthonormality of $x_1, \ldots, x_n$, $V^* x_1 = 0$. Now, using the relation $Ax_1 = \lambda_1 x_1$, we obtain

$$U_1^* A U_1 = \begin{bmatrix} x_1^* \\ V^* \end{bmatrix} A [x_1 V] = \begin{bmatrix} x_1^* \\ V^* \end{bmatrix} [\lambda_1 x_1, AV]$$

$$= \begin{bmatrix} \lambda_1 \|x_1\|^2 & x_1^* AV \\ \lambda_1 V^* x_1 & V^* AV \end{bmatrix} = \begin{bmatrix} \lambda_1 & x_1^* AV \\ 0 & V^* AV \end{bmatrix}$$

Applying the same procedure to the $(n-1) \times (n-1)$ matrix $A_2 \stackrel{\text{def}}{=} V^*AV$, we find an $(n-1) \times (n-1)$ unitary matrix U_2 such that

$$U_2^* A_2 U_2 = \begin{bmatrix} \lambda_2 & * \\ 0 & A_3 \end{bmatrix}$$

for some eigenvalue λ_2 of A_2 and some $(n-2) \times (n-2)$ matrix A_3. Apply the same procedure to A_3 using a suitable $(n-2) \times (n-2)$ unitary matrix U_3, and so on.

Then for the $n \times n$ unitary matrix

$$U = U_1 \begin{bmatrix} 1 & 0 \\ 0 & U_2 \end{bmatrix} \begin{bmatrix} I_2 & 0 \\ 0 & U_3 \end{bmatrix} \cdots \begin{bmatrix} I_{n-1} & 0 \\ 0 & U_n \end{bmatrix}$$

the product U^*AU is upper triangular. Finally, as $U^* = U^{-1}$, we have

$$\det(\lambda I - A) = \det(\lambda I - T) = (\lambda - t_{11}) \cdots (\lambda - t_{nn})$$

so that $t_{11}, \ldots, t_{nn}$ are the eigenvalues of A. □

Let $T = U^*AU$ be a triangular form of the matrix A as in Theorem 1.9.1. Then it follows from Proposition 1.8.4 that there is a maximal chain

$$0 \subset \mathcal{M}_1 \subset \cdots \subset \mathcal{M}_{n-1} \subset \mathcal{M}_n = \mathbb{C}^n$$

where all subspaces $\mathcal{M}_i$ are T invariant. Then Proposition 1.4.2 shows that the maximal chain

$$0 \subset U^*\mathcal{M}_1 \subset \cdots \subset U^*\mathcal{M}_{n-1} \subset \mathcal{M}_n = \mathbb{C}^n$$

consists of A-invariant subspaces. We have obtained the following fact.

Corollary 1.9.2

Any transformation (or $n \times n$ matrix) $A: \mathbb{C}^n \to \mathbb{C}^n$ has a maximal chain of A-invariant subspaces. In particular, for every $i, 1 \le i \le n$, there exists an i-dimensional A-invariant subspace.

In general, a complete chain of A-invariant subspaces is not unique. An extreme case of this situation is provided by $A = \alpha I$, $\alpha \in \mathbb{C}$. For such an A, every complete chain of subspaces is a complete chain of A-invariant subspaces. Clearly, there are many complete chains of subspaces in $\mathbb{C}^n$ (unless $n = 1$).

Let us characterize the matrices A for which there is a unique complete chain of invariant subspaces.

Theorem 1.9.3

An $n \times n$ matrix A has a unique complete chain of invariant subspaces if and only if A has a unique eigenvector (up to multiplication by a scalar).

Proof. We have seen in the proof of Theorem 1.9.1 that for any eigenvector x of A the subspace Span$\{x\}$ appears in some complete chain of A-invariant subspaces. So if a complete chain of invariant subspaces is unique, the matrix A has a unique eigenvector (up to multiplication by a scalar).

The converse part of Theorem 1.9.3 will be proved later using the Jordan normal form of a matrix (see Theorem 2.5.1). $\square$

Theorem 1.9.1 has important consequences for normal transformations. A transformation $A: \mathbb{C}^n \to \mathbb{C}^n$ is called *normal* if $AA^* = A^*A$. Self-adjoint and unitary transformations are normal, of course, but there are also normal transformations that are neither self-adjoint nor unitary.

Theorem 1.9.4

A transformation $A: \mathbb{C}^n \to \mathbb{C}^n$ is normal if and only if there is an orthonormal basis in $\mathbb{C}^n$ consisting of eigenvectors for A.

Proof. Write A as an $n \times n$ matrix. Assuming that A is normal, the matrix T from (1.9.1) is easily seen to be normal as well:

$$TT^* = U^*AUU^*A^*U = U^*AA^*U$$

$$= U^*A^*AU = U^*A^*UU^*AU = T^*T$$

But T is upper triangular:

$$T = \begin{bmatrix} t_{11} & t_{12} & t_{13} & \cdots & t_{1n} \\ 0 & t_{22} & & \cdots & t_{2n} \\ \vdots & \vdots & & & \vdots \\ 0 & 0 & & \cdots & t_{nn} \end{bmatrix}$$

Hence the $(1,1)$ entry in T^*T is $|t_{11}|^2$, whereas this entry in TT^* is $|t_{11}|^2 + |t_{12}|^2 + \cdots + |t_{1n}|^2$. As $T^*T = TT^*$, it follows that $t_{12} = \cdots = t_{1n} = 0$. Comparing the $(2,2)$ entries in T^*T and TT^*, we now find that $t_{23} = \cdots = t_{2n} = 0$, and so on. It turns out that T is diagonal. Now $Ue_1, \ldots, Ue_n$ is an orthonormal basis in $\mathbb{C}^n$ consisting of eigenvectors of A.

Conversely, assume that A has a set of eigenvectors $f_1, \ldots, f_n$ that form an orthonormal basis in $\mathbb{C}^n$. Then the matrix $U = [f_1 f_2 \cdots f_n]$ is unitary and

$$U^*AUe_i = U^*Af_i = \lambda_i U^*f_i = \lambda_i e_i$$

where λ_i is the eigenvalue of A corresponding to f_i. So

$$T \overset{\mathrm{def}}{=} U^* A U = \mathrm{diag}[\lambda_1 \lambda_2 \cdots \lambda_n]$$

As the diagonal matrix T is obviously normal, we find that A is normal as well. $\square$

1.10   EXERCISES

1.1   Prove or disprove the following statements for any linear transformation $A: \mathbb{C}^n \to \mathbb{C}^n$:

(a) $\mathrm{Im}\, A \dotplus \mathrm{Ker}\, A = \mathbb{C}^n$.
(b) $\mathrm{Im}\, A + \mathrm{Ker}\, A = \mathbb{C}^n$ (the sum not necessarily direct).
(c) $\mathrm{Im}\, A \cap \mathrm{Ker}\, A \neq \{0\}$.
(d) $\dim \mathrm{Im}\, A + \dim \mathrm{Ker}\, A = n$.
(e) $\mathrm{Im}\, A$ is the orthogonal complement to $\mathrm{Ker}\, A^*$.

1.2   Prove or disprove statements (d) and (e) in the preceding exercise for a transformation $A: \mathbb{C}^m \to \mathbb{C}^n$, where $m \neq n$.

1.3   Let $A: \mathbb{C}^n \to \mathbb{C}^n$ be the transformation given (in the standard orthonormal basis) by an upper triangular Toeplitz matrix

$$\begin{bmatrix} a_0 & a_1 & a_2 & \cdots & a_{n-1} \\ 0 & a_0 & a_1 & \cdots & a_{n-2} \\ \vdots & \vdots & \vdots & & \vdots \\ 0 & 0 & 0 & \cdots & a_0 \end{bmatrix}$$

where $a_0, \ldots, a_{n-1}$ are complex numbers. Find the subspaces $\mathrm{Im}\, A$ and $\mathrm{Ker}\, A$.

1.4   Given $A: \mathbb{C}^n \to \mathbb{C}^n$ as in Example 1.1.3, identify the A-invariant subspaces $\mathrm{Im}\, A^k$ and $\mathrm{Ker}\, A^k$, $k = 0, 1, \ldots$.

1.5   Identify $\mathrm{Im}\, A^k$ and $\mathrm{Ker}\, A^k$, $k = 0, 1, \ldots$, where

$$A = \begin{bmatrix} a_0 & 0 & \cdots & 0 \\ a_{-1} & a_0 & \cdots & 0 \\ \vdots & \vdots & & \vdots \\ a_{-n+1} & a_{-n+2} & \cdots & a_0 \end{bmatrix}, \qquad a_0, \ldots, a_{-n+1} \in \mathbb{C}$$

is given by a lower triangular Toeplitz matrix.

1.6   Find all one-dimensional invariant subspaces of the following transformations (written as matrices with respect to the standard orthonormal basis):

$$\begin{bmatrix} -2 & 1 & 1 \\ 0 & -1 & 0 \\ 0 & 2 & 1 \end{bmatrix}; \quad \begin{bmatrix} -1 & 2 & -2 \\ 0 & 1 & 0 \\ 0 & 0 & 1 \end{bmatrix}; \quad \begin{bmatrix} 2 & 1 & 0 \\ 0 & 3 & 1 \\ 0 & -1 & 1 \end{bmatrix}$$

1.7 In the preceding exercise, which transformations have a Jordan chain consisting of more than one vector? Find these Jordan chains.

1.8 Show that all invariant subspaces for the projector P on the subspace $\mathcal{N}$ are of the form $\mathcal{M}_1 \dotplus \mathcal{N}_1$, where $\mathcal{M}_1$ (resp. $\mathcal{N}_1$) is a subspace in $\mathcal{M}$ (resp. $\mathcal{N}$). Find the lattice Inv P^*.

1.9 Given P as in Exercise 1.8, find all the invariant subspaces of $\alpha_1 P + \alpha_2(I - P)$, where α_1 and α_2 are complex numbers.

1.10 Let $A: \mathbb{C}^n \to \mathbb{C}^n$ be a transformation with $A^2 = I$. Show that $\text{Im}(I + A)$ and $\text{Im}(I - A)$ are the subspaces consisting of the zero vector and all eigenvectors of A corresponding to the eigenvalues 1 and -1, respectively.

1.11 Find all invariant subspaces of a transformation $A: \mathbb{C}^n \to \mathbb{C}^n$ such that $A^2 = I$.

1.12 Let

$$A\begin{bmatrix} 0 & I \\ I & 0 \end{bmatrix}: \mathbb{C}^n \oplus \mathbb{C}^n \to \mathbb{C}^n \oplus \mathbb{C}^n$$

(a) Show that A is similar to

$$\begin{bmatrix} I & 0 \\ 0 & -I \end{bmatrix}$$

(b) Find all invariant subspaces of A.

1.13 Let

$$A = \begin{bmatrix} 0 & 0 & \cdots & \alpha_k I \\ \vdots & \vdots & & \vdots \\ 0 & \alpha_2 I & \cdots & 0 \\ \alpha_1 I & 0 & \cdots & 0 \end{bmatrix}: \underbrace{\mathbb{C}^n \oplus \cdots \oplus \mathbb{C}^n}_{k \text{ times}} \to \underbrace{\mathbb{C}^n \oplus \cdots \oplus \mathbb{C}^n}_{k \text{ times}}$$

Show that A is similar to a matrix of type

$$\begin{bmatrix} \beta_1 I & 0 & \cdots & 0 \\ 0 & \beta_2 I & \cdots & 0 \\ \vdots & \vdots & & \vdots \\ 0 & 0 & \cdots & \beta_k I \end{bmatrix}$$

and find the lattice $\text{Inv}(A)$. What are the invariant subspaces of A^*?

1.14 Let

$$Q = \begin{bmatrix} 0 & 1 & 0 & \cdots & 0 \\ 0 & 0 & 1 & \cdots & 0 \\ \vdots & \vdots & \vdots & & \vdots \\ 0 & 0 & 0 & \cdots & 1 \\ 1 & 0 & 0 & \cdots & 0 \end{bmatrix}: \mathbb{C}^n \to \mathbb{C}^n$$

Prove that the eigenvalues of Q are $\cos(2\pi k/n) + i\sin(2\pi k/n)$, $k = 0, 1, \ldots, n-1$. Find the corresponding eigenvectors.

1.15 Show that the transformation

$$
\begin{bmatrix}
\alpha & 1 & & & & \\
1 & \alpha & 1 & & 0 & \\
 & 1 & \alpha & 1 & & \\
 & & \ddots & \ddots & \ddots & \\
 & & & & & 1 \\
 & 0 & & & 1 & \alpha
\end{bmatrix} : \mathbb{C}^n \to \mathbb{C}^n, \ \alpha \in \mathbb{C}
$$

has eigenvectors x_p whose jth coordinate is $\sin\{jp\pi/(n+1)\}$, $p = 0, \ldots, n-1$ (independently of α). What are the corresponding eigenvalues?

1.16 Let

$$
A = \begin{bmatrix}
0 & 1 & 0 & \cdots & 0 \\
0 & 0 & 1 & \cdots & 0 \\
\vdots & \vdots & & \ddots & \vdots \\
0 & 0 & & \cdots & 1 \\
a_0 & a_1 & & \cdots & a_{n-1}
\end{bmatrix} : \mathbb{C}^n \to \mathbb{C}^n
$$

where $a_0, \ldots, a_{n-1}$ are complex numbers. Show that λ_0 is an eigenvalue of A if and only if λ_0 is a zero of the equation

$$
\lambda^n - a_{n-1}\lambda^{n-1} - \cdots - a_1\lambda - a_0 = 0
$$

1.17 Let

$$
A = \begin{bmatrix}
0 & 1 & 0 & \cdots & 0 \\
0 & 0 & 1 & \cdots & 0 \\
\vdots & \vdots & \vdots & & \vdots \\
0 & 0 & 0 & \cdots & 1 \\
-1 & -\binom{n}{1} & -\binom{n}{2} & \cdots & -\binom{n}{n-1}
\end{bmatrix} : \mathbb{C}^n \to \mathbb{C}^n
$$

(a) Find all eigenvalues and eigenvectors of A.
(b) Find a longest Jordan chain.
(c) Show that A is similar to a matrix of the form

$$
\begin{bmatrix}
\lambda_0 & 1 & 0 & \cdots & 0 \\
0 & \lambda_0 & 1 & \cdots & 0 \\
\vdots & \vdots & & \ddots & \vdots \\
 & & & & 1 \\
0 & 0 & & & \lambda_0
\end{bmatrix}
$$

and find the similarity matrix.

(d) Find the lattice Inv(A) of all invariant subspaces of A.

(e) Find all invariant subspaces of the transposed matrix A^T.

1.18 Let $A: \mathbb{C}^n \to \mathbb{C}^n$ be a transformation represented as a matrix in the standard orthonormal basis. Show that all invariant subspaces for the transposed matrix A^T are given by the formula $\mathrm{Span}\{\bar{x}_1, \ldots, \bar{x}_k\}$, where $x_1, \ldots, x_k$ is a basis in the orthogonal complement to some A-invariant subspace, and for a vector $y = \langle y_1, \ldots, y_n \rangle \in \mathbb{C}^n$ we denote $\bar{y} = \langle \bar{y}_1, \ldots, \bar{y}_n \rangle$.

1.19 Prove a generalization of Proposition 1.4.4: if $A: \mathbb{C}^n \to \mathbb{C}^n$ is a transformation and $\mathcal{M}$, $\mathcal{N}$ are subspaces in $\mathbb{C}^n$, then $A\mathcal{M} \subset \mathcal{N}$ holds if and only if $A^*\mathcal{N}^\perp \subset \mathcal{M}^\perp$.

1.20 Give an example of a transformation $A: \mathbb{C}^n \to \mathbb{C}^n$ that is not self-adjoint but nevertheless $A\mathcal{M}^\perp \subset \mathcal{M}^\perp$ for every A-invariant subspace $\mathcal{M}$.

1.21 Let

$$\mathcal{M}_1 = \left\{ \begin{bmatrix} x \\ x \end{bmatrix} \in \mathbb{C}^{2n} \mid x \in \mathbb{C}^n \right\}, \qquad \mathcal{M}_2 = \left\{ \begin{bmatrix} x \\ -x \end{bmatrix} \in \mathbb{C}^{2n} \mid x \in \mathbb{C}^n \right\}$$

Find the angular transformations of $\mathcal{M}_1$ and $\mathcal{M}_2$ with respect to the projector on $\mathbb{C}^n \oplus \{0\}$ along $\{0\} \oplus \mathbb{C}^n$.

1.22 Find at least one solution of the quadratic equation

$$RT_{12}R + RT_{11} - T_{22}R - T_{21} = 0$$

where

(a)
$$T_{ij} = \begin{bmatrix} \lambda_{ij} & 1 & 0 & \cdots & 0 \\ 0 & \lambda_{ij} & 1 & \cdots & 0 \\ \vdots & \vdots & \vdots & & \vdots \\ & & & & 1 \\ 0 & 0 & 0 & \cdots & \lambda_{ij} \end{bmatrix}, \qquad \lambda_{ij} \in \mathbb{C}$$

are $n \times n$ matrices.

(b) T_{ij} are $n \times n$ diagonal matrices.

(c) T_{ij} are $n \times n$ circulant matrices.

1.23 Prove that $x_1 + \mathcal{M}, \ldots, x_k + \mathcal{M}$ is a basis in $\mathbb{C}^n/\mathcal{M}$ (where $x_1, \ldots, x_k \in \mathbb{C}$) if and only if for some basis $y_1, \ldots, y_p$ in $\mathcal{M}$ the vectors $x_1, \ldots, x_k, y_1, \ldots, y_p$ form a basis in $\mathbb{C}^n$.

1.24 Let $A = \mathrm{diag}[a_1, \ldots, a_n]: \mathbb{C}^n \to \mathbb{C}^n$, where the numbers $a_1, \ldots, a_n$ are distinct. Show that for any A-invariant subspace $\mathcal{M}$ the induced transformation $A: \mathbb{C}^n/\mathcal{M} \to \mathbb{C}^n/\mathcal{M}$ can also be written in the form $\mathrm{diag}[b_1, \ldots, b_k]$ in some basis in $\mathbb{C}^n/\mathcal{M}$.

1.25 Find all the induced transformations $A: \mathbb{C}^n/\mathcal{M} \to \mathbb{C}^n/\mathcal{M}$, where

$$A = \begin{bmatrix} \lambda_0 & 1 & 0 & \cdots & 0 \\ 0 & \lambda_0 & 1 & \cdots & 0 \\ \vdots & \vdots & & \ddots & \vdots \\ & & & & 1 \\ 0 & 0 & & & \lambda_0 \end{bmatrix}$$

and $\mathcal{M}$ is any A-invariant subspace.

1.26 Show that if P is a projector on $\mathbb{C}^n$ and $\hat{P}$ is the induced transformation on $\mathbb{C}^n/\mathcal{M}$, where $\mathcal{M}$ is a P-invariant subspace, then $\hat{P}$ is a projector as well. Find Im $\hat{P}$ and Ker $\hat{P}$.

1.27 Let

$$A = \begin{bmatrix} 1 & 0 & 3 \\ 0 & 1 & 4 \\ 0 & 0 & 2 \end{bmatrix}$$

be in a triangular form. Show that

$$\begin{bmatrix} 0 & 1 & 0 \\ -1 & 0 & 0 \\ 0 & 0 & 1 \end{bmatrix} A \begin{bmatrix} 0 & -1 & 0 \\ 1 & 0 & 0 \\ 0 & 0 & 1 \end{bmatrix} \neq A$$

is also in a triangular form. Hence the triangular form of a matrix is not unique, in general.

1.28 Find complete chains of invariant subspaces for the transformations given in Exercise 1.6. Check for uniqueness in each case.

1.29 Given a transformation in a matrix form

$$A = \begin{bmatrix} x_{11} & 0 & 0 \\ x_{12} & x_{12} & x_{23} \\ x_{31} & 0 & x_{33} \end{bmatrix}, \qquad x_{ij} \in \mathbb{C}$$

with respect to the basis e_1, e_2, e_3, find a complete chain of A-invariant subspaces. Find a basis in which A has the upper triangular form.

1.30 Let $A: \mathbb{C}^{2n} \to \mathbb{C}^{2n}$ be a transformation. Prove that there exists an orthonormal basis in $\mathbb{C}^{2n}$ such that, with respect to this basis, A has the representation

$$\begin{bmatrix} A_{11} & A_{12} \\ A_{21} & A_{22} \end{bmatrix}$$

where, for each i and j, A_{ij} is an upper triangular matrix.

Chapter Two

The Jordan Form and Invariant Subspaces

We have seen in Section 1.4 and Proposition 1.8.2 that there is a strong relationship between lattices of invariant subspaces of similar transformations, namely

$$S(\mathrm{Inv}(A)) = \mathrm{Inv}(SAS^{-1})$$

for any two tranformations A and S from $\mathbb{C}^n$ into $\mathbb{C}^n$ with S invertible. Thus, for the study of invariant subspaces, it is desirable to use similarity transformations to reduce a given transformation to the simplest form, in the hope that the lattice of invariant subspaces for the simplest form would be more transparent than that for the original transformation. The "simplest form" here is the Jordan form. It is obtained in this chapter and used to study some properties of invariant subspaces. Special insights are obtained into the structure of invariant subspaces and are exploited throughout the book. We examine irreducible invariant subspaces, generators of invariant subspaces, maximal and minimal invariant subspaces, and invariant subspaces of functions of transformations. An interesting class of subspaces is introduced and studied in Section 2.9 that we call "marked." All the subject matter here is well known, although this exposition may be unusual in matters of emphasis and detail that will be useful subsequently.

2.1 ROOT SUBSPACES

In this section we introduce the root subspaces of a transformation. The study of these subspaces is the first step towards an understanding of the Jordan form. At the same time it will be seen that the root subspaces are important examples of invariant subspaces that can be described in terms of Jordan chains.

We consider now some ideas leading up to the definition of root subspaces. Let $A: \mathbb{C}^n \to \mathbb{C}^n$ be a transformation and let λ_0 be an eigenvalue of A. Consider the subspaces $\text{Ker}(A - \lambda_0 I)^i$, $i = 1, 2, \ldots$. For $i = 1$ the subspace $\text{Ker}(A - \lambda_0 I) \neq \{0\}$ is just the subspace spanned by the eigenvectors of A corresponding to λ_0. As $(A - \lambda_0 I)^i x = 0$ implies $(A - \lambda_0 I)^{i+1} x = 0$, we have

$$\text{Ker}(A - \lambda_0 I) \subset \text{Ker}(A - \lambda_0 I)^2 \subset \cdots \subset \text{Ker}(A - \lambda_0 I)^i$$
$$\subset \text{Ker}(A - \lambda_0 I)^{i+1} \subset \cdots \tag{2.1.1}$$

Consequently, $\text{Ker}(A - \lambda_0 I)^{i+1} \neq \text{Ker}(A - \lambda_0 I)^i$ if and only if $\dim \text{Ker}(A - \lambda_0 I)^{i+1} > \dim \text{Ker}(A - \lambda_0 I)^i$. Since the dimensions of the subspaces $\text{Ker}(A - \lambda_0 I)^i$, $i = 1, 2, \ldots$ are bounded above by n, there exists a minimal integer $p \geq 1$ such that

$$\text{Ker}(A - \lambda_0 I)^i = \text{Ker}(A - \lambda_0 I)^p$$

for all integers $i > p$. The subspace $\text{Ker}(A - \lambda_0 I)^p$ is called the *root subspace* of A corresponding to λ_0 and is denoted $\mathcal{R}_{\lambda_0}(A)$.

In other words, $\mathcal{R}_{\lambda_0}(A)$ consists of all vectors $x \in \mathbb{C}^n$ such that $(A - \lambda_0 I)^q x = 0$ for some integer $q \geq 1$. (This integer may depend on x.) Because

$$A(A - \lambda_0 I)^i = (A - \lambda_0 I)^i A, \qquad i = 1, 2, \ldots$$

all subspaces in (2.1.1) are A invariant. In particular, the root subspace $\mathcal{R}_{\lambda_0}(A)$ is A invariant.

By definition, $\mathcal{R}_{\lambda_0}(A) = \text{Ker}(A - \lambda_0 I)^p$ is the biggest subspace in the chain (2.1.1). We see later that, in fact, p is the minimal integer $i \geq 1$ for which the equality $\text{Ker}(A - \lambda_0 I)^i = \text{Ker}(A - \lambda_0 I)^{i+1}$ holds, and that $p \leq n$. Hence we also have

$$\mathcal{R}_{\lambda_0}(A) = \{x \in \mathbb{C}^n \mid (A - \lambda_0 I)^n x = 0\}$$

The nesting of the kernels in (2.1.1) has a dual in the (descending) nesting of images:

$$\text{Im}(A - \lambda_0 I) \supset \text{Im}(A - \lambda_0 I)^2 \supset \cdots \supset \text{Im}(A - \lambda_0 I)^i \supset \cdots$$

But these sequences of inclusions are coupled by the fact that, for any integer $i \geq 0$,

$$\dim \text{Ker}(A - \lambda_0 I)^i + \dim \text{Im}(A - \lambda_0 I)^i = n$$

Consequently, *if p is the least integer for which* $\text{Ker}(A - \lambda_0 I)^{p+1} = \text{Ker}(A - \lambda_0 I)^p$, *it is also the least integer which* $\text{Im}(A - \lambda_0 I)^{p+1} = \text{Im}(a - \lambda_0 I)^p$.

Proposition 2.1.1

The root subspace $\mathcal{R}_{\lambda_0}(A)$ contains the vectors from any Jordan chain of A corresponding to λ_0.

Proof. Let $x_0, \ldots, x_k$ be a Jordan chain of A corresponding to λ_0. Then

$$(A - \lambda_0 I)^{k+1} x_k = (A - \lambda_0 I)^k \cdot (A - \lambda_0 I) x_k = (A - \lambda_0 I)^k x_{k-1}$$
$$= (A - \lambda_0 I)^{k-1} x_{k-2} = \cdots = (A - \lambda_0 I) x_0 = 0$$

Hence all the vectors x_i $(i = 0, \ldots, k)$ belong to $\mathcal{R}_{\lambda_0}(A)$. $\square$

Let us look at the simplest examples. For

$$A = \begin{bmatrix} \lambda_0 & 1 & \cdots & 0 \\ 0 & \lambda_0 & \ddots & \vdots \\ \vdots & & \ddots & 1 \\ 0 & & \cdots & \lambda_0 \end{bmatrix}$$

as well as for $A = \lambda_0 I$, the only eigenvalue is λ_0, and the corresponding root subspace $\mathcal{R}_{\lambda_0}(A)$ is the whole of $\mathbb{C}^n$. If

$$A = \mathrm{diag}[\lambda_1, \lambda_2, \ldots, \lambda_n], \qquad \lambda_i \neq \lambda_j \quad \text{for} \quad i \neq j$$

then the root subspace $\mathcal{R}_{\lambda_i}(A)$ is one-dimensional and is spanned by e_i for $i = 1, 2, \ldots, n$.

Later, we also use the following fact: if $A, S \colon \mathbb{C}^n \to \mathbb{C}^n$ are transformations with S invertible, then

$$\mathcal{R}_{\lambda_0}(SAS)^{-1} = S[\mathcal{R}_{\lambda_0}(A)] \tag{2.1.2}$$

for every eigenvalue λ_0 of A. An analogous property holds also for every member of the chain (2.1.1). The proof of equation (2.1.2) follows the same lines as the proof of Proposition 1.4.3.

The following property of root subspaces is crucial.

Theorem 2.1.2

Let $\lambda_1, \ldots, \lambda_r$ be all the different eigenvalues of a transformation $A \colon \mathbb{C}^n \to \mathbb{C}^n$. Then $\mathbb{C}^n$ decomposes into the direct sum

$$\mathbb{C}^n = \mathcal{R}_{\lambda_1}(A) \dotplus \cdots \dotplus \mathcal{R}_{\lambda_r}(A)$$

We need some preparations to prove this theorem.

Lemma 2.1.3

For every eigenvalue λ_0 of A, the restriction $A_{|\mathcal{R}_{\lambda_0}(A)}$ has the sole eigenvalue λ_0.

Proof. Let $B = A_{|\mathcal{R}_{\lambda_0}(A)}$. We shall show that for every $\lambda_1 \neq \lambda_0$ the transformation $\lambda_1 I - B$ on $\mathcal{R}_{\lambda_0}(A)$ is invertible. Let q be an integer such that

$$\mathcal{R}_{\lambda_0}(A) = \mathrm{Ker}(\lambda_0 I - A)^q$$

Then clearly

$$(\lambda_1 - \lambda_0)^q I = (\lambda_1 - \lambda_0)^q I - (B - \lambda_0 I)^q \qquad (2.1.3)$$

Since this implies that

$$(\lambda_1 - \lambda_0)^q I = (\lambda_1 I - B)$$
$$\times ((\lambda_1 - \lambda_0)^{q-1} I + (\lambda_1 - \lambda_0)^{q-2}(B - \lambda_0 I) + \cdots + (B - \lambda_0 I)^{q-1})$$

and since $\lambda_1 \neq \lambda_0$, the invertibility of $\lambda_1 I - B$ follows. $\square$

Lemma 2.1.4

Given a transformation $A: \mathbb{C}^n \to \mathbb{C}^n$ with an eigenvalue λ_0, let q be a positive integer for which

$$\mathrm{Ker}(A - \lambda_0 I)^q = \mathcal{R}_{\lambda_0}(A) \qquad (2.1.4)$$

Then the subspaces $\mathrm{Ker}(A - \lambda_0 I)^q$ and $\mathrm{Im}(A - \lambda_0 I)^q$ are direct complements to each other in $\mathbb{C}^n$.

Proof. Since

$$\dim \mathrm{Ker}(A - \lambda_0 I)^q + \dim \mathrm{Im}(A - \lambda_0 I)^q = n$$

we have only to check that

$$\mathrm{Ker}(A - \lambda_0 I)^q \cap \mathrm{Im}(A - \lambda_0 I)^q = \{0\} \qquad (2.1.5)$$

Arguing by contradiction, assume that there is an $x \neq 0$ in the left-hand side of equation (2.1.5). Then $x = (A - \lambda_0 I)^q y$ for some y. On the other hand, for some integer $r \geq 1$ we have

$$(A - \lambda_0 I)^r x = 0, \quad \text{and} \quad (A - \lambda_0 I)^{r-1} x \neq 0$$

It follows that

$$(A - \lambda_0 I)^{q+r} y = 0, \quad \text{and} \quad (A - \lambda_0 I)^{q+r-1} y \neq 0$$

Hence

$$\text{Ker}(A - \lambda_0 I)^{q+r} \neq \text{Ker}(A - \lambda_0 I)^{q+r-1}$$

a contradiction with (2.1.4) and the definition of a root subspace. $\square$

Proof of Theorem 2.1.2 Let λ_1 be an eigenvalue of A. Lemma 2.1.4 shows that

$$\text{Ker}(A - \lambda_1 I)^q \dotplus \text{Im}(A - \lambda_1 I)^q = \mathbb{C}^n$$

where q is some positive integer for which

$$\text{Ker}(A - \lambda_1 I)^q = \mathcal{R}_{\lambda_1}(A)$$

By Lemma 2.1.3, the restriction of A to $\text{Ker}(A - \lambda_1 I)^q$ has the sole eigenvalue λ_1. On the other hand, λ_1 is not an eigenvalue of the restriction of A to $\text{Im}(A - \lambda_0 I)^q$.

To see this, observe that we also have

$$\text{Im}(A - \lambda_1 I)^{q+1} = \text{Im}(A - \lambda_1 I)^q$$

Hence $A - \lambda_1 I$ maps $\text{Im}(A - \lambda_1 I)^q$ onto itself. It follows that λ_1 is not an eigenvalue of the restriction of A to the A-invariant subspace $\text{Im}(A - \lambda_1 I)^q$.

So the restrictions of A to the subspaces $\text{Ker}(A - \lambda_1 I)^q = \mathcal{R}_{\lambda_1}(A)$ and $\mathcal{L} \stackrel{\text{def}}{=} \text{Im}(A - \lambda_1 I)^q$ have no common eigenvalues. This property is easily seen to imply that, for any eigenvalue λ_2 of $A_{|\mathcal{L}}$

$$\mathcal{R}_{\lambda_2}(A) = \mathcal{R}_{\lambda_2}(A_{|\mathcal{L}})$$

So we can repeat the previous argument with A replaced by $A_{|\mathcal{L}}$ and with λ_1 replaced by an eigenvalue λ_2 of $A_{|\mathcal{L}}$, to show that

$$\mathcal{R}_{\lambda_1}(A) \dotplus \mathcal{R}_{\lambda_2}(A) \dotplus \mathcal{M} = \mathbb{C}^n$$

for some A-invariant subspace $\mathcal{M}$ such that λ_1 and λ_2 are not eigenvalues of $A_{|\mathcal{M}}$. Continuing this process, we eventually prove Theorem 2.1.2. $\square$

Another approach to the proof of Theorem 2.1.2 is based on the fact that if $q_1(\lambda), \ldots, q_r(\lambda)$ are polynomials (with complex coefficients) with no common zeros, there exist polynomails $p_1(\lambda), \ldots, p_r(\lambda)$ such that

$$p_1(\lambda)q_1(\lambda) + \cdots + p_r(\lambda)q_r(\lambda) \equiv 1 \qquad (2.1.6)$$

(This is easily proved by induction on r, using the Euclidean algorithm for the case $r = 2$.) Now let the characteristic polynomial $\varphi_A(\lambda) = \det(\lambda I - A)$ be factorized in the form

$$\varphi_A(\lambda) = \prod_{i=1}^{r} (\lambda - \lambda_i)^{\nu_i}$$

where $\lambda_1, \ldots, \lambda_r$ are different complex numbers (and are, of course, just the eigenvalues of A) and $\nu_1, \ldots, \nu_r$ are positive integers. Define

$$q_j(\lambda) = \prod_{\substack{i=1 \\ i \neq j}}^{r} (\lambda - \lambda_i)^{\nu_i}$$

for $j = 1, \ldots, r$. Using the fact that $\varphi_A(A) = 0$ (the Cayley–Hamilton theorem) one verifies that actually

$$\mathcal{R}_{\lambda_j}(A) = \operatorname{Im} q_j(A) \qquad (2.1.7)$$

for $j = 1, \ldots, r$. Finally, take advantage of the existence of polynomials $p_1(\lambda), \ldots, p_r(\lambda)$ such that equality (2.1.6) holds, and use equation (2.1.7), to prove Theorem 2.1.2. This approach can be used to prove results analogous to Theorem 2.1.2 for matrices over fields other than $\mathbb{C}$.

Now let $\mathcal{M}$ be an A-invaraint subspace. Consider the restriction $A|_{\mathcal{M}}$ as a linear transformation from $\mathcal{M}$ into $\mathcal{M}$, and note that

$$\mathcal{R}_{\lambda_0}(A|_{\mathcal{M}}) = \{x \in \mathcal{M} \mid (A|_{\mathcal{M}} - \lambda_0 I)^q x = 0\} \quad \text{for some} \quad q \geq 1\}$$

$$= \mathcal{M} \cap \mathcal{R}_{\lambda_0}(A)$$

for every λ_0 that is an eigenvalue of $A|_{\mathcal{M}}$. If λ_0 is an eigenvalue of A but not an eigenvalue of $A|_{\mathcal{M}}$, then $\mathcal{R}_{\lambda_0}(A|_{\mathcal{M}}) = \{0\}$; but also $\mathcal{M} \cap \mathcal{R}_{\lambda_0}(A) = \{0\}$. So the equality $\mathcal{R}_{\lambda_0}(A|_{\mathcal{M}}) = \mathcal{M} \cap \mathcal{R}_{\lambda_0}(A)$ holds for any $\lambda_0 \in \sigma(A)$. Applying Theorem 2.1.2 for the linear transformation $A|_{\mathcal{M}}$ and using the above remark, we obtain the following result.

Theorem 2.1.5

Let $A: \mathbb{C}^n \to \mathbb{C}^n$ be a transformation, and let $\mathcal{M}$ be an A-invariant subspace. Then $\mathcal{M}$ decomposes into a direct sum

$$\mathcal{M} = \mathcal{M} \cap \mathcal{R}_{\lambda_1}(A) \dotplus \cdots \dotplus \mathcal{M} \cap \mathcal{R}_{\lambda_r}(A)$$

where $\lambda_1, \ldots, \lambda_r$ are all the different eigenvalues of A.

Note that Theorem 2.1.2 is actually the particular case of Theorem 2.1.5 with $\mathcal{M} = \mathbb{C}^n$. We consider now some examples in which Theorem 2.1.5 allows us to find all invariant subspaces of a given linear transformation.

EXAMPLE 2.1.1.   Let $A = \mathrm{diag}[\lambda_1, \lambda_2, \ldots, \lambda_n]$ where $\lambda_1, \ldots, \lambda_n$ are different complex numbers (as in Example 1.1.3). Then $\sigma(A) = \{\lambda_1, \ldots, \lambda_n\}$, and

$$\mathcal{R}_{\lambda_i}(A) = \mathrm{Span}\{e_i\}, \qquad i = 1, 2, \ldots, n$$

By Theorem 2.1.5, any A-invariant subspace $\mathcal{M}$ is a direct sum

$$\mathcal{M} = (\mathcal{M} \cap \mathrm{Span}\{e_1\}) \dotplus \cdots \dotplus (\mathcal{M} \cap \mathrm{Span}\{e_n\})$$

As $\mathcal{M} \cap \mathrm{Span}\{e_i\}$ is either $\{0\}$ or $\mathrm{Span}\{e_i\}$, it follows that any A-invariant subspace is of the form

$$\mathcal{M} = \mathrm{Span}\{e_{i_1}\} \dotplus \cdots \dotplus \mathrm{Span}\{e_{i_p}\} = \mathrm{Span}\{e_{i_1}, \ldots, e_{i_p}\}$$

for some indices $1 \le i_1 < i_2 < \cdots < i_p \le n$. This fact was stated without proof in Example 1.1.3. $\square$

EXAMPLE 2.1.2.   Let

$$A = \begin{bmatrix} \lambda_1 & 1 & 0 & 0 \\ 0 & \lambda_1 & 0 & 0 \\ 0 & 0 & \lambda_2 & 0 \\ 0 & 0 & 0 & \lambda_2 \end{bmatrix} : \mathbb{C}^4 \to \mathbb{C}^4$$

where λ_1 and λ_2 are different complex numbers. The matrix A has the eigenvalues λ_1 and λ_2. Further,

$$A - \lambda_1 I = \begin{bmatrix} 0 & 1 & 0 & 0 \\ 0 & 0 & 0 & 0 \\ 0 & 0 & \lambda_2 - \lambda_1 & 0 \\ 0 & 0 & 0 & \lambda_2 - \lambda_1 \end{bmatrix}$$

and thus

$$\mathrm{Ker}(A - \lambda_1 I)^j = \begin{cases} \mathrm{Span}\{e_1\}, & \text{if} \quad j = 1 \\ \mathrm{Span}\{e_1, e_2\}, & \text{if} \quad j > 1 \end{cases}$$

So $\mathcal{R}_{\lambda_1}(A) = \mathrm{Span}\{e_1, e_2\}$. For the eigenvalue λ_2 we have $\mathcal{R}_{\lambda_2}(A) = \mathrm{Span}\{e_3, e_4\}$.

We see (as Theorem 2.1.2 leads us to expect) that $\mathcal{C}^n$ is a direct (even orthogonal) sum of $R_{\lambda_1}(A)$ and $\mathscr{R}_{\lambda_2}(A)$. Let $\mathcal{M}$ be any A-invariant subspace. By Theorem 2.1.5, we obtain

$$\mathcal{M} = \mathcal{M} \cap \operatorname{Span}\{e_1, e_2\} \dotplus \mathcal{M} \cap \operatorname{Span}\{e_3, e_4\}$$

It is easily seen (cf. Example 1.1.1) that the only A-invariant subspaces in $\operatorname{Span}\{e_1, e_2\}$ are $\{0\}$, $\operatorname{Span}\{e_1\}$, and $\operatorname{Span}\{e_1, e_2\}$. On the other hand, any subspace in $\operatorname{Span}\{e_3, e_4\}$ is A invariant.

One can easily describe all subspaces in $\operatorname{Span}\{e_3, e_4\}$ as follows: $\{0\}$; the one-dimensional subspaces $\operatorname{Span}\{e_3 + \alpha e_4\}$, where $\alpha \in \mathcal{C}$ is fixed for each particular subspace; the one-dimensional subspace $\operatorname{Span}\{e_4\}$; and $\operatorname{Span}\{e_3, e_4\}$. Finally, the following is a complete list of A-invariant subspaces:

$\{0\}$, $\operatorname{Span}\{e_1\}$, $\operatorname{Span}\{e_1, e_2\}$
$\operatorname{Span}\{e_3 + \alpha e_4\}$ for a fixed $\alpha \in \mathcal{C}$
$\operatorname{Span}\{e_1, e_3 + \alpha e_4\}$ for a fixed $\alpha \in \mathcal{C}$
$\operatorname{Span}\{e_1, e_2, e_3 + \alpha e_4\}$ for a fixed $\alpha \in \mathcal{C}$
$\operatorname{Span}\{e_4\}$, $\operatorname{Span}\{e_1, e_4\}$, $\operatorname{Span}\{e_1, e_2, e_4\}$
$\operatorname{Span}\{e_3, e_4\}$, $\operatorname{Span}\{e_1, e_3, e_4\}$, $\mathcal{C}^4$. $\square$

2.2   THE JORDAN FORM AND PARTIAL MULTIPLICITIES

Let A be an $n \times n$ matrix. In this section we state one of the most important results in linear algebra—the canonical form of a matrix A under similarity transformations $A \rightarrow S^{-1}AS$, where S is an invertible $n \times n$ matrix.

We start with some notations. The Jordan block of size $k \times k$ with eigenvalue λ_0 is the matrix

$$J_k(\lambda_0) = \begin{bmatrix} \lambda_0 & 1 & 0 & \cdots & 0 \\ 0 & \lambda_0 & 1 & & \vdots \\ \vdots & & & & 0 \\ \vdots & & & & 1 \\ 0 & 0 & & \cdots & \lambda_0 \end{bmatrix}$$

Clearly, $\det(\lambda I - J_k(\lambda_0)) = (\lambda - \lambda_0)^k$, so λ_0 is the only eigenvalue of $J_k(\lambda_0)$. Further

$$\lambda_0 I - J_k(\lambda_0) = \begin{bmatrix} 0 & -1 & 0 & \cdots & 0 \\ 0 & 0 & -1 & \cdots & 0 \\ \vdots & \vdots & & & \vdots \\ & & & & -1 \\ 0 & 0 & 0 & \cdots & 0 \end{bmatrix}$$

so the only eigenvector of $J_k(\lambda_0)$ (up to multiplication by a nonzero complex number) is e_1. The invariant subspaces of $J_k(\lambda_0)$ were described in Example 1.1.1; they form a complete chain of subspaces in $\mathbb{C}^k$:

$$\text{Span}\{e_1\} \subset \text{Span}\{e_1, e_2\} \subset \cdots \subset \text{Span}\{e_1, e_2, \ldots, e_{k-1}\} \subset \mathbb{C}^k$$

It turns out that a similarity transformation can always be found transforming a matrix into a direct sum of Jordan blocks.

Theorem 2.2.1

Let A be an $n \times n$ (complex) matrix. Then there exists an invertible matrix S such that $S^{-1}AS$ is a direct sum of Jordan blocks:

$$S^{-1}AS = J_{k_1}(\lambda_1) \oplus \cdots \oplus J_{k_p}(\lambda_p) \tag{2.2.1}$$

The Jordan blocks $J_{k_j}(\lambda_j)$ in the representation (2.2.1) are uniquely determined by the matrix A (up to permutation) and do not depend on the choice of S.

Since the eigenvalues of a matrix are invariant under similarity, it is clear that the numbers $\lambda_1, \ldots, \lambda_p$ are the eigenvalues of A. Note that they are not necessarily distinct.

We stress that this result holds only for complex matrices. For real matrices there is also a canonical form under similarity with a real similarity matrix. This canonical form is dealt with in Chapter 12.

The right-hand side of equality (2.2.1) is called a *Jordan form* of the matrix (or the linear transformation) A. For a given eigenvalue λ_0 of A, let $J_{k_{i_1}}(\lambda_{i_1}), \ldots, J_{k_{i_m}}(\lambda_{i_m})$ be all the Jordan blocks in the Jordan form of A for which $\lambda_{i_q} = \lambda_0$, $q = 1, \ldots, m$. The positive integer m is called the *geometric multiplicity* of λ_0 as an eigenvalue of A, and the integers $k_{i_1}, \ldots, k_{i_m}$ are called the *partial multiplicities* of λ_0. So the number of partial multiplicities of λ_0 as an eigenvalue of A coincides with the geometric multiplicity of λ_0. In view of Theorem 2.2.1, the geometric multiplicity and the partial multiplicities depend on A and λ_0 only and do not depend on the choice of the invertible matrix S for which (2.2.1) holds. The sum $k_{i_1} + \cdots + k_{i_m}$ of the partial multiplicities of λ_0 is called the *algebraic multiplicity* of λ_0 (as an eigenvalue of A). Obviously, the algebraic multiplicity of λ_0 is not less than its geometric multiplicity.

The following property of the partial multiplicities will be useful in the sequel.

Corollary 2.2.2

If A_1 and A_2 are $n_1 \times n_1$ and $n_2 \times n_2$ matrices with the partial multiplicities $k_1(A_1), \ldots, k_{m_1}(A_1)$ and $k_1(A_2), \ldots, k_{m_2}(A_2)$ of A_1 and A_2, respectively,

all corresponding to the common eigenvalue λ_0, *then* $k_1(A_1), \ldots, k_{m_1}(A_1),$
$k_1(A_2), \ldots, k_{m_2}(A_2)$ *are the partial multiplicities of the matrix*

$$\begin{bmatrix} A_1 & 0 \\ 0 & A_2 \end{bmatrix}$$

corresponding to λ_0. *In particular, the geometric* (*resp. algebraic*) *multiplicity of*

$$\begin{bmatrix} A_1 & 0 \\ 0 & A_2 \end{bmatrix}$$

at λ_0 *is the sum of the algebraic* (*resp. geometric*) *multiplicities of* A_1 *and* A_2
at λ_0.

The proof of this corollary is immediate if one observes that the Jordan form of

$$\begin{bmatrix} A_1 & 0 \\ 0 & A_2 \end{bmatrix}$$

can be obtained as a direct sum of the Jordan forms of A_1 and A_2.

We also need the following property of partial multiplicities.

Corollary 2.2.3

The partial multiplicities of A *at* λ_0 *coincide with the partial multiplicities of the conjugate transpose matrix* A^* *at* $\bar{\lambda}_0$.

Proof. Write $A = SJS^{-1}$, where J is the Jordan form of A and S is a nonsingular matrix. Then $A^* = S^{-1*}J^*S^*$. Now the conjugate transpose J^* of the matrix J is similar to the matrix $\bar{J}$ that is obtained from J by replacing each entry by its complex conjugate. Indeed, if we define the permutation ("rotation") matrix R with elements r_{ij} defined in terms of the Kronecker delta by $r_{ij} = \delta_{i,n+1-j}$, then it is easily verified that $R^{-1} = R$ and

$$RJ_k(\lambda)^*R = J_k(\bar{\lambda})$$

Hence $\bar{J}$ is the Jordan form of A^*, and Corollary 2.2.3 follows from the definition of partial multiplicities. $\quad\square$

To describe the result of Theorem 2.2.1 in terms of linear transformations, let us introduce the following definition. An A-invariant subspace $\mathcal{M}$ is called a *Jordan subspace* corresponding to the eigenvalue λ_0 of A if $\mathcal{M}$ is spanned by the vectors of some Jordan chain of A corresponding to λ_0.

Theorem 2.2.4

Let $A: \mathbb{C}^n \to \mathbb{C}^n$ be a linear transformation. Then there exists a direct sum decomposition

$$\mathbb{C}^n = \mathcal{M}_1 \dotplus \cdots \dotplus \mathcal{M}_p \tag{2.2.2}$$

where $\mathcal{M}_i$ is a Jordan subspace of A corresponding to an eigenvalue λ_i (here $\lambda_1, \ldots, \lambda_p$ are not necessarily different).

If $\mathbb{C}^n = \mathcal{N}_1 \dotplus \cdots \dotplus \mathcal{N}_q$ is another direct sum decomposition with Jordan subspaces $\mathcal{N}_i$ corresponding to eigenvalues μ_i, $i = 1, \ldots, q$, then $q = p$, and (possibly after a permutation of $\mathcal{N}_1, \ldots, \mathcal{N}_q$) dim $\mathcal{M}_i = $ dim $\mathcal{N}_i$ and $\lambda_i = \mu_i$ for $i = 1, \ldots, q$.

Note that in general the decomposition (2.2.2) is not unique. For example, if $A = I$, then one can take $\mathcal{M}_i = \text{Span}\{x_i\}$, where $x_1, \ldots, x_n$ is any basis in $\mathbb{C}^n$.

Theorem 2.2.1 follows easily from Theorem 2.2.2 and vice versa. Indeed, let S be as in Theorem 2.2.1. Then put

$$\mathcal{M}_1 = S(\text{Span}\{e_1, \ldots, e_{k_1}\})$$

$$\mathcal{M}_2 = S(\text{Span}\{e_{k_1+1}, e_{k_1+2}, \ldots, e_{k_1+k_2}\})$$

$$\vdots$$

$$\mathcal{M}_p = S(\text{Span}\{e_{k_1+\cdots+k_{p-1}+1}, \ldots, e_{k_1+\cdots+k_p}\})$$

to satisfy equality (2.2.2).

Conversely, if $\mathcal{M}_i$ are as in (2.2.2), choose a basis $x_1^{(i)}, \ldots, x_{k_i}^{(i)}$ in $\mathcal{M}_i$ whose vectors form a Jordan chain for A. Then put

$$S = [x_1^{(1)} x_2^{(1)} \cdots x_{k_1}^{(1)} x_1^{(2)} \cdots x_{k_2}^{(2)} \cdots x_1^{(p)} \cdots x_{k_p}^{(p)}]$$

The direct sum decomposition (2.2.2) ensures that S is an $n \times n$ nonsingular matrix, and the definition of a Jordan chain ensures that $S^{-1}AS$ has the form (2.2.1).

Theorem 2.2.1 (or Theorem 2.2.4) is proved in the next section. Note that because of Theorem 2.1.2 one has to prove Theorem 2.2.1 only for the case when $\mathcal{R}_{\lambda_0}(A) = \mathbb{C}^n$, that is, A has only one eigenvalue λ_0. In this sense the property of root subspaces described in Theorem 2.1.2 is the first step toward a proof of the Jordan form.

In view of Proposition 1.4.2, there are many cases in which the Jordan form allows us to reduce the consideration of invariant subspaces of a general linear transformation to the consideration of invariant subspaces of a linear transformation that is given by the Jordan normal form in the standard orthonormal basis. This reduction is used many times in the sequel.

As a first example of such a reduction we note the following simple fact.

Proposition 2.2.5

Let $A: \mathbb{C}^n \to \mathbb{C}^n$ be a linear transformation. Then the geometric multiplicity of any $\lambda_0 \in \sigma(A)$ coincides with $\dim \operatorname{Ker}(A - \lambda_0 I)$, and the algebraic multiplicity of λ_0 coincides with the dimension of $\mathscr{R}_{\lambda_0}(A)$, the root subspace of λ_0 [i.e., with the dimension of $\operatorname{Ker}(A - \lambda_0 I)^n$].

Proof. By (2.1.2) and Theorem 2.2.1 we can assume without loss of generality that

$$A = J_{k_1}(\lambda_1) \oplus \cdots \oplus J_{k_p}(\lambda_p)$$

Then for any $\lambda_0 \in \mathbb{C}$ we have

$$A - \lambda_0 I = J_{k_1}(\lambda_1 - \lambda_0) \oplus \cdots \oplus J_{k_p}(\lambda_p - \lambda_0)$$

From the definition of the Jordan block it is easily seen that

$$\operatorname{Ker} J_{k_i}(\lambda_i - \lambda_0) = \begin{cases} \{0\}, & \text{if } \lambda_0 \neq \lambda_i \\ \operatorname{Span}\{e_1\}, & \text{if } \lambda_0 = \lambda_i \end{cases}$$

Hence

$$\dim \operatorname{Ker}(A - \lambda_0 I) = \sum_{j=1}^{p} \dim \operatorname{Ker} J_{k_j}(\lambda_j - \lambda_0)$$

is the number of indices j for which $\lambda_0 = \lambda_j$, and, by definition, this number coincides [in case $\lambda_0 \in \sigma(A)$] with the geometric multiplicity of λ_0.

Similarly

$$\operatorname{Ker}[J_{k_i}(\lambda_i - \lambda_0)]^q$$

$$= \begin{cases} \{0\}, & \text{if } \lambda_0 \neq \lambda_i \\ \operatorname{Span}\{e_1, \ldots, e_q\}, & \text{if } \lambda_0 = \lambda_i \text{ and } q = 1, \ldots, k_i - 1 \\ \mathbb{C}^{k_i}, & \text{if } \lambda_0 = \lambda_i \text{ and } q \geq k_i \end{cases}$$

So for $q = 1, 2, \ldots$ and $\lambda_0 \in \mathbb{C}$ we have

$$\dim[\operatorname{Ker}(A - \lambda_0 I)^q] = \sum_{j=1}^{p} \dim[\operatorname{Ker}[J_{k_j}(\lambda_j - \lambda_0)]^q]$$

$$= \sum_{\lambda_0 = \lambda_j} \min(k_j, q) \tag{2.2.3}$$

As $\mathscr{R}_{\lambda_0}(A)$ is the maximal subspace of the type $\operatorname{Ker}(A - \lambda_0 I)^q$, $q = 1, 2, \ldots$, we obtain

$$\dim[\operatorname{Ker} \mathscr{R}_{\lambda_0}(A)] = \sum_{\lambda_0 = \lambda_j} k_j$$

which, by definition, is just the algebraic multiplicity of λ_0. $\square$

Proposition 2.2.5 is actually a particular case of the following general proposition.

Proposition 2.2.6

Let $A: \mathbb{C}^n \to \mathbb{C}^n$ be a transformation with partial multiplicities $k_1, \ldots, k_m$ corresponding to the eigenvalue λ_0 of A. Then

$$\dim[\operatorname{Ker}(A - \lambda_0 I)^q] = \sum_{i=1}^{q} \{j \mid 1 \le j \le m, \quad k_j \ge i\}^{\#}, \quad q = 1, 2, \ldots$$

where $\Omega^{\#}$ represents the number of different elements in a finite set Ω.

Proof. In view of formula (2.2.3) we have only to show that

$$\sum_{i=1}^{m} \min\{k_i, q\} = \sum_{i=1}^{q} \{j \mid 1 \le j \le m, \quad k_j \ge i\}^{\#}, \quad q = 1, 2, \ldots \quad (2.2.4)$$

This equality is certainly true for $q = 1$ (for then both sides are equal to m). Assume that the equality is true for $q - 1$. We have

$$\sum_{i=1}^{m} \min\{k_i, q\} - \sum_{i=1}^{m} \min\{k_i, q-1\} = \sum_{i=1}^{m} [\min\{k_i, q\} - \min\{k_i, q-1\}]$$

$$= \{j \mid 1 \le j \le m, \quad k_j \ge q\}^{\#}$$

Adding the relation

$$\sum_{i=1}^{m} \min\{k_i, q-1\} = \sum_{i=1}^{q-1} \{j \mid 1 \le j \le m, \quad k_j \ge i\}^{\#}$$

(which is just the induction hypothesis) we verify (2.2.4). $\square$

It follows from Proposition 2.2.6 that if

$$\operatorname{Ker}(A - \lambda_0 I)^q = \operatorname{Ker}(A - \lambda_0 I)^{q+1}$$

for some positive integer q, then actually

$$\operatorname{Ker}(A - \lambda_0 I)^q = \operatorname{Ker}(A - \lambda_0 I)^p$$

for all $p \ge q$, that is

$$\operatorname{Ker}(A - \lambda_0 I)^q = \mathscr{R}_{\lambda_0}(A)$$

2.3 PROOF OF THE JORDAN FORM

In this section we prove Theorem 2.2.4. In view of Theorem 2.1.5, it is sufficient to consider $A_{|\mathcal{R}_{\lambda_0}(A)}$, where $\lambda_0 \in \sigma(A)$ is fixed, in place of A. In other words, we can assume that A has only one eigenvalue λ_0, possibly with several partial multiplicities.

Let $\mathcal{S}_j = \mathrm{Ker}(A - \lambda_0 I)^j$, $j = 1, 2, \ldots, m$, where m is chosen so that $\mathcal{S}_m = \mathcal{R}_{\lambda_0}(A)$ but $\mathcal{S}_{m-1} \neq \mathcal{R}_{\lambda_0}(A)$. Note that $\mathcal{S}_1 \subset \mathcal{S}_2 \subset \cdots \subset \mathcal{S}_m$. Let $x_m^{(1)}, \ldots, x_m^{(t_m)}$ be a basis in $\mathcal{S}_m$ modulo $\mathcal{S}_{m-1}$, that is, a linearly independent set in $\mathcal{S}_m$ such that

$$\mathcal{S}_{m-1} \dotplus \mathrm{Span}\{x_m^{(1)}, \ldots, x_m^{(t_m)}\} = \mathcal{S}_m \tag{2.3.1}$$

(the sum here is direct). We claim that the mt_m vectors

$$(A - \lambda_0 I)^k x_m^{(1)}, \ldots, (A - \lambda_0 I)^k x_m^{(t_m)}, \qquad k = 0, \ldots, m-1$$

are linearly independent. Indeed, assume

$$\sum_{k=0}^{m-1} \sum_{i=1}^{t_m} \alpha_{ik} (A - \lambda_0 I)^k x_m^{(i)} = 0, \qquad \alpha_{ik} \in \mathbb{C} \tag{2.3.2}$$

Applying $(a - \lambda_0 I)^{m-1}$ to the left-hand side and using the property that $(a - \lambda_0 I)^m x_m^{(i)} = 0$ for $i = 1, \ldots, t_m$, we find that

$$(A - \lambda_0 I)^{m-1}\left\{\sum_{i=1}^{t_m} \alpha_{i0} x_m^{(i)}\right\} = 0$$

Hence $\sum_{i=1}^{t_m} \alpha_{i0} x_m^{(i)} \in \mathcal{S}_{m-1}$ and because of (2.3.1), $\alpha_{10} = \cdots = \alpha_{t_m 0} = 0$. Applying $(A - \lambda_0 I)^{m-2}$ to the left-hand side of (2.3.2) we show similarly that $\alpha_{11} = \cdots = \alpha_{t_m 1} = 0$, and so on, We put

$$\mathcal{M}_1 = \mathrm{Span}\{(A - \lambda_0 I)^k x_m^{(1)}, \qquad k = 0, \ldots, m-1\}$$

$$\mathcal{M}_2 = \mathrm{Span}\{(A - \lambda_0 I)^k x_m^{(2)}, \qquad k = 0, \ldots, m-1\}$$

$$\vdots$$

$$\mathcal{M}_{t_m} = \mathrm{Span}\{(A - \lambda_0 I)^k x_m^{(t_m)}, \qquad k = 0, \ldots, m-1\}$$

As we have just seen, the sum $\mathcal{M}_1 \dotplus \mathcal{M}_2 \dotplus \cdots \dotplus \mathcal{M}_{t_m}$ is direct.

Consider now the vectors

$$x_{m-1}^{(i)} = (A - \lambda_0 I) x_m^{(i)}, \qquad i = 1, \ldots, t_m$$

We claim that

$$\mathscr{S}_{m-2} \cap \text{Span}\{x_{m-1}^{(1)}, x_{m-1}^{(2)}, \ldots, x_{m-1}^{(t_m)}\} = \{0\} \qquad (2.3.3)$$

Indeed, assume

$$\sum_{i=1}^{t_m} \alpha_i x_{m-1}^{(i)} \in \mathscr{S}_{m-2}, \qquad \alpha_i \in \mathbb{C}$$

Applying $(A - \lambda_0 I)^{m-2}$ to the left-hand side, we get

$$(A - \lambda_0 I)^{m-1} \sum_{i=1}^{t_m} \alpha_i x_m^{(i)} = 0$$

which implies $\alpha_1 = \cdots = \alpha_{t_m} = 0$ in view of equality (2.3.1). So equation (2.3.3) follows.

Assume first that $\mathscr{S}_{m-2} \dotplus \text{Span}\{x_{m-1}^{(1)}, \ldots, x_{m-1}^{(t_m)}\}$ does not coincide with $\mathscr{S}_{m-1}$. Then there exist vectors $x_{m-1}^{(t_m+1)}, \ldots, x_{m-1}^{(t_m+t_{m-1})}$ in $\mathscr{S}_{m-1}$ such that the set $\{x_{m-1}^{(i)}\}_{i=1}^{t_m+t_{m-1}}$ is linearly independent and

$$\mathscr{S}_{m-2} \dotplus \text{Span}\{x_{m-1}^{(1)}, \ldots, x_{m-1}^{(t_m+t_{m-1})}\} = \mathscr{S}_{m-1} \qquad (2.3.4)$$

Applying the previous argument to (2.3.4) as with (2.3.1), we find that the vectors

$$(A - \lambda_0 I)^k x_{m-1}^{(1)}, \ldots, (A - \lambda_0 I)^k x_{m-1}^{(t_m-t_{m-1})}, \qquad k = 0, \ldots, m-2$$

are linearly independent. Now put

$$\mathscr{M}_{t_m+1} = \text{Span}\{(A - \lambda_0 I)^k x_{m-1}^{(t_m+1)}, \qquad k = 0, \ldots, m-2\}$$
$$\vdots$$
$$\mathscr{M}_{t_m+t_{m-1}} = \text{Span}\{(A - \lambda_0 I)^k x_{m-1}^{(t_m+t_{m-1})}, \qquad k = 0, \ldots, m-2\}$$

If it happens that

$$\mathscr{S}_{m-2} \dotplus \text{Span}\{x_{m-1}^{(i)}, \qquad i = 1, \ldots, t_m\} = \mathscr{S}_{m-1}$$

then put formally $t_{m-1} = 0$.

At the next step put

$$x_{m-2}^{(i)} = (A - \lambda_0 I) x_{m-1}^{(i)}, \qquad i = 1, \ldots, t_m + t_{m-1}$$

and show similarly that

$$\mathscr{S}_{m-3} \cap \text{Span}\{x_{m-2}^{(i)}, \qquad i = 1, \ldots, t_m + t_{m-1}\} = \{0\}$$

Assuming that $\mathscr{S}_{m-3} \dotplus \text{Span}\{x_{m-2}^{(i)}, i = 1, \ldots, t_m + t_{m-1}\} \neq \mathscr{S}_{m-2}$, choose

$x_{m-2}^{(i)}$, $i = t_m + t_{m-1} + 1, \ldots, t_m + t_{m-1} + t_{m-2}$ in such a way that the vectors $x_{m-2}^{(i)}$, $i = 1, \ldots, t_m + t_{m-1} + t_{m-2}$ are linearly independent and the linear span of these vectors is a direct complement to $\mathscr{S}_{m-3}$ in $\mathscr{S}_{m-2}$. Then put

$$\mathscr{M}_{t_m+t_{m-1}+j} = \text{Span}\{(A - \lambda_0 I)^k x_{m-2}^{(t_m+t_{m-1}+j)}, \qquad k = 0, \ldots, m-3\}$$

for $j = 1, \ldots, t_{m-2}$. We continue this process of construction of $\mathscr{M}_i$, $i = 1, \ldots, p$, where $p = t_m + t_{m-1} + \cdots + t_1$. The construction shows that each $\mathscr{M}_i$ is a Jordan subspace of A and the sume $\mathscr{M}_1 \dotplus \cdots \dotplus \mathscr{M}_p$ is a direct sum. Also

$$\mathscr{M}_1 \dotplus \cdots \dotplus \mathscr{M}_p = \mathscr{R}_{\lambda_0}(A) = \mathbb{C}^n$$

because of our assumption that $\sigma(A) = \{\sigma_0\}$. Hence (2.2.2) holds.

Let us prove the uniqueness part of Theorem 2.2.4. Assume that (2.2.2) holds, and let $\hat{\lambda}_1, \ldots, \hat{\lambda}_k$ be all the different eigenvalues of A. Denoting by E_j the set of all integers i, $1 \le i \le p$, such that $\lambda_i = \hat{\lambda}_j$, we have for $t = 0, 1, 2, \ldots$:

$$\dim \text{Ker}(A_{|\mathscr{M}_i} - \hat{\lambda}_j I)^t = \begin{cases} 0, & \text{if } i \notin E_j \\ \min(t, \dim \mathscr{M}_i), & \text{if } i \in E_j \end{cases}$$

Consequently

$$\dim \text{Ker}(A - \hat{\lambda}_j I)^t = \sum_{i \in E_j} \min(t, \dim \mathscr{M}_i) \qquad (2.3.5)$$

In particular (taking $t = 1$), the number of elements in E_j coincides with $\dim \text{Ker}(A - \hat{\lambda}_j I)$. This proves that for a direct sum decomposition $\mathbb{C}^n = \mathscr{N}_1 \dotplus \cdots \dotplus \mathscr{N}_q$ as in Theorem 2.2.4 we have $q = p$ and for a fixed j the number of μ_i values that are equal to $\hat{\lambda}_j$ coincides with the numbers of λ_i values that are equal to $\hat{\lambda}_j$. Hence we can assume $\mu_i = \lambda_i$, $i = 1, \ldots, p$. Further, (2.3.5) implies that (for fixed $\hat{\lambda}_j$) the number

$$\dim \text{Ker}(A - \hat{\lambda}_j I)^t - \dim \text{Ker}(A - \hat{\lambda}_j I)^{t-1}$$

coincides with the number of indices $i \in E_j$ such that $\dim \mathscr{M}_i \ge t$ ($t = 1, 2, \ldots$), and thus it also coincides with the number of indices $i \in E_j$ such that $\dim \mathscr{N}_i \ge t$. This implies the uniqueness part of Theorem 2.2.4.

2.4   SPECTRAL SUBSPACES

Let $A: \mathbb{C}^n \to \mathbb{C}^n$ be a transformation. A subspace $\mathscr{M} \subset \mathbb{C}^n$ is called a *spectral subspace* for A if $\mathscr{M}$ is a sum of root subspaces for A. The zero subspace is also considered spectral. Since root subspaces are A invariant, a spectral subspace for A is A invariant. It is easily seen that the total number of spectral subspaces for A is 2^r, where r is the number of distinct eigenvalues of A.

By Theorem 2.1.5, for every A invariant subspace $\mathcal{M}$, we have

$$\mathcal{M} = [\mathcal{M} \cap \mathcal{R}_{\lambda_1}(A)] \dotplus \cdots \dotplus [\mathcal{M} \cap \mathcal{R}_{\lambda_p}(A)] \tag{2.4.1}$$

where $\lambda_1, \ldots, \lambda_p$ are all the distinct eigenvalues of A. From this formula it is clear that $\mathcal{M}$ *is spectral if and only if for every* λ_j *either* $\mathcal{M} \cap \mathcal{R}_{\lambda_j}(A) = \{0\}$ *or the inclusion* $\mathcal{R}_{\lambda_j}(A) \subset \mathcal{M}$ *holds*. Another consequence of formula (2.4.1) is that, for any nonzero spectral subspace $\mathcal{M}$ of A,

$$\mathcal{M} = \mathcal{R}_{\mu_1}(A) \dotplus \cdots \dotplus \mathcal{R}_{\mu_s}(A)$$

where $\mu_1, \ldots, \mu_s$ are all the distinct eigenvalues of the restriction $A|_{\mathcal{M}}$.

A useful characterization of spectral subspaces is given by their maximality property.

Proposition 2.4.1

An A-invariant subspace $\mathcal{M} \neq \{0\}$ is spectral if and only if any A-invariant subspace $\mathcal{L}$ with the property $\sigma(A|_{\mathcal{L}}) \subset \sigma(A|_{\mathcal{M}})$ is contained in $\mathcal{M}$.

Proof. Assume that $\mathcal{M}$ is not spectral so that, in particular, $\{0\} \neq \mathcal{M} \cap \mathcal{R}_{\lambda_0}(A) \neq \mathcal{R}_{\lambda_0}(A)$ for some $\lambda_0 \in \sigma(A)$. Define the A-invariant subspace $\mathcal{L}$ by the equalities

$$\mathcal{L} \cap \mathcal{R}_{\lambda_0}(A) = \mathcal{R}_{\lambda_0}(A), \qquad \mathcal{L} \cap \mathcal{R}_{\lambda_i}(A) = \mathcal{M} \cap \mathcal{R}_{\lambda_i}(A)$$

for all eigenvalues λ_i of A different from λ_0. Obviously, $\sigma(A|_{\mathcal{L}}) = \sigma(A|_{\mathcal{M}})$ but $\mathcal{L}$ is not contained in $\mathcal{M}$ (actually, $\mathcal{L}$ contains $\mathcal{M}$ properly).

On the other hand, assume that $\mathcal{M}$ is spectral. If $\mathcal{L}$ is A invariant with $\sigma(A|_{\mathcal{L}}) \subset \sigma(A|_{\mathcal{M}})$, then the equality

$$\mathcal{L} = [\mathcal{L} \cap \mathcal{R}_{\lambda_1}(A)] \dotplus \cdots \dotplus [\mathcal{L} \cap \mathcal{R}_{\lambda_p}(A)] \tag{2.4.2}$$

(where $\lambda_1, \ldots, \lambda_p$ are the distinct eigenvalues of A) implies that $\mathcal{L} \cap \mathcal{R}_{\lambda_0}(A) = \emptyset$ for every $\lambda_0 \in \sigma(A)$ not belonging to the spectrum of $A|_{\mathcal{M}}$. It follows then from (2.4.2) that

$$\mathcal{L} \subset \mathcal{R}_{\mu_1}(A) \dotplus \cdots \dotplus \mathcal{R}_{\mu_s}(A) \tag{2.4.3}$$

where $\mu_1, \ldots, \mu_s$ are the distinct eigenvalues of $A|_{\mathcal{M}}$. As the right-hand side of (2.4.3) is equal to $\mathcal{M}$, the inclusion $\mathcal{L} \subset \mathcal{M}$ follows. $\square$

Another characterization of spectral subspaces can be given in terms of direct complements.

Theorem 2.4.2

The following statements are equivalent for an A-invariant subspace $\mathcal{M}$: (a) $\mathcal{M}$ is spectral for A; (b) there exists a direct complement $\mathcal{N}$ to $\mathcal{M}$ such that $\mathcal{N}$ is A invariant and

$$\sigma(A|_{\mathcal{M}}) \cap \sigma(A|_{\mathcal{N}}) = \emptyset \qquad (2.4.4)$$

(c) there exists a unique A-invariant direct complement $\mathcal{N}$ to $\mathcal{M}$; (d) for any A-invariant subspace $\mathcal{L}$ that contains $\mathcal{M}$ properly, $\sigma(A|_{\mathcal{L}})$ contains $\sigma(A|_{\mathcal{M}})$ properly.

To accommodate the cases $\mathcal{M} = \{0\}$ and $\mathcal{M} = \mathcal{C}^n$ in Theorem 2.4.2 we adopt the convention that the spectrum of the restriction of A to the zero subspace is empty.

Proof. The equivalence of (a) and (d) follows immediately from Theorem 2.4.1. By Theorem 2.1.5 (considering each root subspace of A separately) we can assume that $\mathcal{R}_{\lambda_0}(A) = \mathcal{C}^n$, that is, A has the single eigenvalue λ_0. Then the only spectral subspaces of A are $\{0\}$ and $\mathcal{C}^n$. Further, since $\sigma(A|_{\mathcal{L}}) = \{\lambda_0\}$ for every nonzero A-invariant subspace $\mathcal{L}$, equation (2.4.4) implies that either $\sigma(A|_{\mathcal{M}})$ or $\sigma(A|_{\mathcal{N}})$ is empty; in other words, either $\mathcal{M} = \{0\}$ or $\mathcal{N} = \{0\}$. But if the latter case holds, then obviously $\mathcal{M} = \mathcal{C}^n$. Thus $\mathcal{M} = \{0\}$ and $\mathcal{M} = \mathcal{C}^n$ are the only subspaces satisfying (b), and (a) and (b) are equivalent.

Obviously, (a) implies (c). So it remains to prove that (c) implies (a).

Let $\mathcal{M}$ be a nontrivial A-invariant subspace (i.e., different from $\{0\}$ and $\mathcal{C}^n$) that has an A-invariant direct complement $\mathcal{N}$. Then $\mathcal{N}$ is nontrivial as well. We now use the Jordan form (Theorem 2.2.4) for the restriction $A|_{\mathcal{N}}$:

$$\mathcal{N} = \operatorname{Span}\{x_1^{(1)}, \ldots, x_{k_1}^{(1)}\} \dotplus \operatorname{Span}\{x_1^{(2)}, \ldots, x_{k_2}^{(2)}\}$$
$$\dotplus \cdots \dotplus \operatorname{Span}\{x_1^{(q)}, \ldots, x_{k_q}^{(q)}\}$$

where $x_1^{(i)}, \ldots, x_{k_i}^{(i)}$ is a Jordan chain (necessarily with eigenvalue λ_0) of A, $i = 1, \ldots, q$. It is easily seen (cf. Proposition 1.3.4) that the vectors $x_j^{(i)}$, $j = 1, \ldots, k_i$, $i = 1, \ldots, q$ are linearly independent and hence form a basis in $\mathcal{N}$.

We now construct another direct complement for $\mathcal{M}$ that is A invariant. Let $y \ (\neq 0)$ be an eigenvector of A in $\mathcal{M}$, and put

$$\mathcal{N}' = \operatorname{Span}\{x_1^{(1)}, \ldots, x_{k_1-1}^{(1)}, x_{k_1}^{(1)} + y\} \dotplus \operatorname{Span}\{x_1^{(2)}, \ldots, x_{k_2}^{(2)}\}$$
$$\dotplus \cdots \dotplus \operatorname{Span}\{x_1^{(q)}, \ldots, x_{k_q}^{(q)}\} .$$

As $Ay = \lambda_0 y$, one checks easily that $\mathcal{N}'$ is A invariant. Also, $\mathcal{N}' \neq \mathcal{N}$, because otherwise y would belong to $\mathcal{N}$, a contradiction with the direct sum $\mathcal{M} \dotplus \mathcal{N} = \mathbb{C}^n$. We verify that $\mathcal{N}'$ is a direct complement to $\mathcal{M}$. Indeed, observe that the vectors $x_1^{(1)}; \ldots; x_{k_1-1}^{(1)}; x_{k_1}^{(1)} + y; x_j^{(i)}, j = 1, \ldots, k_i, i = 2, \ldots, q$ are linearly independent and hence $\dim \mathcal{N}' = \dim \mathcal{N}$. So we must only check that $\mathcal{M} \cap \mathcal{N}' = \{0\}$. Let

$$z = \sum_{i=2}^{q} \sum_{j=1}^{k_i} \alpha_{ij} x_j^{(i)} + \sum_{j=1}^{k_1-1} \alpha_{ij} x_j^{(1)} + \alpha_{1k_1}(x_{k_1}^{(1)} + y) \in \mathcal{M} \cap \mathcal{N}' \quad (2.4.5)$$

where α_{ij} are complex numbers. The condition $\mathcal{M} \cap \mathcal{N} = \{0\}$ implies

$$z - \alpha_{1k_1} y = 0 \qquad (2.4.6)$$

which in turn implies

$$\sum_{i=1}^{q} \sum_{j=1}^{k_j} \alpha_{ij} x_j^{(i)} = 0$$

and, because of the linear independence of $x_j^{(i)}$, all the coefficients α_{ij} are zeros. In particular, $\alpha_{1k_1} = 0$, and $z = 0$ in view of equation (2.4.6).

We have proved that (when $\sigma(A) = \{\lambda_0\}$) any nontrivial A-invariant subspace either does not have A-invariant direct complements or has at least two of them. This means that (c) implies (a). $\quad \square$

We deduce immediately from Theorem 2.4.2 that the unique A-invariant direct complement $\mathcal{N}$ to a spectral subspace $\mathcal{M}$ is spectral as well: if $\mathcal{M} = \mathcal{R}_{\mu_1}(A) \dotplus \cdots \dotplus \mathcal{R}_{\mu_s}(A)$, then $\mathcal{N} = \mathcal{R}_{\nu_1}(A) \dotplus \cdots \dotplus \mathcal{R}_{\nu_t}(A)$, where $\mu_1, \ldots, \mu_s, \nu_1, \ldots, \nu_t$ is a complete list of all the distinct eigenvalues of A.

We say that the spectral subspace $\mathcal{M}$ for A corresponds to the part Λ of the spectrum of A if $\sigma(A|_{\mathcal{M}}) = \Lambda$. Obviously, there is a unique spectral subspace corresponding to any given subset Λ of $\sigma(A)$ [with the understanding that $\sigma(A|_{\{0\}}) = \emptyset$]. This spectral subspace can easily be described in case A is given by an $n \times n$ matrix in Jordan form as in equation (2.2.1). Indeed, using the notation of that equation, if $\Lambda \subset \sigma(A)$, define the $k_i \times k_i$ matrix K_i by $K_i = I$ if $\lambda_i \in \Lambda$ and $K_i = 0$ if $\lambda_i \notin \Lambda$. Then the subspace

$$\text{Im}(K_1 \oplus \cdots \oplus K_p)$$

is the spectral subspace for A corresponding to Λ. Its only A-invariant direct complement is

$$\text{Im}[(I - K_1) \oplus \cdots \oplus (I - K_p)]$$

We conclude this section with a description of spectral subspaces in terms of contour integrals. (Actually, this description is a particular case of the properties of functions of transformations that are studied in more detail in Section 2.10.) Let Γ be a simple, closed, rectifiable, and positively oriented contour in the complex plane. In fact, for our purposes polygonal contours will suffice. Given an $n \times n$ matrix $B(\lambda) = [b_{ij}(\lambda)]_{i,j=1}^n$, that depends continuously on the variable $\lambda \in \Gamma$ (this means that each entry $b_{ij}(\lambda)$ in $B(\lambda)$ is a continuous function of λ on Γ) the integral

$$\int_\Gamma B(\lambda) \, d\lambda = [c_{ij}]_{i,j=1}^n$$

is defined naturally as the $n \times n$ matrix whose entries are the integrals of the entries of $B(\lambda)$:

$$c_{ij} = \int_\Gamma b_{ij}(\lambda) \, d\lambda \; ; \qquad i, j = 1, \dots, n$$

The same definition of a contour integral applies also for transformations $B(\lambda) : \mathbb{C}^n \to \mathbb{C}^n$ that are continuous functions of λ on Γ. We have only to write $B(\lambda)$ as a matrix $[b_{ij}(\lambda)]_{i,j=1}^n$ in a fixed basis, and then interpret $\int_\Gamma B(\lambda) \, d\lambda$ as a transformation represented by the matrix $[\int_\Gamma b_{ij}(\lambda) \, d\lambda]_{i,j=1}^n$ in the same basis. One checks easily that this defintion is independent of the chosen basis.

Proposition 2.4.3

Let Λ be a subset of $\sigma(A)$ where A is a transformation on $\mathbb{C}^n$, and let Γ be a closed contour having Λ in its interior and $\sigma(A) \smallsetminus \Lambda$ outside Γ. Then the transformation

$$\frac{1}{2\pi i} \int_\Gamma (\lambda I - A)^{-1} \, d\lambda$$

is a projector (known as a *Riesz* projector) *onto the spectral subspace associated with Λ and along the spectral subspace associated with $\sigma(A) \smallsetminus \Lambda$.*

Proof. Using the relation $S(\lambda I - A)^{-1}S^{-1} = (\lambda I - SAS^{-1})^{-1}$, equation (2.1.2), and the Jordan form, we can assume that A is an $n \times n$ matrix given by

$$A = J_{k_1}(\lambda_1) \oplus \cdots \oplus J_{k_p}(\lambda_p)$$

where $J_{k_i}(\lambda_i)$ is the $k_i \times k_i$ Jordan block with λ_i on the main diagonal. One easily verifies that

$$[\lambda I - J_{k_i}(\lambda_i)]^{-1} = \begin{bmatrix} (\lambda - \lambda_i)^{-1} & (\lambda - \lambda_i)^{-2} & \cdots & (\lambda - \lambda_i)^{-k_i} \\ 0 & (\lambda - \lambda_i)^{-1} & & \vdots \\ \vdots & \vdots & & (\lambda - \lambda_i)^{-2} \\ 0 & 0 & \cdots & (\lambda - \lambda_i)^{-1} \end{bmatrix}$$

As a first consequenc of this formula we see immediately that, because $\sigma(A) \cap \Gamma = \emptyset$, $(\lambda I - A)^{-1}$ is indeed continuous on Γ. Further, the Cauchy formula gives

$$\int_\Gamma (\lambda - \lambda_i)^{-m} \, d\lambda = \begin{cases} 2\pi i, & \text{if } m = 1 \text{ and } \lambda_i \text{ is inside } \Gamma \\ 0, & \text{otherwise} \end{cases}$$

Thus

$$\frac{1}{2\pi i} \int_\Gamma (\lambda I - A)^{-1} \, d\lambda = S(K_1 \oplus \cdots \oplus K_p)S^{-1} \qquad (2.4.7)$$

where $K_i = I$ if $\lambda \in \Lambda$ and $K_i = 0$ if $\lambda \notin \Lambda$. Thus the matrix (2.4.7) is indeed a projector with image and kernel as prescribed by the theorem. $\square$

2.5 IRREDUCIBLE INVARIANT SUBSPACES AND UNICELLULAR TRANSFORMATIONS

In this section we use the Jordan form to study irreducible invariant subspaces. An invariant subspace $\mathcal{M}$ of a transformation $A: \mathbb{C}^n \to \mathbb{C}^n$ is called *reducible* if $\mathcal{M}$ can be represented as a direct sum of nonzero A-invariant subspaces $\mathcal{M}_1$ and $\mathcal{M}_2$; otherwise $\mathcal{M}$ is called *irreducible*.

Let us consider some examples.

EXAMPLE 2.5.1. Let A be a Jordan block. Then, as Example 1.1.1 shows, each nonzero A-invariant subspace (including $\mathbb{C}^n$ itself) is irreducible.

EXAMPLE 2.5.2. Let $A = \lambda_0 I$, $\lambda_0 \in \mathbb{C}$. Then an A-invariant subspace is irreducible if and only if it is one-dimensional.

EXAMPLE 2.5.3. Let

$$A = \begin{bmatrix} 0 & 1 & 0 \\ 0 & 0 & 0 \\ 0 & 0 & 0 \end{bmatrix}$$

According to Theorem 2.1.5, the A-invariant subspaces are as follows: $\{0\}$; $\text{Span}\{\alpha e_1 + \beta e_3\}$ for fixed numbers $\alpha, \beta \in \mathbb{C}$ with at least one of them different from zero; $\text{Span}\{e_1, e_2\}$; $\text{Span}\{e_1, e_3\}$; $\mathbb{C}^3$. Among these subspaces $\text{Span}\{e_1, e_3\}$ and $\mathbb{C}^3$ are reducible and the rest are irreducible. $\square$

The following theorem gives various characterizations of irreducible invariant subspaces.

Theorem 2.5.1

The following statements are equivalent for an A-invariant subspace $\mathcal{M}$: (a) $\mathcal{M}$ is irreducible; (b) each A-invariant subspace $\mathcal{N}$ contained in $\mathcal{M}$ is irreducible; (c) $\mathcal{M}$ is Jordan, that is, has a basis consisting of vectors that form a Jordan chain of A; (d) there is a unique eigenvector (up to multiplication by a scalar) of A in $\mathcal{M}$; (e) the lattice of invariant subspaces of $A|_{\mathcal{M}}$ is a chain—that is, for any A-invariant subspaces $\mathcal{L}_1$, $\mathcal{L}_2 \subset \mathcal{M}$ either $\mathcal{L}_1 \subset \mathcal{L}_2$ or $\mathcal{L}_2 \subset \mathcal{L}_1$ holds; (f) every nonzero A-invariant subspace that is contained in $\mathcal{M}$ is Jordan; (g) the spectrum of $A|_{\mathcal{M}}$ is a singleton $\{\lambda_0\}$, and

$$\operatorname{rank}[(A|_{\mathcal{M}} - \lambda_0 I)^i] = \max\{0, (\dim \mathcal{M}) - i\}, \qquad i = 0, 1, \ldots$$

(h) the Jordan form of the linear transformation $A|_{\mathcal{M}}$ consists of a single Jordan block.

Proof. The definition of a Jordan block and the description of its invariant subspaces (Example 1.1.1) show that (h) implies all the other statements in Theorem 2.5.1.

The implications (f) $\rightarrow$ (c) and (b) $\rightarrow$ (a) are obvious. Let us show that (c) $\rightarrow$ (d). Let $x_1, \ldots, x_k$ be a basis in $\mathcal{M}$ such that $Ax_1 = \lambda_0 x_1$; $Ax_2 - \lambda_0 x_2 = x_1$; $\ldots$; $Ax_k - \lambda_0 x_k = x_{k-1}$. The matrix of $A|_{\mathcal{M}}$ in this basis is the $k \times k$ Jordan block with λ_0 on the main diagonal, so the spectrum of $A|_{\mathcal{M}}$ is the singleton $\{\lambda_0\}$. If $x = \Sigma_{i=1}^{k} \alpha_i x_i$ is an eigenvector of A (necessarily corresponding to λ_0), then $(A - \lambda_0 I)x = 0$, which implies $\Sigma_{i=2}^{k} \alpha_i x_{i-1} = 0$. As $x_1, \ldots, x_k$ are linearly independent, $\alpha_2 = \cdots = \alpha_k = 0$, and x is a scalar multiple of x_1. So (d) holds.

If x and y are two eigenvectors of $A|_{\mathcal{M}}$ such that $\operatorname{Span}\{x\} \neq \operatorname{Span}\{y\}$, then for the A-invariant subspaces $\mathcal{L}_1 = \operatorname{Span}\{x\}$ and $\mathcal{L}_2 = \operatorname{Span}\{y\}$ we have $\mathcal{L}_1 \not\subset \mathcal{L}_2$ and $\mathcal{L}_2 \not\subset \mathcal{L}_1$. So (e) implies (d).

It remains, therefore, to show that (d) $\rightarrow$ (h), (a) $\rightarrow$ (h), and (g) $\rightarrow$ (h). To this end we can assume that $A|_{\mathcal{M}}$ is in Jordan form (written as a matrix in a suitable basis in λ_0):

$$A|_{\mathcal{M}} = J_{k_1}(\lambda_1) \oplus \cdots \oplus J_{k_p}(\lambda_p) \tag{2.5.1}$$

If $p > 1$, then e_1 and e_{k_1+1} are two eigenvectors of A in $\mathcal{M}$ that are not scalar multiples of each other; so (d) $\rightarrow$ (h). Further, if $p > 1$, then

$$\mathcal{M} = \operatorname{Span}\{e_1, \ldots, e_{k_1}\} \dotplus \operatorname{Span}\{e_{k_1+1}, \ldots, e_{k_1+k_2+\cdots+k_p}\}$$

is a direct sum of two nonzero A-invariant subspaces. Hence (a) $\rightarrow$ (h).

Finally, assume that (g) holds. Then we have $\lambda_1 = \lambda_2 = \cdots = \lambda_p = \lambda_0$ in equation (2.5.1), and this equation implies

$$\operatorname{rank}(A|_{\mathcal{M}} - \lambda_0 I)^i = \sum_{j=1}^{p} \max\{0, k_j - i\}, \qquad i = 0, 1, 2, \ldots$$

On the other hand, the statement (g) implies that the left-hand side of this equation is also equal to $\max\{0, k_1 + \cdots + k_p - i\}$. In particular (for $i = 1$), we have

$$\sum_{j=1}^{p} (k_j - 1) = k_1 + \cdots + k_p - 1$$

which implies $p = 1$. So (h) holds, and Theorem 2.5.1 is proved. $\square$

Observe that with $\mathcal{M} = \mathbb{C}^n$, Theorem 1.9.3 is just the equivalence (d) $\Leftrightarrow$ (e). Thus the proof of that theorem is now complete.

A transformation $A: \mathbb{C}^n \to \mathbb{C}^n$ is called *unicellular* if the Jordan form of A consists of a single Jordan block. Comparing statements (a) and (h) of Theorem 2.5.1, we obtain another characterization of a unicellular transformation.

Proposition 2.5.2

A transformation $A: \mathbb{C}^n \to \mathbb{C}^n$ is unicellular if and only if the whole space $\mathbb{C}^n$ is irreducible as an A-invariant subspace.

Indeed, rewriting Theorem 2.5.1 for the particular case $\mathcal{M} = \mathbb{C}^n$, one obtains various characterizations of unicellular transformations.

Another important property of a unicellular transformation is the "near" uniqueness of an orthonormal basis in which this transformation has upper triangular form (see Section 1.9).

Theorem 2.5.3

A transformation $A: \mathbb{C}^n \to \mathbb{C}^n$ is unicellular if and only if for any two orthonormal bases $x_1, \ldots, x_n$ and $y_1, \ldots, y_n$ in which A has an upper triangular form we have

$$x_j = \theta_j y_j, \qquad j = 1, \ldots, n \tag{2.5.2}$$

where $\theta_j \in \mathbb{C}$ and $|\theta_j| = 1$.

Proof. Assume that A is not unicellular. By Theorem 2.5.1 there exist two eigenvectors x_1 and y_1 (which can be assumed to have norm 1) such that $\mathrm{Span}\{x_1\} \neq \mathrm{Span}\{y_1\}$. The proof of Theorem 1.9.1 shows that there exists an orthonormal basis whose first vector is x_1 and in which A has a triangular form. Similarly, there exists such a basis whose first vector is y_1. So equation (2.5.2) does not hold for $j = 1$.

Assume now that A is unicellular, and let $z_1, \ldots, z_n$ be a Jordan basis for A in $\mathbb{C}^n$. So

$$Az_1 = \lambda_0 z_1; \qquad Az_i - \lambda_0 z_i = z_{i-1}, \qquad i = 2, \ldots, n$$

For $i = 1, 2, \ldots, n$ define x_i to be a vector in $\mathrm{Span}\{z_1, \ldots, z_i\}$ that is orthogonal to $\mathrm{Span}\{z_1, \ldots, z_{i-1}\}$ and has norm 1. (By definition, $x_1 = \alpha z_1 / \|z_1\|$ for some $\alpha \in \mathbb{C}$ with $|\alpha| = 1$.) Then

$$\mathrm{Span}\{x_1, \ldots, x_i\} = \mathrm{Span}\{z_1, \ldots, z_i\}, \qquad i = 1, \ldots, n$$

and these subspaces are A invariant. By Proposition 1.8.4, A has an upper triangular form with respect to the orthonormal basis $x_1, \ldots, x_n$.

If A also has an upper triangular form in an orthonormal basis $y_1, \ldots, y_n$, then

$$\mathrm{Span}\{y_1\} \subset \mathrm{Span}\{y_1, y_2\} \subset \cdots \subset \mathrm{Span}\{y_1, \ldots, y_n\} \qquad (2.5.3)$$

is a chain of A-invariant subspaces. But the lattice of all A-invariant subspaces is a chain (Example 1.1.1); therefore, (2.5.3) is a unique complete chain of A-invariant subspaces. Hence the chain (2.5.3) coincides with

$$\mathrm{Span}\{z_1\} \subset \mathrm{Span}\{z_1, z_2\} \subset \cdots \subset \mathrm{Span}\{z_1, z_2, \ldots, z_n\}$$

Hence $\mathrm{Span}\{y_1, \ldots, y_i\} = \mathrm{Span}\{z_1, \ldots, z_i\}$ for $i = 1, 2, \ldots, n$, and the orthonormality of $y_1, \ldots, y_n$ implies that (2.5.2) holds. $\square$

We conclude this section with a proposition that was promised in Section 1.1.

Proposition 2.5.4

The set $\mathrm{Inv}(A)$ *of all invariant subspaces of a fixed transformation* $A: \mathbb{C}^n \to \mathbb{C}^n$ *is either a continuum* [*i.e., there exists a bijection* $\varphi: \mathrm{Inv}(A) \to \mathbb{R}$] *or a finite set.*

Proof. In view of Theorem 2.1.5 we can assume that A has only one eigenvalue λ_0, that is, $\mathcal{R}_{\lambda_0}(A) = \mathbb{C}^n$. If A is unicellular, then by Example 2.1.1 the set $\mathrm{Inv}(A)$ is finite (namely, there are exactly $n + 1$ A-invariant subspaces). If A is not unicellular, then by the equivalence (c) $\Leftrightarrow$ (d) in Theorem 2.5.1 there exist two linearly independent eigenvectors x and y of A: $Ax = \lambda_0 x$, and $Ay = \lambda_0 y$. Then $\{\mathrm{Span}\{x + \alpha y\} \mid \alpha \in \mathbb{R}\}$ is a set of A-invariant subspaces which is a continuum. On the other hand, let ψ be the map from the set of all n-tuples $(x_1, \ldots, x_n)$ of n – dimensional vectors $x_1, \ldots, x_n$ onto $\mathrm{Inv}(A)$ defined by $\psi(x_1, \ldots, x_n) = \mathrm{Span}\{x_1, \ldots, x_n\}$ if the subspace $\mathrm{Span}\{x_1, \ldots, x_n\}$ is A invariant and $\psi(x_1, \ldots, x_n) = \{0\}$ otherwise. As the set of all n-tuples $(x_1, \ldots, x_n)$, $x_i \in \mathbb{C}^n$ is a continuum, by an elementary result in set theory it follows that $\mathrm{Inv}(A)$ is a continuum as well. $\square$

2.6 GENERATORS OF INVARIANT SUBSPACES

Let $\mathcal{M}$ be an invariant subspace for the transformation $A: \mathbb{C}^n \to \mathbb{C}^n$. The vectors $x_1, \ldots, x_m \in \mathbb{C}^n$ are called *generators* for $\mathcal{M}$ if

$$\mathcal{M} = \text{Span}\{x_1, \ldots, x_m, Ax_1, \ldots, Ax_m, A^2x_1, \ldots, A^2x_m, \ldots\}$$

For example, any basis for $\mathcal{M}$ forms a set of generators for $\mathcal{M}$. In connection with this definition note that for any vectors $y_1, \ldots, y_p \in \mathbb{C}^n$ the subspace $\text{Span}\{y_1, \ldots, y_p, Ay_1, \ldots, Ay_p, A^2y_1, \ldots, A^2y_p, \ldots\}$ is A invariant. The particular case when $\mathcal{M}$ has one generator is of special interest (see also Section 1.1), that is, when $\mathcal{M} = \text{Span}\{x, Ax, A^2x, \ldots\}$ for some $x \in \mathbb{C}^n$. In this case we call $\mathcal{M}$ a *cyclic* invariant subspace (and is frequently referred to as a "Krylov subspace" in the literature on numerical analysis).

The notion of generators behaves well with respect to similarity. That is, if $\mathcal{M}$ is an A-invariant subspace with generators $x_1, \ldots, x_m$, then $S\mathcal{M}$ is an SAS^{-1}-invariant subspace with generators $Sx_1, \ldots, Sx_m$ (here S is any invertible transformation). So the study of generators of A-invariant subspaces can be reduced to the study of generators of J-invariant subspaces, where J is a Jordan form for A. Let us give some examples.

EXAMPLE 2.6.1. Let $A = I$ (or, more generally, $A = \alpha I$, where $\alpha \in \mathbb{C}$). Then a k-dimensional subspace $\mathcal{M}$ in $\mathbb{C}^n$ (which is obviously A-invariant) has not less than k generators. Any set of vectors that span $\mathcal{M}$ is a set of generators.

EXAMPLE 2.6.2. Let $A = J_n(\lambda)$ be the $n \times n$ Jordan block with eigenvalue λ. An A-invariant subspace $\mathcal{M}_k = \text{Span}\{e_1, \ldots, e_k\}$ is cyclic with the generator e_k. $\square$

The generators $x_1, \ldots, x_m$ of $\mathcal{M}$ are called *minimal generators* for $\mathcal{M}$ if m is the smallest number of generators of $\mathcal{M}$. Obviously, any set of minimal generators is a minimal set of generators. (A set of generators $x_1, \ldots, x_p$ for the A-invariant subspace $\mathcal{M}$ is called *minimal* if any proper subset of $\{x_1, \ldots, x_p\}$ does not constitute a set of generators for $\mathcal{M}$.) However, not every minimal set of generators is a set of minimal generators. Let us demonstrate this in an example.

EXAMPLE 2.6.3. Let

$$A = \begin{bmatrix} 1 & 0 \\ 0 & 0 \end{bmatrix}$$

and let $\mathcal{M} = \mathbb{C}^2$ be the A-invariant subspace. The vector $\langle 1, 1 \rangle$ is obviously a generator for $\mathcal{M}$, so a set of minimal generators must consist of a single vector.

On the other hand, the set of two vectors $\{e_1, e_2\}$ is a set of generators of $\mathbb{C}^2$ that is minimal. Indeed, neither of the vectors e_1 and e_2 is a generator of $\mathbb{C}^2$. $\square$

The number of vectors in a set of minimal generators admits an intrinsic characterization as follows.

Theorem 2.6.1

Let $\mathcal{M}$ be an A-invariant subspace. Then the number of vectors in a set of minimal generators coincides with the maximal dimension m of $\operatorname{Ker}(A - \lambda_0 I)|_{\mathcal{M}}$, where λ_0 is any eigenvalue of $A|_{\mathcal{M}}$.

Proof. We can assume that

$$A|_{\mathcal{M}} = J_{k_1}(\lambda_1) \oplus \cdots \oplus J_{k_p}(\lambda_p) \qquad (2.6.1)$$

a matrix in Jordan form (with respect to a certain basis in $\mathcal{M}$). Further, we can assume that $\lambda_1 = \cdots = \lambda_m$ where $m \leq p$ (recall that m is the maximal number of Jordan blocks corresponding to any eigenvalue). Let $x_1, \ldots, x_q$ be generators of $\mathcal{M}$. Let y_i be the m-dimensional vector formed by the k_1th, $(k_1 + k_2)$th, $\ldots$, $(k_1 + k_2 + \cdots + k_m)$th coordinates of x_i $(i = 1, \ldots, q)$. Now

$$e_{k_1} \in \operatorname{Span}\{x_1, \ldots, x_q, Ax_1, \ldots, Ax_q, A^2 x_1, \ldots, A^2 x_q, \ldots\}$$

Examining the k_1th, $\ldots$, $(k_1 + k_2 + \cdots + k_m)$th coordinates of x_i and using the condition $\lambda_1 = \cdots = \lambda_m$, we see that $e_1 \in \operatorname{Span}\{y_1, \ldots, y_q\}$. Similarly, the condition

$$e_{k_1 + k_2} \in \operatorname{Span}\{x_1, \ldots, x_q, Ax_1, \ldots, Ax_q, A^2 x_1, \ldots, A^2 x_q, \ldots\}$$

gives rise to the conclusion that $e_2 \in \operatorname{Span}\{y_1, \ldots, y_q\}$. Continuing in this way, we eventually find that $e_i \in \operatorname{Span}\{y_1, \ldots, y_q\}$, $i = 1, 2, \ldots, m$. So $y_1, \ldots, y_q$ span the whole space $\mathbb{C}^m$, and thus $q \geq m$.

We now prove that there is a set of m generators for $\mathcal{M}$. We proceed by induction on m.

Suppose first that $m = 1$, that is, the eigenvalues $\lambda_1, \ldots, \lambda_p$ are all different. Then the vector $x = e_{k_1} + e_{k_1 + k_2} + \cdots + e_{k_1 + \cdots + k_p}$ is a generator for $\mathcal{M}$. Indeed

$$(A - \lambda_2 I)^{k_2} \cdots (A - \lambda_p I)^{k_p} x = (A - \lambda_2 I)^{k_2} \cdots (A - \lambda_p I)^{k_p} e_{k_1} \overset{\text{def}}{=} f_1$$

Because of the form (2.6.1) of $A|_{\mathcal{M}}$ the matrix $(A - \lambda_2 I)^{k_2} \cdots (A - \lambda_p I)^{k_p}|_{\mathcal{M}}$ has the form $T_1 \oplus 0_{k_1} \oplus \cdots \oplus 0_{k_p}$, where T_1 is an upper triangular nonsingular matrix. Hence the k_1th coordinate f_{1,k_1} of f_1 is nonzero. Now $(A - \lambda_1 I)^j f_1$ has $(k_1 - j)$th coordinate equal to f_{1,k_1} (and thus nonzero) and all the coordinates of $(A - \lambda_1 I)^j f_1$ below the $(k_1 - j)$th coordinate are zeros $(j = 1, \ldots, k_1 - 1)$. Consequently, the vectors $e_1, \ldots, e_k$ belong to the span of $f_1, (A - \lambda_1 I)f_1, \ldots, (A - \lambda_1 I)^{k_1-1} f_1$. Similarly, one shows that the span of vectors

$$f_2, (A - \lambda_2 I)f_2, \ldots, (A - \lambda_2 I)^{k_2-1} f_2$$

where

$$f_2 = (A - \lambda_1 I)^{k_1}(A - \lambda_3 I)^{k_3} \cdots (A - \lambda_p)^{k_p} x$$

contains vectors $e_{k_1+1}, \ldots, e_{k_1+k_2}$. Proceeding in this way we find eventually that all the vectors e_i, $i = 1, \ldots, k_1 + \cdots + k_p$ belong to $\mathrm{Span}\{x, Ax, A^2 x, \ldots\}$.

Assume now that $m > 1$. Suppose that for any transformation B and any B-invariant subspace $\mathscr{L}$ such that

$$\max_{\lambda \in \sigma(B|_\mathscr{L})} \dim \mathrm{Ker}(B - \lambda I)|_\mathscr{L} = m - 1$$

there exists a set of $m - 1$ generators in $\mathscr{L}$. Given the transformation $A: \mathbb{C}^n \to \mathbb{C}^n$, write

$$A|_\mathscr{M} = A|_{\mathscr{M}_1} \oplus A|_{\mathscr{M}_2}$$

where $\mathscr{M}_1$ and $\mathscr{M}_2$ are some A-invariant subspaces such that

$$\max_{\lambda \in \sigma(A|_{\mathscr{M}_1})} \dim \mathrm{Ker}(A|_{\mathscr{M}_1} - \lambda I) = m - 1$$

$$\max_{\lambda \in \sigma(A|_{\mathscr{M}_2})} \dim \mathrm{Ker}(A|_{\mathscr{M}_2} - \lambda I) = 1$$

(Such subspaces $\mathscr{M}_1$ and $\mathscr{M}_2$ are easily found by using the Jordan form of A.) By the induction hypothesis we have a set of $m - 1$ generators $x_1, \ldots, x_{m-1}$ for the A-invariant subspace $\mathscr{M}_1$. Also, we have proved that there is a generator x_m for the A-invariant subspace $\mathscr{M}_2$. Then, obviously, $x_1, \ldots, x_m$ is a set of generators for $\mathscr{M}$. $\square$

In particular, an A-invariant subspace $\mathscr{M}$ is cyclic if and only if there is only one eigenvector (up to multiplication by a nonzero number) in $\mathscr{M}$ corresponding to any eigenvalue of the restricton $A|_\mathscr{M}$.

We conclude this section with an example.

EXAMPLE 2.6.4.   Let $A = \text{diag}[\lambda_1 \lambda_2 \cdots \lambda_n]$ where $\lambda_1, \ldots, \lambda_n$ are different complex numbers. Then $\mathbb{C}^n$ is a cyclic subspace for A. A vector $x = \langle x_1, \ldots, x_n \rangle \in \mathbb{C}^n$ is cyclic, that is

$$\mathbb{C}^n = \text{Span}\{x, Ax, A^2x, \ldots\}$$

if and only if all the coordinates x_i are different from zero. Indeed, if $x_i = 0$ for some i, then e_i does not belong to $\text{Span}\{x, Ax, A^2x, \ldots\}$. On the other hand, if $x_i \neq 0$ for $i = 1, \ldots, n$, then

$$\det[x, Ax, \ldots, A^{n-1}x] = \det \begin{bmatrix} x_1 & \lambda_1 x_1 & \cdots & \lambda_1^{n-1} x_1 \\ x_2 & \lambda_2 x_2 & \cdots & \lambda_2^{n-1} x_2 \\ \vdots & \vdots & & \vdots \\ x_n & \lambda_n x_n & \cdots & \lambda_n^{n-1} x_n \end{bmatrix}$$

$$= x_1 x_2 \cdots x_n \det \begin{bmatrix} 1 & \lambda_1 & \cdots & \lambda_1^{n-1} \\ 1 & \lambda_2 & \cdots & \lambda_2^{n-1} \\ \vdots & \vdots & & \vdots \\ 1 & \lambda_n & \cdots & \lambda_n^{n-1} \end{bmatrix}$$

The determinant on the right-hand side is known as the *Vandermonde determinant*, and it is well known that it is equal to $\Pi_{i<j}(\lambda_j - \lambda_i) \neq 0$. So $\det[x, Ax, \ldots, A^{n-1}x] \neq 0$. It follows that the vectors $x, Ax, \ldots, A^{n-1}x$ are linearly independent and thus span $\mathbb{C}^n$. $\square$

2.7   MAXIMAL INVARIANT SUBSPACE IN A GIVEN SUBSPACE

Given a transformation $A: \mathbb{C}^n \to \mathbb{C}^n$ and a subspace $\mathcal{N} \subset \mathbb{C}^n$, we say that an A-invariant subspace $\mathcal{M}$ is *maximal* in $\mathcal{N}$ if $\mathcal{M} \subset \mathcal{N}$ and there is no A-invariant subspace that is contained in $\mathcal{N}$ and contains $\mathcal{M}$ properly.

Proposition 2.7.1

A maximal A-invaraint subspace in $\mathcal{N}$ exists, is unique and is equal to the sum of all A-invariant subspaces that are contained in $\mathcal{N}$.

Note that, because the dimension of $\mathcal{N}$ is finite, $\mathcal{M}$ can actually be expressed as the sum of a finite number of A-invariant subspaces.

Proof.   Clearly, $\mathcal{M}$ is A-invariant and contained in $\mathcal{N}$. Also, $\mathcal{M}$ is maximal in $\mathcal{N}$. This follows from the definition of $\mathcal{M}$ that implies that every A-invariant subspace in $\mathcal{N}$ is contained in $\mathcal{M}$.

For the uniqueness, assume that there are two different maximal A-invariant subspaces in $\mathcal{N}$, say, $\mathcal{M}_1$ and $\mathcal{M}_2$. Then $\mathcal{M}_1 + \mathcal{M}_2$ is an A-invariant subspace in $\mathcal{N}$ that contains $\mathcal{M}_1$ properly, a contradiction with the definition of a maximal A-invariant subspace in $\mathcal{N}$. $\square$

Observe that if $\mathcal{N}$ is A invariant, the maximal A-invariant subspace in $\mathcal{N}$ coincides with $\mathcal{N}$ itself. At the other extreme, assume that $\mathcal{N}$ does not contain any eigenvector, it follows that $\mathcal{N}$ does not contain nonzero A-invariant subspaces. Hence the maximal A-invariant subspace in $\mathcal{N}$ is the zero subspace.

Let us consider some examples.

EXAMPLES 2.7.1. Let $A = \mathrm{diag}[\lambda_1, \lambda_2, \ldots, \lambda_n]$, where $\lambda_1, \ldots, \lambda_n$ are different complex numbers. Then the maximal A-invariant subspace in $\mathcal{N}$ is $\mathrm{Span}\{e_{j_1}, \ldots, e_{j_k}\}$, where e_{j_p}, $p = 1, \ldots, k$ are all the vectors among $e_1, \ldots, e_n$ that are contained in $\mathcal{N}$ (by definition, $\mathrm{Span}\{e_{j_1}, \ldots, e_{j_k}\} = \{0\}$ if none of the vectors $e_1, \ldots, e_n$ belongs to $\mathcal{N}$).

EXAMPLE 2.7.2. Let $A = J_n(\lambda_0)$, the $n \times n$ Jordan block with eigenvalue λ_0. Then the maximal A-invariant subspace in $\mathcal{N}$ is $\mathrm{Span}\{e_1, \ldots, e_{p-1}\}$, where p is the minimal index such that $e_p \not\in \mathcal{N}$ (again, we put $\mathrm{Span}\{e_1, \ldots, e_{p-1}\} = \{0\}$ if $\mathcal{N}$ does not contain e_1). $\square$

The following more explicit description of maximal A-invariant subspaces is sometimes useful.

Theorem 2.7.2

The maximal A-invariant subspace in $\mathcal{N}$ coincides with

$$\mathcal{M} \stackrel{\mathrm{def}}{=} \bigcap_{j=0}^{\infty} \mathcal{N}_j$$

where $\mathcal{N}_j = \{x \in \mathbb{C}^n \mid A^j x \in \mathcal{N}\}$ (in particular, $\mathcal{N}_0 = \mathcal{N}$).

Proof. We have $\mathcal{M} \subset \mathcal{N}_0 = \mathcal{N}$. Further, $\mathcal{M}$ is A-invariant. For, if $x \in \mathcal{M}$, then $A^j x = y_j$ for some $y_j \in \mathcal{N}$ ($j = 0, 1, \ldots$) and

$$A^j(Ax) = y_{j+1}, \qquad j = 0, 1, \ldots$$

Hence $Ax \in \mathcal{M}$. It remains to verify that $\mathcal{M}$ is maximal in $\mathcal{N}$. Let $\mathcal{L}$ be an A-invariant subspace contained in $\mathcal{N}$. Then for $j = 0, 1, \ldots$,

$$\{x \in \mathbb{C}^n \mid A^j x \in \mathcal{L}\} \subset \{x \in \mathbb{C}^n \mid A^j x \in \mathcal{N}\}$$

and (because $\mathcal{L}$ is A invariant)

$$\mathcal{L} \subset \{x \in \mathbb{C}^n \mid A^j x \in \mathcal{L}\}$$

Combining these inclusions, we have

$$\mathscr{L} \subset \bigcap_{j=0}^{\infty} \{x \in \mathbb{C}^n \mid A^j x \in \mathscr{L}\} \subset \bigcap_{j=0}^{\infty} \{x \in \mathbb{C}^n \mid A^j x \in \mathscr{N}\} = \mathscr{M}$$

and $\mathscr{M}$ is indeed maximal in $\mathscr{N}$. $\square$

In connection with Theorem 2.7.2 observe that $\mathscr{N}_j = A^{-j}\mathscr{N}$ with A is invertible.

Given a transformation $A: \mathbb{C}^n \to \mathbb{C}^n$, it is well known that there are scalar polynomials $f(\lambda)$ such that, if $f(\lambda) = \sum_{i=0}^{l} a_i \lambda^i$, then

$$f(A) \overset{\text{def}}{=} \sum_{i=0}^{l} a_i A^i = 0$$

Indeed, the characteristic polynomial of A has this property (the Cayley–Hamilton theorem). A nonzero polynomial $g(\lambda)$ of least degree—say, p—for which $g(A) = 0$ is called a *minimal polynomial* for A (it can be shown that p is uniquely defined). Then it is clear that for any integer $j \geq p$, we can equate A^j to a polynomial in A of degree less than p. Thus (in the notation of Theorem 2.7.2)

$$\bigcap_{j=0}^{\infty} \mathscr{N}_j = \bigcap_{j=0}^{p-1} \mathscr{N}_j \qquad (2.7.1)$$

where p is the degree of a minimal polynomial of A. Indeed, the inclusion $\subset$ in equation (2.7.1) is obvious. To prove the opposite inclusion, let $q(\lambda) = \lambda^p + \sum_{j=0}^{p-1} \alpha_j \lambda^j$ be a minimal polynomial of A, so

$$q(A) = A^p + \sum_{j=0}^{p-1} \alpha_j A^j = 0$$

Let $x \in \bigcap_{j=0}^{p-1} \mathscr{N}_j$, so $A^j x \in \mathscr{N}$ for $j = 0, \ldots, p-1$. Then

$$A^p x = -\sum_{j=0}^{p-1} \alpha_j A^j x \in \mathscr{N}$$

and $x \in \mathscr{N}_p$. Assume inductively that we have already proved that $x \in \mathscr{N}_j$, $j = 0, \ldots, q-1$ for some $q \geq p$. Then

$$A^q x = -\sum_{j=0}^{p-1} \alpha_j A^{q-p+j} x \in \mathscr{N}$$

and $x \in \mathscr{N}_q$. So actually $x \in \bigcap_{j=0}^{\infty} \mathscr{N}_j$, and equation (2.7.1) is proved.
Observe that (2.7.1) implies

$$\bigcap_{j=0}^{\infty} \mathscr{N}_j = \bigcap_{j=0}^{q} \mathscr{N}_j \qquad (2.7.2)$$

for every $q \geq p - 1$. In particular, equation (2.7.2) holds with $q = n - 1$.

The case when $\mathcal{N} = \operatorname{Ker} C$ and $C \colon \mathbb{C}^n \to \mathbb{C}^r$ is a transformation is of particular interest. In this case one can describe the maximal A-invariant subspaces in $\operatorname{Ker} C$ in terms of the kernels of transformations CA^j, $j = 0, 1, \ldots$.

Theorem 2.7.3

Given linear transformations $A \colon \mathbb{C}^n \to \mathbb{C}^n$ *and* $C \colon \mathbb{C}^n \to \mathbb{C}^r$, *the maximal* A-*invariant subspace in* $\operatorname{Ker} C$ *is*

$$\mathcal{K}(C, A) \overset{\text{def}}{=} \bigcap_{j=0}^{\infty} \operatorname{Ker}(CA^j)$$

Moreover, the subspace $\mathcal{K}(C, A)$ *coincides with* $\bigcap_{j=0}^{q-1} \operatorname{Ker}(CA^j)$ *for every integer q greater than or equal to the degree of a minimal polynomial of A.*

Proof. In view of Theorem 2.7.2 and equality (2.7.2), we have only to show that

$$\operatorname{Ker}(CA^j) = \{x \in \mathbb{C}^n \mid A^j x \in \operatorname{Ker} C\}, \qquad j = 0, 1, \ldots$$

However, this equality is immediately verified using the defintions of $\operatorname{Ker} C$ and $\operatorname{Ker}(CA^j)$. $\square$

We say that a pair of linear transformations (C, A) where $A \colon \mathbb{C}^n \to \mathbb{C}^n$ and $C \colon \mathbb{C}^n \to \mathbb{C}^r$ is a *null kernel pair* if the maximal A-invariant subspace in $\operatorname{Ker} C$ is the zero subspace, or, equivalently, if

$$\bigcap_{j=0}^{n-1} \operatorname{Ker}(CA^j) = \{0\}$$

It is easily seen that, also, the pair (C, A) is a null kernel pair if and only if

$$\operatorname{rank} \begin{bmatrix} C \\ CA \\ \vdots \\ CA^{n-1} \end{bmatrix} = n$$

EXAMPLE 2.7.3.   Let $C = [c_1 \cdots c_n] \colon \mathbb{C}^n \to \mathbb{C}$, and

$$A = \begin{bmatrix} 0 & 1 & 0 & \cdots & 0 \\ 0 & 0 & 1 & \cdots & 0 \\ \vdots & \vdots & \vdots & & \vdots \\ & & & & 1 \\ 0 & 0 & 0 & \cdots & 0 \end{bmatrix}$$

For $j = 0, 1, \ldots, n - 1$ we have

$$CA^n = [0 \cdots 0c_1 \cdots c_{n-j}]$$

and hence

$$\bigcap_{j=0}^{n-1} \mathrm{Ker}(CA^j) = \mathrm{Span}\{e_1, \ldots, e_{k-1}\}$$

where k is the smallest index such that $c_k \neq 0$. In particular, (C, A) is a null kernel pair if and only if $c_1 \neq 0$. □

The notion of null kernel pairs plays important roles in realization theory for rational matrix functions and in linear systems theory, as we see in Chapters 7 and 8. Here we prove that every pair of transformations has a naturally defined null kernel part.

Theorem 2.7.4

Let $C: \mathbb{C}^n \to \mathbb{C}^r$ *and* $A: \mathbb{C}^n \to \mathbb{C}^n$ *be transformations, and let* $\mathcal{M}_1$ *be the maximal A-invariant subspace in* $\mathrm{Ker}\, C$. *Then for every direct complement* $\mathcal{M}_2$ *to* $\mathcal{M}_1$ *in* $\mathbb{C}^n$, *C and A have the following block matrix form with respect to the direct sum decomposition* $\mathbb{C}^n = \mathcal{M}_1 \dotplus \mathcal{M}_2$:

$$C = [0, C_2], \qquad A = \begin{bmatrix} A_{11} & A_{12} \\ 0 & A_{22} \end{bmatrix} \qquad (2.7.3)$$

where the pair $C_2: \mathcal{M}_2 \to \mathbb{C}^r$, $A_{22}: \mathcal{M}_2 \to \mathcal{M}_2$ *is a null kernel pair. If* $\mathbb{C}^n = \mathcal{M}_1' \dotplus \mathcal{M}_2'$ *is another direct sum with respect to which C and A have the form*

$$C = [0, C_2'], \qquad A = \begin{bmatrix} A_{11}' & A_{12}' \\ 0 & A_{22}' \end{bmatrix} \qquad (2.7.4)$$

where the pair (C_2', A_{22}') *is a null kernel pair, then* $\mathcal{M}_1'$ *is the maximal A-invariant subspace in* $\mathrm{Ker}\, C$ *and there exists an invertible linear transformation* $S: \mathcal{M}_2 \to \mathcal{M}_2'$ *such that*

$$C_2 = C_2' S, \qquad A_{22} = S^{-1} A_{22}' S \qquad (2.7.5)$$

Proof. As $\mathcal{M}_1$ is A invariant and $Cx = 0$ for every $x \in \mathcal{M}_1$, the transformations C and A indeed have the form of equality (2.7.3). Let us show that the pair (C_2, A_{22}) is null kernel. Assume $x \in \cap_{j=0}^{\infty} \mathrm{Ker}(C_2 A_{22}^j)$. As

$$A^j = \begin{bmatrix} A_{11}^j & * \\ 0 & A_{22}^j \end{bmatrix}, \qquad j = 0, 1, \ldots$$

where by * we denote a transformation of no immediate interest, we have

$$CA^j = [0, C_{22}A_{22}^j]$$

and hence

$$x \in \bigcap_{j=0}^{\infty} \mathrm{Ker}(CA^j) \in \mathcal{M}_1$$

On the other hand, x belongs to the domain of definition of A_{22}, that is, $x \in \mathcal{M}_2$. Since $\mathcal{M}_1 \cap \mathcal{M}_2 = \{0\}$, the vector x must be the zero vector. Consequently, (C_2, A_{22}) is a null kernel pair.

Now consider a direct sum decomposition $\mathbb{C}^n = \mathcal{M}_1' \dotplus \mathcal{M}_2'$, with respect to which C and A have the form of equality (2.7.4) with the null kernel pair (C_2', A_{22}'). As

$$CA^j = [0, C_{22}'(A_{22}')^j], \qquad j = 0, 1, \ldots$$

we have

$$\bigcap_{j=0}^{\infty} \mathrm{Ker}(CA^j) = \mathcal{M}_1' + \left(\bigcap_{j=0}^{\infty} \mathrm{Ker}[C_{22}'(A_{22}')^j] \right) = \mathcal{M}_1'$$

where the last equality follows from the null kernel property of (C_2', A_{22}'). Hence $\mathcal{M}_1'$ actually coincides with $\mathcal{M}_1$. Further, write the identity transformation $I: \mathbb{C}^n \to \mathbb{C}^n$ as a 2×2 block matrix

$$I = \begin{bmatrix} I & * \\ 0 & S \end{bmatrix} : \mathcal{M}_1 \dotplus \mathcal{M}_2 \to \mathcal{M}_1 \dotplus \mathcal{M}_2'$$

Here $S: \mathcal{M}_2 \to \mathcal{M}_2'$ is a linear transformation that must be invertible in view of the invertibility of I. The inverse of I (which is I itself) written as a 2×2 block matrix with respect to the direct sum decompositions $\mathbb{C}^n = \mathcal{M}_1 \dotplus \mathcal{M}_2' = \mathcal{M}_1 \dotplus \mathcal{M}_2$ has the form

$$I = \begin{bmatrix} I & * \\ 0 & S^{-1} \end{bmatrix} : \mathcal{M}_1 \dotplus \mathcal{M}_2' \to \mathcal{M}_1 \dotplus \mathcal{M}_2$$

We obtain the equalities

$$\begin{bmatrix} A_{11} & A_{12} \\ 0 & A_{22} \end{bmatrix} = \begin{bmatrix} I & * \\ 0 & S^{-1} \end{bmatrix} \begin{bmatrix} A_{11}' & A_{12}' \\ 0 & A_{22}' \end{bmatrix} \begin{bmatrix} I & * \\ 0 & S \end{bmatrix}$$

$$[0 \quad C_2] = [0 \quad C_2'] \begin{bmatrix} I & * \\ 0 & S \end{bmatrix}$$

which imply equality (2.7.5). $\square$

Observe that if (2.7.5) holds, one can identify both $\mathcal{M}_2$ and $\mathcal{M}_2'$ with $\mathbb{C}^m$, for some integer m. Write C_2 and A_{22} as $r \times m$ and $m \times m$ matrices, respectively, with respect to a fixed basis in $\mathbb{C}^r$ and some basis in $\mathbb{C}^m$. Then C_2' and A_{22}' are transformations represented by the matrices C_2 and A_{22}, respectively, with respect to the same basis in $\mathbb{C}^r$ and a possibly different basis in $\mathbb{C}^m$. So the pairs (C_2, A_{22}) and (C_2', A_{22}') are essentially the same.

We conclude this section with an example.

EXAMPLE 2.7.4. Let C and A be as in Example 2.7.3, and assume that $c_1 = \cdots = c_{k-1} = 0$, $c_k \neq 0$ $(k > 1)$. Then (C, A) is not a null kernel pair. The null kernel part (C_2, A_{22}) of (C, A) (as in Theorem 2.7.3) is given by

$$C_2 = [c_k, c_{k+1}, \ldots, c_n], \qquad A_{22} = J_{n-k+1}(0)$$

2.8   MINIMAL INVARIANT SUBSPACES OVER A GIVEN SUBSPACE

Here we present properties of invariant subspaces that contain a given subspace and are minimal with respect to this property. It turns out that such subspaces are in a certain sense dual to the maximal subspaces studied in the preceding section. We also see a connection with generators of invariant subspaces, as studied in Section 2.6.

Given a transformation $A: \mathbb{C}^n \rightarrow \mathbb{C}^n$ and a subspace $\mathcal{N} \subset \mathbb{C}^n$, we say that an A-invariant subspace $\mathcal{M}$ is *minimal over* $\mathcal{N}$ if $\mathcal{M} \supset \mathcal{N}$ and there is no A-invariant subspace that contains $\mathcal{N}$ and is contained properly in $\mathcal{M}$. As an analog of Proposition 2.7.1 we see that *a minimal A-invariant subspace over $\mathcal{N}$ exists, is unique, and is equal to the intersection of all A-invariant subspaces that contain $\mathcal{N}$.* The proof of this statement is left to the reader.

If $\mathcal{N}$ is A invariant, then the minimal A-invariant subspace over $\mathcal{N}$ coincides with $\mathcal{N}$ itself. On the other hand, it can happen that $\mathbb{C}^n$ is the minimal A-invariant subspace over $\mathcal{N}$, even when $\mathcal{N}$ is one-dimensional.

EXAMPLE 2.8.1. Let $A = \text{diag}[\lambda_1, \lambda_2, \ldots, \lambda_n]$, with different complex numbers $\lambda_1, \ldots, \lambda_n$. Let $\mathcal{N} = \text{Span} \sum_{i=1}^{n} \alpha_i e_i$ be a one-dimensional subspace. Then the minimal A-invariant subspace over $\mathcal{N}$ is $\text{Span}\{e_j \mid \alpha_j \neq 0\}$. In particular, if all α_j are different from zero, then the minimal A-invariant subspace over $\mathcal{N}$ is $\mathbb{C}^n$. $\square$

Our next result expresses the duality between minimal and maximal invariant subspaces in a precise form. (Recall that by Proposition 1.4.4 the subspace $\mathcal{M}$ is A invariant if and only if its orthogonal complement $\mathcal{M}^\perp$ is A^* invariant.)

Proposition 2.8.1

An A-invariant subspace $\mathcal{M}$ is minimal over $\mathcal{N}$ if and only if the A^*-invariant subspace $\mathcal{M}^\perp$ is maximal in $\mathcal{N}^\perp$.

Proof. Assume that the A-invariant subspace $\mathcal{M}$ is minimal over $\mathcal{N}$. In particular, $\mathcal{M} \supset \mathcal{N}$, so $\mathcal{M}^{\perp} \subset \mathcal{N}^{\perp}$. If there were an A^{*}-invariant subspace $\mathcal{L}$ such that $\mathcal{M}^{\perp} \subset \mathcal{L} \subset \mathcal{N}^{\perp}$ and $\mathcal{M}^{\perp} \neq \mathcal{L}$, the subspace $\mathcal{L}^{\perp}$ would be A invariant and $\mathcal{M} \supset \mathcal{L}^{\perp} \supset \mathcal{N}$, $\mathcal{M} \neq \mathcal{L}^{\perp}$. This contradicts the definition of $\mathcal{M}$ as a minimal A-invariant subspace over $\mathcal{N}$. Hence $\mathcal{M}^{\perp}$ is a maximal A^{*}-invariant subspace in $\mathcal{N}^{\perp}$. Reversing the argument, we find that, if the A^{*}-invariant subspace $\mathcal{M}^{\perp}$ is a maximal in $\mathcal{N}^{\perp}$, the A-invariant subspace $\mathcal{M}$ is minimal over $\mathcal{N}$. $\square$

Proposition 2.8.1 allows us to obtain many properties of minimal invariant subspaces from the corresponding properties of maximal invariant subspaces proved in the preceding section. For example, let us prove an analogue of Theorem 2.7.2 in this way.

Theorem 2.8.2

The minimal A-invariant subspace $\mathcal{M}$ over $\mathcal{N}$ coincides with $\sum_{j=0}^{\infty} A^{j}\mathcal{N}$.

Proof. By Proposition 2.8.1 and Theorem 2.7.2, we have

$$\mathcal{M}^{\perp} = \bigcap_{j=0}^{\infty} \mathcal{N}_{j} \qquad (2.8.1)$$

where $\mathcal{N}_{j} = \{x \in \mathbb{C}^{n} \mid A^{*j}x \in \mathcal{N}^{\perp}\}$. It is not difficult to check that for $j = 0, 1, \ldots$

$$\mathcal{N}_{j}^{\perp} = A^{j}\mathcal{N} \qquad (2.8.2)$$

Indeed, let $y \in A^{j}\mathcal{N}$, so that $y = A^{j}z$ for some $z \in \mathcal{N}$. Then for every $x \in \mathbb{C}^{n}$ such that $A^{*j}x \in \mathcal{N}^{\perp}$ we have

$$(y, x) = (A^{j}z, x) = (z, A^{*j}x) = 0$$

Hence $y \in \mathcal{N}_{j}^{\perp}$. If the equality (2.8.2) were not true, there would exist a nonzero $y_{0} \in \mathcal{N}_{j}^{\perp}$ such that y_{0} would be orthogonal to $A^{j}\mathcal{N}$. Hence for every $z \in \mathcal{N}$ we have

$$0 = (y_{0}, A^{j}z) = (A^{*j}y_{0}, z)$$

which implies $y_{0} \in \mathcal{N}_{j}$, a contradiction with $y_{0} \in \mathcal{N}_{j}^{\perp}$.
Now (2.8.1) and (2.8.2) give

$$\mathcal{M} = (\mathcal{M}^{\perp})^{\perp} = \left(\bigcap_{j=0}^{\infty} \mathcal{N}_{j} \right)^{\perp} = \sum_{j=0}^{\infty} \mathcal{N}_{j}^{\perp} = \sum_{j=0}^{\infty} A^{j}\mathcal{N} \qquad \square$$

Note that the equality $\mathcal{M} = \sum_{j=0}^{\infty} A^{j}\mathcal{N}$ can also be verified directly without difficulty. To this end, observe that the subspace $\mathcal{M}_0 \overset{\text{def}}{=} \Sigma_{j=0}^{\infty} A^j \mathcal{N}$ is A invariant: if $x = A^j z$ for some $z \in \mathcal{N}$, then $Ax = A^{j+1}z$ belongs to $\mathcal{M}_0$. Obviously, $\mathcal{M}_0$ contains $\mathcal{N}$. If $\mathcal{M}'$ is an A-invariant subspace that contains $\mathcal{N}$, then

$$\mathcal{M}' = \sum_{n=0}^{\infty} A^j \mathcal{M}' \supset \sum_{n=0}^{\infty} A^j \mathcal{N} = \mathcal{M}_0$$

So $\mathcal{M}_0$ is indeed the minimal A-invariant subspace over $\mathcal{N}$.

As all subspaces under consideration are finite dimensional, the sum $\Sigma_{j=0}^{\infty} A^j \mathcal{N}$ is actually the sum of a finite number of subspaces $A^j \mathcal{N}$ ($j = 0, 1, \ldots$). In fact

$$\sum_{j=0}^{\infty} A^j \mathcal{N} = \sum_{j=0}^{q-1} A^j \mathcal{N} \tag{2.8.3}$$

where q is any integer greater than or equal to the degree p of a minimal polynomial for A. Indeed, it is sufficient to verify equation (2.8.3) for $q = p$. Let $r(\lambda) = \lambda^p + \Sigma_{j=0}^{p-1} \alpha_j \lambda^j$ be a minimal polynomial of A, so

$$A^p + \sum_{j=0}^{p-1} \alpha_j A^j = 0$$

Assuming by induction that we have already proved the inclusion

$$\sum_{j=0}^{s-1} A^j \mathcal{N} \subset \sum_{j=0}^{p-1} A^j \mathcal{N}$$

for some $s \geq p$, for $x = A^s y$, $y \in \mathcal{N}$ we have

$$x = A^s y = -\sum_{j=0}^{p-1} \alpha_j A^{s-p+j} y \in \sum_{j=0}^{p-1} A^j \mathcal{N}$$

So the inclusion $A^s \mathcal{N} \subset \Sigma_{j=0}^{p-1} A^j \mathcal{N}$ follows, and by induction we have proved the inclusion $\subset$ in (2.8.3) (with $q = p$). As the opposite inclusion is obvious, (2.8.3) is proved.

Going back to Theorem 2.8.2, observe that

$$\mathcal{M} = \text{Span}\{ A^j f_1, \ldots, A^j f_k \mid j = 0, 1, \ldots \}$$

where $f_1, \ldots, f_k$ is a basis in $\mathcal{N}$. In other words, the A-invariant subspace $\mathcal{M}$ has a set of k generators, where $k = \dim \mathcal{N}$. Combining this observation with Theorem 2.6.1, we obtain the following fact.

Theorem 2.8.3

If M is the minimal A-invariant subspace over N, and $k = \dim N$, then for any eigenvalue λ_0 of $A_{|M}$ we have

$$\dim \mathrm{Ker}(A - \lambda_0 I)_{|M} \leq k \tag{2.8.4}$$

In particular, the theorem implies that if N is one-dimensional, then M is cyclic.

It is easy to produce examples when the inequality in (2.8.4) is strict. For instance, in the extreme case when $N = \mathbb{C}^n$ and A has n distinct eigenvalues, we have $M = \mathbb{C}^n$ and

$$\max_{\lambda_0 \in \sigma(A)} \dim \mathrm{Ker}(A - \lambda_0 I) = 1$$

The case when $N = \mathrm{Im}\, B$, and $B: \mathbb{C}^s \to \mathbb{C}^n$ is a transformation is of special interest. Noting that $A^j(\mathrm{Im}\, B) = \mathrm{Im}(A^j B)$, Theorem 2.8.2 together with (2.8.3) gives the following.

Theorem 2.8.4

Let $B: \mathbb{C}^s \to \mathbb{C}^n$ and $A: \mathbb{C}^n \to \mathbb{C}^n$ be transformations. Then the minimal A-invariant subspace over $\mathrm{Im}\, B$ coincides with

$$\mathscr{I}(A, B) \overset{\mathrm{def}}{=} \sum_{j=0}^{\infty} \mathrm{Im}(A^j B) = \sum_{j=0}^{q-1} \mathrm{Im}(A^j B)$$

for every integer q greater than or equal to the degree of a minimal polynomial for A. [In particular, $\mathscr{I}(A, B) = \sum_{j=0}^{n-1} \mathrm{Im}(A^j B)$.]

We say that a pair of transformations (A, B), where $A: \mathbb{C}^n \to \mathbb{C}^n$ and $B: \mathbb{C}^s \to \mathbb{C}^n$, is a *full-range pair* if the minimal A-invariant subspace over $\mathrm{Im}\, B$ coincides with $\mathbb{C}^n$, or, equivalently, if

$$\sum_{j=0}^{n-1} \mathrm{Im}(A^j B) = \mathbb{C}^n$$

It is easy to see that, also, (A, B) is a full-range pair if and only if

$$\mathrm{rank}[B \quad AB \quad \cdots \quad A^{n-1}B] = n$$

The duality generated by Proposition 2.8.1 now takes the form: *the pair (A, B) is a full-range pair if and only if the adjoint pair (B^*, A^*) is a null kernel pair.* This follows from the orthogonal decomposition

$$\mathbb{C}^n = \text{Im}[B \quad AB \quad \cdots \quad A^{n-1}B] \oplus \text{Ker} \begin{bmatrix} B^* \\ A^*B^* \\ \vdots \\ \vdots \\ (A^*)^{n-1}B^* \end{bmatrix}$$

which is obtained directly from Proposition 1.4.4.

EXAMPLE 2.8.2.   Let

$$A = \begin{bmatrix} 0 & 1 & 0 & \cdots & 0 \\ 0 & 0 & 1 & \cdots & 0 \\ \vdots & \vdots & \vdots & & \vdots \\ & & & & 1 \\ 0 & 0 & 0 & \cdots & 0 \end{bmatrix}, \qquad B = \begin{bmatrix} b_1 \\ b_2 \\ \vdots \\ b_n \end{bmatrix} : \mathbb{C} \to \mathbb{C}^n$$

Then

$$\mathcal{I}(A, B) = \text{Span}\{e_1, \ldots, e_m\}$$

where m is the index determined by the properties that $b_m \neq 0$, $b_{m+1} = \cdots = b_n = 0$. In particular, $\mathcal{I}(A, B) = \{0\}$ if and only if $B = 0$, and the pair (A, B) is full range if and only if $b_n \neq 0$. $\square$

As with null kernel pairs, full-range pairs will be important in realization theory for rational matrix functions and in linear systems theory (see Chapters 7 and 8).

We conclude this section with an analog of Theorem 2.7.4 concerning the full-range part of a pair of transformations.

Theorem 2.8.5

Given transformations $A: \mathbb{C}^n \to \mathbb{C}^n$, $B: \mathbb{C}^s \to \mathbb{C}^n$, let $\mathcal{N}_1$ be the minimal A-invariant subspace over $\text{Im } B$. Then for every direct complement $\mathcal{N}_2$ to $\mathcal{N}_1$ in $\mathbb{C}^n$, and with respect to the decomposition $\mathbb{C}^n = \mathcal{N}_1 \dotplus \mathcal{N}_2$, the transformations A and B have the block matrix form

$$A = \begin{bmatrix} A_{11} & A_{12} \\ 0 & A_{22} \end{bmatrix}, \qquad B = \begin{bmatrix} B_1 \\ 0 \end{bmatrix} \tag{2.8.5}$$

where the pair $A_{11}: \mathcal{N}_1 \to \mathcal{N}$, $B_1: \mathbb{C}^s \to \mathcal{N}_1$ is full-range. If $\mathbb{C}^n = \mathcal{N}_1' \dotplus \mathcal{N}_2'$ is another direct sum decomposition with respect to which A and B have the form

$$A = \begin{bmatrix} A_{11}' & A_{12}' \\ 0 & A_{22}' \end{bmatrix}, \qquad B = \begin{bmatrix} B_1' \\ 0 \end{bmatrix} \tag{2.8.6}$$

with full-range pair (A_{11}', B_1'), then $\mathcal{N}_1' = \mathcal{N}_1$ and $A_{11}' = A_{11}$, $B_1' = B_1$.

Proof. Equality (2.8.5) holds because $\mathcal{N}_1$ is A invariant and $\mathcal{N}_1 \supset \text{Im } B$. Further, in view of (2.8.5) we have

$$\sum_{j=0}^{\infty} \text{Im}(A_{11}^j B_1) = \sum_{j=0}^{\infty} \text{Im}(A^j B) = \mathcal{N}_1$$

so (A_{11}, B_1) is indeed a full-range pair. If (2.8.6) holds for a direct sum decomposition $\mathbb{C}^n = \mathcal{N}_1' + \mathcal{N}_2'$, then

$$\sum_{j=0}^{\infty} \text{Im}(A^j B) = \sum_{j=0}^{\infty} \text{Im}((A_{11}')^j B_1')$$

which is equal to $\mathcal{N}_1'$ in view of the full-range property of (A_{11}', B_1). Hence $\mathcal{N}_1'$ is the minimal A-invariant subspace over $\text{Im } B$ and thus $\mathcal{N}_1' = \mathcal{N}_1$. Now clearly $A_{11}' = A_{11}$ (which is the restriction of A to $\mathcal{N}_1' = \mathcal{N}_1$) and $B_1' = B_1$. $\square$

2.9 MARKED INVARIANT SUBSPACES

Let $A: \mathbb{C}^n \to \mathbb{C}^n$ be a transformation, and let

$$f_{11}, \ldots, f_{1k_1}; f_{21}, \ldots, f_{2k_2}; \ldots; f_{p1}, \ldots, f_{pk_p}$$

be a basis in which A has the Jordan form

$$A = J_{k_1}(\lambda_1) \oplus \cdots \oplus J_{k_p}(\lambda_p)$$

Obviously, any subspace of the form

$$\text{Span}\{f_{11}, \ldots, f_{1,m_1}, f_{21}, \ldots, f_{2,m_2}, \ldots, f_{p1}, \ldots, f_{pm_p}\} \quad (2.9.1)$$

for some choice of integers m_i, $0 \le m_i \le k_i$, is A invariant. [Here $m_i = 0$ is interpreted in the sense that the vectors $f_{i1}, \ldots, f_{i,k_i}$ do not appear in (2.9.1) at all.] Such A-invariant subspaces are called *marked* (with respect to the given basis f_{ij} in which A is in the Jordan form).

The following example shows that, in general, not every A-invariant subspace is marked (with respect to some Jordan basis for A).

EXAMPLE 2.9.1. Let

$$A = \begin{bmatrix} 0 & 1 & 1 & 0 \\ 0 & 0 & 0 & 0 \\ 0 & 0 & 0 & 1 \\ 0 & 0 & 0 & 0 \end{bmatrix} : \mathbb{C}^4 \to \mathbb{C}^4$$

We shall verify that the A-invariant subspace $\mathcal{M} = \mathrm{Span}\{e_1, e_2\}$ is not marked in any Jordan basis for A. Indeed, it is easy to see (because $A^2 \neq 0$ and rank $A = 2$) that the Jordan form of A is

$$
J = \begin{bmatrix} 0 & 1 & 0 & 0 \\ 0 & 0 & 1 & 0 \\ 0 & 0 & 0 & 0 \\ 0 & 0 & 0 & 0 \end{bmatrix}
$$

So any Jordan basis of A is of the form f_1, f_2, f_3, g, where $Af_1 = Ag = 0$, $Af_2 = f_1$, $Af_3 = f_2$. If $\mathcal{M}$ were marked with respect to this basis, we would have either $\mathcal{M} = \mathrm{Span}\{f_1, g\}$ or $\mathcal{M} = \mathrm{Span}\{f_1, f_2\}$. The former case is impossible because $A|_{\mathcal{M}} \neq 0$, and the latter case is impossible because it implies $\mathcal{M} \subset \mathrm{Im}\, A$, which is not true ($e_2 \notin \mathrm{Im}\, A$). $\quad \square$

The description of marked invariant subspaces can be reduced to the description of invariant subspaces which are marked with respect to a fixed Jordan basis. This reduction is achieved with the use of matrices commuting with J.

Theorem 2.9.1

Let J be an $n \times n$ matrix in Jordan form. Then every marked J-invariant subspace $\mathcal{L}$ can be represented in the form $\mathcal{L} = B\mathcal{M}$, where $\mathcal{M}$ is marked (with respect to the standard basis, $e_1, \ldots, e_n$ in $\mathbb{C}^n$) and B is an $n \times n$ matrix commuting with J.

Proof. Assume $\mathcal{L} = B\mathcal{M}$, where $\mathcal{M}$ is marked (with respect to the standard basis) and $BJ = JB$. Denoting by $f_1, \ldots, f_n$ the columns of B, we find that $\mathcal{L}$ is a marked J-invariant subspace in the basis $f_1, \ldots, f_n$. (In view of the equality $BJ = JB$, the matrix J has the same Jordan form in the basis $f_1, \ldots, f_n$.)

Conversely, if $\mathcal{L}$ is marked with respect to some Jordan basis $f_1, \ldots, f_n$ of J, then, denoting $B = [f_1 f_2 \cdots f_n]$ and $\mathcal{M} = B^{-1}\mathcal{L}$, we obtain L in the required form $\mathcal{L} = B\mathcal{M}$. $\quad \square$

Note that the characteristic property of a marked invariant subspace depends only on the parts of this subspace corresponding to each eigenvalue: an A-invariant subspace $\mathcal{M}$ is marked if and only if for every eigenvalue λ_0 of A the $A|_{\mathcal{R}_{\lambda_0}(A)}$-invariant subspace $\mathcal{M} \cap \mathcal{R}_{\lambda_0}(A)$ is marked. This follows immediately from the definition of a marked subspace.

In view of Example 2.9.1 it is of interest to find transformations for which every invariant subspace is marked. We have the following result.

Theorem 2.9.2

Let $A: \mathbb{C}^n \to \mathbb{C}^n$ be a transformation such that, for every eigenvalue λ_0 of A, at least one of the following holds: (a) the geometric multiplicity of λ_0 is equal to its algebraic multiplicity; (b) $\dim \mathrm{Ker}(A - \lambda_0 I) = 1$. Then every A-invariant subspace is marked.

Proof. Considering $\mathcal{M} \cap \mathcal{R}_{\lambda_0}(A)$ and $A|_{\mathcal{R}_{\lambda_0}}(A)$ in place of $\mathcal{M}$ and A, respectively, we can assume that A has a single eigenvalue λ_0.

If $\dim \mathrm{Ker}(A - \lambda_0 I) = 1$, then there is a unique maximal chain of A-invariant subspaces:

$$\{0\}, \mathrm{Span}\{f_1, f_2, \dots, f_k\}, \qquad k = 1, \dots, n$$

where $f_1, f_2, \dots, f_n$ is any Jordan chain for A. So obviously every A-invariant subspace is marked.

Assume now that the geometric multiplicity of the eigenvalue λ_0 of A is equal to its algebraic multiplicity. Then $A|_{\mathcal{R}_{\lambda_0}}(A) = \lambda_0 I$, and since every nonzero vector in $\mathcal{R}_{\lambda_0}(A)$ is an eigenvector for A, again every A-invariant subspace is marked. $\square$

It is easy to produce examples of transforamtions for which the hypotheses of Theorem 2.9.2 fail, but nevertheless every invariant subspace is marked; for example

$$\begin{bmatrix} 0 & 1 & 0 \\ 0 & 0 & 0 \\ 0 & 0 & 0 \end{bmatrix}$$

2.10 FUNCTIONS OF TRANSFORMATIONS

We recall the definition of functions of matrices. Let $f(\lambda) = \Sigma_{i=0}^{l} \lambda^i f_i$ be a scalar polynomial of the complex variable λ, and let $A: \mathbb{C}^n \to \mathbb{C}^n$ be a transformation written as a matrix in the standard basis. Then $f(A)$ is defined as $f(A) = \Sigma_{i=0}^{l} f_i A^i$. Letting $J_k(\lambda)$ be the Jordan block of size k with eigenvalue λ, define

$$J = S^{-1} A S = J_{k_1}(\lambda_1) \oplus \cdots \oplus J_{k_p}(\lambda_p)$$

a Jordan form for A. Then

$$f(A) = \sum_{i=0}^{l} f_i A^i = \sum_{i=0}^{l} f_i (SJS^{-1})^i = S \left[\sum_{i=0}^{l} f_i J^i \right] S^{-1}$$

$$= S \left\{ \sum_{i=0}^{l} f_i [J_{k_1}(\lambda_1)^i \oplus \cdots \oplus J_{k_p}(\lambda_p)^i] \right\} S^{-1}$$

A computation shows that

$$
J_k(\lambda)^2 =
\begin{bmatrix}
\lambda^2 & 2\lambda & 1 & 0 & \cdots & 0 \\
0 & \lambda^2 & 2\lambda & 1 & & \vdots \\
\vdots & & & & & \\
& & & & & 2\lambda \\
0 & 0 & & \cdots & & \lambda^2
\end{bmatrix}
$$

and in general the (s, q) entry of $J_k(\lambda)^i$ is $\begin{pmatrix} i \\ q - s \end{pmatrix} \lambda^{i-(q-s)}$ if $q \geq s$ and zero otherwise (here $\begin{pmatrix} i \\ q - s \end{pmatrix} = i!/[(q-s)!(i-(q-s))!]$ if $i \geq q - s$ and $\begin{pmatrix} i \\ q - s \end{pmatrix} = 0$ if $i < q - s$). It follows that

$$
\sum_{i=0}^{l} f_i[J_k(\lambda)]^i =
\begin{bmatrix}
f(\lambda) & \dfrac{1}{1!}f'(\lambda) & \dfrac{1}{2!}f''(\lambda) & \cdots & \dfrac{1}{(k-1)!}f^{(k-1)}(\lambda) \\
0 & f(\lambda) & \dfrac{1}{1!}f'(\lambda) & \cdots & \dfrac{1}{(k-2)!}f^{(k-2)}(\lambda) \\
& & \ddots & & \vdots \\
0 & 0 & \cdots & 0 & f(\lambda)
\end{bmatrix}
$$

$$
f(A) = S \left\{ \sum_{j=1}^{p} \oplus
\begin{bmatrix}
f(\lambda_j) & \dfrac{1}{1!}f'(\lambda_j) & \dfrac{1}{2!}f''(\lambda_j) & \cdots & \dfrac{1}{(k_j-1)!}f^{(k_j-1)}(\lambda_j) \\
0 & f(\lambda_j) & \dfrac{1}{1!}f'(\lambda_j) & \cdots & \dfrac{1}{(k_j-2)!}f^{(k_j-2)}(\lambda_j) \\
& & \ddots & & \\
0 & 0 & \cdots & 0 & f(\lambda_j)
\end{bmatrix}
\right\} S^{-1}
$$

$$(2.10.1)$$

Hence for fixed A the matrix $f(A)$ depends only on the values of the derivatives

$$
f(\mu_j), \ldots, f^{(m_j-1)}(\mu_j), \qquad j = 1, \ldots, r
$$

where $\mu_1, \ldots, \mu_r$ are all the different eigenvalues of A and m_j is the *height* of μ_j, that is, the maximal size of Jordan blocks with eigenvalue μ_j in a Jordan form of A. Equivalently, the height of μ_j is the minimal integer m such that $\mathrm{Ker}(A - \mu_j I)^m = R_{\mu_j}(A)$. This observation allows us to define $f(A)$ by equality (2.10.1) not only for polynomials $f(\lambda)$, but also for complex-valued functions that are analytic in a neighbourhood of each eigenvalue of A.

Note that for a fixed A the correspondence $f(\lambda) \to f(A)$ is an algebraic homomorphism. This means that for any two functions $f(\lambda)$ and $g(\lambda)$ that are analytic in a neighbourhood of each eigenvalue of A the following holds:

$$(\alpha f + \beta g)(A) = \alpha f(A) + \beta g(A) \; ; \qquad \alpha, \beta \in \mathbb{C} \; ; \qquad (fg)(\dot{A}) = f(A)g(A)$$

On the left-hand side the function $\alpha f + \beta g$ (which is analytic in a neighbourhood of each eigenvalue of A) is naturally defined by

$$(\alpha f + \beta g)(\lambda) = \alpha f(\lambda) + \beta g(\lambda)$$

Also, we define

$$(fg)(\lambda) = f(\lambda)g(\lambda) \tag{2.10.2}$$

These properties can be verified by a straightforward computation using (2.10.1). For example:

$$
\begin{bmatrix}
f(\lambda_0) & \dfrac{1}{1!}f'(\lambda_0) & \cdots & \dfrac{1}{(k-1)!}f^{(k-1)}(\lambda_0) \\[2mm]
0 & f(\lambda_0) & \cdots & \dfrac{1}{(k-2)!}f^{(k-2)}(\lambda_0) \\[2mm]
\vdots & \vdots & \ddots & \vdots \\[2mm]
0 & 0 & \cdots & f(\lambda_0)
\end{bmatrix}
$$

$$
\times
\begin{bmatrix}
g(\lambda_0) & \dfrac{1}{1!}g'(\lambda_0) & \cdots & \dfrac{1}{(k-1)!}g^{(k-2)}(\lambda_0) \\[2mm]
0 & g(\lambda_0) & \cdots & \dfrac{1}{(k-2)!}g^{(k-2)}(\lambda_0) \\[2mm]
\vdots & \vdots & \ddots & \vdots \\[2mm]
0 & 0 & \cdots & g(\lambda_0)
\end{bmatrix}
$$

$$
=
\begin{bmatrix}
h(\lambda_0) & \dfrac{1}{1!}h'(\lambda_0) & \cdots & \dfrac{1}{(k-1)!}h^{(k-1)}(\lambda_0) \\[2mm]
0 & h(\lambda_0) & \cdots & \dfrac{1}{(k-2)!}h^{(k-2)}(\lambda_0) \\[2mm]
\vdots & \vdots & \ddots & \vdots \\[2mm]
0 & 0 & \cdots & h(\lambda_0)
\end{bmatrix}
$$

where $f(\lambda)$ and $g(\lambda)$ are analytic functions in a neighbourhood of λ_0 and $h(\lambda) = f(\lambda)g(\lambda)$. In particular, the property (2.10.2) ensures that $f(A)g(A) = g(A)f(A)$ for any functions $f(\lambda)$ and $g(\lambda)$ that are analytic in a neighbourhood of each eigenvalue of A.

In the sequel we need integral formulas for functions of matrices. Let A

be an $n \times n$ matrix, and let Γ be any simple rectifiable contour in the complex plane with the property that all eigenvalues of A are inside Γ. For instance, one can take Γ to be a circle with center 0 and radius greater than $\|A\|$ (here and elsewhere the norm $\|A\|$ of a transformation $A: \mathscr{C}^n \to \mathscr{C}^n$ is defined by

$$\|A\| = \max_{\|x\|=1} \|Ax\|$$

where $\|x\| = (|x_1|)^2 + \cdots + (|x_n|^2)^{1/2}$ for a vector $x = \langle x_1, \ldots, x_n \rangle \in \mathscr{C}^n$).

Proposition 2.10.1

$$\frac{1}{2\pi i} \int_\Gamma \lambda^j (I\lambda - A)^{-1} \, d\lambda = A^j, \qquad j = 0, 1, \ldots \qquad (2.10.3)$$

Proof. Suppose first that T is a Jordan block with eigenvalue $\lambda = 0$:

$$T = \begin{bmatrix} 0 & 1 & & \cdots & 0 \\ 0 & 0 & 1 & \cdots & 0 \\ \vdots & \vdots & & \ddots & \vdots \\ & & & & 1 \\ 0 & 0 & & & 0 \end{bmatrix} \qquad (2.10.4)$$

Then

$$(\lambda I - T)^{-1} = \begin{bmatrix} \lambda^{-1} & \lambda^{-2} & \cdots & \lambda^{-n} \\ 0 & \lambda^{-1} & \cdots & \lambda^{-n+1} \\ \vdots & \vdots & \ddots & \vdots \\ & & & \lambda^{-2} \\ 0 & 0 & \cdots & \lambda^{-1} \end{bmatrix} \qquad (2.10.5)$$

(recall that n is the size of T). So

$$\frac{1}{2\pi i} \int_\Gamma \lambda^j (\lambda I - T)^{-1} \, d\lambda = \frac{1}{2\pi i} \int_\Gamma \begin{bmatrix} \lambda^{j-1} & \lambda^{j-2} & \cdots & \lambda^{j-n} \\ 0 & \lambda^{j-1} & & \lambda^{j-n+1} \\ \vdots & \vdots & \ddots & \vdots \\ 0 & 0 & \cdots & \lambda^{j-1} \end{bmatrix} d\lambda$$

$$= \begin{bmatrix} 0 & \cdots & \overset{\displaystyle \overset{j+1}{\downarrow}}{1} & \cdots & 0 \\ & & & \ddots & \vdots \\ \vdots & & & & 1 \\ 0 & \cdots & & \cdots & 0 \end{bmatrix} = T^j$$

It is then easy to verify (2.10.3) for a Jordan block T with eigenvalue λ_0 (not necessarily 0). Indeed, $T - \lambda_0 I$ has an eigenvalue 0, so by the case already considered

$$\frac{1}{2\pi i} \int_{\Gamma_0} \lambda^j [I\lambda - (T - \lambda_0 I)]^{-1} \, d\lambda = (T - \lambda_0 I)^j, \qquad j = 0, 1, \ldots$$

where $\Gamma_0 = \{\lambda - \lambda_0 \mid \lambda \in \Gamma\}$. The change of variables $\mu = \lambda + \lambda_0$ on the left-hand side leads to

$$\frac{1}{2\pi i} \int_{\Gamma} (\mu - \lambda_0)^j (I\mu - T)^{-1} \, d\mu = (T - \lambda_0 I)^j, \qquad j = 0, 1, \ldots \qquad (2.10.6)$$

Now

$$\frac{1}{2\pi i} \int_{\Gamma} \mu^j (I\mu - T)^{-1} \, d\mu = \sum_{p=0}^{j} \binom{j}{p} \lambda_0^{j-p} \frac{1}{2\pi i} \int_{\Gamma} (\mu - \lambda_0)^p (I\mu - T)^{-1} \, d\mu$$

$$= \sum_{p=0}^{j} \binom{j}{p} \lambda_0^{j-1} (T - \lambda_0 I)^p = T^j$$

so (2.10.3) holds for the block T.

Applying (2.10.3) separately for each Jordan block, we can carry the result further for arbitrary Jordan matrices J. Finally, for a given matrix A there exists a Jordan matrix J and an invertible matrix S such that $T = S^{-1}JS$. Since (2.10.3) is already proved for J, we have

$$\frac{1}{2\pi i} \int_{\Gamma} \lambda^j (I\lambda - A)^{-1} \, d\lambda = S^{-1} \frac{1}{2\pi i} \int_{\Gamma} \lambda^j (I\lambda - J)^{-1} \, d\lambda \cdot S = S^{-1}J^jS = A^j \qquad \square$$

As a consequence of Proposition 2.10.1 we see that for a scalar polynomial $f(\lambda)$ the formula

$$\frac{1}{2\pi i} \int_{\Gamma} f(\lambda)(I\lambda - A)^{-1} \, d\lambda = f(A)$$

holds. Note that here Γ can be replaced by a composite contour that consists of a small circle around each eigenvalue of A. (Indeed, the matrix function $(I\lambda - A)^{-1}$ is analytic outside the spectrum of A.) Using this observation and formula (2.10.1) we see that for any function that is analytic in a neighbourhood of each eigenvalue of A, the formula

$$f(A) = \frac{1}{2\pi i} \int_{\Gamma} f(\lambda)(I\lambda - A)^{-1} \, d\lambda$$

holds, where Γ consists of a sufficiently small circle around each eigenvalue of A [so that $f(\lambda)$ is analytic inside and on Γ].

A transformation $A: \mathbb{C}^n \to \mathbb{C}^n$ (or an $n \times n$ matrix A) is called *diagonable* if there exist eigenvectors $x_1, \ldots, x_n$ of A that form a basis in $\mathbb{C}^n$. Equivalently, an $n \times n$ matrix A is diagonable if for some nonsingular matrix S the matrix $S^{-1}AS$ has a diagonal form:

$$S^{-1}AS = \text{diag}[\alpha_1 \quad \alpha_2 \quad \cdots \quad \alpha_n]$$

So a diagonable matrix has n Jordan blocks in its Jordan form with each block of size 1. If one knows that A is diagonable, then $f(A)$ can be given a meaning [by the same formula (2.10.1)] for every function $f(\lambda)$ that is defined on the set of all eigenvalues of A. So, given a diagonable A, there is an S such that

$$S^{-1}AS = \text{diag}[\alpha_1 \quad \cdots \quad \alpha_n]$$

For any function $f(\lambda)$ that is defined for $\lambda = \alpha_1, \ldots, \lambda = \alpha_n$, put

$$f(A) = S\,\text{diag}[f(\alpha_1) \quad \cdots \quad f(\alpha_n)]S^{-1}$$

In particular, $f(A)$ is defined for a hermitian A and any function f defined on $\mathbb{R}$. Also, for a unitary A and any function f defined on the unit circle, the matrix $f(A)$ is well defined in this way.

Consider now the application of these ideas to the exponential function. This is subsequently used in connection with the solution of systems of differential equations with constant coefficients. As $f(\lambda) = e^\lambda$ is analytic on the whole complex plane, the linear transformation $f(A) = e^A$ is defined for every linear transfomation $A: \mathbb{C}^n \to \mathbb{C}^n$. In fact

$$e^A = I + A + \frac{A^2}{2!} + \frac{A^3}{3!} + \cdots \tag{2.10.7}$$

is given by the same power series as e^λ. In order to verify (2.10.7), we can assume that A is in the Jordan form:

$$A = J_{k_1}(\lambda_1) \oplus \cdots \oplus J_{k_p}(\lambda_p)$$

Then, by definition

$$e^A = \bigoplus_{i=1}^{p} \begin{bmatrix} e^{\lambda_i} & \frac{1}{1!}e^{\lambda_i} & \cdots & \frac{1}{(k_i-1)!}e^{\lambda_i} \\ 0 & e^{\lambda_i} & \cdots & \frac{1}{(k_i-2)!}e^{\lambda_i} \\ \vdots & \vdots & \ddots & \vdots \\ 0 & 0 & \cdots & e^{\lambda_i} \end{bmatrix}$$

On the other hand

$$(J_{k_i}(\lambda_i))^j = \begin{bmatrix} \lambda_i^j & \binom{j}{1}\lambda_i^{j-1} & \binom{j}{2}\lambda_i^{j-2} & \cdots & \binom{j}{k_i-1}\lambda_i^{j-k_i+1} \\ 0 & \lambda_i^j & \binom{j}{1}\lambda_i^{j-1} & \cdots & \binom{j}{k_i-2}\lambda_i^{j-k_i+2} \\ \vdots & \vdots & & \ddots & \vdots \\ 0 & 0 & \cdots & 0 & \lambda_i^j \end{bmatrix}$$

$j = 0, 1, \ldots$. So the (s, q) $(q \geq s)$ entry in the matrix

$$I + J_{k_i}(\lambda_i) + \frac{(J_{k_i}(\lambda_i))^2}{2!} + \cdots$$

is

$$\sum_{t=0}^{\infty} \frac{1}{t!}\binom{t}{q-s}\lambda_i^{t-(q-s)} = \sum_{t=q-s}^{\infty} \frac{1}{(q-s)!(t-(q-s))!}\lambda_i^{t-(q-s)} = \frac{1}{(q-s)!}e^{\lambda_i}$$

Hence formula (2.10.7) follows.

This argument shows that the series (2.10.7) converges for every transformation A. Actually, it converges absolutely in the sense that the series

$$I + \|A\| + \frac{\|A\|^2}{2!} + \frac{\|A\|^3}{3!} + \cdots$$

converges as well.

The exponential function appears naturally in the solution of systems of differential equations of the type

$$\begin{cases} \dfrac{dx_1(t)}{dt} = \displaystyle\sum_{j=1}^{n} a_{1j}x_j(t) \\[2mm] \dfrac{dx_2(t)}{dt} = \displaystyle\sum_{j=1}^{n} a_{2j}x_j(t) \\[1mm] \vdots \\[1mm] \dfrac{dx_n(t)}{dt} = \displaystyle\sum_{j=1}^{n} a_{nj}x_j(t) \end{cases}$$

Here a_{kj} are fixed (i.e., independent of t) complex numbers, and $x_1(t), \ldots, x_n(t)$ are functions of the real variable t to be found. Denoting $A = [a_{kj}]_{k,j=1}^{n}$ and $x(t) = \langle x_1(t), \ldots, x_n(t) \rangle$, we rewrite this system in the form

$$\frac{dx(t)}{dt} = Ax(t)$$

A general solution is given by the formula

$$x(t) = e^{tA}x_0, \qquad -\infty < t < \infty \qquad (2.10.8)$$

where $x_0 = x(0)$ is the initial value of $x(t)$.

In connection with this formula observe that $e^{(t+s)A} = e^{tA}e^{sA}$, as follows, for instance, from (2.10.7). In fact, $e^{A+B} = e^A e^B$ provided A and B commute. However, e^{A+B} is not equal to $e^A \cdot e^B$ in general.

2.11   PARTIAL MULTIPLICITIES AND INVARIANT SUBSPACES OF FUNCTIONS OF TRANSFORMATIONS

From the definition of a function of a transformation $A: \mathbb{C}^n \to \mathbb{C}^n$ it follows immediately that if $\lambda_1, \ldots, \lambda_n$ are the eigenvalues of A (not necessarily distinct), then $f(\lambda_1), \ldots, f(\lambda_n)$ are the eigenvalues of $f(A)$. Moreover we compute the partial multiplicities of $f(A)$, as follows.

Theorem 2.11.1

Let $A: \mathbb{C}^n \to \mathbb{C}^n$ be a transformation with distinct eigenvalues $\mu_1, \ldots, \mu_r$ and partial multiplicities $m_{i1}, \ldots, m_{ik_i}$ corresponding to μ_i, $i = 1, \ldots, r$. Let $f(\lambda)$ be an analytic function in a neighbourhood of each μ_i (if all m_{ij} are 1, it is sufficient to require that $f(\mu_i)$ be defined for $i = 1, \ldots, r$). For each m_{ij} define a positive integer s_{ij} as follows: $s_{ij} = m_{ij}$ if $m_{ij} = 1$ or if $f^{(k)}(\mu_i) = 0$ for $k = 1, \ldots, m_{ij} - 1$; otherwise $f^{(s_{ij})}(\mu_i)$ is the first nonvanishing derivative of $f(\lambda)$ at μ_i. Then the partial multiplicities of $f(A)$ corresponding to the eigenvalue λ are as follows:

$$\left[\frac{m_{ij}}{s_{ij}}\right] \text{ repeated } s_{ij}\left[\frac{m_{ij}}{s_{ij}}\right] - m_{ij} + s_{ij} \text{ times}, \qquad j = 1, \ldots, k_i$$

$$\left[\frac{m_{ij}}{s_{ij}}\right] + 1 \text{ repeated } m_{ij} - s_{ij}\left[\frac{m_{ij}}{s_{ij}}\right] \text{ times}, \qquad j = 1, \ldots, k_i$$

for all indices i such that $f(\mu_i) = \lambda$.

Proof.   By Corollary 2.2.2, if suffices to consider the case when $A = J_m(\mu)$ is a Jordan block. Using equations (2.10.1), we see that

$$\dim \text{Ker}[f(A) - f(\mu)I] = s$$

where $f^{(s)}(\mu)$ is the first nonvanishing derivative of $f(\lambda)$ at μ. [If $m = 1$ or if $f^{(k)}(\mu) = 0$ for $k = 1, \ldots, m$, we put $s = m$.] More generally

$$\dim \text{Ker}[f(A) - f(\mu)I]^j = \min(m, js), \qquad j = 1, 2, \ldots \quad (2.11.1)$$

Denoting the left-hand side of this relation by t_j, note that the sizes of Jordan blocks of $f(A)$ are uniquely determined by the sequence $t_1, \ldots, t_m$. Indeed, the number of Jordan blocks of $f(A)$ with size not less than j is just $t_j - t_{j-1}$, where $j = 1, \ldots, m$ and t_0 is zero by definition. This observation, together with (2.11.1), leads to the conclusion of the theorem. $\square$

Let us give an illustrative example for Theorem 2.11.1.

EXAMPLE 2.11.1.   Let A be a 23×23 matrix with only two distinct eigenvalues 0 and 1, and with partial multiplicities 1,4,9 corresponding to the eigenvalue 0, and with partial multiplicities 2,7 corresponding to the eigenvalue 1. Let $f(\lambda) = \lambda^2(\lambda - 1)^4$. Then $f(A)$ has the unique eigenvalue 0, and the different partial multiplicities of A have the following contribution to the partial multiplicity (PM) of $f(A)$, according to Theorem 2.7.1:

The PM 1 for A gives rise to the PM 1 of $f(A)$.
The PM 4 of A gives rise to the PM values 2,2 of $f(A)$.
The PM 9 of A gives rise to the PM values 4,5 of $f(A)$.
The PM 2 of A gives rise to the PM values 1,1 of $f(A)$.
The PM 7 of A gives rise to the PM values 1,2,2,2 of $f(A)$.

Hence a Jordan form for the transformation $A^2(A - I)^4$ has four Jordan blocks of size 1,5 Jordan blocks of size 2, one Jordan block of size 4 and one Jordan block of size 5, all corresponding to the eigenvalue zero. $\square$

Note that for a given transformation $A: \mathcal{C}^n \to \mathcal{C}^n$ and a function $f(\lambda)$ such that $f(A)$ can be defined as above, there exists a polynomial $p(\lambda)$ such that $p(A) = f(A)$. Indeed, take $p(\lambda)$ such that

$$p(\mu_j) = f(\mu_j), \ldots, p^{(m_j-1)}(\mu_j) = f^{(m_j-1)}(\mu_j), \qquad j = 1, \ldots, r$$

where $\mu_1, \ldots, \mu_r$ are all the different eigenvalues of A and m_j is the height of μ_j.

Consider now the connections between invariant subspaces of A and the invariant subspaces of a function of A.

Proposition 2.11.2

If $\mathcal{M}$ is an invariant subspace of a transformation A, then $\mathcal{M}$ is also invariant for every transformation $f(A)$, where $f(\lambda)$ is a function for which $f(A)$ is defined.

The proof is immediate:

$$f(A) = p(A) = \sum_{j=0}^{m} p_j A^j$$

for some polynomial $p(\lambda) = \sum_{j=0}^{m} p_j \lambda^j$; so for every $x \in \mathcal{M}$ we have $A^j x \in \mathcal{M}$, $j = 0, \ldots, m$, and thus

$$\left(\sum_{j=0}^{m} p_j A^j \right) x \in \mathcal{M}$$

Note that in general the linear transformation $f(A)$ may have more invariant subspaces than A, as the following example shows.

EXAMPLE 2.11.2.   Let

$$A = \begin{bmatrix} 0 & 1 \\ 0 & 0 \end{bmatrix}$$

The invariant subspaces of A are $\{0\}$, $\text{Span}\{e_1\}$, $\mathbb{C}^2$, but the invariant subspaces of $A^2 = 0$ are all the subspaces in $\mathbb{C}^2$.   □

We characterize the cases when $f(A)$ has exactly the same invariant subspaces as A.

Theorem 2.11.3

(a) *Assume that $f(\lambda)$ is an analytic function in a neighbourhood of each eigenvalue $\mu_1, \ldots, \mu_r$ of A ($\mu_1, \ldots, \mu_r$ are assumed to be distinct). Then $f(A)$ has exactly the same invariant subspaces as A if and only if the following conditions hold: (i) $f(\mu_i) \neq f(\mu_j)$ if $\mu_i \neq \mu_j$; (ii) $f'(\mu_i) \neq 0$ for every eigenvalue μ_j with height greater than 1. (b) If A is diagonable and $f(\lambda)$ is a function defined at each eigenvalue of A, then $f(A)$ has exactly the same invariant subspaces as A if and only if condition (i) of part (a) holds.*

Proof.   We shall assume that A has the Jordan form

$$A = J_{k_1}(\lambda_1) \oplus \cdots \oplus J_{k_p}(\lambda_p)$$

where each λ_i coincides with some μ_j, $1 \leq j \leq r$.

Suppose that (i) does not hold, and suppose, for instance, that $\lambda_1 \neq \lambda_2$ but $f(\lambda_1) = f(\lambda_2)$. Formula (2.10.1) shows that $e_1 + e_{k_1+1}$ is an eigenvector of $f(A)$ corresponding to the eigenvalue $f(\lambda_1)$. Hence $\text{Span}\{e_1 + e_{k_1+1}\}$ is $f(A)$ invariant; but this subspace is easily seen not to be A invariant.

Suppose that (ii) does not hold; say, $k_1 > 1$ and $f'(\lambda_1) = 0$. Formula (2.6.1) implies that $e_1 + e_2$ is an eigenvector of $f(A)$ corresponding to $f(\lambda_1)$. So $\text{Span}\{e_1 + e_2\}$ is an $f(A)$-invariant subspace that is not A invariant.

Assume now that (i) and (ii) hold. As $f(A) = p(A)$ for some polynomial $p(\lambda)$, we can assume that $f(\lambda)$ is itself a polynomial. Condition (i) imposed on the polynomial f ensures that the root subspace of A corresponding to some eigenvalue λ_0 is also a root subspace of $f(A)$ corresponding to the eigenvalue $f(\lambda_0)$. Since every A-invariant [resp. $f(A)$-invariant] subspace is a direct sum of A-invariant [resp. $f(A)$-invariant] subspaces, each summand belonging to a root subspace, we can assume that $\sigma(A)$ consists of a single point; say, $\sigma(A) = \{0\}$. Replacing, if necessary, $f(\lambda)$ by $\alpha f(\lambda) + \beta$, where $\alpha, \beta \in \mathbb{C}$ are constants and $\alpha \neq 0$—such a replacement does not alter the set $\mathrm{Inv}(f(A))$ of all $f(A)$-invariant subspaces—we can assume that $f(0) = 0$, $f'(0) = 1$. In this case

$$f(A) = A + \sum_{i=2}^{p} \alpha_i A^i, \qquad \alpha_i \in \mathbb{C}$$

But then $f(A) = AF$, where $F = I + \sum_{i=1}^{p-1} \alpha_{i+1} A^i$ is an invertible matrix. Clearly, every A-invariant subspace is also AF invariant. Note that F^{-1} is a polynomial in AF (this can be checked, for instance, by direct computation in each Jordan block of A, using the fact that A is a Jordan matrix and $\sigma(A) = \{0\}$); so every AF-invariant subspace is also $(AF \cdot F^{-1})$ invariant, that is, A invariant. Thus we have proved that $\mathrm{Inv}(f(A)) = \mathrm{Inv}\, A$. $\square$

2.12   EXERCISES

2.1   Let

$$A = \begin{bmatrix} A_1 & 0 \\ 0 & A_2 \end{bmatrix} \colon \mathbb{C}^{m+n} \to \mathbb{C}^{m+n}$$

where $A_1 \colon \mathbb{C}^m \to \mathbb{C}^m$ and $A_2 \colon \mathbb{C}^n \to \mathbb{C}^n$ are transformations.

(a)   Prove or disprove the following statement: every A-invariant subspace is a direct sum of an A_1-invariant subspace and an A_2-invariant subspace.

(b)   Prove or disprove the preceding statement under the additional condition that the spectra of A_1 and A_2 do not intersect.

(c)   Prove or disprove the preceding statement under the additional condition that A_1 and A_2 are unicellular with the same eigenvalue.

2.2   Let $A \colon \mathbb{C}^n \to \mathbb{C}^n$ be a transformation with $A^2 = I$. Describe the root subspaces of A.

2.3   Describe the root subspaces of a transformation A such that $A^3 = I$. How many spectral A-invariant subspaces are there?

2.4   Find the root subspaces of the transformation

$$A\begin{bmatrix} \lambda_1 I & B \\ 0 & \lambda_2 I \end{bmatrix} \colon \mathbb{C}^n \oplus \mathbb{C}^n \to \mathbb{C}^n \oplus \mathbb{C}^n$$

where $B \colon \mathbb{C}^n \to \mathbb{C}^n$ is some transformation and $\lambda_1 \neq \lambda_2$. Is it true that $\mathcal{R}_{\lambda_i}(A) = \mathrm{Ker}(\lambda_i I - A)$, $i = 1, 2$?

2.5 Find the Jordan form for the following matrices A:

$$\begin{bmatrix} -2 & -1 & -1 \\ 1 & -1 & 0 \\ 0 & 2 & 1 \end{bmatrix}; \quad \begin{bmatrix} -1 & -2 & 3 \\ 0 & 1 & 0 \\ 0 & 0 & 1 \end{bmatrix}; \quad \begin{bmatrix} 2 & 1 & -1 \\ 0 & 3 & 1 \\ 0 & -1 & 0 \end{bmatrix}$$

For each one of the matrices A and each eigenvalue λ_0 of A, check whether $\mathcal{R}_{\lambda_0}(A) = \text{Ker}(\lambda_0 I - A)$.

2.6 Find all possible Jordan forms of transformations $A: \math{C}^n \to \math{C}^n$ satisfying $A^2 = 0$. Express the number of Jordan blocks of size 2 in terms of A.

2.7 Find the Jordan form of the transformation

$$Q = \begin{bmatrix} 0 & 1 & 0 & \cdots & 0 \\ 0 & 0 & 1 & \cdots & 0 \\ \vdots & \vdots & \vdots & & \vdots \\ 0 & 0 & 0 & \cdots & 1 \\ 1 & 0 & 0 & \cdots & 0 \end{bmatrix} : \math{C}^n \to \math{C}^n$$

2.8 What is the Jordan form of Q^k, $k = 2, 3, \ldots$, where Q is given in Exercise 2.7.]

2.9 Describe the Jordan form of a *circulant matrix*

$$A = \begin{bmatrix} a_1 & a_2 & \cdots & a_{n-1} & a_n \\ a_n & a_1 & \cdots & a_{n-2} & a_{n-1} \\ \vdots & \vdots & & \vdots & \vdots \\ a_2 & a_3 & \cdots & a_n & a_1 \end{bmatrix}$$

where $a_1, \ldots, a_n$ are complex numbers. Prove that there exists an invertible matrix S independent of $a_1, \ldots, a_n$ such that SAS^{-1} is diagonal. [*Hint: A is a polynomial in Q, where Q is defined in Exercise 2.7*].

2.10 What is the Jordan form of the transformation

$$Q_2 = \begin{bmatrix} 0 & I_2 & 0 & \cdots & 0 \\ 0 & 0 & I_2 & \cdots & 0 \\ \vdots & \vdots & \vdots & & \vdots \\ 0 & 0 & 0 & \cdots & I_2 \\ I_2 & 0 & 0 & \cdots & 0 \end{bmatrix} : \math{C}^{2n} \to \math{C}^{2n}?$$

2.11 Find the Jordan form of the transformation

$$Q_k = \begin{bmatrix} 0 & I_k & 0 & \cdots & 0 \\ 0 & 0 & I_k & \cdots & 0 \\ \vdots & \vdots & \vdots & & \vdots \\ 0 & 0 & 0 & \cdots & I_k \\ I_k & 0 & 0 & \cdots & 0 \end{bmatrix} : \mathbb{C}^{nk} \to \mathbb{C}^{nk}$$

2.12 Let $A_1, A_2, \ldots, A_n$ be transformations on $\mathbb{C}^2$, and define

$$A = \begin{bmatrix} A_1 & A_2 & A_3 & \cdots & A_n \\ A_n & A_1 & A_2 & \cdots & A_{n-1} \\ \vdots & \vdots & \vdots & & \vdots \\ A_2 & A_3 & A_4 & \cdots & A_1 \end{bmatrix} : \mathbb{C}^{2n} \to \mathbb{C}^{2n}$$

(a) Show that A is similar to a block diagonal matrix with 2×2 blocks on the main diagonal. [*Hint*: On writing

$$A_j = \begin{bmatrix} b_j & c_j \\ d_j & f_j \end{bmatrix}; \qquad b_j, c_j, d_j, f_j \in \mathbb{C}$$

for $j = 1, \ldots, n$, A is similar to

$$\begin{bmatrix} B & C \\ D & F \end{bmatrix}$$

where B is the circulant matrix

$$\begin{bmatrix} b_1 & b_2 & \cdots & b_n \\ b_n & b_1 & \cdots & b_{n-1} \\ \vdots & \vdots & & \vdots \\ b_2 & b_3 & \cdots & b_1 \end{bmatrix}$$

and analogously for C, D, and F. Now use the existence of one similarity transformation that takes B, C, D, and F to the Jordan form (Exercise 2.9).]

(b) Prove that in the Jordan form of A only Jordan blocks of size 2 or 1 may appear.

(c) Show that if all A_j, $j = 1, \ldots, n$ are diagonal matrices, then A is *diagonable*, that is, the Jordan form of A is a diagonal matrix. Give an example of nondiagonal $A_1, \ldots, A_n$ for which A is diagonable nevertheless.

2.13 Prove that the block circulant matrix

$$
\begin{bmatrix}
A_1 & A_2 & \cdots & A_n \\
A_n & A_1 & \cdots & A_{n-1} \\
\vdots & \vdots & & \vdots \\
A_2 & A_3 & \cdots & A_1
\end{bmatrix} : \mathbb{C}^{nk} \to \mathbb{C}^{nk}
$$

where $A_1, \ldots, A_n$ are $k \times k$ matrices, has Jordan blocks of sizes less than or equal to k in its Jordan form.

2.14 Find the Jordan form for the transformation

$$
A = \begin{bmatrix}
0 & 1 & 0 & \cdots & 0 \\
0 & 0 & 1 & \cdots & 0 \\
\vdots & \vdots & \vdots & & \vdots \\
0 & 0 & 0 & \cdots & 1 \\
a_0 & a_1 & a_2 & \cdots & a_{n-1}
\end{bmatrix}
$$

where $a_0, \ldots, a_{n-1} \in \mathbb{C}$ and the polynomial $\lambda^n - \sum_{j=0}^{n-1} a_j \lambda^j$ has n distinct zeros. Show that a similarity that takes A to its Jordan form is given by the Vandermonde matrix of type

$$
\begin{bmatrix}
1 & 1 & \cdots & 1 \\
\lambda_1 & \lambda_2 & \cdots & \lambda_n \\
\vdots & \vdots & & \vdots \\
\lambda_1^{n-1} & \lambda_2^{n-1} & \cdots & \lambda_n^{n-1}
\end{bmatrix}
$$

2.15 Let

$$
A = \begin{bmatrix}
0 & 1 & 0 & \cdots & 0 \\
0 & 0 & 1 & \cdots & 0 \\
\vdots & \vdots & \vdots & & \vdots \\
0 & 0 & 0 & \cdots & 1 \\
a_0 & a_1 & a_2 & \cdots & a_{n-1}
\end{bmatrix}
$$

(a) Prove that, for each eigenvalue, A has only one Jordan block in its Jordan form. (*Hint*: Use the description of partial multiplicities of A in terms of the matrix polynomial $\lambda I - A$; see the appendix.)

(b) Find the Jordan form of A.

2.16 Show that any matrix of the type

$$
\begin{bmatrix}
0 & I_k & 0 & \cdots & 0 \\
0 & 0 & I_k & \cdots & 0 \\
\vdots & \vdots & \vdots & & \vdots \\
0 & 0 & 0 & \cdots & I_k \\
A_0 & A_1 & A_2 & \cdots & A_{n-1}
\end{bmatrix} : \mathbb{C}^{nk} \to \mathbb{C}^{nk}
$$

where A_j are $k \times k$ matrices, has not more than k Jordan blocks corresponding to each eigenvalue in its Jordan form.

2.17 What is the Jordan form of the *upper triangular Toeplitz matrix*

$$
\begin{bmatrix}
a_0 & a_1 & \cdots & a_{n-1} \\
0 & a_0 & \ddots & \vdots \\
\vdots & \vdots & & a_1 \\
0 & 0 & \cdots & a_0
\end{bmatrix}
$$

where $a_0, \ldots, a_{n-1}$ are complex numbers with $a_1 \neq 0$?

2.18 Find the Jordan form of $(J_n(\lambda_0))^k$, $k = 2, 3, \ldots$. Show that $[J_n(0)]^k$ has infinitely many invariant subspaces if $k \geq 2$.

2.19 Describe the Jordan form of the matrix in Exercise 2.17 without the restriction $a_1 \neq 0$. When does this matrix have infinitely many invariant subspaces? [*Hint*: Observe that the matrix is a polynomial in $J_n(0)$ and use Theorem 2.11.1.]

2.20 Prove that an $n \times n$ matrix A is similar to its transpose A^T.

2.21 Let $A: \mathbb{C}^n \to \mathbb{C}^n$ be a transformation such that $p(A) = 0$, where $p(\lambda)$ is a polynomial of degree k with k distinct zeros $\lambda_1, \ldots, \lambda_k$.

(a) Show that $\operatorname{Ker}(\lambda_j I - A) \neq \{0\}$, $j = 1, \ldots, k$.
(b) Verify the direct sum decomposition

$$
\mathbb{C}^n = \operatorname{Ker}(\lambda_1 I - A) \dotplus \cdots \dotplus \operatorname{Ker}(\lambda_k I - A)
$$

(c) Prove that A is diagonable.

2.22 Assume that the transformation $A: \mathbb{C}^n \to \mathbb{C}^n$ satisfies the equation $p(A) = 0$, where $p(\lambda)$ is a polynomial. Let λ_0 be a zero of $p(\lambda)$, and let k be its multiplicity. Show that the A-invariant subspace $\operatorname{Im} q(A)$, where $q(\lambda) = p(\lambda)(\lambda - \lambda_0)^{-k}$, is spectral.

2.23 Prove that for any transformation $A: \mathbb{C}^n \to \mathbb{C}^n$ the inequalities

$$
\dim \operatorname{Ker} A^{s+1} + \dim \operatorname{Ker} A^{s-1} \leq 2 \dim \operatorname{Ker} A^s, \qquad s = 1, 2, \ldots
$$

hold.

2.24 Prove that a transformation $A: \mathbb{C}^n \to \mathbb{C}^n$ has the property that $A\mathcal{M}^\perp \subset \mathcal{M}^\perp$ for every A-invariant subspace $\mathcal{M}$ if and only if A is normal.

2.25 Show that a transformation has only one-dimensional irreducible subspaces if and only if A is diagonable.

2.26 Find the minimal number of generators in $\mathbb{C}^n$ of the following transformations:

(a) The circulant

$$\begin{bmatrix} a_1 & a_2 & \cdots & a_n \\ a_n & a_1 & \cdots & a_{n-1} \\ \vdots & \vdots & & \vdots \\ a_2 & a_3 & \cdots & a_1 \end{bmatrix}, \qquad a_j \in \mathbb{C}$$

(b) The lower triangular Toeplitz matrix

$$\begin{bmatrix} a_0 & 0 & \cdots & 0 \\ a_1 & a_0 & \cdots & 0 \\ \vdots & \vdots & & \vdots \\ a_{n-1} & a_{n-2} & \cdots & a_0 \end{bmatrix}$$

(c) The companion matrix

$$\begin{bmatrix} 0 & 1 & 0 & \cdots & 0 \\ 0 & 0 & 1 & \cdots & 0 \\ \vdots & \vdots & \vdots & & \vdots \\ & & & & 1 \\ a_0 & a_1 & a_2 & \cdots & a_{n-1} \end{bmatrix}$$

2.27 Prove that if $A: \mathbb{C}^n \to \mathbb{C}^n$ has one-dimensional image, the minimal number of generators of any A-invariant subspace is less than or equal to $n-1$. Show that Ker A is the only nontrivial A-invariant subspace whose minimal number of generators is precisely $n-1$.

2.28 For a given transformation A, denote by $g(\mathcal{M})$ the minimal number of generators in an A-invariant subspace $\mathcal{M}$. Prove that

$$g(\mathcal{M}) = \max g(\mathcal{M} \cap \mathcal{R}_{\lambda_0}(A)]$$

where the maximum is taken over all eigenvalues λ_0 of A [$g(\{0\})$ is interpreted as zero].

2.29 Let

$$A = \begin{bmatrix} A_1 & 0 \\ 0 & A_2 \end{bmatrix}$$

where A_1 and A_2 are transformations such that every invariant subspace of each of them is cyclic. Prove or disprove the following statements:

(a) Every A-invariant subspace is cyclic.

(b) Every A-invariant subspace has not more than two minimal generators.

2.30 Show that the vector $\langle 0, 0, \ldots, 0, 1 \rangle \in \mathcal{C}^n$ is a generator of $\mathcal{C}^n$ as an invariant subspace of a companion matrix.

2.31 Find the minimal A-invariant subspace over Im B for the following pairs of transformations:

(a)

$$A = \begin{bmatrix} a_1 & & a_2 & \cdots & a_n \\ & & a_1 & \cdots & a_{n-1} \\ & 0 & & & \\ & & & \ddots & \\ & & & & a_1 \end{bmatrix}, \qquad B = \begin{bmatrix} 0 \\ 0 \\ \vdots \\ 0 \\ 1 \end{bmatrix}$$

(b)

$$A = \begin{bmatrix} a_1 & 0 & \cdots & 0 \\ a_2 & a_1 & \cdots & 0 \\ \vdots & \vdots & & \vdots \\ a_n & a_{n-1} & \cdots & a_1 \end{bmatrix}, \qquad B = \begin{bmatrix} 0 & 0 \\ \vdots & \vdots \\ 0 & 0 \\ 1 & 0 \\ 0 & 1 \end{bmatrix}$$

(c)

$$A = \begin{bmatrix} & & & a_n I_k \\ & 0 & & \\ & & \ddots & \\ & a_2 I_k & & \\ a_1 I_k & & & 0 \end{bmatrix}, \qquad B = \begin{bmatrix} 0 \\ 0 \\ \vdots \\ 0 \\ I_k \end{bmatrix}$$

Here $a_1, \ldots, a_n$ are complex numbers.

2.32 Find the maximal A-invariant subspace in Ker C for the following pairs of transformations:

(a) $C = [1 \quad 0 \quad \cdots \quad 0]$; A is a companion matrix.

(b) $C = [1 \quad 0 \quad \cdots \quad 0]$; A is an upper triangular Toeplitz matrix.

(c) $C = [I_k \quad 0 \quad \cdots \quad 0]$; A is an in Exercise 2.31, (c).

2.33 Prove or disprove the following statements:

(a) If M_1 is the maximal A-invariant subspace in V_1 and M_2 is the maximal A-invariant subspace in V_2, then $M_1 + M_2$ is the maximal A-invariant subspace in $V_1 + V_2$.

(b) If M_i and V_i ($i = 1, 2$) are as in (a), then $M_1 \cap M_2$ is the maximal A-invariant subspace in $V_1 \cap V_2$.

(c) The analog of (a) for the case of minimal A-invariant subspaces M_i over V_i, $i = 1, 2$.

(d) The analog of (b) for the case of minimal A-invariant subspaces M_i over V_i, $i = 1, 2$.

2.34 Find when the following pairs of matrices are full-range pairs:

(a)
$$\left([J_n(\lambda_0)], \begin{bmatrix} b_1 \\ b_2 \\ \vdots \\ b_n \end{bmatrix} \right), \quad \text{where} \quad b_1, b_2, \ldots, b_n \in \mathbb{C}$$

(b) (A, B), where A is an $n \times n$ matrix with $A^n = 0$ and B is an $n \times 1$ matrix.

2.35 Find when the following pairs of matrices are null kernel pairs:

(a) $([c_1, \ldots, c_n], J_n(\lambda_0)^k)$, where $k \geq 1$ is a fixed integer and $c_1, \ldots, c_n \in \mathbb{C}$.

(b) (C, A), where C is an $1 \times n$ matrix and A is an $n \times n$ upper triangular matrix with zeros on the main diagonal.

2.36 Given a full-range pair $A: \mathbb{C}^n \to \mathbb{C}^n$, $B: \mathbb{C}^m \to \mathbb{C}^n$, prove that if $A': \mathbb{C}^n \to \mathbb{C}^n$, $B': \mathbb{C}^m \to \mathbb{C}^n$ are transformations sufficiently close to A and B, respectively, (i.e., $\|A' - A\| < \epsilon$, $\|B' - B\| < \epsilon$, where $\epsilon > 0$ depends on A and B only), then (A', B') is a full-range pair as well.

2.37 Prove that for every pair of transformations $A: \mathbb{C}^n \to \mathbb{C}^n$, $B: \mathbb{C}^m \to \mathbb{C}^n$ there exists a sequence of full-range pairs (A_p, B_p), $p = 1, 2, \ldots$ such that $\lim_{p \to \infty} \|A_p - A\| = 0$ and $\lim_{p \to \infty} \|B_p - B\| = 0$.

2.38 State and prove the analogs of Exercises 2.36 and 2.37 for null kernel pairs.

2.39 Let A and B be transformations on $\mathbb{C}^n$. Show that the biggest A-invariant (or, equivalently, B-invariant) subspace M for which $A_{|M} = B_{|M}$ consists of all vectors $x \in \mathbb{C}^n$ such that $A^j x = B^j x$, $j = 1, 2, \ldots$.

2.40 Let $A_1, \ldots, A_k$ be transformations on $\mathbb{C}^n$. Show that the biggest A_i-invariant subspace M for which $A_{i|M} = A_{p|M}$ for $p = 1, \ldots, k$, consists of all $x \in \mathbb{C}^n$ such that $A_i^j x = A_p^j x$ for $p = 1, \ldots, k$ and $j = 1, 2, \ldots$.

2.41 Show that the transformation e^A is nonsingular for every transformation $A: \mathbb{C}^n \to \mathbb{C}^n$. Find the eigenvalues and the partial multiplicities of e^A in terms of the eigenvalues and the partial multiplicities of A.

2.42 Give an example of a transformation A such that $\text{Inv}(A)$ is finite but $\text{Inv}(e^A)$ is infinite.

2.43 Show that for a transformation A the series

$$f(A) = I - A + \frac{1}{2}A^2 - \frac{1}{3}A^3 + \cdots$$

converges provided all eigenvalues of A are less than 1 in absolute value. For such an A prove that $A = e^{f(A)} - I$, so one can write $f(A) = \ln(I + A)$. Prove that A and $\ln(I + A)$ have exactly the same invariant subspaces.

2.44 Find all marked A-invariant subspaces for the transformation A of Example 2.9.1.

2.45 Show that for any transformation A, all A-hyperinvariant subspaces are marked.

2.46 For which of the following classes of $n \times n$ matrices are all invariant subspaces marked?

(a) Companion matrices
(b) Block companion matrices

$$\begin{bmatrix} 0 & I & 0 & \cdots & 0 \\ 0 & 0 & I & \cdots & 0 \\ \vdots & \vdots & \vdots & & \vdots \\ A_0 & A_1 & A_2 & \cdots & A_{p-1} \end{bmatrix}$$

with 2×2 blocks A_j ($p = n/2$)
(c) Upper triangular Toeplitz matrices
(d) Circulant matrices
(e) Block circulant matrices

$$\begin{bmatrix} A_1 & A_2 & \cdots & A_p \\ A_p & A_1 & \cdots & A_{p-1} \\ \vdots & \vdots & & \vdots \\ A_2 & A_3 & \cdots & A_1 \end{bmatrix}$$

with 2×2 blocks A_j
(f) Matrices A such that $A^2 = 0$

2.47 Prove that every invariant subspace of a matrix of type

$$
\begin{bmatrix}
\alpha_1 & 0 & 0 & \cdots & 0 & \beta_1 \\
0 & \alpha_2 & 0 & \cdots & \beta_2 & 0 \\
\vdots & \vdots & \vdots & & \vdots & \vdots \\
0 & \beta_{n-1} & 0 & \cdots & \alpha_{n-1} & 0 \\
\beta_n & 0 & 0 & \cdots & 0 & \alpha_n
\end{bmatrix}
$$

is marked.

2.48 Prove that for any transformation $A: \mathbb{C}^3 \to \mathbb{C}^3$ every invariant subspace is marked.

2.49 Find all Jordan forms of transformations $A: \mathbb{C}^4 \to \mathbb{C}^4$ for which there exists a nonmarked invariant subspace.

Chapter Three

Coinvariant and Semiinvariant Subspaces

In this chapter we study two classes of subspaces closely related to invariant ones; namely, coinvariant and semiinvariant subspaces. A subspace is called *coinvariant* if it is a direct complement to an invariant subspace. A subspace is called *semiinvariant* if it is a coinvariant part of an invariant subspace. Also, we introduce here the related notion of a triinvariant decomposition for a transformation. This requires a decomposition of the whole space into a direct sum of three subspaces with respect to which the transformation has a block upper triangular form. It follows that the first, second, and third subspace are invariant, semiinvariant, and coinvariant, respectively. The triinvariant decomposition will play an important role in subsequent applications.

3.1 COINVARIANT SUBSPACES

A subspace $\mathcal{M} \subset \mathbb{C}^n$ is called *coinvariant* for the transformation $A: \mathbb{C}^n \to \mathbb{C}^n$ (or, in short, A coinvariant) if there is an A-invariant direct complement to $\mathcal{M}$ in $\mathbb{C}^n$. Consider some simple examples.

EXAMPLE 3.1.1. Let A be an $n \times n$ Jordan block. Then for each i ($1 \le i \le n$) $\mathrm{Span}\{e_i, e_{i+1}, \ldots, e_n\}$ is an A-coinvariant subspace (although there are many other A-coinvariant subspaces). For this subspace there is a unique A-invariant subspace that is its direct complement, namely, $\mathrm{Span}\{e_1, e_2, \ldots, e_{i-1}\}$ ($\{0\}$ if $i = 1$). Note that, in this case, the only subspaces that are simultaneously A invariant and A coinvariant are the trivial ones $\{0\}$ and $\mathbb{C}^n$. $\square$

EXAMPLE 3.1.2.   Let $A = \text{diag}[\lambda_1, \ldots, \lambda_n]$, where all λ_i are different. As we have seen in Example 1.1.3, the only A-invariant subspaces are $\{0\}$, $\mathbb{C}^n$, and $\text{Span}\{e_{i_1}, \ldots, e_{i_k}\}$, $k = 1, \ldots, n - 1$, for any choice of $i_1 < i_2 < \cdots < i_k$. In contrast, *every* subspace in $\mathbb{C}^n$ is A coinvariant. Indeed, let $\mathcal{M} = \text{Span}\{x_1, \ldots, x_q\}$, where $x_1, \ldots, x_q$ are linearly independent vectors in $\mathbb{C}^n$. Then the columns of the $n \times q$ matrix $X = [x_1 x_2 \cdots x_q]$ are linearly independent. So there exist q rows of X, say, the i_1th, $\ldots$, i_qth rows, which are also linearly independent. Put $\{j_1, \ldots, j_{n-q}\} = \{1, \ldots, n\} \setminus \{i_1, \ldots, i_q\}$ and $\mathcal{N} = \text{Span}\{e_{j_1}, \ldots, e_{j_{n-q}}\}$ so that $\mathcal{N}$ is an A-invariant subspace. As, by construction, the $n \times n$ matrix

$$[x_1 \ x_2 \cdots x_q \ e_{j_1} \ e_{j_2} \cdots e_{j_{n-q}}]$$

is nonsingular, $\mathcal{N}$ is a direct complement to $\mathcal{M}$ in $\mathbb{C}^n$. Thus $\mathcal{M}$ is A coinvariant. $\square$

EXAMPLE 3.1.3.   If $A = \alpha I$, $a \in \mathbb{C}$, then every subspace in $\mathbb{C}^n$ is obviously A coinvariant. For every A-coinvariant subspace $\mathcal{M}$ there is a continuum of A-invariant subspaces that are direct complements to $\mathcal{M}$ in $\mathbb{C}^n$. $\square$

For an A-coinvariant subspace $\mathcal{M}$ and any projector P onto $\mathcal{M}$ such that $\text{Ker } P$ is A invariant, we have $PAP = PA$. This follows, for instance, when equation (1.5.5) is applied to $I - P$, or else it can be proved directly. Conversely, if $PAP = PA$ for some projector P onto a subspace $\mathcal{M} \subset \mathbb{C}^n$, then $\mathcal{M}$ is A coinvariant and $\text{Ker } P$ is an A-invariant direct complement to $\mathcal{M}$ in $\mathbb{C}^n$.

Given an A-coinvariant subspace $\mathcal{M}$ and a projector P onto $\mathcal{M}$ such that $\text{Ker } P$ is A invariant, the linear transformation A has the following block triangular form:

$$A = \begin{bmatrix} A_{11} & 0 \\ A_{21} & A_{22} \end{bmatrix} \qquad (3.1.1)$$

with respect to the decomposition $\mathbb{C}^n = \text{Im } P \dotplus \text{Ker } P$. In particular, we find that every eigenvalue of the *compression* $PA|_{\mathcal{M}} : \mathcal{M} \to \mathcal{M}$ of A to its coinvariant subspace $\mathcal{M}$, is also an eigenvalue of A. Indeed, in the representation (3.1.1) the compression $PA|_{\mathcal{M}}$ coincides with A_{11}, and this immediately implies that $\sigma(PA|_{\mathcal{M}}) \subset \sigma(A)$.

We note that, essentially, the compression to a coinvariant subspace depends on the invariant direct complement only. (Actually, we have encountered this property already in Theorem 2.7.4 and its proof.)

Proposition 3.1.1

Let $\mathcal{M}_1$ and $\mathcal{M}_2$ be A-coinvariant subspaces with a common A-invariant direct complement $\mathcal{N}$. Then the compressions $P_1 A|_{\mathcal{M}_1} : \mathcal{M}_1 \to \mathcal{M}_1$ and

$P_2 A_{|\mathcal{M}_2}: \mathcal{M}_2 \to \mathcal{M}_2$ (where P_j is the projector on $\mathcal{M}_j$ along $\mathcal{N}$ for $j = 1, 2$) are similar.

Proof. Write

$$A = \begin{bmatrix} A_{11} & 0 \\ A_{21} & A_{22} \end{bmatrix}: \mathcal{M}_1 \dotplus \mathcal{N} \to \mathcal{M}_1 \dotplus \mathcal{N}$$

$$A = \begin{bmatrix} A'_{11} & 0 \\ A'_{21} & A'_{22} \end{bmatrix}: \mathcal{M}_2 \dotplus \mathcal{N} \to \mathcal{M}_2 \dotplus \mathcal{N}$$

with respect to the direct sum decompositions $\mathcal{C}^n = \mathcal{M}_1 \dotplus \mathcal{N}$ and $\mathcal{C}^n = \mathcal{M}_2 \dotplus \mathcal{N}$, respectively. Also, write the identity transformation $I: \mathcal{C}^n \to \mathcal{C}^n$ in the 2×2 block matrix form

$$I = \begin{bmatrix} S_{11} & S_{12} \\ S_{21} & S_{22} \end{bmatrix}: \mathcal{M}_1 \dotplus \mathcal{N} \to \mathcal{M}_2 \dotplus \mathcal{N}$$

(so $S_{11}: \mathcal{M}_1 \to \mathcal{M}_2$, $S_{12}: \mathcal{N} \to \mathcal{M}_2$, $S_{21}: \mathcal{M}_1 \to \mathcal{N}$, $S_{22}: \mathcal{N} \to \mathcal{N}$). It is easily seen that $S_{12} = 0$ and $S_{22} = I_N$, the identity transformation on $\mathcal{N}$. As I is invertible, the transformation S_{11} must be invertible as well, and

$$I^{-1} = \begin{bmatrix} S_{11}^{-1} & 0 \\ -S_{21}S_{11}^{-1} & I_{\mathcal{N}} \end{bmatrix}: \mathcal{M}_2 \dotplus \mathcal{N} \to \mathcal{M}_1 \dotplus \mathcal{N}$$

Now

$$\begin{bmatrix} A_{11} & 0 \\ A_{21} & A_{22} \end{bmatrix} = \begin{bmatrix} S_{11}^{-1} & 0 \\ -S_{21}S_{11}^{-1} & I_{\mathcal{N}} \end{bmatrix} \begin{bmatrix} A'_{11} & 0 \\ A'_{21} & A'_{22} \end{bmatrix} \begin{bmatrix} S_{11} & 0 \\ S_{21} & I_{\mathcal{N}} \end{bmatrix}$$

which gives, in particular, $A_{11} = S_{11}^{-1} A'_{11} S_{11}$. It remains to observe that $P_1 A_{|\mathcal{M}_1} = A_{11}$, $P_2 A_{|\mathcal{M}_2} = A'_{11}$. $\square$

The following property of coinvariant subspaces is analogous to the property of A-invariant subspaces proved in Section 1.4.

Proposition 3.1.2

A subspace $\mathcal{M}$ is A coinvariant if and only if its orthogonal complement $\mathcal{M}^\perp$ is A^* coinvariant.

Proof. Assume that $\mathcal{M}$ is A coinvariant, and let $\mathcal{N}$ be an A-invariant direct complement to $\mathcal{M}$ in $\mathcal{C}^n$. Then $\mathcal{M}^\perp \cap \mathcal{N}^\perp = (\mathcal{M} + \mathcal{N})^\perp = (\mathcal{C}^n)^\perp = \{0\}$, and since $\dim \mathcal{M}^\perp + \dim \mathcal{N}^\perp = (n - \dim \mathcal{M}) + (n - \dim \mathcal{N}) = n$, we have $\mathcal{M}^\perp + \mathcal{N}^\perp = \mathcal{C}^n$. As $\mathcal{N}^\perp$ is A^* invariant (see Section 1.4) it follows that $\mathcal{M}^\perp$ is A^* coinvariant. Conversely, if $\mathcal{M}^\perp$ is A^* coinvariant, then by the part of this proposition already proved, the subspace $(\mathcal{M}^\perp)^\perp = \mathcal{M}$ is $(A^*)^*$ coinvariant, that is, $\mathcal{M}$ is A coinvariant. $\square$

A subspace $\mathcal{M} \subset \mathbb{C}^n$ is called *orthogonally coinvariant* for the transformation $A: \mathbb{C}^n \to \mathbb{C}^n$ (in short, orthogonally A coinvariant) if the orthogonal complement $\mathcal{M}^\perp$ of $\mathcal{M}$ is A invariant.

Proposition 3.1.3

A subspace $\mathcal{M}$ is orthogonally A-coinvariant if and only if $\mathcal{M}$ is invariant for the adjoint linear transformation A^*.

Proof. Assume that $\mathcal{M}$ is orthogonally A coinvariant. So $Ax \in \mathcal{M}^\perp$. Then we have

$$(Ax, y) = 0 \qquad (3.1.2)$$

for all $y \in \mathcal{M}$. But the left-hand side of (3.1.2) is just (x, A^*y). Hence $A^*y \in (\mathcal{M}^\perp)^\perp = \mathcal{M}$ for all $y \in \mathcal{M}$ and $\mathcal{M}$ is A^* invariant. Reversing this argument we find that if $\mathcal{M}$ is A^* invariant, then $Ax \in \mathcal{M}^\perp$ for every $x \in \mathcal{M}^\perp$, that is, $\mathcal{M}$ is orthogonally A coinvariant. $\square$

We observe that, in general, A-coinvariant subspaces do not form a lattice, that is, the sum and intersection of A-coinvariant subspaces need not be A coinvariant. This is illustrated in the following example.

EXAMPLE 3.1.4.   Let

$$A = \begin{bmatrix} 0 & 1 & 0 \\ 0 & 0 & 1 \\ 0 & 0 & 0 \end{bmatrix}: \mathbb{C}^3 \to \mathbb{C}^3$$

The only A-invariant subspaces are $\{0\}$, $\mathrm{Span}\{e_1\}$, $\mathrm{Span}\{e_1, e_2\}$, $\mathbb{C}^3$. Consequently, all A-coinvariant subspaces are as follows:

$$\{0\} ; \qquad \mathbb{C}^3;$$

$$\mathrm{Span}\{\langle x, y, 1 \rangle\}, \qquad x, y \in \mathbb{C}$$

$$\mathrm{Span}\{\langle x, 1, 0 \rangle, \qquad \langle y, 0, 1 \rangle\}, \qquad x, y \in \mathbb{C}$$

Indeed, assume $\mathrm{Span}\{u, v\}$ is a two-dimensional subspace for which $\mathrm{Span}\{e_1\}$ is a direct complement. Writing $u = \langle u_1, u_2, u_3 \rangle$, $v = \langle v_1, v_2, v_3 \rangle$ we see that $\det \begin{bmatrix} u_2 & v_2 \\ u_3 & v_3 \end{bmatrix} \neq 0$. Hence replacing u and v with their linear combinations, if necessary, we see that

$$\mathrm{Span}\{u, v\} = \mathrm{Span}\{\langle x, 1, 0 \rangle, \langle y, 0, 1 \rangle\}$$

for some $x, y \in \mathbb{C}$. Now $\mathrm{Span}\{e_2, e_3\}$ and $\mathrm{Span}\{e_2, \langle 1, 0, 1 \rangle\}$ are A-coinvariant subspaces but their intersection (which is equal to $\mathrm{Span}\{e_2\}$) is not. Also, $\mathrm{Span}\{e_3\}$ and $\mathrm{Span}\{\langle 1, 0, 1 \rangle\}$ are A-coinvariant subspaces but their sum (which is equal to $\mathrm{Span}\{e_3, \langle 1, 0, 1 \rangle\}$) is not. $\square$

In contrast, it follows immediately from Proposition 3.1.3 that the set of all orthogonally A-coinvariant subspaces is a lattice. Note also the following property of orthogonally coinvariant subspaces.

Proposition 3.1.4

Any transformation has a complete chain of orthogonally coinvariant subspaces.

Proof. Let $A: \mathbb{C}^n \to \mathbb{C}^n$ be a transformation. As we have seen in Section 1.9, there is an orthonormal basis $x_1, \ldots, x_n$ for $\mathbb{C}^n$ in which A has the upper triangular form:

$$
A = \begin{bmatrix}
a_{11} & a_{12} & \cdots & a_{1n} \\
0 & a_{22} & \cdots & a_{2n} \\
\vdots & \vdots & & \vdots \\
0 & 0 & \cdots & a_{nn}
\end{bmatrix}
$$

Clearly, the subspaces $\mathrm{Span}\{x_k, \ldots, x_n\}$, $k = 1, \ldots, n$ are orthogonally A coinvariant and form a complete chain. $\square$

3.2 REDUCING SUBSPACES

An invariant subspace $\mathcal{L}$ of a transformation $A: \mathbb{C}^n \to \mathbb{C}^n$ is called *reducing* for A if $\mathcal{L} \dotplus \mathcal{M} = \mathbb{C}^n$ for some other A-invariant subspace $\mathcal{M}$. In other words, a subspace $\mathcal{L} \subset \mathbb{C}^n$ is reducing for A if it is simultaneously A invariant and A coinvariant. In particular, $\{0\}$ and $\mathbb{C}^n$ are trivially reducing. A more important example follows from Theorem 2.1.2. This shows that the root subspaces $\mathcal{R}_{\lambda_0}(A)$ are reducing for A. A unicellular linear transformation is an example in which the only reducing subspaces are the trivial ones $\{0\}$ and $\mathbb{C}^n$. On the other hand, $A = I$ is a linear transformation for which every subspace in $\mathbb{C}^n$ is invariant and reducing.

As a transformation on $\mathbb{C}^n$ with only one Jordan block (i.e., a unicellular transformation) has the smallest possible number of reducing subspaces, one might expect that a transformation with the most Jordan blocks has the most reducing subspaces. This is indeed so. Recall that a transformation is called *diagonable* if its Jordan form is a diagonal matrix.

Theorem 3.2.1

If A is diagonable, then each invariant subspace of A is reducing. Conversely, if each invariant subspace of A is reducing, then A is diagonable.

Proof. Assume that A is diagonable. Using Proposition 1.4.2, it is easily seen that each invariant subspace of A is reducing if and only if the same is true for $S^{-1}AS$, for any nonsingular matrix S. So we can assume that $A = \text{diag}[\alpha_1 \cdots \alpha_n]$ for some $\alpha_1, \ldots, \alpha_n \in \mathbb{C}$. Let $\lambda_1, \ldots, \lambda_p$ be all the different numbers among the α_i values, and for notational convenience assume that

$$\alpha_1 = \cdots = \alpha_{k_1} = \lambda_1 \, ; \qquad \alpha_{k_1+1} = \alpha_{k_1+2} = \cdots = \alpha_{k_2} = \lambda_2 \, ; \qquad \cdots$$

$$\alpha_{k_{p-1}+1} = \cdots = \alpha_{k_p} = \lambda_p$$

where

$$1 \le k_1 < k_2 < \cdots < k_{p-1} < k_p = n$$

are integers. Obviously, the eigenvalues of A are $\lambda_1, \ldots, \lambda_p$, and the root subspaces of A are

$$\mathcal{R}_{\lambda_i}(A) = \text{Span}\{e_{k_{i-1}+1}, e_{k_{i-1}+2}, \ldots, e_{k_i}\} \, , \qquad i = 1, \ldots, p$$

(by definition we put $k_0 = 0$). By Theorem 2.1.5 any A-invariant subspace $\mathcal{M}$ has the form

$$\mathcal{M} = \mathcal{M}_i \dotplus \cdots \dotplus \mathcal{M}_p$$

where $\mathcal{M}_i \subset \mathcal{R}_{\lambda_i}(A)$. Let $\mathcal{N}_i$ be any direct complement for $\mathcal{M}_i$ in $\mathcal{R}_{\lambda_i}(A)$. As $Ax = \lambda_i x$ for every $x \in \mathcal{R}_{\lambda_i}(A)$, the subspace $\mathcal{N}_i$ is obviously A invariant. Hence the subspace $\mathcal{N} = \mathcal{N}_1 \dotplus \cdots \dotplus \mathcal{N}_p$, which is a direct complement to $\mathcal{M}$ in $\mathbb{C}^n$, is also A invariant. This means, by definition, that $\mathcal{M}$ is reducing.

Conversely, assume that A is not diagonable. Let $\mathcal{M}$ be the A-invariant subspace of A spanned by its eigenvectors. As A is not diagonable, $\mathcal{M} \ne \mathbb{C}^n$. If $\mathcal{N}$ is any other A-invariant subspace and x is an eigenvector of $A|_{\mathcal{N}}$, then x is also an eigenvector of A, and thus $x \in \mathcal{M}$. So $\mathcal{M} \cap \mathcal{N} \ne \{0\}$ for every A-invariant $\mathcal{N}$. Consequently, $\mathcal{M}$ is not reducing. $\square$

An important class of diagonable transformations $A: \mathbb{C}^n \to \mathbb{C}^n$ are those that have n *distinct* eigenvalues $\lambda_1, \ldots, \lambda_n$. Indeed, the corresponding eigenvectors $x_1, \ldots, x_n$ are linearly independent (and, therefore, form a basis in $\mathbb{C}^n$) because $x_i \in \mathcal{R}_{\lambda_i}(A)$ and the subspaces $\mathcal{R}_{\lambda_1}(A), \ldots, \mathcal{R}_{\lambda_n}(A)$ form a direct sum. We have the following.

Corollary 3.2.2

If a transformation $A: \mathbb{C}^n \to \mathbb{C}^n$ has n distinct eigenvalues, then every A-invariant subspace is reducing.

Consider now the situation in which an A-invariant subspace is reducing and is orthogonal to its A-invariant comlementary subspace. An invariant subspace $\mathcal{M}$ of a transformation $A: \mathbb{C}^n \to \mathbb{C}^n$ is called *orthogonally reducing* if its orthogonal complement $\mathcal{M}^\perp$ is also A invariant.

Theorem 3.2.3

Every invariant subspace of A is orthogonally reducing if and only if A is normal.

Proof.   Recall first (Theorem 1.9.4) that A is normal if and only if there is an orthonormal basis of eigenvectors $x_1, \ldots, x_n$ of A.

Assume that A is normal, and let $x_1, \ldots, x_n$ be an orthonormal basis of eigenvectors of A that is ordered in such a way that

$$x_1, \ldots, x_{k_1} \text{ correspond to the eigenvalue } \lambda_1$$

$$x_{k_1+1}, \ldots, x_{k_2} \text{ correspond to the eigenvalue } \lambda_2$$

$$\vdots$$

$$x_{k_{p-1}}, \ldots, x_{k_p} \text{ correspond to the eigenvalue } \lambda_p$$

Here $\lambda_1, \ldots, \lambda_p$ are all the different eigenvalues of A. Arguing as in the proof of Theorem 3.2.1 we see that any A-invariant subspace is of the form

$$\mathcal{M} = \mathcal{M}_i \dotplus \cdots \dotplus \mathcal{M}_p$$

where $\mathcal{M}_i \subset \text{Span}\{x_{k_{i-1}}, \ldots, x_{k_i}\}$, $i = 1, \ldots, p$ (by definition $k_0 = 0$) and its orthogonal complement

$$\mathcal{M}^\perp = \mathcal{M}_i^\perp \dotplus \cdots \dotplus \mathcal{M}_p^\perp$$

in $\mathbb{C}^n$ is also A invariant. Here $\mathcal{M}_i^\perp$ is the orthogonal complement to $\mathcal{M}_i$ in the space $\mathbb{C}^{k_i - k_{i-1}}$.

Conversely, assume that every A-invariant subspace is orthogonally reducing. In particular, every A-invariant subspace is reducing, and by Theorem 3.2.1, $A = \text{diag}[\alpha_1, \ldots, \alpha_n]$ in a certain basis in $\mathbb{C}^n$.

Denoting by $\lambda_1, \ldots, \lambda_p$ all the different eigenvalues of A, it follows that $\mathcal{R}_{\lambda_i}(A)$ is spanned by the eigenvectors of A corresponding to λ_i. Now for each i_0, $1 \le i_0 \le p$, the subspace $\mathcal{R}_{\lambda_{i_0}}(A)$ is the unique A-invariant subspace that is a direct complement to $\Sigma_{i \ne i_0} \mathcal{R}_{\lambda_i}(A)$ in $\mathbb{C}^n$. [This follows from the fact that any A-invariant subspace $\mathcal{M}$ has the form $\mathcal{M} = \Sigma_{i=1}^{p} \mathcal{M} \cap \mathcal{R}_{\lambda_0}(A)$.] The orthogonal reducing property of $\mathcal{R}_{\lambda_{i_0}}(A)$ implies that the subspaces $\mathcal{R}_{\lambda_1}(A), \ldots, \mathcal{R}_{\lambda_p}(A)$ are orthogonal to each other. Taking an orthonormal basis in each $\mathcal{R}_{\lambda_j}(A)$ (which necessarily consists of eigenvectors of A corresponding to λ_i), we obtain an orthonormal basis in $\mathbb{C}^n$ in which A has a diagonal form. Hence A is normal. $\square$

The proof of Theorem 3.2.3 shows that if every A-invariant subspace is reducing and every root subspace for A is orthogonally reducing, then every A-invariant subspace is orthogonally reducing.

Note also the important special cases of Theorem 3.2.3: every invariant subspace of a hermitian or unitary transformation is orthogonally reducing.

3.3 SEMIINVARIANT SUBSPACES

A subspace $\mathcal{M} \subset \mathbb{C}^n$ is called *semiinvariant* for a transformation $A: \mathbb{C}^n \to \mathbb{C}^n$ (or, in short, A semiinvariant) if there exists an A-invariant subspace $\mathcal{N}$ such that $\mathcal{N} \cap \mathcal{M} = \{0\}$ and the sum $\mathcal{M} \dotplus \mathcal{N}$ is again A invariant. By taking $\mathcal{N} = \{0\}$ we see that any A-invariant subspace is also A semiinvariant.

If $\mathcal{M}$ is an A-coinvariant subspace, then there is an A-invariant direct complement $\mathcal{N}$ to $\mathcal{M}$ in $\mathbb{C}^n$ (so the conditions that $\mathcal{N} \cap \mathcal{M} = \{0\}$ and that $\mathcal{M} \dotplus \mathcal{N}$ is A invariant are automatically satisfied). Thus we see that any A-coinvariant subspace is also A semiinvariant. In general, a subspace $\mathcal{M} \subset \mathbb{C}^n$ is A semiinvariant if and only if $\mathcal{M}$ is $A|_L$-coinvariant for some A-invariant subspace $\mathcal{L}$ containing $\mathcal{M}$.

EXAMPLE 3.3.1.   Let A be an $n \times n$ Jordan block. Then it is easily seen that the subspaces $\text{Span}\{e_i, e_{i+1}, \ldots, e_j\}$, where $1 \le i \le j \le n$, are A semiinvariant (but there are many other A semiinvariant subspaces). This example shows that in general there exist semiinvariant subspaces that are neither invariant nor coinvariant. $\square$

Consider now the A-semiinvariant subspace $\mathcal{M}$, and let $\mathcal{N}$ be an A-invariant subspace such that $\mathcal{N} \cap \mathcal{M} = \{0\}$ and $\mathcal{M} \dotplus \mathcal{N}$ is A invariant. Then we have a direct sum decomposition

$$\mathbb{C}^n = \mathcal{N} \dotplus \mathcal{M} \dotplus \mathcal{L} \tag{3.3.1}$$

where $\mathcal{L}$ is a direct complement to $\mathcal{M} \dotplus \mathcal{N}$ in $\mathbb{C}^n$. To emphasize the fact that this is a decomposition of $\mathbb{C}^n$ into the sum of invariant, semiinvariant, and coinvariant subspaces, respectively, we call equation (3.3.1) a *triinvariant decomposition* associated with the A-semiinvariant subspace $\mathcal{M}$. Triinvariant decompositions play an important role in the applications of Chapters 5 and 7.

Note that in general a triinvariant decomposition associated with a given $\mathcal{M}$ is not unique. With respect to the triinvariant decomposition (3.3.1), the transformation A has the following 3×3 block form:

$$A = \begin{bmatrix} A_{11} & A_{12} & A_{13} \\ 0 & A_{22} & A_{23} \\ 0 & 0 & A_{33} \end{bmatrix} \tag{3.3.2}$$

Here $A_{11}: \mathcal{N} \to \mathcal{N}$, $A_{22}: \mathcal{M} \to \mathcal{M}$, $A_{33}: \mathcal{L} \to \mathcal{L}$, $A_{12}: \mathcal{M} \to \mathcal{N}$, $A_{23}: \mathcal{L} \to \mathcal{M}$, $A_{13}: \mathcal{L} \to \mathcal{N}$. The presence of zeros in (3.3.2) follows from the A invariance of $\mathcal{N}$ and $\mathcal{M} \dotplus \mathcal{N}$ (see Section 1.5). The converse is also true: if A is a transformation from $\mathbb{C}^n$ into $\mathbb{C}^n$, and A has the form (3.3.2) with respect to some direct sum decomposition (3.3.1), then $\mathcal{M}$ is A semiinvariant, and the A-invariant subspace $\mathcal{N}$ is such that $\mathcal{M} \dotplus \mathcal{N}$ is A invariant as well.

In particular, it follows from the formula (3.3.2) that the spectrum of the *compression* $PA|_{\mathcal{M}}$ (where $P: \mathcal{N} \dotplus \mathcal{M} \to \mathcal{N} \dotplus \mathcal{M}$ is the projector on $\mathcal{M}$ along $\mathcal{N}$) of A to its semiinvariant subspace $\mathcal{M}$ is contained in the spectrum of A.

We characterize A-semiinvariant subspaces in terms of functions of A, as follows.

Theorem 3.3.1

Let $A: \mathbb{C}^n \to \mathbb{C}^n$ be a transformation. The following statements are equivalent for a subspace $\mathcal{M} \subset \mathbb{C}^n$: (a) $\mathcal{M}$ is semiinvariant for A; (b) for a suitable projector P mapping $\mathbb{C}^n$ onto $\mathcal{M}$, we have

$$PA^m|_{\mathcal{M}} = (PA|_{\mathcal{M}})^m, \qquad m = 0, 1, 2, \ldots$$

(c) for any function $f(\lambda)$ such that $f(A)$ is defined we have

$$Pf(A)|_{\mathcal{M}} = f(PA|_{\mathcal{M}}) \tag{3.3.3}$$

where P is a suitable projector with $\operatorname{Im} P = \mathcal{M}$.

In (b) $PA^m|_{\mathcal{M}}$ is understood as a transformation from $\mathcal{M}$ into $\mathcal{M}$. Recall that $f(A)$ is certainly defined for a function $f(\lambda)$ that is analytic on the spectrum of A and, if A is diagonable, for any function $f(\lambda)$ that is merely defined on the spectrum of A. As the spectrum of $PA|_{\mathcal{M}}$ is contained in the spectrum of A (provided $\mathcal{M}$ is A semiinvariant), it follows that $f(PA|_{\mathcal{M}})$ is well defined if $f(\lambda)$ is analytic on the spectrum of A. We shall see in Section 4.1 that if A is diagonable, so is $PA|_{\mathcal{M}}$ (provided $\mathcal{M}$ is A semiinvariant), and thus $f(PA|_{\mathcal{M}})$ is well defined in the case when A is diagonable and $f(\lambda)$ is defined on the spectrum of A.

Proof. Assume that $\mathcal{M}$ is A semiinvariant, and write A as in (3.3.2), with respect to the triinvariant decomposition (3.3.1). Let P be the projector on $\mathcal{M}$ along $\mathcal{N} \dotplus \mathcal{L}$. Then $PA|_{\mathcal{M}} = A_{22}$. Now a straightforward calculation shows that

$$A^m = \begin{bmatrix} A_{11}^m & * & * \\ 0 & A_{22}^m & * \\ 0 & 0 & A_{33}^m \end{bmatrix}, \qquad m = 0, 1, 2, \ldots$$

so that $PA^m|_{\mathcal{M}} = A_{22}^m = (PA|_{\mathcal{M}})^m$.

Now assume that (b) holds. Let $\mathcal{L}$ be the smallest A-invariant subspace containing $\mathcal{M}$. (In other words, $\mathcal{L}$ is the intersection of all A-invariant subspaces that contain $\mathcal{M}$.) Equivalently, $\mathcal{L}$ is the span of all vectors of type $A^j x$, where $x \in \mathcal{M}$ and $j = 0, 1, \ldots$. In particular, $\mathcal{L} \supset \mathcal{M}$. Let Q be a projector on $\mathcal{L}$ such that $\operatorname{Ker} Q \subset \operatorname{Ker} P$ (e.g., take any direct complement $\mathcal{N}'$ to $\mathcal{L} \cap \operatorname{Ker} P$ in $\operatorname{Ker} P$, so that $\operatorname{Ker} P = \mathcal{N}' \dotplus (\mathcal{L} \cap \operatorname{Ker} P)$, and let Q be the projector on $\mathcal{L}$ along $\mathcal{N}'$). Then $\operatorname{Im}(I - Q) \subset \operatorname{Ker} P$ or, equivalently, $P(I - Q) = 0$, that is, $PQ = P$. As $\mathcal{L} \supset \mathcal{M}$, the equality $QP = P$ obviously holds. Now

$$(Q - P)(Q - P) = Q^2 - PQ - QP + P^2 = Q - P$$

so $Q - P$ is a projector, and $\operatorname{Im}(Q - P)$ is a direct complement to $\mathcal{M}$ in $\mathcal{L}$.

We shall prove that $\operatorname{Im}(Q - P)$ is A invariant, which shows that $\mathcal{M}$ is semiinvariant for A. Clearly, $QAQ = AQ$ (because $\operatorname{Im} Q = \mathcal{L}$ is A invariant) and $QAP = AP$ (because for every vector $x \in \operatorname{Im} P = \mathcal{M}$, the vector Ax belongs to $\mathcal{L}$ and thus $QAx = Ax$). Let us show that

$$PAP = PAQ \qquad (3.3.4)$$

For every $x \in \mathcal{M}$ and for any $j = 0, 1, 2, \ldots$ we have

$$PAPA^j x = PA|_{\mathcal{M}} PA^j|_{\mathcal{M}} x = PA|_{\mathcal{M}} \cdot (PA|_{\mathcal{M}})^j x$$

$$= (PA|_{\mathcal{M}})^{j+1} x = PA^{j+1} x = PA \cdot A^j x$$

where we have used the property (b) twice. As the subspace $\mathcal{L}$ is spanned by $A^j x$, $x \in \mathcal{M}$, $j = 0, 1, \ldots$, we conclude that $PAPy = PAy$ for every $y \in \mathcal{L}$, which amounts to the equality $PAPQ = PAQ$, and (3.3.4) follows. Using the equalities $QAQ = AQ$, $QAP = AP$, $PAP = PAQ$, we easily verify that $(Q - P)A(Q - P) = A(Q - P)$. This means that $\operatorname{Im}(Q - P)$ is A invariant.

Finally, let $f(\lambda)$ be a function such that $f(A)$ is defined. Then $f(A) = p(A)$, where $p(\lambda)$ is a polynomial such that

$$p^{(j)}(\lambda_k) = f^{(j)}(\lambda_k), \qquad j = 0, \ldots, m_k - 1; \qquad k = 1, \ldots, s$$

where $\lambda_1, \ldots, \lambda_s$ are all the distinct eigenvalues of A, and m_k is the height of λ_k $(k = 1, \ldots, s)$. Such a polynomial $p(\lambda)$ always exists. For example, the Lagrange–Sylvester interpolation polynomial, which is given by the formula

$$p(\lambda) = \sum_{k=1}^{s} [\alpha_{k1} + \alpha_{k2}(\lambda - \lambda_k) + \cdots + \alpha_{k,m_k}(\lambda - \lambda_k)^{m_k - 1}] \psi_k(\lambda)$$

where

$$\alpha_{kj} = \frac{1}{(j-1)!} \left[\frac{f(\lambda)}{\psi_k(\lambda)} \right]^{(j-1)}_{\lambda = \lambda_k}, \qquad j = 1, \ldots, m_k; \qquad k = 1, \ldots, s$$

and $\psi_k(\lambda) = (\lambda - \lambda_k)^{-m_k} \prod_{i=1}^{s} (\lambda - \lambda_i)^{m_i}$, $k = 1, \ldots, s$ [see, e.g. Chapter V of Gantmacher (1959)]. As the eigenvalues of $PA|_{\mathscr{M}}$ are also eigenvalues of A, and the height of $\lambda_0 \in \sigma(PA|_{\mathscr{M}})$ does not exceed the height of λ_0 as an eigenvalue of A (see Section 4.1), we obtain $f(PA|_{\mathscr{M}}) = p(PA|_{\mathscr{M}})$. Now equality (3.3.3) follows from (b). Conversely, (c) obviously implies (b). $\square$

Given an A-semiinvariant subspace $\mathscr{M}$ with an associated triinvariant decomposition $\mathbb{C}^n = \mathscr{N} \dotplus \mathscr{M} \dotplus \mathscr{L}$, the proof of Theorem 3.1 shows that (b) holds with P being the projector on $\mathscr{M}$ along $\mathscr{N} \dotplus \mathscr{L}$. And conversely, if a projector P satisfies (b), then $\mathrm{Ker}\, P = \mathscr{N} \dotplus \mathscr{L}$, where $\mathscr{N}$ and $\mathscr{L}$ are A-invariant and A-coinvariant subspaces, respectively, taken from some tri-invariant decomposition associated with $\mathscr{M}$.

Extending the notion of orthogonally coinvariant subspaces, we introduce the notion of orthogonally semiinvariant subspaces as follows. A subspace $\mathscr{M} \subset \mathbb{C}^n$ is called *orthogonally semiinvariant* for a transformation $A: \mathbb{C}^n \to \mathbb{C}^n$ if there exists an A-invariant subspace $\mathscr{N}$ such that $\mathscr{M} + \mathscr{N}$ is again A invariant and $\mathscr{M}$ is the orthogonal complement to $\mathscr{N}$ in $\mathscr{M} + \mathscr{N}$. Clearly, an orthogonally semiinvariant subspace is semiinvariant. For an orthogonally A-semiinvariant subspace $\mathscr{M}$ there exists an orthogonal decomposition

$$\mathbb{C}^n = \mathscr{N} \oplus \mathscr{M} \oplus \mathscr{L} \tag{3.3.5}$$

where $\mathscr{L} = (\mathscr{M} + \mathscr{N})^\perp$. Decomposition (3.3.5) will be called an *orthogonal triinvariant decomposition* associated with $\mathscr{M}$. Again, for a given $\mathscr{M}$ there are generally many associated orthogonal triinvariant decompositions. (The extreme case of this situation appears for $A = 0$.)

Consider the orthogonal triinvariant decomposition (3.3.5), and choose orthonormal bases in $\mathscr{N}$, $\mathscr{M}$, and $\mathscr{L}$. Then we represent A as the 3×3 block matrix

$$A = \begin{bmatrix} A_{11} & A_{12} & A_{13} \\ 0 & A_{22} & A_{23} \\ 0 & 0 & A_{33} \end{bmatrix} \tag{3.3.6}$$

in the orthonormal basis for $\mathbb{C}^n$ obtained by putting together the orthonormal bases in $\mathcal{N}$, $\mathcal{M}$, and $\mathcal{L}$. As the representation (3.3.6) is in an orthonormal basis, we have

$$A^* = \begin{bmatrix} A_{11}^* & 0 & 0 \\ A_{12}^* & A_{22}^* & 0 \\ A_{13}^* & A_{23}^* & A_{33}^* \end{bmatrix}$$

This leads to the following conclusion.

Proposition 3.3.2

An orthogonally A-semiinvariant subspace is also orthogonally A^* semiinvariant.

Indeed, if equation (3.3.5) holds, then $\mathcal{L}$ is A^* invariant, and $\mathcal{M}$ is the orthogonal complement to $\mathcal{L}$ in the A^*-invariant subspace $\mathcal{N}^\perp = \mathcal{M} \oplus \mathcal{L}$.

An analog of Theorem 3.3.1 holds for orthogonally semiinvariant subspaces.

Theorem 3.3.3

The following statements are equivalent for a transformation $A: \mathbb{C}^n \to \mathbb{C}^n$ and a subspace $\mathcal{M} \subset \mathbb{C}^n$: (a) $\mathcal{M}$ is orthogonally semiinvariant for A; (b) we have

$$P_{\mathcal{M}} A^m |_{\mathcal{M}} = (P_{\mathcal{M}} A |_{\mathcal{M}})^m , \qquad m = 0, 1, 2, \ldots$$

where $P_{\mathcal{M}}$ is the orthogonal projector on $\mathcal{M}$; (c) for any function $f(\lambda)$ such that $f(A)$ is defined we have

$$P_{\mathcal{M}} f(A) |_{\mathcal{M}} = f(P_{\mathcal{M}} A |_{\mathcal{M}})$$

The proof is like the proof of Theorem 3.3.1, with the only difference that an orthogonal triinvariant decomposition is used and the projector Q is taken to be orthogonal.

3.4 SPECIAL CLASSES OF TRANSFORMATIONS

In this section we shall describe coinvariant and semiinvariant subspaces for certain classes of transformations. We start with the relatively simple case of unicellular transformations.

Proposition 3.4.1

Let $A: \mathbb{C}^n \to \mathbb{C}^n$ be a unicellular transformation that is represented as a Jordan block in some basis $x_1, \ldots, x_n$. Then a k-dimensional subspace

$\mathcal{M} \subset \mathbb{C}^n$ *is A-coinvariant if and only if $\mathcal{M}$ is spanned by a set of vectors $y_1, \ldots, y_k$ with the property that $x_1, \ldots, x_{n-k}, y_1, \ldots, y_k$ is a basis in $\mathbb{C}^n$.*

A k-dimensional subspace $\mathcal{M}$ is A semiinvariant if and only if $\mathcal{M} = \mathrm{Span}\{y_1, \ldots, y_k\}$ where the vectors $y_1, \ldots, y_k$ are such that, for some index l with $k \leq l \leq n$, we have $y_i \in \mathrm{Span}\{x_1, \ldots, x_l\}$, $i = 1, \ldots, k$ and $x_1, \ldots, x_{l-k}, y_1, \ldots, y_k$ is a basis in $\mathrm{Span}\{x_1, \ldots, x_l\}$.

The proof follows easily from the definitions of coinvariant and semiinvariant subspaces and from the fact that the only A-invariant subspaces are $\{0\}$ and $\mathrm{Span}\{x_1, \ldots, x_l\}$, $l = 1, \ldots, n$.

Consider now a diagonable transformation $A: \mathbb{C}^n \to \mathbb{C}^n$, so that $A = \mathrm{diag}[\lambda_1, \ldots, \lambda_n]$ in some basis in $\mathbb{C}^n$. As we have seen in Example 1.2, if all λ_i are different, then every subspace in $\mathbb{C}^n$ is A coinvariant and hence also A semiinvariant. In fact, this conclusion holds for any diagonable transformation (not necessarily with all eigenvalues distinct). Indeed, consider the transformation B given by the matrix $\mathrm{diag}[\mu_1, \ldots, \mu_n]$ with different μ_i values in the same basis in which A is given by $\mathrm{diag}[\lambda_1, \ldots, \lambda_n]$. As every B-invariant subspace is also A invariant, it follows that every B-coinvariant subspace is also A coinvariant. But we have already seen that every subspace is B-coinvariant.

We consider now the orthogonally coinvariant and semiinvariant subspaces. We say that a transformation $A: \mathbb{C}^n \to \mathbb{C}^n$ is *orthogonally unicellular* if there exists a Jordan chain $x_1, \ldots, x_n$ of A such that the vectors $x_1, \ldots, x_n$ form an orthogonal basis in $\mathbb{C}^n$. Clearly, any orthogonally unicellular transformation is unicellular.

Proposition 3.4.2

Let $A: \mathbb{C}^n \to \mathbb{C}^n$ be an orthogonally unicellular transformation, and let $x_1, \ldots, x_n$ be its orthogonal Jordan chain. Then the only orthogonally A-coinvariant subspaces are $\mathrm{Span}\{x_k, x_{k+1}, \ldots, x_n\}$; $k = 1, \ldots, n$; and $\{0\}$. The only orthogonally A-semiinvariant subspaces are $\mathrm{Span}\{x_k, \ldots, x_l\}$, $1 \leq k \leq l \leq n$ and $\{0\}$.

Again, Proposition 3.4.2 follows from the description of all A-invariant subspaces.

Consider a normal transformation $A: \mathbb{C}^n \to \mathbb{C}^n$: $AA^* = A^*A$. By Theorem 1.9.4, A has an orthonormal basis of eigenvectors (and conversely, if a transformation has an orthonormal basis of eigenvectors, it is normal). It turns out that normal transformations are exactly those for which the classes of invariant subspaces and of orthogonally semiinvariant subspaces coincide.

Theorem 3.4.3

The following statements are equivalent for a transformation: (a) A is normal; (b) every A-invariant subspace is orthogonally A coinvariant; (c) every orthogonally A-coinvariant subspace is A invariant; (d) every orthogonally A-semiinvariant subspace is A invariant.

Proof. Obviously, (d) implies (c). Assume that A is normal, and let $\lambda_1, \ldots, \lambda_k$ be all the different eigenvalues of A. Then

$$\mathbb{C}^n = \mathcal{R}_{\lambda_i(A)} \oplus \cdots \oplus \mathcal{R}_{\lambda_k}(A)$$

is an orthogonal sum, and $A|_{R_{\lambda_i}}(A) = \lambda_i I$. Let $\mathcal{M}$ be an orthogonally A-semiinvariant subspace, so that $\mathcal{M}$ is the orthogonal complement to an A-invariant subspace $\mathcal{N}$ in another A-invariant subspace $\mathcal{L}$. We have

$$\mathcal{N} = \mathcal{N}_1 \oplus \cdots \oplus \mathcal{N}_k , \qquad \mathcal{L} = \mathcal{L}_1 \oplus \cdots \mathcal{L}_k$$

where $\mathcal{N}_i \subset \mathcal{L}_i \subset \mathcal{R}_{\lambda_i}(A)$, $i = 1, \ldots, k$. Denoting by $\mathcal{M}_i$ the orthogonal complement of $\mathcal{N}_i$ in $\mathcal{L}_i$, the definition of $\mathcal{M}$ implies that

$$\mathcal{M} = \mathcal{M}_1 \oplus \cdots \oplus \mathcal{M}_k .$$

It follows that $\mathcal{M}$ is A invariant. So (a) implies (d). One sees easily that (a) implies (b) also.

It remains to show that (c) $\Rightarrow$ (a) and (b) $\Rightarrow$ (a). Assume (c) holds, that is (cf. Proposition 3.1.2) every A^*-invariant subspace is A-invariant. Write A^* in an upper triangular form with respect to some orthonormal basis $x_1, \ldots, x_n$:

$$A^* = \begin{bmatrix} a_{11} & a_{12} & \cdots & a_{1n} \\ 0 & a_{22} & \cdots & a_{2n} \\ \vdots & \vdots & & \vdots \\ 0 & 0 & \cdots & a_{nn} \end{bmatrix} \qquad (3.4.1)$$

As $\mathrm{Span}\{x_1, \ldots, x_k\}$, $k = 1, \ldots, n$ are A^*-invariant subspaces, they are also A invariant. Hence (Proposition 1.8.4) A also has an upper triangular form in the same basis:

$$A = \begin{bmatrix} b_{11} & b_{12} & \cdots & b_{1n} \\ 0 & b_{22} & \cdots & b_{2n} \\ \vdots & \vdots & & \vdots \\ 0 & 0 & \cdots & b_{nn} \end{bmatrix} \qquad (3.4.2)$$

On the other hand, equality (3.4.1) implies

$$A = \begin{bmatrix} \bar{a}_{11} & 0 & \cdots & 0 \\ \bar{a}_{12} & \bar{a}_{22} & \cdots & 0 \\ \vdots & \vdots & & \vdots \\ \bar{a}_{1n} & \bar{a}_{2n} & \cdots & \bar{a}_{nn} \end{bmatrix} \qquad (3.4.3)$$

Comparison of (3.4.2) and (3.4.3) reveals that $b_{ij} = 0$ for $i < j$, and A is normal.

Assume now that (b) holds, and write

$$A = \begin{bmatrix} b_{11} & b_{12} & \cdots & b_{1n} \\ 0 & b_{22} & \cdots & b_{2n} \\ \vdots & \vdots & & \vdots \\ 0 & 0 & \cdots & b_{nn} \end{bmatrix} \qquad (3.4.4)$$

in some orthonormal basis $x_1, \ldots, x_n$ in $\mathbb{C}^n$. The subspaces Span$\{x_1, \ldots, x_k\}$, $k = 1, \ldots, n$ are A invariant and, by (b), orthogonally A coinvariant. Hence Span$\{x_{k+1}, \ldots, x_n\}$, $k = 1, \ldots, n-1$ are A-invariant subspaces, which means that A has a lower triangular form

$$A = \begin{bmatrix} c_{11} & 0 & \cdots & 0 \\ c_{21} & c_{22} & \cdots & 0 \\ \vdots & \vdots & & \vdots \\ c_{n1} & c_{n2} & \cdots & c_{nn} \end{bmatrix} \qquad (3.4.5)$$

Comparing equations (3.4.4) and (3.4.5), we find that A is normal. $\square$

As a corollary of Theorem 3.4.3 we obtain the following characterization of a normal transformation in terms of its invariant subspaces.

Corollary 3.4.4

A transformation $A: \mathbb{C}^n \to \mathbb{C}^n$ is normal if and only if a subspace $\mathcal{M}$ is A invariant exactly when its orthogonal complement is A invariant.

Indeed, it follows from the definition that the subspace $\mathcal{M}^\perp$ is A invariant if and only if $\mathcal{M}$ is orthogonally A coinvariant.

3.5 EXERCISES

3.1 Prove that, in Example 3.1.2, there is a unique A-invariant direct complement to the A-coinvariant subspace $\mathcal{M}$ if and only if $\mathcal{M}$ itself is A invariant.

3.2 Prove that a subspace $\mathcal{M}$ is A coinvariant (resp. A semiinvariant) if and only if $\mathcal{M}$ is $(\alpha A + \beta I)$ coinvariant [resp. $(\alpha A + \beta I)$ semiinvariant]. Here α, β are complex numbers and $\alpha \neq 0$.

3.3 Show that a subspace $\mathcal{M}$ is A coinvariant (resp. A semiinvariant) if and only if $\mathcal{S}\mathcal{M}$ is SAS^{-1} coinvariant (resp. SAS^{-1} semiinvariant), where S is an invertible transformation.

3.4 Let $A: \mathfrak{C}^n \to \mathfrak{C}^n$ $(n \geq 3)$ be a unicellular transformation. Give an example of a subspace $\mathcal{M} \subset \mathfrak{C}^n$ that is not A semiinvariant. List all such subspaces when $n = 3$.

3.5 Show that every subspace in $\mathfrak{C}^n$ is A coinvariant if and only if A is diagonable (i.e., it is similar to a diagonal matrix).

3.6 Prove that every subspace in $\mathfrak{C}^n$ is coinvariant for any $n \times n$ circulant matrix.

3.7 Give an example of a nondiagonable transformation $A: \mathfrak{C}^n \to \mathfrak{C}^n$ such that every subspace in $\mathfrak{C}^n$ is A semiinvariant.

3.8 Find all the coinvariant subspaces for the matrices

$$\begin{bmatrix} 0 & 1 & 0 \\ 0 & 0 & 1 \\ 0 & -4 & -4 \end{bmatrix}, \quad \begin{bmatrix} 0 & 1 & 0 \\ 0 & 0 & 1 \\ i & 3 & -3i \end{bmatrix}$$

3.9 Find all coinvariant and semiinvariant subspaces for the matrix

$$\begin{bmatrix} 0 & 1 & -1 \\ 0 & 0 & 1 \\ 0 & 0 & 1 \end{bmatrix}$$

3.10 Prove that every reducing A-invariant subspace is reducing also for $f(A)$, where $f(\lambda)$ is any function such that $f(A)$ is defined. Is the converse true?

3.11 If J is a Jordan block, for which positive integers k does the matrix J^k have a nontrivial reducing invariant subspace? Is the reducing subspace unique?

3.12 Prove that an A-invariant subspace $\mathcal{M}$ is reducing if and only if $\mathcal{M} \cap \mathcal{R}_{\lambda_0}(A)$ is reducing for every eigenvalue λ_0 of A.

3.13 Find all the triinvariant decompositions $\mathfrak{C}^3 = \mathcal{N} \dotplus \mathcal{M} \dotplus \mathcal{L}$ with $\dim \mathcal{N} = \dim \mathcal{M} = \dim \mathcal{L} = 1$ for the following matrices:

$$\begin{bmatrix} 0 & 1 & 0 \\ 0 & 0 & 1 \\ 0 & 2 & 1 \end{bmatrix}; \quad \begin{bmatrix} 0 & 1 & 0 \\ 0 & 0 & 1 \\ -i & 3 & 3i \end{bmatrix}$$

Chapter Four

Jordan Forms for Extensions and Completions

Consider a transformation $A: \mathbb{C}^n \to \mathbb{C}^n$ and an A-coinvariant subspace $\mathcal{M}$. Thus there is an A-invariant subspace $\mathcal{N}$ such that $\mathbb{C}^n = \mathcal{M} \dotplus \mathcal{N}$ and there is a projector P onto $\mathcal{M}$ along $\mathcal{N}$. The main problems of this chapter are: given Jordan normal forms for $A|_{\mathcal{N}}$ and $PA|_{\mathcal{M}}$, what are the possible Jordan forms for A itself? In general, this problem is open. Here, we present partial results and important inequalities.

4.1 EXTENSIONS FROM AN INVARIANT SUBSPACE

Let $\mathcal{M} \subset \mathbb{C}^n$ be a subspace, and consider a transformation $A_0: \mathcal{M} \to \mathcal{M}$. A linear transformation $A: \mathbb{C}^n \to \mathbb{C}^n$ is called an *extension* of A_0 if $Ax = A_0 x$ for every $x \in \mathcal{M}$. Then, in particular, $\mathcal{M}$ is A invariant. Also, A_0 is called the *restriction* of A to $\mathcal{M}$. We are interested in the Jordan form (or, equivalently, the partial multiplicities) of A_0 and its extensions.

We start with a relatively simple but important case in which A as well as its extension are in the Jordan form and have special spectral properties. These spectral properties ensure that the partial multiplicities corresponding to a particular eigenvalue λ_0 are the same for A_0 and its extension A.

Theorem 4.1.1

Let J_1 and J_2 be matrices in Jordan normal form with sizes $p \times p$ and $q \times q$, respectively. Let B be a $p \times q$ matrix and

$$J = \begin{bmatrix} J_1 & B \\ 0 & J_2 \end{bmatrix}$$

Denote by J_{10} and J_{20} the Jordan submatrices of J_1 and J_2, respectively, formed by those Jordan blocks with the same eigenvalue λ_0.

Then the partial multiplicities of J corresponding to λ_0 coincide with the partial multiplicities of the submatrix

$$\begin{bmatrix} J_{10} & B_0 \\ 0 & J_{20} \end{bmatrix}$$

of J, where B_0 is the submatrix of J formed by the rows that belong to the rows of J_{10} and by the columns that belong to the columns of J_{20} (so actually B_0 is a submatrix of B).

Theorem 4.1.1 is used later to reduce problems concerning the Jordan form of an extension to the case when the transformations involved have only one eigenvalue. The proof of Theorem 4.1.1 is based on two lemmas, which are also independently important.

Lemma 4.1.2

Let A, B, C be given matrices of sizes $n \times n$, $m \times m$, and $n \times m$, respectively. Consider the equation

$$AX - BX = C \tag{4.1.1}$$

where X is an $n \times m$ matrix to be found. Equation (4.1.1) has a unique solution X for every C if and only if $\sigma(A) \cap \sigma(B) = \emptyset$.

This lemma follows immediately from the fact that, for the linear transformation $L: \mathbb{C}^{n \times m} \to \mathbb{C}^{n \times m}$ defined by $L(X) = AX - XB$, $\sigma(L) = \{\lambda - \mu \mid \lambda \in \alpha(A) \text{ and } \mu \in \sigma(B)\}$. [See Chapter 12 of Lancaster and Tismenetsky (1985), for example.] Here we give a direct proof based on the Jordan decompositions of A and B.

Proof. Equation (4.1.1) may be regarded as a system of linear equations in the rs variables x_{ij} $(i = 1, \ldots, r; j = 1, \ldots, s)$ that form the entries in the matrix X. Thus it is sufficient to prove that the homogeneous equation

$$AX - XB = 0 \tag{4.1.2}$$

has only the trivial solution $X = 0$ if and only if $\sigma(A) \cap \sigma(B) = \emptyset$.

Let J_A and J_B be the Jordan forms of A and B, respectively; so $A = S_A J_A S_A^{-1}$, $B = S_B J_B S_B^{-1}$ for some invertible matrices S_A and S_B. It follows that X is a solution of (4.1.2) if and only if $Z = S_A^{-1} X S_B$ is a solution of

$$J_A Z - Z J_B = 0 \tag{4.1.3}$$

Thus we can restrict ourselves to equation (4.1.3). Let us write down J_A and J_B explicitly:

$$J_A = \text{diag}[J_{A,1}, \ldots, J_{A,\mu}]; \qquad J_B = \text{diag}[J_{B,1}, \ldots, J_{B,\nu}]$$

where $J_{A,i}$ (resp. $J_{B,j}$) is a Jordan block of size $m_{A,i}$ (resp. $m_{B,j}$) with eigenvalue $\lambda_{A,i}$ (resp. $\lambda_{B,j}$). The matrix Z from (4.1.3) is decomposed into blocks accordingly:

$$Z = [Z_{ij}]_{i,j=1}^{\mu,\nu} \tag{4.1.4}$$

where Z_{ij} is of size $m_{A,i} \times m_{B,j}$.

Suppose first that $\sigma(A) \cap \sigma(B) \neq \emptyset$. Without loss of generality we can assume that $\lambda_{A,1} = \lambda_{B,1}$. Then we can construct a nonzero solution Z of equation (4.1.3) as follows. In the representation equation (4.1.4) put $Z_{ij} = 0$, except for the case that $i = j = 1$; and let

$$Z_{11} = \begin{bmatrix} I \\ 0 \end{bmatrix} \quad \text{or} \quad [I \quad 0]$$

(according as $m_{A,1} \geq m_{B,1}$ or $m_{A,1} \leq m_{B,1}$). Direct examination shows that such a matrix Z satisfies (4.1.3).

Suppose now that $\sigma(A) \cap \sigma(B) = \emptyset$. Let Z be given by (4.1.4) and suppose that Z satisfies (4.1.3). We have to prove that $Z = 0$.

Equation (4.1.3) means that

$$J_{A,i} Z_{ij} = Z_{ij} J_{B,j} \quad \text{for} \quad i = 1, \ldots, \mu; \quad j = 1, \ldots, \nu \tag{4.1.5}$$

Write

$$J_{A,i} = \lambda_{A,i} I + H; \qquad J_{B,j} = \lambda_{B,j} I + G$$

where H and G are the nilpotent matrices [i.e., $\sigma(H) = \sigma(G) = \{0\}$] having 1 on the first superdiagonal and zeros elsewhere. Rewrite equation (4.1.5) in the form

$$(\lambda_{A,i} - \lambda_{B,j}) Z_{ij} = Z_{ij} G - H Z_{ij}$$

Multiply the left-hand side by $\lambda_{A,i} - \lambda_{B,j}$, and in each term on the right-hand side replace $(\lambda_{A,i} - \lambda_{B,j}) Z_{ij}$ by $Z_{ij} G - H Z_{ij}$. We obtain

$$(\lambda_{A,i} - \lambda_{B,j})^2 Z_{ij} = Z_{ij} G^2 - 2 H Z_{ij} G + H^2 Z_{ij}$$

Repeating this process, we obtain for every $p = 1, 2, \ldots$

$$(\lambda_{A,i} - \lambda_{B,j})^p Z_{ij} = \sum_{q=0}^{p} (-1)^q \binom{p}{q} H^q Z_{ij} G^{p-q} \tag{4.1.6}$$

Choose p large enough so that either $H^q = 0$ or $G^{p-q} = 0$ for every $q = 0, \dots, p$. Then the right-hand side of equation (4.1.6) is zero, and since $\lambda_{A,i} \neq \lambda_{B,j}$, we find that $Z_{ij} = 0$. Thus $Z = 0$. $\square$

Lemma 4.1.3

If A and B are $n \times n$ and $m \times m$ matrices, respectively, with $\sigma(A) \cap \sigma(B) = \emptyset$, then for every $n \times m$ matrix C the $(m + n) \times (m + n)$ matrices

$$\begin{bmatrix} A & C \\ 0 & B \end{bmatrix} \quad and \quad \begin{bmatrix} A & 0 \\ 0 & B \end{bmatrix}$$

are similar.

Proof. By Lemma 4.1.2, for every $n \times m$ matrix C there is a unique $n \times m$ matrix X such that $AX - XB = -C$. With this X, one verifies that

$$\begin{bmatrix} I & -X \\ 0 & I \end{bmatrix} \begin{bmatrix} A & C \\ 0 & B \end{bmatrix} \begin{bmatrix} I & X \\ 0 & I \end{bmatrix} = \begin{bmatrix} A & 0 \\ 0 & B \end{bmatrix}$$

As

$$\begin{bmatrix} I & -X \\ 0 & I \end{bmatrix} = \begin{bmatrix} I & X \\ 0 & I \end{bmatrix}^{-1}$$

the lemma follows. $\square$

Proof of Theorem 1.1. For notational simplicity assume that

$$J = \begin{bmatrix} J_{10} & B_{12} & B_0 & B_{14} \\ 0 & J_{11} & B_{23} & B_{24} \\ 0 & 0 & J_{20} & B_{34} \\ 0 & 0 & 0 & J_{21} \end{bmatrix}$$

where J_{11} (resp. J_{21}) are the Jordan blocks from J_1 (resp. J_2) with eigenvalues different from λ_0, and B_{ij} are the corresponding submatrices in J.

Applying Lemma 4.1.3 twice, we see that J is similar to

$$\begin{bmatrix} J_{10} & B_{12} & B_0 & 0 \\ 0 & J_{11} & 0 & B_{24} \\ 0 & 0 & J_{20} & B_{34} \\ 0 & 0 & 0 & J_{21} \end{bmatrix}$$

which after interchanging the second and third block rows and columns (this is a similarity operation) becomes

$$\begin{bmatrix} J_{10} & B_0 & B_{12} & 0 \\ 0 & J_{20} & 0 & B_{34} \\ 0 & 0 & J_{11} & B_{24} \\ 0 & 0 & 0 & J_{21} \end{bmatrix}$$

It remains to apply Lemma 4.1.3 once more to prove that J is similar to

$$\begin{bmatrix} J_{10} & B_0 \\ 0 & J_{20} \end{bmatrix} \oplus \begin{bmatrix} J_{11} & B_{24} \\ 0 & J_{21} \end{bmatrix} \quad \square$$

It is convenient to describe the partial multiplicities of a transformation $A: \mathbb{C}^n \to \mathbb{C}^n$ at an eigenvalue λ_0 as a nonincreasing sequence of nonnegative integers $\alpha_1(A; \lambda_0) \geq \alpha_2(A; \lambda_0) \geq \alpha_3(A; \lambda_0) \geq \cdots$, where the nonzero members of this sequence are exactly the partial multiplicities of A at λ_0. In particular, not more than n of the numbers $\alpha_j(A; \lambda_0)$ are different from zero. Also, if λ_0 is not an eigenvalue of A, we define $\alpha_j(A; \lambda_0) = 0$ for $j = 1, 2, \ldots$. Thus the nonnegative integers $\alpha_j(A; \lambda_0)$ are defined for all $\lambda_0 \in \mathbb{C}$, and we have

$$\sum_{\lambda_0 \in \mathbb{C}} \sum_{j=1}^{\infty} \alpha_j(A; \lambda_0) = n$$

The following result describes the connections between the partial multiplicities of a transformation and those of its extension.

Theorem 4.1.4

Let $\mathcal{M} \subset \mathbb{C}^n$ be a subspace and let $A_0: \mathcal{M} \to \mathcal{M}$ be a transformation. Then for every extension $A: \mathbb{C}^n \to \mathbb{C}^n$ of A_0 we have

$$\alpha_j(A; \lambda_0) \geq \alpha_j(A_0; \lambda_0), \qquad j = 1, 2, \ldots \tag{4.1.7}$$

for every $\lambda_0 \in \mathbb{C}$. Conversely, let $\beta_1 \geq \beta_2 \geq \cdots$ be a nonincreasing sequence of nonnegative integers such that

$$\sum_{i=1}^{\infty} \beta_i \leq n \tag{4.1.8}$$

and

$$\beta_j \geq \alpha_j(A_0; \lambda_0), \qquad j = 1, 2, \ldots \tag{4.1.9}$$

for a fixed complex number λ_0. Then there is an extension A of A_0 such that $\alpha_j(A; \lambda_0) = \beta_j$, $j = 1, 2, \ldots$.

Proof. We prove (4.1.7) for an extension A of A_0. In view of Theorem 4.1.1, we may restrict ourselves to the case when $\alpha(A) = \{\lambda_0\}$. (Indeed, without loss of generality it can be assumed that A_0 is in the Jordan form. Furthermore, the transformation $PA|_{\mathcal{N}}: \mathcal{N} \to \mathcal{N}$, where $\mathcal{N}$ is a direct complement to $\mathcal{M}$ and P is the projector on $\mathcal{N}$ along $\mathcal{M}$, may also be assumed to have Jordan normal form.) There exists a chain of A-invariant subspaces

$$\mathcal{M} = \mathcal{M}_0 \subset \mathcal{M}_1 \subset \cdots \subset \mathcal{M}_{n-m} = \mathbb{C}^n \qquad (4.1.10)$$

where $\dim \mathcal{M}_i = m + i$, $i = 0, 1, \ldots, n - m$ (so $m = \dim \mathcal{M}$). This can be seen by considering the transformation $\hat{A}: \mathbb{C}^n/\mathcal{M} \to \mathbb{C}^n/\mathcal{M}$ induced by A and using the existence of a complete chain of $\hat{A}$-invariant subspaces.

In view of the chain (4.1.10) and using induction on the index i of $\mathcal{M}_i$, it will suffice to prove inequalities (4.1.7) for the case $\dim \mathcal{M} = n - 1$. Writing A_0 in a basis for $\mathcal{M}$ in which A_0 has a Jordan form, we can assume

$$A = \begin{bmatrix} J & B \\ 0 & \lambda_0 \end{bmatrix}$$

where $J = J_{k_1}(\lambda_0) \oplus \cdots \oplus J_{k_p}(\lambda_0)$, $k_1 \geq \cdots \geq k_p$ is the Jordan form of A_0 and B is an $(n-1)$-dimensional vector.

Let j be the first index $(1 \leq j \leq p)$ for which the $(k_1 + k_2 + \cdots + k_j)$th coordinate of B is nonzero (if such a j exists). Let S be the $(n-1) \times (n-1)$ matrix

$$S = \begin{bmatrix} I_{k_1} & 0 & \cdots & 0 & 0 & \cdots & 0 \\ 0 & I_{k_2} & & \vdots & \vdots & & 0 \\ 0 & \cdots & & I_{k_j} & 0 & \cdots & 0 \\ 0 & \cdots & & \alpha_{j+1}Q_{j+1} & I_{k_j+1} & & 0 \\ \vdots & & & \vdots & & \ddots & \vdots \\ 0 & \cdots & & \alpha_p Q_p & 0 & \cdots & I_{k_p} \end{bmatrix}$$

where Q_m is the $k_m \times k_j$ matrix of the form $[0 \quad I_{k_j}]$ and $\alpha_{j+1}, \ldots, \alpha_p$ are complex numbers chosen so that the $(k_1 + k_2 + \cdots + k_m)$th coordinates of SB are zeros for $m = j + 1, \ldots, p$. If all coordinates $k_1, k_1 + k_2, \ldots,$ $k_1 + \cdots k_p$ of B are zeros, put $S = I_{n-1}$. It is easy to see that $SJ = JS$ and S is nonsingular. Moreover, the k_1th, $(k_1 + k_2)$th, $\ldots, (k_1 + k_2 + \cdots + k_p)$th coordinates of SB are all zero except for at most one of them. Further, let X be an $(n-1)$-dimensional vector such that the nonzero coordinates of the vector

$$Y = (\lambda_0 I - J)X + SB$$

can appear only in the places $k_1, k_1 + k_2, \ldots, k_1 + k_2 + \cdots + k_p$ (this is possible because

$$\text{Im}(\lambda_0 I - J) = \text{Span}\{e_j \mid j \neq k, k_1 + k_2, \ldots, k_1 + k_2 + \cdots + k_p\})$$

Now a computation shows that

$$\begin{bmatrix} S & X \\ 0 & 1 \end{bmatrix} \begin{bmatrix} J & B \\ 0 & \lambda_0 \end{bmatrix} \begin{bmatrix} S^{-1} & -S^{-1}X \\ 0 & 1 \end{bmatrix} = \begin{bmatrix} J & Y \\ 0 & \lambda_0 \end{bmatrix}$$

As $\begin{bmatrix} S^{-1} & -S^{-1}X \\ 0 & 1 \end{bmatrix}$ is the inverse of $\begin{bmatrix} S & X \\ 0 & 1 \end{bmatrix}$, it follows that $\begin{bmatrix} J & B \\ 0 & \lambda_0 \end{bmatrix}$ and $\begin{bmatrix} J & Y \\ 0 & \lambda_0 \end{bmatrix}$ have the same partial multiplicities. Now the partial multiplicities of $\begin{bmatrix} J & Y \\ 0 & \lambda_0 \end{bmatrix}$ are easy to discover: they are $k_1, \ldots, k_p, 1$ if $Y = 0$, and $k_1, \ldots, k_{j-1}, k_j + 1, k_{j+1}, \ldots, k_p$ if $Y \neq 0$ and the nonzero coordinate of Y (by construction of Y there is exactly one) appears in the place $k_1 + \cdots + k_j$. So the inequalities (4.1.7) are satisfied. If $B = 0$, then (4.1.7) is obviously satisfied.

Now let β_i be a sequence with the properties described in the theorem. Let $x_1, \ldots, x_k$ be a basis in $\mathcal{M}$ in which A_0 has the Jordan form. We assume also that the first p Jordan blocks in the Jordan form have eigenvalues λ_0 and sizes $\alpha_1(A_0; \lambda_0), \ldots, \alpha_p(A_0; \lambda_0)$, respectively. (Here, $\alpha_1(A_0; \lambda_0), \ldots, \alpha_p(A_0; \lambda_0)$ are all the nonzero integers in the sequence $\{\alpha_j(A_0; \lambda_0)\}_{j=1}^{\infty}$). So in the basis $x_1, \ldots, x_k$ we have

$$A_0 = J_{\alpha_1}(\lambda_0) \oplus \cdots \oplus J_{\alpha_p}(\lambda_0) \oplus J_{m_1}(\lambda_1) \oplus \cdots \oplus J_{m_u}(\lambda_u)$$

where $\lambda_1, \ldots, \lambda_u$ are different from λ_0, and $\alpha_j = \alpha_j(A_0; \lambda_0)$. Now let $y_1, \ldots, y_{n-k}$ be vectors in $\mathbb{C}^n$ such that $x_1, \ldots, x_k, y_1, \ldots, y_{n-k}$ is a basis in $\mathbb{C}^n$. Put

$$z_1 = x_1, \ldots, z_{\alpha_1} = x_{\alpha_1}, z_{\alpha_1+1} = y_1, \ldots, z_{\beta_1} = y_{\beta_1 - \alpha_1}$$
$$z_{\beta_1+1} = x_{\alpha_1+1}, \ldots, z_{\beta_1+\alpha_2} = x_{\alpha_1+\alpha_2}$$

$$z_{\beta_1+\alpha_2+1} = y_{\beta_1 - \alpha_1+1}, \ldots, z_{\beta_1+\beta_2} = y_{\beta_1 - \alpha_1 + \beta_2 - \alpha_2}, \ldots, z_s = y_r$$

where $s = \Sigma_{i=1}^q \beta_i$, $r = \Sigma_{i=1}^q (\beta_i - \alpha_i)$, and q is the number of positive β_i values. Further, setting $t = \Sigma_{i=1}^p \alpha_i$, put

$$z_{s+1} = x_{t+1}, \ldots, z_{k+s-t} = x_k; \qquad z_{k+s-t+1} = y_{r+1}, \ldots, z_n = y_{n-k}$$

Now let $A: \mathbb{C}^n \to \mathbb{C}^n$ be a transformation that is given in the basis $z_1, \ldots, z_n$ by the matrix

$$A = J_{\beta_1}(\lambda_0) \oplus \cdots \oplus J_{\beta_q}(\lambda_0) \oplus J_{m_1}(\lambda_1) \oplus \cdots \oplus J_{m_u}(\lambda_u) \oplus J$$

where J is any $(n - k - r) \times (n - k - r)$ matrix in the Jordan form with the property that λ_0 is not an eigenvalue of J. From the construction of A it is clear that $\beta_1, \ldots, \beta_q$ are the partial multiplicities of A corresponding to λ_0 and that A is an extension of A_0. $\square$

In particular, the theorem shows that if A is diagonable, then so is the restriction of A to any A-invariant subspace.

For coinvariant subspaces the notions of coextension and corestriction become natural. Let $\mathcal{M} \subset \mathbb{C}^n$ be a subspace, and let $A_0 : \mathcal{M} \to \mathcal{M}$ be a linear transformation. A transformation $A : \mathbb{C}^n \to \mathbb{C}^n$ is called a *coextension* of A_0 if there exists an A-invariant direct complement $\mathcal{N}$ to $\mathcal{M}$ in $\mathbb{C}^n$ such that $PA|_{\mathcal{M}} = A_0$, where P is the projector on $\mathcal{M}$ along $\mathcal{N}$. Clearly, in this case $\mathcal{M}$ is an A-coinvariant subspace. There is a connection between the partial multiplicities of a transformation and those of a coextension of the kind described in Theorem 4.1.4.

Theorem 4.1.5

Let $\mathcal{M} \subset \mathbb{C}^n$ be a subspace and $A_0 : \mathcal{M} \to \mathcal{M}$ be a transformation. Then for every coextension A of A_0 we have $\alpha_j(A; \lambda_0) \geq \alpha_j(A_0; \lambda_0)$, $j = 1, 2, \ldots$ for every $\lambda_0 \in \mathbb{C}$. Conversely, let $\beta_j \geq \beta_2 \geq \cdots$ be a nonincreasing sequence of nonnegative integers such that equations (4.1.8) and (4.1.9) hold. Then there is a coextension A of A_0 such that $\alpha_j(A; \lambda_0) = \beta_j$, $j = 1, 2, \ldots$.

The proof of Theorem 4.1.5 is similar to the proof of theorem 4.1.4.

Given a transformation $A_0 : \mathcal{M} \to \mathcal{M}$, where $\mathcal{M} \subset \mathbb{C}^n$, we say that a transformation $A : \mathbb{C}^n \to \mathbb{C}^n$ is a *dilation* of A_0 if there exists an A-invariant subspace $\mathcal{N}$ for which $\mathcal{N} \cap \mathcal{M} = \{0\}$, $\mathcal{M} \dotplus \mathcal{N}$ is A invariant as well, and $PA|_{\mathcal{M}} = A_0$, where P is some projector on $\mathcal{M}$ with $\mathcal{N} \subset \operatorname{Ker} P$. (The term "semiextension" would be more logical in the context of our terminology; however, "dilation" is widely used in the literature.) In this case $\mathcal{M}$ is an A-semiinvariant subspace and A_0 is the *reduction* of A (again the term "semirestriction" would be consistent with our terminology, but "reduction" is already widely used.) Thus there is a subspace $\mathcal{L}$ of $\mathbb{C}^n$ for which the decomposition (3.3.1) holds, and this decomposition determines a triangular representation such as (3.3.2) for A in which $A_{22} = A_0$. A result similar to theorems 4.1.4 and 4.1.5 also holds for dilations, and it can be proved by first applying one of these theorems and then applying the second. In particular, if A is diagonable, so is any reduction of A.

4.2   COMPLETIONS FROM A PAIR OF INVARIANT AND COVARIANT SUBSPACES

Let $A : \mathcal{M} \to \mathcal{M}$ and $B : \mathcal{N} \to \mathcal{N}$ be transformations, where $\mathcal{M}$ and $\mathcal{N}$ are subspaces in $\mathbb{C}^n$ which are direct complements to each other. A transformation $C : \mathbb{C}^n \to \mathbb{C}^n$ is called a *completion* of A and B if $\mathcal{M}$ is C invariant and

$C|_{\mathcal{M}} = A$, $PC|_{\mathcal{N}} = B$, where P is the projector on $\mathcal{N}$ along $\mathcal{M}$. So with respect to the direct sum decomposition $\mathbb{C}^n = \mathcal{M} \dotplus \mathcal{N}$, C has the form

$$C = \begin{bmatrix} A & D \\ 0 & B \end{bmatrix} \tag{4.2.1}$$

for some matrix D.

Let $\alpha_1 \geq \alpha_2 \geq \cdots$ (resp. $\beta_1 \geq \beta_2 \geq \cdots$) be a sequence of nonnegative integers whose nonzero elements are exactly the partial multiplicities of A (resp. B) corresponding to a fixed point $\lambda_0 \in \mathbb{C}$. Assuming that C is a completion of A and B, let $\gamma_1 \geq \gamma_2 \geq \cdots$ be a sequence of nonnegative integers such that the nonzero γ_i values are the partial multiplicities of C at λ_0. In this section we study the connections between α_i, β_i, and γ_i. In view of Theorem 4.1.1, these connections describe the Jordan form of C in terms of the Jordan forms of A and B.

Some such connections are easily seen. We have

$$\det(C - \lambda I) = \det(A - \lambda I)\det(B - \lambda I) \tag{4.2.2}$$

for every $\lambda \in \mathbb{C}$. Now the algebraic multiplicity of an eigenvalue λ_0 of a matrix X coincides with the multiplicity of λ_0 as a zero of the polynomial $\det(X - \lambda I)$. (When λ_0 is not an eigenvalue of X this statement is also true if we accept the convention that, in this case, the algebraic multiplicity of λ_0 is zero.) It follows from equation (4.2.2) that the algebraic multiplicity* of C at λ_0 is equal to the sum of the algebraic multiplicities of A and B at λ_0. In other words

$$\sum_{i=1}^{\infty} \gamma_i = \sum_{i=1}^{\infty} \alpha_i + \sum_{i=1}^{\infty} \beta_i \tag{4.2.3}$$

Further, as C is an extension of A and a coextension of B, Theorems 4.1.4 and 4.1.5 imply that

$$\gamma_i \geq \max(\alpha_i, \beta_i), \qquad i = 1, 2, \ldots \tag{4.2.4}$$

The following inequality between $\{\alpha_j\}_{j=1}^{\infty}$, $\{\beta_j\}_{j=1}^{\infty}$, and $\{\gamma_j\}_{j=1}^{\infty}$ is deeper.

Proposition 4.2.1

Let C be a completion of A and B, with the partial multiplicities of A, B, and C at a fixed $\lambda_0 \in \mathbb{C}$ given by the nonincreasing sequences of nonnegative integers $\{\alpha_i\}_{i=1}^{\infty}$, $\{\beta_i\}_{i=1}^{\infty}$, and $\{\gamma_i\}_{i=1}^{\infty}$, respectively. Then

---

*It is convenient here to talk about the "algebraic multiplicity of C at λ_0" rather than the "algebraic multiplicity of λ_0" as an eigenvalue of C.

$$\sum_{j=1}^{m} \{k \mid \gamma_k \geq j\}^{\#} \leq \sum_{j=1}^{m} \{k \mid \beta_k \geq j\}^{\#} + \sum_{j=1}^{m} \{k \mid \alpha_k \geq j\}^{\#}, \qquad m = 1, 2, \ldots$$

$$(4.2.5)$$

As usual in this book, the symbol $\Omega^{\#}$ represents the number of different elements in a finite set Ω.

Proof. First we prove the following inequalities:

$$\dim \operatorname{Ker}(C - \lambda_0 I)^i \leq \dim \operatorname{Ker}(A - \lambda_0 I)^i + \dim \operatorname{Ker}(B - \lambda_0 I)^i, \qquad i = 1, 2, \ldots$$

$$(4.2.6)$$

Indeed, for every $\epsilon \neq 0$ we have [using formula (4.2.1)]

$$\begin{bmatrix} I & 0 \\ 0 & \epsilon^{-1}I \end{bmatrix} \begin{bmatrix} A - \lambda_0 I & D \\ 0 & B - \lambda_0 I \end{bmatrix} \begin{bmatrix} I & 0 \\ 0 & \epsilon I \end{bmatrix} = \begin{bmatrix} A - \lambda_0 I & \epsilon D \\ 0 & B - \lambda_0 I \end{bmatrix}$$

and thus

$$\dim \operatorname{Ker}(C - \lambda_0 I)^i = \dim \operatorname{Ker} \begin{bmatrix} A - \lambda_0 I & \epsilon D \\ 0 & B - \lambda_0 I \end{bmatrix}^i, \qquad i = 1, 2, \ldots$$

$$(4.2.7)$$

Fix some i, and let

$$m = \operatorname{rank} \begin{bmatrix} A - \lambda_0 I & 0 \\ 0 & B - \lambda_0 I \end{bmatrix}^i$$

So there exists an $m \times m$ nonsingular submatrix Q in $(A - \lambda_0 I)^i \oplus (B - \lambda_0 I)^i$. Consider the $m \times m$ submatrix $Q(\epsilon)$ of

$$\begin{bmatrix} A - \lambda_0 I & \epsilon D \\ 0 & B - \lambda_0 I \end{bmatrix}$$

which is formed by the same rows and columns as Q itself. Now $Q(\epsilon)$ is as close as we wish to Q provided ϵ is sufficiently close to 0. Take ϵ so small that the matrix $Q(\epsilon)$ is also nonsingular. For such an ϵ

$$\operatorname{rank} \begin{bmatrix} A - \lambda_0 I & \epsilon D \\ 0 & B - \lambda_0 I \end{bmatrix}^i \geq m$$

Comparing with (4.2.7), we obtain the desired inequality (4.2.6). Now use Proposition 2.2.6 to obtain the inequalities (4.2.5). $\square$

In connection with inequalities (4.2.5), note that

$$\sum_{j=1}^{\infty} \{k \mid \gamma_k \geq j\}^{\#} = \sum_{j=1}^{\infty} \{k \mid \alpha_k \geq j\}^{\#} + \sum_{j=1}^{\infty} \{k \mid \beta_k \geq j\}^{\#} \qquad (4.2.8)$$

Indeed, as $\{k \mid \gamma_k \geq j\}^{\#} = 0$ for $j > \gamma_1$, and similarly for $\{\alpha_k\}_{k=1}^{\infty}$ and $\{\beta_k\}_{k=1}^{\infty}$, all the sums in equation (4.2.8) are finite, so (4.2.8) makes sense. Further, for any nonincreasing sequence of nonnegative integers $\{\delta_i\}_{i=1}^{\infty}$ with finite sum $\Sigma_{i=1}^{\infty} \delta_i$ we have

$$\sum_{i=1}^{\infty} \delta_i = \sum_{i-1}^{\infty} \{k \mid \delta_k \geq i\}^{\#} \qquad (4.2.9)$$

The easiest way to verify (4.2.9) is by representing each nonzero δ_i as the rectangle with height δ_i and width 1 and putting these rectangles one next to another. The result is a ladderlike figure Φ. For instance, if $\delta_1 = 5$, $\delta_2 = \delta_3 = 4$, $\delta_4 = 1$, $\delta_j = 0$ for $j > 4$, then Φ is the following figure:

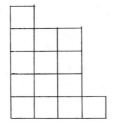

Obviously, the area of Φ is just the left-hand side of equation (4.2.9). On the other hand, the right-hand side of (4.2.9) is also the area of Φ calculated by the rows of Φ (indeed, $\{k \mid \delta_k \geq i\}^{\#}$ is the area of the ith row in Φ counting from the bottom); hence equality holds in (4.2.9). Now appeal to (4.2.3) and (4.2.8) follows. We need a completely different line of argument to prove the following proposition.

Proposition 4.2.2

With $\{\alpha_i\}_{i=1}^{\infty}$, $\{\beta_i\}_{i=1}^{\infty}$ and $\{\gamma_i\}_{i=1}^{\infty}$ as in Proposition 4.2.1, we have

$$\sum_{j=1}^{m} \gamma_j \leq \sum_{j=1}^{m} \alpha_j + \sum_{j=1}^{m} \beta_j, \qquad m = 1, 2, \ldots \qquad (4.2.10)$$

Proof. Assuming that C is given by (4.2.1), one easily obtains

$$C - \lambda I = \begin{bmatrix} I & 0 \\ 0 & B - \lambda I \end{bmatrix} \begin{bmatrix} I & D \\ 0 & I \end{bmatrix} \begin{bmatrix} A - \lambda I & 0 \\ 0 & I \end{bmatrix}$$

Using Theorem A.4.3 of the appendix, pick a $p \times p$ submatrix $C_0(\lambda)$ in $C - \lambda I$ such that λ_0 is a zero of det $C_0(\lambda)$ of multiplicity $\gamma_n + \cdots + \gamma_{n-p+1}$

(here $n \times n$ is the size of C). The integer p is assumed to be greater than $\max(n_A, n_B)$ where $n_A \times n_A$ is the size of A and $n_B \times n_B$ is the size of B (so $n = n_A + n_B$). By the Binet–Cauchy formula (Theorem A.2.1 of the appendix) we have

$$\det C_0(\lambda) = \sum_{i,j,k} \det B_i(\lambda) \cdot \det D_j \cdot \det A_k(\lambda) \qquad (4.2.11)$$

where $B_i(\lambda)$, D_j, $A_k(\lambda)$ are $p \times p$ submatrices of $\begin{bmatrix} I & 0 \\ 0 & B - \lambda I \end{bmatrix}$, $\begin{bmatrix} I & D \\ 0 & I \end{bmatrix}$, and $\begin{bmatrix} A - \lambda I & 0 \\ 0 & I \end{bmatrix}$, respectively, and the summation is taken over certain triples i, j, k. Note that $\det B_i(\lambda) = 0$ unless $B_i(\lambda)$ is of the form $I_s \oplus \tilde{B}_i(\lambda)$, where $\tilde{B}_i(\lambda)$ is a $(p - s) \times (p - s)$ submatrix of $B - \lambda I$ (here s is an integer that may depend on i and for which $0 \le s \le n_A$). Similarly, $\det A_k(\lambda) = 0$ unless $A_k(\lambda)$ is of the form $I_t \oplus \tilde{A}_k(\lambda)$, where $\tilde{A}_k(\lambda)$ is a $(p - t) \times (p - t)$ submatrix of $A - \lambda I$ $(0 \le t \le n_B)$. Taking these observations into account, rewrite equation (4.2.11) as follows:

$$\det C_0(\lambda) = \sum_{i,j,k} \det \tilde{B}_i(\lambda) \cdot \det D_j \cdot \det \tilde{A}_k(\lambda)$$

Now the size of $\tilde{B}_i(\lambda)$ is at least $(p - n_A) \times (p - n_A)$, so by the same theorem, Theorem A.4.3, the multiplicity of λ_0 as a zero of $\det \tilde{B}_i(\lambda)$ is at least

$$\beta_{n_B} + \beta_{n_B - 1} + \cdots + \beta_{n_B - (p - n_A) + 1} = \beta_n + \beta_{n-1} + \cdots + \beta_{n - p + 1}$$

(here we use $n_B + n_A = n$ and $\beta_i = 0$ for $i > n_B$). Similarly, the multiplicity of λ_0 as a zero of $\det \tilde{A}_k(\lambda)$ is at least $\alpha_n + \alpha_{n-1} + \cdots + \alpha_{n-p+1}$. We find that the multiplicity $\sum_{j=n-p+1}^{n} \gamma_j$ of λ_0 as a zero of $\det C_0(\lambda)$ is at least $\sum_{j=n-p+1}^{n} (\alpha_j + \beta_j)$. It follows from equation (4.2.3) that

$$\sum_{j=1}^{n-p} \gamma_j \le \sum_{j=1}^{n-p} \alpha_j + \sum_{j=1}^{n-p} \beta_j \qquad (4.2.12)$$

If it happens that $p \le n_A$, then the inequality

$$\sum_{j=n-p+1}^{n} \gamma_j \ge \sum_{j=n-p+1}^{n} \alpha_j + \sum_{j=n-p+1}^{n} \beta_j$$

and hence also the relation (4.2.12), follows from (4.2.4) because in this case $\beta_j = 0$ for $j \ge n - p + 1$. Similarly, (4.2.12) holds for $p \le n_B$. We have proved (4.2.10) for $m = 1, \ldots, n$. For $m \ge n$ the inequality (4.2.10) coincides with (4.2.3), so the proof of (4.2.10) is complete. $\square$

We have proved various inequalities and equalities relating the sequences

$\{\alpha_i\}_{i=1}^{\infty}$, $\{\beta_i\}_{i=1}^{\infty}$, and $\{\gamma_i\}_{i=1}^{\infty}$ [relations (4.2.3), (4.2.5), (4.2.8), (4.2.10)]. These relations are by no means the only connections between these sequences. More specifically, there exist nonincreasing sequences of non-negative integers $\{\alpha_i\}_{i=1}^{\infty}$, $\{\beta_i\}_{i=1}^{\infty}$ and $\{\gamma_i\}_{i=1}^{\infty}$, only a finite number of them nonzero, that satisfy equations (4.2.3), (4.2.5), (4.2.8), and (4.2.10), but for which there is *no* completion C of A and B with the property that for some $\lambda_0 \in \mathbb{C}$ the sequences $\{\alpha_i\}_{i=1}^{\infty}$, $\{\beta_i\}_{i=1}^{\infty}$, and $\{\gamma_i\}_{i=1}^{\infty}$ give the partial multiplicities of A, B, and C, respectively, corresponding to λ_0. In the next section we see more general inequalities, but even they do not completely describe the connections between the partial multiplicities of extensions of A and B and the partial multiplicities of A and B. The problem of describing all such connections is open.

4.3   THE SIGAL INEQUALITIES

The main result in this section is the following generalization of Proposition 4.2.2.

Theorem 4.3.1

Let $\{\alpha_i\}_{i=1}^{\infty}$, $\{\beta_i\}_{i=1}^{\infty}$, and $\{\gamma_i\}_{i=1}^{\infty}$ be as in Proposition 4.2.1. Then for every sequence $r_1 < r_2 < \cdots < r_m$ of positive integers we have

$$\sum_{i=1}^{m} \gamma_{r_i} \leq \sum_{i=1}^{m} \alpha_{r_i} + \sum_{i=1}^{m} \beta_i \tag{4.3.1}$$

and

$$\sum_{i=1}^{m} \gamma_{r_i} \leq \sum_{i=1}^{m} \alpha_i + \sum_{i=1}^{m} \beta_{r_i} \tag{4.3.2}$$

Proposition 4.2.2 is obtained from this theorem by putting $r_j = j$, $j = 1, \ldots, m$. It will be convenient to prove a lemma (which is actually a particular case of Theorem 4.3.1) before proving the theorem itself.

Lemma 4.3.2

Let

$$C = \begin{bmatrix} 0_{k \times k} & 0 \\ X & B \end{bmatrix}$$

where B is $(n-k) \times (n-k)$ with $\sigma(B) = \{0\}$. If $\{\gamma_i\}_{i=1}^{\infty}$ and $\{\beta_i\}_{i=1}^{\infty}$ are the nonincreasing sequences of partial multiplicities of C and B, respectively, then $\gamma_i = \beta_i + \delta_i$, $i = 1, 2, \ldots$, where δ_i is zero or one, and $\sum_{i=1}^{\infty} \delta_i = k$.

Proof. Let $x_1, \ldots, x_l$ $(l \geq 2)$ *be a Jordan chain for* C:

$$Cx_{i+1} = x_i, \qquad i = 1, \ldots, l-1; \qquad x_1 \neq 0 \qquad (4.3.3)$$

Write $x_i = \begin{bmatrix} y_i \\ z_i \end{bmatrix}$, where y_i is a k-dimensional vector and z_i is $(n-k)$-dimensional. Equalities (4.3.3) then imply $y_1 = \cdots = y_{l-1} = 0$ and $Bz_{i+1} = z_i$, $i = 1, \ldots, l-2, z_1 \neq 0$. In other words, $z_1, \ldots, z_{l-1}$ is a Jordan chain for B. Moreover, if $Xy_l = 0$, then $z_1, \ldots, z_l$ is also a Jordan chain for B. Now let

$$x_{11}, \ldots, x_{1,\gamma_1}; \ldots; x_{q1}, \ldots, x_{q,\gamma_q} \qquad (4.3.4)$$

be a basis in $\mathbb{C}^n$ consisting of Jordan chains for C (so q is the maximal index such that $\gamma_q > 0$). Denoting by p the maximal index such that $\gamma_p \geq 2$, let Z be the subspace spanned by the Jordan chains $z_{11}, \ldots, z_{1,l_1}; \ldots; z_{p1}, \ldots, z_{p,l_p}$ for B constructed as in the preceding paragraph from the Jordan chains $x_{j1}, \ldots, x_{j,\gamma_j}, j = 1, \ldots, p$ of C. Here l_j is either $\gamma_j - 1$ or γ_j. The order of Jordan chains in equation (4.3.4) of the same length can be adjusted so that $l_1 \geq \cdots \geq l_p$. Since Z is B invariant, Theorem 4.1.4 gives $\beta_i \geq l_i$, $i = 1, \ldots, p$. On the other hand, by Theorem 4.1.5 $\gamma_i \geq \beta_i$, $i = 1, 2, \ldots$. So we obtain $\gamma_i - \beta_i \leq \delta_i$, $i = 1, 2, \ldots$, where each δ_i is either zero or one. The equality $\Sigma_{i=1}^{\infty} \delta_i = k$ follows from the fact that the sum of the partial multiplicities of C (resp. of B) is n (resp. $n - k$). $\square$

Proof of Theorem 4.3.1. Let $\mathbb{C}^n = \mathcal{M} \dotplus \mathcal{N}$ and let $A: \mathcal{M} \to \mathcal{M}$, $B: \mathcal{N} \to \mathcal{N}$ be transformations such that $\{\alpha_i\}_{i=1}^{\infty}$, $\{\beta_i\}_{i=1}^{\infty}$, and $\{\gamma_i\}_{i=1}^{\infty}$ are the nonincreasing sequences of nonnegative integers representing the partial multiplicities of A, B, and

$$C = \begin{bmatrix} A & D \\ 0 & B \end{bmatrix}$$

respectively, corresponding to the eigenvalue λ_0 (here D is some transformation from $\mathcal{N}$ into $\mathcal{M}$). Applying a similarity transformation, if necessary, we can assume that $\mathcal{N} = \mathcal{M}^{\perp}$.

Without loss of generality (Theorem 4.1.1) we can assume also that $\lambda_0 = 0$ and $\sigma(A) = \sigma(B) = \{0\}$ (then also $\sigma(C) = \{0\}$). We can assume also that A is in the Jordan form:

$$A = \text{diag}[J_{\alpha_1}(0), \ldots, J_{\alpha_l}(0)], \qquad (\alpha_j = 0 \text{ for } j > l)$$

We use induction on the size α_1 of the biggest Jordan block in A. If

$\alpha_1 = 1$, then $A = 0$ and by Lemma 4.3.2 (applied to B^* and C^* in place of B and C, respectively) we have

$$\sum_{i=1}^{m} \gamma_{r_i} = \sum_{i=1}^{m} (\beta_{r_i} + \delta_{r_i}) \leq \sum_{i=1}^{m} \beta_{r_i} + \min(m, l) = \sum_{i=1}^{m} \beta_{r_i} + \sum_{i=1}^{m} \alpha_i$$

Assume that inequality (4.3.2) is proved for all A with the property that the size of the biggest Jordan block is less than α_1. Using a matrix similar to A in place of A, we can assume that

$$A = \begin{bmatrix} 0_{l \times l} & A_1 \\ 0 & A_2 \end{bmatrix}$$

where A_2 is a Jordan matrix with partial multiplicities $\{\alpha_i'\}_{i=1}^{\infty}$ satisfying

$$\alpha_1' = \alpha_1 - 1, \ldots, \alpha_l' = \alpha_l - 1; \alpha_j' = 0 \quad \text{for} \quad j > l \qquad (4.3.5)$$

With the corresponding partition $D = \begin{bmatrix} X_1 \\ X_2 \end{bmatrix}$, and using the induction hypothesis the partial multiplicities $\{\gamma_i'\}_{i=1}^{\infty}$ of the matrix $C' = \begin{bmatrix} A_2 & X_2 \\ 0 & B \end{bmatrix}$ satisfy the inequalities:

$$\sum_{i=1}^{m} \gamma_{r_i}' \leq \sum_{i=1}^{m} \alpha_i' + \sum_{i=1}^{m} \beta_{r_i} \qquad (4.3.6)$$

But in view of Lemma 4.3.2 (applied with C'^* and C^* in place of B and C, respectively)

$$\sum_{i=1}^{m} \gamma_{r_i} \leq \sum_{i=1}^{m} \gamma_{r_i}' + \min(m, l) \qquad (4.3.7)$$

Now combine relations (4.3.5), (4.3.6), and (4.3.7) to obtain the inequality (4.3.2). The inequalities (4.3.1) are obtained from (4.3.2) applied to the transformation C^* written as the 2×2 block matrix with respect to the direct sum decomposition $\mathbb{C}^n = \mathcal{N} \dotplus \mathcal{M}$. $\square$

Inequalities (4.3.1) and (4.3.2) admit the following geometric interpretation. Let q be any index such that $\gamma_i = 0$ for $i > q$ (e.g., $q = \sum_{i=1}^{\infty} \alpha_i + \sum_{i=1}^{\infty} \beta_i$). Denote by $K_1 \subset \mathbb{R}^q$ the convex hull of the points

$$\alpha_1 + \beta_{\pi(1)}, \alpha_2 + \beta_{\pi(2)}, \ldots, \alpha_q + \beta_{\pi(q)}$$

where π is any permutation of $\{1, 2, \ldots, q\}$, that is

$$K_1 = \left\{ \sum_{\pi} k_{\pi}(\alpha_1 + \beta_{\pi(1)}, \ldots, \alpha_q + \beta_{\pi(q)}) \mid k \geq 0, \quad \sum_{\pi} k_{\pi} = 1 \right\}$$

Also let

$$K_2 = \left\{ \sum_\pi k_\pi(\alpha_{\pi(1)} + \beta_1, \ldots, \alpha_{\pi(q)} + \beta_q) \mid k_\pi \geq 0, \quad \sum_\pi k_\pi = 1 \right\}$$

Then inequalities (4.3.1) and (4.3.2) imply

$$(\gamma_1, \ldots, \gamma_q) \in K_1 \cap K_2 \tag{4.3.8}$$

Actually, the inclusion (4.3.8) in turn implies (4.3.1) and (4.3.2). The proof of these statements would take us too far afield; we only mention that it is essentially the same as the proof of Theorem 10 of Lidskii (1966). It is interesting that the geometric interpretation of inequalities (4.3.1) and (4.3.2) is completely analogous to the geometric interpretation of the inequalities for the eigenvalues of the sum of two hermitian matrices in terms of the eigenvalues of each hermitian matrix [see Lidskii (1966)].

Inequalities (4.3.1) and (4.3.2) can be generalized. In fact, for any sequence $r_1 < r_2 < \cdots < r_m$ of positive integers and any nonnegative integer $k < r_1$ the following inequalities hold [see Thijsse (1984)]:

$$\sum_{i=1}^m \gamma_{r_i} \leq \sum_{i=1}^m \alpha_{r_i - k} + \sum_{i=1}^m \beta_{i+k} ; \quad \sum_{i=1}^m \gamma_{r_i} \leq \sum_{i=1}^m \alpha_{i+k} + \sum_{i=1}^m \beta_{r_i - k} \tag{4.3.9}$$

Theorem 4.3.1 is a particular case of (4.3.9) with $k = 0$.

We have seen that, given the sequences $\{\alpha_i\}_{i=1}^\infty$ and $\{\beta_i\}_{i=1}^\infty$ of partial multiplicities of A and B, respectively, corresponding to λ_0, the sequence $\{\gamma_i\}_{i=1}^\infty$ of partial multiplicities corresponding to λ_0 of any completion C of A and B satisfies the properties of (4.2.3), (4.2.4), (4.2.5), (4.3.1), and (4.3.2); moreover, (4.3.9) is satisfied as well. However, the following example shows that, in general, these properties do not characterize the partial multiplicities of completions.

EXAMPLE 4.3.1.   Let $\alpha_1 = \alpha_2 = 3$, $\alpha_i = 0$ for $i > 2$; $\beta_1 = \beta_2 = 5$; $\beta_3 = 4$; $\beta_i = 0$ for $i > 3$; $\gamma_1 = 7$, $\gamma_2 = 6$, $\gamma_3 = 4$, $\gamma_4 = 3$, $\gamma_i = 0$ for $i > 4$. One verifies that relations (4.2.3), (4.2.4), (4.2.5), and (4.3.9) hold [the verification of (4.3.9) is lengthy because of the many possibilities involved]. However, Theorem 7 of Rodman and Schaps (1979) implies that there is no completion C of A and B such that the partial multiplicities of A, B, and C corresponding to some λ_0 are given by $\{\alpha_i\}_{i=1}^\infty$, $\{\beta_i\}_{i=1}^\infty$, and $\{\gamma_i\}_{i=1}^\infty$, respectively.

4.4   SPECIAL CASE OF COMPLETIONS

In this section we describe all the possible sequences of partial multiplicities corresponding to λ_0 for completions of A and B in case at least one of A and B has only one partial multiplicity at λ_0. First, we establish some general observations on partial multiplicities of completions that are used in this description.

It is convenient to introduce the set Ω of all nondecreasing sequences of nonnegative integers such that, in each sequence, only a finite number of integers is different from zero. For $\alpha = (\alpha_1, \alpha_2, \ldots)$, $\beta = (\beta_1, \beta_2, \ldots) \in \Omega$ denote by $\Gamma(\alpha, \beta)$ the set of all sequences $\gamma = (\gamma_1, \gamma_2, \ldots) \in \Omega$ with the following properties: (a) there is a transformation $C: \mathbb{C}^n \to \mathbb{C}^n$ (for some n) and a C-invariant subspace $\mathcal{M}$ such that the restriction $C|_{\mathcal{M}}$ has partial multiplicities $\alpha_1, \alpha_2, \ldots$ corresponding to a certain eigenvalue λ_0; (b) the compression of C to a coinvariant subspace that is a complement to $\mathcal{M}$ has partial multiplicities $\beta_1, \beta_2, \ldots$ corresponding to λ_0, and (c) C itself has partial multiplicities $\gamma_1, \gamma_2, \ldots$ corresponding to the same λ_0.

Proposition 4.4.1

Let $\alpha = (\alpha_1, \alpha_2, \ldots) \in \Omega$, $\beta = (\beta_1, \beta_2, \ldots) \in \Omega$, *and put* $m = \Sigma_{i=1}^{\infty} \alpha_i$, $n = \Sigma_{i=1}^{\infty} \beta_i$. *Then a sequence* $\gamma = (\gamma_1, \gamma_2, \ldots) \in \Omega$ *belongs to* $\Gamma(\alpha, \beta)$ *if and only if there is an* $m \times n$ *matrix* A *such that the partial multiplicities of the matrix*

$$J = \begin{bmatrix} J_1 & A \\ 0 & J_2 \end{bmatrix} \tag{4.4.1}$$

where $J_1 = J_{\alpha_1}(0) \oplus \cdots \oplus J_{\alpha_{n_1}}(0)$, $J_2 = J_{\beta_1}(0) \oplus \cdots \oplus J_{\beta_{n_2}}(0)$ $[n_1 \text{ (resp. } n_2)]$ *is the largest index such that* $\alpha_{n_1} \neq 0$ $[resp. \ \beta_{n_2} \neq 0]$ *are* $\gamma_1, \gamma_2, \ldots$.

Proof. As the part "if" follows from the definition of $\Gamma(\alpha, \beta)$, we have only to prove the "only if" part. Assume $\gamma \in \Gamma(\alpha, \beta)$. By definition, there is a matrix C partitioned as follows:

$$C = \begin{bmatrix} C_{11} & C_{12} \\ 0 & C_{22} \end{bmatrix}$$

where for some eigenvalue λ_0 of C the partial multiplicities of C (resp. C_{11}, C_{22}) at λ_0 are given by γ (resp. α, β). Replacing C by $C - \lambda_0 I$, we can assume $\lambda_0 = 0$. Furthermore, we can assume that C_{11} and C_{22} are matrices in the Jordan form. It remains to appeal to Theorem 4.4.1. $\square$

It follows immediately from Proposition 4.1.1 that $\Gamma(\alpha, \beta) = \Gamma(\beta, \alpha)$. Indeed, in the notation of Proposition 4.4.1 we have

$$\begin{bmatrix} 0 & I \\ I & 0 \end{bmatrix} \begin{bmatrix} J_1 & A \\ 0 & J_2 \end{bmatrix} \begin{bmatrix} 0 & I \\ I & 0 \end{bmatrix} = \begin{bmatrix} J_2 & 0 \\ A & J_1 \end{bmatrix}$$

so the matrices

$$\begin{bmatrix} J_1 & A \\ 0 & J_2 \end{bmatrix} \quad \text{and} \quad \begin{bmatrix} J_2 & 0 \\ A & J_1 \end{bmatrix}$$

have the same Jordan form. But then (in view of Corollary 2.2.3) this is also true for the matrices

$$\begin{bmatrix} J_1 & A \\ 0 & J_2 \end{bmatrix} \quad \text{and} \quad \begin{bmatrix} J_2 & 0 \\ A & J_1 \end{bmatrix}^* = \begin{bmatrix} J_2^* & A^* \\ 0 & J_1^* \end{bmatrix}$$

As J_2^* and J_1^* are similar to J_2 and J_1, respectively, the conclusion $\Gamma(\alpha, \beta) = \Gamma(\beta, \alpha)$ follows.

In view of Proposition 4.4.1, in order to determine $\Gamma(\alpha, \beta)$, we have to find the partial multiplicities $\gamma_1 \geq \gamma_2 \geq \cdots$ (or, what is the same, the Jordan form) of matrices J of type (4.4.1). As

$$\{k \mid \gamma_k \geq i + 1\}^{\#} = \operatorname{rank} J^i - \operatorname{rank} J^{i+1}, \qquad i = 0, 1, \ldots$$

(by definition, $J^0 = I$), we focus on a formula for computation of the ranks of J^i, $i = 1, 2, \ldots$.

Divide the matrix A into blocks A_{ij}, $i = 1, \ldots, n_1; j = 1, \ldots, n_2$ according to the sizes of Jordan blocks in J_1 and J_2 (so the size of A_{ij} is $\alpha_i \times \beta_j$). For fixed i and j, write $A_{ij} = \sum_{q=1}^{\beta_j} \sum_{p=1}^{\alpha_i} u_{pq} E_{pq}$, where E_{pq} is an $\alpha_i \times \beta_j$ matrix with 1 in the intersection of the $(\alpha_i - p + 1)$th row and qth column and zero in all other places. Let

$$d_{ij}^{(t)} = u_{1t} + u_{2,t-1} + \cdots + u_{t1}$$

(we put $u_{pq} = 0$ if $p > \alpha_i$ or $q > \beta_j$). Define

$$B_{ij}^{(k)} = \sum d_{ij}^{(p+q-k)} E_{pq}, \qquad k = 1, 2, \ldots \qquad (4.4.2)$$

where the sum is over all the pairs p, q such that $p \leq \min(k, \alpha_i)$, $q \leq \min(k, \beta_j)$, and $p + q > k$. For example, $B_{ij}^{(1)}$ has u_{11} in the lower left corner and zeros elsewhere, $B_{ij}^{(2)}$ has $\begin{bmatrix} u_{11} & u_{12} + u_{21} \\ 0 & u_{11} \end{bmatrix}$ in the lower left corner and zero elsewhere, $B_{ij}^{(3)}$ has

$$\begin{bmatrix} u_{11} & u_{12} + u_{21} & u_{13} + u_{22} + u_{31} \\ 0 & u_{11} & u_{12} + u_{21} \\ 0 & 0 & u_{11} \end{bmatrix}$$

in the lower left corner and zeros elsewhere (provided $\alpha_i, \beta_j \geq 3$). Let $B^{(k)}$ be the $m \times n$ matrix with blocks $B_{ij}^{(k)} (i = 1, \ldots, n_1; j = 1, \ldots, n_2)$.

Lemma 4.4.2

In the preceding notation we have

$$\operatorname{rank} J^k = \operatorname{rank} J_1^k + \operatorname{rank} J_2^k + \operatorname{rank} B^{(k)}, \qquad k = 1, 2, \ldots$$

Proof. Let $A^{(k)}$ be defined by $J^k = \begin{bmatrix} J_1^k & A^{(k)} \\ 0 & J_2^k \end{bmatrix}$. An easy induction argument on k shows that

$$A^{(k)} = \sum_{s=0}^{k-1} J_1^s A J_2^{k-1-s}, \qquad k = 1, 2, \ldots$$

and hence

$$A_{ij}^{(k)} = \sum_{s=0}^{k-1} J_{1i}^s A_{ij} J_{2j}^{k-1-s} = \sum_{s=0}^{k-1} \sum_{p,q} u_{pq} J_{1i}^s E_{pq} J_{2j}^{k-s-1}$$

$$= \sum_{p,q} \sum_{s=0}^{k-1} u_{pq} E_{p+s, q+k-s-1}$$

$$= \sum_{p'+q' \geq k+1} \sum_{s=0}^{k-1} u_{p'-s, q'-k+s+1} E_{p'q'},$$

where $E_{ab} = 0$ whenever at least one of the inequalities $1 \leq a \leq \alpha_i$; $1 \leq b \leq \beta_j$ is violated, and $u_{ab} = 0$ for $a < 1$ or $b < 1$.

It follows that

$$A_{ij}^{(k)} = B_{ij}^{(k)} + (\text{terms with } E_{p'q'} \text{ such that } p' > k \text{ or } q' > k)$$

By column operations from J_1^k and row operations from J_2^k, we can eliminate all terms of $A_{ij}^{(k)}$ except those in the block $B_{ij}^{(k)}$. Permuting the rows and columns of the resulting matrix, we obtain the following matrix that has the same rank as J^k:

$$\begin{bmatrix} 0 & I_{a_k} & 0 & 0 \\ 0 & 0 & B^{(k)} & 0 \\ 0 & 0 & 0 & I_{b_k} \\ 0 & 0 & 0 & 0 \end{bmatrix}$$

where $a_k = \text{rank } J_1^k$ and $b_k = \text{rank } J_2^k$. Lemma 4.4.2 follows. $\square$

It is an immediate consequence of the lemma that the sequence $\{\gamma_j\}_{j=1}^{\infty}$ depends only on the diagonal sums $d_{ij}^{(t)}$, for $t \leq \min(\alpha_i, \beta_i)$. Thus we can replace each A_{ij} by a matrix in which only the first column can contain nonzero entries. Alternatively, we can presume that only the bottom row of A_{ij} can contain nonzero entries.

For illustration of Lemma 4.4.2, consider the following example.

EXAMPLE 4.4.1. Let $\alpha = (\alpha_1, 0, 0, \ldots)$, $\beta = (\beta_1, 0, 0, \ldots)$, where $\alpha_1, \beta_1 > 0$. We suppose for definiteness that $\alpha_1 \geq \beta_1$. If $d^{(1)} \neq 0$, it is easily seen that

$$\text{rank } B_{11}^{(k)} = \min(k, \alpha_1) + \min(k, \beta_1) - k, \qquad k \geq 1$$

In general, we have

$$\text{rank } B_{11}^{(k)} = \begin{cases} \min(k, \alpha_1) + \min(k, \beta_1) - k - t_0 + 1 & \text{for} \quad k \geq t_0 \\ 0 & \text{for} \quad k < t_0 \end{cases}$$

$$(4.4.3)$$

where t_0 is the smallest t such that $d_{11}^{(t)} \neq 0$, or $t_0 = \beta_1 + 1$ if all $d_{11}^{(t)}$ are zeros. It is now clear that $\gamma = (\gamma_1, \gamma_2, \ldots) \in \Gamma(\alpha, \beta)$ is determined completely by the value of t_0. Further, using formula (4.4.3) and Lemma 4.4.2, we compute

$$\{k \mid \gamma_k \geq i + 1\}^\# = \text{rank } J^i - \text{rank } J^{i+1}$$

Computation shows that

$$\Gamma(\alpha, \beta) = \{(\alpha_1 + \beta_1, 0), (\alpha_1 + \beta_1 - 1, 1), \ldots, (\alpha_1 + 1, \beta_1 - 1), (\alpha_1, \beta_1)\}$$

(In every γ sequence we write only the first members; the others are zeros.) The γ sequence $(\alpha_1 + \beta_1 - \rho, \rho)$ corresponds to the value $t_0 = \rho + 1$.

The possibility of $\gamma = (\alpha_1 + \beta_1 - \rho, \rho)$, $\rho = 0, \ldots, \beta_1$, is realized for the matrix

$$J^{(\rho)} = \begin{bmatrix} J_1 & A_\rho \\ 0 & J_2 \end{bmatrix}$$

where A_ρ is an $\alpha_1 \times \beta_1$ matrix with all but the $(\alpha_1 - \rho, 1)$th entry equal to zero, and this exceptional entry is equal to 1 (for $\rho = \beta_1$ we put $A_\rho = 0$). It is not difficult to construct two independent Jordan chains of $\lambda I - J^{(\rho)}$ of lengths $\alpha_1 + \beta_1 - \rho$ and ρ. Namely, the Jordan chain of length $\alpha_1 + \beta_1 - \rho$ is $e_{\alpha_1 + \beta_1}, e_{\alpha_1 + \beta_1 - 1}, \ldots, e_{\alpha_1 + 1}, e_{\alpha_1 - \rho}, e_{\alpha_1 - \rho - 1}, \ldots, e_1$. The Jordan chain of length ρ is $e_{\alpha_1} - e_{\alpha_1 + \rho}, e_{\alpha_1 - 1} - e_{\alpha_1 + \rho - 1}, \ldots, e_{\alpha_1 - \rho + 1} - e_{\alpha_1 + 1}$. $\quad \square$

Using Lemma 4.4.2, we shall now give a complete description of the set $\Gamma(\alpha, \beta)$ in the case that $\alpha = (\alpha_1, \alpha_2, \ldots, \alpha_n, 0, \ldots)$ and $\beta = (\beta_1, 0, 0, \ldots)$ where α_n and β_1 are positive.

Introduce the set Ω_0 of all n-tuples $(\omega_1, \omega_2, \ldots, \omega_n)$, where ω_j are integers such that $1 \leq \omega_j \leq \lambda_j + 1$ and $\lambda_j = \min(\alpha_j, \beta_1)$. For a given sequence $\omega = (\omega_1, \omega_2, \ldots, \omega_n) \in \Omega_0$ and $i = 1, 2, \ldots$, define integers $c_{ij}^{(\omega)}$ as follows:

$$c_{ij}^{(\omega)} = \begin{cases} i - \min(\omega_j - 1, i) & \text{for} \quad 1 \leq i \leq \lambda_j \\ \lambda_j - \min(\omega_j - 1, \lambda_j) & \text{for} \quad \lambda_j \leq i \leq \mu_j \\ \lambda_j + \mu_j - i - \min(\omega_j - 1, \lambda_j + \mu_j - i) & \text{for} \quad i \geq \mu_j \end{cases}$$

where $\mu_j = \max(\alpha_j, \beta_1)$. Now let $\gamma = (\gamma_1, \gamma_2, \ldots)$ be the nonincreasing sequence of nonnegative integers defined by the equalities

$$\{j \mid \gamma_j \geq k + 1\}^{\#} = \{j \mid \alpha_j \geq k + 1\}^{\#} + \max(\beta_1 - k, 0)$$
$$- \max(\beta_1 - k - 1, 0) + f_k - f_{k+1}$$

for $k = 0, 1, 2, \ldots$, where $f_0 = 0$ and

$$f_k = \max(c_{k1}^{(\omega)}, c_{k2}^{(\omega)}, \ldots, c_{kn}^{(\omega)}) \quad \text{for} \quad k > 0 \tag{4.4.4}$$

Thus for every $\omega \in \Omega_0$ we have constructed a sequence γ. Let us denote this sequence by $F(\omega)$.

Theorem 4.4.3

For every $\omega \in \Omega_0$ the sequence $F(\omega)$ belongs to $\Gamma(\alpha, \beta)$. Conversely, if $\gamma \in \Gamma(\alpha, \beta)$, there exists $\omega \in \Omega_0$ such that $\gamma = F(\omega)$.

Proof. Recall that

$$\{j \mid \gamma_j \geq k + 1\}^{\#} = \text{rank } J^k - \text{rank } J^{k+1}$$

In view of Lemma 4.4.2, we find that

$$\text{rank } J^k - \text{rank } J^{k+1} = \text{rank } J_1^k - \text{rank } J_1^{k+1} + \text{rank } J_2^k - \text{rank } J_2^{k+1}$$
$$+ \text{rank } B^{(k)} - \text{rank } B^{(k+1)}$$

$$= \{j \mid \alpha_j \geq k + 1\}^{\#} + \max(\beta_1 - k, 0)$$
$$- \max(\beta_1 - k - 1, 0) + \text{rank } B^{(k)} - \text{rank } B^{(k+1)}$$

It remains to check, therefore, that for every $\omega \in \Omega$ it is possible to pick the complex numbers $d_{j1}^{(t)}$ ($1 \leq j \leq n$, $t = 1, 2, \ldots$) in such a way that $f_k = \text{rank } B^{(k)}$ for $k = 1, 2, \ldots$, where f_k is defined by equation (4.4.4) and $B^{(k)}$ is defined as in Lemma 4.4.2; and conversely, for every choice of $d_{j1}^{(t)}$ it is possible to find an $\omega \in \Omega_0$ such that $f_k = \text{rank } B^{(k)}$.

Note that $B^{(k)}$ depends on $d_{j1}^{(t)}$ with $t \leq \lambda_j$, so we restrict ourselves only to these values of t.

Given $\omega = (\omega_1, \ldots, \omega_n) \in \Omega_0$, choose $d_{j1}^{(t)}$ in such a way that ω_j is the smallest index t with the property that $d_{j1}^{(t)} \neq 0$ [if $\omega_j = \lambda_j + 1$, put $d_{j1}^{(t)} = 0$ for all t]. It is easy to see that $c_{kj}^{(\omega)}$ is just the rank of the matrix $B_{j1}^{(k)}$ [defined by (4.4.2)]. Observe that after crossing out some zero columns and rows, if necessary, $B_{j1}^{(k)}$ is an upper triangular Toeplitz matrix with $\min(k, \beta_1)$ columns. Thus the rank of

$$B^{(k)} = \begin{bmatrix} B^{(k)}_{11} \\ \vdots \\ B^{(k)}_{n1} \end{bmatrix}$$

is just the maximum of the ranks of $B^{(k)}_{11}$, $B^{(k)}_{21}$, ..., $B^{(k)}_{n1}$, that is, f_k.

Conversely, if $d^{(t)}_{j1}$ are given, define ω_j as the minimal $t(1 \le t \le \lambda_j)$ such that $d^{(t)}_{j1} \ne 0$; and if $d^{(t)}_{j1} = 0$ for every t, $1 \le t \le \lambda_j$, put $\omega_j = \lambda_j + 1$. $\square$

4.5   EXERCISES

4.1 Supply a proof of Theorem 4.1.5.

4.2 State and prove a result for dilations analogous to Theorems 4.1.4 and 4.1.5.

4.3 Prove that the maximal dimension of an irreducible A-invariant subspace coincides with the maximal dimension of a Jordan block in the Jordan form of A.

4.4 Find all possibilities for the partial multiplicities of matrices of type

$$\begin{bmatrix} 0 & X \\ 0 & 0 \end{bmatrix}$$

where X is any $n \times m$ matrix.

4.5 What is the answer to the preceding exercise under the restriction that rank $X \le k$, where k is a fixed positive integer?

4.6 Find all possibilities for the partial multiplicities of matrices of the following types:

(a)
$$\begin{bmatrix} J_n(0) & X \\ 0 & 0 \end{bmatrix}$$

where X is any $n \times m$ matrix

(b)
$$\begin{bmatrix} J_n(0) & X \\ 0 & 0 \end{bmatrix}$$

where X is any $n \times m$ matrix of rank 1. (*Hint:* Prove that there exists an $n \times m$ matrix X_0 with exactly one nonzero entry such that

$$\begin{bmatrix} J_n(0) & X \\ 0 & 0 \end{bmatrix} \quad \text{and} \quad \begin{bmatrix} J_n(0) & X_0 \\ 0 & 0 \end{bmatrix}$$

are similar.)

(c) What happens if we allow matrices X of rank 2?

4.7 Find all possibilities for partial multiplicities of matrices of type

$$\begin{bmatrix} J_n(0) & X \\ 0 & J_m(0) \end{bmatrix}$$

where X is any $n \times m$ matrix.

4.8 Let

$$C_1 = \begin{bmatrix} a_1 & a_2 & \cdots & a_n \\ a_n & a_1 & \cdots & a_{n-1} \\ \vdots & \vdots & & \vdots \\ a_2 & a_3 & \cdots & a_1 \end{bmatrix} \quad \text{and} \quad C_2 = \begin{bmatrix} b_1 & b_2 & \cdots & b_n \\ b_n & b_1 & \cdots & b_{n-1} \\ \vdots & \vdots & & \ddots \\ b_2 & b_3 & \cdots & b_1 \end{bmatrix}$$

be circulant matrices. Find all possibilities for the partial multiplicities of matrices of type

$$\begin{bmatrix} C_1 & X \\ 0 & C_2 \end{bmatrix}$$

where X is an $n \times n$ matrix.

Chapter Five

Applications to Matrix Polynomials

Let $A_0, A_1, \ldots, A_{l-1}$ be complex $n \times n$ matrices. We call the matrix-valued function $L(\lambda) = I\lambda^l + \Sigma_{j=0}^{l-1} A_j \lambda^j$ a monic matrix polynomial of degree l. It will be seen that there are $ln \times ln$ matrices C such that

$$\begin{bmatrix} L(\lambda) & 0 \\ 0 & I_{(l-1)n} \end{bmatrix} \quad \text{and} \quad \lambda I - C$$

are equivalent. (See the appendix for the notion of equivalence.) In this case C is said to be a linearization of $L(\lambda)$. The invariant, coinvariant, and semiinvariant subspaces for C play a special role in the study of the matrix polynomial $L(\lambda)$. For example, certain invariant subspaces of C are related to factorizations of $L(\lambda)$. More precisely, certain invariant subspaces determine monic right divisors of $L(\lambda)$, certain coinvariant subspaces determine monic left divisors, and certain semiinvariant subspaces determine three monic factors of $L(\lambda)$. In this chapter we explore these and similar connections and study the behavior of solutions of differential and difference equations with constant coefficients.

5.1 LINEARIZATIONS, STANDARD TRIPLES, AND REPRESENTATIONS OF MONIC MATRIX POLYNOMIALS

In this section we introduce the main tools required for the study of monic matrix polynomials. These tools are freely used in subsequent sections.

Let $L(\lambda) = I\lambda^l + \Sigma_{j=0}^{l-1} A_j \lambda^j$ be a monic matrix polynomial of degree l, where the A_j are $n \times n$ matrices with complex entries. Note that det $L(\lambda)$ is a polynomial of degree nl. A linear matrix polynomial $I\lambda - A$ of size $(n + p) \times (n + p)$ is called a *linearization* of $L(\lambda)$ if

$$I\lambda - A = E(\lambda)\begin{bmatrix} L(\lambda) & 0 \\ 0 & I_p \end{bmatrix} F(\lambda) \qquad (5.1.1)$$

where $E(\lambda)$ and $F(\lambda)$ are $(n + p) \times (n + p)$ matrix polynomials with constant nonzero determinants. Admitting a small abuse of language, we also call matrix A from equation (5.1.1) a linearization of $L(\lambda)$. Comparing determinants on both sides of (5.1.1), we conclude that $\det(I\lambda - A)$ is a polynomial of degree nl, where l is the degree of $L(\lambda)$. So the size of a linearization A of $L(\lambda)$ is necessarily nl.

As an illustration of the notion of linearization, consider the linearizations of a scalar polynomial ($n = 1$). Let $L(\lambda) = \Pi_{i=1}^{k} (\lambda - \lambda_i)^{\alpha_i}$ be a scalar polynomial having different zeros $\lambda_1, \ldots, \lambda_k$ with multiplicities $\alpha_1, \ldots, \alpha_k$, respectively. To construct a linearization of $L(\lambda)$, let J_i $(i = 1, \ldots, k)$ be the Jordan block of size α_i with eigenvalue λ_i, and consider the linear polynomial $\lambda I - J$ of size $\Sigma_{i=1}^{k} \alpha_i$, where $J = \text{diag}[J_i]_{i=1}^{k}$. Then J is a linearization of $L9\lambda)$. Indeed, $I\lambda - J$ and $\begin{bmatrix} L(\lambda) & 0 \\ 0 & I \end{bmatrix}$ have the same elementary divisors; so using Theorem A.3.1, we find that J is a linearization of $L(\lambda)$.

The following theorem describes a linearization of a monic matrix polynomial directly in terms of the coefficients of the polynomial.

Theorem 5.1.1

For a monic matrix polynomial $L(\lambda) = I\lambda^l + \Sigma_{j=0}^{l-1} A_j \lambda^j$ of size $n \times n$, define the $nl \times nl$ matrix

$$C_1 = \begin{bmatrix} 0 & I & 0 & \cdots & 0 \\ 0 & 0 & I & \cdots & 0 \\ \vdots & \vdots & \vdots & & \vdots \\ & & & \cdots & I \\ -A_0 & -A_i & & \cdots & -A_{l-1} \end{bmatrix}$$

Then C_1 is a linearization of $L(\lambda)$.

Proof. Define $nl \times nl$ matrix polynomials $E(\lambda)$ and $F(\lambda)$ as follows:

$$F(\lambda) = \begin{bmatrix} I & 0 & \cdots & 0 & 0 \\ -\lambda I & I & \cdots & 0 & 0 \\ \vdots & \vdots & \cdots & \vdots & \vdots \\ 0 & 0 & \cdots & I & 0 \\ 0 & 0 & \cdots & -\lambda I & I \end{bmatrix}$$

$$E(\lambda) = \begin{bmatrix} B_{l-1}(\lambda) & B_{l-2}(\lambda) & \cdots & & B_0(\lambda) \\ -I & 0 & \cdots & & 0 \\ 0 & -I & \cdots & & \\ \vdots & \vdots & & & \vdots \\ 0 & & & -I & 0 \end{bmatrix}$$

where $B_0(\lambda) = I$ and $B_{r+1}(\lambda) = \lambda B_r(\lambda) + A_{l-r-1}$ for $r = 0, 1, \ldots, l-2$. It is immediately seen that $\det F(\lambda) \equiv 1$ and $\det E(\lambda) \equiv \pm 1$. Direct multiplication on both sides shows that

$$E(\lambda)(\lambda I - C_1) = \begin{bmatrix} L(\lambda) & 0 \\ 0 & I \end{bmatrix} F(\lambda) \tag{5.1.2}$$

and Theorem 5.1.1 follows. □

The matrix C_1 from Theorem 5.1.1 will be called the (first) *companion matrix* of $L(\lambda)$, and will play an important role in the sequel. From the definition of C_1 it is clear that

$$\det(I\lambda - C_1) = \det L(\lambda)$$

In particular, the *eigenvalues* of $L(\lambda)$, that is, zeros of the scalar polynomial $\det L(\lambda)$, and the eigenvalues of $I\lambda - C_1$ are the same. In fact, we can say more: since C_1 is a linearization of $L(\lambda)$, it follows that the elementary divisors (and thus also the partial multiplicities of every eigenvalue) of $I\lambda - C_1$ and $L(\lambda)$ are the same.

Now we prove an important result connecting the rational matrix function $L(\lambda)^{-1}$ with the resolvent function for the linearization C_1.

Proposition 5.1.2

For every $\lambda \in \mathbb{C}$ that is not an eigenvalue of $L(\lambda)$, the following equality holds:

$$[L(\lambda)]^{-1} = P_1(I\lambda - C_1)^{-1}R_1 \tag{5.1.3}$$

where

$$P_1 = [I \quad 0 \quad \cdots \quad 0]$$

is an $n \times nl$ matrix and

$$R_1 = \begin{bmatrix} 0 \\ \vdots \\ 0 \\ I \end{bmatrix} \tag{5.1.4}$$

is an $n \times nl$ matrix.

Proof. Consider the equality (5.1.2) used in the proof of Theorem 5.1.1. We have

$$\begin{bmatrix} [L(\lambda)]^{-1} & 0 \\ 0 & I \end{bmatrix} = F(\lambda)(I\lambda - C_1)^{-1}[E(\lambda)]^{-1} \qquad (5.1.5)$$

It is easy to see that the first n columns of the matrix $[E(\lambda)]^{-1}$ have the form (5.1.4). Now, multiplying equation (5.1.5) on the left by P_1 and on the right by P_1^T and using the relation

$$P_1 F(\lambda) = [I \quad 0 \quad \cdots \quad 0] = P_1$$

we obtain the desired formula (5.1.3). $\square$

Formula (5.1.3) is referred to as a *resolvent form* of the monic matrix polynomial $L(\lambda)$. The following result follows directly from the definition of a linearization and Theorem A.4.1.

Proposition 5.1.3

Any two linearizations of a monic matrix polynomial $L(\lambda)$ are similar. Conversely, if a matrix T is a linearization of $L(\lambda)$ and matrix S is similar to T, then S is also a linearization of $L(\lambda)$.

This proposition and the resolvent form (5.1.3) suggest the following important definition: a triple of matrices (X, T, Y), where T is $nl \times nl$, X is $n \times nl$, and Y is $nl \times n$, is called a *standard triple* of $L(\lambda)$ if

$$L(\lambda)^{-1} = X(I\lambda - T)^{-1}Y$$

For example, Proposition 5.1.2 shows that (P_1, C_1, R_1) is a standard triple of $L(\lambda)$.

It is evident from the definition that, if (X, T, Y) is a standard triple for $L(\lambda)$, then so is any other triple $(\tilde{X}, \tilde{T}, \tilde{Y})$ that is *similar* to (X, T, Y), that is, such that

$$X = \tilde{X}S, \qquad T = S^{-1}\tilde{T}S, \qquad Y = S^{-1}\tilde{Y}$$

for some nonsingular matrix S. As we see in Theorem 5.1.5, this is the only freedom in the choice of standard triples.

We start with some useful properties of standard triples. Here and in the sequel we adopt the notation $\operatorname{col}[Z_i]_{i=0}^p$ for the column matrix

$$\begin{bmatrix} Z_0 \\ Z_1 \\ \vdots \\ Z_p \end{bmatrix}$$

Proposition 5.1.4

If (X, T, Y) is a standard triple of a monic $n \times n$ matrix polynomial $I\lambda^l + \Sigma_{j=0}^{l-1} A_j \lambda^j$, then the $nl \times nl$ matrices

$$\text{col}[XT^i]_{i=0}^{l-1} \quad \text{and} \quad [Y, TY, \ldots, T^{l-1}Y]$$

are nonsingular. Further, the equalities

$$A_0 X + A_1 XT + \cdots + A_{l-1} XT^{l-1} + XT^l = 0 \tag{5.1.6}$$

and

$$YA_0 + TYA_1 + \cdots + T^{l-1}YA_{l-1} + T^l Y = 0 \tag{5.1.7}$$

hold.

Proof. We have

$$L(\lambda)^{-1} = X(I\lambda - T)^{-1} Y$$

and by Proposition 2.10.1,

$$\frac{1}{2\pi i} \int_\Gamma \lambda^j L(\lambda)^{-1} \, dy = \frac{1}{2\pi i} \int_\Gamma \lambda^j X(I\lambda - T)^{-1} Y \, d\lambda = XT^j Y , \quad j = 0, 1, \ldots \tag{5.1.8}$$

where Γ is a circle with centre 0 and sufficiently large radius so that $\sigma(T)$ and the eigenvalues of $L(\lambda)$ are inside Γ. On the other hand, since $L(\lambda)$ is a monic polynomial of degree l, the matrix function $\tilde{L}(\lambda) = \lambda^{-l} L(\lambda)$ is analytic and invertible in a neighbourhood of infinity and takes the value I at infinity. In fact, $\tilde{L}(\lambda)$ is analytic outside and on Γ. Hence

$$\frac{1}{2\pi i} \int_\Gamma \lambda^j L(\lambda)^{-1} \, d\lambda = \frac{1}{2\pi i} \int_\Gamma \lambda^{j-l} \tilde{L}(\lambda)^{-1} \, d\lambda$$

and representing $\tilde{L}(\lambda)^{-1}$ as a power series $I + \Sigma_{k=1}^\infty \lambda^{-k} \tilde{L}_k$, we see that

$$\frac{1}{2\pi i} \int_\Gamma \lambda^j L(\lambda)^{-1} \, d\lambda = \begin{cases} 0 & \text{for} \quad j = 0, \ldots, l-2 \\ I & \text{for} \quad j = l-1 \end{cases}$$

Combining this with (5.1.8), we have

$$\frac{1}{2\pi i} \int_\Gamma \begin{bmatrix} 1 & \lambda & \cdots & \lambda^{l-1} \\ \lambda & \lambda^2 & \cdots & \lambda^l \\ \vdots & & & \vdots \\ \lambda^{l-1} & & \cdots & \lambda^{2l-2} \end{bmatrix} L(\lambda)^{-1}\, d\lambda = \begin{bmatrix} 0 & \cdots & 0 & I \\ \vdots & & I & \\ 0 & I & & * \\ I & & & \end{bmatrix}, \begin{bmatrix} X \\ XT \\ \vdots \\ XT^{l-1} \end{bmatrix}$$

$$\times [Y \quad TY \quad \cdots \quad T^{l-1}Y] \qquad (5.1.9)$$

As the right-hand side in equation (5.1.9) is nonsingular, the $nl \times nl$ matrices $\mathrm{col}[XT^i]_{i=0}^{l-1}$ and $[Y\ TY\ \cdots\ T^{l-1}Y]$ are both nonsingular.

Now use equation (5.1.8) again and we find that, for $i = 0, 1, \ldots, l-1$,

$$0 = \frac{1}{2\pi i} \int_\Gamma \lambda^i L(\lambda) L(\lambda)^{-1}\, d\lambda = \frac{1}{2\pi i} \int_\Gamma \lambda^i L(\lambda) X(I\lambda - T)^{-1} Y\, d\lambda$$

$$= (XT^l + \cdots + A_1 XT + A_0 X) T^i Y$$

It follows that

$$(XT^l + \cdots + A_1 XT + A_0 X)[Y, YT, \ldots, T^{l-1}Y] = 0$$

and since the second factor is nonsingular, formula (5.1.6) follows.

Similarly, starting with the equality

$$0 = \frac{1}{2\pi i} \int_\Gamma \lambda^i L(\lambda)^{-1} L(\lambda)\, d\lambda$$

formula (5.1.7) can be verified.   □

We are now ready to state and prove the basic result that the standard triple for a monic matrix polynomial is essentially unique (up to similarity).

Theorem 5.1.5

Let (X_1, T_1, Y_1) and (X_2, T_2, Y_2) be two standard triples of the monic matrix polynomial $L(\lambda)$ of degree l. Then there exists a unique nonsingular matrix S such that

$$X_1 = X_2 S, \qquad T_1 = S^{-1} T_2 S, \qquad Y_1 = S^{-1} Y_2 \qquad (5.1.10)$$

The matrix S is given by the formula

$$S = (\mathrm{col}[X_2 T_2^i]_{i=0}^{l-1})^{-1} \cdot \mathrm{col}[X_1 T_1^i]_{i=0}^{l-1}$$

$$= [Y_2, T_2 Y_2, \ldots, T_2^{l-1} Y_2] \cdot [Y_1, T_1 Y_1, \ldots, T_1^{l-1} Y_1]^{-1} \qquad (5.1.11)$$

where the invertibility of the matrices involved is ensured by Proposition

5.1.4. *In particular, if (X, T, Y) is a standard triple of $L(\lambda)$, then T is a linearization of $L(\lambda)$.*

Proof. Assume we have already found a nonsingular S such that (5.1.10) holds. Then

$$\text{col}[X_1 T_1^i]_{i=0}^{l-1} = \text{col}[X_2 T_2^i]_{i=0}^{l-1} S$$

and

$$[Y_1, T_1 Y_1, \ldots, T_1^{l-1} Y_1] = S^{-1}[Y_2, T_2 Y_2, \ldots, T_2^{l-1} Y_2]$$

Thus formulas (5.1.11) hold and consequently S is unique.

Now we prove the existence of an S such that (5.1.10) holds. Without loss of generality, and taking advantage of Proposition 5.1.2, we can assume that $X_2 = P_1$, $T_2 = C_1$, $Y_2 = R_1$. Using (5.1.6) [with (X, T, Y) replaced by (X_1, T_1, Y_1)], the equality

$$\text{col}[X_1 T_1^i]_{i=0}^{l-1} T_1 = C_1 \, \text{col}[X_1 T_1^i]_{i=0}^{l-1}$$

where C_1 is the companion matrix of $L(\lambda)$, is easily verified. Also, (5.1.9) implies

$$\text{col}[X_1 T_1^i]_{i=0}^{l-1} Y_1 = \text{col}[\delta_{il} I]_{i=1}^{l}$$

where δ_{ij} is the Kronecker index ($\delta_{ij} = 0$ if $i \neq j$; $\delta_{ij} = 1$ if $i = j$). Obviously

$$X_1 = [I \quad 0 \quad \cdots \quad 0] \, \text{col}[X_1 T_1^i]_{i=0}^{l-1}$$

and equations (5.1.10) hold with $S = \text{col}[X_1 T_1^i]_{i=0}^{l-1}$.

Finally, if (X, T, Y) is a standard triple of $L(\lambda)$, then, by the part of Theorem 5.1.4 already proved, T is similar to the companion matrix C_1 of $L(\lambda)$, and thus T is also a linearization of $L(\lambda)$. $\square$

Proposition 5.1.2 gives an example of a standard triple based on the companion matrix of $L(\lambda)$. Another useful example of a standard triple is

$$([0 \quad \cdots \quad 0 \quad I], C_2, \text{col}[\delta_{i1} I]_{i=1}^{l}) \tag{5.1.12}$$

where

$$C_2 = \begin{bmatrix} 0 & \cdots & 0 & -A_0 \\ I & \cdots & 0 & -A_1 \\ \vdots & & \vdots & \vdots \\ 0 & \cdots & I & -A_{l-1} \end{bmatrix}$$

and is called the *second companion matrix* of $L(\lambda)$. Indeed, if we define

$$B = \begin{bmatrix} A_1 & A_2 & \cdots & A_{l-1} & I \\ A_2 & & & I & 0 \\ \vdots & & \ddots & & \vdots \\ I & 0 & & \cdots & 0 \end{bmatrix}$$

then we have

$$[0 \quad \cdots \quad 0 \quad I] = [I \quad 0 \quad \cdots \quad 0] B^{-1}, \quad C_2 = BC_1 B^{-1}$$

$$\operatorname{col}[\delta_{i1} I]_{i=1}^l = B \operatorname{col}[\delta_{il} I]_{i=1}^l$$

Thus the triple (5.1.10) is similar to the standard triple given in Proposition 5.1.2. The notion of a standard triple is the main tool in the following representation theorem.

Theorem 5.1.6

Let $L(\lambda) = I\lambda^l + \sum_{j=0}^{i-1} A_j \lambda^j$ be a monic matrix polynomial of degree l with standard triple (X, T, Y). Then $L(\lambda)$ admits the following representations:
(a) *Right canonical form*:

$$L(\lambda) = I\lambda^l - XT^l(V_1 + V_2\lambda + \cdots + V_l\lambda^{l-1}) \tag{5.1.13}$$

where V_i are $nl \times n$ matrices such that

$$[V_1 \quad \cdots \quad V_l] = \{\operatorname{col}[XT^i]_{i=0}^{l-1}\}^{-1}$$

(b) *Left canonical form*:

$$L(\lambda) = \lambda^l I - (W_1 + \lambda W_2 + \cdots + \lambda^{l-1} W_l) T^l Y \tag{5.1.14}$$

where W_i are $n \times nl$ matrices such that

$$\operatorname{col}[W_i]_{i=1}^l = [Y, TY, \ldots, T^{l-1}Y]^{-1}$$

Note that only X and T appear in the right canonical form of $L(\lambda)$, whereas only T and Y appear in the left canonical form.

Proof. Observe that the forms (5.1.13) and (5.1.14) are independent of the choice of the standard triple (X, T, Y). Let us check this for (5.1.13), for example. We have to prove that if (X, T, Y) and (X', T', Y') are standard triples of $L(\lambda)$, then

$$XT^l[V_1 \quad \cdots \quad V_l] = X'(T')^l[V_1' \quad \cdots \quad V_l'] \tag{5.1.15}$$

where

$$[V_1 \quad \cdots \quad V_l] = \{\text{col}[XT^i]_{i=0}^{l-1}\}^{-1} , \qquad [V_1' \quad \cdots \quad V_l'] = \{\text{col}[X'T'^i]_{i=0}^{l-1}\}^{-1}$$

But these standard triples are similar:

$$X' = XS , \qquad T' = S^{-1}TS , \qquad Y' = S^{-1}Y$$

Therefore

$$[V_1' \quad \cdots \quad V_l'] = \{\text{col}[X'T'^i]_{i=0}^{l-1}\}^{-1} = \{\text{col}[XT^i]_{i=0}^{l-1}S\}^{-1}$$
$$= S^{-1}[V_1 \quad \cdots \quad V_l]$$

and (5.1.15) follows.

Thus it suffices to check equation (5.1.13) only for the special standard triple

$$X = [I \quad 0 \quad \cdots \quad 0] , \qquad T = C_1 , \qquad Y = \text{col}[\delta_{il}I]_{i=1}^{l}$$

and for checking (5.1.14), we choose the standard triple defined by (5.1.12). To prove (5.1.13), observe that

$$[I \quad 0 \quad \cdots \quad 0]C_1^l = [-A_0 \quad -A_1 \quad \cdots \quad -A_{l-1}]$$

and

$$[V_1 \quad \cdots \quad V_l] = \{\text{col}[[I \quad 0 \quad \cdots \quad 0]C_1^i]_{i=0}^{l-1}\}^{-1} = I$$

so

$$[I \quad 0 \quad \cdots \quad 0]C_1^l[V_1 \quad V_2 \quad \cdots \quad V_l] = [-A_0 \quad -A_1 \quad \cdots \quad -A_{l-1}]$$

and (5.1.13) becomes evident. To prove (5.1.14), note that by direct computation one easily checks that for the standard triple (5.1.12)

$$C_2^j \text{col}[\delta_{i1}I]_{i=1}^{l} = \text{col}[\delta_{i,j+1}I]_{i=1}^{l} , \qquad j = 0, \ldots, l-1$$

and

$$C_2^l \text{col}[\delta_{i1}I]_{i=1}^{l} = \text{col}[-A_i]_{i=0}^{l-1}$$

So

$$[\text{col}[\delta_{i1}]_{i=1}^{l} , \qquad C_2 \text{col}[\delta_{i1}I]_{i=1}^{l}, \ldots, C_2^{l-1} \text{col}[\delta_{i1}I]_{i=1}^{l}] = I$$

Thus

$$\text{col}[W_i]_{i=1}^l = I$$

and

$$W_i C_2^l \text{col}[\delta_{i1} I]_{i=1}^l = -A_{i-1}, \qquad i = 1, \ldots, l$$

So equations (5.1.14) follows. $\square$

5.2 MULTIPLICATION OF MONIC MATRIX POLYNOMIALS AND PARTIAL MULTIPLICITIES OF A PRODUCT

In this section we describe multiplication of monic matrix polynomials in terms of their standard triples. First we compute the inverse $L^{-1}(\lambda)$ of the product $L(\lambda) = L_2(\lambda)L_1(\lambda)$ of two monic matrix polynomials $L_1(\lambda)$ and $L_2(\lambda)$.

Theorem 5.2.1

Let $L_i(\lambda)$ be a matrix polynomial with standard triple (X_i, T_i, Y_i) for $i = 1, 2$, and let $L(\lambda) = L_2(\lambda)L_1(\lambda)$. Then

$$L^{-1}(\lambda) = [X_1\ 0](I\lambda - T)^{-1}\begin{bmatrix} 0 \\ Y_2 \end{bmatrix} \tag{5.2.1}$$

where

$$T = \begin{bmatrix} T_1 & Y_1 X_2 \\ 0 & T_2 \end{bmatrix}$$

Proof. It is easily verified that

$$(I\lambda - T)^{-1} = \begin{bmatrix} (I\lambda - T_1)^{-1} & (I\lambda - T_1)^{-1}Y_1 X_2 (I\lambda - T_2)^{-1} \\ 0 & (I\lambda - T_2)^{-1} \end{bmatrix}$$

The product on the right of equation (5.2.1) is then found to be

$$X_1(I\lambda - T_1)^{-1}Y_1 X_2 (I\lambda - T_2)^{-1}Y_2$$

But, using the definition of standard triples, this is just $L_1^{-1}(\lambda)L_2^{-1}(\lambda)$, and the theorem follows immediately. $\square$

Corollary 5.2.2

If $L_i(\lambda)$ are monic matrix polynomials with standard triples (X_i, T_i, Y_i) for $i = 1, 2$, then $L(\lambda) = L_2(\lambda)L_1(\lambda)$ has a standard triple (X, T, Y) with the representations

$$X = [X_1 \ 0], \qquad T = \begin{bmatrix} T_1 & Y_1 X_2 \\ 0 & T_2 \end{bmatrix}, \qquad Y = \begin{bmatrix} 0 \\ Y_2 \end{bmatrix}$$

Proof.   Combine Theorem 5.1.5 with Theorem 5.2.1.   □

Corollary 5.2.2 allows us to describe the partial multiplicities of a product of monic matrix polynomials. We first give some necessary definitions. For a monic matrix polynomial $L(\lambda)$ and its eigenvalue λ_0 [i.e., det $L(\lambda_0) = 0$], let $\alpha_1 \geq \alpha_2 \geq \cdots > \alpha_r$ be the degrees of the elementary divisors of $L(\lambda)$ corresponding to λ_0. The integers α_i are called the *partial multiplicities* of $L(\lambda)$ corresponding to λ_0. It is convenient to augment the α_i values by zeros and call the sequence $\alpha = (\alpha_1, \alpha_2, \ldots, \alpha_r, 0, \ldots)$ the sequence of partial multiplicities of $L(\lambda)$ at λ_0. Thus $\alpha \in \Omega$ (see Section 4.4 for the definition of Ω). Also, we shall say formally that the partial multiplicities of $L(\lambda)$ corresponding to a complex number that is not an eigenvalue of $L(\lambda)$ are all zeros. Recall also the definition of the set $\Gamma(\alpha, \beta)$ given in Section 4.4.

Theorem 5.2.3

Let $L_1(\lambda)$ and $L_2(\lambda)$ be $n \times n$ monic matrix polynomials. Let α, β and γ be the sequences of partial multiplicities of $L_1(\lambda)$, $L_2(\lambda)$, and $L_2(\lambda)L_1(\lambda)$, respectively, at λ_0. Then $\gamma \in \Gamma(\alpha, \beta)$. Conversely, if $\gamma \in \Gamma(\alpha, \beta)$, then for n sufficiently large there exist $n \times n$ monic matrix polynomials $L_1(\lambda)$ and $L_2(\lambda)$, such that the sequence of their partial multiplicities at λ_0 are α and β, respectively, and the sequence of partial multiplicities of $L_2(\lambda)L_1(\lambda)$ is γ.

Proof.   Let (X_i, T_i, Y_i) be a standard triple for $L_i(\lambda)$ and $i = 1, 2$. By the multiplication formula (Corollary 5.2.2), the matrix

$$T = \begin{bmatrix} T_1 & Y_1 X_2 \\ 0 & T_2 \end{bmatrix}$$

is a linearization of $L_2(\lambda)L_1(\lambda)$. From the properties of a linearization it follows that γ is also the sequence of partial multiplicities of T at λ_0. Now from the structure of T it is clear that $\gamma \in \Gamma(\alpha, \beta)$, and the first part of the theorem follows.

To prove the second part of Theorem 5.2.3, we first prove the following assertion: let A be an $r_1 \times r_2$ matrix. Then for n sufficiently large there exist an $r_1 \times n$ matrix Y and an $n \times r_2$ matrix X such that $YX = A$, the rows of Y are linearly independent, and the columns of X are linearly independent. Indeed, multiplying A by invertible matrices from the left and the right (if necessary), we can suppose that

$$A = \begin{bmatrix} I & 0 \\ 0 & 0 \end{bmatrix}$$

where I is the unit $r \times r$ matrix (for some $r \leq \min(r_1, r_2)$). Then we can take

$$Y = \begin{bmatrix} I & 0 & 0 \\ 0 & Y_1 & 0 \end{bmatrix}, \qquad X = \begin{bmatrix} I & 0 \\ 0 & 0 \\ 0 & X_1 \end{bmatrix}$$

where Y_1 is an $(r_1 - r) \times r_1$ matrix with linearly independent rows and X_1 is an $r_2 \times (r_2 - r)$ matrix with linearly independent columns. Then $n = r + r_1 + r_2$, of course.

Now let $\gamma \in \Gamma(\alpha, \beta)$, so that γ is the sequence of partial multiplicities of

$$T_0 = \begin{bmatrix} T_1 & A \\ 0 & T_2 \end{bmatrix}$$

for some T_1, T_2, A, and the partial multiplicities of T_1 (resp. T_2) corresponding to λ_0 are given by the sequence α (resp. β). Applying a similarity to T_0, if necessary, we can assume that T_1 and T_2 are in Jordan form. Further, in view of Theorem 4.1.1 we can assume that $\sigma(T_1) = \sigma(T_2) = \{\lambda_0\}$.

According to the assertion proved in the preceding paragraph, for n sufficiently large there exist matrices X_0 and Y_0 of sizes $n \times r_2$ and $r_1 \times n$, respectively (where $r_1 = \Sigma_{j=1}^{\infty} \alpha_j$, $r_2 = \Sigma_{j=1}^{\infty} \beta_j$) such that $Y_0 X_0 = A$, the rows of Y_0 are linearly independent, and so are the columns of X_0. Choose an $n \times (n - r_2)$ matrix X_1 such that the matrix $[X_0 \, X_1]$ (of size $n \times n$) is invertible, and put

$$L_2(\lambda) = \lambda I - [X_0 \, X_1] \begin{bmatrix} T_2 & 0 \\ 0 & zI \end{bmatrix} [X_0 \, X_1]^{-1}$$

where z is some complex number different from λ_0. Similarly, choose an $(n - r_1) \times n$ matrix Y_1 such that $\begin{bmatrix} Y_0 \\ Y_1 \end{bmatrix}$ is nonsingular, and put

$$L_1(\lambda) = \lambda I - \begin{bmatrix} Y_0 \\ Y_1 \end{bmatrix}^{-1} \begin{bmatrix} T_1 & 0 \\ 0 & zI \end{bmatrix} \begin{bmatrix} Y_0 \\ Y_1 \end{bmatrix}$$

As $T_2 \oplus zI$ (resp. $T_1 \oplus zI$) is a linearization of $L_2(\lambda)$ [resp. of $L_1(\lambda)$], it follows that the partial multiplicities of $L_2(\lambda)$ [resp. of $L_1(\lambda)$] corresponding to λ_0 are given by the sequence β (resp. α). Further

$$\left([X_0 \, X_1], \begin{bmatrix} T_2 & 0 \\ 0 & zI \end{bmatrix}, [X_0 \, X_1]^{-1} \right) \quad \text{and} \quad \left(\begin{bmatrix} Y_0 \\ Y_1 \end{bmatrix}^{-1}, \begin{bmatrix} T_1 & 0 \\ 0 & zI \end{bmatrix}, \begin{bmatrix} Y_0 \\ Y_1 \end{bmatrix} \right)$$

are the standard triples for $L_2(\lambda)$ and $L_1(\lambda)$, respectively. By Corollary 5.2.2 the matrix

$$T = \begin{bmatrix} T_1 & 0 & Y_0X_0 & Y_0X_1 \\ 0 & zI & Y_1X_0 & Y_1X_1 \\ 0 & 0 & T_2 & 0 \\ 0 & 0 & 0 & zI \end{bmatrix} = \begin{bmatrix} T_1 & 0 & A & Y_0X_1 \\ 0 & zI & Y_1X_0 & Y_1X_1 \\ 0 & 0 & T_2 & 0 \\ 0 & 0 & 0 & zI \end{bmatrix}$$

is a linearization of $L_2(\lambda)L_1(\lambda)$. Now Theorem 4.1.1 ensures that the partial multiplicities of T corresponding to λ_0 are exactly those for T_0; that is, they are given by the sequence γ. $\square$

The proof of the converse statement of Theorem 5.2.3 shows that for a $\gamma \in \Gamma(\alpha, \beta)$ there exist *linear* monic matrix polynomials $L_1(\lambda)$ and $L_2(\lambda)$ with the desired properties and with the size not exceeding $\min(r_1, r_2) + r_1 + r_2$, where r_1 (resp. r_2) is the sum of all integers in α (resp. β).

Our analysis of partial multiplicities of completions in sections 4.2–4.4, combined with Theorem 5.2.3, allows us to deduce various connections between the partial multiplicities of monic matrix polynomials and the partial multiplicities of their product, as indicated, for instance, in the following corollary.

Corollary 5.2.4

Let $L_1(\lambda)$ and $L_2(\lambda)$ be $n \times n$ monic matrix polynomials. Let $\alpha = (\alpha_1, \alpha_2, \ldots)$, $\beta = (\beta_1, \beta_2, \ldots)$, and $\gamma = (\gamma_1, \gamma_2, \ldots)$ be sequences of partial multiplicities of $L_1(\lambda)$, $L_2(\lambda)$, and $L_2(\lambda)L_1(\lambda)$, respectively, at λ_0. Then

$$\sum_{i=1}^{m} \gamma_{r_i} \le \min\left(\sum_{i=1}^{m} \alpha_{r_i} + \sum_{i=1}^{m} \beta_i, \sum_{i=1}^{m} \alpha_i + \sum_{i=1}^{m} \beta_{r_i} \right)$$

for any sequence $r_1 < \cdots < r_m$ of positive integers.

The corollary follows from Theorems 4.3.1 and 5.2.3.

5.3   DIVISIBILITY OF MONIC MATRIX POLYNOMIALS

Let $L(\lambda)$ be an $n \times n$ monic matrix polynomial of degree l, and let (X, T, Y) be a standard triple for $L(\lambda)$. Consider a T-semiinvariant subspace $\mathcal{M}$. Thus there exists a *triinvariant decomposition* (see Section 3.3) associated with $\mathcal{M}$:

$$\mathbb{C}^{nl} = \mathcal{L} \dotplus \mathcal{M} \dotplus \mathcal{N} \tag{5.3.1}$$

where the subspaces $\mathcal{L}$ and $\mathcal{L} \dotplus \mathcal{M}$ are T invariant. The triinvariant decomposition (5.3.1) is called *supporting* [with respect to (X, T, Y)] if, for some integers p and q, the transformations

$$\begin{bmatrix} X \\ XT \\ \vdots \\ XT^{p-1} \end{bmatrix}_{|\mathcal{L}+\mathcal{M}} : \mathcal{L} \dotplus \mathcal{M} \to \mathbb{C}^{np} \tag{5.3.2}$$

and

$$\begin{bmatrix} X \\ XT \\ \\ XT^{q-1} \end{bmatrix}_{|\mathcal{L}} : \mathcal{L} \to \mathbb{C}^{nq} \tag{5.3.3}$$

are invertible (in particular, this implies that $\dim(\mathcal{L} \dotplus \mathcal{M}) = np$, $\dim \mathcal{L} = nq$).

Cases in which $\mathcal{L} = \{0\}$ are of particular interest; then $\mathcal{M}$ is T invariant and condition (5.3.3) is vacuous. Also, if $\mathcal{N} = \{0\}$, then $\mathcal{M}$ is T coinvariant and the condition (5.3.2) is satisfied automatically with $p = l$. (Indeed, we have seen in Proposition 5.1.4 that the matrix $\mathrm{col}[XT^i]_{i=0}^{l-1}$ is nonsingular.)

The definition of a supporting triinvariant decomposition is given in terms of (X, T) only. However, if $P_{\mathcal{N}}$ is a projector with $\mathrm{Ker}\, P_{\mathcal{N}} = \mathcal{N}$, the following lemma shows that the invertibility of (5.3.2) is equivalent to the invertibility of the transformation from $C^{n(l-p)}$ into $\mathrm{Im}\, P_{\mathcal{N}}$ defined by

$$P_{\mathcal{N}}[T^{l-p-1}Y, \ldots, TY, Y] = [P_{\mathcal{N}}T^{l-p-1}P_{\mathcal{N}}Y, \ldots, P_{\mathcal{N}}TP_{\mathcal{N}}Y, P_{\mathcal{N}}Y]$$

Similarly, (5.3.3) is invertible if and only if

$$P_{\mathcal{L}}[T^{l-q-1}Y, \ldots, TY, Y] = [P_{\mathcal{L}}T^{l-q-1}P_{\mathcal{L}}Y, \ldots, P_{\mathcal{L}}TP_{\mathcal{L}}Y, P_{\mathcal{L}}Y]$$

is invertible, where $P_{\mathcal{L}}$ is a projector with $\mathrm{Ker}\, P_{\mathcal{L}} = \mathcal{L}$ (note that because of the T invariance of $\mathcal{L}$ and $\mathcal{N}$ we have $P_{\mathcal{N}}T^j = P_{\mathcal{N}}T^jP_{\mathcal{N}}$, $P_{\mathcal{L}}T^j = P_{\mathcal{L}}T^jP_{\mathcal{L}}$, $j = 1, 2, \ldots$).

Lemma 5.3.1

Let $L(\lambda)$ be a monic matrix polynomial of degree l with standard triple (X, T, Y), and let P be a projector in $\mathbb{C}^{nl}$. Then the transformation

$$\mathrm{col}[XT^{i-1}]_{i=1}^k|_{\mathrm{Im}\, P} : \mathrm{Im}\, P \to \mathbb{C}^{nk} \tag{5.3.4}$$

(where $k < l$) is invertible if and only if the transformation

$$(I - P)[T^{l-k-1}Y, \quad T^{l-k-2}Y, \ldots, Y] : \mathbb{C}^{n(l-k)} \to \mathrm{Ker}\, P \tag{5.3.5}$$

is invertible.

Proof. Put $A = \mathrm{col}[XT^{i-1}]_{i=1}^{l}$ and $B = [T^{l-1}Y, \ldots, TY, Y]$. With respect to the decompositions $\mathbb{C}^{nl} = \mathrm{Im}\, P \dotplus \mathrm{Ker}\, P$ and $\mathbb{C}^{nl} = \mathbb{C}^{nk} \oplus \mathbb{C}^{n(l-k)}$ write

$$A = \begin{bmatrix} A_1 & A_2 \\ A_3 & A_4 \end{bmatrix}, \qquad B = \begin{bmatrix} B_1 & B_2 \\ B_3 & B_4 \end{bmatrix}$$

Thus the A_i are transformations with the following domains and ranges:

$$A_1 : \mathrm{Im}\, P \to \mathbb{C}^{nk};$$
$$A_2 : \mathrm{Ker}\, P \to \mathbb{C}^{nk};$$
$$A_3 : \mathrm{Im}\, P \to \mathbb{C}^{n(l-k)};$$
$$A_4 : \mathrm{Ker}\, P \to \mathbb{C}^{n(l-k)};$$

and similarly for the B_i.

Observe that A_1 and B_4 coincide with the transformations (5.3.4) and (5.3.5), respectively. By formula (5.1.9) the product AB has the form

$$AB = \begin{bmatrix} D_1 & 0 \\ * & D_2 \end{bmatrix}$$

where D_1 and D_2 are nonsingular matrices. Recall that A and B are also nonsingular by Proposition 5.1.4. But then A_1 is invertible if and only if B_4 is invertible. This may be seen as follows.

Suppose that B_4 is invertible. Then

$$B \begin{bmatrix} I & 0 \\ -B_4^{-1}B_3 & B_4^{-1} \end{bmatrix} = \begin{bmatrix} B_1 & B_2 \\ B_3 & B_4 \end{bmatrix} \begin{bmatrix} I & 0 \\ -B_4^{-1}B_3 & B_4^{-1} \end{bmatrix}$$

$$= \begin{bmatrix} B_1 - B_2 B_4^{-1} B_3 & B_2 B_4^{-1} \\ 0 & I \end{bmatrix}$$

is invertible in view of the invertibility of B, and then also $B_1 - B_2 B_4^{-1} B_3$ is invertible. The special form of AB implies $A_1 B_2 + A_2 B_4 = 0$. Hence $D_1 = A_1 B_1 + A_2 B_3 = A_1 B_1 - A_1 B_2 B_4^{-1} B_3 = A_1 (B_1 - B_2 B_4^{-1} B_3)$ and it follows that A_1 is invertible. A similar argument shows that invertibility of A_1 implies the invertibility of B_4. This proves the lemma. $\square$

The importance of supporting triinvariant decompositions stems from the following result describing factorizations of a monic matrix polynomial $L(\lambda)$ in terms of supporting triinvariant decompositions associated with a linearization of $L(\lambda)$.

Theorem 5.3.2

Let $\mathcal{L}(\lambda)$ be an $n \times n$ monic matrix polynomial with standard triple (X, T, Y), and let $\mathbb{C}^{nl} = \mathcal{L} \dotplus \mathcal{M} \dotplus \mathcal{N}$ be a supporting triinvariant decomposition associated with a T-semiinvariant subspace $\mathcal{M}$. Then $L(\lambda)$ admits a factorization

$$L(\lambda) = L_1(\lambda) L_2(\lambda) L_3(\lambda) \tag{5.3.6}$$

where $L_i(\lambda)$, $i = 1, 2, 3$ are monic matrix polynomials with the following property: (a) $(X_{|\mathcal{L}}, T_{|\mathcal{L}}, \tilde{Y})$ is a standard triple of $L_3(\lambda)$, where

$$\tilde{Y} = \left(\begin{bmatrix} X \\ XT \\ \vdots \\ XT^{q-1} \end{bmatrix}_{|\mathcal{L}} \right)^{-1} \begin{bmatrix} 0 \\ 0 \\ \vdots \\ 0 \\ I \end{bmatrix} \tag{5.3.7}$$

(b) $(\tilde{X}, P_{\mathcal{N}} T_{|\mathrm{Im}\, P_{\mathcal{N}}}, P_{\mathcal{N}} Y)$ is a standard triple for $L_1(\lambda)$, where $P_{\mathcal{N}}$ is a projector with $\mathrm{Ker}\, P_{\mathcal{N}} = \mathcal{L} \dotplus \mathcal{M}$ and

$$\tilde{X} = [0 \quad 0 \quad \cdots \quad I](P_{\mathcal{N}}[Y, TY, \ldots, T^{l-p-1}Y])^{-1} \tag{5.3.8}$$

(c) $(Z_{|\mathcal{M}}, P_{\mathcal{M}} T_{|\mathcal{M}}, \hat{Y})$ is a standard triple for $L_2(\lambda)$, where $P_{\mathcal{M}}$ is the projector on $\mathcal{M}$ along $\mathcal{L} \dotplus \mathrm{Im}\, P_{\mathcal{N}}$,

$$Z = [0 \quad \cdots \quad 0 \quad I]\{(P_{\mathcal{M}} + P_{\mathcal{N}})[Y, TY, \ldots, T^{l-q-1}Y]\}^{-1} : \mathcal{M} \dotplus \mathrm{Im}\, P_{\mathcal{N}} \to \mathbb{C}^n \tag{5.3.9}$$

and

$$\hat{Y} = \left(\begin{bmatrix} Z \\ ZP_{\mathcal{M}} T \\ \vdots \\ Z(P_{\mathcal{M}} T)^{p-q-1} \end{bmatrix}_{|\mathcal{M}} \right)^{-1} \begin{bmatrix} 0 \\ \vdots \\ 0 \\ I \end{bmatrix} \tag{5.3.10}$$

[Here $q \le l$ and $l - p \le l$ are the unique nonnegative integers such that the linear transformations $\mathrm{col}(XT^i)_{i=0|\mathcal{L}}^{q-1} : \mathcal{L} \to \mathbb{C}^{nq}$ and $P_{\mathcal{N}}[Y, \ldots, T^{l-p-2}Y, T^{l-p-1}Y] : \mathbb{C}^{n(l-p)} \to \mathcal{L} \dotplus \mathcal{M}$ are invertible.]

Conversely, if equation (5.3.6) is a factorization of $L(\lambda)$ into a product of three monic matrix polynomials $L_1(\lambda)$, $L_2(\lambda)$, and $L_3(\lambda)$, there exists a supporting triinvariant decomposition

$$\mathbb{C}^{nl} = \mathcal{L} \dotplus \mathcal{M} \dotplus \mathcal{N} \tag{5.3.11}$$

*associated with a T-semiinvariant subspace $\mathcal{M}$ such that the standard triples
of $L_1(\lambda)$, $L_2(\lambda)$, and $L_3(\lambda)$ are $(\tilde{X}, P_{\mathcal{N}} T_{|\mathrm{Im}\, P_{\mathcal{N}}}, P_{\mathcal{N}} Y)$, $(Z_{|\mathcal{M}}, P_{\mathcal{M}} T_{|\mathcal{M}}, P_{\mathcal{M}} \hat{Y})$,
and $(X_{|\mathcal{L}}, T_{|\mathcal{L}}, \tilde{Y})$, respectively, where $P_{\mathcal{N}}$ is a projector with $\mathrm{Ker}\, P_{\mathcal{N}} =
\mathcal{L} \dotplus \mathcal{M}$, $P_{\mathcal{M}}$ is the projector on $\mathcal{M}$ along $\mathcal{L} \dotplus \mathrm{Im}\, P_{\mathcal{N}}$, and $\tilde{X}, \tilde{Y}, Z, \hat{Y}$ are given
by (5.3.8), (5.3.7), (5.3.9) and (5.3.10), respectively.*

*Moreover, the T-invariant subspaces $\mathcal{L}$ and $\mathcal{L} \dotplus \mathcal{M}$ in (5.3.11) are unique-
ly determined by the factors $L_1(\lambda)$, $L_2(\lambda)$, and $L_3(\lambda)$.*

It is assumed in Theorem 5.3.2 that $P_{\mathcal{M}} \colon \mathbb{C}^{nl} \to \mathcal{M}$, $P_{\mathcal{N}} \colon \mathbb{C}^{nl} \to \mathcal{N}$, where
l is the degree of $L(\lambda)$.

As a monic matrix polynomial $M(\lambda)$ and its inverse are uniquely determined by any standard triple (see Theorem 5.1.6 and the definition of a
standard triple), Theorem 5.3.2 provides an explicit description of the
factors $L_i(\lambda)$ in (5.3.6) in terms of supporting triinvariant decompositions.
For instance, if $\mathbb{C}^{nl} = \mathcal{L} \dotplus \mathcal{M} \dotplus \mathcal{N}$ is a supporting triinvariant decomposition (associated with a T-semiinvariant subspace $\mathcal{M}$) and $L(\lambda) =
L_1(\lambda) L_2(\lambda) L_3(\lambda)$ is the corresponding factorization of $L(\lambda)$, then (in the
notation of Theorem 5.3.2) we have

$$L_1(\lambda)^{-1} = \tilde{X}(\lambda I - P_{\mathcal{N}} T_{|\mathrm{Im}\, P_{\mathcal{N}}})^{-1} P_{\mathcal{N}} Y$$

$$L_2(\lambda)^{-1} = Z_{|\mathcal{M}}(\lambda I - P_{\mathcal{M}} T_{|\mathrm{Im}\, P_{\mathcal{M}}})^{-1} \hat{Y}$$

$$L_3(\lambda)^{-1} = X_{|\mathcal{L}}(\lambda I - T_{|\mathcal{L}})^{-1} \tilde{Y}$$

Similarly, using Theorem 5.3.2, one can produce the formulas for $L_1(\lambda)$,
$L_2(\lambda)$, and $L_3(\lambda)$ themselves. The proof of Theorem 5.3.2 is quite lengthy
and is relegated to the next section.

The following particular case of Theorem 5.3.2 is especially important.
We assume that $L(\lambda)$ and (X, T, Y) are as in Theorem 5.3.2.

Corollary 5.3.3

*Let $\mathbb{C}^{nl} = \mathcal{L} \dotplus \mathcal{M} \dotplus \mathcal{N}$ be a supporting triinvariant decomposition associated
with a T-semiinvariant subspace $\mathcal{M}$ such that $\mathcal{L} \dotplus \mathcal{M} = \mathbb{C}^{nl}$ (so $\mathcal{N} = \{0\}$ and $\mathcal{M}$
is actually T coinvariant). Then $L(\lambda)$ admits a factorization*

$$L(\lambda) = L_2(\lambda) L_3(\lambda) \tag{5.3.12}$$

*where $L_3(\lambda)$ is a monic matrix polynomial of degree q with a standard triple
of the form $(X_{|\mathcal{L}}, T_{|\mathcal{L}}, \tilde{Y})$, where*

$$\tilde{Y} = \left(\begin{bmatrix} X \\ XT \\ \vdots \\ XT^{q-1} \end{bmatrix}_{|\mathcal{L}} \right)^{-1} \begin{bmatrix} 0 \\ 0 \\ \vdots \\ 0 \\ I \end{bmatrix}$$

Also, $L_2(\lambda)$ is a monic matrix polynomial of degree $l - p$ with a standard triple of the form $(X_{|\mathcal{M}}, P_{\mathcal{M}} T_{|\mathcal{M}}, P_{\mathcal{M}} Y)$ where

$$\tilde{X} = [0 \quad \cdots \quad 0 \quad I](P_{\mathcal{M}}[Y, TY, \dots, T^{l-p-1}Y])^{-1}$$

and $P_{\mathcal{M}}$ is the projector on $\mathcal{M}$ along $\mathcal{L}$.

Conversely, if equation (5.3.12) is a factorization of $L(\lambda)$ into a product of two monic matrix polynomials $L_2(\lambda)$ and $L_3(\lambda)$, there exists a unique T-invariant subspace $\mathcal{L}$ such that the triinvariant decomposition $\mathbb{C}^{nl} = \mathcal{L} \dot{+} \mathcal{M} \dot{+} \{0\}$ (where $\mathcal{M}$ is a direct complement to $\mathcal{L}$) is supporting and the standard triples of $L_2(\lambda)$ and $L_3(\lambda)$ are as described above.

Note that under the conditions of Corollary 5.3.3 we have $q = l - p$ (cf. Lemma 5.3.1).

Again, as in Theorem 5.3.2, one can write down explicit formulas for the factors in (5.3.12) and their inverses using the triinvariant decomposition $\mathbb{C}^{nl} = \mathcal{L} \dot{+} \mathcal{M} \dot{+} \mathcal{N}$ with $\mathcal{N} = \{0\}$. For example

$$L_2(\lambda)^{-1} = X_{|\mathcal{M}}(\lambda I - P_{\mathcal{M}} T_{|\mathcal{M}})^{-1} P_{\mathcal{M}} Y$$

in the notation of Corollary 5.3.3.

5.4 PROOF OF THEOREM 5.3.2

We need the following fact.

Proposition 5.4.1

Let $L(\lambda) = \sum_{j=0}^{l} A_j \lambda^i$ be an $n \times n$ matrix polynomial (not necessarily monic) and let $L_1(\lambda)$ be an $n \times n$ monic matrix polynomial with standard triple (X_1, T_1, Y_1). Then (a) $L(\lambda) = L_2(\lambda) L_1(\lambda)$ for some matrix polynomial $L_2(\lambda)$ if and only if the equality

$$\sum_{j=0}^{l} A_j X_1 T_1^j = 0 \tag{5.4.1}$$

holds; (b) $L(\lambda) = L_1(\lambda) L_3(\lambda)$ for some matrix polynomial $L_3(\lambda)$ if and only if the equality

$$\sum_{j=0}^{l} T_1^j Y_1 A_j = 0$$

holds.

Proof. Let us prove (a). We have $L_1(\lambda)^{-1} = X_1(\lambda I - T_1)^{-1} Y_1$. Therefore

$$L(\lambda)L_1(\lambda)^{-1} = \left(\sum_{j=0}^{l} A_j\lambda^j\right)X_1(\lambda I - T_1)^{-1}Y_1$$

and for $|\lambda|$ large enough (e.g., for $|\lambda| > \|T_1\|$) we have

$$L(\lambda)L_1(\lambda)^{-1} = \left(\sum_{j=0}^{l} A_j\lambda^j\right)X_1\left(\sum_{j=0}^{\infty} T^j\lambda^{-j-1}\right)Y_1 \qquad (5.4.2)$$

Now assume $L(\lambda)L_1(\lambda)^{-1}$ is a polynomial. Then in formula (5.4.2) the coefficients of negative powers of λ are zeros. But the coefficient of $\lambda^{-j-1}(j = 0, 1, \ldots)$ in (5.4.2) is

$$A_0X_1T_1^jY_1 + A_1X_1T_1^{j+1}Y_1 + \cdots + A_lX_1T_1^{j+l}Y_1$$

which is zero. So

$$\left(\sum_{i=0}^{l} A_iX_1T_1^i\right)T_1^jY_1 = 0, \qquad j = 0, 1, \ldots$$

As $[Y_1, T_1, Y_1, \ldots, T_1^{k-1}Y_1]$ is nonsingular [where k is the degree of $L_1(\lambda)$; see Proposition 5.1.4], we obtain equality (5.4.1).

Conversely, if (5.4.1) holds, then

$$A_0X_1T_1^jY_1 + A_1X_1T_1^{j+1}Y_1 + \cdots + A_lX_1T_1^{j+l}Y_1 = 0, \qquad j = 0, 1, \ldots$$

which means that all coefficients of negative powers of λ in (5.4.2) are zeros, that is, $L(\lambda)L_1(\lambda)^{-1}$ is a polynomial.

Statement (b) of Proposition 5.4.1 follows from the (already proved) statement (a) when applied to the matrix polynomials $\tilde{L}(\lambda) \stackrel{\text{def}}{=} (L(\bar{\lambda}))^* = \Sigma_{j=0}^{l} A_j^*\lambda^j$ and $\tilde{L}_1(\lambda) \stackrel{\text{def}}{=} (L_1(\bar{\lambda}))^*$ in place of $L(\lambda)$ and $L_1(\lambda)$, respectively. {Note that (Y_1^*, T_1^*, X_1^*) is a standard triple for $\tilde{L}_1(\lambda)$, and that $L(\lambda) = L_1(\lambda)L_3(\lambda)$ if and only if $\tilde{L}(\lambda) = \tilde{L}_3(\lambda)\tilde{L}_1(\lambda)$, where $\tilde{L}_3(\lambda) = [L_3(\bar{\lambda})]^*$ is a matrix polynomial together with $L_3(\lambda)$.} $\square$

Assume now that $\mathbb{C}^{nl} = \mathcal{L} \dotplus \mathcal{M} \dotplus \mathcal{N}$ is a supporting triinvariant decomposition associated with T-semiinvariant subspace $\mathcal{M}$, as in Theorem 5.3.2. As $[\text{col}[XT^i]_{i=0}^{q-1}]_{|\mathcal{L}} : \mathcal{L} \to \mathbb{C}^{nq}$ is an invertible transformation, we can define the $n \times n$ monic matrix polynomial $L_3(\lambda)$ by the formula

$$L_3(\lambda) = I\lambda^q - X_{|\mathcal{L}}(T_{|\mathcal{L}})^q(V_1 + V_2\lambda + \cdots + V_q\lambda^{q-1})$$

where

$$[V_1 \quad V_2 \quad \cdots \quad V_q] = [\text{col}[XT^i]_{i=0}^{q-1}]^{-1} : \mathbb{C}^{nq} \to \mathcal{L}$$

(so $V_i: \mathbb{C}^n \to \mathcal{L}$, $i = 1, \ldots, q$). It turns out that $(X_{|\mathcal{L}}, T_{|\mathcal{L}}, V_q)$ is a standard triple of $L_3(\lambda)$. Indeed, we note that the following equalities hold:

$$[I \quad 0 \quad \cdots \quad 0][\text{col}[XT^i]_{i=0}^{q-1}]_{|\mathcal{L}} = X_{|\mathcal{L}}$$

$$C_3[\text{col}[XT^i]_{i=0}^{q-1}]_{|\mathcal{L}} = [\text{col}[XT^i]_{i=1}^{q}]_{|\mathcal{L}} T_{|\mathcal{L}}$$

where C_3 is the companion matrix for $L_3(\lambda)$, and

$$\text{col}[\delta_{iq}I]_{i=1}^{q} = [\text{col}[XT^i]_{i=0}^{q-1}]_{|\mathcal{L}} \cdot V_q$$

(The second equality is obtained from

$$[V_1 \quad V_2 \quad \cdots \quad V_q][\text{col}[XT^i]_{i=0}^{q-1}]_{|\mathcal{L}} = V_q X_{|\mathcal{L}} + V_2 XT_{|\mathcal{L}} + \cdots + V_q XT_{|\mathcal{L}}^{q-1} = I$$

on premultiplication by $XT_{|\mathcal{L}}^q$.) Hence $(X_{|\mathcal{L}}, T_{|\mathcal{L}}, V_q)$ is similar to the standard triple (P_1, C_1, R_1) for the matrix polynomial $L_3(\lambda)$ (in the notation of Proposition 5.1.2), so $(X_{|\mathcal{L}}, T_{|\mathcal{L}}, V_q)$ is itself a standard triple for $L_3(\lambda)$.

Because of the equality

$$A_0 X + A_1 XT + \cdots + A_{l-1} XT^{l-1} + XT^l = 0$$

where l is the degree of $L(\lambda)$ and A_j is the coefficient of λ^j in $L(\lambda)$ [see formula (5.1.6)], Proposition 5.4.1 ensures that there exists a matrix polynomial $L_4(\lambda)$ such that $L(\lambda) = L_4(\lambda)L_3(\lambda)$. The matrix polynomial $L_4(\lambda)$ is necessarily monic and of degree $l - q$. Let us find its standard triple. First note that the transformation $Q \underset{\text{def}}{=} P_{\mathcal{M}} + P_{\mathcal{N}}$ is a projector on $\mathcal{M} \dotplus \text{Im } P_{\mathcal{N}}$ along $\mathcal{L}$. Indeed, for every $x \in \mathcal{L}$ we have $Qx = P_{\mathcal{M}}x + P_{\mathcal{N}}x = 0 + 0 = 0$, and for every $y \in \mathcal{M}$ (resp. $y \in \text{Im } P_{\mathcal{N}}$) we have $Qy = P_{\mathcal{M}}y + P_{\mathcal{N}}y = y + 0 = 0$ (resp. $Qy = P_{\mathcal{N}}y = y$). Then by Lemma 5.3.1, the transformation

$$Q[Y, TY, \ldots, T^{l-q-1}Y]: \mathbb{C}^{n(l-q)} \to \mathcal{M} \dotplus \text{Im } P_{\mathcal{N}}$$

is invertible.

Now we check that

$$L_4(\lambda)^{-1} = Z(I\lambda - QTQ)^{-1}QY \tag{5.4.3}$$

where

$$Z = [0 \quad \cdots \quad 0 \quad I]\{Q[Y, TY, \ldots, T^{l-q-1}Y]\}^{-1}: \text{Im } Q \to \mathbb{C}^n$$

and QTQ is considered as a transformation from $\text{Im } Q$ into itself. In view of the multiplication theorem (Theorem 5.2.1) it will suffice to check that the triple (X, T, Y) is similar to the triple

$$\left([X_{|\mathscr{L}} \quad 0], \begin{bmatrix} T_{|\mathscr{L}} & V_q Z \\ 0 & QTQ \end{bmatrix}, \begin{bmatrix} 0 \\ QY \end{bmatrix} \right)$$

For then we have $L(\lambda)^{-1} = L_3(\lambda)^{-1} \tilde{L}_4(\lambda)^{-1}$, where $\tilde{L}_4(\lambda)$ is the right-hand side of (5.4.3) and thus $\tilde{L}_4(\lambda) \equiv L_4(\lambda)$. To this end define

$$P' = [\text{col}[X_1 T_1^{i-1}]_{i=1}^q]^{-1} \text{col}[XT^{i-1}]_{i=1}^q : \mathbb{C}^{nl} \to \mathbb{C}^{nl}$$

where $X_1 = X_{|\mathscr{L}}$, $T_1 = T_{|\mathscr{L}}$. Then P' is a projector and $\text{Im } P' = \mathscr{L}$. Indeed, we obviously have $P'y = y$ for every $y \in \mathscr{L}$. Further, formula (5.1.9) implies that

$$\text{Ker } P' \supset \text{Im}[Y, TY, \ldots, T^{l-q-1}Y]$$

In fact, we have the equality:

$$\text{Ker } P' = \text{Im}[Y, TY, \ldots, T^{l-q-1}Y] \tag{5.4.4}$$

To check this, let $y \in \text{Ker } P'$. As $[Y, TY, \ldots, T^{l-q-1}Y]$ is invertible, we have $y = \sum_{i=0}^{l-1} T^i Y x_i$ for some $x_0, \ldots, x_{l-1} \in \mathbb{C}^n$. Now

$$0 = \text{col}[XT^{i-1}]_{i=1}^q y = \begin{bmatrix} X \\ XT \\ \vdots \\ XT^{q-1} \end{bmatrix} [Y, TY, \ldots, T^{l-1}Y] \begin{bmatrix} x_0 \\ x_1 \\ \vdots \\ x_{l-1} \end{bmatrix}$$

and formula (5.1.9) easily implies that $x_{l-q} = \cdots = x_{l-1} = 0$. Hence (5.4.4) follows.

In view of Lemma 5.3.1 the transformation $[Y, TY, \ldots, T^{l-q-1}Y]$ is one-to-one; therefore

$$\dim \text{Im}[Y, TY, \ldots, T^{l-q-1}Y] = (l - q)n$$

Using (5.4.4) and the fact that $P'_{|\mathscr{L}} = I$, it follows that $\mathscr{L}$ and $\text{Im}[Y, TY, \ldots, T^{l-q-1}Y]$ are direct complements to each other in $\mathbb{C}^{nl}$. Thus P' is indeed a projector.

Define $S: \mathbb{C}^{nl} \to \text{Im } P \dotplus \text{Im } Q$ by

$$S = \begin{bmatrix} P' \\ Q \end{bmatrix}$$

where P' and Q are considered as transformations from $\mathbb{C}^{nl}$ into $\text{Im } P'$ and

Im Q, respectively. One verifies easily that S is invertible. We show that

$$[X_1 \ 0]S = X, \qquad ST = \begin{bmatrix} T_1 & V_q Z \\ 0 & QTQ \end{bmatrix} S, \qquad SY = \begin{bmatrix} 0 \\ QY \end{bmatrix}$$

$$(5.4.5)$$

Take $y \in \mathcal{C}^{nl}$. Then $P'y \in \mathcal{L}$ and $\text{col}[X_1 T_1^{i-1} P'y]_{i=1}^k = \text{col}[XT^{i-1}y]_{i=1}^k$. In particular, $X_1 P'y = Xy$. This proves that $[X_1 \ 0]S = X$. The second equality in (5.4.5) is equivalent to the relations

$$P'T = T_1 P' + V_q ZQ \qquad (5.4.6)$$

and $QT = QTQ$. The last follows immediately from the fact that Ker Q is an invariant subspace for T. To prove (5.4.6), take $y \in \mathcal{C}^{nl}$. The case when $y \in \text{Ker } Q = \text{Im } P'$ is trivial. Therefore, assume that $y \in \text{Ker } P'$. We then have to demonstrate that $P'Ty = V_1 Z_2 Qy$. Since $y \in \text{Ker } P'$, there exist $x_0, \ldots, x_{l-q-1} \in \mathcal{C}^n$ such that $v = \Sigma_{i=1}^{l-q} T^{l-q-i} Y x_{i-1}$. Hence

$$Ty = T^{l-q} Y x_0 + T^{l-q-1} Y x_1 + \cdots + TY x_{l-q-1} = T^{l-q} Y x_0 + u$$

with $u \in \text{Ker } P'$ and, as a consequence, $P'Ty = P'T^{l-q} Y x_0$. But then it follows from the definition of P' that

$$P'Ty = [T_1^{q-1} V_q, \ldots, T_1 V_q, V_q] \text{col}[0, \ldots, 0, x_0] = V_q x_0$$

On the other hand, putting $x = \text{col}[x_{i-1}]_{i=1}^{l-q}$, we obtain

$$Qy = Q[T^{l-q-1}Y, \ldots, TY, Y]x$$
$$= [(QTQ)^{l-q-1} QY, \ldots, (QTQ) \cdot QY, QY]x$$

and so $V_q ZQy$ is also equal to $V_q x_0$. This completes the proof of equation (5.4.6). Finally, the last equality in (5.4.5) is obvious because $P'Y = 0$.

We have now proved equality (5.4.3), from which it follows that (Q, QTQ, QY) is a standard triple for $L_4(\lambda)$.

Now define the monic matrix polynomial

$$L_1(\lambda) = \lambda^{l-p} I - (U_1 + U_2 \lambda + \cdots + U_{l-p} \lambda^{l-p-1})(P_{\mathcal{N}} T_{|\text{Im } P_{\mathcal{N}}})^{l-p} P_{\mathcal{N}} Y$$

where $\text{col}[U_i]_{i=1}^{l-p} = A^{-1}$ and

$$A = [P_{\mathcal{N}} Y, P_{\mathcal{N}} T_{|\text{Im } P_{\mathcal{N}}} \cdot P_{\mathcal{N}} Y, \ldots, (P_{\mathcal{N}} T_{|\text{Im } P_{\mathcal{N}}})^{l-p-1} P_{\mathcal{N}} Y]$$

Then $(\tilde{X}, P_{\mathcal{N}} T_{|\text{Im } P_{\mathcal{N}}}, P_{\mathcal{N}} Y)$ is a standard triple for $L_1(\lambda)$. Indeed, this follows from the equalities

$$P_{\mathcal{N}}Y = A \operatorname{col}[\delta_{i1}]_{i=1}^{l-p}, \ P_{\mathcal{N}}T_{|\operatorname{Im} P_{\mathcal{N}}}A = AC_2, \ \check{X}A = [0 \cdots 0 \ I] \quad (5.4.7)$$

where C_2 is the second companion matrix of $L_1(\lambda)$. The first and third equations of (5.4.7) follow from the definitions; the second equality follows from the structure of C_2 using the fact that $A \operatorname{col}[U_i]_{i=1}^{l-p} = I$.

Now Proposition 5.4.1, (b) implies that $L_4(\lambda) = L_1(\lambda)L_2(\lambda)$ for some (necessarily monic) matrix polynomial $L_2(\lambda)$. So in order to prove the direct statement of Theorem 5.3.2 we have only to verify that $(Z_{|\mathcal{M}}, P_{\mathcal{M}}T_{|\mathcal{M}}, \hat{Y})$ is indeed a standard triple for $L_2(\lambda)$. To this end, put

$$\hat{L}_2(\lambda) = I\lambda^{p-q} - Z_{|\mathcal{M}}(P_{\mathcal{M}}T_{|\mathcal{M}})^{p-q}(\hat{V}_1 + \hat{V}_2\lambda + \cdots + \hat{V}_{p-q}\lambda^{p-q-1})$$

where

$$[\hat{V}_1 \ \hat{V}_2 \ \cdots \ \hat{V}_{p-q}] = [\operatorname{col}[Z_{|\mathcal{M}}(P_{\mathcal{M}}T_{|\mathcal{M}})^i]_{i-0}^{p-q-1}]^{-1} \quad (5.4.8)$$

Note that, in view of Lemma 5.3.1, the invertibility of the transformation on the right-hand side of (5.4.8) follows from the invertibility of A. As shown earlier in this section, $(Z_{|\mathcal{M}}, P_{\mathcal{M}}T_{|\mathcal{M}}, \hat{Y})$ is a standard triple for $\hat{L}_2(\lambda)$, and $L_4(\lambda) = \hat{L}_1(\lambda)\hat{L}_2(\lambda)$ for some monic matrix polynomial $\hat{L}_1(\lambda)$ with standard triple $(\check{X}, P_{\mathcal{N}}T_{|\operatorname{Im} P_{\mathcal{N}}}, P_{\mathcal{N}}Y)$. Hence $\hat{L}_1(\lambda) = L_1(\lambda)$, and thus $\hat{L}_2(\lambda) = L_2(\lambda)$.

Consider now the proof of the converse statement of Theorem 5.3.2. This statement amounts to the following: if $L(\lambda) = L_4(\lambda)L_3(\lambda)$ for some monic matrix polynomials $L_4(\lambda)$ and $L_3(\lambda)$, then there is a unique T-invariant subspace $\mathcal{L}$ such that $(X_{|\mathcal{L}}, T_{|\mathcal{L}}, \tilde{Y})$, with

$$\tilde{Y} = [\operatorname{col}[XT^i]_{i=0|\mathcal{L}}^{q-1}]^{-1} \operatorname{col}[\delta_{iq}I]_{i=1}^q,$$

is a standard triple for $L_3(\lambda)$. Here q is the degree of $L_3(\lambda)$. Let C be the first companion matrix of $L(\lambda)$. Proposition 5.4.1 implies that

$$C \operatorname{col}[X_1T_1^i]_{i=0}^{l-1} = \operatorname{col}[X_1T_1^i]_{i=0}^{l-1}T_1 \quad (5.4.9)$$

where (X_1, T_1, Y_1) is a standard triple for $L_3(\lambda)$. Also

$$C \operatorname{col}[XT^i]_{i=0}^{l-1} = \operatorname{col}[XT^i]_{i=0}^{l-1}T \quad (5.4.10)$$

Eliminating C from (5.4.9) and (5.4.10), we obtain

$$T[\operatorname{col}[XT^i]_{i=0}^{l-1}]^{-1}[\operatorname{col}[X_1T_1^i]_{i=1}^{l-1}] = [\operatorname{col}[XT^i]_{i=0}^{l-1}]^{-1}[\operatorname{col}[X_1T_1^i]_{i=0}^{l-1}]T \quad (5.4.11)$$

This readily implies that the subspace

$$\mathcal{L} = \operatorname{Im}([\operatorname{col}[XT^i]_{i=0}^{l-1}]^{-1}[\operatorname{col}[X_1T_1^i]_{i=0}^{l-1}])$$

Example   167

is T invariant. Moreover, it is easily seen that the columns of $[\text{col}[XT^i]_{i=0}^{l-1}]^{-1}[\text{col}[X_1 T_1^i]_{i=0}^{l-1}]$ are linearly independent; equation (5.4.11) implies that in the basis of $\mathcal{L}$ formed by these columns, $T_{|\mathcal{L}}$ is represented by the matrix T_1.

Further

$$X[\text{col}[XT^i]_{i=0}^{l-1}]^{-1}[\text{col}[X_1 T_1^i]_{i=0}^{l-1}] = X_1$$

so $X_{|\mathcal{L}}$ is represented in the same basis in $\mathcal{L}$ by the matrix X. Now it is clear that $(X_{|\mathcal{L}}, T_{|\mathcal{L}}, \hat{Y})$ is similar to (X_1, T_1, Y_1), and thus $(X_{|\mathcal{L}}, T_{|\mathcal{L}}, \hat{Y})$ is also a standard triple for $L_3(\lambda)$.

It remains to prove the uniqueness of $\mathcal{L}$. Assume that $\mathcal{L}'$ is also a T-invariant subspace such that $(X_{|\mathcal{L}'}, T_{|\mathcal{L}'}, \hat{Y})$ is a standard triple for $L_3(\lambda)$ (for some admissible $\hat{Y}$). As any two standard triples of $L_3(\lambda)$ are similar, there exists an invertible transformation $S: \mathcal{L}' \to \mathcal{L}$ such that $X_{|\mathcal{L}'} = X_{|\mathcal{L}}S$, $T_{|\mathcal{L}'} = S^{-1}T_{|\mathcal{L}}S$. Then

$$\text{col}[XT^i]_{i=0 \, | \, \mathcal{L}'}^{l-1} = \text{col}[XT^i]_{i=0 \, | \, \mathcal{L}}^{l-1} S$$

In particular

$$\text{Im}(\text{col}[XT^i]_{i=0 \, | \, \mathcal{L}'}^{l-1}) = \text{Im}(\text{col}[XT^i]_{i=0 \, | \, \mathcal{L}}^{l-1})$$

But the matrix $\text{col}[XT^i]_{i=0}^{l-1}$ is invertible, so $\mathcal{L}' = \mathcal{L}$.

Theorem 5.3.2 is proved completely.

5.5   EXAMPLE

We illustrate Theorem 5.3.2 with an example. Let

$$L(\lambda) = \begin{bmatrix} \lambda(\lambda-1)^2 & \lambda \\ 0 & \lambda^2(\lambda-2) \end{bmatrix}$$

Then

$$\left(\begin{bmatrix} 1 & 0 & 0 & 0 & 0 & 0 \\ 0 & 1 & 0 & 0 & 0 & 0 \end{bmatrix}, \quad C, \quad \begin{bmatrix} 0 & 0 \\ 0 & 0 \\ 0 & 0 \\ 0 & 0 \\ 1 & 0 \\ 0 & 1 \end{bmatrix} \right)$$

is the standard triple for $L(\lambda)$ of Proposition 5.1.2, where

$$C = \begin{bmatrix} 0 & 0 & 1 & 0 & 0 & 0 \\ 0 & 0 & 0 & 1 & 0 & 0 \\ 0 & 0 & 0 & 0 & 1 & 0 \\ 0 & 0 & 0 & 0 & 0 & 1 \\ 0 & 0 & -1 & -1 & 2 & 0 \\ 0 & 0 & 0 & 0 & 0 & 2 \end{bmatrix}$$

is the companion matrix for $L(\lambda)$. As we are concerned with semiinvariant subspaces for C, it is more convenient to use a Jordan form for C in place of C itself. The only eigenvalues of $L(\lambda)$ (and thus also of C) are 0, 1, and 2. A calculation shows that the vectors

$$x_1 = \langle -1, 1, 0, 0, 0, 0 \rangle, \qquad x_2 = \langle 0, 0, -1, 1, 0, 0 \rangle$$

form a Jordan chain of C corresponding to 0; the vector $x_3 = \langle 1, 0, 0, 0, 0, 0 \rangle$ is an eigenvector of C corresponding to 0; the vectors

$$x_4 = \langle 1, 0, 1, 0, 1, 0 \rangle, \qquad x_5 = \langle 0, 0, 1, 0, 2, 0 \rangle$$

form a Jordan chain of C corresponding to 1; and the vector $x_6 = \langle -1, 1, -2, 2, -4, 4 \rangle$ is an eigenvector of C corresponding to 2. The vectors $x_1, \ldots, x_6$ are easily seen to be linearly independent. Denoting by S the invertible 6×6 matrix with columns $x_1, \ldots, x_6$, let

$$X = \begin{bmatrix} 1 & 0 & 0 & 0 & 0 & 0 \\ 0 & 1 & 0 & 0 & 0 & 0 \end{bmatrix} S = \begin{bmatrix} -1 & 0 & 1 & 1 & 0 & -1 \\ 1 & 0 & 0 & 0 & 0 & 1 \end{bmatrix}$$

$$J = S^{-1}CS = \begin{bmatrix} 0 & 1 \\ 0 & 0 \end{bmatrix} \oplus [0] \oplus \begin{bmatrix} 1 & 1 \\ 0 & 1 \end{bmatrix} \oplus [2] \qquad (5.5.1)$$

(J is the Jordan form of C); and

$$Y = S^{-1} \begin{bmatrix} 0 & 0 \\ 0 & 0 \\ 0 & 0 \\ 0 & 0 \\ 1 & 0 \\ 0 & 1 \end{bmatrix} = \begin{bmatrix} 0 & -\frac{1}{4} \\ 0 & -\frac{1}{2} \\ 1 & 1 \\ -1 & -1 \\ 1 & 1 \\ 0 & \frac{1}{4} \end{bmatrix}$$

Clearly, (X, J, Y) is a standard triple for $L(\lambda)$.

We now find some factorizations

$$L(\lambda) = L_1(\lambda)L_2(\lambda)L_3(\lambda) \qquad (5.5.2)$$

where $L_i(\lambda)$, $i = 1, 2, 3$ are monic matrix polynomials of the first degree. As

Example **169**

in Theorem 5.3.2, we express these factorizations in terms of the supporting triinvariant decompositions

$$\mathbb{C}^6 = \mathcal{L} \dotplus \mathcal{M} \dotplus \mathcal{N} \tag{5.5.3}$$

with respect to the standard triple (X, J, Y). So we are looking for J-semiinvariant subspace $\mathcal{M}$ with $\mathcal{L}$ and $\mathcal{L} \dotplus \mathcal{M}$ J invariant, such that the transformations

$$X_{|\mathcal{L}} \colon \mathcal{L} \to \mathbb{C}^2$$

and

$$\begin{bmatrix} X \\ XJ \end{bmatrix}_{|\mathcal{L} \dotplus \mathcal{M}} \colon \mathcal{L} \dotplus \mathcal{M} \to \mathbb{C}^4$$

are invertible. In particular, $\dim \mathcal{L} = \dim \mathcal{M} = \dim \mathcal{N} = 2$. As $\mathcal{L}$ and $\mathcal{L} \dotplus \mathcal{M}$ are J invariant, we have

$$\mathcal{L} = (\mathcal{L} \cap \mathcal{R}_0(J)) \dotplus (\mathcal{L} \cap \mathcal{R}_1(J)) \dotplus (\mathcal{L} \cap \mathcal{R}_2(J))$$

and

$$\mathcal{L} \dotplus \mathcal{M} = ((\mathcal{L} \dotplus \mathcal{M}) \cap \mathcal{R}_0(J)) \dotplus ((\mathcal{L} \dotplus \mathcal{M}) \cap \mathcal{R}_1(J)) \dotplus ((\mathcal{L} \dotplus \mathcal{M}) \cap \mathcal{R}_2(J))$$

where $\mathcal{R}_{\lambda_0}(J)$ is the root subspace of J corresponding to the eigenvalue λ_0. We consider only those supporting triinvariant decomposition (5.5.3) for which

$$\dim(\mathcal{L} \cap \mathcal{R}_0(J)) = \dim(\mathcal{L} \cap \mathcal{R}_1(J)) = 1, \qquad \dim(\mathcal{L} \cap \mathcal{R}_2(J)) = 0 \tag{5.5.4}$$

and

$$\dim((\mathcal{L} \dotplus \mathcal{M}) \cap \mathcal{R}_0(J)) = 2,$$
$$\dim((\mathcal{L} \dotplus \mathcal{M}) \cap \mathcal{R}_1(J)) = \dim((\mathcal{L} \dotplus \mathcal{M}) \cap \mathcal{R}_2(J)) = 1. \tag{5.5.5}$$

In other words, we consider only those factorizations (5.5.1) for which $\det L_3(\lambda) = \lambda(\lambda - 1)$ and $\det(L_2(\lambda)L_3(\lambda)) = \lambda^2(\lambda - 1)(\lambda - 2)$, or, equivalently

$$\det L_1(\lambda) = \lambda(\lambda - 1) ; \qquad \det L_2(\lambda) = \lambda(\lambda - 2) ; \qquad \det L_3(\lambda) = \lambda(\lambda - 1)$$

One could consider all other factorizations (5.5.2) of $L(\lambda)$ in a similar way.

First, we find all pairs of J-invariant subspaces $(\mathcal{L}, \mathcal{L} \dotplus \mathcal{M})$ with the properties (5.5.4) and (5.5.5). Using the Jordan form (5.5.1), it is not difficult to see that all such pairs are given by the following formulas:

(a) $\mathcal{L} = \text{Span}\{e_1 + \alpha e_3, e_4\}$; $\mathcal{L} \dotplus \mathcal{M} = \text{Span}\{e_1, e_3, e_4, e_6\}$, where $\alpha \in \mathbb{C}$ is arbitrary.

(b) $\mathcal{L} = \text{Span}\{e_3, e_4\}$; $\mathcal{L} \dotplus \mathcal{M} = \text{Span}\{e_1, e_3, e_4, e_6\}$.

(c) $\mathcal{L} = \text{Span}\{e_1, e_4\}$; $\mathcal{L} \dotplus \mathcal{M} = \text{Span}\{e_1, e_2 + \beta e_3, e_4, e_6\}$, where $\beta \in \mathbb{C}$ is arbitrary.

Let us check which of these pairs $(\mathcal{L}, \mathcal{L} \dotplus \mathcal{M})$ give rise to supporting triinvariant decompositions, that is, for which pairs the transformations

$$X_{|\mathcal{L}} : \mathcal{L} \to \mathbb{C}^2, \quad \begin{bmatrix} X \\ XJ \end{bmatrix}_{|\mathcal{L} + \mathcal{M}} : \mathcal{L} \dotplus \mathcal{M} \to \mathbb{C}^4$$

are invertible. We have for $\mathcal{L} = \text{Span}\{e_1 + \alpha e_3, e_4\}$:

$$X_{|\mathcal{L}} = \begin{bmatrix} -1 + \alpha & 1 \\ 1 & 0 \end{bmatrix}$$

(in the basis $e_1 + \alpha e_3, e_4$ in $\mathcal{L}$ and the standard basis in $\mathbb{C}^2$), and this matrix is invertible for all $\alpha \in \mathbb{C}$. For $\mathcal{L} = \text{Span}\{e_3, e_4\}$

$$X_{|\mathcal{L}} = \begin{bmatrix} 1 & 1 \\ 0 & 0 \end{bmatrix}$$

which is not invertible. For $\mathcal{L} \dotplus \mathcal{M} = \text{Span}\{e_1, e_3, e_4, e_6\}$

$$\begin{bmatrix} X \\ XJ \end{bmatrix}_{|\mathcal{L} + \mathcal{M}} = \begin{bmatrix} -1 & 1 & 1 & -1 \\ 1 & 0 & 0 & 1 \\ 0 & 0 & 1 & -2 \\ 0 & 0 & 0 & 2 \end{bmatrix}$$

which is invertible. For $\mathcal{L} \dotplus \mathcal{M} = \text{Span}\{e_1, e_2 + \beta e_3, e_4, e_6\}$

$$\begin{bmatrix} X \\ XJ \end{bmatrix}_{|\mathcal{L} + \mathcal{M}} = \begin{bmatrix} -1 & \beta & 1 & -1 \\ 1 & 0 & 0 & 1 \\ 0 & -1 & 1 & -2 \\ 0 & 1 & 0 & 2 \end{bmatrix}$$

which is invertible if and only if $\beta \neq 0$. (In this calculation we have used the formula

$$\begin{bmatrix} X \\ XJ \end{bmatrix} = \begin{bmatrix} -1 & 0 & 1 & 1 & 0 & -1 \\ 1 & 0 & 0 & 0 & 0 & 1 \\ 0 & -1 & 0 & 1 & 1 & -2 \\ 0 & 1 & 0 & 0 & 0 & 2 \end{bmatrix})$$

Summarizing, one obtains all the supporting triinvariant decompositions (5.5.3) with the properties (5.5.4) and (5.5.5) where *either* $\mathscr{L} = \mathrm{Span}\{e_1 + \alpha e_3, e_4\}$ for some $\alpha \in \mathbb{C}$, $\mathcal{M}$ is a direct complement to $\mathrm{Span}\{e_1, e_3, e_4, e_6\}$ in $\mathbb{C}^6$, *or*, for some nonzero $\beta \in \mathbb{C}$ we have $\mathscr{L} = \mathrm{Span}\{e_1, e_4\}$, $\mathcal{M}$ is a direct complement to $\mathscr{L}$ in $\mathrm{Span}\{e_1, e_2 + \beta e_3, e_4, e_6\}$, and $\mathcal{N}$ is a direct complement to $\mathrm{Span}\{e_1, e_2 + \beta e_3, e_4, e_6\}$ in $\mathbb{C}^6$.

Using the formulas given in Theorem 5.3.2, one finds all the factorizations (5.5.2) corresponding to the supporting triinvariant decomposition with properties (5.5.4) and (5.5.5) (here $\alpha \in \mathbb{C}$ and $\beta \in \mathbb{C}$, $\beta \neq 0$ are as above):

$$L(\lambda) = \begin{bmatrix} \lambda - 1 & \lambda - 1 \\ 0 & \lambda \end{bmatrix}\begin{bmatrix} \lambda & -\lambda + 2 \\ 0 & \lambda - 2 \end{bmatrix}\begin{bmatrix} \lambda - 1 & \lambda - 1 \\ 0 & \lambda \end{bmatrix}$$

$$L(\lambda) = \begin{bmatrix} \lambda - \dfrac{\beta + 2}{\beta} & -\dfrac{\beta + 2}{\beta} \\ \dfrac{2}{\beta} & \lambda + \dfrac{2}{\beta} \end{bmatrix}\begin{bmatrix} \lambda + \dfrac{2}{\beta} & \dfrac{2(\beta + 1)}{\beta} \\ -\dfrac{2}{\beta} & \lambda - \dfrac{2(\beta + 1)}{\beta} \end{bmatrix}\begin{bmatrix} \lambda - 1 & -1 \\ 0 & \lambda \end{bmatrix}$$

5.6 FACTORIZATION INTO SEVERAL FACTORS AND CHAINS OF INVARIANT SUBSPACES

In this section we study factorizations of the monic $n \times n$ matrix polynomial $L(\lambda)$ of degree l into the product of several factors:

$$L(\lambda) = L_1(\lambda)L_2(\lambda) \cdots L_k(\lambda) \tag{5.6.1}$$

where $L_1(\lambda), \ldots, L_k(\lambda)$ are monic $n \times n$ matrix polynomials of positive degrees $l_1, \ldots, l_k$, respectively (of course, $l_1 + \cdots + l_k = l$). We have already encountered particular cases of factorizations (5.6.1) in Theorem 5.3.2 (with $k = 3$) and in Corollary 5.3.3 (with $k = 2$). In Theorem 5.3.2 factorizations (5.6.1) with $k = 3$ were described in terms of supporting triinvariant decompositions associated with semiinvariant subspaces of a linearization of $L(\lambda)$. In contrast, the description of (5.6.1) is to be given in terms of chains of invariant subspaces for a linearization of $L(\lambda)$.

The following main result can be regarded as a generalization of Corollary 5.3.3.

Theorem 5.6.1

Let (X, T, Y) be a standard triple for $L(\lambda)$. Then for every chain of T-invariant subspaces

$$\{0\} \subset \mathscr{L}_k \subset \mathscr{L}_{k-1} \subset \cdots \subset \mathscr{L}_2 \subset \mathbb{C}^{nl} \tag{5.6.2}$$

satisfying the property that the transformations

$$\begin{bmatrix} X \\ XT \\ \vdots \\ XT^{m_j-1} \end{bmatrix}_{|\mathscr{L}_j} : \mathscr{L}_j \to \mathbb{C}^{nm_j}, \qquad j = 2, \ldots, k$$

are invertible (for some positive integers $m_k < m_{k-1} < \cdots < m_2 < l$) there exists a factorization (5.6.1) of $L(\lambda)$, with the factors $L_j(\lambda)$ uniquely determined by the chain (5.6.2), as follows. For $j = 1, 2, \ldots, k-1$, let $\mathcal{M}_j$ be a direct complement to $\mathscr{L}_{j+1}$ in $\mathscr{L}_j$ (by definition, $\mathscr{L}_1 = \mathbb{C}^{nl}$) and let $P_{\mathcal{M}_j} : \mathscr{L}_j \to \mathcal{M}_j$ be the projector on $\mathcal{M}_j$ along $\mathscr{L}_{j+1}$. Then for $j = 1, 2, \ldots, k-1$,

$$L_j(\lambda) = \lambda^{l_j} I - (W_{j1} + \lambda W_{j2} + \cdots + \lambda^{l_j-1} W_{j,l_j})(P_{\mathcal{M}_j} T_{|\mathcal{M}_j})^{l_j} P_{\mathcal{M}_j} \tilde{Y}_j \tag{5.6.3}$$

where $l_j = m_j - m_{j+1}$ (by definition, $m_1 = l$) and

$$\tilde{Y}_j = (\mathrm{col}[XT^i]_{i=0|\mathscr{L}_j}^{m_j-1})^{-1} \mathrm{col}[\delta_{im_j} I]_{i=1}^{m_j} \tag{5.6.4}$$

and the transformations $W_{ji} : \mathcal{M}_j \to \mathbb{C}^n$ ($i = 1, \ldots, l_j$) are determined by

$$\mathrm{col}[W_{ji}]_{i=1}^{l_j} = [P_{\mathcal{M}_j} \tilde{Y}_j, \ P_{\mathcal{M}_j} T_{|\mathcal{M}_j} P_{\mathcal{M}_j} \tilde{Y}_j, \ldots, (P_{\mathcal{M}_j} T_{|\mathcal{M}_j} P_{\mathcal{M}_j})^{l_j-1} P_{\mathcal{M}_j} \tilde{Y}_j]$$

Further

$$L_k(\lambda) = \lambda^{m_k} I - X_{|\mathscr{L}_k}(T_{|\mathscr{L}_k})^{m_k}(V_{k1} + V_{k2}\lambda + \cdots + V_{k,m_k}\lambda^{m_k-1}) \tag{5.6.5}$$

where

$$[V_{k1} \quad V_{k2} \quad \cdots \quad V_{km_k}] = (\mathrm{col}[XT^i]_{i=0|\mathscr{L}_k}^{m_k-1})^{-1} : \mathbb{C}^{nm_k} \to \mathscr{L}_k$$

so V_{kq} are transformations from $\mathbb{C}^n$ into $\mathscr{L}_k$ for $q = 1, \ldots, m_k$. Conversely, for every factorization (5.6.1) of $L(\lambda)$ there is a unique chain of T-invariant subspaces (5.6.2) such that for $j = 2, 3, \ldots, k$ the transformations

$$\mathrm{col}[XT^i]_{i=0|\mathscr{L}_j}^{m_j-1} : \mathscr{L}_j \to \mathbb{C}^{nm_j}$$

where $m_j = l_j + l_{j+1} + \cdots + l_k$ (l_j is the degree of L_j), are invertible and formulas (5.6.3) and (5.6.5) hold.

Observe that in view of Proposition 3.1.1 formulas (5.6.3) do not depend on the choice of $\mathcal{M}_j$.

Proof. Apply Corollary 5.3.3 several times to see that factorization

(5.6.1) holds for the monic matrix polynomial $L_j(\lambda)L_{j+1}(\lambda)\cdots L_k(\lambda)$ having the standard triple $(X_{|\mathscr{L}_j}, T_{|\mathscr{L}_j}, \tilde{Y}_j)$, where $\tilde{Y}_j$ is given by (5.6.4) $(j = 2, \ldots, k)$. Now use Theorem 5.1.6 to produce the formulas

$$L_j(\lambda)L_{j+1}(\lambda)\cdots L_k(\lambda)$$
$$= \lambda^{m_j}I - X_{|\mathscr{L}_j}(T_{|\mathscr{L}_j})^{m_j}(V_{j1} + \cdots + V_{j,m_j}\lambda^{m_j-1}), \qquad j = 2, \ldots, k$$

where

$$[V_{j1} \quad V_{j2} \quad \cdots \quad V_{jm_j}] = [\mathrm{col}[XT^i]_{i=0|\mathscr{L}_j}^{m_j-1}]^{-1} \colon \mathbb{C}^{nm_j} \to \mathscr{L}_j$$

(so V_{jq} are transformations from $\mathbb{C}^n$ into $\mathscr{L}_j$ for $q = 1, \ldots, m_j$). In particular (with $j = k$), formula (5.6.5) follows. Further, using the formulas for the standard triple of the factor $L_2(\lambda)$ in Corollary 5.3.3, one easily obtains the desired formulas [equation (5.6.3)]. The converse statement also follows by repeated application of the converse statement of Corollary 5.3.3. □

A "dual" version of Theorem 5.6.1 can be obtained if one uses the left canonical form [equation (5.1.14) instead of the right canonical form equation (5.1.13)] to produce formulas for $L_j(\lambda)L_{j+1}(\lambda)\cdots L_k(\lambda)$. Then one uses (5.1.13) [instead of (5.1.14)] to derive the formulas for $L_1(\lambda), \ldots, L_{k-1}(\lambda)$. We omit an explicit formulation of these results.

We are interested particularly in factorizations (5.6.1) with linear factors $L_j(\lambda)$: $L_j(\lambda) = \lambda I + A_j$ for some $n \times n$ matrices A_j $(j = 1, \ldots, k)$. Note that in contrast to the scalar case, not every monic matrix polynomial admits such a factorization:

EXAMPLE 5.6.1. Let

$$L(\lambda) = \begin{bmatrix} \lambda^2 & -1 \\ 0 & \lambda^2 \end{bmatrix}$$

We claim that $L(\lambda)$ cannot be factorized into the product of (two) linear factors. Indeed, assume the contrary:

$$\begin{bmatrix} \lambda^2 & -1 \\ 0 & \lambda^2 \end{bmatrix} = \begin{bmatrix} \lambda + a_1 & b_1 \\ c_1 & \lambda + d_1 \end{bmatrix}\begin{bmatrix} \lambda + a_2 & b_2 \\ c_2 & \lambda + d_2 \end{bmatrix} \tag{5.6.6}$$

for some complex numbers $a_i, b_i, c_i, d_i, i = 1, 2$. Multiplying the factors on the right-hand side and comparing entries, we obtain

$$a_1 + a_2 = 0; \qquad b_2 + b_1 = 0$$

$$c_1 + c_2 = 0; \qquad d_1 + d_2 = 0$$

Letting

$$A = \begin{bmatrix} a_1 & b_1 \\ c_1 & d_1 \end{bmatrix}$$

we can rewrite equality (5.6.6) in the form

$$\begin{bmatrix} \lambda^2 & -1 \\ 0 & \lambda^2 \end{bmatrix} = (\lambda I + A)(\lambda I - A)$$

which implies $A^2 = \begin{bmatrix} 0 & 1 \\ 0 & 0 \end{bmatrix}$. However, there is no 2×2 matrix A with this property (indeed, such an A must have only the zero eigenvalue, but then inevitably $A^2 = 0$). $\square$

As we shall see in the next theorem, a necessary (but not sufficient) condition for a monic matrix polynomial $L(\lambda)$ *not* to be decomposable into a product of monic linear factors is that the linearization of $L(\lambda)$ is not diagonable. Indeed, in Example 5.6.1 the linearization of $\begin{bmatrix} \lambda^2 & -1 \\ 0 & \lambda^2 \end{bmatrix}$ has only one Jordan block $J_4(0)$ in its Jordan form.

Theorem 5.6.2

Let $L(\lambda)$ be an $n \times n$ monic matrix polynomial of degree l for which the companion matrix is diagonable. Then there exist $n \times n$ matrices $A_1, \ldots, A_l$ such that

$$L(\lambda) = (\lambda I + A_1)(\lambda I + A_2) \cdots (\lambda I + A_l)$$

Proof. Let (X, T, Y) be a standard triple for $L(\lambda)$, and let

$$\mathcal{N}_j = \mathrm{Ker} \begin{bmatrix} X \\ XT \\ \vdots \\ XT^{j-1} \end{bmatrix}, \qquad j = 1, \ldots, l-1$$

Obviously, the $\mathcal{N}_j$ are subspaces in $\mathbb{C}^{nl}$ and

$$\mathcal{N}_1 \supset \mathcal{N}_2 \supset \cdots \supset \mathcal{N}_{l-1}$$

By Theorem 1.8.5 there exist T-invariant subspaces $\mathcal{M}_1 \subset \mathcal{M}_2 \subset \cdots \subset \mathcal{M}_{l-1}$ such that $\mathcal{M}_j$ is a direct complement to $\mathcal{N}_j$ in $\mathbb{C}^{nl}$. The transformations

$$\mathrm{col}[XT^i]_{i=0|\mathcal{M}_j}^{j-1} : \mathcal{M}_j \to \mathbb{C}^{nj}, \qquad j = 1, \ldots, l-1 \qquad (5.6.7)$$

are invertible. Indeed, by the choice of $\mathcal{M}_j$ we have $\mathrm{Ker}(\mathrm{col}[XT^i]_{i=0|\mathcal{M}_j}^{j-1}) = \{0\}$. As the matrix $\mathrm{col}[XT^i]_{i=0}^{l-1}$ is invertible, the matrix $\mathrm{col}[XT^i]_{i=0}^{j-1}$ has linearly independent rows and thus $\mathrm{Im}(\mathrm{col}[XT^i]_{i=0|\mathcal{M}_j}^{j-1}) = \mathrm{Im}(\mathrm{col}[XT^i]_{i=0}^{j-1}) = \mathbb{C}^{nj}$, $j = 1, \ldots, l-1$. Invertibility of (5.6.7) now follows. The proof is completed by applying Theorem 5.6.1. $\square$

5.7   DIFFERENTIAL EQUATIONS

Consider the homogeneous system of differential equations with constant coefficients:

$$\frac{d^l x(t)}{dt^l} + \sum_{j=0}^{l-1} A_j \frac{d^j x(t)}{dt^j} = 0, \qquad t \in [a, \infty) \qquad (5.7.1)$$

where $A_0, \ldots, A_{l-1}$ are $n \times n$ (complex) matrices, and $x(t)$ is an n-dimensional vector function of t to be found. The behaviour of solutions of equation (5.7.1) as $t \to \infty$ is an important question in applications to physical systems. We look for solutions with prescribed growth (or decay) at infinity. It will turn out that such solutions depend on certain invariant subspaces of a linearization of the monic matrix polynomial

$$L(\lambda) = I\lambda^l + \sum_{j=0}^{l-1} A_j \lambda^j$$

connected with (5.7.1).

First we observe that a solution of (5.7.1) is uniquely defined by the initial data $x^{(j)}(a) = x_j$, $j = 0, \ldots, l - 1$, with given initial vectors $x_0, \ldots, x_{l-1}$. Indeed, denoting by $y(t)$ the nl-dimensional vector

$$\begin{bmatrix} x(t) \\ x'(t) \\ \vdots \\ x^{(l-1)}(t) \end{bmatrix}$$

equation (5.7.1) is equivalent to the following equation:

$$\frac{dy(t)}{dt} = \begin{bmatrix} 0 & I & 0 & \cdots & 0 \\ 0 & 0 & I & \cdots & 0 \\ & & & \ddots & \\ 0 & 0 & 0 & \cdots & \\ -A_0 & -A_1 & -A_2 & \cdots & -A_{l-1} \end{bmatrix} y(t), \qquad t \in [a, \infty) \qquad (5.7.2)$$

As it is well known [cf. Section 2.10, especially formula (2.10.8)], a solution of equation (5.7.2) is uniquely defined by the initial data $y(a)$, which amounts to the initial data $x^{(j)}(a), j = 0, \ldots, l - 1$ for equation (5.7.1). In particular, the dimension of the set of all solutions of (5.7.1) (this set obviously is a linear space) is nl, the number of (complex) parameters in the n-dimensional vectors $x_0, \ldots, x_{l-1}$ that determine the initial data of a solution and thus the solution itself.

It will be convenient to describe the general solution of (5.7.1) in terms of a standard triple (X, T, Y) of the monic matrix polynomial $L(\lambda)$.

Lemma 5.7.1

A function $x(t)$ is a solution of (5.7.1) *if and only if it has the form*

$$x(t) = Xe^{tT}c, \qquad t \in [a, \infty) \tag{5.7.3}$$

for some vector $c \in \mathbb{C}^{nl}$.

   Proof.   Differentiating (5.7.3), we obtain

$$x^{(j)}(t) = XT^j e^{tT}c, \qquad j = 0, 1, \dots$$

so

$$\frac{d^l x(t)}{dt^l} + \sum_{j=0}^{l-1} A_j \frac{d^j x(t)}{dt^j} = XT^l e^{tT}c + \sum_{j=0}^{l-1} A_j XT^j e^{tT}c = \left(XT^l + \sum_{j=0}^{l-1} A_j XT^j \right) e^{tT}c$$

which is equal to zero in view of Proposition 5.1.4. It remains to show that every solution of (5.7.1) is of the type (5.7.3) for some $c \in \mathbb{C}^{nl}$. As the linear space of all solutions of (5.7.1) has dimension nl it will suffice to show that the solutions $Xe^{tT}c_1, \dots, Xe^{tT}c_{nl}$ that correspond to a basis $c_1, \dots, c_{nl}$ in $\mathbb{C}^{nl}$ are linearly independent. In other words, we should prove that $Xe^{tT}c = 0$ for all $t \geq a$ implies $c = 0$. Indeed, differentiating the relation $Xe^{tT}c \equiv 0$ j times, we obtain $XT^j e^{tT}c \equiv 0$ for $j = 0, 1, 2, \dots$. In particular

$$\begin{bmatrix} X \\ XT \\ \vdots \\ XT^{l-1} \end{bmatrix} e^{tT}c \equiv 0$$

As the matrices e^{tT} and $\mathrm{col}(XT^i)_{i=0}^{l-1}$ are nonsingular (Proposition 5.1.4), it follows that $c = 0$. $\square$

   Now let us introduce some T-invariant subspaces: $\mathcal{R}_+(T)$ [resp. $\mathcal{R}_-(T)$] is the sum of all root subspaces of T corresponding to its eigenvalues with positive real part (resp. with negative real part); $\mathcal{R}_0(T)$ is the sum of all root subspaces of T corresponding to its pure imaginary eigenvalues (including zero); and

$$\mathcal{K}_0(T) = \sum_{\substack{\lambda_0 \in \sigma(T) \\ \lambda_0 + \bar{\lambda}_0 = 0}} \mathrm{Ker}(T - \lambda_0 I)$$

Obviously, $\mathcal{K}_0(T)$ is a T-invariant subspace contained in $\mathcal{R}_0(T)$. If it happens that T has no eigenvalues with positive real part, we set $\mathcal{R}_+(T) = \{0\}$. A similar convention will apply for $\mathcal{R}_-(T)$, $\mathcal{R}_0(T)$, and $\mathcal{K}_0(T)$.

Let $\mathcal{K}_1(T)$ be a fixed direct complement to $\mathcal{K}_0(T)$ in $\mathcal{R}_0(T)$ and note that $\mathcal{K}_1(T)$ is never T invariant [unless $\mathcal{K}_1(T) = \{0\}$]. Otherwise $\mathcal{K}_1(T)$ would contain an eigenvector of T that, by definition, should belong to $\mathcal{K}_0(T)$.

We now have the direct sum $\mathbb{C}^{nl} = \mathcal{R}_-(T) \dotplus \mathcal{K}_0(T) \dotplus \mathcal{K}_1(T) \dotplus \mathcal{R}_+(T)$. For a given vector $c \in \mathbb{C}^{nl}$, let

$$c = c_- + c_0 + c_1 + c_+ \tag{5.7.4}$$

where $c_- \in \mathcal{R}_-(T)$, $c_0 \in \mathcal{K}_0(T)$, $c_1 \in \mathcal{K}_1(T)$, $c_+ \in \mathcal{R}_+(T)$. We describe the qualitative behaviour of solutions of (5.7.1) in terms of this decomposition of the initial value of the solution $x(t)$.

A solution $x(t)$ of (5.7.1) is said to be *exponentially increasing* if for some positive number μ

$$0 < \overline{\lim_{t \to \infty}} \|e^{-\mu t} x(t)\| \leq \infty \tag{5.7.5}$$

but

$$\lim_{t \to \infty} \|e^{-(\mu + \epsilon)t} x(t)\| = 0 \tag{5.7.6}$$

for every $\epsilon > 0$. Obviously, such a positive number μ is unique and is called the *exponent* of the exponentially increasing solution $x(t)$. A solution $x(t)$ of (5.7.1) is *exponentially decreasing* if (5.7.5) and (5.7.6) hold for some negative number μ [which is unique and is called again the *exponent* of $x(t)$]. We say that a solution $x(t)$ is *polynomially increasing* if

$$0 < \overline{\lim_{t \to \infty}} \|t^{-m} x(t)\| < \infty$$

for some positive integer m. Finally, we say that a solution $x(t)$ is *oscillatory* if

$$0 < \overline{\lim_{t \to \infty}} \|x(t)\| < \infty$$

These classes of solutions of (5.7.1) can be distinguished according to the decomposition (5.7.4) of the vector c, as follows.

Theorem 5.7.2

Let $x(t) = X e^{tT} c$ be a solution of (5.7.1). Then (a) $x(t)$ is exponentially increasing if and only if $c_+ \neq 0$; (b) $x(t)$ is polynomially increasing if and only if $c_+ = 0$, $c_1 \neq 0$; (c) $x(t)$ is oscillatory if and only if $c_+ = c_1 = 0$, $c_0 \neq 0$; (d) $x(t)$ is exponentially decreasing if and only if $c_+ = c_1 = c_0 = 0$, $c_- \neq 0$. In cases (a) and (d), the exponent of $x(t)$ is equal to the maximum of the real parts of the eigenvalues λ_0 of T with the property that $P_{\lambda_0} c \neq 0$, where P_{λ_0} is the projector on $\mathcal{R}_{\lambda_0}(T)$ along $\sum_{\substack{\lambda \in \sigma(T) \\ \lambda \neq \lambda_0}} \mathcal{R}_\lambda(T)$.

Proof. We have

$$x(t) = Xe^{tT_-}c_- + Xe^{tT_0}(c_0 + c_1) + Xe^{tT_+}c_+ \tag{5.7.7}$$

where $T_- = T_{|\mathcal{R}_-(T)}$, $T_0 = T_{|\mathcal{R}_0(T)}$, $T_+ = T_{|\mathcal{R}_+(T)}$. Without loss of generality [passing to a similar triple (X, T, Y), if necessary] we can assume that T_-, T_0, T_+ are matrices in Jordan form.

Note that for the Jordan block $J_k(\lambda)$ we have (according to Section 2.10)

$$e^{tJ_k(\lambda)} = \begin{bmatrix} e^{t\lambda} & \dfrac{1}{1!}te^{t\lambda} & \cdots & \dfrac{t^{k-1}}{(k-1)!}e^{t\lambda} \\ 0 & e^{t\lambda} & \ddots & \vdots \\ \vdots & \ddots & \ddots & \dfrac{1}{1!}te^{t\lambda} \\ 0 & \cdots & 0 & e^{t\lambda} \end{bmatrix}$$

So every entry in $Xe^{tT_+}c_+$ is a function of the type

$$\sum_{\substack{\lambda_i \in \sigma(T) \\ \mathcal{R}e\,\lambda_i > 0}} e^{t\lambda_i}p_i(t) \tag{5.7.8}$$

for some polynomials $p_i(t)$. Also, every entry in $Xe^{tT_0}(c_0 + c_1)$ is of the type

$$\sum_{\substack{\lambda_i \in \sigma(T) \\ \mathcal{R}e\,\lambda_i = 0}} e^{t\lambda_i}p_i(t) \tag{5.7.9}$$

whereas every entry in $Xe^{tT_0}c_0$ is of the type (5.7.9) with all polynomials $p_i(t)$ constant. Finally, every entry in $Xe^{tT_-}c_-$ is of the type

$$\sum_{\substack{\lambda_i \in \sigma(T) \\ \mathcal{R}e\,\lambda_i < 0}} e^{t\lambda_i}p_i(t) \tag{5.7.10}$$

Further, note that

$$Xe^{tT_\pm}c_\pm = 0 \quad \text{for all} \quad t \geq a \tag{5.7.11}$$

if and only if $c_\pm = 0$. Indeed, if equality (5.7.11) holds, then successive differentiation gives $XT_\pm^j e^{tT_\pm}c_\pm = 0$, $j = 0, 1, \ldots$. In particular

$$[\text{col}[XT^i]_{i=0}^{l-1}]_{|\mathcal{R}_\pm(T)}e^{tT_\pm}c_\pm = 0 \tag{5.7.12}$$

As $\text{col}[XT^i]_{i=0}^{l-1}$ is a nonsingular matrix, the transformation

$$[\mathrm{col}[XT^i]_{i=0}^{l-1}]_{|\mathcal{R}_\pm(T)} \colon \mathcal{R}_\pm(T) \to \mathbb{C}^{nl}$$

has zero kernel, and equation (5.7.12) implies $c_\pm = 0$. Also, the equality

$$Xe^{tT_0}(c_0 + c_1) = 0 , \qquad t \geq a$$

holds if and only if $c_0 + c_1 = 0$. Also

$$Xe^{tT_0}c_0 = 0 , \qquad t \geq a$$

if and only if $c_0 = 0$. According to the observation made in this and the preceding paragraphs, statements (a)–(d) follow easily from formula (5.7.7). For instance, assume that $x(t)$ is exponentially increasing. In view of (5.7.8)–(5.7.10), this means that $Xe^{tT_+}c_+ \neq 0$ (since $|e^z| = e^{\mathscr{R}ez}$ for any complex number z), and this is equivalent to the inequality $c_+ \neq 0$. $\square$

A special case with $X = [I \quad 0 \quad \cdots \quad 0]$ and T the companion matrix of $L(\lambda)$ deserves special attention. In this case the matrix $\mathrm{col}[XT^i]_{i=0}^{l-1}$ is just the identity, and thus

$$c = \begin{bmatrix} x(0) \\ x'(0) \\ \vdots \\ x^{(l-1)}(0) \end{bmatrix}$$

Exponentially decreasing solutions of (5.7.1) are of particular interest. We present one result on existence and uniqueness of exponentially decreasing solutions in which only partial initial data are prescribed.

Theorem 5.7.3

For every set of k vectors $x_0, \ldots, x_{k-1}$ in $\mathbb{C}^n$ there exists a unique exponentially decreasing solution $x(t)$ of (5.7.1) such that

$$x^{(i)}(a) = x_i , \qquad i = 0, \ldots, k-1$$

if and only if the matrix polynomial $L(\lambda)$ admits a factorization $L(\lambda) = L_2(\lambda)L_1(\lambda)$, where $L_1(\lambda)$ and $L_2(\lambda)$ are monic matrix polynomials of degrees k and $l-k$, respectively, such that $\mathrm{Re}\ \lambda < 0$ for all $\lambda \in \sigma(L_1)$ and $\mathscr{R}e\ \lambda \geq 0$ for all $\lambda \in \sigma(L_2)$.

Proof. In the notation of Theorem 5.7.2 the solution $x(t)$ is exponentially decreasing if and only if

$$x(t) = Xe^{tT}c_- \tag{5.7.13}$$

where $c_- \in \mathcal{R}_-(T)$. When $x(t)$ is given by (5.7.10) we have

$$x^{(i)}(a) = XT^i e^{aT} c_-, \qquad i = 0, 1, 2, \ldots$$

It follows that for every set $x_0, \ldots, x_{k-1} \in \mathbb{C}^n$ there exists a unique exponentially decreasing solution $x(t)$ of (5.7.1) with $x^{(i)}(a) = x_i$, $i = 0, \ldots, k-1$ if and only if the transformation

$$\mathrm{col}[XT^i]_{i=0}^{k-1} \cdot [e^{aT}]_{|\mathscr{R}_-(T)} : \mathscr{R}_-(T) \to \mathbb{C}^{nk}$$

is one-to-one and onto. This amounts to the invertibility of $\mathrm{col}[X(T_{|\mathscr{R}_-(T)})^i]_{i=0}^{k-1}$, which in turn is equivalent (by Corollary 5.3.3) to the existence of a factorization $L(\lambda) = L_2(\lambda) L_1(\lambda)$. Moreover, in this factorization $[X_{|\mathscr{R}_-(T)}, T_{|\mathscr{R}_-(T)}, \tilde{Y}]$ is a standard triple for $L_1(\lambda)$ (for a suitable $\tilde{Y}$), whereas $(\tilde{X}, PT_{|\mathrm{Im}\,P}, PY)$ is a standard triple for $L_2(\lambda)$ for a suitable $\tilde{X}$, where P is the projector on $\mathscr{R}_0(T) + \mathscr{R}_+(T)$ along $\mathscr{R}_-(T)$. As $T_{|\mathscr{R}_-(T)}$ and $PT_{|\mathrm{Im}\,P}$ are linearizations of $L_1(\lambda)$ and $L_2(\lambda)$, respectively (Theorem 5.1.5), it follows that indeed $\mathscr{R}e\,\lambda < 0$ for all $\lambda \in \sigma(L_1)$, and $\mathscr{R}e\,\lambda \geq 0$ for all $\lambda \in \sigma(L_2)$. $\quad\square$

5.8   DIFFERENCE EQUATIONS

In this section we consider the system of difference equations

$$x_{j+l} + A_{l-1} x_{j+l-1} + \cdots + A_0 x_j = 0, \qquad j = 0, 1, \ldots \qquad (5.8.1)$$

where $A_0, \ldots, A_{l-1}$ are given $n \times n$ matrices, and $\{x_j\}_{j=0}^{\infty}$ is a sequence of n-dimensional vectors to be found. Clearly, given l initial vectors $x_0, \ldots, x_{l-1}$, the vectors x_l, x_{l+1}, and so on are determined uniquely from (5.8.1). Hence, a solution $\{x_j\}_{j=0}^{\infty}$ of equation (5.8.1) is determined by its first l vectors.

Again, it will turn out that the asymptotic behaviour of solutions of (5.8.1) can be described in terms of certain invariant subspaces of a linearization of the associated monic matrix polynomial

$$L(\lambda) = I\lambda^l + \sum_{j=0}^{l-1} A_j \lambda^j$$

Let (X, T, Y) be a standard triple for $L(\lambda)$. The general solution of (5.8.1) is then

$$\{XT^i c\}_{i=0}^{\infty} \qquad (5.8.2)$$

where $c \in \mathbb{C}^{nl}$ is an arbitrary vector. Indeed, putting $x_j = XT^j c$, $j = 0, 1, \ldots$, we have

$$x_{j+l} + A_{l-1}x_{j+l-1} + \cdots + A_0 x_j = XT^{l+j}c + A_{l-1}XT^{j+l-1}c + \cdots + A_0 XT^j c$$
$$= (XT^l + A_{l-1}XT^{l-1} + \cdots + A_0 X)T^j c$$

which is zero in view of Proposition 5.1.4. If the first l vectors in (5.8.2) are zeros, that is

$$Xc = XTc = \cdots = XT^{l-1}c = 0$$

then by the nonsingularity of $\mathrm{col}[XT^i]_{i=0}^{l-1}$ we obtain $c = 0$. This means that the solutions (5.8.2) are indeed *all* the solution of (5.8.1).

The solutions of (5.8.1) are now to be classified according to the rate of growth of the sequence $\{x_j\}_{j=0}^{\infty}$. We say that the solution $\{x_j\}_{j=0}^{\infty}$ is of *geometric growth* (resp. *geometric decay*) if there exists a number $q > 1$ (resp. a positive number $q < 1$) such that

$$0 < \overline{\lim_{m \to \infty}} \| q^{-m}x_m \| \leq \infty$$

but

$$\overline{\lim_{m \to \infty}} \| (q + \epsilon)^{-m}x_m \| = 0$$

for every positive number ϵ. The number q is called the *multiplier* of the geometrically growing (or decaying) solution $\{x_j\}_{j=0}^{\infty}$. The solution $\{x_j\}_{j=0}^{\infty}$ is said to be of *arithmetic growth* if for some positive integer k the inequalities

$$0 < \overline{\lim_{m \to \infty}} \| m^{-k}x_m \| < \infty$$

holds. Finally, $\{x_j\}_{j=0}^{\infty}$ is *oscillatory* if

$$0 < \overline{\lim_{m \to \infty}} \| x_m \| < \infty$$

The classification of the solution $x_j = XT^j c$, $j = 0, 1, \ldots$ of (5.8.1) in terms of $c \in \mathbb{C}^{nl}$ is based on certain T-invariant subspaces. Let us introduce these subspaces. Denote by $\mathcal{R}^+(T)$ [resp. $\mathcal{R}^-(T)$] the sum of all root subspaces of T corresponding to the eigenvalues λ_0 of T with $|\lambda_0| > 1$ (resp. with $|\lambda_0| < 1$), and let $\mathcal{K}^1(T)$ be a direct complement to the subspace

$$\mathcal{K}^0(T) = \sum_{\substack{|\lambda_0| = 1 \\ \lambda_0 \in \sigma(T)}} \mathrm{Ker}(T - \lambda_0 I)$$

in the sum of all root substances of T corresponding to the eigenvalues λ_0 with $|\lambda_0| = 1$. Observe that $\mathcal{R}^+(T)$, $\mathcal{R}^-(T)$, and $\mathcal{K}^0(T)$ are T invariant. We have a direct sum decomposition

$$\mathbf{C}^{nl} = \mathcal{R}^{+}(T) \dotplus \mathcal{K}^{0}(T) \dotplus \mathcal{K}^{1}(T) \dotplus \mathcal{R}^{-}(T)$$

according to which every vector $c \in \mathbf{C}^{nl}$ will be represented as

$$c = c^{+} + c^{0} + c^{1} + c^{-}$$

Theorem 5.8.1

Let $\{x_i = XT^i c\}_{i=0}^{\infty}$ be a solution of (5.8.1). Then the solution is (a) of geometric growth if and only if $c^{+} \neq 0$; (b) of arithmetic growth if and only if $c^{+} = 0$, $c^{1} \neq 0$; (c) oscillatory if and only if $c^{+} = 0$, $c^{1} = 0$, $c^{0} \neq 0$; (d) of geometric decay if and only if $c^{+} = c^{1} = c^{0} = 0$, $c^{-} \neq 0$. In cases (a) and (d) the multiplier of $\{x_i\}_{i=0}^{\infty}$ is equal to the maximum of the absolute values of the eigenvalues λ_0 of T with the property that $P_{\lambda_0} c \neq 0$, where P_{λ_0} is the projector on $\mathcal{R}_{\lambda_0}(T)$ along $\Sigma_{\substack{\lambda \in \sigma(T) \\ \lambda \neq \lambda_0}} \mathcal{R}_{\lambda}(T)$.

The proof of Theorem 5.8.1 is similar to the proof of Theorem 5.7.2 if we first observe that the mth power of the Jordan block of size $k \times k$ with eigenvalue λ is

$$[J_k(\lambda)]^m = \begin{bmatrix} \lambda^m & \binom{m}{1}\lambda^{m-1} & \binom{m}{2}\lambda^{m-2} & \cdots & \binom{m}{k-1}\lambda^{m-k+1} \\ 0 & \lambda^m & \binom{m}{1}\lambda^{m-1} & \cdots & \binom{m}{k-2}\lambda^{m-k+2} \\ 0 & 0 & \lambda^m & \cdots & \binom{m}{k-3}\lambda^{m-k+3} \\ \vdots & \vdots & \vdots & & \vdots \\ 0 & 0 & 0 & \cdots & \lambda^m \end{bmatrix} \quad (5.8.3)$$

(It is assumed here that $\binom{m}{j} = 0$ if $j > m$.) This formula can be easily verified by induction on m.

The following result on existence of geometrically decaying solutions of equation (5.8.1) can be established using a proof similar to that of Theorem 5.7.3.

Theorem 5.8.2

For every set of k vectors $y_0, \ldots, y_{k-1}$ in $\mathbf{C}^n$ there exists a unique geometrically decaying solution $\{x_i\}_{i=0}^{\infty}$ with $x_0 = y_0, \ldots, x_{k-1} = y_{k-1}$ if and only if $L(\lambda)$ admits a factorization $L(\lambda) = L_2(\lambda)L_1(\lambda)$, where $L_2(\lambda)$ and $L_1(\lambda)$ are monic matrix polynomials of degrees $l - k$ and k, respectively, such that $|\lambda_0| < 1$ for every $\lambda_0 \in \sigma(L_1)$ and $|\lambda_0| \geq 1$ for every $\lambda_0 \in \sigma(L_2)$.

5.9   EXERCISES

5.1   For a monic $n \times n$ matrix polynomial $L(\lambda)$ of degree l, the pair of matrices (X, T), where X and T have sizes $n \times nl$ and $nl \times nl$, respectively, is called a *right standard pair* for $L(\lambda)$ if (X, T, Y) is a standard triple of $L(\lambda)$, for some $n \times n$ matrix Y.

(a)   Prove that a pair of matrices (X, T) of sizes $n \times nl$ and $nl \times nl$, respectively, is a right standard pair for a monic matrix polynomial $L(\lambda) = I\lambda^l + \sum_{j=0}^{l-1} A_j \lambda^j$ if and only if $\mathrm{col}[XT^i]_{i=0}^{l-1}$ is invertible and

$$A_0 X + \cdots + A_{l-1} XT^{l-1} + XT^l = 0$$

[*Hint*: The necessity follows from Proposition 5.1.4. To prove sufficiency, define

$$Y = (\mathrm{col}[XT^i]_{i=0}^{l-1})^{-1} \mathrm{col}[\delta_{il} I]_{i=1}^{l} \qquad (1)$$

and verify that (X, T, Y) is similar to the triple (P_1, C_1, R_1) from Proposition 5.1.2 with the similarity matrix $\mathrm{col}[XT^i]_{i=0}^{l-1}$.]

(b)   Show that given a right standard pair (X, T) of $L(\lambda)$, there exists a unique Y such that (X, T, Y) is a standard triple for $L(\lambda)$, and in fact Y is given by formula (1). [*Hint*: Use formula (5.1.11) for the similarity between the standard triple (X, T, Y) and the standard triple (P_1, C_1, R_1) from Proposition 5.1.2.]

5.2   A pair of matrices (T, Y) of sizes $nl \times nl$ and $nl \times n$, respectively, is called a *left standard pair* for the monic $n \times n$ matrix polynomial $L(\lambda)$ if for some $n \times nl$ matrix X the triple (X, T, Y) is a standard triple of $L(\lambda)$.

(a)   Prove that a pair of matrices (T, Y) of sizes $nl \times nl$ and $nl \times n$, respectively, is a left standard pair for $L(\lambda) = I\lambda^l + \sum_{j=0}^{l-1} A_j \lambda^j$ if and only if $[Y, TY, \ldots, T^{l-1}Y]$ is invertible and

$$YA_0 + TYA_1 + \cdots + T^{l-1}YA_{l-1} + T^l Y = 0$$

(b)   Show that given a left standard pair (T, Y) of $L(\lambda)$, there exists a unique X such that (X, T, Y) is a standard triple of $L(\lambda)$, and in fact

$$X = [0 \quad \cdots \quad 0 \quad I][Y, TY, \ldots, T^{l-1}Y]^{-1}$$

(c)   Prove that (T, Y) is a left standard pair for $L(\lambda) = I\lambda^l + \sum_{j=0}^{l-1} A_j \lambda^j$ if and only if (Y^*, T^*) is a right standard pair for the monic matrix polynomial $I\lambda^l + \sum_{j=0}^{l-1} A_j^* \lambda^j$.

5.3   Let $L(\lambda) = \lambda^l + \Sigma_{i=0}^{l-1} a_i \lambda^i$ be a scalar polynomial with l distinct zeros $\lambda_1, \ldots, \lambda_l$.

(a)   Show that

$$(X, T) = ([1 \quad 1 \quad \cdots \quad 1], \text{ diag}[\lambda_1, \ldots, \lambda_l])$$

is a right standard pair for $L(\lambda)$. Find Y such that (X, T, Y) is a standard triple for $L(\lambda)$.

(b)   Show that

$$(T, Y) = \text{ diag}[\lambda_1, \ldots, \lambda_l], \quad \begin{bmatrix} 1 \\ 1 \\ \vdots \\ 1 \end{bmatrix}$$

is a left standard pair for $L(\lambda)$, and find X such that (X, T, Y) is a standard triple for $L(\lambda)$.

5.4   Let $L(\lambda) = (\lambda - \lambda_0)^l$ be a scalar polynomial. Show that $([1 \quad 0 \quad \cdots \quad 0], J_l(\lambda_0))$ is a right standard pair of $L(\lambda)$ and that $(J_l(\lambda_0), \text{col}[\delta_{il}]_{i=1}^l)$ is a left standard pair for $L(\lambda)$. Find X and Y such that $([1 \quad 0 \quad \cdots \quad 0], J_l(\lambda_0), Y)$ and $(X, J_l(\lambda_0), \text{col}[\delta_{il}]_{i=1}^l)$ are standard triples for $L(\lambda)$.

5.5   Let $L(\lambda) = (\lambda - \lambda_1)^{l_1} \cdots (\lambda - \lambda_k)^{l_k}$ be a scalar polynomial, where $\lambda_1, \ldots, \lambda_k$ are distinct complex numbers. Show that

$$([X_1, \ldots, X_k], J_{l_1}(\lambda_1) \oplus \cdots \oplus J_{l_k}(\lambda_k))$$

and

$$\left(J_{l_1}(\lambda_1) \oplus \cdots \oplus J_{l_k}(\lambda_k), \quad \begin{bmatrix} Y_1 \\ Y_2 \\ \vdots \\ Y_k \end{bmatrix} \right)$$

are right and left standard pairs, respectively, of $L(\lambda)$, where $X_i = [1 \quad 0 \quad \cdots \quad 0]$ is an $1 \times l_i$ matrix and

$$Y_i = \begin{bmatrix} 0 \\ 0 \\ \vdots \\ 1 \end{bmatrix}$$

is an $l_i \times 1$ matrix.

5.6   Let

$$L(\lambda) = \begin{bmatrix} L_1(\lambda) & 0 \\ 0 & L_2(\lambda) \end{bmatrix}$$

be a monic matrix polynomial, and let (X_1, T_1, Y_1) and (X_2, T_2, Y_2) be standard triples for the polynomials $L_1(\lambda)$ and $L_2(\lambda)$, respectively. Find a standard triple for the polynomial $L(\lambda)$.

5.7   Given a standard triple for the polynomial $L(\lambda)$, find a standard triple for the polynomial $S^{-1}L(\lambda + \alpha)S$, where S is an invertible matrix, and α is a complex number.

5.8   Let (X, T, Y) be a standard triple for $L(\lambda)$. Show that

$$\left([X, 0], \begin{bmatrix} 0 & I \\ T & 0 \end{bmatrix}, \begin{bmatrix} 0 \\ Y \end{bmatrix} \right)$$

is a standard triple for the matrix polynomial $L(\lambda^2)$.

5.9   Given a standard triple for the matrix polynomial $L(\lambda)$, find a standard triple for the polynomial $L(p(\lambda))$, where $p(\lambda) = \lambda^m + \sum_{j=0}^{m-1} \lambda^j \alpha_j$ is a scalar polynomial.

5.10   Let

$$L(\lambda) = I\lambda^l + \sum_{j=0}^{l-1} A_j\lambda^j$$

be a 3×3 matrix polynomial whose coefficients are circulants:

$$A_k = \begin{bmatrix} a_k & b_k & c_k \\ c_k & a_k & b_k \\ b_k & c_k & a_k \end{bmatrix}, \qquad k = 0, 1, \ldots, l-1$$

(a_k, b_k, and c_k are complex numbers). Describe right and left standard pairs of $L(\lambda)$. [*Hint*: Find an invertible S such that $S^{-1}L(\lambda)S$ is diagonal and use the results of Exercises 5.5–5.7.]

5.11   Identify right and left standard pairs of a monic $n \times n$ matrix polynomial with circulant coefficients.

5.12   Using the right standard pair of a scalar polynomial given in Exercise 5.5, describe:

(a)   The solutions of differential equation

$$x^{(l)}(t) + \sum_{j=0}^{l-1} a_j x^{(j)}(t) = 0$$

where $a_0, \ldots, a_{l-1}$ are complex numbers;

(b)   The solutions of difference equations

$$x_{j+l} + a_{l-1}x_{j+l-1} + \cdots + a_1 x_{j+1} + a_0 x_j = 0, \qquad j = 0, 1, \ldots$$

5.13   Find the solution of the system of differential equations

$$\begin{cases} x^{(l)}(t) + \sum_{j=0}^{l-1} (a_j x^{(j)}(t) + b_j y^{(j)}(t)) = 0 \\ y^{(l)}(t) + \sum_{j=0}^{l-1} (b_j x^{(j)}(t) + a_j y^{(j)}(t)) = 0 \end{cases}$$

where $a_0, \ldots, a_{l-1}$ and $b_0, \ldots, b_{l-1}$ are complex numbers. When are all solutions exponentially decreasing? When does there exist a nonzero oscillatory solution?

5.14   Find the solutions of the system of difference equations

$$\begin{cases} x_{j+l} + \sum_{k=0}^{l-1} (a_k x_{j+k} + b_k y_{j+k}) = 0 \\ y_{j+l} + \sum_{k=0}^{l-1} (b_k y_{j+k} + a_k x_{j+k}) = 0 ; \qquad j = 0, 1, 2, \ldots \end{cases}$$

When do all nonzero solutions have geometric growth?

5.15   Find the supporting triinvariant decomposition $\mathcal{L} \dotplus \mathcal{M} \dotplus \{0\} = \mathbb{C}^l$ corresponding to the divisor $(\lambda - \lambda_1)^{\alpha_1} \cdots (\lambda - \lambda_k)^{\alpha_k}$ of the scalar polynomial $(\lambda - \lambda_1)^{\beta_1} \cdots (\lambda - \lambda_k)^{\beta_k}$ (here $\alpha_j \le \beta_j$, $j = 1, \ldots, k$, and α_j are nonnegative integers). Use the standard triple determined by the right standard pair described in Exercise 5.5.

5.16   Let $\lambda I - X_1$ and $\lambda I - X_2$ be linear $n \times n$ matrix polynomials such that the matrix $X_1 - X_2$ is invertible. Construct a monic $n \times n$ matrix polynomial of second degree with right divisors $\lambda I - X_1$ and $\lambda I - X_2$. [*Hint*: Look for a matrix polynomial with the standard pair $([I \ I], X_1 \oplus X_2)$.]

5.17   Let $L_1(\lambda)$ and $L_2(\lambda)$ be monic matrix polynomials with no partial multiplicities greater than 1. Show that the product $L_1(\lambda)L_2(\lambda)$ has no partial multiplicities greater than 2.

5.18   State and prove a generalization of the preceding exercise for the product of k monic matrix polynomials with no partial multiplicities greater than 1.

5.19   Show that a monic $n \times n$ matrix polynomial has not more than n partial multiplicities corresponding to any zero of its determinant. (*Hint*: Use Exercise 2.16.)

5.20   Prove that a monic $n \times n$ matrix polynomial of degree l with circulant coefficients has not more than l partial multiplicities corresponding to any zero of its determinant.

5.21   Describe all supporting triinvariant decompositions for the scalar polynomial $(\lambda - \lambda_0)^n$.

5.22   Given an $n \times n$ monic matrix polynomial $L(\lambda)$ of degree l, a C_L-invariant subspace $\mathscr{L}$ is called *supporting* if the direct sum decomposition $\mathscr{L} \dot{+} \mathscr{M} \dot{+} \{0\} = \mathbb{C}^{nl}$ is a supporting triinvariant decomposition with respect to the standard triple

$$[I \quad 0 \cdots \quad 0], \quad C_L, \quad \begin{bmatrix} 0 \\ \vdots \\ 0 \\ I \end{bmatrix}$$

Find all supporting subspaces for the scalar polynomial

$$(\lambda - \lambda_i)^{k_1}(\lambda - \lambda_2)^{k_2}$$

5.23   Find all supporting subspaces for the scalar polynomial

$$(\lambda - \lambda_1)^{k_1} \cdots (\lambda - \lambda_r)^{k_r}$$

5.24   Prove that for a scalar monic polynomial $L(\lambda)$, every C_L-invariant subspace is supporting.

5.25   Describe all supporting subspaces for a monic matrix polynomial whose coefficients are circulant matrices, that is, matrices of type

$$\begin{bmatrix} a_1 & a_2 & \cdots & a_n \\ a_n & a_1 & \cdots & a_{n-1} \\ \vdots & \vdots & & \vdots \\ a_2 & a_3 & \cdots & a_1 \end{bmatrix}, \quad a_j \in \mathbb{C}$$

5.26   Give an example of a monic matrix polynomial of second degree with nondiagonable companion matrix that admits factorization into linear factors.

5.27   Prove the following extension of Theorem 5.6.2 for polynomials of second degree. Let $L(\lambda)$ be a monic $n \times n$ matrix polynomial of second degree such that its companion matrix has at least $2n - 1$ blocks in its Jordan form. Then $L(\lambda)$ admits a factorization into linear factors $(\lambda I - A_1)(\lambda I - A_2)$. [*Hint*: Let (X, J) be a right standard pair of $L(\lambda)$ with J in the Jordan form. Arguing by contradiction, assume that every n columns of X formed by the eigenvectors of $L(\lambda)$ are linearly dependent. Then the columns in $\begin{bmatrix} X \\ XJ \end{bmatrix}$ that correspond to the eigenvectors of $L(\lambda)$ are linearly dependent, and this contradicts the invertibility of $\begin{bmatrix} X \\ XJ \end{bmatrix}$.]

5.28   A factorization $L(\lambda) = L_2(\lambda)L_3(\lambda)$ of a monic matrix polynomial $L(\lambda)$ is called *spectral* if $\det L_2(\lambda)$ and $\det L_3(\lambda)$ have no common zeros. Show that the factorization is spectral if and only if in the corresponding triinvariant decomposition $\mathscr{L} \dotplus \mathscr{M} \dotplus \{0\} = \math{C}^{nl}$ (Corollary 5.3.3) the T-invariant subspace $\mathscr{L}$ is spectral.

5.29   Prove or disprove the following statement: each monic matrix polynomial $L(\lambda)$ has a spectral factorization corresponding to every triinvariant decomposition $\mathscr{L} \dotplus \mathscr{M} \dotplus \{0\} = \math{C}^n$ with spectral T-invariant subspaces $\mathscr{L}$ and $\mathscr{M}$, where T is a linearization for $L(\lambda)$.

5.30   Let a_1, a_2, a_3, a_4 be distinct complex numbers, and let

$$L(\lambda) = \begin{bmatrix} (\lambda - a_1)(\lambda - a_2) & \lambda - a_1 \\ 0 & (\lambda - a_3)(\lambda - a_4) \end{bmatrix}$$

(a)   Show that

$$(X, T) = \left(\begin{bmatrix} 1 & 1 & (a_2 - a_3)^{-1} & (a_2 - a_4)^{-1} \\ 0 & 0 & 1 & 1 \end{bmatrix}, \ \mathrm{diag}[a_1, a_2, a_3, a_4] \right)$$

is a right standard pair for $L(\lambda)$.

(b)   Find Y such that (X, T, Y) is a standard triple for $L(\lambda)$.

(c)   Using the supporting triinvariant decomposition $\mathscr{L} \dotplus \mathscr{M} \dotplus \{0\} = \math{C}^4$ with spectral T-invariant subspace $\mathscr{L}$, find all spectral factorizations of $L(\lambda)$.

5.31   Let $M(\lambda)$ and $N(\lambda)$ be a monic matrix polynomials of sizes $n \times n$ and $m \times m$, respectively, and of the same degree l, and let

$$L(\lambda) = \begin{bmatrix} M(\lambda) & 0 \\ 0 & N(\lambda) \end{bmatrix}$$

be a direct sum of $M(\lambda)$ and $N(\lambda)$. Prove or disprove each of the following statements: (a) the monic matrix polynomials $L_1(\lambda)$ and $L_2(\lambda)$ in every factorization $L(\lambda) = L_1(\lambda)L_2(\lambda)$ are also direct sums; (b) same as (a) with the extra assumption that $M(\lambda)$ and $N(\lambda)$ do not have common eigenvalues.

5.32   Verify formula (5.8.3).

5.33   Supply the details for the proof of Theorem 5.8.1.

5.34   Prove Theorem 5.8.2.

Chapter Six

Invariant Subspaces For Transformations Between Different Spaces

We are now to generalize the notion of an invariant subspace for transformations from $\mathbb{C}^n$ into $\mathbb{C}^n$ in such a way that it will apply to transformations from $\mathbb{C}^{m+n}$ into $\mathbb{C}^n$, or from $\mathbb{C}^n$ into $\mathbb{C}^{m+n}$. The definitions introduced will have associated with them a natural generalization of similarity, called "block similarity", that will apply to transformations between different spaces. This will form an equivalence relation on the class of transformations between two given (generally different) spaces. A canonical form is developed for this similarity that is a generalization of the Jordan normal form. These ideas and results are then applied to the resolution of two spectral assignment problems. This really means analysis of the changes in spectra brought about by block similarity transformations.

Although this material is based on the theory of feedback in time-invariant linear systems, the presentation here is in the framework of linear algebra.

6.1 [A B]-INVARIANT SUBSPACES

Consider a transformation from $\mathbb{C}^{m+n}$ into $\mathbb{C}^n$. Our objective in this section is to develop and investigate a generalization of the notion of an invariant subspace that will apply to such transformations and that reduces to the familiar concept when $m = 0$. Let P be the projector on $\mathbb{C}^{m+n}$ that maps each vector onto the corresponding vector with zeros in the last m positions. We treat vectors of $\mathbb{C}^{m+n}$ in terms of their components in $\text{Im } P$ and

$\text{Im}(I - P)$, respectively, and, for $x = \langle x_1, \ldots, x_{m+n} \rangle \in \mathbb{C}^{m+n}$ we identify $Px = \langle x_1, \ldots, x_n, 0, \ldots, 0 \rangle$ with $\langle x_1, \ldots, x_n \rangle \in \mathbb{C}^n$. Then we may represent any $x \in \mathbb{C}^{m+n}$ as an ordered pair $(Px, (I - P)x)$ and, with respect to this decomposition, a transformation from $\mathbb{C}^{m+n}$ into $\mathbb{C}^n$ can be written in the block form $[A \ B]$ where $A: \mathbb{C}^n \to \mathbb{C}^n$ and $B: \mathbb{C}^m \to \mathbb{C}^n$. We also write $[A \ B]: \mathbb{C}^n \dotplus \mathbb{C}^m \to \mathbb{C}^n$.

A subspace $\mathcal{M}$ of $\mathbb{C}^n$ will be said to be $[A \ B]$ *invariant* if there is a subspace $\mathcal{S}$ of $\mathbb{C}^{m+n}$ with $\mathcal{M} = P\mathcal{S}$ and $[A \ B]\mathcal{S} \subset P\mathcal{S} = \mathcal{M}$. Of course, when $m = 0$, $P = I$, and this is interpreted as the familiar definition $A\mathcal{M} \subset \mathcal{M}$ for A invariance.

We now characterize this concept in different ways and, for this purpose, introduce another definition. Given a transformation $[A \ B]: \mathbb{C}^n \dotplus \mathbb{C}^m \to \mathbb{C}^n$, a transformation $T: \mathbb{C}^{m+n} \to \mathbb{C}^{m+n}$ is called an *extension* of $[A \ B]$ if it has the form

$$T = \begin{bmatrix} A & B \\ C & D \end{bmatrix}$$

for some transformations $C: \mathbb{C}^n \to \mathbb{C}^m$ and $D: \mathbb{C}^m \to \mathbb{C}^m$.

Theorem 6.1.1

Let $\mathcal{M}$ be a subspace of $\mathbb{C}^n$ and $[A \ B]$ be a transformation from $\mathbb{C}^{m+n}$ into $\mathbb{C}^n$. Then the following are equivalent: (a) $\mathcal{M}$ is $[A \ B]$ invariant; (b) there exists a subspace $\mathcal{S}$ of $\mathbb{C}^{m+n}$ with $\mathcal{M} = P\mathcal{S}$ and an extension of $[A \ B]$ under which $\mathcal{S}$ is invariant; (c) the subspace $\mathcal{M}$ satisfies

$$A\mathcal{M} \subset \mathcal{M} + \text{Im } B \qquad (6.1.1)$$

(d) there is a transformation $F: \mathbb{C}^n \to \mathbb{C}^m$ such that

$$(A + BF)\mathcal{M} \subset \mathcal{M} \qquad (6.1.2)$$

Proof. The theorem will be proved by verifying the implications (a) $\Rightarrow$ (d) $\Rightarrow$ (c) $\Rightarrow$ (b) $\Rightarrow$ (a).

(a) $\Rightarrow$ (d): Since $\mathcal{M}$ is $[A \ B]$ invariant, there is a subspace $\mathcal{S}$ of $\mathbb{C}^{m+n}$ with $\mathcal{M} = P\mathcal{S}$ and $[A \ B]\mathcal{S} \subset \mathcal{M}$. Let $x_1, \ldots, x_k$ be a basis for $\mathcal{M}$. Then there exist $z_1, \ldots, z_k \in \mathcal{S}$ such that $x_j = Pz_j$, $j = 1, 2, \ldots, k$. Define $y_j = (I - P)z_j \in \mathbb{C}^m$, $j = 1, 2, \ldots, k$ and then, since $\begin{bmatrix} x_j \\ y_j \end{bmatrix} \in \mathcal{S}$, $[A \ B]\mathcal{S} \subset \mathcal{M}$ implies that, for $j = 1, 2, \ldots, k$, $Ax_j + By_j \in \mathcal{M}$. Now define a transformation $F: \mathbb{C}^n \to \mathbb{C}^m$ by setting $Fx_j = y_j$ for $j = 1, \ldots, k$ and letting F be arbitrary on some direct complement to $\mathcal{M}$ in $\mathbb{C}^n$. Then for any $m = \sum_{j=1}^k \alpha_j x_j \in \mathcal{M}$ we have

$$(A + BF)m = \sum_{j=1}^k \alpha_j (Ax_j + By_j) \in \mathcal{M}$$

as required.

(d) $\Rightarrow$ (c): Given condition (6.1.2) we have, for any $x \in \mathcal{M}$

$$Ax = (A + BF)x - BFx \in \mathcal{M} + \text{Im } B$$

and (6.1.1) follows.

(c) $\Rightarrow$ (b): Let $x_1, \ldots, x_k$ be a basis for $\mathcal{M}$ and, using formula (6.1.1), let $y_1, \ldots, y_k$ be vectors in $\mathbb{C}^m$ for which $Ax_j + By_j \in \mathcal{M}$ for $j = 1, 2, \ldots, k$. Define a transformation $H: \mathbb{C}^n \to \mathbb{C}^m$ by means of the relation $Hx_j = y_j$, $j = 1, 2, \ldots, k$ and letting H be arbitrary on some direct complement to $\mathcal{M}$ in $\mathbb{C}^n$. Then define the subspace $\mathcal{S}$ of $\mathbb{C}^{m+n}$ by

$$\mathcal{S} = \left\{ \begin{bmatrix} m \\ Hm \end{bmatrix} \,\middle|\, m \in \mathcal{M} \right\}$$

and note our construction ensures that $(A + BH)m \in \mathcal{M}$ for any $m \in \mathcal{M}$. Consider the extension $\begin{bmatrix} A & B \\ HA & HB \end{bmatrix}$ of $[A\ B]$. It is easily verified that $\mathcal{S}$ is invariant under this extension.

(b) $\Rightarrow$ (a): This follows immediately from the definitions. $\square$

We will find the next simple corollary useful.

Corollary 6.1.2

With the notation of Theorem 6.1.1, if $\mathcal{M}$ is $[A\ B]$ invariant, then for any transformation $F: \mathbb{C}^n \to \mathbb{C}^m$, $\mathcal{M}$ is $[A + BF\ B]$ invariant.

Proof. We use the equivalence of statements (a) and (d) of the theorem. The fact that $\mathcal{M}$ is $[A\ B]$ invariant implies the existence of an $F_0: \mathbb{C}^m \to \mathbb{C}^m$ such that $\mathcal{M}$ is $(A + BF_0)$ invariant. Thus, for any $F: \mathbb{C}^n \to \mathbb{C}^m$, $\mathcal{M}$ is invariant under $A + BF + B(F_0 - F)$. Consequently, $\mathcal{M}$ is $[A + BF\ B]$ invariant. $\square$

Subspaces characterized by equation (6.1.1) are described in more geometric terms by replacement of Im B by some subspace $\mathcal{V}$ of $\mathbb{C}^n$. In this context it is useful to describe a subspace $\mathcal{M}$ as A *invariant* (mod $\mathcal{V}$) if

$$A\mathcal{M} \subset \mathcal{M} + \mathcal{V}$$

When $\mathcal{V} = \{0\}$ a subspace is A invariant (mod $\mathcal{V}$) if and only if it is A invariant. At the other extreme, when $\mathcal{V} = \mathbb{C}^n$, every subspace is A invariant (mod $\mathcal{V}$).

For a given transformation $A: \mathbb{C}^n \to \mathbb{C}^n$ and a subspace $\mathcal{V}$ of $\mathbb{C}^n$, consider the class of all subspaces that are A invariant (mod $\mathcal{V}$). It is easy to see that this class is closed under addition of subspaces, but is *not* closed under intersection. This is illustrated in the next example. We observe that (reverting to the language of transformations), although the set of all A-invariant subspaces form a lattice, the same is not generally true for the set of all $[A \ B]$-invariant subspaces.

EXAMPLE 6.1.1.   Let $A: \mathbb{C}^3 \to \mathbb{C}^3$ be defined by linearity and the equalities $Ae_1 = e_2$, $Ae_2 = e_1$, $Ae_3 = e_1$. Let $\mathcal{V} = \operatorname{Span}\{e_2 + e_3\}$. The subspaces $\operatorname{Span}\{e_1, e_2\}$ and $\operatorname{Span}\{e_1, e_3\}$ are both A invariant (mod $\mathcal{V}$). (The subspace $\operatorname{Span}\{e_1, e_2\}$ is actually A invariant.) However, their intersection $\operatorname{Span}\{e_1\}$ is not A invariant (mod $\mathcal{V}$). Indeed, $Ae_1 = e_2 \not\subseteq \operatorname{Span}\{e_1\} + \operatorname{Span}\{e_2 + e_3\}$.   □

Given A and $\mathcal{V}$ as above, it is natural to look for a "largest" subspace among all of those that are A invariant (mod $\mathcal{V}$). More generally (cf. Section 2.7), given a subspace $\mathcal{M}$ of $\mathbb{C}^n$, a subspace $\mathcal{U}$ of $\mathcal{M}$ that is A invariant (mod $\mathcal{V}$), is said to be *maximal in $\mathcal{M}$* if $\mathcal{U}$ contains all other subspaces of $\mathcal{M}$ that are A invariant (mod $\mathcal{V}$).

Proposition 6.1.3

For every subspace $\mathcal{M} \subset \mathbb{C}^n$ there is a unique subspace of $\mathbb{C}^n$ that is A invariant (mod $\mathcal{V}$) and maximal in $\mathcal{M}$.

Proof.   Let $\mathcal{U}$ be the sum of all subspaces that are A invariant (mod $\mathcal{V}$) and are contained in $\mathcal{M}$. Because of the finite dimension of $\mathcal{M}$, $\mathcal{U}$ is in fact the sum of a finite number of such subspaces. Consequently, $\mathcal{U}$ is itself A invariant (mod $\mathcal{V}$) and thus maximal in $\mathcal{M}$. The uniqueness is clear from the definition.   □

6.2   BLOCK SIMILARITY

In the preceding section the idea of $[A \ B]$-invariant subspaces has been developed where $[A \ B]$ is viewed as a transformation from $\mathbb{C}^n \dotplus \mathbb{C}^m$ into $\mathbb{C}^n$. We must also consider transformations of the other kind, namely, those acting from $\mathbb{C}^n$ to $\mathbb{C}^n \dotplus \mathbb{C}^m$. Such transformations can be written in the form $\begin{bmatrix} A \\ C \end{bmatrix}$ where $A: \mathbb{C}^n \to \mathbb{C}^n$ and $C: \mathbb{C}^n \to \mathbb{C}^m$. For these transformations, we need a dual concept of $\begin{bmatrix} A \\ C \end{bmatrix}$-invariant subspaces where $\begin{bmatrix} A \\ C \end{bmatrix}$ is viewed as a transformation from $\mathbb{C}^n$ into $\mathbb{C}^n \dotplus \mathbb{C}^m$. Thus, guided by Proposition 1.4.4, it is natural to define a subspace $\mathcal{M}$ of $\mathbb{C}^n$ is $\begin{bmatrix} A \\ C \end{bmatrix}$ invariant if and only if $\mathcal{M}^\perp$ is $[A^* \ C^*]$ invariant in the sense of Section 6.1. We develop this idea in Section 6.6. The purpose of this section is to generalize the notion of similarity to transformations $[A \ B]$ and $\begin{bmatrix} A \\ C \end{bmatrix}$ in a way that will be consistent with the definitions of these generalized invariant subspaces.

Let us begin with similarity for transformations $\begin{bmatrix} A \\ C \end{bmatrix}$ from $\mathbb{C}^n$ into $\mathbb{C}^n \dotplus \mathbb{C}^m$. In this case it is natural to say that a transformation $\begin{bmatrix} A_1 \\ C_1 \end{bmatrix}$ is similar to $\begin{bmatrix} A \\ C \end{bmatrix}$ if there is an invertible transformation S on $\mathbb{C}^n \dotplus \mathbb{C}^m$ such that

$$\begin{bmatrix} A_1 \\ C_1 \end{bmatrix} = S \begin{bmatrix} A \\ C \end{bmatrix} (S_{|\mathbb{C}^n})^{-1} \tag{6.2.1}$$

and the additional assumption that $\mathbb{C}^n$ is S invariant. Thus $S|_{\mathbb{C}^n}$ defines an invertible transformation on $\mathbb{C}^n$—the space on which $\begin{bmatrix} A \\ C \end{bmatrix}$ acts. This means that, with respect to the decomposition $\mathbb{C}^{m+n} = \mathbb{C}^n \dotplus \mathbb{C}^m$, S has the representation

$$S = \begin{bmatrix} X & Z \\ 0 & Y \end{bmatrix}$$

where X, Y are invertible transformations on $\mathbb{C}^n$ and $\mathbb{C}^m$, respectively. The formal definition is thus as follows: transformations $\begin{bmatrix} A_1 \\ C_1 \end{bmatrix}$ and $\begin{bmatrix} A_2 \\ C_2 \end{bmatrix}$ from $\mathbb{C}^n$ into $\mathbb{C}^n \dotplus \mathbb{C}^m$ are said to be *block similar* if there is an invertible transformation

$$S = \begin{bmatrix} X & Z \\ 0 & Y \end{bmatrix} : \mathbb{C}^n \dotplus \mathbb{C}^m \rightarrow \mathbb{C}^n \dotplus \mathbb{C}^m$$

such that

$$\begin{bmatrix} A_2 \\ C_2 \end{bmatrix} = S \begin{bmatrix} A_1 \\ C_1 \end{bmatrix} X^{-1} \tag{6.2.2}$$

Going to the adjoint transformations, this leads us to the dual definition: transformations $[A_1 \ B_1]$ and $[A_2 \ B_2]$ from $\mathbb{C}^n \dotplus \mathbb{C}^m$ into $\mathbb{C}^n$ are said to be *block similar* if there is an invertible transformation

$$S = \begin{bmatrix} N & 0 \\ L & M \end{bmatrix} : \mathbb{C}^n \dotplus \mathbb{C}^m \rightarrow \mathbb{C}^n \dotplus \mathbb{C}^m \tag{6.2.3}$$

such that

$$[A_2 \ B_2] = N^{-1}[A_1 \ B_1] \begin{bmatrix} N & 0 \\ L & M \end{bmatrix} \tag{6.2.4}$$

Now let us describe block-similar pairs $[A_1 \ B_1]$ and $[A_2 \ B_2]$ in two other ways.

Theorem 6.2.1

Let $[A_1 \ B_1]$ and $[A_2 \ B_2]$ be transformations from $\mathbb{C}^n \dotplus \mathbb{C}^m$ into $\mathbb{C}^n$. Then the following statements are equivalent: (a) $[A_1 \ B_1]$ and $[A_2 \ B_2]$ are block similar; (b) there exist invertible transformations N and M on $\mathbb{C}^n$ and $\mathbb{C}^m$, respectively, and a transformation $F: \mathbb{C}^n \to \mathbb{C}^m$ such that

$$A_2 = N^{-1}(A_1 + B_1F)N \qquad \text{and} \qquad B_2 = N^{-1}B_1M \qquad (6.2.5)$$

(c) for any extension T_1 of $[A_1 \ B_1]$ there is an extension T_2 of $[A_2 \ B_2]$ and a triangular invertible transformation S of the form (6.2.3) for which $T_1 = ST_2S^{-1}$.

Proof. Given statement (a) and, hence, equation (6.2.4), let $F = LN^{-1}$, and it is found immediately that equation (6.2.4) implies the relations (6.2.5). So (a) $\Rightarrow$ (b).

Given statement (b), define S as in (6.2.3), let $L = FN$, and let

$$T_1 = \begin{bmatrix} A_1 & B_1 \\ C_1 & D_1 \end{bmatrix} \qquad (6.2.6)$$

be an extension of $[A_1 \ B_1]$. Then it is easily verified that $S^{-1}T_1S$ is an extension of $[A_2 \ B_2]$, and statement (c) follows.

Finally, statement (c) implies that for any extension T_1 of $[A_1 \ B_1]$ [as in (6.2.6)] there is an extension T_2 of $[A_2 \ B_2]$ such that $T_2 = S^{-1}T_1S$ with S as in (6.2.3). This immediately implies equation (6.2.4). Thus (c) $\Rightarrow$ (a). $\square$

Corollary 6.2.2

Let $[A_1 \ B_1]$ and $[A_2 \ B_2]$ be block-similar transformations with transforming matrix S given by [6.2.3]. Then $\mathcal{M}$ is an $[A_1 \ B_1]$-invariant subspace if and only if $N^{-1}\mathcal{M}$ is an $[A_2 \ B_2]$-invariant subspace.

Proof. Assume that $\mathcal{M}$ is $[A_1 \ B_1]$ invariant. By Theorem 6.1.1 there is an extension T_1 of $[A_1 \ B_1]$ and a subspace $\mathcal{S}$ such that $\mathcal{M} = P\mathcal{S}$ and $T_1\mathcal{S} \subset \mathcal{S}$. Since $T_1 = ST_2S^{-1}$, $T_2(S^{-1}\mathcal{S}) \subset (S^{-1}\mathcal{S})$. But also, using (6.2.3), $P(S^{-1}\mathcal{S}) = N^{-1}P\mathcal{S} = N^{-1}\mathcal{M}$. Hence $[A_2 \ B_2](S^{-1}\mathcal{S}) \subset N^{-1}\mathcal{M}$ and, by definition, $N^{-1}\mathcal{M}$ is $[A_2 \ B_2]$ invariant.

If we are given that $N^{-1}\mathcal{M}$ is $[A_2 \ B_2]$ invariant, it follows from $T_2 = S^{-1}T_1S$ that $\mathcal{M}$ is $[A_2 \ B_1]$ invariant. $\square$

Corollary 6.2.3

If transformations $[A_1 \ B_1]$ and $[A_2 \ B_2]$ are block similar, they have the same rank.

Proof. Let $[A_1 \ B_1]$ and $[A_2 \ B_2]$ be block similar. Then Theorem 6.2.1 implies that

$$[A_2 \quad B_2] = N^{-1}[(A_1 + B_1F)N \quad B_1M]$$

Writing $G = FNM^{-1}$, we see that

$$\mathrm{rank}[A_2 \quad B_2] = \mathrm{rank}[A_1 + B_1G \quad B_1]$$

But it is easily verified that $\mathrm{Im}[A_1 + B_1G \quad B_1] = \mathrm{Im}[A_1 \quad B_1]$, and so $\mathrm{rank}[A_2 \quad B_2] = \mathrm{rank}[A_1 \quad B_1]$. □

By use of the characterizations of block-similar transformations developed in Theorem 6.2.1, it is easily verified that block similarity determines an equivalence relation on the class of all transformations from $\mathbb{C}^n \dotplus \mathbb{C}^m$ into $\mathbb{C}^n$. This immediately raises the problem of finding a canonical form for representations of the transformations in the equivalence classes determined by this relation. The rest of this section is devoted to the derivation of such a form. It will, of course, be a generalization of (and so be more complicated than) the Jordan normal form, which is associated with similarity of transformations in the usual sense, and which appears herein as Theorem 2.2.1.

Our argument will make use of the Kronecker canonical form for linear matrix polynomials under strict equivalence, as developed in the appendix. The following proposition is an important step in the argument. Note that it is convenient to work with matrices here. The previous analysis applies, of course, when they are viewed as transformations in the natural way.

Proposition 6.2.4

Let A_1 and A_2 be $n \times n$ matrices and B_1 and B_2 be $n \times m$ matrices. Then $[A_1 \quad B_1]$ and $[A_2 \quad B_2]$ are block similar if and only if the linear matrix polynomials $[I\lambda + A_1 \quad B_1]$ and $[I\lambda + A_2 \quad B_2]$ are strictly equivalent, that is, there exist invertible matrices S and T such that

$$S[I\lambda + A_1 \quad B_1]T = [I\lambda + A_2 \quad B_2] \tag{6.2.7}$$

Proof. Assume that (6.2.7) holds and write

$$T = \begin{bmatrix} T_{11} & T_{12} \\ T_{21} & T_{22} \end{bmatrix}$$

where T_{11} is $n \times n$. Then

$$S(I\lambda + A_1)T_{11} + SB_1T_{21} = I\lambda + A_2$$

Hence $T_{11} = S^{-1}$, and

$$S(A_1 + B_1T_{21}S)S^{-1} = A_2 \tag{6.2.8}$$

Equation (6.2.7) also implies that

$$S(I\lambda + A_1)T_{12} + SB_1T_{22} = B_2$$

It follows that $T_{12} = 0$ and then that $SB_1T_{22} = B_2$. Combining this relation with (6.2.5), it follows from Theorem 6.2.1 that $[A_1 \ B_1]$ and $[A_2 \ B_2]$ are block-similar.

Conversely, suppose that the relations (6.2.5) hold for appropriate N, M and F. Then (6.2.7) holds with $S = N^{-1}$, $T_{11} = N$, $T_{12} = 0$, $T_{21} = FN^{-1}$, and $T_{22} = M$. $\square$

Now we are ready to state and prove a result giving a canonical form for block-similar transformations and known as the *Brunovsky canonical form*. In the statement of the theorem $J_k(\lambda)$ will, as usual, denote the $k \times k$ Jordan block with eigenvalue λ.

Theorem 6.2.5

Given a transformation $[A \ B]: \mathbb{C}^n \dotplus \mathbb{C}^m \to \mathbb{C}^n$, there is a block-similar transformation $[A_0 \ B_0]$ that (in some bases for $\mathbb{C}^n$ and $\mathbb{C}^m$) has the representation

$$A_0 = J_{k_1}(0) \oplus \cdots \oplus J_{k_p}(0) \oplus J_{l_1}(\lambda_1) \oplus \cdots \oplus J_{l_q}(\lambda_q) \qquad (6.2.9)$$

for some integers $k_1 \ge \cdots \ge k_p > 0$ and all entries in B_0 are zero except for those in positions $(k_1, 1)$, $(k_1 + k_2, 2)$, ..., $(k_1 + \cdots + k_p, p)$, and these exceptional entries are equal to one. Moreover, the matrices A_0 and B_0 defined in this way are uniquely determined by $[A \ B]$, apart from a permutation of the blocks $J_{l_1}(\lambda_1)$, ..., $J_{l_q}(\lambda_q)$ in (6.2.9).

Thus the pair of matrices A_0, B_0 or the block matrix $[A_0 \ B_0]$ may be seen as making up the Brunovsky canonical form for the transformation $[A \ B]$. It will be convenient to call the matrix $J_{k_1}(0) \oplus \cdots \oplus J_{k_p}(0)$ the *Kronecker part* of A_0 and the integers $k_1, \ldots, k_p$ the *Kronecker indices* of $[A \ B]$. Similarly, we call $J_{l_1}(\lambda_1) \oplus \cdots \oplus J_{l_q}(\lambda_q)$ the *Jordan part* of A_0 and $l_1, \ldots, l_q$ the *Jordan indices* of $[A \ B]$.

Proof. We use the terminology and results of the appendix to this book. We may consider A and B to be $n \times n$ and $n \times m$ matrices, respectively. Consider the linear matrix polynomial

$$C(\lambda) = [\lambda I + A, \ B]$$

of size $n \times (n + m)$. As the equation

$$x(\lambda)^T C(\lambda) = 0^T, \qquad \lambda \in \mathbb{C}$$

has no nontrivial polynomial solution $x(\lambda)$, the minimal row indices of $C(\lambda)$

are absent. Further, the polynomial $\lambda C(\lambda^{-1}) = [I + \lambda A, \lambda B]$ obviously has no elementary divisors at zero, so $C(\lambda)$ has no elementary divisors at infinity. Let $k_1, \ldots, k_p$ be the minimal column indices of $C(\lambda)$ and $(\lambda + \lambda_1)^{l_1}, \ldots, (\lambda + \lambda_q)^{l_q}$ be the elementary divisors of $C(\lambda)$. Then Theorem A.7.3 ensures that $C(\lambda)$ is strictly equivalent to the linear matrix polynomial

$$[L_{k_1} \oplus \cdots \oplus L_{k_p} \oplus (\lambda I + J_{l_1}(\lambda_1)) \oplus \cdots \oplus (\lambda I + J_{l_q}(\lambda_q)), 0_{n \times s}] \qquad (6.2.10)$$

where L_k is the $k \times (k + 1)$ matrix

$$\begin{bmatrix} \lambda & 1 & 0 & \cdots & 0 & 0 \\ 0 & \lambda & 1 & \cdots & 0 & 0 \\ \vdots & \vdots & \vdots & & \vdots & \vdots \\ 0 & 0 & 0 & \cdots & \lambda & 1 \end{bmatrix}$$

and $s = \max_{\lambda \in \mathbb{C}} (\text{rank } C(\lambda)) - n$ [and we have used the elementary fact that $-J_l(\lambda_0)$ and $J_l(-\lambda_0)$ are similar]. After a permutation of columns the polynomial (6.2.10) becomes $[I\lambda + A_0 \; B_0]$ with A_0 and B_0 as defined in the statement of the theorem. The theorem itself now follows in view of Proposition 6.2.4. □

6.3 ANALYSIS OF THE BRUNOVSKY CANONICAL FORM

We first draw attention to an important special case of Theorem 6.2.5. This concerns transformations $[A \; B]: \mathbb{C}^{m+n} \to \mathbb{C}^n$ in which the pair (A, B) is a full-range pair in the sense defined in Section 2.8. That is, when

$$\sum_{j=0}^{\infty} \text{Im}(A^j B) = \sum_{j=0}^{p-1} \text{Im}(A^j B) = \mathbb{C}^n$$

where p is the degree of a minimal polynomial for A.

The following lemma will be useful.

Lemma 6.3.1

Consider any transformations $[A \; B]: \mathbb{C}^{m+n}$ and $F: \mathbb{C}^n \to \mathbb{C}^m$. For $s = 0, 1, 2, \ldots$ we have

$$\sum_{j=0}^{s} \text{Im}(A^j B) = \sum_{j=0}^{s} \text{Im}(A + BF)^j B \qquad (6.3.1)$$

Proof. The proof is by induction on s. When $s = 0$, equation (6.3.1) is trivially true. Using a binomial expansion it is found that

$$\text{Im } A(A + BF)^{r-1}B = A \text{ Im}((A + BF)^{r-1}B)$$
$$\subset A \text{ Im}[B \quad AB \quad \cdots \quad A^{r-1}B]$$
$$\subset \text{Im}[AB \quad A^2B \quad \cdots \quad A^rB]$$

Hence

$$\text{Im}(A + BF)^rB \subset \text{Im } A(A + BF)^{r-1}B + \text{Im } BF(A + BF)^{r-1}B$$
$$\subset \text{Im}[AB \quad \cdots \quad A^rB] + \text{Im } B$$
$$= \text{Im}[B \quad AB \quad \cdots \quad A^rB]$$

Assuming that the relation (6.3.1) holds when $s = r - 1$, this implies that the right-hand side of (6.3.1) is contained in the left-hand side. But the opposite inclusion follows from that already proved on replacing A by $A - BF$. $\square$

We now formulate other characterizations of full-range pairs (A, B).

Theorem 6.3.2

For a transformation $[A \quad B]: \mathbb{C}^{m+n} \to \mathbb{C}^n$ the following statements are equivalent: (a) the pair (A, B) is a full-range pair; (b) there is a full-range pair (A_1, B_1) for which $[A_1 \quad B_1]$ and $[A \quad B]$ are block-similar; (c) in the Brunovsky form $[A_0 \quad B_0]$ for $[A \quad B]$, the matrix A_0 has no Jordan part; (d) the rank of the transformation $[I\lambda + A \quad B]$ does not depend on the complex parameter λ.

Proof. Consider statement (b). If $[A_1 \quad B_1]$ and $[A \quad B]$ are block-similar, then, by Theorem 6.2.1, there are invertible transformations N, M and a transformation F such that

$$A_1 = N^{-1}(A + BF)N, \qquad B_1 = N^{-1}BM$$

Thus $A_1^j B_1 = N^{-1}(A + BF)^j M$. From the definition of full-range pairs and Lemma 6.3.1 it follows that (A, B) is a full-range pair. So (a) and (b) are equivalent.

Now consider a canonical pair (A_0, B_0) as defined in Theorem 6.2.5. It is easily verified that such a pair is a full-range pair if and only if the Jordan part of A_0 is absent. Since $[A \quad B]$ is block-similar to a canonical pair $[A_0 \quad B_0]$ (by Theorem 6.2.5), the equivalence of (a) and (c) follows from the equivalence of (a) and (b).

Consider condition (d). It follows from Corollary 6.2.3 that the rank of $[I\lambda + A \quad B]$ for any $\lambda \in \mathbb{C}$ is just that of $[I\lambda + A_0 \quad B_0]$ where $[A_0 \quad B_0]$ is a Brunovsky form for $[A \quad B]$. A moment's examination of A_0 and B_0 convinces us that the rank of $[I\lambda + A_0 \ B_0]$ takes the same numerical value, except at the points $\lambda = -\lambda_j, j = 1, \ldots, q$, where there is a reduction in rank. Thus the rank of $[I\lambda + A \ B]$ is independent of λ if and only if there is no Jordan part in A_0, and the equivalence of (c) and (d) is proved. $\square$

So far, the discussion of this section has focussed on cases in which the matrix A_0 of a canonical pair (A_0, B_0) has no Jordan part. This can be described as the case $q = 0$ in equation (6.2.9). It is also possible that A_0 has no Kronecker part; the case $p = 0$ in equation (6.2.9). In this case $B_0 = 0$ as well. We return to this case in Section 6.6.

We conclude this section by showing that the Kronecker indices of the Brunovsky form can be determined directly from geometric properties of the transformation $[A \ B]$ without resort to the computation of the minimal column indices of $[I\lambda + A \ B]$.

Proposition 6.3.3

Let $[A \ B]$ be a transformation from $\mathbb{C}^{m+n}$ into $\mathbb{C}^n$ and define the sequence $d_{-1}, d_0, d_1, \ldots$ by $d_{-1} = 0$ and, for $s = 0, 1, \ldots$

$$d_s = \dim \sum_{j=0}^{s} \operatorname{Im}(A^j B) \tag{6.3.2}$$

Then the Kronecker indices $k_1, \ldots, k_p$ of $[A \ B]$ are determined by the relations

$$\{k_j \mid k_j > s\}^{\#} = d_s - d_{s-1} \tag{6.3.3}$$

Note that the sequence $d_{-1}, d_0, \ldots$ is ultimately constant and (if $B \neq 0$), is initially strictly increasing (see Section 2.8).

Proof. Use Theorems 6.2.1 and 6.2.5 to write

$$A^j B = N^{-1}(A_0 + B_0 F)^j B_0 M$$

where M and N are invertible and $[A_0 \ B_0]$ is block similar to $[A \ B]$. Now Lemma 6.3.1 implies

$$\sum_{j=0}^{s} \operatorname{Im}(A^j B) = N^{-1}\left(\sum_{j=0}^{s} \operatorname{Im}(A_0 + B_0 F)^j B_0\right) M = N^{-1}\left(\sum_{j=0}^{s} \operatorname{Im}(A_0^j B_0)\right) M$$

Consequently, the integers d_s defined by formula (6.3.2) are invariant under block similarity. Now formula (6.3.3) is easily verified for a canonical pair (A_0, B_0). $\square$

Note that the number of Kronecker indices p is given by equation (6.3.3) in the case $s = 0$. Thus

$$d_0 = p = \dim(\operatorname{Im} B) = \operatorname{rank} B$$

6.4   DESCRIPTION OF $[A\ B]$-INVARIANT SUBSPACES

In some special cases Theorem 6.2.5 can be used to describe explicitly all $[A\ B]$-invariant subspaces. We consider a primitive but important "full-range" case in this section.

Theorem 6.4.1

Let $[A\ B]$ be a transformation from $\mathbb{C}^{n+1}$ into $\mathbb{C}^n$ for which (A, B) is a full-range pair. Then there exists a basis $f_1, \dots, f_n$ in $\mathbb{C}^n$ such that every m-dimensional $[A\ B]$-invariant subspace $\mathcal{M} \neq \{0\}$ admits the description:

$$\mathcal{M} = \operatorname{Span}\left\{ \sum_{k=1}^{n} \lambda_j^{k-1} f_k, \sum_{k=1}^{n} \binom{k-1}{1} \lambda_j^{k-2} f_k, \dots \right.$$

$$\left. \dots, \sum_{k=1}^{n} \binom{k-1}{r_j-1} \lambda_j^{k-r_j} f_k ; \qquad j = 1, \dots, l \right\} \qquad (6.4.1)$$

where $r_1, \dots, r_l$ are positive integers with $r_1 + \cdots + r_l = m$ and $\lambda_1, \dots, \lambda_l$ are distinct complex numbers (as usual, $\binom{m}{p} = m! / [p!(m-p)]$ with the understanding that $0! = 1$ and that $\binom{m}{p} = 0$ for $m < p$). Conversely, every subspace $\mathcal{M} \subset \mathbb{C}^n$ of the form (6.4.1) is $[A\ B]$ invariant.

Proof. Taking advantage of the equivalence (a) ⇔ (b) in Theorem 6.3.2, we can assume that

$$A = \begin{bmatrix} 0 & 1 & 0 & \cdots & 0 \\ 0 & 0 & 1 & \cdots & 0 \\ \vdots & \vdots & \vdots & & \vdots \\ 0 & 0 & 0 & \cdots & 0 \end{bmatrix}, \qquad B = \begin{bmatrix} 0 \\ \vdots \\ 0 \\ 1 \end{bmatrix}$$

Let $\mathcal{M} \neq \{0\}$ be an $[A\ B]$-invariant subspace. Then, by Theorem 6.1.1, there exists a $1 \times n$ matrix $F = [a_0, \dots, a_{n-1}]$ such that $\mathcal{M}$ is invariant for the matrix

$$A + BF = \begin{bmatrix} 0 & 1 & 0 & \cdots & 0 \\ 0 & 0 & 1 & \cdots & 0 \\ \vdots & \vdots & \vdots & & \vdots \\ 0 & 0 & 0 & \cdots & 1 \\ a_0 & a_1 & a_2 & \cdots & a_{n-1} \end{bmatrix}$$

Let $r_1, \ldots, r_l$ be all the partial multiplicities of $(A + BF)|_{\mathcal{M}}$ (so $r_1 + \cdots + r_l = \dim \mathcal{M}$), and let $\lambda_1, \ldots, \lambda_l$ be the corresponding eigenvalues. For every $\lambda_0 \in \mathbb{C}$ the matrix $\lambda_0 I - (A + BF)$ has a nonsingular $(n - 1) \times (n - 1)$ submatrix (namely, that formed by the rows $1, 2, \ldots, n - 1$ and columns $2, 3, \ldots, n$). It follows that $\dim \operatorname{Ker}(\lambda_0 I - (A + BF)) = 1$ for every $\lambda_0 \in \sigma(A)$. So there is exactly one Jordan block in the Jordan form of $A + BF$ corresponding to each $\lambda_0 \in \sigma(A + BF)$. Hence the same property holds for $(A + BF)|_{\mathcal{M}}$, and the eigenvalues $\lambda_1, \ldots, \lambda_l$ must be distinct. It follows that in order to prove that $\mathcal{M}$ has the form (6.4.1), it will suffice to verify that for any Jordan chain $g_1, \ldots, g_r$ of $(A + BF)|_{\mathcal{M}}$ corresponding to λ_j we have

$$\operatorname{Span}\{g_1, \ldots, g_r\} = \operatorname{Span}\left\{ \sum_{k=1}^{n} \lambda_j^{k-1} e_k, \ldots, \sum_{k=1}^{n} \binom{k-1}{r-1} \lambda_j^{k-r} e_k \right\} \qquad (6.4.2)$$

Observe first that

$$\det(\lambda I - (A + BF)) = \lambda^n - a_{n-1}\lambda^{n-1} - \cdots - a_1 \lambda - a_0$$

and consequently λ_j is a zero of the polynomial $\lambda^n - a_{n-1}\lambda^{n-1} - \cdots - a_1\lambda - a_0$ of multiplicity at least r. Further, for $t = 1, 2, \ldots, r$

$$[(A + BF) - \lambda_j I] \sum_{k=1}^{n} \binom{k-1}{t-1} \lambda_j^{k-t} e_k = \sum_{k=1}^{n} \binom{k-1}{t-2} \lambda_j^{k+1-t} e_k \qquad (6.4.3)$$

(and the right-hand side is interpreted as zero for $t = 1$). Indeed, equality in the sth place ($s = 1, \ldots, n - 1$) on both sides of (6.4.3) follows from the easily verified combinatorial identity:

$$\binom{s}{t-1} - \binom{s-1}{t-1} = \binom{s-1}{t-2}$$

Equality in the nth place on both sides of (6.4.3) amounts to

$$\sum_{k=1}^{n} a_{k-1} \binom{k-1}{t-1} \lambda_j^{k-t} - \lambda_j \binom{n-1}{t-1} \lambda_j^{n-1} = \binom{n-1}{t-2} \lambda_j^{n+1-t}$$

or

$$-\sum_{k=1}^{n} a_{k-1} \binom{k-1}{t-1} \lambda_j^{k-t} + \binom{n}{t-1} \lambda_j^{n+1-t} = 0 \qquad (6.4.4)$$

but the left-hand side of this equation is just the $(t - 1)$th derivative of the polynomial $\lambda^n - a_{n-1}\lambda^{n-1} - \cdots - a_0$ evaluated at λ_j; so equation (6.4.4), and hence (6.4.3), is confirmed.

We have verified that the vectors $\sum_{k=1}^{n} \lambda_j^{k-1} e_k, \ldots, \sum_{k=1}^{n} \binom{k-1}{r-1} \lambda_j^{k-r} e_k$ form a Jordan chain of $A + BF$ corresponding to λ_j. As the restriction $(A + BF)|_{\mathcal{R}_{\lambda_j}(A+BF)}$ is unicellular, there exists a unique $(A + BF)$-

invariant subspace in $\mathcal{R}_{\lambda_j}(A + BF)$ of dimension r, and this subspace is spanned by the vectors in any Jordan chain of $(A + BF)$ of length r corresponding to λ_j. So (6.4.2) follows.

Conversely, let $\mathcal{M}$ be given by (6.4.1) (with f_k replaced by e_k, $k = 1, \ldots, n$). Let $f(\lambda) = \lambda^n - a_{n-1}\lambda^{n-1} - \cdots - a_0$ be a polynomial such that λ_j is a zero of $f(\lambda)$ of multiplicity of at least r_j, $j = 1, \ldots, l$. As we have seen above, the vectors $\sum_{k=1}^{n} \lambda_j^{k-1} e_k, \ldots, \sum_{k=1}^{n} \binom{k-1}{r_j - 1} \lambda_j^{k-r_j}$ form a Jordan chain of $A + BF$ corresponding to λ_j for $j = 1, \ldots, l$ (here $F = [a_0, a_1, \ldots, a_{n-1}]$). So by Theorem 6.1.1, $\mathcal{M}$ is $[A\ B]$ invariant. $\square$

The case $l = m$ in Theorem 6.4.1 deserves special attention.

Corollary 6.4.2

Let $[A\ B]$ be as in Theorem 6.4.1. Then there exists a basis $f_1, \ldots, f_n$ in $\mathbb{C}^n$ such that, for every m-tuple of distinct complex numbers $\lambda_1, \ldots, \lambda_m$, the m-dimensional subspace

$$\text{Span}\left\{ \sum_{k=1}^{n} \lambda_j^{k-1} f_k ; \qquad j = 1, \ldots, m \right\}$$

is $[A\ B]$ invariant.

This corollary shows that (at least in the case of a full-range pair $A: \mathbb{C}^n \to \mathbb{C}^n$ and $B: \mathbb{C} \to \mathbb{C}^n$) there are a lot of $[A\ B]$-invariant subspaces. Indeed, Corollary 6.4.2 shows the existence of a family of $[A\ B]$-invariant m-dimensional subspaces that depends on m complex parameters (namely, $\lambda_1, \ldots, \lambda_m$).

For the general case of a full-range pair we have the following partial description of $[A\ B]$-invariant subspaces.

Theorem 6.4.3

Let (A, B) be a full-range pair with Kronecker indices $k_1 \geq \cdots \geq k_r$. Then there exists a basis $f_{i1}, \ldots, f_{i,k_i}$, $i = 1, \ldots, r$ in $\mathbb{C}^n$ such that for every r-tuple of nonnegative integers $l_i, \ldots, l_r$ satisfying $l_i \leq k_i$, $i = 1, \ldots, r$, and for every collection $\{\lambda_{i1}, \ldots, \lambda_{i,l_i}; i = 1, \ldots, r\}$ of complex numbers the subspace

$$\sum_{i=1}^{r} \text{Span}\left\{ \sum_{p=1}^{n} \lambda_{ij}^{p-1} f_{ip} \mid j = 1, \ldots, l_i \right\}$$

is $[A\ B]$ invariant.

The proof of Theorem 6.4.3 is obtained by combining Theorem 6.3.2 and Corollary 6.4.2.

6.5 THE SPECTRAL ASSIGNMENT PROBLEM

For a transformation A on $\mathbb{C}^n$ the eigenvalues are invariant under similarity transformations. More generally, if A is defined by a transformation $[A \ B]\colon \mathbb{C}^n \dotplus \mathbb{C}^m \to \mathbb{C}^n$, then, by Theorem 6.2.1, block similarity transforms A into $N^{-1}(A + BF)N$ for some invertible N. Thus the eigenvalues of A are no longer invariant, but are transformed to those of $A + BF$, where F depends on the similarity. Now we ask, for given $[A \ B]$, what are the attainable eigenvalues of $A + BF$? We do not answer this question directly, but we present solutions to two closely related problems.

First, suppose that we are given n complex numbers $\lambda_1, \ldots, \lambda_n$ (possibly with repetitions) that are candidates for the eigenvalues of $A + BF$. Under what conditions on the transformation $[A \ B]$ does a transformation $F\colon \mathbb{C}^n \to \mathbb{C}^m$ exist such that the numbers $\lambda_1, \ldots, \lambda_n$ are just the eigenvalues of $A + BF$, counting algebraic multiplicities? This is known as the *spectral assignment problem*. It is important in its own right and is also relevant to our discussion of the stability of $[A \ B]$-invariant subspaces.

Clearly, when $B = 0$, the problem is not generally solvable. Another extreme case arises if $B = I$ when it is easily seen that a solution can always be found by using diagonable matrices F. We show first that the problem is always solvable as long as (A, B) is a full-range pair.

Theorem 6.5.1

Let $A\colon \mathbb{C}^n \to \mathbb{C}^n$, $B\colon \mathbb{C}^m \to \mathbb{C}^n$ be a full-range pair of transformations. Then for every n-tuple of complex numbers $\lambda_1, \ldots, \lambda_n$ there exists a transformation $F\colon \mathbb{C}^n \to \mathbb{C}^m$ such that $A + BF$ has eigenvalues $\lambda_1, \ldots, \lambda_n$.

Proof. With the use of Theorem 6.2.1 it is easily seen that we can assume, without loss of generality, that A and B are in Brunovsky canonical form. Furthermore, by Theorem 6.3.2, it follows that the Jordan part of A is absent [see equation (6.2.9)]. So the Kronecker indices $k_1, \ldots, k_p$ of $[A \ B]$ satisfy the condition $k_1 + \cdots + k_p = n$. For $j = 1, \ldots, p$ let

$$a_j(\lambda) = \lambda^{k_j} + \sum_{q=0}^{k_j - 1} c_{jq} \lambda^q$$

be the scalar polynomial with zeros $\lambda_{l_{j-1}+1}, \ldots, \lambda_{l_j}$, where $l_j = k_1 + \cdots + k_j$ (and we define $l_0 = 0$). Let

$$F = [F_1 \quad F_2 \quad \cdots \quad F_p]$$

where F_i is the $m \times k_i$ matrix whose ith row is $[-c_{i0}, -c_{i1}, \ldots, -c_{i,k_i-1}]$ and the other rows are zeros. Then

$$A + BF = \text{diag}[A_1, A_2, \ldots, A_p]$$

where

$$A_i = \begin{bmatrix} 0 & 1 & 0 & \cdots & 0 \\ 0 & 0 & 1 & \cdots & 0 \\ \vdots & \vdots & \vdots & & \vdots \\ 0 & 0 & 0 & & 1 \\ -c_{i0} & -c_{i1} & -c_{12} & \cdots & -c_{i,k_i-1} \end{bmatrix}$$

is a $k_i \times k_i$ matrix for $i = 1, \ldots, p$ [the companion matrix of $a_i(\cdot)$]. It is well known that the eigenvalues of A_i are exactly $\lambda_{l_{j-1}+1}, \ldots, \lambda_{l_j}$. This proves the theorem. $\square$

The argument used in proving Theorem 6.5.1 can also be utilized to obtain a full description of the solvable cases of the spectral assignment problem. We omit the details of the proof.

Theorem 6.5.2

Let $A: \mathbb{C}^n \to \mathbb{C}^n$ and $B: \mathbb{C}^m \to \mathbb{C}^n$ be a pair of transformations, and let the $l \times l$ matrix $J = J_{l_1}(\lambda_1) \oplus \cdots \oplus J_{l_q}(\lambda_q)$ be the Jordan part of the Brunovsky form for $[A\ B]$. Then, given an n-tuple of (not necessarily distinct) complex numbers $\mu_1, \ldots, \mu_n$, there exists a transformation $F: \mathbb{C}^n \to \mathbb{C}^m$ such that $A + BF$ has eigenvalues $\mu_1, \ldots, \mu_n$ if and only if at least l numbers among $\mu_1, \ldots, \mu_n$ coincide with the eigenvalues of J (counting multiplicities).

We need another version of the spectral assignment problem, known as the *spectral shifting problem*. Given a transformation $[A\ B]: \mathbb{C}^{m+n} \to \mathbb{C}^n$ and a nonempty set $\Omega \subset \mathbb{C}$, when does there exist a transformation $F: \mathbb{C}^n \to \mathbb{C}^m$ such that $\sigma(A + BF) \subset \Omega$? When (A, B) is a full-range pair, such an F always exists in view of Theorem 6.3.2. In general, the answer depends on the relationship between the root subspaces of A and the minimal A-invariant subspace over Im B:

$$\langle A \mid \text{Im } B \rangle \overset{\text{def}}{=} \text{Im } B + A(\text{Im } B) + \cdots + A^{n-1}(\text{Im } B) = \sum_{i=0}^{n-1} \text{Im}(A^i B)$$

[known as the "controllable subspace" of the pair (A, B) in the systems theory literature; see also Proposition 8.4.1]. Observe first that the subspace $\langle A \mid \text{Im } B \rangle$ is the minimal A-invariant subspace over Im B (see Theorem 2.8.4). In particular, $\langle A \mid \text{Im } B \rangle$ is A invariant. Also, equation (6.3.1) can be expressed in the form

$$\langle A \mid \text{Im } B \rangle = \langle A + BF \mid \text{Im } B \rangle \tag{6.5.1}$$

for any transformation $F: \mathbb{C}^n \to \mathbb{C}^m$.

Theorem 6.5.3

Given a nonempty set $\Omega \subset \mathbb{C}$ and a transformation $[A \; B]: \mathbb{C}^{m+n} \to \mathbb{C}^n$, there exists a transformation $F: \mathbb{C}^n \to \mathbb{C}^m$ such that $\sigma(A + BF) \subset \Omega$ if and only if

$$\mathcal{R}_{\lambda_0}(A) \subset \langle A \mid \operatorname{Im} B \rangle \tag{6.5.2}$$

for every eigenvalue λ_0 of A that does not belong to Ω.

Recall that $\mathcal{R}_{\lambda_0}(A) = \operatorname{Ker}(\lambda_0 I - A)^n$ is the root subspace of A corresponding to the eigenvalue λ_0 and, by definition, $\mathcal{R}_{\lambda_0}(A) = \{0\}$ if $\lambda_0 \notin \sigma(A)$.

In the proof we use the following basic fact about induced transformations in factor spaces. (Recall the definition of the induced transformation given in Section 1.7.)

Lemma 6.5.4

Let $X: \mathbb{C}^n \to \mathbb{C}^n$ be a transformation with an invariant subspace $\mathcal{L}$, and let $\bar{X}: \mathbb{C}^n/\mathcal{L} \to \mathbb{C}^n/\mathcal{L}$ be the induced transformation. Then for every $\lambda_0 \in \mathbb{C}$ we have

$$\mathcal{R}_{\lambda_0}(\bar{X}) = P\mathcal{R}_{\lambda_0}(X) \tag{6.5.3}$$

where $P: \mathbb{C}^n \to \mathbb{C}^n/\mathcal{L}$ is the canonical transformation: $Px = x + \mathcal{L}$, $x \in \mathbb{C}^n$. In particular, every eigenvalue of $\bar{X}$ is also an eigenvalue of X.

Proof. Let $p(\lambda) = (\lambda_0 - \lambda)^n$. Then for every $x \in \mathbb{C}^n$ with $Px \in \mathcal{R}_{\lambda_0}(\bar{X})$ we have

$$P(p(X)x) = p(\bar{X})(Px) = 0$$

So $p(X)x \in \mathcal{L}$. Let $q(\lambda) = \Pi_{j=1}^r (\lambda_j - \lambda)^n$, where $\lambda_1, \ldots, \lambda_r$ are all the eigenvalues of X different from λ_0. As $p(\lambda)$ and $q(\lambda)$ are polynomials with no common zeros, there exist polynomials $g(\lambda)$ and $h(\lambda)$ such that $g(\lambda)p(\lambda) + h(\lambda)q(\lambda) = 1$. (This is well known and is easily deduced from the Euclidean algorithm.) Hence

$$x = g(X)p(X)x + h(X)q(X)x \tag{6.5.4}$$

Since $p(X)x \in \mathcal{L}$, we also have $g(X)p(X)x \in \mathcal{L}$. On the other hand, the Cayley–Hamilton theorem ensures that $p(X)h(X)g(X)x = 0$, that is, the vector $u = h(X)g(X)x$ belongs to $\mathcal{R}_{\lambda_0}(X)$. Now equation (6.5.4) implies

$$Px = Pu \in P\mathcal{R}_{\lambda_0}(X)$$

We have proved the inclusion $\subset$ in equality (6.5.3). The opposite inclusion follows from the relation

$$P(p(X)y) = p(\bar{X})(Py)$$

for every vector $y \in \mathcomplex^n$. $\square$

Proof of Theorem 6.5.3.   First consider a pair (A_0, B_0) in the Brunovsky canonical form, as described in Theorem 6.2.5. Then

$$\langle A_0 \mid \operatorname{Im} B_0 \rangle = \operatorname{Span}\{e_j \mid j = 1, \ldots, k_1 + \cdots + k_p\}$$

The condition $\mathscr{R}_{\lambda_0}(A_0) \subset \langle A_0 \mid \operatorname{Im} B_0 \rangle$ for every $\lambda_0 \in \mathcomplex \smallsetminus \Omega$ means that [in the notation of equation (6.2.9)] $\lambda_1, \ldots, \lambda_q \in \Omega$. It remains to apply Theorem 6.5.2.

Now consider the general case, and let

$$A_0 = N^{-1}(A + BF_0)N; \qquad B_0 = N^{-1}BM$$

where $[A_0 \quad B_0]$ is in Brunovsky canonical form. It is easily seen that there exists a transformation F_1 such that $\sigma(A_0 + B_0F_1) \subset \Omega$ if and only if there exists an F_2 with $\sigma(A + BF_2) \subset \Omega$ (indeed, one can take $F_2 = F_0 + MF_1N^{-1}$). Further, using equation (6.5.1), we have

$$\langle A \mid \operatorname{Im} B \rangle = \langle A + BF_0 \mid \operatorname{Im} B \rangle = N\langle A_0 \mid \operatorname{Im} B_0 \rangle$$

and obviously, for any $\lambda_0 \in \mathcomplex$

$$\mathscr{R}_{\lambda_0}(A + BF_0) = \mathscr{N}\mathscr{R}_{\lambda_0}(A_0)$$

So it remains to show that (6.5.2) holds if and only if

$$\mathscr{R}_{\lambda_0}(A + BF_0) \subset \langle A \mid \operatorname{Im} B \rangle, \qquad \lambda_0 \in \mathcomplex \smallsetminus \Omega \tag{6.5.5}$$

This is done by using Lemma 6.5.4. Denote by $P \colon \mathcomplex^n \to \mathcomplex^n / \langle A \mid \operatorname{Im} B \rangle$ the canonical transformation

$$Px = x + \langle A \mid \operatorname{Im} B \rangle, \qquad x \in \mathcomplex^n$$

For a transformation $X \colon \mathcomplex^n \to \mathcomplex^n$ with invariant subspace $\langle A \mid \operatorname{Im} B \rangle$, let

$$\bar{X} \colon \mathcomplex^n / \langle A \mid \operatorname{Im} B \rangle \to \mathcomplex^n / \langle A \mid \operatorname{Im} B \rangle$$

be the induced transformation. Using (6.5.1), we see that $\bar{A}$ and $\overline{A + BF}$ are well defined. Further, for every $x \in \mathcomplex^n$

$$Ax = (A + BF_0)x - BFx \in (A + BF_0)x + \langle A \mid \mathrm{Im}\ B \rangle$$

so $\bar{A} = \overline{A + BF_0}$. Now, assuming that (6.5.5) holds, and in view of Lemma 6.5.4, we find that for every $\lambda_0 \in \mathbb{C} \smallsetminus \Omega$:

$$P\mathcal{R}_{\lambda_0}(A) = \mathcal{R}_{\lambda_0}(\bar{A}) = \mathcal{R}_{\lambda_0}(\overline{A + BF_0}) = P\mathcal{R}_{\lambda_0}(A + BF) \subset P\langle A \mid \mathrm{Im}\ B \rangle = \{0\}$$

Hence $\mathcal{R}_{\lambda_0}(A) \subset \langle A \mid \mathrm{Im}\ B \rangle$. A similar argument shows that (6.5.2) implies (6.5.5). □

6.6   SOME DUAL CONCEPTS

The definitions and analysis of this chapter have primarily concerned transformations $[A\ B]$: $\mathbb{C}^n \dotplus \mathbb{C}^m \to \mathbb{C}^n$. Questions arise concerning analogs for transformations $\begin{bmatrix} A \\ C \end{bmatrix}$: $\mathbb{C}^n \to \mathbb{C}^n \dotplus \mathbb{C}^m$. In this section we quickly review some notions and results in this direction. Recall first that a subspace $\mathcal{M}$ of $\mathbb{C}^n$ will be called $\begin{bmatrix} A \\ C \end{bmatrix}$ invariant if and only if $\mathcal{M}^\perp$ is $[A^*\ C^*]$ invariant. Thus, with the characterization (d) of Theorem 6.1.1 for $[A^*\ C^*]$-invariant subspaces, there is a transformation G^* such that

$$(A^* + C^*G^*)\mathcal{M}^\perp \subset \mathcal{M}^\perp \tag{6.6.1}$$

if and only if $\mathcal{M}$ is $\begin{bmatrix} A \\ C \end{bmatrix}$ invariant. Using Proposition 1.4.4, we see that this is equivalent to

$$(A + GC)\mathcal{M} \subset \mathcal{M}$$

We include this discussion as part of the following statement.

Theorem 6.6.1

Let $\mathcal{M}$ be a subspace of $\mathbb{C}^n$ and $\begin{bmatrix} A \\ C \end{bmatrix}$ be a transformation from $\mathbb{C}^n$ into $\mathbb{C}^n \dotplus \mathbb{C}^m$. Then the following are equivalent: (a) $\mathcal{M}$ is $\begin{bmatrix} A \\ C \end{bmatrix}$ invariant;

(b) $$A(\mathcal{M} \cap \mathrm{Ker}\ C) \subset \mathcal{M} \tag{6.6.2}$$

(c) there is a transformation G: $\mathbb{C}^m \to \mathbb{C}^n$ such that

$$(A + GC)\mathcal{M} \subset \mathcal{M} \tag{6.6.3}$$

Proof. It remains only to establish the equivalence of (a) and (b). This is done by using the equivalence of statements (a) and (c) in Theorem 6.1.1. Thus $\mathcal{M}$ is $\begin{bmatrix} A \\ C \end{bmatrix}$ invariant if and only if

$$A^* \mathcal{M}^\perp \subset \mathcal{M}^\perp + \text{Im } C^* = \mathcal{M}^\perp + (\text{Ker } C)^\perp \qquad (6.6.4)$$

Now it is easily verified that, for subspaces $\mathcal{U}$, $\mathcal{V}$, and a transformation A, the relations $A\mathcal{V} \subset \mathcal{U}$ and $A^*\mathcal{U}^\perp \subset \mathcal{V}^\perp$ are equivalent. Thus equation (6.6.4) is equivalent to

$$A(\mathcal{M}^\perp + (\text{Ker } C)^\perp)^\perp \subset (\mathcal{M}^\perp)^\perp$$

or

$$A(\mathcal{M} \cap \text{Ker } C) \subset \mathcal{M}$$

which is condition (6.6.2). $\quad\square$

It is useful to have a terminology involving an arbitrary subspace in the place of Ker C in (6.6.2). Thus, if A is a transformation on $\mathbb{C}^n$ and $\mathcal{V}$ is a subspace of $\mathbb{C}^n$, we say that a subspace $\mathcal{M}$ is A *invariant intersect* $\mathcal{V}$, or A invariant (int $\mathcal{V}$), if $A(\mathcal{M} \cap \mathcal{V}) \subset \mathcal{M}$.

Through extension of the terminology of Section 2.8 for any given subspace $\mathcal{M}$, a subspace $\mathcal{U}$ that is A invariant (int $\mathcal{V}$) is said to be *minimal over* $\mathcal{M}$ if $\mathcal{U} \supset \mathcal{M}$ and there is no other A-invariant (int $\mathcal{V}$) subspace that contains $\mathcal{M}$ and is contained in $\mathcal{U}$.

Now consider a generalization of similarity for transformations from $\mathbb{C}^n$ to $\mathbb{C}^n \dotplus \mathbb{C}^m$. If $\begin{bmatrix} A \\ C \end{bmatrix}$ is such a transformation, an *extension* of $\begin{bmatrix} A \\ C \end{bmatrix}$ is a transformation T on $\mathbb{C}^n \dotplus \mathbb{C}^m$ of the form

$$T = \begin{bmatrix} A & B \\ C & D \end{bmatrix}$$

Then we say that transformations $\begin{bmatrix} A_1 \\ C_1 \end{bmatrix}$ and $\begin{bmatrix} A_2 \\ C_2 \end{bmatrix}$ from $\mathbb{C}^n$ into $\mathbb{C}^n \dotplus \mathbb{C}^m$ are *block similar* if, given any extension T_1 of $\begin{bmatrix} A_1 \\ C_1 \end{bmatrix}$, there is an extension T_2 of $\begin{bmatrix} A_2 \\ C_2 \end{bmatrix}$ such that T_1 and T_2 are similar. Comparing this with the corresponding definition of Section 6.2, we see that this is equivalent to the block similarity of $[A_1^* \; C_1^*]$ and $[A_2^* \; C_2^*]$. We may thus apply Theorem 6.2.1 to obtain the following theorem.

Theorem 6.6.2

The transformations $\begin{bmatrix} A_1 \\ C_1 \end{bmatrix}$ *and* $\begin{bmatrix} A_2 \\ C_2 \end{bmatrix}$ *from* $\mathbb{C}^n$ *to* $\mathbb{C}^{m+n}$ *are block similar if and only if there exist invertible transformations N on $\mathbb{C}^n$ and M on $\mathbb{C}^m$, and a transformation $G: \mathbb{C}^m \rightarrow \mathbb{C}^n$ such that*

$$A_2 = N(A_1 + GC_1)N^{-1}, \qquad C_2 = MC_1N^{-1} \qquad (6.6.5)$$

Once again, it is found that block similarity determines an equivalence relation on all transformations from $\mathbb{C}^n$ into $\mathbb{C}^n \dotplus \mathbb{C}^m$. Furthermore, the canonical forms in the corresponding equivalence follow immediately from the Brunovsky form of Theorem 6.2.5 by duality.

Theorem 6.6.3

Given a transformation $\begin{bmatrix} A \\ C \end{bmatrix}: C^n \to \mathbb{C}^n \dotplus \mathbb{C}^m$ there is a block-similar transformation $\begin{bmatrix} A_0 \\ C_0 \end{bmatrix}$ that (in some bases for $\mathbb{C}^m$ and $\mathbb{C}^n$) has the representation

$$A_0 = J_{k_1}(0) \oplus \cdots \oplus J_{k_p}(0) \oplus J_{l_1}(\lambda_1) \oplus \cdots \oplus J_{l_q}(\lambda_q) \qquad (6.6.6)$$

for some integers $k_1 \geq k_2 \geq \cdots \geq k_p$, and al entries in C_0 are zero except for those in positions $(1, 1), (2, k_1 + 1), \ldots, (p, k_1 + \cdots + k_{p-1} + 1)$, and those exceptional entries are equal to one. Moreover, the matrices A_0 and C_0 defined in this way are uniquely determined by A and C, apart from a permutation of the blocks $J_{l_1}(\lambda_1), \ldots, J_{l_q}(\lambda_q)$ in equation (6.6.6).

The case of full-range pairs (A, B), which was one of our concerns in Section 6.3, is now replaced by the dual case in which (C, A) is a null kernel pair (see the definition in Section 2.7 and Theorem 2.8.2). The dual of Theorem 6.3.2 is now as follows.

Theorem 6.6.4

For a transformation $\begin{bmatrix} A \\ C \end{bmatrix}: \mathbb{C}^n \to \mathbb{C}^n \dotplus \mathbb{C}^m$ the following statements are equivalent: (a) the pair (A, C) is a null kernel pair; (b) there is a null kernel pair (A_1, C_1) for which $\begin{bmatrix} A \\ C \end{bmatrix}$ and $\begin{bmatrix} A_1 \\ C_1 \end{bmatrix}$ are block similar; (c) in the Brunovsky form $\begin{bmatrix} A_0 \\ C_0 \end{bmatrix}$ for $\begin{bmatrix} A \\ C \end{bmatrix}$, the matrix A_0 has no Jordan part; (d) the rank of the transformation $\begin{bmatrix} I\lambda + A \\ C \end{bmatrix}$ does not depend on the complex parameter λ.

6.7 EXERCISES

6.1 Let $A: \mathbb{C}^n \to \mathbb{C}^n$ be a transformation. A chain

$$\mathcal{M}_1 \subset \cdots \subset \mathcal{M}_k \qquad (1)$$

of subspaces in $\mathbb{C}^n$ will be called *almost A invariant* if $A\mathcal{M}_i \subset \mathcal{M}_{i+1}$, $i = 1, \ldots, k - 1$. Show that the chain (1) is almost A invariant if and only if A has the block matrix form $A = [A_{ij}]_{i,j=1}^{k+1}$ with $A_{ij} = 0$ for $i - j > 1$, with respect to the direct sum decomposition $C^n = \mathcal{L}_1 \dotplus \cdots \dotplus \mathcal{L}_{k+1}$, where $\mathcal{L}_i$ is a direct complement to $\mathcal{M}_{i-1}$ in $\mathcal{M}_i$ (by definition, $\mathcal{M}_0 = \{0\}$ and $\mathcal{M}_{k+1} = \mathbb{C}^n$).

6.2   Prove that every transformation $A: \mathbb{C}^n \to \mathbb{C}^n$ has an almost A-invariant chain

$$\{0\} \subset \mathcal{M}_1 \subset \cdots \subset \mathcal{M}_{n-1} \subset \mathbb{C}^n$$

consisting of $n+1$ distinct subspaces, where $\mathcal{M}_1$ is any given one-dimensional subspace. (*Hint:* For a given $\mathcal{M}_1 = \mathrm{Span}\{x\}$, $x \neq 0$, put $\mathcal{M}_2 = \mathrm{Span}\{x, Ax\}, \ldots, \mathcal{M}_k = \mathrm{Span}\{x, Ax, \ldots, A^{k-1}x\}$, where k is the least positive integer such that the vectors $x, Ax, \ldots, A^kx$ are linearly dependent. Use the preceding exercise.)

6.3   A block matrix $A = [A_{ij}]_{i,j=1}^p$ is called *tridiagonal* if $A_{ij} = 0$ for $|i - j| > 1$. Show that a transformation A has tridiagonal block matrix form with respect to a direct sum decomposition $\mathbb{C}^n = \mathcal{L}_1 \dotplus \cdots \dotplus \mathcal{L}_p$ if and only if the chains

$$\mathcal{L}_1 \subset \mathcal{L}_1 \dotplus \mathcal{L}_2 \subset \cdots \subset \mathcal{L}_1 \dotplus \cdots \dotplus \mathcal{L}_{p-1}$$

and

$$\mathcal{L}_p \subset \mathcal{L}_{p-1} \dotplus \mathcal{L}_p \subset \cdots \subset \mathcal{L}_2 \dotplus \cdots \dotplus \mathcal{L}_p$$

are almost A invariant.

6.4   Let $A: \mathbb{C}^n \to \mathbb{C}^n$ be a self-adjoint transformation. Prove that for any vector $x_1 \in \mathbb{C}^n$ with norm 1 there exists an orthonormal basis $x_1, \ldots, x_n$ in $\mathbb{C}^n$ such that the chains

$$\mathrm{Span}\{x_1\} \subset \mathrm{Span}\{x_1, x_2\} \subset \cdots \subset \mathrm{Span}\{x_1, x_2, \ldots, x_{n-1}\}$$

and

$$\mathrm{Span}\{x_n\} \subset \mathrm{Span}\{x_{n-1}, x_n\} \subset \cdots \subset \mathrm{Span}\{x_2, \ldots, x_n\}$$

are almost A invariant (so A has a tridiagonal form with respect to the basis $x_1, \ldots, x_n$). [*Hint:* Apply Gram–Schmidt orthogonalization to a basis $x_1, y_2, \ldots, y_n$ in $\mathbb{C}^n$ such that the chain

$$\mathrm{Span}\{x_1\} \subset \mathrm{Span}\{x_1, y_2\} \subset \cdots \subset \mathrm{Span}\{x_1, y_2, \ldots, y_{n-1}\}$$

is almost A invariant (Exercise 6.2) and use the self-adjointness of A.]

6.5   Let $A: \mathbb{C}^n \to \mathbb{C}^n$ and $B: \mathbb{C}^m \to \mathbb{C}^n$ be transformations.
    (a)   Show that

$$\dim \mathrm{Im}[\lambda I - A, B] = n \qquad (2)$$

for every $\lambda \in \mathbb{C}$ with the possible exception of not more than $n - (\dim \mathrm{Im}\, B)$ points.

    (b)   Show that if equation (2) holds at k eigenvalues of A (counting multiplicities) then for every k-tuple $\mu_1, \ldots, \mu_k$ there exists a transformation $F: \mathbb{C}^n \to \mathbb{C}^m$ such that $\mu_1, \ldots, \mu_k \in \sigma(A + BF)$.

6.6 State and prove the analogs of Exercises 6.5 (a) and (b) for a pair of transformations $C: \mathbb{C}^q \to \mathbb{C}^n$, $A: \mathbb{C}^n \to \mathbb{C}^n$.

6.7 Let (A, B) be a full-range pair of transformations. Show that for any F the transformation $A + BF$ has not more than dim Im B Jordan blocks corresponding to each eigenvalue in its Jordan form.

6.8 Let

$$A = \begin{bmatrix} 0 & 1 & 0 \\ 0 & 0 & 1 \\ 1 & 0 & 2 \end{bmatrix}, \qquad B = \begin{bmatrix} -1 \\ 0 \\ 1 \end{bmatrix}$$

(a) Show that (A, B) is a full-range pair.

(b) Find matrices N, M and F, where N and M are invertible, such that the pair $N^{-1}(A + BF)N$, $N^{-1}BM$ is in the Brunovsky canonical form.

(c) Find G such that $A + BG$ has the eigenvalues $0, 2, -1$.

6.9 Let $A: \mathbb{C}^n \to \mathbb{C}^n$ be a transformation, and let $x \in \mathbb{C}^n$ be a cyclic vector in $\mathbb{C}^n$ for A (i.e., $\mathbb{C}^n = \text{Span}\{x, Ax, A^2x, \ldots\}$). Show that for any n-tuple of not necessarily distinct complex numbers $\lambda_1, \ldots, \lambda_n$ there exists a transformation $B: \mathbb{C}^n \to \mathbb{C}^n$ with Im $B \subset \text{Span}\{x\}$ such that $A + B$ has the eigenvalues $\lambda_1, \ldots, \lambda_n$.

6.10 Let $A: \mathbb{C}^n \to \mathbb{C}^n$ be a transformation, and let $\mathcal{M} \subset \mathbb{C}^n$ be a subspace such that $\mathbb{C}^n$ is the minimal A-invariant subspace that contains $\mathcal{M}$. Show that for n-tuple $\lambda_1, \ldots, \lambda_n$ of not necessarily distinct complex numbers there exists a transformation $B: \mathbb{C}^n \to \mathbb{C}^n$ with Im $B \subset \mathcal{M}$ such that $A + B$ has eigenvalues $\lambda_1, \ldots, \lambda_n$.

6.11 Let

$$C = [1, -1, 0]; \qquad A = \begin{bmatrix} 0 & 0 & 1 \\ 1 & 0 & 0 \\ 0 & 1 & -2 \end{bmatrix}$$

(a) Show that (C, A) is a null kernel pair.

(b) Find matrices N, M and F, where N and M are invertible such that MCN^{-1}, $N(A + FC)N^{-1}$ are in the canonical form as described in Theorem 6.6.3.

(c) Find G such that $A + GC$ has the eigenvalues $1, -1, 0$.

6.12 Let $A: \mathbb{C}^n \to \mathbb{C}^n$ be a transformation, and let $\mathcal{M} \subset \mathbb{C}^n$ be a subspace such that $\{0\}$ is the maximal A-invariant subspace in $\mathcal{M}$. Prove that for any n-tuple of not necessarily distinct complex numbers $\lambda_1, \ldots, \lambda_n$ there exists a transformation $C: \mathbb{C}^n \to \mathbb{C}^n$ with Ker $C \subset \mathcal{M}$ such that $A + C$ has eigenvalues $\lambda_1, \ldots, \lambda_n$.

Chapter Seven

Rational Matrix
Functions

In this chapter we study $r \times n$ matrices $W(\lambda)$ whose elements are rational functions of a complex variable λ. Thus we may write

$$W(\lambda) = \left[\frac{p_{ij}(\lambda)}{q_{ij}(\lambda)} \right]_{i,j=1}^{r,n}$$

where $p_{ij}(\lambda)$ and $q_{ij}(\lambda)$ are scalar polynomials and $q_{ij}(\lambda)$ are not identically zero. Such functions $W(\lambda)$ are called *rational matrix functions*.

We focus on problems for rational matrix functions in which different types of invariant subspaces and triinvariant decompositions play a decisive role. All these problems are motivated mostly by linear systems theory, and their solutions are used in Chapter 8. The problems we have in mind are the following: (1) the realization problem, which concerns representations of a rational matrix function in the form $D + C(\lambda I - A)^{-1}B$ with constant matrices A, B, C, D; (2) the problem of minimal factorization; and (3) the problem of linear fractional decomposition.

7.1 REALIZATIONS OF RATIONAL MATRIX FUNCTIONS

Let $W(\lambda)$ be an $r \times n$ rational matrix function. We assume that $W(\lambda)$ is finite at infinity; that is, in each entry $p_{ij}(\lambda)/q_{ij}(\lambda)$ of $W(\lambda)$ the degree of the polynomial $p_{ij}(\lambda)$ is less than or equal to the degree of $q_{ij}(\lambda)$.

A *realization* of the rational matrix function $W(\lambda)$ is a representation of the form

$$W(\lambda) = D + C(\lambda I_m - A)^{-1}B , \qquad \lambda \notin \sigma(A) \qquad (7.1.1)$$

where A, B, C, D are matrices of sizes $m \times m$, $m \times n$, $r \times m$, $r \times n$, respectively. Observe that $\lim_{\lambda \to \infty} (\lambda I - A)^{-1} = 0$. [To verify this, assume that A is in the Jordan form and use formula (1.9.5).] So if there exists a realization (7.1.1), then necessarily $D = W(\infty)$. We may thus identify such a realization with the triple (A, B, C). The following lemma is useful in the proof of existence of a realization.

Lemma 7.1.1

Let $H(\lambda) = \Sigma_{j=0}^{l-1} \lambda^j H_j$ and $L(\lambda) = \lambda^l I + \Sigma_{j=0}^{l-1} \lambda^j A_j$ be $r \times n$ and $n \times n$ matrix polynomials, respectively. Put

$$B = \begin{bmatrix} 0 \\ \vdots \\ 0 \\ I \end{bmatrix}, \quad A = \begin{bmatrix} 0 & I & \cdots & 0 \\ \vdots & & \ddots & \\ 0 & 0 & \cdots & I \\ -A_0 & -A_1 & \cdots & -A_{l-1} \end{bmatrix}, \quad C = [H_0 \quad \cdots \quad H_{l-1}]$$

Then

$$H(\lambda)L(\lambda)^{-1} = C(\lambda I - A)^{-1}B$$

Proof. We know already (see Section 5.1) that for $\lambda \notin \sigma(A)$

$$L(\lambda)^{-1} = Q(\lambda I - A)^{-1}B \tag{7.1.2}$$

where $Q = [I \quad 0 \quad \cdots \quad 0]$. We may define $C_1(\lambda), \ldots, C_l(\lambda)$ for all $\lambda \notin \sigma(A)$ by

$$\mathrm{col}[C_j(\lambda)]_{j=1}^{l} = (\lambda I - A)^{-1}B$$

From equation (7.1.2) we see that $C_1(\lambda) = L(\lambda)^{-1}$. As

$$(\lambda I - A)[\mathrm{col}[C_j(\lambda)]_{j=1}^{l}] = B$$

the special form of A yields

$$C_i(\lambda) = \lambda^{i-1}C_1(\lambda), \quad 1 \le i \le l$$

It follows that $C(\lambda I - A)^{-1}B = \Sigma_{j=0}^{l-1} H_j C_{j+1}(\lambda) = H(\lambda)L(\lambda)^{-1}$, and the proof is complete. $\square$

Theorem 7.1.2

Every $r \times n$ rational matrix function that is finite at infinity has a realization.

Proof. Let $W(\lambda)$ be an $r \times n$ rational matrix function with finite value at infinity. There exists a monic scalar polynomial $l(\lambda)$ such that $l(\lambda)W(\lambda)$ is a (matrix) polynomial. For instance, take $l(\lambda)$ to be a least common multiple of the denominators of entries in $W(\lambda)$. Put $H(\lambda) = l(\lambda)(W(\lambda) - W(\infty))$. Then $H(\lambda)$ is an $r \times n$ matrix polynomial. Clearly, $L(\lambda) = l(\lambda)I_n$ is monic and $W(\lambda) = W(\infty) + H(\lambda)L(\lambda)^{-1}$. Further

$$\lim_{\lambda \to \infty} H(\lambda)L(\lambda)^{-1} = \lim_{\lambda \to \infty} [W(\lambda) - W(\infty)] = 0$$

So the degree of $H(\lambda)$ is strictly less than the degree of $L(\lambda)$. We can apply Lemma 7.1.1 to find A, B, C for which

$$W(\lambda) = W(\infty) + H(\lambda)L(\lambda)^{-1} = W(\infty) + C(\lambda I - A)^{-1}B$$

This is a realization of $W(\lambda)$. $\square$

A realization for $W(\lambda)$ is far from being unique. This can be seen from our construction of a realization because there are many choices for $l(\lambda)$. In general, if (A, B, C) is a realization of $W(\lambda)$, then so is $(\tilde{A}, \tilde{B}, \tilde{C})$, where

$$\tilde{A} = \begin{bmatrix} A_{11} & A_{12} & A_{13} \\ 0 & A & A_{23} \\ 0 & 0 & A_{33} \end{bmatrix}, \qquad \tilde{B} = \begin{bmatrix} B_1 \\ B \\ 0 \end{bmatrix}, \qquad \tilde{C} = [0 \ C \ C_1]$$

$$(7.1.3)$$

for any matrices A_{ij}, B_1, and C_1 with suitable sizes (in other words, the matrices $\tilde{A}$, $\tilde{B}$, $\tilde{C}$ are of size $s \times s$, $s \times n$, $r \times s$, respectively, and partitioned with respect to the orthogonal sum $\mathbb{C}^s = \mathbb{C}^p \oplus \mathbb{C}^m \oplus \mathbb{C}^q$, where m is the size of A; for instance, A_{13} is a $p \times q$ matrix). Indeed, for every $\lambda \notin \sigma(\tilde{A})$ we have

$$(\lambda I - \tilde{A})^{-1} = \begin{bmatrix} (\lambda I - A_{11})^{-1} & * & * \\ 0 & (\lambda I - A)^{-1} & * \\ 0 & 0 & (\lambda I - A_{33})^{-1} \end{bmatrix}$$

and thus

$$\tilde{B}(\lambda I - \tilde{A})^{-1}\tilde{C} = B(\lambda I - A)^{-1}C = W(\lambda) - W(\infty)$$

Among all the realizations of $W(\lambda)$ those with the properties that (C, A) is a null kernel pair and (A, B) is a full-range pair will be of special interest. That is, for which

$$\bigcap_{j=0}^{\infty} \text{Ker } CA^j = \{0\} \qquad \text{and} \qquad \sum_{j=0}^{\infty} \text{Im } A^jB = \mathbb{C}^m \qquad (7.1.4)$$

The next result shows that any realization "contains" a realization with those properties. To make this precise it is convenient to introduce another definition. Let (A, B, C) be a realization of $W(\lambda)$, and let $m \times m$ be the size of A. Given a triinvariant decomposition $\mathbb{C}^m = \mathcal{L} \dotplus \mathcal{M} \dotplus \mathcal{N}$ associated with an A-semiinvariant subspace $\mathcal{M}$ (so that the subspaces $\mathcal{L}$ and $\mathcal{L} \dotplus \mathcal{M}$ are A invariant) with the property that $C|_{\mathcal{L}} = 0$ and $\operatorname{Im} B \subset \mathcal{L} + \mathcal{M}$, a realization $(P_{\mathcal{M}} A|_{\mathcal{M}}, P_{\mathcal{M}} B, C|_{\mathcal{M}})$, where $P_{\mathcal{M}} \colon \mathbb{C}^m \to \mathcal{M}$ is a projector on $\mathcal{M}$ with $\operatorname{Ker} P_{\mathcal{M}} \supset \mathcal{L}$, is called a *reduction* of (A, B, C). Note that $(P_{\mathcal{M}} A|_{\mathcal{M}}, P_{\mathcal{M}} B, C|_{\mathcal{M}})$ is again a realization for the same $W(\lambda)$. [See the proof that (7.1.3) is a realization of $W(\lambda)$ if (A, B, C) is.] We shall also say that (A, B, C) is a *dilation* of $(P_{\mathcal{M}} A|_{\mathcal{M}}, P_{\mathcal{M}} B, C|_{\mathcal{M}})$ (in a natural extension of the terminology introduced at the end of Section 4.1).

Theorem 7.1.3

Any realization (A, B, C) of $W(\lambda)$ is the dilation of a realization (A_0, B_0, C_0) of $W(\lambda)$ with null kernel pair (C_0, A_0) and full-range pair (A_0, B_0).

Proof. Let $\mathcal{K}(C, A) = \bigcap_{j=0}^{\infty} \operatorname{Ker}(CA^j)$ and $\mathcal{J}(A, B) = \sum_{j=0}^{\infty} \operatorname{Im}(A^j B)$ be the maximal A-invariant subspace in $\operatorname{Ker} C$ and the minimal A-invariant subspace over $\operatorname{Im} B$, respectively.

Put $\mathcal{L} = \mathcal{K}(C, A)$, let $\mathcal{M}$ be a direct complement of $\mathcal{L} \cap \mathcal{J}(A, B)$ in $\mathcal{J}(A, B)$, and choose $\mathcal{N}$ so that

$$\mathbb{C}^m = \mathcal{L} \dotplus \mathcal{M} \dotplus \mathcal{N} \tag{7.1.5}$$

and we recall that m is the size of A. Let us verify that equality (7.1.5) is a triinvariant decomposition associated with an A-semiinvariant subspace $\mathcal{M}$, and that the realization

$$(P_{\mathcal{M}} A|_{\mathcal{M}}, P_{\mathcal{M}} B, C|_{\mathcal{M}}) \tag{7.1.6}$$

where $P \colon \mathbb{C}^M \to \mathcal{M}$ is the projector on $\mathcal{M}$ along $\mathcal{L} \dotplus \mathcal{N}$, is a reduction of (A, B, C), and has the required properties. Indeed, $\mathcal{L}$ and $\mathcal{L} \dotplus \mathcal{M} = \mathcal{K}(C, A) + \mathcal{J}(A, B)$ are A-invariant subspaces, so (7.1.5) is indeed a triinvariant decomposition. Further, $C|_{\mathcal{L}} = 0$ and

$$\operatorname{Im} B \subset \mathcal{J}(A, B) \subset \mathcal{L} \dotplus \mathcal{M}$$

so (7.1.6) is a reduction of (A, B, C). It remains only to prove that the realization (7.1.6) of $W(\lambda)$ has the null kernel and full-range properties. Indeed

$$\operatorname{Ker} C|_{\mathcal{M}} (P_{\mathcal{M}} A|_{\mathcal{M}})^j = (\operatorname{Ker} CA^j) \cap \mathcal{M}, \qquad j = 0, 1, \ldots$$

So

$$\bigcap_{j=0}^{\infty} \text{Ker } C|_{\mathcal{M}}(P_{\mathcal{M}}A|_{\mathcal{M}})^j = \bigcap_{j=0}^{\infty} \text{Ker } CA^j \cap \mathcal{M} = \mathcal{L} \cap \mathcal{M} = \{0\}$$

Also

$$\text{Im}(P_{\mathcal{M}}A|_{\mathcal{M}})^j P_{\mathcal{M}}B = P_{\mathcal{M}}(\text{Im } A^jB)$$

Hence

$$\sum_{j=0}^{\infty} \text{Im}(P_{\mathcal{M}}A|_{\mathcal{M}})^j P_{\mathcal{M}}B = P_{\mathcal{M}}\left(\sum_{j=0}^{\infty} \text{Im } A^jB\right) = P_{\mathcal{M}}(\mathcal{J}(A, B)) = \mathcal{M}$$

because by construction $\mathcal{M} \subset \mathcal{J}(A, B)$. $\square$

It turns out that a realization (A, B, C) for which conditions (7.1.4) are satisfied is essentially unique. To state this result precisely and to prove it, we need some observations concerning one-sided invertibility of matrices. By Theorems 2.7.3 and 2.8.4 we have

$$\text{Ker col}[CA^j]_{j=0}^{p-1} = \{0\} , \qquad \text{Im}[B, AB, \ldots, A^{p-1}B] = \mathbb{C}^m$$

where p is any integer not smaller than the degree of the minimal polynomial for A. Hence there exists a *left inverse* $[\text{col}[CA^j]_{j=0}^{p-1}]^{-L}$. Thus

$$[\text{col}[CA^j]_{j=0}^{p-1}]^{-L} \text{ col}[CA^j]_{j=0}^{p-1} = I$$

Also, there exists a *right inverse* $[B, AB, \ldots, A^{p-1}B]^{-R}$:

$$[B, AB, \ldots, A^{p-1}B][B, AB, \ldots, A^{p-1}B]^{-R} = I$$

Note that in general the left and right inverses involved are not unique.

Theorem 7.1.4

Let (A_1, B_1, C_1) and (A_2, B_2, C_2) be realizations for a rational matrix function $W(\lambda)$ for which (C_1, A_1) and (C_2, A_2) are null kernel pairs and (A_1, B_1), (A_2, B_2) are full-range pairs. Then the sizes of A_1 and A_2 coincide, and there exists a nonsingular matrix S such that

$$A_1 = S^{-1}A_2S , \qquad B_1 = S^{-1}B_2 , \qquad C_1 = C_2S \tag{7.1.7}$$

Moreover, the matrix S is unique and is given by

$$S = [\text{col}[C_2A_2^j]_{j=0}^{p-1}]^{-L}[\text{col}[C_1A_1^j]_{j=0}^{p-1}]$$
$$= [B_2, A_2B_2, \ldots, A_2^{p-1}B_2][B_1, A_1B_1, \ldots, A_1^{p-1}B_1]^{-R} \tag{7.1.8}$$

Here p is any integer greater than or equal to the maximum of the degrees of minimal polynomials for A_1 and A_2, and the superscript $-L$ (resp. $-R$) indicates left (resp. right) inverse.

Proof. We have

$$W(\lambda) = D + C_1(\lambda I - A_1)^{-1}B_1 = D + C_2(\lambda I - A_2)^{-1}B_2$$

For $|\lambda| > \max\{\|A_1\|, \|A_2\|\}$ the matrices $\lambda I - A_1$ and $\lambda I - A_2$ are nonsingular and for $i = 1, 2$

$$(\lambda I - A_i)^{-1} = \lambda^{-1}(I - \lambda^{-1}A_i)^{-1} = \sum_{j=0}^{\infty} \lambda^{-j-1}A_i^j$$

Consequently, we have

$$C_1\left(\sum_{j=0}^{\infty} \lambda^{-j-1}A_1^j\right)B_1 = C_2\left(\sum_{j=0}^{\infty} \lambda^{-j-1}A_2^j\right)B_2$$

for any λ with $|\lambda| > \max\{\|A_1\|, \|A_2\|\}$. Comparing coefficients, we see that $C_1A_1^jB_1 = C_2A_2^jB_2$, $j = 0, 1, \ldots$. This implies $\Omega_1\Delta_1 = \Omega_2\Delta_2$, where, for $k = 1, 2$ we write

$$\Omega_k = \mathrm{col}[C_kA_k^j]_{j=0}^{p-1}, \qquad \Delta_k = [B_k, A_kB_k, \ldots, A_k^{p-1}B_k]$$

Premultiplying by a left inverse of Ω_2 and postmultiplying by a right inverse of Δ_2, we find that the second equality in (7.1.8) holds. Now define S as in (7.1.8). Let us check first that S is (two-sided) invertible. Indeed, we can verify the relations

$$(\Omega_1^{-L}\Omega_2)S = I, \qquad S(\Delta_1\Delta_2^{-R}) = I$$

Since $C_1A_1^jB_1 = C_2A_2^jB_2$, for $j = 0, 1, \ldots$, we have $\Omega_2^{-L}\Omega_1\Delta_1\Delta_2^{-R} = \Omega_2^{-L}\Omega_2\Delta_2\Delta_2^{-R} = I$. Similarly, one checks that $\Omega_1^{-L}\Omega_2\Delta_2\Delta_1^{-R} = I$. Because S is invertible, the sizes of A_1 and A_2 must coincide.

It remains to check equations (7.1.7). Write

$$\Omega_2A_2\Delta_2 = \Omega_1A_1\Delta_1 = \Omega_1\Delta_1\Delta_1^{-R}A_1\Delta_1 = \Omega_2\Delta_2\Delta_1^{-R}A_1\Delta_1$$

Premultiply by Ω_2^{-L} and postmultiply by Δ_2^{-R} to obtain $A_2S = SA_1$. Now

$$SB_1 = \Omega_2^{-L}\Omega_1B_1 = \Omega_2^{-L}\Omega_2B_2 = B_2$$

and

$$C_2S = C_2\Delta_2\Delta_1^{-R} = C_1\Delta_1\Delta_1^{-R} = C_1 \qquad \square$$

Theorems 7.1.3 and 7.1.4 allow us to deduce the following important fact.

Theorem 7.1.5

In a realization (A, B, C) *of* $W(\lambda)$, (C, A) *and* (A, B) *are null kernel pairs and full-range pairs, respectively, if and only if the size of* A *is minimal among all possible realizations of* $W(\lambda)$.

Proof. Assume that the size m of A is minimal. By Theorem 7.1.3, there is a reduction (A', B', C') of (A, B, C) that is a realization for $W(\lambda)$ and satisfies conditions (7.1.4). But because of the minimality of m the realizations (A', B', C') and (A, B, C) must be similar, and this implies that (A, B, C) also satisfies condition (7.1.4).

Conversely, assume that (A, B, C) satisfies conditions (7.1.4). Arguing by contradiction, suppose that there is a realization (A', B', C') with A' of smaller size than A. By Theorem 7.1.3, there is a reduction (A'', B'', C'') of (A', B', C') that satisfies conditions (7.1.4). But then the size of A'' is smaller than that of A, which contradicts Theorem 7.1.4. $\square$

Realizations of the kind described in this theorem are, naturally, called *minimal realizations* of $W(\lambda)$. That is, they are those realizations for which the dimension of the space on which A acts is as small as possible.

7.2   PARTIAL MULTIPLICITIES AND MULTIPLICATION

In this section we study multiplication and partial multiplicities of rational matrix functions. To facilitate the presentation, it is assumed that the functions take values in the *square* matrices and that the determinant function is not identically zero.

Let $W(\lambda)$ by an $n \times n$ rational matrix function with $\det W(\lambda) \not\equiv 0$. In a neighbourhood of each point $\lambda_0 \in \mathbb{C}$ the function $W(\lambda)$ admits the representation, called the *local Smith form* of $W(\lambda)$, at λ_0:

$$W(\lambda) = E_1(\lambda) \operatorname{diag}[(\lambda - \lambda_0)^{\nu_1}, \ldots, (\lambda - \lambda_0)^{\nu_n}]E_2(\lambda) \qquad (7.2.1)$$

where $E_1(\lambda)$ and $E_2(\lambda)$ are rational matrix functions that are defined and invertible at λ_0, and $\nu_1, \ldots, \nu_n$ are integers. Indeed, for matrix polynomials equation (7.2.1) follows from Theorem A.3.4 in the appendix. In the general case write $W(\lambda) = p(\lambda)^{-1}\tilde{W}(\lambda)$, where $\tilde{W}(\lambda)$ and $p(\lambda)$ are matrix and scalar polynomials, respectively. Since we have a representation (7.2.1) for $\tilde{W}(\lambda)$, it immediately follows that a similar representation holds for $W(\lambda)$.

The integers $\nu_1, \ldots, \nu_n$ in (7.2.1) are uniquely determined by $W(\lambda)$ and

λ_0 up to permutation and do not depend on the particular choice of the local Smith form (7.2.1). To see this, assume that $\nu_1 \leq \cdots \leq \nu_n$, and define the multiplicity of a scalar rational function $g(\lambda) \not\equiv 0$ at λ_0 as the integer ν such that the function $g(\lambda)(\lambda - \lambda_0)^{-\nu}$ is analytic and nonzero at λ_0. Then, using the Cauchy–Binet formula (Theorem A.2.1 in the appendix), we see that $\nu_1 + \cdots + \nu_i$ is the minimal multiplicity at λ_0 of the not identically zero minors of size $i \times i$ of $W(\lambda)$, $i = 1, \ldots, n$. Thus the numbers $\nu_1 + \cdots + \nu_i$, $i = 1, \ldots, n$, and, consequently, $\nu_1, \ldots, \nu_n$ are uniquely determined by $W(\lambda)$.

The integers $\nu_1, \ldots, \nu_n$ from the local Smith form (7.2.1) of $W(\lambda)$ are called the *partial multiplicities* of $W(\lambda)$ at λ_0.

Note that $\lambda_0 \in \mathbb{C}$ is a *pole* of $W(\lambda)$ [i.e., a pole of at least one entry in $W(\lambda)$] if and only if $W(\lambda)$ has a negative partial multiplicity at λ_0. Indeed, the minimal partial multiplicity of $W(\lambda)$ at λ_0 coincides with the minimal multiplicity at λ_0 of the not identically zero entries of $W(\lambda)$. Also, $\lambda_0 \in \mathbb{C}$ is a *zero* of $W(\lambda)$ [by definition, this means that λ_0 is a pole of $W(\lambda)^{-1}$] if and only if $W(\lambda)$ has a positive partial multiplicity. In particular, for every $\lambda_0 \in \mathbb{C}$, except for a finite number of points, all partial multiplicities are zeros.

There is a close relationship between the partial multiplicities of $W(\lambda)$ and the minimal realization of $W(\lambda)$. Namely, let $W(\lambda)$ be a rational $n \times n$ matrix function with determinant not identically zero. Let

$$W(\lambda) = \sum_{j=-\infty}^{q} \lambda^j W_j \qquad (7.2.2)$$

be the Laurent series of $W(\lambda)$ at infinity (here q is some nonnegative integer and the coefficients W_i are $n \times n$ matrices): write $U(\lambda) = \sum_{j=0}^{q} \lambda^j W_j$ for the polynomial part of $W(\lambda)$. Thus $W(\lambda) - U(\lambda)$ takes the value 0 at infinity, and we may write

$$W(\lambda) = C(\lambda I - A)^{-1} B + U(\lambda) \qquad (7.2.3)$$

where $C(\lambda I - A)^{-1} B$ is a minimal realization of the rational matrix function $W(\lambda) - U(\lambda)$. We say that (7.2.3) is a *minimal realization* of $W(\lambda)$. We see later (Theorem 7.2.3) that $\lambda_0 \in \mathbb{C}$ is a pole of $W(\lambda)$ if and only if λ_0 is an eigenvalue of A. Moreover, for a fixed pole of $W(\lambda)$ the number of negative partial multiplicities of $W(\lambda)$ at λ_0 coincides with the number of Jordan blocks with eigenvalue λ_0 in the Jordan normal form of A, and the absolute values of these partial multiplicities coincide with the sizes of these Jordan blocks. A similar statement holds for the zeros of $W(\lambda)$.

An analytic n-dimensional vector function

$$\psi(\lambda) = \sum_{j=0}^{\infty} (\lambda - \lambda_0)^j \psi_j$$

defined on a neighbourhood of $\lambda_0 \in \mathbb{C}$ is said to be a *null function* of a rational matrix function $W(\lambda)$ at λ_0 if $\psi_0 \neq 0$, $W(\lambda)\psi(\lambda)$ is analytic in a neighbourhood of λ_0, and $[W(\lambda)\psi(\lambda)]_{\lambda=\lambda_0} = 0$. The multiplicity of λ_0 as a zero of the vector function $W(\lambda)\psi(\lambda)$ is the *order* of $\psi(\lambda)$, and ψ_0 is the *null vector* of $\psi(\lambda)$. From this definition it follows immediately that for $n \times n$ matrix-valued functions $U(\lambda)$ and $V(\lambda)$ that are rational and invertible in a neighbourhood of λ_0, $\psi(\lambda)$ is a null function of $V(\lambda)W(\lambda)U(\lambda)$ at λ_0 of order k if and only if $U(\lambda)\psi(\lambda)$ is a null function of $W(\lambda)$ at λ_0 of order k. A set of null functions $\psi_1(\lambda), \ldots, \psi_p(\lambda)$ of $W(\lambda)$ at λ_0 with orders $k_1, \ldots, k_p$, respectively, is said to be *canonical* if the null vectors $\psi_1(\lambda_0), \ldots, \psi_p(\lambda_0)$ are linearly independent and the sum $k_1 + k_2 + \cdots + k_p$ is maximal among all sets of null functions with linearly independent null vectors.

Proposition 7.2.1

Let $W(\lambda)$ be as defined above and $\psi_1(\lambda), \ldots, \psi_p(\lambda)$ be a canonical set of null functions of $W(\lambda)$ (resp. $W(\lambda)^{-1}$) at λ_0. Then the number p is the number of positive (resp. negative) partial multiplicities of $W(\lambda)$ at λ_0, and the corresponding orders $k_1, \ldots, k_p$ are the positive (resp. absolute values of the negative) partial multiplicities of $W(\lambda)$ at λ_0.

Proof. Briefly, reduce $W(\lambda)$ to local Smith form as described above and apply the observation made in the paragraph preceding Proposition 7.2.1. $\square$

Now we fix an $n \times n$ rational matrix function $W(\lambda)$ with $\det W(\lambda) \not\equiv 0$. Let

$$W(\lambda) = C(\lambda I - A)^{-1}B + U(\lambda) \tag{7.2.4}$$

be its minimal realization, and fix an eigenvalue λ_0 of A. Replacing (7.2.4), if necessary, by a similar realization, we can assume that

$$A = \begin{bmatrix} A_p & 0 \\ 0 & A_p' \end{bmatrix}, \qquad C = [C_p, C_p'], \qquad B = \begin{bmatrix} B_p \\ B_p' \end{bmatrix}$$

where $\sigma(A_p) = \{\lambda_0\}$ and $\lambda_0 \notin \sigma(A_p')$. Note also that if λ_0 is a pole of $W(\lambda)$, then equation (7.2.4) implies that λ_0 is an eigenvalue of A.

Proposition 7.2.2

Let $W(\lambda)$, A_p, and B_p be defined as above. Let λ_0 be a pole of $W(\lambda)$, let $\psi(\lambda)$ be a null function of $W(\lambda)^{-1}$ at λ_0 of order k, and let φ_ν be the coefficients of $\varphi(\lambda) \stackrel{\text{def}}{=} W(\lambda)^{-1}\psi(\lambda)$:

$$\varphi(\lambda) = \sum_{j=k}^{\infty} (\lambda - \lambda_0)^j \varphi_j \qquad (7.2.5)$$

Then

$$x_j = \sum_{\nu=k}^{\infty} (A_p - \lambda_0 I)^{\nu-j-1} B_p \varphi_\nu , \qquad j = 0, \ldots, k-1 \qquad (7.2.6)$$

is a Jordan chain for A at λ_0. Conversely, if λ_0 is an eigenvalue of A and $x_0, \ldots, x_{k-1}$ is a Jordan chain of A at λ_0, there is a null function $\psi(\lambda)$ of $W(\lambda)^{-1}$ at λ_0 with order not less than k for which (7.2.6) holds [in particular, λ_0 is a pole of $W(\lambda)$].

Note that as $\sigma(A_p) = \{\lambda_0\}$, the series in (7.2.6) is actually finite.

Proof. By definition, vectors (7.2.6) form the Jordan chain for A_p at λ_0 if

$$\begin{cases} (A_p - \lambda_0 I)x_0 = 0, & x_0 \neq 0 \\ (A_p - \lambda_0 I)x_j = x_j - 1, & j = 1, 2, \ldots, k-1 \end{cases}$$

The last $k-1$ statements follow immediately from (7.2.6). Also

$$(A_p - \lambda_0 I)x_0 = \sum_{\nu=k}^{\infty} (A_p - \lambda_0 I)^\nu B_p \varphi_\nu$$

Now the Laurent series for $W(\lambda)$ at λ_0, say, $W(\lambda) = \sum_{j=-q}^{\infty} (\lambda - \lambda_0)^j W_j$, has the following coefficients of negative powers of $(\lambda - \lambda_0)$:

$$W_{-j} = C_p (A_p - \lambda_0 I)^{j-1} B_p , \qquad j = 1, 2, \ldots, q$$

and it is easily seen that q is the least positive integer for which $(A_p - \lambda_0 I)^q = 0$. (One checks this by passing to the Jordan form of A_p.) Now recall that $\psi(\lambda) = W(\lambda)\varphi(\lambda)$ is analytic near λ_0; so equating coefficients of negative powers of $(\lambda - \lambda_0)$ to zero and using the fact that $(A_p - \lambda_0 I)^q = 0$, we obtain for $j = 1, 2, \ldots$

$$0 = \sum_{\nu=k}^{q-j} W_{-\nu-j}\varphi_\nu = \sum_{\nu=k}^{q-j} C_p(A_p - \lambda_0 I)^{\nu+j-1} B_p \varphi_\nu$$

$$= \sum_{\nu=k}^{\infty} C_p(A_p - \lambda_0 I)^{\nu+j-1} B_p \varphi_\nu$$

$$= C_p(A_p - \lambda_0 I)^{j-1}(A_p - \lambda_0 I)x_0$$

Since $\cap_{j=0}^{\infty} \operatorname{Ker} CA^j = \{0\}$, it follows that $\cap_{j=0}^{\infty} \operatorname{Ker} C_p A_p^j = \{0\}$ or, what is the same, that $\operatorname{col}[C_p A_p^i]_{i=0}^{r-1}$ is left invertible for some integer r. As

$$
\begin{bmatrix} C_p \\ C_p(A_p - \lambda_0 I) \\ \vdots \\ C_p(A_p - \lambda_0 I)^{r-1} \end{bmatrix} = \begin{bmatrix} I & & & 0 \\ -\lambda_0 & I & & \\ \vdots & \vdots & & \\ (-\lambda_0)^{r-1} & \binom{r-1}{1}(-\lambda_0)^{r-2} & \cdots & I \end{bmatrix} \begin{bmatrix} C_p \\ C_p A_p \\ \vdots \\ C_p A_p^{r-1} \end{bmatrix}
$$

the matrix $\operatorname{col}[C_p(A_p - \lambda_0 I)^i]_{i=0}^{r-1}$ is left invertible as well, and since $(A_p - \lambda_0 I)^s = 0$ for $s \geq q$, we obtain the left invertibility of $\operatorname{col}[C_p(A_p - \lambda_0 I)^{j-1}]_{j=1}^q$. It now follows that $(A_p - \lambda_0 I)x_0 = 0$ as required. Finally, since $\psi(\lambda_0) = x_0$, it is also true that $x_0 \neq 0$. Thus, as asserted, equations (7.2.6) do associate a Jordan chain for A_p with the null function $\psi(\lambda)$.

Conversely, let $x_0, x_1, \ldots, x_{k-1}$ be a Jordan chain of A_p at λ_0. From the definition of a minimal realization it follows that the matrix

$$
[B_p, (A_p - \lambda_0 I)B_p, \ldots, (A_p - \lambda_0 I)^{m-1} B_p]
$$

is right invertible for some integer m. Consequently, there exist vectors $\varphi_k, \varphi_{k+1}, \ldots$, with only finitely many nonzero, such that

$$
x_{k-1} = \sum_{j=k}^{\infty} (A_p - \lambda_0 I)^{j-k} B_p \varphi_j \tag{7.2.7}
$$

The definition of a Jordan chain includes $(A_p - \lambda_0 I)x_j = x_{j-1}$ for $j = 1, 2, \ldots, k-1$, and so equations (7.2.6) follow immediately from (7.2.7). It remains only to check that $W(\lambda)\varphi(\lambda)$ is now a null function of $W(\lambda)^{-1}$ at λ_0, where $\varphi(\lambda) = \sum_{j=k}^{\infty} (\lambda - \lambda_0)^j \varphi_j$.

Observe first that $x_0 \neq 0$ and that $C_p A_p^j x_0 = \lambda^j C_p x_0$ for $j = 0, 1, 2, \ldots$. As the matrix $\operatorname{col}[C_p(A_p - \lambda_0 I)^j]_{j=0}^{m-1}$ is left invertible for some integer m, so is $\operatorname{col}[C_p A_p^j]_{j=0}^{m-1}$, and it follows that $C_p x_0 \neq 0$. But using (7.2.6), we obtain

$$
0 \neq C_p x_0 = \sum_{\nu=k}^{\infty} C_p(A_p - \lambda_0 I)^{\nu-1} B_p \varphi_\nu = \lim_{\lambda \to \lambda_0} W(\lambda)\varphi(\lambda) \quad \square
$$

If the Jordan chain $x_0, \ldots, x_{k-1}$ of A_p at λ_0 cannot be prolonged, then $x_{k-1} \not\in \operatorname{Im}(A_p - \lambda_0 I)$, and it follows from (7.2.7) that $\varphi_k \neq 0$. Thus a maximal Jordan chain of length k determines, by means of (7.2.6), an *associated* null function $\psi(\lambda)$ of $W(\lambda)^{-1}$ of order k.

Propositions 7.2.1 and 7.2.2 prove the following result. [The second part of Theorem 7.2.3 concerning zeros of $W(\lambda)$ is obtained by applying the first part to $W(\lambda)^{-1}$.]

Theorem 7.2.3

Let $W(\lambda)$ be a rational $n \times n$ matrix function with $\det W(\lambda) \not\equiv 0$, and let its minimal realization be given by equation (7.2.4). A complex number λ_0 is a pole of $W(\lambda)$ if and only if λ_0 is an eigenvalue of A, and then the absolute values of negative partial multiplicities of $W(\lambda)$ at λ_0 coincide with the sizes of Jordan blocks with eigenvalue λ_0 in the Jordan form of A, that is, with the partial multiplicities of λ_0 as an eigenvalue of A.

A complex number λ_0 is a zero of $W(\lambda)$ if and only if λ_0 is an eigenvalue of A_1, where A_1 is taken from a minimal realization for $W(\lambda)^{-1}$:

$$W(\lambda)^{-1} = C_1(\lambda I - A_1)B_1 + V(\lambda)$$

with matrix polynomial $V(\lambda)$. In this case the positive partial multiplicities of $W(\lambda)$ at λ_0 coincide with the partial multiplicities of λ_0 as an eigenvalue of A_1.

Now we apply Theorem 7.2.3 to study the partial multiplicities of a product of two rational matrix functions. Let $W_1(\lambda)$ and $W_2(\lambda)$ be rational $n \times n$ matrix functions with realizations

$$W_i(\lambda) = D_i + C_i(\lambda I - A_i)^{-1}B_i \qquad (7.2.8)$$

for $i = 1$ and 2. [Of course, the existence of realizations (7.2.8) presumes that $W_1(\lambda)$ and $W_2(\lambda)$ are finite at infinity.] Then the product $W_1(\lambda)W_2(\lambda)$ has a realization

$$W_1(\lambda)W_2(\lambda) = D_1D_2 + [C_1, D_1C_2]\left(\lambda I - \begin{bmatrix} A_1 & B_1C_2 \\ 0 & A_2 \end{bmatrix}\right)^{-1}\begin{bmatrix} B_1D_2 \\ B_2 \end{bmatrix} \quad (7.2.9)$$

Indeed, the following formula is easily verified by multiplication:

$$\left(\lambda I - \begin{bmatrix} A_1 & B_1C_2 \\ 0 & A_2 \end{bmatrix}\right)^{-1} = \begin{bmatrix} (\lambda I - A_1)^{-1} & (\lambda I - A_1)^{-1}B_1C_2(\lambda I - A_2)^{-1} \\ 0 & (\lambda I - A_2)^{-1} \end{bmatrix}$$

so the right-hand side of (7.2.9) is equal to

$$
\begin{aligned}
D_1D_2 &+ C_1(\lambda I - A_1)^{-1}B_1D_2 \\
&+ [C_1(\lambda I - A_1)^{-1}B_1C_2(\lambda I - A_2)^{-1} + D_1C_2(\lambda I - A_2)^{-1}]B_2 \\
&= D_1D_2 + C_1(\lambda I - A_1)^{-1}B_1D_2 \\
&\quad + C_1(\lambda I - A_1)^{-1}B_1C_2(\lambda I - A_2)^{-1}B_2 + D_1C_2(\lambda I - A_2)^{-1}B_2 \\
&= W_1(\lambda)D_2 + (W_1(\lambda) - D_1)(W_2(\lambda) - D_2) + D_1(W_2(\lambda) - D_2) \\
&= W_1(\lambda)W_2(\lambda)
\end{aligned}
$$

So formula (7.2.9) produces a realization for the product $W_1 W_2$ in terms of the realizations for each factor. Easy examples show that (7.2.9) is not necessarily minimal even if the realizations (7.2.8) are minimal. See the following example, for instance.

EXAMPLE 2.1.   Let

$$W_1(\lambda) = \frac{\lambda}{\lambda - 1}, \qquad W_2(\lambda) = \frac{\lambda - 1}{\lambda}$$

Minimal realizations for $W_i(\lambda)$, $i = 1, 2$ are not difficult to obtain:

$$W_1(\lambda) = 1 + 1 \cdot (\lambda - 1)^{-1} \cdot 1 ; \qquad W_2(\lambda) = 1 - 1 \cdot \lambda^{-1} \cdot 1$$

Formula (7.2.9) gives

$$I = W_1(\lambda) W_2(\lambda) = 1 + [1, -1] \left[\lambda I - \begin{bmatrix} 1 & -1 \\ 0 & 0 \end{bmatrix} \right]^{-1} \begin{bmatrix} 1 \\ 1 \end{bmatrix}$$

which is a realization of the rational matrix function I, but not a minimal one. More generally, if $W_2(\lambda) = W_1(\lambda)^{-1}$, then the realization (7.2.9) is not minimal [unless $W_1(\lambda)$ is a constant]. $\square$

Let $W(\lambda)$ be an $n \times n$ rational matrix function with determinant not identically zero. For $\lambda_0 \in \mathbb{C}$, denote by $\pi(W; \lambda_0) = \{\pi_j\}_{j=1}^{\infty}$ the nonincreasing sequence of absolute values of negative partial multiplicities of $W(\lambda)$ at λ_0. This means that $\pi_1 \geq \pi_2 \geq \cdots$ are nonnegative integers with only a finite number of them nonzero (say, $\pi_k > \pi_{k+1} = 0$), and $-\pi_1, -\pi_2, \ldots, -\pi_k$ are the negative partial multiplicities of $W(\lambda)$ at λ_0.

Consider nonincreasing sequences $\alpha = \{\alpha_j\}_{j=1}^{\infty}$ and $\beta = \{\beta_j\}_{j=1}^{\infty}$ of nonnegative integers such that only finitely many of them are nonzero, and recall the definition of the set $\Gamma(\alpha, \beta)$ given in Section 4.4.

Theorem 7.2.4

Let $W_1(\lambda)$ and $W_2(\lambda)$ be $n \times n$ rational matrix functions with determinant not identically zero and that take finite value at infinity. Then for every $\lambda_0 \in \mathbb{C}$ and $j = 1, 2, \ldots$ we take $\pi_j \leq \delta_j$, where $\{\pi_j\}_{j=1}^{\infty} = \pi(W_1 W_2; \lambda_0)$ and $\{\delta_j\}_{j=1}^{\infty}$ is some sequence from $\Gamma(\pi(W_1; \lambda_0), \pi(W_2; \lambda_0))$. If, in addition, $W_1(\lambda)$ and $W_2(\lambda)$ admit minimal realizations (7.2.8) for which the realization (7.2.9) of $W_1(\lambda) W_2(\lambda)$ is minimal as well, then actually

$$\pi(W_1 W_2; \lambda_0) \in \Gamma(\pi(W_1, \lambda_0), \pi(W_2; \lambda_0))$$

Proof. Let $W_1(\lambda)$ and $W_2(\lambda)$ have minimal realizations as in equation (7.2.8). Using Theorem 7.2.3 and the definition of $\Gamma(\pi(W_1; \lambda_0)$,

$\pi(W_2; \lambda_0))$, we see that the nonincreasing sequence $\delta = \{\delta_j\}_{j=1}^{\infty}$ of partial multiplicities of the matrix

$$A = \begin{bmatrix} A_1 & B_1 C_2 \\ 0 & A_2 \end{bmatrix}$$

belongs to $\Gamma(\pi(W_1; \lambda_0), \pi(W_2; \lambda_0))$. Now (7.2.9) is a realization (not necessarily minimal) of $W_1(\lambda)W_2(\lambda)$. Theorem 7.1.2 shows that there is a restriction (A_0, B_0, C_0) of $(A, \begin{bmatrix} B_1 \\ B_2 \end{bmatrix}, [C_1 C_2])$ to some A-semiinvariant subspace such that the realization

$$W_1(\lambda)W_2(\lambda) = I + C_0(\lambda I - A_0)^{-1}B_0$$

is minimal. Then $\{\pi_j\}_{j=1}^{\infty}$ is the sequence of partial multiplicities of A_0 at λ_0. But as A_0 is a restriction of A to $\mathcal{M}$, we have $\pi_j \leq \delta_j$, for $j = 1, 2, \ldots$ (see Section 4.1). $\square$

The assumption that both $W_1(\lambda)$ and $W_2(\lambda)$ take finite values at infinity is not essential in Theorem 7.2.4. However, we do not pursue this generalization.

The condition that the realization (7.2.9) is minimal for some minimal realizations (7.2.8) is important in the theory of rational matrix functions and in the theory of linear systems. It leads to the notion of minimal factorization and is studied in detail in the following sections.

7.3 MINIMAL FACTORIZATIONS OF RATIONAL MATRIX FUNCTIONS

In this section we describe the minimal factorizations of a rational matrix function in terms of certain invariant subspaces. To make the presentation more transparent, we restrict ourselves to the case when the rational matrix functions involved are $n \times n$ and have value I at infinity. (The same analysis applies to the case when the matrix function has invertible value at infinity.)

We start with a definition. The *McMillan degree* of a rational $n \times n$ matrix function $W(\lambda)$ [with $W(\infty) = I$], denoted $\delta(W)$, is the size of the matrix A in a minimal realization

$$W(\lambda)^{-1} = I - C(\lambda I - A)^{-1}B \tag{7.3.1}$$

It is easily verified that

$$W(\lambda)^{-1} = I - C(\lambda I - A^{\times})^{-1}B \tag{7.3.2}$$

where $A^{\times} = A - BC$. Moreover, if realization (7.3.1) is minimal, so is equation (7.3.2). Indeed, equation (6.3.1) shows that the pair $(A - BC, B)$ is a full-range pair [because (A, B) is so]. Further, (C, A) is a null kernel pair, or, equivalently, (A^*, C^*) is a full-range pair. By the same argument, the pair $(A^* - C^*B^*, C^*)$ is also a full-range pair. Hence $(C, A - BC)$ is a null kernel pair, and therefore realization (7.3.2) is minimal. In particular, $\delta(W^{-1}) = \delta(W)$.

Consider the factorization

$$W(\lambda) = W_1(\lambda)W_2(\lambda) \cdots W_p(\lambda) \tag{7.3.3}$$

where, for $j = 1, \ldots, p$, $W_j(\lambda)$ are $n \times n$ rational matrix functions with minimal realizations

$$W_j(\lambda) = I + C_j(\lambda I - A_j)^{-1}B_j$$

Formula (7.2.9) applied several times yields a realization for $W(\lambda)$:

$$W(\lambda) = I + [C_1 \quad C_2 \quad \cdots \quad C_p]\left(\lambda I - \begin{bmatrix} A_1 & B_1C_2 & \cdots & B_1C_p \\ 0 & A_2 & \cdots & B_2C_p \\ 0 & 0 & & \vdots \\ \vdots & \vdots & & \vdots \\ 0 & 0 & \cdots & A_p \end{bmatrix}\right)^{-1}\begin{bmatrix} B_1 \\ B_2 \\ \vdots \\ B_p \end{bmatrix}$$

$$\tag{7.3.4}$$

This realization is not necessarily minimal, so we have (in view of Theorem 7.1.2)

$$\delta(W) \leq \delta(W_1) + \cdots + \delta(W_p)$$

We say that the *factorization* (7.3.3) *is minimal* if actually $\delta(W) = \delta(W_1) + \cdots + \delta(W_p)$, that is, realization (7.3.4) is minimal as well. In informal terms, minimality of (7.3.3) means that zero-pole cancellation does not occur between the factors $W_j(\lambda)$. Because the McMillan degrees of a rational matrix function (with value I at infinity) and of its inverse are the same, (7.3.3) is minimal if and only if the corresponding factorization for the inverse matrix function

$$W(\lambda)^{-1} = W_p(\lambda)^{-1}W_{p-1}(\lambda)^{-1} \cdots W_1(\lambda)^{-1}$$

is minimal.

Let us focus on minimal factorizations (7.3.3) with three factors ($p = 3$). A description of all such factorizations in terms of certain triinvariant decompositions associated with A-semiinvariant subspaces is given. Here A is taken from a minimal realization $W(\lambda) = I + C(\lambda I - A)^{-1}B$. Write $A^\times = A - BC$, and let A and $A^\times$ be of size m.

We say that a direct sum decomposition

$$\mathbb{C}^m = \mathcal{L} \dotplus \mathcal{M} \dotplus \mathcal{N} \tag{7.3.5}$$

is a *supporting triinvariant decomposition* for $W(\lambda)$ if (7.3.5) is a triinvariant decomposition associated with an A-semiinvariant subspace $\mathcal{M}$ (so $\mathcal{L}$ and $\mathcal{L} \dotplus \mathcal{M}$ are A invariant) and at the same time $\mathcal{M}$ is $A^\times$ semiinvariant with associated triinvariant decomposition $\mathbb{C}^m = \mathcal{N} \dotplus \mathcal{M} \dotplus \mathcal{L}$ (i.e., $\mathcal{N}$ and $\mathcal{N} \dotplus \mathcal{M}$ are $A^\times$ invariant). Note that a supporting triinvariant decomposition for $W(\lambda)$ depends on the choice of minimal realization. We assume, however, that the minimal realization of $W(\lambda)$ is fixed and thereby suppresses the dependence of supporting triinvariant decompositions on this choice. (In view of Theorems 7.1.4 and 7.1.5, there is no loss of generality in making this assumption.)

The role of supporting triinvariant decompositions in the minimal factorization problem is revealed in the next theorem.

Theorem 7.3.1

Let (7.3.5) be a supporting triinvariant decomposition for $W(\lambda)$. Then $W(\lambda)$ admits a minimal factorization

$$
\begin{aligned}
W(\lambda) &= [I + C\pi_{\mathcal{L}}(\lambda I - A)^{-1}\pi_{\mathcal{L}}B][I + C\pi_{\mathcal{M}}(\lambda I - A)^{-1}\pi_{\mathcal{M}}B] \\
&\quad \times [I + C\pi_{\mathcal{N}}(\lambda I - A)^{-1}\pi_{\mathcal{N}}B] \\
&= [I + C(\lambda I - A)^{-1}\pi_{\mathcal{L}}B][I + C\pi_{\mathcal{M}}(\lambda I - A)^{-1}\pi_{\mathcal{M}}B] \\
&\quad \times [I + C\pi_{\mathcal{N}}(\lambda I - A)^{-1}B] \tag{7.3.6}
\end{aligned}
$$

where $\pi_{\mathcal{L}}$ is the projector on $\mathcal{L}$ along $\mathcal{M} \dotplus \mathcal{N}$, and $\pi_{\mathcal{M}}$ and $\pi_{\mathcal{N}}$ are defined similarly.

Conversely, for every minimal factorization $W(\lambda) = W_1(\lambda)W_2(\lambda)W_3(\lambda)$ where the factors are rational matrix functions with value I at infinity there exists a unique supporting triinvariant decomposition $\mathbb{C}^m = \mathcal{L} \dotplus \mathcal{M} \dotplus \mathcal{N}$ such that

$$W_1(\lambda) = I + C\pi_{\mathcal{L}}(\lambda I - A)^{-1}\pi_{\mathcal{L}}B$$

$$W_2(\lambda) = I + C\pi_{\mathcal{M}}(\lambda I - A)^{-1}\pi_{\mathcal{M}}B \tag{7.3.7}$$

$$W_3(\lambda) = I + C\pi_{\mathcal{N}}(\lambda I - A)^{-1}\pi_{\mathcal{N}}B$$

Note that the second equality in (7.3.6) follows from the relations $\pi_{\mathcal{L}}A\pi_{\mathcal{L}} = A\pi_{\mathcal{L}}$ and $\pi_{\mathcal{N}}A\pi_{\mathcal{N}} = \pi_{\mathcal{N}}A$, which express the A invariance of $\mathcal{L}$ and $\mathcal{L} \dotplus \mathcal{M}$, respectively (see Section 1.5).

Proof. With respect to the direct sum decomposition (7.3.5), write

$$A = \begin{bmatrix} A_{11} & A_{12} & A_{13} \\ 0 & A_{22} & A_{23} \\ 0 & 0 & A_{33} \end{bmatrix}, \qquad A^\times = \begin{bmatrix} A_{11}^\times & 0 & 0 \\ A_{21}^\times & A_{22}^\times & 0 \\ A_{31}^\times & A_{32}^\times & A_{33}^\times \end{bmatrix}$$

$$C = [C_1 \quad C_2 \quad C_3], \qquad B = \begin{bmatrix} B_1 \\ B_2 \\ B_3 \end{bmatrix}$$

Note, in particular, that the triangular form of $A^\times$ implies $A_{12} = B_1 C_2$, $A_{13} = B_1 C_3$, and $A_{23} = B_2 C_3$. Applying formula (7.2.9) twice, we now see that the product on the right-hand side of (7.3.6) is indeed $W(\lambda)$. Further, denoting $W_{\mathcal{K}}(\lambda) = I + C\pi_{\mathcal{K}}(\lambda I - A)^{-1}\pi_{\mathcal{K}}B$, for $\mathcal{K} = \mathcal{L}$, $\mathcal{M}$, or $\mathcal{N}$, we obviously have $\delta(W_{\mathcal{K}}) \leq \dim \mathcal{K}$. Hence

$$\delta(W) \leq \delta(W_{\mathcal{L}}) + \delta(W_{\mathcal{M}}) + \delta(W_{\mathcal{N}}) \leq \dim \mathcal{L} + \dim \mathcal{M} + \dim \mathcal{N} = m$$

Since, by definition, $m = \delta(W)$, it follows that

$$\delta(W_{\mathcal{L}}) + \delta(W_{\mathcal{M}}) + \delta(W_{\mathcal{N}}) = m = \delta(W)$$

and the factorization (7.3.6) is minimal.

Next assume that $W = W_1 W_2 W_3$ is a minimal factorization of W, and for $i = 1, 2, 3$ let

$$W_i(\lambda) = I + C_i(\lambda I - A_i)^{-1}B_i$$

be a minimal realization of $W_i(\lambda)$. By the multiplication formula (7.2.9)

$$W(\lambda) = I + \tilde{C}(\lambda I - \tilde{A})^{-1}\tilde{B} \tag{7.3.8}$$

where

$$\tilde{C} = [C_1 \quad C_2 \quad C_3], \qquad \tilde{A} = \begin{bmatrix} A_1 & B_1 C_2 & B_1 C_3 \\ 0 & A_2 & B_2 C_3 \\ 0 & 0 & A_3 \end{bmatrix}, \qquad \tilde{B} = \begin{bmatrix} B_1 \\ B_2 \\ B_3 \end{bmatrix}$$

Note that

$$\tilde{A} - \tilde{B}\tilde{C} = \begin{bmatrix} A_1 - B_1 C_1 & 0 & 0 \\ -B_2 C_1 & A_2 - B_2 C_2 & 0 \\ -B_3 C_1 & -B_3 C_2 & A_3 - B_3 C_3 \end{bmatrix}$$

As the factorization $W = W_1 W_2 W_3$ is minimal, the realization (7.3.8) is minimal. Hence, by Theorem 7.1.4, for some invertible matrix S we have

$$\tilde{C} = CS , \qquad \tilde{A} = S^{-1}AS , \qquad \tilde{B} = S^{-1}B$$

To satisfy (7.3.7), put $\mathscr{L} = S\tilde{\mathscr{L}}$, $\mathscr{M} = S\tilde{\mathscr{M}}$, and $\mathscr{N} = S\tilde{\mathscr{N}}$, where

$$\tilde{\mathscr{L}} = \mathrm{Span}\{e_1, \ldots, e_{p_1}\} , \qquad \tilde{\mathscr{M}} = \mathrm{Span}\{e_{p_1+1}, \ldots, e_{p_1+p_2}\}$$
$$\tilde{\mathscr{N}} = \mathrm{Span}\{e_{p_1+p_2+1}, \ldots, e_{p_1+p_2+p_3}\}$$

and A_i has size p_i for $i = 1, 2, 3$.

It remains to prove the uniqueness of $\mathscr{L}$, $\mathscr{M}$, and $\mathscr{N}$. Assume that $\mathbb{C}^m = \mathscr{L}' \dotplus \mathscr{M}' \dotplus \mathscr{N}'$ is also a supporting triinvariant decomposition such that

$$W_1(\lambda) = I + C\pi_{\mathscr{L}'}(\lambda I - A)^{-1}\pi_{\mathscr{L}'}B$$
$$W_2(\lambda) = I + C\pi_{\mathscr{M}'}(\lambda I - A)^{-1}\pi_{\mathscr{M}'}B \qquad (7.3.9)$$
$$W_3(\lambda) = I + C\pi_{\mathscr{N}'}(\lambda I - A)^{-1}\pi_{\mathscr{N}'}B$$

As the realizations (7.3.7) and (7.3.9) are minimal (see the first part of the proof), there exist invertible transformations $T_{\mathscr{L}}: \mathscr{L}' \to \mathscr{L}$, $T_{\mathscr{M}}: \mathscr{M}' \to \mathscr{M}$, $T_{\mathscr{N}}: \mathscr{N}' \to \mathscr{N}$ such that

$$C\pi_{\mathscr{H}'} = C\pi_{\mathscr{H}}T_{\mathscr{H}} , \qquad \pi_{\mathscr{H}'}A\pi_{\mathscr{H}'} = (T_{\mathscr{H}})^{-1}\pi_{\mathscr{H}}A\pi_{\mathscr{H}}T_{\mathscr{H}}$$
$$\pi_{\mathscr{H}'}B = (T_{\mathscr{H}})^{-1}\pi_{\mathscr{H}}B , \qquad \mathscr{H} = \mathscr{L}, \mathscr{M}, \mathscr{N}$$

Therefore, the invertible transformation $T: \mathbb{C}^m \to \mathbb{C}^m$ defined by $T_{|\mathscr{H}'} = T_{\mathscr{H}}$ for $\mathscr{H} = \mathscr{L}, \mathscr{M}, \mathscr{N}$ is a similarity between the minimal realization $W(\lambda) = I + C(\lambda I - A)^{-1}B$ and itself:

$$C = CT ; \qquad A = T^{-1}AT ; \qquad B = T^{-1}B$$

Because of the uniqueness of such a similarity (Theorem 7.1.4), we must have $T = I$. So $\mathscr{L}' = \mathscr{L}$, $\mathscr{M}' = \mathscr{M}$, $\mathscr{N}' = \mathscr{N}$. $\square$

Using formula (7.3.2), we can rewrite the minimal factorization (7.3.6) in terms of the minimal factorization of the inverse matrix function:

$$W(\lambda)^{-1} = [I - C\pi_{\mathscr{N}}(\lambda I - A^{\times})^{-1}\pi_{\mathscr{N}}B][I - C\pi_{\mathscr{M}}(\lambda I - A^{\times})^{-1}\pi_{\mathscr{M}}B]$$
$$\times [I - C\pi_{\mathscr{L}}(\lambda I - A^{\times})^{-1}\pi_{\mathscr{L}}B]$$
$$= [I - C(\lambda I - A^{\times})^{-1}\pi_{\mathscr{N}}B][I - C\pi_{\mathscr{M}}(\lambda I - A^{\times})^{-1}\pi_{\mathscr{M}}B]$$
$$\times [I - C\pi_{\mathscr{L}}(\lambda I - A^{\times})^{-1}B]$$

where the second equality follows from $\pi_{\mathcal{N}} A^{\times} \pi_{\mathcal{N}} = A^{\times} \pi_{\mathcal{N}}$ and $\pi_{\mathcal{L}} A^{\times} \pi_{\mathcal{L}} = \pi_{\mathcal{L}} A^{\times}$, expressing the $A^{\times}$ invariance of $\mathcal{N}$ and $\mathcal{M} \dotplus \mathcal{N}$.

An important particular case of Theorem 7.3.1 appears when $\mathcal{N} = \{0\}$ in the supporting triinvariant decomposition (7.3.5). This corresponds to the minimal factorization of $W(\lambda)$ into the product of two factors, as follows.

Corollary 7.3.2

Let $\mathcal{L}$ and $\mathcal{M}$ be subspaces in $\mathbb{C}^m$ that are direct complements of each other. Assume that $\mathcal{L}$ is A invariant and $\mathcal{M}$ is $A^{\times}$ invariant. Then $W(\lambda)$ admits a minimal factorization

$$W(\lambda) = [I + C(\lambda I - A)^{-1} \pi_{\mathcal{L}} B][I + C(I - \pi_{\mathcal{L}})(\lambda I - A)^{-1} B]$$

where $\pi_{\mathcal{L}}$ is the projector on $\mathcal{L}$ along $\mathcal{M}$. Conversely, if $W(\lambda) = W_1(\lambda) W_2(\lambda)$ is a minimal factorization with $W_1(\infty) = W_2(\infty) = I$, then there exists a unique direct sum decomposition $\mathbb{C}^m = \mathcal{L} \dotplus \mathcal{M}$, where $\mathcal{L}$ is A invariant, $\mathcal{M}$ is $A^{\times}$ invariant, and such that $W_1(\lambda) = I + C(\lambda I - A)^{-1} \pi_{\mathcal{L}} B$, $W_2(\lambda) = I + C(I - \pi_{\mathcal{L}})(\lambda I - A)^{-1} B$.

7.4   EXAMPLE

Let us illustrate the description of minimal factorizations obtained in Theorem 7.3.1. The rational matrix function

$$W(\lambda) = \begin{bmatrix} 1 + \lambda^{-1}(\lambda - 1)^{-1} & \lambda^{-1} \\ 0 & 1 + \lambda^{-1} \end{bmatrix}$$

has a realization

$$W(\lambda) = I + C(\lambda I - A)^{-1} B \qquad\qquad (7.4.1)$$

where

$$A = \begin{bmatrix} 1 & 0 & 0 \\ 1 & 0 & 0 \\ 0 & 0 & 0 \end{bmatrix}, \qquad B = \begin{bmatrix} 1 & 0 \\ 0 & 1 \\ 0 & 1 \end{bmatrix}, \qquad C = \begin{bmatrix} 0 & 1 & 0 \\ 0 & 0 & 1 \end{bmatrix}$$

This realization is minimal. Indeed, the matrix

$$\begin{bmatrix} C \\ CA \end{bmatrix} = \begin{bmatrix} 0 & 1 & 0 \\ 0 & 0 & 1 \\ 1 & 0 & 0 \\ 1 & 0 & 0 \end{bmatrix}$$

has rank 3 and hence zero kernel. The matrix

Example 231

$$[B, AB] = \begin{bmatrix} 1 & 0 & 1 & 0 \\ 0 & 1 & 1 & 0 \\ 0 & 1 & 0 & 0 \end{bmatrix}$$

has rank 3, and hence its image is $\mathbb{C}^3$. Further

$$A^\times = A - BC = \begin{bmatrix} 1 & -1 & 0 \\ 1 & 0 & -1 \\ 0 & 0 & -1 \end{bmatrix}$$

Let us find all invariant subspaces for A and $A^\times$. It is easy to see that $\langle 1, 1, 0 \rangle$ is an eigenvector of A corresponding to the eigenvalue 1, whereas the vectors $\langle 0, 0, 1 \rangle$, $\langle 0, 1, 0 \rangle$ are the eigenvectors of A corresponding to the eigenvalue 0. Hence all one-dimensional A-invariant subspaces are of the form $\mathrm{Span}\{\langle 1, 1, 0 \rangle\}$; $\mathrm{Span}\{\langle 0, 1, 0 \rangle\}$; $\mathrm{Span}\{\langle 0, \alpha, 1 \rangle\}$, $\alpha \in \mathbb{C}$. All two-dimensional A-invariant subspaces are of the form

$$\mathrm{Span}\{\langle 1, 0, 0 \rangle, \langle 0, 1, 0 \rangle\} \, ; \qquad \mathrm{Span}\{\langle 1, 1, 0 \rangle, \langle 0, \alpha, 1 \rangle\} \, , \quad \alpha \in \mathbb{C}$$

$$\mathrm{Span}\{\langle 0, 1, 0 \rangle, \langle 0, 0, 1 \rangle\}$$

Passing to $A^\times$, we find that $A^\times$ has three eigenvalues -1, $\gamma = \frac{1}{2}(1 + i\sqrt{3})$, and $\bar{\gamma}$ with corresponding eigenvectors $\langle 1, 2, 3 \rangle$, $\langle 1, \bar{\gamma}, 0 \rangle$, and $\langle 1, \gamma, 0 \rangle$, respectively. There are three one-dimensional $A^\times$-invariant subspaces $\mathrm{Span}\{\langle 1, 2, 3 \rangle\}$, $\mathrm{Span}\{\langle 1, \bar{\gamma}, 0 \rangle\}$, $\mathrm{Span}\{\langle 1, \gamma, 0 \rangle\}$, and three two-dimensional $A^\times$-invariant subspaces $\mathrm{Span} \{\langle 1, 2, 3 \rangle, \langle 1, \bar{\gamma}, 0 \rangle\}$, $\mathrm{Span}\{\langle 1, 2, 3 \rangle, \langle 1, \gamma, 0 \rangle\}$, and $\mathrm{Span}\{\langle 1, 0, 0 \rangle, \langle 0, 1, 0 \rangle\}$.

Now we describe supporing triinvariant decompositions

$$\mathbb{C}^3 = \mathscr{L} \dotplus \mathscr{M} \dotplus \mathscr{N} \tag{7.4.2}$$

of $W(\lambda)$ with $\mathscr{N} = \mathrm{Span}\{\langle 1, 2, 3 \rangle\}$, $\mathscr{L} = \mathrm{Span}\{\langle 1, 1, 0 \rangle\}$. If we let $\mathscr{M} = \mathrm{Span}\{\langle x, y, z \rangle\}$, we easily see that

$$\mathbb{C}^3 = \mathrm{Span}\{\langle 1, 1, 0 \rangle\} \dotplus \mathscr{M} \dotplus \mathrm{Span}\{\langle 1, 2, 3 \rangle\}$$

if and only if $z \neq 3(y - x)$. Further, one of the following four cases appears:

(a) $\mathscr{M} = \mathrm{Span}\{\langle 1, 0, 0 \rangle, \langle 0, 1, 0 \rangle\} \cap \mathrm{Span}\{\langle 1, 2, 3 \rangle, \langle 1, \bar{\gamma}, 0 \rangle\}$.
(b) $\mathscr{M} = \mathrm{Span}\{\langle 1, 1, 0 \rangle, \langle 0, \alpha, 1 \rangle\} \cap \mathrm{Span}\{\langle 1, 2, 3 \rangle, \langle 1, \bar{\gamma}, 0 \rangle\}$, $\quad \alpha \in \mathbb{C}$.
(c) $\mathscr{M} = \mathrm{Span}\{\langle 1, 0, 0 \rangle, \langle 0, 1, 0 \rangle\} \cap \mathrm{Span}\langle\{1, 2, 3 \rangle, \langle 1, \gamma, 0 \rangle\}$.
(d) $\mathscr{M} = \mathrm{Span}\{\langle 1, 1, 0 \rangle, \langle 0, \alpha, 1 \rangle\} \cap \mathrm{Span}\{\langle 1, 2, 3 \rangle, \langle 1, \gamma, 0 \rangle\}$, $\quad \alpha \in \mathbb{C}$.

In cases (a) and (c) we obtain $\mathscr{M} = \mathrm{Span}\{\langle 1, \bar{\gamma}, 0 \rangle\}$ and $\mathscr{M} = \mathrm{Span}\{\langle 1, \gamma, 0 \rangle\}$, respectively. In case (b) we have

$$\begin{bmatrix} x \\ y \\ z \end{bmatrix} = p \begin{bmatrix} 1 \\ 1 \\ 0 \end{bmatrix} + q \begin{bmatrix} 0 \\ \alpha \\ 1 \end{bmatrix} = r \begin{bmatrix} 1 \\ 2 \\ 3 \end{bmatrix} + s \begin{bmatrix} 1 \\ \gamma \\ 0 \end{bmatrix} \tag{7.4.3}$$

for some complex numbers p, q, r, s. Consider the second equality in (7.4.3) as an equation with unknowns p, q, r, s. Solving this equation and putting $r = 1 - \bar{\gamma}$, we get $q = 3 - 3\bar{\gamma}, s = 1 - 3\alpha, p = 2 - 3\alpha - \bar{\gamma}$, and M is spanned by $\langle 2 - 3\alpha - \bar{\gamma}, 2 - 3\alpha\bar{\gamma} - \bar{\gamma}, 3 - 3\bar{\gamma} \rangle$, where $\alpha \neq \frac{1}{3}$. [This condition reflects the inequality $z \neq 3(y - x)$.] Similarly, in case (d) we obtain $M = \mathrm{Span}\{\langle 2 - 3\alpha - \gamma, 2 - 3\alpha\gamma - \gamma, 3 - 3\gamma \rangle\}$, where $\alpha \neq \frac{1}{3}$. To summarize, the subspaces M for which

$$\mathbb{C}^3 = \mathrm{Span}\{\langle 1, 1, 0 \rangle\} \dotplus M + \mathrm{Span}\{\langle 1, 2, 3 \rangle\}$$

is a supporting triinvariant decomposition for $W(\lambda)$ are exactly the following: $\mathrm{Span}\{\langle 1, \gamma, 0 \rangle\}$; $\mathrm{Span}\{\langle 1, \bar{\gamma}, 0 \rangle\}$; $\mathrm{Span}\{\langle 2 - 3\alpha - \bar{\gamma}, 2 - 3\alpha\bar{\gamma} - \bar{\gamma}, 3 - 3\bar{\gamma} \rangle\}$, $\alpha \neq \frac{1}{3}$; and $\mathrm{Span}\{\langle 2 - 3\alpha - \gamma, 2 - 3\alpha\gamma - \gamma, 3 - 3\gamma \rangle\}$, $\alpha \neq \frac{1}{3}$.

To compute the corresponding minimal factorizations according to formula (7.3.6), write the matrices A, B, C (understood as transformations in the standard orthonormal bases in $\mathbb{C}^2$ and $\mathbb{C}^3$) with respect to the basis $\langle 1, 1, 0 \rangle, \langle 1, \gamma, 0 \rangle, \langle 1, 2, 3 \rangle$ in $\mathbb{C}^3$ and the standard basis $\langle 1, 0 \rangle, \langle 0, 1 \rangle$ in $\mathbb{C}^2$:

$$A = \begin{bmatrix} 1 & 1 & 1 \\ 0 & 0 & 0 \\ 0 & 0 & 0 \end{bmatrix}; \qquad C = \begin{bmatrix} 1 & \gamma & 2 \\ 0 & 0 & 3 \end{bmatrix};$$

$$B = \begin{bmatrix} -\gamma(1 - \gamma)^{-1} & -\frac{1}{3}(\gamma + 1)(\gamma - 1)^{-1} \\ (1 - \gamma)^{-1} & \frac{2}{3}(\gamma - 1)^{-1} \\ 0 & \frac{1}{3} \end{bmatrix}$$

So the minimal factorization corresponding to the supporting triinvariant decomposition (7.4.3) with $M = \mathrm{Span}\{\langle 1, \gamma, 0 \rangle\}$ is $W(\lambda) = W_1(\lambda)W_2(\lambda)W_3(\lambda)$, where

$$W_1(\lambda) = I + \begin{bmatrix} 1 \\ 0 \end{bmatrix}(\lambda - 1)^{-1}\left[-\gamma(1 - \gamma)^{-1}, -\frac{1}{3}(\gamma + 1)(\gamma - 1)^{-1} \right]$$

$$= \begin{bmatrix} 1 - \dfrac{\gamma}{(1 - \gamma)(\lambda - 1)} & \dfrac{-(\gamma + 1)}{3(\gamma - 1)(\lambda - 1)} \\ 0 & 1 \end{bmatrix}$$

Example 233

$$W_2(\lambda) = I + \begin{bmatrix} \gamma \\ 0 \end{bmatrix} \lambda^{-1} \Big[(1-\gamma)^{-1}, \frac{2}{3}(\gamma-1)^{-1} \Big] = \begin{bmatrix} 1 + \dfrac{\gamma}{(1-\gamma)\lambda} & \dfrac{2\gamma}{3(\gamma-1)\lambda} \\ 0 & 1 \end{bmatrix}$$

$$W_3(\lambda) = I + \begin{bmatrix} 2 \\ 3 \end{bmatrix} \lambda^{-1} \Big[0, \frac{1}{3} \Big] = \begin{bmatrix} 1 & \dfrac{2}{3\lambda} \\ 0 & \dfrac{1+\lambda}{\lambda} \end{bmatrix}$$

Replacing γ by $\bar{\gamma}$ in these expressions we obtain the minimal factorization corresponding to (7.4.3) with $\mathcal{M} = \text{Span}\{\langle 1, \bar{\gamma}, 0\rangle\}$.

Now for $\alpha \neq \frac{1}{3}$ write A, B, and C in the basis $\langle 1, 1, 0\rangle$, $\langle 2 - 3\alpha - \gamma,$ $2 - 3\alpha\gamma - \gamma, 3 - 3\gamma\rangle$, $\langle 1, 2, 3\rangle$:

$$A = \begin{bmatrix} 1 & 2-3\alpha-\gamma & 1 \\ 0 & 0 & 0 \\ 0 & 0 & 0 \end{bmatrix}, \quad C = \begin{bmatrix} 1 & 2-3\alpha\gamma-\gamma & 2 \\ 0 & 3-3\gamma & 3 \end{bmatrix}$$

$$B = \begin{bmatrix} \gamma(\gamma-1)^{-1} & \frac{1}{3}(\gamma+1)(1-\gamma)^{-1} \\ (3\alpha-1)^{-1}(\gamma-1)^{-1} & \frac{2}{3}(3\alpha-1)^{-1}(1-\gamma)^{-1} \\ (3\alpha-1)^{-1} & (\alpha-1)(2\alpha-1)^{-1} \end{bmatrix}$$

The corresponding minimal factorization is given by

$$W_1(\lambda) = \begin{bmatrix} 1 + \dfrac{\gamma}{(\gamma-1)(\lambda-1)} & \dfrac{\gamma+1}{3(1-\gamma)(\lambda-1)} \\ 0 & 1 \end{bmatrix}$$

$$W_2(\lambda) = \begin{bmatrix} 1 + \dfrac{2-3\alpha\gamma-\gamma}{(3\alpha-1)(\gamma-1)\lambda} & \dfrac{2(2-3\alpha\gamma-\gamma)}{3(3\alpha-1)(1-\gamma)\lambda} \\ \dfrac{3-3\gamma}{(3\alpha-1)(\gamma-1)\lambda} & 1 + \dfrac{2(3-3\gamma)}{3(3\alpha-1)(1-\gamma)\lambda} \end{bmatrix}$$

$$W_3(\lambda) = \begin{bmatrix} 1 + \dfrac{2}{(3\alpha-1)\lambda} & \dfrac{2(\alpha-1)}{(3\alpha-1)\lambda} \\ \dfrac{3}{(3\alpha-1)\lambda} & 1 + \dfrac{3(\alpha-1)}{(3\alpha-1)\lambda} \end{bmatrix}$$

Taking $\bar{\gamma}$ in the place of γ in these expressions, we obtain the minimal factorization corresponding to (7.4.3) with $\mathcal{M} = \mathrm{Span}\{\langle 2 - 3\alpha - \bar{\gamma},$ $2 - 3\alpha\bar{\gamma} - \bar{\gamma}, 3 - 3\bar{\gamma}\rangle\}$.

Note that these four factorizations exhaust all minimal factorizations

$$\begin{bmatrix} 1 + \lambda^{-1}(\lambda - 1)^{-1} & \lambda^{-1} \\ 0 & 1 + \lambda^{-1} \end{bmatrix} = W_1(\lambda)W_2(\lambda)W_3(\lambda)$$

with not identically constant rational 2×2 matrix functions $W_i(\lambda)$ with value I at infinity and for which $W_1(\lambda)$ has a pole at $\lambda_0 = 1$ and $W_3(\lambda)$ has a zero at $\lambda_0 = -1$ [i.e., $W_3(\lambda)^{-1}$ has a pole at $\lambda_0 = -1$].

7.5 MINIMAL FACTORIZATIONS INTO SEVERAL FACTORS AND CHAINS OF INVARIANT SUBSPACES

Let $W(\lambda)$ be an $n \times n$ rational matrix function with minimal realization

$$W(\lambda) = I + C(\lambda I - A)^{-1}B \tag{7.5.1}$$

so that, in particular, $W(\infty) = I$. We study minimal factorizations of $W(\lambda)$ by means of the realization (7.5.1), and in terms of chains of invariant subspaces for A and $A^\times = A - BC$. We state the main theorem of this section.

Theorem 7.5.1

Let m be the size of A in equation (7.5.1), and let

$$\mathbb{C}^m = \mathcal{L}_1 \dotplus \cdots \dotplus \mathcal{L}_p \tag{7.5.2}$$

where the chain

$$\mathcal{L}_1 \subset \mathcal{L}_1 \dotplus \mathcal{L}_2 \subset \cdots \subset \mathcal{L}_1 \dotplus \mathcal{L}_2 \dotplus \cdots \dotplus \mathcal{L}_{p-1} \tag{7.5.3}$$

consists of A-invariant subspaces, whereas the chain

$$\mathcal{L}_p \subset \mathcal{L}_p \dotplus \mathcal{L}_{p-1} \subset \cdots \subset \mathcal{L}_p \dotplus \mathcal{L}_{p-1} \dotplus \cdots \dotplus \mathcal{L}_2 \tag{7.5.4}$$

consists of $A^\times$-invariant subspaces. Then $W(\lambda)$ admits the minimal factorization

$$W(\lambda) = [I + C\pi_1(\lambda I - A)^{-1}\pi_1 B] \cdots [I + C\pi_p(\lambda I - A)^{-1}\pi_p B] \tag{7.5.5}$$

where π_j is the projector on $\mathcal{L}_j$ along $\mathcal{L}_j \dotplus \cdots \dotplus \mathcal{L}_{j-1} \dotplus \mathcal{L}_{j+1} \dotplus \cdots \dotplus \mathcal{L}_p$. Conversely, for every minimal factorization

$$W(\lambda) = W_1(\lambda) \cdots W_p(\lambda) \qquad (7.5.6)$$

where $W_j(\lambda)$ are rational $n \times n$ matrix functions with $W_j(\infty) = I$, there exists a unique direct sum decomposition (7.5.2) with the property that the chains (7.5.3) and (7.5.4) consist of invariant subspaces for A and $A^\times$, respectively, such that

$$W_j(\lambda) = I + C\pi_j(\lambda I - A)^{-1}\pi_j B, \qquad j = 1, \ldots, p$$

The proof is obtained by $p - 1$ consecutive applications of Corollary 7.3.2.

As in the remark following the proof of Theorem 7.3.1, the factorization (7.5.5) implies the minimal factorization for $W(\lambda)^{-1}$:

$$W(\lambda)^{-1} = [I - C\pi_p(\lambda I - A^\times)^{-1}\pi_p B]$$
$$\times [I - C\pi_{p-1}(\lambda I - A^\times)^{-1}\pi_{p-1}B] \cdot \ \cdots \ \cdot [I - C\pi_1(\lambda I - A^\times)^{-1}\pi_1 B]$$

We are interested in the case when p, the number of factors in the minimal factorization (7.5.6), is maximal [of course, we exclude the case when some of the $W_j(\lambda)$ values are identically equal to I]. Obviously, p cannot exceed the McMillan degree of $W(\lambda)$, $\delta(W)$, for then each factor $W_j(\lambda)$ must have McMillan degree 1. It is not difficult to find a general form of rational $n \times n$ matrix functions $V(\lambda)$ with $V(\infty) = I$ and $\delta(V) = 1$; namely

$$V(\lambda) = I + (\lambda - \lambda_0)^{-1}R \qquad (7.5.7)$$

where λ_0 is a complex number and R is an $n \times n$ matrix of rank 1. Indeed, if $V(\lambda)$ has the form (7.5.7), then by writing $R = C_0 B_0$, where C_0 is an $n \times 1$ matrix and B_0 is a $1 \times n$ matrix, we obtain a realization $I + C_0(\lambda - \lambda_0)^{-1}B_0$ of $V(\lambda)$ that is obviously minimal. So $\delta(V) = 1$. Conversely, if $\delta(V) = 1$, then we take a minimal realization $I + C_0(\lambda - \lambda_0)^{-1}B_0$ of $V(\lambda)$ and put $R = C_0 B_0$ to obtain (7.5.7).

Note that if $V(\lambda)$ has the form (7.5.7), then so does $V(\lambda)^{-1}$ [because $\delta(V^{-1}) = \delta(V) = 1$]. Indeed, by equation (7.3.2)

$$V(\lambda)^{-1} = I - (\lambda - (\lambda_0 - \operatorname{tr} R))^{-1}R$$

where $\operatorname{tr} R$ is the trace of R (the sum of its diagonal entries).

We arrive at the following problem: study minimal factorizations

$$W(\lambda) = V_1(\lambda) \cdots V_m(\lambda) \qquad (7.5.8)$$

of $W(\lambda)$, where each $V_j(\lambda)$ has the form (7.5.7) for some λ_0 and R. First let us see an example showing that not every $W(\lambda)$ admits a minimal factorization of this type.

EXAMPLE 5.1.   Let

$$W(\lambda) = I + \begin{bmatrix} 1 & 0 \\ 0 & 0 \end{bmatrix}\left(\lambda I - \begin{bmatrix} 0 & 1 \\ 0 & 0 \end{bmatrix}\right)^{-1}\begin{bmatrix} 0 & 0 \\ 0 & 1 \end{bmatrix} = \begin{bmatrix} 1 & \lambda^{-2} \\ 0 & 1 \end{bmatrix}$$

This realization (with

$$A = \begin{bmatrix} 0 & 1 \\ 0 & 0 \end{bmatrix}, \qquad B = \begin{bmatrix} 0 & 0 \\ 0 & 1 \end{bmatrix}, \qquad C = \begin{bmatrix} 1 & 0 \\ 0 & 0 \end{bmatrix})$$

is easily seen to be minimal. As $BC = 0$, we have

$$A^{\times} = A = \begin{bmatrix} 0 & 1 \\ 0 & 0 \end{bmatrix}$$

Obviously, there is no (nontrivial) direct sum decomposition $\mathbb{C}^2 = \mathscr{L}_1 \dot{+} \mathscr{L}_2$, where $\mathscr{L}_1$ and $\mathscr{L}_2$ are A invariant. So by Theorem 7.5.1 (or Corollary 7.3.2) $W(\lambda)$ does not admit minimal factorizations, except for the trivial ones $W(\lambda) = W(\lambda)I = IW(\lambda)$.   □

We give a sufficient condition for the existence of a minimal factorization (7.5.8). This condition is based on the following independently interesting property of chains of invariant subspaces.

Lemma 7.5.2

Let A_1, A_2: $\mathbb{C}^n \to \mathbb{C}^n$ be transformations and assume that at least one of them is diagonable. Then there exists a direct sum decomposition $\mathbb{C}^n = \mathscr{L}_1 \dot{+} \cdots \dot{+} \mathscr{L}_n$ with one-dimensional subspaces $\mathscr{L}_j$, $j = 1, \ldots, n$, such that the complete chains

$$\mathscr{L}_1 \subset \mathscr{L}_1 \dot{+} \mathscr{L}_2 \subset \cdots \subset \mathscr{L}_1 \dot{+} \cdots \dot{+} \mathscr{L}_{n-1}$$

and

$$\mathscr{L}_n \subset \mathscr{L}_n \dot{+} \mathscr{L}_{n-1} \subset \cdots \subset \mathscr{L}_n \dot{+} \mathscr{L}_{n-1} \dot{+} \cdots \dot{+} \mathscr{L}_2$$

consist of A_1-invariant and A_2-invariant subspaces, respectively.

Proof.   It is sufficient to prove the existence of a direct sum decomposition

$$\mathbb{C}^n = \mathscr{M} \dot{+} \mathscr{L} \tag{7.5.9}$$

where dim $\mathscr{M} = n - 1$, dim $\mathscr{L} = 1$, $\mathscr{M}$ is A_1 invariant, and $\mathscr{L}$ is A_2 invariant.

Indeed, we can then use induction on n and assume that Lemma 7.5.2 is already proved for $A_{1|\mathcal{M}}$ and $P_{\mathcal{M}}A_{2|\mathcal{M}}$ in place of A_1 and A_2, respectively, where $P_{\mathcal{M}}$ is projector on $\mathcal{M}$ along $\mathcal{L}$. (Remember that if at least one of A_1 and A_2 is diagonable, the same is true for $A_{1|\mathcal{M}}$ and $P_{\mathcal{M}}A_{2|\mathcal{M}}$; see Theorems 4.1.4 and 4.1.5.) Combining (7.5.9) with the result of Lemma 7.5.2 for $A_{1|\mathcal{M}}$ and $P_{\mathcal{M}}A_{2|\mathcal{M}}$, we prove the lemma for A_1 and A_2.

To establish the existence of the decomposition (7.5.9), assume first that A_1 is diagonable, and let $f_1, \ldots, f_n$ be a basis for $\mathbb{C}^n$ consisting of eigenvectors of A_1. If g is an eigenvector of A_2, (7.5.9) is satisfied with $\mathcal{L} = \operatorname{Span}\{f_{i_1}, \ldots, f_{i_{n-1}}\}$, where the indices $i_1, \ldots, i_{n-1}$ are such that $f_{i_1}, \ldots, f_{i_{n-1}}, g$ form a basis in $\mathbb{C}^n$.

If A_2 is diagonable but A_1 is not, then use the part of the theorem already proved with A_2^* and A_1^* in place of A_1 and A_2, respectively. We obtain an $(n-1)$-dimensional A_2^*-invariant subspace $\hat{\mathcal{M}}$ and a one-dimensional A_1^*-invariant subspace $\hat{\mathcal{L}}$ that are direct complements of each other. Then put $\mathcal{M} = (\hat{\mathcal{L}})^{\perp}$ and $\mathcal{L} = (\hat{\mathcal{M}})^{\perp}$ to satisfy (7.5.9). $\square$

We can now state and prove the following sufficient condition for minimal factorization of a rational matrix function $W(\lambda)$ into the product of $\delta(W)$ nontrivial factors.

Theorem 7.5.3

Let $W(\lambda)$ be a rational $n \times n$ matrix function with a minimal realization

$$W(\lambda) = I + C(\lambda I - A)^{-1}B \qquad (7.5.10)$$

and assume that at least one of the matrices A and $A - BC$ is diagonable. Then $W(\lambda)$ admits a minimal factorization of the form

$$W(\lambda) = [I + (\lambda - \lambda_1)^{-1}R_1] \cdots [I + (\lambda - \lambda_m)^{-1}R_m] \qquad (7.5.11)$$

where $\lambda_1, \ldots, \lambda_m$ are complex numbers and $R_1, \ldots, R_m$ are $n \times n$ matrices of rank 1.

The proof of Theorem 7.5.3 is obtained by combining Theorem 7.5.1 and Corollary 7.5.2. In Example 7.5.1 the hypothesis of Theorem 7.5.3 is obviously violated. Indeed, the matrix $\begin{bmatrix} 0 & 1 \\ 0 & 0 \end{bmatrix}$ is not diagonable. The following form of Theorem 7.5.3 may be more easily applied in many cases.

Theorem 7.5.4

Let $W(\lambda)$ be a rational $n \times n$ matrix function with $W(\infty) = I$. Assume that either in $W(\lambda)$, or in $W(\lambda)^{-1}$, all the poles (if any) of each entry are of the first order. Then $W(\lambda)$ admits a factorization (7.5.11).

Recall that the order of a pole λ_0 of a scalar rational matrix $f(\lambda)$ is defined as the minimal positive integer r such that $\lim_{\lambda \to \lambda_0}[(\lambda - \lambda_0)^r f(\lambda)]$ is finite.

Proof. Assume that all the poles of each entry in $W(\lambda)$ are of the first order. The local Smith form (7.2.1) implies that all the negative partial multiplicities (if any) of $W(\lambda)$ at each point λ_0 are -1s. By Theorem 7.2.3, all the partial multiplicities of the matrix A from the minimal realization (7.5.10) are 1s. Hence A is diagonable and Theorem 7.5.3 applies. If all poles of $W(\lambda)^{-1}$ are of the first order, apply the above reasoning to $W(\lambda)^{-1}$, using its realization $W(\lambda)^{-1} = I - C(\lambda I - (A - BC))^{-1}B$, which is minimal if (7.5.10) is minimal. $\square$

7.6 LINEAR FRACTIONAL TRANSFORMATIONS

In this and the next sections we study linear fractional transformations and decompositions of general (nonsquare) rational matrix functions. We deviate here from our custom and denote certain matrices by lower case Latin and Greek letters.

Let $W(\lambda)$ be a rational matrix function of size $r \times m$ written in a 2×2 block matrix form as follows:

$$W(\lambda) = \begin{bmatrix} W_{11}(\lambda) & W_{12}(\lambda) \\ W_{21}(\lambda) & W_{22}(\lambda) \end{bmatrix} : \mathbb{C}^{m_1} \dotplus \mathbb{C}^{m_2} \to \mathbb{C}^{r_1} \dotplus \mathbb{C}^{r_2} \qquad (7.6.1)$$

Here m_i and r_i ($i = 1, 2$) are positive integers such that $m = m_1 + m_2$, $r = r_1 + r_2$. Let $V(\lambda)$ be a rational $m_2 \times r_1$ matrix function for which $\det(I - W_{12}(\lambda)V(\lambda)) \neq 0$, and define matrix function

$$U(\lambda) = W_{21}(\lambda) + W_{21}(\lambda)V(\lambda)(I - W_{12}(\lambda)V(\lambda))^{-1}W_{11}(\lambda) \qquad (7.6.2)$$

So $U(\lambda)$ is a rational matrix function of size $r_2 \times m_1$. It is called the *linear fractional transformation* of $V(\lambda)$ by $W(\lambda)$ [with respect to the block matrix form of (7.6.1)] and is denoted by $\mathcal{F}_W(V)$. It is easily seen that when $m_1 = r_1$ and $\det W_{11}(\lambda) \neq 0$, (7.6.2) can be rewritten in the form

$$U(\lambda) = (R_1(\lambda) - R_2(\lambda)V(\lambda))(R_3(\lambda) - R_4(\lambda)V(\lambda))^{-1} \qquad (7.6.3)$$

where

$$R_1(\lambda) = W_{21}(\lambda)W_{11}(\lambda)^{-1}, \qquad R_2(\lambda) = W_{21}(\lambda)W_{11}(\lambda)^{-1}W_{12}(\lambda) - W_{22}(\lambda)$$

$$R_3(\lambda) = W_{11}(\lambda)^{-1}, \qquad R_4(\lambda) = W_{11}(\lambda)^{-1}W_{12}(\lambda)$$

Conversely, if (7.6.3) holds, then we have (7.6.2) with

$$W_{11}(\lambda) = R_3(\lambda)^{-1}, \qquad\qquad W_{12}(\lambda) = R_3(\lambda)^{-1}R_4(\lambda),$$

$$W_{21}(\lambda) = R_1(\lambda)R_3(\lambda)^{-1}, \qquad W_{22}(\lambda) = R_1(\lambda)R_3(\lambda)^{-1}R_4(\lambda) - R_2(\lambda)$$

The form (7.6.3) justifies the terminology "linear fractional transformation", however, the form (7.6.2) will be more convenient for our analysis.

Observe that multiplication of rational matrix functions is a particular case of the linear fractional transformation, which is obtained in case $W_{21}(\lambda) \equiv 0$, $W_{12}(\lambda) \equiv 0$, and either $W_{22}(\lambda) \equiv I$ or $W_{11}(\lambda) \equiv I$.

Assume now that both $W(\lambda)$ and $V(\lambda)$ take finite values at infinity. Then (see Section 7.1) there exist realizations

$$W(\lambda) = D + C(\lambda I - A)^{-1}B \qquad (7.6.4)$$

where A, B, C, and D are matrices of sizes $n \times n$, $n \times m$, $r \times n$, and $r \times m$, respectively, and

$$V(\lambda) = d + c(\lambda I - a)^{-1}b \qquad (7.6.5)$$

with matrices a, b, c, d of size $p \times p$, $p \times r_1$, $m_2 \times p$, and $m_2 \times r_1$, respectively. At this point we do not require that the realizations (7.6.4) and (7.6.5) be minimal. We are to find a realization of $\mathcal{F}_W(V)$ in terms of the realizations (7.6.4) and (7.6.5) of $W(\lambda)$ and $V(\lambda)$.

With respect to the direct sum decompositions $\mathbb{C}^m = \mathbb{C}^{m_1} \dotplus \mathbb{C}^{m_2}$ and $\mathbb{C}^r = \mathbb{C}^{r_1} \dotplus \mathbb{C}^{r_2}$, we write B, C, and D as block matrices

$$B = [B_1 \quad B_2], \qquad C = \begin{bmatrix} C_1 \\ C_2 \end{bmatrix}, \qquad D = \begin{bmatrix} D_{11} & D_{12} \\ D_{21} & D_{22} \end{bmatrix}$$

As $D = W(\infty)$, formula (7.6.2) shows that $\mathcal{F}_W(V)$ is analytic at infinity (i.e., has no poles there) provided the matrix $I - D_{12}d$ is invertible; in this case

$$\mathcal{F}_W(V)(\infty) = D_{21} + D_{22}d(I - D_{12}d)^{-1}D_{11}$$

We restrict our attention to rational matrix functions that are analytic at infinity, so it will be assumed that $I - D_{12}d$ is invertible. Then $I - dD_{12}$ is invertible as well and

$$(I - dD_{12})^{-1} = I + d(I - D_{12}d)^{-1}D_{12} \qquad (7.6.6)$$

Indeed, multiplication gives

$$(I - dD_{12})[I + d(I - D_{12}d)^{-1}D_{12}]$$

$$= I - dD_{12} + (d - dD_{12}d)(I - D_{12}d)^{-1}D_{12}$$

$$= I + d[-I + (I - D_{12}d)(I - D_{12}d)^{-1}]D_{12} = I$$

Define transformations:

$$\alpha = \begin{bmatrix} A + B_2 d(I - D_{12}d)^{-1}C_1 & B_2(I - dD_{12})^{-1}c \\ b(I - D_{12}d)^{-1}C_1 & a + b(I - D_{12}d)^{-1}D_{12}c \end{bmatrix}:$$

$$\mathbb{C}^n \dotplus \mathbb{C}^p \to \mathbb{C}^n \dotplus \mathbb{C}^p \qquad (7.6.7)$$

$$\beta = \begin{bmatrix} \beta_1 \\ \beta_2 \end{bmatrix} = \begin{bmatrix} B_1 + B_2 d(I - D_{12}d)^{-1}D_{11} \\ b(I - D_{12}d)^{-1}D_{11} \end{bmatrix}: \mathbb{C}^{m_1} \to \mathbb{C}^n \dotplus \mathbb{C}^p \qquad (7.6.8)$$

$$\gamma = [\gamma_1, \gamma_2] = [C_2 + D_{22}d(I - D_{12}d)^{-1}C_1, \, D_{22}(I - dD_{12})^{-1}c]: \mathbb{C}^n \dotplus \mathbb{C}^p \to \mathbb{C}^{r_2}$$

$$(7.6.9)$$

$$\delta = D_{21} + D_{22}d(I - D_{12}d)^{-1}D_{11}: \mathbb{C}^{m_1} \to \mathbb{C}^{r_2} \qquad (7.6.10)$$

Theorem 7.6.1

We have

$$\mathscr{F}_W(V)(\lambda) = \delta + \gamma(\lambda I - \alpha)^{-1}\beta \qquad (7.6.11)$$

Further, if this realization of $\mathscr{F}_W(V)$ is minimal, then the realizations (7.6.4) and (7.6.5) of $W(\lambda)$ and $V(\lambda)$, respectively, are minimal as well.

Proof.   Write

$$W(\lambda) = \begin{bmatrix} D_{11} & D_{12} \\ D_{21} & D_{22} \end{bmatrix} + \begin{bmatrix} C_1 \\ C_2 \end{bmatrix}(\lambda I - A)^{-1}[B_1 \quad B_2]$$

$$= \begin{bmatrix} D_{11} + C_1(\lambda I - A)^{-1}B_1 & D_{12} + C_1(\lambda I - A)^{-1}B_2 \\ D_{21} + C_2(\lambda I - A)^{-1}B_1 & D_{22} + C_2(\lambda I - A)^{-1}B_2 \end{bmatrix}$$

So

$$W_{ij}(\lambda) = D_{ij} + C_i(\lambda I - A)^{-1}B_j, \qquad i, j = 1, 2$$

We use a step-by-step procedure to compute a realization for

$$\mathscr{F}_W(V) = W_{21}(\lambda) + W_{22}(\lambda)V(\lambda)(I - W_{12}(\lambda)V(\lambda))^{-1}W_{11}(\lambda)$$

using these realizations for $W_{ij}(\lambda)$ and the realization (7.6.5) for $V(\lambda)$ by the following rules: given two rational matrix functions $X_1(\lambda)$ and $X_2(\lambda)$ with finite values at infinity and realizations

$$X_i(\lambda) = \hat{D}_i + \hat{C}_i(\lambda I - \hat{A}_i)^{-1}\hat{B}_i, \qquad i = 1, 2$$

realizations for $X_1(\lambda) + X_2(\lambda)$, $X_1(\lambda)X_2(\lambda)$, and $X_1(\lambda)^{-1}$ can be found as follows [cf. formulas (7.2.9) and (7.3.2)]:

$$X_1(\lambda) + X_2(\lambda) = \hat{D}_1 + \hat{D}_2 + [\hat{C}_1 \ \hat{C}_2]\left(\lambda I - \begin{bmatrix} \hat{A}_1 & 0 \\ 0 & \hat{A}_2 \end{bmatrix}\right)^{-1} \begin{bmatrix} \hat{B}_1 \\ \hat{B}_2 \end{bmatrix}$$

$$X_1(\lambda) X_2(\lambda) = \hat{D}_1 \hat{D}_2 + [\hat{C}_1, \ \hat{D}_1 \hat{C}_2]\left(\lambda I - \begin{bmatrix} \hat{A}_1 & \hat{B}_1 \hat{C}_2 \\ 0 & \hat{A}_2 \end{bmatrix}\right)^{-1} \begin{bmatrix} \hat{B}_1 \hat{D}_2 \\ \hat{B}_2 \end{bmatrix}$$

$$X_1(\lambda)^{-1} = \hat{D}_1^{-1} - \hat{D}_1^{-1} \hat{C}_1 (\lambda I - (\hat{A}_1 - \hat{B}_1 \hat{D}_1^{-1} \hat{C}_1))^{-1} \hat{B}_1 \hat{D}_1^{-1}$$

(it is assumed in the last formula that $\hat{D}_1$ is invertible). A computation shows that

$$\mathscr{F}_W(V) = \tilde{\delta} + \tilde{\gamma}(\lambda I - \tilde{\alpha})^{-1}\tilde{\beta} \tag{7.6.12}$$

where

$$\tilde{\delta} = D_{21} + D_{22}d(I - D_{12}d)^{-1}D_{11}$$

$$\tilde{\gamma} = [C_2, \ C_2, \ D_{22}c, \ D_{22}d(I - D_{12}d)^{-1}C_1, \ D_{22}d(I - D_{12}d)^{-1}D_{12}c,$$
$$D_{22}d(I - D_{12}d)^{-1}C_1]$$

$$\tilde{\alpha} = \begin{bmatrix} A & 0 & 0 & 0 & 0 & 0 \\ 0 & A & B_2 c & XC_1 & XD_{12}c & XC_1 \\ 0 & 0 & a & yC_1 & yD_{12}c & yC_1 \\ 0 & 0 & 0 & A + XC_1 & (B_2 + XD_{12})c & XC_1 \\ 0 & 0 & 0 & yC_1 & a + yD_{12}c & yC_1 \\ 0 & 0 & 0 & 0 & 0 & A \end{bmatrix}$$

and $X = B_2 d(I - D_{12}d)^{-1}$, $y = b(I - D_{12}d)^{-1}$;

$$\tilde{\beta} = \begin{bmatrix} B_1 \\ XD_{11} \\ yD_{11} \\ XD_{11} \\ yD_{11} \\ B_1 \end{bmatrix}$$

Let

$$S = \begin{bmatrix} I_n & 0 & 0 & 0 & 0 & I_n \\ 0 & 0 & I_n & I_n & 0 & -I_n \\ 0 & I_p & 0 & 0 & I_p & 0 \\ 0 & 0 & 0 & I_n & 0 & -I_n \\ 0 & 0 & 0 & 0 & I_p & 0 \\ 0 & 0 & 0 & 0 & 0 & I_n \end{bmatrix}$$

where $n \times n$ and $p \times p$ are the sizes of A and a, respectively. Then

$$S^{-1} = \begin{bmatrix} I_n & 0 & 0 & 0 & 0 & -I_n \\ 0 & 0 & I_p & 0 & -I_p & 0 \\ 0 & I_n & 0 & -I_n & 0 & 0 \\ 0 & 0 & 0 & I_n & 0 & I_n \\ 0 & 0 & 0 & 0 & I_p & 0 \\ 0 & 0 & 0 & 0 & 0 & I_n \end{bmatrix}$$

and

$$\tilde{\gamma} S = [c_2, D_{22}c, C_2, C_2 + D_{22}d(I - D_{12}d)^{-1}C_1, D_{22}(I - dD_{12})^{-1}c, 0]$$

$$S^{-1}\tilde{\alpha} S = \begin{bmatrix} A & 0 & 0 \\ 0 & a & 0 \\ 0 & B_2c & A \end{bmatrix} \oplus \begin{bmatrix} A + XC_1 & B_2(I - dD_{12})^{-1}c \\ yC_1 & a + yD_{12}c \end{bmatrix} \oplus A$$

$$S^{-1}\tilde{\beta} = \begin{bmatrix} 0 \\ 0 \\ 0 \\ B_1 + XD_{11} \\ yD_{11} \\ B_1 \end{bmatrix}$$

Writing (7.6.12) in the form

$$\mathcal{F}_W(V) = \tilde{\delta} + \tilde{\gamma}S(\lambda I - S^{-1}\tilde{\alpha}S)^{-1}S^{-1}\tilde{\beta}$$

we see that formula (7.6.11) follows.

Assume now that (7.6.11) is a minimal realization. Let $x \in \mathbb{C}^n$ be such that

$$\begin{bmatrix} C_1 \\ C_2 \end{bmatrix} A^k x = 0 \qquad (7.6.13)$$

for all nonnegative integers k. Using formula (7.6.7), one proves by induction on k that

$$\alpha^k \begin{bmatrix} x \\ 0 \end{bmatrix} = \begin{bmatrix} A^k x \\ 0 \end{bmatrix}, \qquad k = 0, \ldots \qquad (7.6.14)$$

Indeed, (7.6.14) holds for $k = 0$. Assuming that (7.6.14) is true for $k - 1$, we have

$$\alpha^k \begin{bmatrix} x \\ 0 \end{bmatrix} = \alpha \cdot \alpha^{k-1} \begin{bmatrix} x \\ 0 \end{bmatrix} = \alpha \begin{bmatrix} A^k x \\ 0 \end{bmatrix} = \begin{bmatrix} A^{k+1} x \\ 0 \end{bmatrix}$$

where the last equality follows in view of (7.6.13). Now

$$\gamma \alpha^k \begin{bmatrix} x \\ 0 \end{bmatrix} = (C_2 + D_{22}d(I - D_{12}d)^{-1}C_1)A^k x = 0, \qquad k = 0, 1, \ldots$$

and $x = 0$ because (γ, α) is a null kernel pair. [This follows from the minimality of (7.6.11).] So the pair (C, A) is also a null kernel pair.

To prove that (A, B) is a full-range pair, observe that α^k can be written in the form

$$\alpha^k = \begin{bmatrix} A^k + \sum_{l=0}^{k-1} A^l B_2 Y_{lk} & \sum_{l=0}^{k-1} A^l B_2 Z_{lk} \\ * & * \end{bmatrix}, \qquad k = 0, 1, \ldots \qquad (7.6.15)$$

where Y_{lk} and Z_{lk} are certain matrices and the stars denote matrices of no immediate interest. Formula (7.6.15) can be proved by induction on k by means of formula (7.6.7). From the minimality of (7.6.11) it follows that for every $x \in \mathbb{C}^n$ there exist vectors $v_0, \ldots, v_q \in \mathbb{C}^{m_1}$ such that

$$\begin{bmatrix} x \\ 0 \end{bmatrix} = \sum_{k=0}^{q} \alpha^k \begin{bmatrix} \beta_1 \\ \beta_2 \end{bmatrix} v_k$$

But then, using (7.6.15) and (7.6.8), we have

$$x = \sum_{k=0}^{q} \left\{ \left[A^k + \sum_{l=0}^{k-1} A^l B_2 Y_{l,k} \right][B_1 + B_2 d(I - D_{12}d)^{-1} D_{11}]v_k \right.$$
$$\left. + \sum_{l=0}^{k-1} A^l B_2 Z_{lk} b(I - D_{12}d)^{-1} D_{11} v_k \right\} = \sum_{k=0}^{q} A^k [B_1, B_2] \begin{bmatrix} v_k \\ * \end{bmatrix}$$

and (A, B) is a full-range pair. So the realization (7.6.4) is minimal.

Now consider the realization (7.6.5). Let $x \in \mathbb{C}^p$ be such that $ca^k x = 0$, $k = 0, 1, \ldots$. One proves that $\alpha^k \begin{bmatrix} 0 \\ x \end{bmatrix} = \begin{bmatrix} 0 \\ a^k x \end{bmatrix}$, $k = 0, 1, \ldots$ using an argument analogous to that used in obtaining (7.6.14). Hence

$$[\gamma_1, \gamma_2]\alpha^k \begin{bmatrix} 0 \\ x \end{bmatrix} = [\gamma_1, \gamma_2] \begin{bmatrix} 0 \\ a^k x \end{bmatrix} = D_{22}(I - dD_{12})^{-1} ca^k x = 0$$

In view of the minimality of (7.6.11) we obtain $x = 0$, and (c, a) is a null kernel pair. Finally, write α^k in the form

$$\alpha^k = \begin{bmatrix} * & * \\ \sum_{l=0}^{k-1} a^l b z_{lk} & a^k + \sum_{l=0}^{k-1} a^l b y_{lk} \end{bmatrix}, \qquad k = 0, 1, \ldots \qquad (7.6.16)$$

for some matrices z_{lk} and y_{lk}. [Again, equation (7.6.16) can be proved by induction on k using (7.6.7).] For every $x \in \mathbb{C}^p$ by the minimality of (7.6.11) there exist vectors $u_0, \ldots, u_q \in \mathbb{C}^{m_1}$ such that

$$\begin{bmatrix} 0 \\ x \end{bmatrix} = \sum_{k=0}^{q} a^k \begin{bmatrix} \beta_1 \\ \beta_2 \end{bmatrix} u_k$$

From (7.6.16), it follows that

$$x = \sum_{k=0}^{q} a^k b w_k$$

for some vectors $w_0, \ldots, w_q$, and the full-range property of (a, b) is proved. Hence the realization (7.6.5) is minimal as well. □

Observe that if $D_{12} = 0$, $D_{21} = 0$, $D_{11} = I$, $C_1 = 0$, $B_1 = 0$, we have $W_{21}(\lambda) \equiv 0$, $W_{12}(\lambda) \equiv 0$, $W_{11}(\lambda) \equiv I$, and so

$$\mathscr{F}_W(V)(\lambda) = W_{22}(\lambda)V(\lambda)$$

On the other hand, formulas (7.6.7)–(7.6.10) take the form

$$\alpha = \begin{bmatrix} A & B_2 c \\ 0 & a \end{bmatrix}, \qquad \beta = \begin{bmatrix} B_2 d \\ d \end{bmatrix}, \qquad \gamma = [C_2, D_{22}c], \qquad \delta = D_{22}d$$

which coincides with formula (7.2.9) for the realization of a product of rational matrix functions. So (7.6.11) is a generalization of (7.2.9). On the other hand, putting $D_{12} = 0$, $D_{21} = 0$, $D_{22} = I$, $C_2 = 0$, $B_2 = 0$, we have

$$\mathscr{F}_W(V)(\lambda) = V(\lambda)W_{11}(\lambda)$$

and formula (7.6.11) gives another version for the realization of the product of two rational matrix functions:

$$\alpha = \begin{bmatrix} A & 0 \\ bC_1 & a \end{bmatrix}, \qquad \beta = \begin{bmatrix} B_1 \\ bD_{11} \end{bmatrix}, \qquad \gamma = [dC_1, c], \qquad \delta = dD_{11}$$

7.7   LINEAR FRACTIONAL DECOMPOSITIONS AND INVARIANT SUBSPACES FOR NONSQUARE MATRICES

Let $U(\lambda)$ be a rational matrix function of size $q \times s$ with finite value at infinity. A *linear fractional decomposition* of $U(\lambda)$ is a representation of $U(\lambda)$ in the form

$$U(\lambda) = \mathscr{F}_W(\lambda) \tag{7.7.1}$$

for some rational matrix functions $W(\lambda)$ and $V(\lambda)$ that take finite values at infinity. In this section we describe linear fractional decompositions of $U(\lambda)$

in terms of certain invariant subspaces for nonsquare matrices related to a realization of $U(\lambda)$.

Minimal linear fractional decompositions (7.7.1) are of particular interest. First observe that the definition of the McMillan degree of a rational matrix function with value I at infinity (given in Section 7.3) extends verbatim to a (possibly rectangular) rational matrix function $W(\lambda)$ with finite value at infinity: namely, $\delta(W)$ is the size of the matrix A taken from any minimal realization

$$W(\lambda) = D + C(\lambda I - A)^{-1}B$$

of $W(\lambda)$. In any linear fractional decomposition (7.7.1) of $U(\lambda)$ for which the rational functions $W(\lambda)$ and $V(\lambda)$ take finite values at infinity, we have

$$\delta(U) \le \delta(W) + \delta(V) \qquad (7.7.2)$$

Indeed, assuming that (7.6.4) and (7.6.5) are minimal realizations of $W(\lambda)$ and $V(\lambda)$, respectively, then by Theorem 7.6.1 $U(\lambda)$ has a realization (not necessarily minimal) $\delta + \gamma(\lambda I - \alpha)^{-1}\beta$, where the size of α is $t \times t$, with $t = \delta(W) + \delta(V)$. Hence (7.7.2) follows.

The linear fractional decomposition (7.7.1) is called *minimal* if equality holds in (7.7.2), that is, $\delta(U) = \delta(W) + \delta(V)$. As in the preceding paragraph, Theorem 7.6.1 implies that (7.7.1) is minimal if and only if for some (and hence for any) minimal realizations (7.6.4) and (7.6.5) of $W(\lambda)$ and $U(\lambda)$, respectively, the realization (7.6.11) of $U(\lambda) = \mathcal{F}_W(V)$ is again minimal.

Let

$$U(\lambda) = \delta + \gamma(\lambda I - \alpha)^{-1}\beta \qquad (7.7.3)$$

be a realization (not necessarily minimal) of $U(\lambda)$, where α, β, γ, and δ are matrices of sizes $l \times l$, $l \times s$, $q \times l$, and $q \times s$, respectively. Recall from Theorem 6.1.1 that a subspace $\mathcal{M} \subset \mathbb{C}^l$ is $[\alpha \ \beta]$ invariant if and only if there exists an $s \times l$ matrix F such that $\mathcal{M}$ is invariant for $\alpha + \beta F$. Also (see Theorem 6.6.1), a subspace $\mathcal{N} \subset \mathbb{C}^l$ is $\begin{bmatrix} \alpha \\ \gamma \end{bmatrix}$ invariant if and only if there exists an $l \times q$ matrix G such that $(\alpha + G\gamma)\mathcal{N} \subset \mathcal{N}$. For the purpose of this section we can accept these properties as definitions of $[\alpha \ \beta]$-invariant and $\begin{bmatrix} \alpha \\ \gamma \end{bmatrix}$-invariant subspaces, respectively.

A pair of subspaces $(\mathcal{M}_1, \mathcal{M}_2)$ of $\mathbb{C}^l$ will be called *reducing* with respect to realization (7.7.3) if $\mathcal{M}_1$ is $[\alpha \ \beta]$ invariant, $\mathcal{M}_2$ is $\begin{bmatrix} \alpha \\ \gamma \end{bmatrix}$ invariant, and $\mathcal{M}_1$ and $\mathcal{M}_2$ are direct complements to each other in $\mathbb{C}^l$.

The following theorem provides a geometrical characterization of minimal linear fractional decompositions of $U(\lambda)$ in terms of its realization (7.7.3).

Theorem 7.7.1

Assume that $(\mathcal{M}_1, \mathcal{M}_2)$ is a reducing pair with respect to the realization (7.7.3) of $U(\lambda)$. The following recipe may be used to construct realizations of rational matrix functions $W(\lambda)$ and $V(\lambda)$ such that

$$U(\lambda) = \mathscr{F}_W(V) \tag{7.7.4}$$

and

$$W(\lambda) = \begin{bmatrix} D_{11} & D_{12} \\ D_{21} & D_{22} \end{bmatrix} + \begin{bmatrix} C_1 \\ C_2 \end{bmatrix}(\lambda I - A)^{-1}[B_1\ B_2]$$

$$= \begin{bmatrix} D_{11} + C_1(\lambda I - A)^{-1}B_1 & D_{12} + C_1(\lambda I - A)^{-1}B_2 \\ D_{21} + C_2(\lambda I - A)^{-1}B_1 & D_{22} + C_2(\lambda I - A)^{-1}B_2 \end{bmatrix}:$$

$$\mathbb{C}^s \dotplus \mathbb{C}^q \to \mathbb{C}^s \dotplus \mathbb{C}^q \tag{7.7.5}$$

with a transformation $A: \mathcal{M}_1 \to \mathcal{M}_1$, and

$$V(\lambda) = d + c(\lambda I - a)^{-1}b: \mathbb{C}^s \to \mathbb{C}^q \tag{7.7.6}$$

with a transformation $a: \mathcal{M}_2 \to \mathcal{M}_2$: (a) choose any transformation

$$D = \begin{bmatrix} D_{11} & D_{12} \\ D_{21} & D_{22} \end{bmatrix}: \mathbb{C}^s \dotplus \mathbb{C}^q \to \mathbb{C}^s \dotplus \mathbb{C}^q$$

and any transformation $d: \mathbb{C}^s \to \mathbb{C}^q$ such that the transformations D_{11}, D_{22} and $I - D_{12}d$ are invertible and

$$\delta = D_{21} + D_{22}d(I - D_{12}d)^{-1}D_{11} \tag{7.7.7}$$

(b) choose any transformations $F: \mathbb{C}^l \to \mathbb{C}^s$ and $G: \mathbb{C}^q \to \mathbb{C}^l$ for which $(\alpha + \beta F)\mathcal{M}_1 \subset \mathcal{M}_1$ and $(\alpha + G\gamma)\mathcal{M}_2 \subset \mathcal{M}_2$; (c) let

$$\alpha = \begin{bmatrix} \alpha_{11} & \alpha_{12} \\ \alpha_{21} & \alpha_{22} \end{bmatrix}: \mathcal{M}_1 \dotplus \mathcal{M}_2 \to \mathcal{M}_1 \dotplus \mathcal{M}_2$$

$$\beta = \begin{bmatrix} \beta_1 \\ \beta_2 \end{bmatrix}: \mathbb{C}^s \to \mathcal{M}_1 \dotplus \mathcal{M}_2; \qquad \gamma = [\gamma_1\ \ \gamma_2]: \mathcal{M}_1 \dotplus \mathcal{M}_2 \to \mathbb{C}^q \tag{7.7.8}$$

$$F = [F_1\ \ F_2]: \mathcal{M}_1 \dotplus \mathcal{M}_2 \to \mathbb{C}^s; \qquad G = \begin{bmatrix} G_1 \\ G_2 \end{bmatrix}: \mathbb{C}^q \to \mathcal{M}_1 \dotplus \mathcal{M}_2$$

be block matrix representations with respect to the direct sum decomposition $\mathbb{C}^l = \mathcal{M}_1 \dotplus \mathcal{M}_2$. Then, defining

$$A = \alpha_{11} - G_1(\delta - D_{21})F_1$$
$$B_1 = \beta_1 + G_1(\delta - D_{21})$$
$$B_2 = -G_1 D_{22} \qquad\qquad (7.7.9)$$
$$C_1 = -D_{11}F_1$$
$$C_2 = \gamma_1 + (\delta - D_{21})F_1$$

and

$$a = \alpha_{22} - \beta_2 D_{11}^{-1} D_{12}(I - dD_{12})D_{22}^{-1}\gamma_2$$
$$b = \beta_2 D_{11}^{-1}(I - D_{12}d) \qquad\qquad (7.7.10)$$
$$c = (I - dD_{12})D_{22}^{-1}\gamma_2$$

equation (7.7.4) holds. Moreover, if, in addition, the realization (7.7.3) is minimal, the linear fractional decomposition (7.7.4) is minimal as well; and conversely, any minimal linear fractional decomposition

$$U(\lambda) = W_{21}(\lambda) + W_{22}(\lambda)V(\lambda)(I - W_{21}(\lambda)V(\lambda))^{-1}W_{11}(\lambda) \quad (7.7.11)$$

of $U(\lambda)$ where the rational matrix functions

$$W(\lambda) = \begin{bmatrix} W_{11}(\lambda) & W_{12}(\lambda) \\ W_{21}(\lambda) & W_{22}(\lambda) \end{bmatrix}$$

and $V(\lambda)$ take finite values at infinity and the matrices $W_{11}(\infty)$ and $W_{22}(\infty)$ are invertible, can be obtained by this recipe.

Proof. Let A, B_j, C_j, D_{ij} and a, b, c, d be defined as in the recipe. Then, using the relationships (7.7.7), (7.7.9), and (7.7.10) and the equalities $\alpha_{21} + \beta_2 F_1 = 0$ and $\alpha_{12} + G_1\gamma_2 = 0$ (which follow from the invariance of $\mathcal{M}_1$ and $\mathcal{M}_2$ under the transformations $\alpha + \beta F$ and $\alpha + G\gamma$, respectively), one checks that the equalities (7.6.7)–(7.6.10) hold. Now by Theorem 7.6.1. we obtain the linear fractional decomposition (7.7.4).

Assume now that (7.7.3) is a minimal realization of $U(\lambda)$; hence $\delta(U) = l$. By Theorem 7.6.1 the realizations (7.7.5) and (7.7.6) are minimal, so

$$\delta(W) = \dim \mathcal{M}_1, \qquad \delta(V) = \dim \mathcal{M}_2$$

As $\mathcal{M}_1$ and $\mathcal{M}_2$ are direct complements to each other in $\mathbb{C}^l$, we have $\delta(U) = \delta(W) + \delta(V)$, and the minimality of the linear fractional decomposition (7.7.4) follows.

Conversely, assume that (7.7.3) is a minimal realization of $U(\lambda)$, and let (7.7.10) be a minimal linear fractional decomposition of $U(\lambda)$, where the rational functions $W(\lambda)$ and $V(\lambda)$ are finite at infinity and $W_{11}(\infty)$, $W_{22}(\infty)$ are invertible. Here

$$W(\lambda) = \begin{bmatrix} W_{11}(\lambda) & W_{12}(\lambda) \\ W_{21}(\lambda) & W_{22}(\lambda) \end{bmatrix} : \mathbb{C}^s \dotplus \mathbb{C}^q \rightarrow \mathbb{C}^s \dotplus \mathbb{C}^q \qquad (7.7.12)$$

and $V(\lambda)$ is of size $q \times s$. [The sizes of $W(\lambda)$ and $V(\lambda)$ are dictated by formula (7.7.11) and by the invertibility of $W_{11}(\infty)$ and $W_{22}(\infty)$; in particular, the matrix functions $W_{11}(\lambda)$ and $W_{22}(\lambda)$ must be square.] Let

$$W(\lambda) = \begin{bmatrix} D_{11} & D_{12} \\ D_{21} & D_{22} \end{bmatrix} + \begin{bmatrix} C_1 \\ C_2 \end{bmatrix} (\lambda I - A)^{-1} [B_1 \quad B_2] \qquad (7.7.13)$$

be a minimal realization of $W(\lambda)$ partitioned as in (7.7.12), where the matrix A has size $n \times n$, $n = \delta(W)$. Let

$$V(\lambda) = d + c(\lambda I - a)^{-1}b \qquad (7.7.14)$$

be a minimal realization of $V(\lambda)$ in which a is $p \times p$, $p = \delta(V)$. By Theorem 7.6.1, form a realization

$$U(\lambda) = \delta' + \gamma'(\lambda I - \alpha')^{-1}\beta' \qquad (7.7.15)$$

where α', β', γ', and δ' are given by formulas (7.6.7), (7.6.8), (7.6.9), and (7.6.10), respectively, using the realizations (7.7.13) and (7.7.14). As (7.7.11) is a minimal linear fractional decomposition, the realization (7.7.15) is minimal. [The size of α' is $(n + p) \times (n + p)$.] Comparing the minimal realizations (7.7.3) and (7.7.15) we find, in view of Theorem 7.1.4, that $\delta = \delta'$ and there exists an invertible transformation $S: \mathbb{C}^n \dotplus \mathbb{C}^p \rightarrow \mathbb{C}^l$ such that

$$\alpha = S\alpha'S^{-1}, \qquad \beta = S\beta', \qquad \gamma = \gamma'S^{-1}$$

Putting $\mathcal{M}_1 = S(\mathbb{C}^n \dotplus \{0\})$, $\mathcal{M}_2 = S(\{0\} \dotplus \mathbb{C}^p)$,

$$F = [-D_{11}^{-1}C_1S^{-1}|_{\mathcal{M}_1}, \; *], \qquad G = \begin{bmatrix} -S|_{\mathbb{C}^N + \{0\}}B_2D_{22}^{-1} \\ * \end{bmatrix}$$

one verifies that $(\alpha + \beta F)\mathcal{M}_1 \subset \mathcal{M}_1$, $(\alpha + G\gamma)\mathcal{M}_2 \subset \mathcal{M}_2$, and the minimal linear fractional decomposition (7.7.11) is given by our recipe. $\square$

Observe that the linear fractional decomposition of $U(\lambda)$ described in the recipe of Theorem 7.7.1 depends on the reducing pair $(\mathcal{M}_1, \mathcal{M}_2)$, on the choice of D and d such that condition (a) holds, and on the choice of F and

G such that $(\alpha + \beta F)\mathcal{M}_1 \subset \mathcal{M}_1$, $(\alpha + G\gamma)\mathcal{M}_2 \subset \mathcal{M}_2$. [We assume that the realization (7.7.3) of $U(\lambda)$ is fixed in advance.] We determine the parts of this information that are uniquely defined by the linear fractional decomposition. Let us introduce the following definition. Let $(\mathcal{M}_1, \mathcal{M}_2)$ be a reducing pair [with respect to the realization (7.7.3)] and $F: \mathbb{C}^l \to \mathbb{C}^s$, $G: \mathbb{C}^q \to \mathbb{C}^l$ be transformations such that $(\alpha + \beta F)\mathcal{M}_1 \subset \mathcal{M}_1$, $(\alpha + G\gamma)\mathcal{M}_2 \subset \mathcal{M}_2$, and write

$$F = [F_1, F_2], \qquad G = \begin{bmatrix} G_1 \\ G_2 \end{bmatrix}$$

with respect to the direct sum decomposition $\mathbb{C}^l = \mathcal{M}_1 \dot{+} \mathcal{M}_2$. The quadruple $(\mathcal{M}_1, \mathcal{M}_2; F_1, G_1)$ will be called a *supporting quadruple* [with respect to the realization (7.7.3)]. Given a supporting quadruple, for every choice of D and d satisfying condition (a) of Theorem 7.7.1, the recipe produces a linear fractional decomposition of $U(\lambda)$. We now have the following important addition to Theorem 7.7.1.

Theorem 7.7.2

Assume that the realization (7.7.3) is minimal, and let (7.7.11) be a minimal linear fractional decomposition of $U(\lambda)$ such that $W_{ij}(\lambda)$ and $V(\lambda)$ take finite values at infinity and the matrices $W_{11}(\infty)$ and $W_{22}(\infty)$ are invertible. Then there exists a unique supporting quadruple $Q = (\mathcal{M}_1, \mathcal{M}_2; F_1, G_1)$ that produces, together with some choice of D and d satisfying condition (a), the decomposition (7.7.11) according to the recipe of Theorem 7.7.1.

Proof. The existence of Q is ensured by Theorem 7.7.1. To prove the uniqueness of Q, assume that $Q' = (\mathcal{M}'_1, \mathcal{M}'_2; F'_1, G'_1)$ is another supporting quadruple that gives rise (with some choice of D and d) to the same decomposition (7.7.11). As $D = W(\infty)$, $d = V(\infty)$, we see that actually the matrices

$$D = \begin{bmatrix} D_{11} & D_{12} \\ D_{21} & D_{22} \end{bmatrix}$$

and d, which, together with Q', give rise to the decomposition (7.7.11) are the same matrices chosen to produce (7.7.11), together with Q. Further, let (7.7.8) be the block matrix representations of α, β, and γ with respect to the direct sum decomposition $\mathbb{C}^l = \mathcal{M}_1 \dot{+} \mathcal{M}_2$, and let

$$\alpha = \begin{bmatrix} \alpha'_{11} & \alpha'_{12} \\ \alpha'_{21} & \alpha'_{22} \end{bmatrix} \qquad \beta = \begin{bmatrix} \beta'_1 \\ \beta'_2 \end{bmatrix}, \qquad \gamma = [\gamma'_1, \gamma'_2]$$

be the corresponding representations with respect to the direct sum $\mathbb{C}^l = \mathcal{M}'_1 \dot{+} \mathcal{M}'_2$. We now have two realizations for $W(\lambda)$:

$$W(\lambda) = \begin{bmatrix} D_{11} & D_{12} \\ D_{21} & D_{22} \end{bmatrix} + \begin{bmatrix} C_1 \\ C_2 \end{bmatrix} (\lambda I - A)^{-1} [B_1 \quad B_2]$$

$$= \begin{bmatrix} D_{11} & D_{12} \\ D_{21} & D_{22} \end{bmatrix} + \begin{bmatrix} C_1' \\ C_2' \end{bmatrix} (\lambda I - A')^{-1} [B_1' \quad B_2'] \qquad (7.7.16)$$

where A, B_i, and C_j are given by formulas (7.7.9) and A', B_i', and C_j' are given by (7.7.9) with α_{11}, G_1, F_1, β_1, γ_1 replaced by α_{11}', G_1', F_1', β_1', γ_1', respectively. By Theorem 7.6.1, both realizations (7.7.16) are minimal, so in view of Theorem 7.1.3 there exists an invertible transformation $S: \mathcal{M}_1 \to \mathcal{M}_1'$ such that

$$A = S^{-1}A'S, \begin{bmatrix} C_1 \\ C_2 \end{bmatrix} = \begin{bmatrix} C_1' \\ C_2' \end{bmatrix} S, [B_1 \quad B_2] = S^{-1}[B_1' \quad B_2'] \qquad (7.7.17)$$

Similarly, we have

$$V(\lambda) = d + c(\lambda I - a)^{-1}b = d + c'(\lambda I - a')^{-1}b' \qquad (7.7.18)$$

where a, b, and c are given by (7.7.10) and a', b', and c' are given by (7.7.10) with α_{22}, β_2, γ_2 replaced by α_{22}', β_2', γ_2', respectively. Since both realizations (7.7.18) are minimal, we have

$$a = T^{-1}a'T, \qquad c = c'T, \qquad b = T^{-1}b' \qquad (7.7.19)$$

for some invertible transformation $T: \mathcal{M}_2 \to \mathcal{M}_2'$.

We now verify that

$$\begin{bmatrix} S^{-1} & 0 \\ 0 & T^{-1} \end{bmatrix} \begin{bmatrix} \alpha_{11}' & \alpha_{12}' \\ \alpha_{21}' & \alpha_{22}' \end{bmatrix} \begin{bmatrix} S & 0 \\ 0 & T \end{bmatrix} = \begin{bmatrix} \alpha_{11} & \alpha_{12} \\ \alpha_{21} & \alpha_{22} \end{bmatrix}$$

$$\begin{bmatrix} S^{-1} & 0 \\ 0 & T^{-1} \end{bmatrix} \begin{bmatrix} \beta_1' \\ \beta_2' \end{bmatrix} = \begin{bmatrix} \beta_1 \\ \beta_2 \end{bmatrix}; \qquad [\gamma_1' \quad \gamma_2'] \begin{bmatrix} S & 0 \\ 0 & T \end{bmatrix} = [\gamma_1 \quad \gamma_2]$$

$$(7.7.20)$$

Indeed, formulas (7.7.9) together with (7.7.17) give

$$F_1 = F_1'S, \qquad G_1 = S^{-1}G_1', \qquad \beta_1 = S^{-1}\beta_1', \qquad \gamma_1 = \gamma_1'S$$

and

$$\alpha_{11} - G_1(\delta - D_{21})F_1 = S^{-1}(\alpha_{11}' - G_1'(\delta - D_{21})F_1')S$$

$$= S^{-1}\alpha_{11}'S - G_1(\delta - D_{21})F_1$$

so $\alpha_{11} = S^{-1}\alpha_{11}'S$. From formulas (7.7.10) and (7.7.19) one obtains

$$\beta_2 = T^{-1}\beta_2', \qquad \gamma_2 = \gamma_2'T$$

and

$$\alpha_{22} = a - \beta_2 D_{11}^{-1}D_{12}(I - dD_{12})D_{22}^{-1}\gamma_2$$
$$= T^{-1}(a' - \beta_2'D_{11}^{-1}D_{12}(I - dD_{12})D_{22}^{-1}\gamma_2')T = T^{-1}\alpha_{22}'T$$

Further, the definition of the supporting quadruples Q and Q' implies

$$\alpha_{12} = -G_1\gamma_2, \qquad \alpha_{21} = -\beta_2 F_1, \qquad \alpha_{12}' = -G_1'\gamma_2', \qquad \alpha_{21}' = -\beta_2'F_1'$$

so

$$S^{-1}\alpha_{12}'T = -S^{-1}G_1'\gamma_2'T = -G_1\gamma_2 = \alpha_{12}$$

and

$$T^{-1}\alpha_{21}'S = -T^{-1}\beta_2'F_1'S = -\beta_2 F_1 = \alpha_{21}$$

All the established relationships verify the equalities (7.7.20).

It remains to observe that the transformation $V = \begin{bmatrix} S & 0 \\ 0 & T \end{bmatrix}$ is a similarity of the minimal realization (7.7.3) with itself. Since such a similarity must be unique (Theorem 7.1.4), it follows that $V = I$ and hence $\mathcal{M}_1 = \mathcal{M}_1'$, $\mathcal{M}_2 = \mathcal{M}_2'$, $F_1 = F_1'$, and $G_1 = G_1'$. $\square$

7.8. LINEAR FRACTIONAL DECOMPOSITIONS: FURTHER DEDUCTIONS

We consider here some deductions, examples, and results on linear fractional decompositions that follow from the main theorems, Theorems 7.7.1 and 7.7.2.

The particular case when $\delta = I$, $D = I$, and $d = I$ in Theorems 7.7.1 and 7.7.2 is of special interest. In this case condition (a) of Theorem 7.7.1 is satisfied automatically, and we have the following.

Theorem 7.8.1

Let

$$U(\lambda) = I + \gamma(\lambda I - \alpha)^{-1}\beta \tag{7.8.1}$$

be a minimal realization of the rational $q \times q$ matrix function $U(\lambda)$. Let $(\mathcal{M}_1, \mathcal{M}_2)$ be a reducing pair for the realization (7.8.1), and write

$$\alpha = \begin{bmatrix} \alpha_{11} & \alpha_{12} \\ \alpha_{21} & \alpha_{22} \end{bmatrix} : \mathcal{M}_1 \dotplus \mathcal{M}_2 \to \mathcal{M}_1 \dotplus \mathcal{M}_2$$

$$\beta = \begin{bmatrix} \beta_1 \\ \beta_2 \end{bmatrix} : \mathbb{C}^q \to \mathcal{M}_1 \dotplus \mathcal{M}_2, \qquad \gamma = [\gamma_1 \quad \gamma_2] : \mathcal{M}_1 \dotplus \mathcal{M}_2 \to \mathbb{C}^q$$

Choose any transformations

$$F = [F_1 \quad F_2] : \mathcal{M}_1 \dotplus \mathcal{M}_2 \to \mathbb{C}^q, \qquad G = \begin{bmatrix} G_1 \\ G_2 \end{bmatrix} : \mathbb{C}^q \to \mathcal{M}_1 \dotplus \mathcal{M}_2$$

in such a way that $(\alpha + \beta F)\mathcal{M}_1 \subset \mathcal{M}_1$, $(\alpha + G\gamma)\mathcal{M}_2 \subset \mathcal{M}_2$. *Then*

$$W(\lambda) = I + \begin{bmatrix} -F_1 \\ \gamma_1 + F_1 \end{bmatrix} (\lambda I - (\alpha_{11} - G_1 F_1))^{-1} [\beta_1 + G_1, -G_1] \qquad (7.8.2)$$

and

$$V(\lambda) = I + \gamma_2 (\lambda I - \alpha_{22})^{-1} \beta_2 \qquad (7.8.3)$$

produce a minimal linear fractional decomposition $U(\lambda) = \mathcal{F}_W(V)$. *Conversely, every minimal linear fractional decomposition* $U(\lambda) = \mathcal{F}_W(V)$ *with* $W(\infty) = I$ *and* $V(\infty) = I$ *can be obtained in this way, and the quadruple* $(\mathcal{M}_1, \mathcal{M}_2; F_1, G_1)$ *is determined uniquely by* $W(\lambda)$ *and* $V(\lambda)$.

Let us give a simple example illustrating Theorem 7.8.1.

EXAMPLE 7.8.1. Let

$$U(\lambda) = \begin{bmatrix} (\lambda + 1)\lambda^{-1} & \epsilon \lambda^{-2} \\ 0 & (\lambda + 1)\lambda^{-1} \end{bmatrix}, \qquad \epsilon \neq 0$$

A minimal realization for $U(\lambda)$ is easy to find:

$$U(\lambda) = \delta + \gamma(\lambda I - \alpha)^{-1}\beta$$

with $\delta = \gamma = \beta = I$, $\alpha = \begin{bmatrix} 0 & \epsilon \\ 0 & 0 \end{bmatrix}$. We find all nontrivial [i.e., such that $W(\lambda) \neq I$, $V(\lambda) \neq I$] minimal linear fractional decompositions $U(\lambda) = \mathcal{F}_W(V)$ such that $W(\infty) = I$, $V(\infty) = I$. Every subspace in $\mathbb{C}^2$ is $[\alpha \ \beta]$ invariant, as well as $\begin{bmatrix} \alpha \\ \gamma \end{bmatrix}$ invariant. We consider the case when the one-dimensional subspaces $\mathcal{M}_1$ and $\mathcal{M}_2$ and $\mathbb{C}^2$ that are direct complements to each other are of the form

$$\mathcal{M}_1 = \text{Span}\begin{bmatrix} 1 \\ x \end{bmatrix}, \qquad \mathcal{M}_2 = \text{Span}\begin{bmatrix} 1 \\ y \end{bmatrix}, \qquad x \neq y$$

Then one computes

$$\alpha = \begin{bmatrix} \dfrac{\epsilon xy}{y-x} & \dfrac{\epsilon y^2}{y-x} \\[2ex] \dfrac{-\epsilon x^2}{y-x} & \dfrac{-\epsilon xy}{y-x} \end{bmatrix}, \qquad \beta = \begin{bmatrix} \dfrac{y}{y-x} & \dfrac{-1}{y-x} \\[2ex] \dfrac{-x}{y-x} & \dfrac{1}{y-x} \end{bmatrix}, \qquad \gamma = \begin{bmatrix} 1 & 1 \\ x & y \end{bmatrix}$$

with respect to the direct sum decomposition $\mathbb{C}^2 = \mathcal{M}_1 \dot{+} \mathcal{M}_2$, where $\langle 1, x \rangle$ and $\langle 1, y \rangle$ are chosen as bases in $\mathcal{M}_1$ and $\mathcal{M}_2$, respectively. Further, $F = [F_1 \quad F_2]$ is such that $(\alpha + \beta F)\mathcal{M}_1 \subset \mathcal{M}_1$ if and only if the transformation $F_1 = \begin{bmatrix} f_1 \\ f_2 \end{bmatrix}$ satisfies

$$-xf_1 + f_2 = \epsilon x^2 \tag{7.8.4}$$

The transformation $G = \begin{bmatrix} G_1 \\ G_2 \end{bmatrix}$ is such that $(\alpha + G\gamma)\mathcal{M}_2 \subset \mathcal{M}_2$ if and only if for $G_1 = [g_1 \quad g_2]$ we have

$$g_1 + yg_2 = \frac{\epsilon y^2}{y-x} \tag{7.8.5}$$

Now formulas (7.8.2) and (7.8.3) give

$$W(\lambda) = I + \left(\lambda - \left(\frac{\epsilon xy}{y-x} - g_1 f_1 - g_2 f_2 \right) \right)^{-1} \begin{bmatrix} -f_1 \\ -f_2 \\ f_1 \\ x + f_2 \end{bmatrix}$$

$$\times \left[\frac{y}{y-x} + g_1, \frac{-1}{y-x} + g_2, -g_1, -g_2 \right] \tag{7.8.6}$$

$$V(\lambda) = I + \left(\lambda + \frac{\epsilon xy}{y-x} \right)^{-1} \begin{bmatrix} 1 \\ y \end{bmatrix} \left[\frac{-x}{y-x}, \frac{1}{y-x} \right] \tag{7.8.7}$$

We conclude that for every six-tuple of complex numbers $(x, y, f_1, f_2; g_1, g_2)$ such that $x \neq y$ and (7.8.4) and (7.8.5) hold, there is a minimal linear fractional decomposition $U(\lambda) = \mathcal{F}_W(V)$ where $W(\lambda)$ and $V(\lambda)$ are given by equalities (7.8.6) and (7.8.7), respectively. $\square$

As an application of Theorem 7.7.1, let us consider linear fractional decompositions with several factors.

Theorem 7.8.2

Let $U(\lambda)$ be a rational matrix function that has no pole at infinity, and let $m = \delta(U)$. Then $U(\lambda)$ admits a linear fractional decomposition

$$U(\lambda) = \mathcal{F}_{W_1}(\mathcal{F}_{W_2}(\cdots (\mathcal{F}_{W_{m-1}}(W_m))\cdots) \qquad (7.8.8)$$

where for $j = 1, \ldots, m$ $W_j(\lambda)$ is a rational matrix function that is finite at infinity with McMillan degree 1. Moreover, $W_j(\lambda)$ can be chosen in such a way that

$$\mathcal{F}_{W_j}(V) = W_{j2}(\lambda) + V(\lambda)W_{j1}(\lambda), \qquad j = 1, \ldots, m-1 \qquad (7.8.9)$$

for any rational matrix function $V(\lambda)$ of suitable size, where $W_{j1}(\lambda)$ and $W_{j2}(\lambda)$ are rational matrix functions of appropriate sizes with $W_{j2}(\infty) = 0$, $W_{j1}(\infty) = I$.

Observe that the decomposition (7.8.8) is minimal in the sense that $\delta(U) = \delta(W_1) + \cdots + \delta(W_m)$. So, in contrast with the factorization of rational matrix functions (Example 7.5.1), nontrivial minimal linear fractional decompositions always exist.

Proof. Choose a minimal realization

$$U(\lambda) = \delta + \gamma(\lambda I - \alpha)^{-1}\beta$$

By the pole assignment theorem (Theorem 6.5.1), there exists a transformation F such that $\sigma(\alpha + \beta F) = \{\lambda_1, \ldots, \lambda_l\}$ with distinct numbers $\lambda_1, \ldots, \lambda_l$ (here $l \times l$ is the size of α). So there is a basis $g_1, \ldots g_l$ in $\mathbb{C}^l$ such that $(\alpha + \beta F)g_j = \lambda_j g_j$, $j = 1, \ldots, l$. On the other hand, for any transformation $G: \mathbb{C}^q \to \mathbb{C}^l$ there is a basis $f_1, \ldots, f_l$ in $\mathbb{C}^l$ in which the matrix of $\alpha + G\gamma$ has a lower triangular form (Theorem 1.9.1). Choose g_j in such a way that $g_j, f_2, f_3, \ldots, f_l$ are linearly independent and put $\mathcal{M}_1 = \text{Span}\{g_j\}$, $\mathcal{M}_2 = \text{Span}\{f_2, f_3, \ldots, f_l\}$. Then $(\mathcal{M}_1, \mathcal{M}_2)$ is a reducing pair and the recipe of Theorem 7.7.1 (with $D = I$, $d = \delta$) produces a minimal linear fractional decomposition $U(\lambda) = \mathcal{F}_W(U_1)$, where $\delta(W) = 1$ and $W(\infty) = I$. Moreover, taking $G = 0$ it follows that $W(\lambda)$ has the form

$$W(\lambda) = \begin{bmatrix} W_{11}(\lambda) & 0 \\ W_{12}(\lambda) & I \end{bmatrix}$$

Hence $\mathcal{F}_W(V)$ has the form (7.8.9). Now apply the preceding argument to $U_1(\lambda)$, and so on. Eventually we obtain the desired linear fractional decomposition (7.8.8). $\square$

Observe that, because $\delta(W_j) = 1$, each function $W_j(\lambda)$ from Theorem 7.8.2 has only one pole μ_j, and the multiplicity of this pole is 1. The proof of

Theorem 7.8.2, together with formula (7.7.8) for the transformation A, shows that the functions $W_j(\lambda)$ can be chosen with the additional property that $\mu_1, \ldots, \mu_m$ are the eigenvalues (counted with multiplicities) of the transformation α taken from a minimal realization (7.8.3) of $U(\lambda)$.

7.9 EXERCISES

7.1 Find realizations for the following rational matrix functions:

(a)
$$\begin{bmatrix} 1 & \lambda^{-1} \\ \lambda^{-1} & 1 \end{bmatrix}$$

(b)
$$\begin{bmatrix} 1 & \lambda^{-1} \\ (\lambda+1)^{-1} & 1+\lambda^{-1} \end{bmatrix}$$

(c)
$$\begin{bmatrix} 1+(\lambda-1)^{-1} & \lambda^{-1} \\ 0 & 1+(\lambda+1)^{-1} \end{bmatrix}$$

Determine whether these realizations are minimal.

7.2 Find the McMillan degree and a minimal realization for the following rational matrix functions:

(a)
$$\begin{bmatrix} \dfrac{\lambda^2+3\lambda+2}{\lambda^2+2\lambda+1} & \dfrac{\lambda}{\lambda^2+1} \\ \dfrac{1}{\lambda+2} & \dfrac{\lambda^2+3\lambda+4}{\lambda^2+3\lambda+2} \end{bmatrix}$$

(b)
$$\begin{bmatrix} \dfrac{\lambda^2}{(\lambda-2)^2} & \dfrac{1}{\lambda} \\ \dfrac{1}{\lambda^2} & \dfrac{\lambda}{\lambda+2} \end{bmatrix}$$

7.3 Reduce the following realizations to minimal realizations

(a)
$$1 + \begin{bmatrix} 0 & 1 & 0 & \cdots & 0 \end{bmatrix} (\lambda I - J_n(\lambda_0))^{-1} \begin{bmatrix} 0 \\ 0 \\ \vdots \\ 1 \end{bmatrix}$$

(b)
$$1 + C_p(\lambda I - J_n(\lambda_0))^{-1} C_q^T$$

where C_p is the $1 \times n$ matrix with 1 in the pth place and zeros elsewhere;

(c)
$$1 + \begin{bmatrix} 1 & 0 & 0 \end{bmatrix} \left(\lambda I - \begin{bmatrix} 1 & 1 & 1 \\ 1 & 1 & 1 \\ 1 & 1 & 1 \end{bmatrix} \right)^{-1} \begin{bmatrix} 0 \\ 1 \\ 0 \end{bmatrix}$$

7.4 Find minimal realizations for the following scalar rational functions:

(a)
$$\frac{\lambda - \lambda_1}{\lambda - \lambda_2}, \qquad \lambda_1 \neq \lambda_2$$

(b)
$$\frac{(\lambda - \lambda_1)^k}{(\lambda - \lambda_2)^k}; \qquad \lambda_1 \neq \lambda_2, \quad \text{where } k \text{ is a positive integer}$$

[*Hint*: In the minimal realization $I + C(\lambda I - A)^{-1}B$ the matrix A is the Jordan block of size k with eigenvalue λ_2.]

(c)
$$\sum_{j=0}^{k} a_j(\lambda - \lambda_0)^{-j}$$

7.5 Find a minimal realization for the scalar rational function with finite value at infinity, assuming that its representation as a sum of simple fractions is known, that is, of the form $\sum_{l=1}^{r} \sum_{j=0}^{k} a_{jl}(\lambda - \lambda_l)^{-j}$. [*Hint*: Use Exercise 7.4 (c) and Exercise 7.11.]

7.6 Show that if

$$W_1(\lambda) = I_n + C_1(\lambda I - A_1)^{-1}B_1, \qquad W_2(\lambda) = I_m + C_2(\lambda I - A_2)^{-1}B_2 \tag{1}$$

are realizations for $n \times n$ and $m \times m$ rational matrix functions $W_1(\lambda)$ and $W_2(\lambda)$, then the $(n + m) \times (n + m)$ rational matrix function $W_1(\lambda) \oplus W_2(\lambda)$ has realization

$$W_1(\lambda) \oplus W_2(\lambda) = I + \begin{bmatrix} C_1 & 0 \\ 0 & C_2 \end{bmatrix} \left(\lambda I - \begin{bmatrix} A_1 & 0 \\ 0 & A_2 \end{bmatrix} \right)^{-1} \begin{bmatrix} B_1 & 0 \\ 0 & B_2 \end{bmatrix} \tag{2}$$

Show, furthermore, that (2) is minimal if and only if each realization (1) is minimal.

7.7 Describe a minimal realization for the 2×2 circulant rational matrix function

$$W(\lambda) = \begin{bmatrix} a_1(\lambda) & a_2(\lambda) \\ a_2(\lambda) & a_1(\lambda) \end{bmatrix}$$

where $a_1(\lambda)$ and $a_2(\lambda)$ are scalar rational functions with finite value at infinity.

7.8 Describe a minimal realization for the $n \times n$ circulant rational matrix function

$$W = \begin{bmatrix} a_1(\lambda) & a_2(\lambda) & \cdots & a_n(\lambda) \\ a_n(\lambda) & a_1(\lambda) & \cdots & a_{n-1}(\lambda) \\ \vdots & \vdots & & \vdots \\ a_2(\lambda) & a_3(\lambda) & \cdots & a_1(\lambda) \end{bmatrix}$$

[As usual, assume that $W(\infty)$ is finite at infinity.]

7.9 Let $W_1(\lambda)$ and $W_2(\lambda)$ be rational matrix functions with realizations

$$W_j(\lambda) = D_j + C_j(\lambda I - A_j)^{-1} B_j, \qquad j = 1, 2 \qquad (3)$$

Show that the sum $W_1(\lambda) + W_2(\lambda)$ has the realization

$$W_1(\lambda) + W_2(\lambda) = D_1 + D_2 + [C_1 \quad C_2]\left(\lambda I - \begin{bmatrix} A_1 & 0 \\ 0 & A_2 \end{bmatrix}\right)^{-1} \begin{bmatrix} B_1 \\ B_2 \end{bmatrix}$$
$$(4)$$

7.10 Give an example of rational matrix functions $W_1(\lambda)$ and $W_2(\lambda)$ with minimal realizations (3) for which the realization (4) is not minimal.

7.11 Assume that the realizations (3) are minimal and A_1 and A_2 do not have common eigenvalues. Prove that (4) is minimal as well. [*Hint*: We have to show that $\left([C_1, \ C_2], \begin{bmatrix} A_1 & 0 \\ 0 & A_2 \end{bmatrix}\right)$ is a null kernel pair and $\left(\begin{bmatrix} A_1 & 0 \\ 0 & A_2 \end{bmatrix}, \begin{bmatrix} B_1 \\ b_2 \end{bmatrix}\right)$ is a full-range pair. Suppose that x and y are such that

$$C_1 A_1^k x + C_2 A_2^k y = 0$$

for $k = 0, 1, \ldots$. Because $\sigma(A_1) \cap \sigma(A_2) = \emptyset$, for $k = 0, 1, \ldots$ there exists a polynomial $p_k(\lambda)$ such that $p_k(A_1) = 0$, $p_k(A_2) = A_2^k$. Then

$$0 = C_1 p_k(A_1) x + C_2 p_k(A_2) y = C_2 A_2^k y$$

Hence $y = 0$. Similarly, one proves that $x = 0$.]

7.12 Let

$$W(\lambda) = D + \sum_{j=1}^{k} Z_j (\lambda - \lambda_j)^{-1}$$

be a rational $n \times n$ matrix function, where $\lambda_1, \ldots, \lambda_k$ are distinct complex numbers. Show that $W(\lambda)$ admits a realization

$$W(\lambda) = D + [I \quad \cdots \quad I] \, \mathrm{diag}[(\lambda - \lambda_1)^{-1}I, \ldots, (\lambda - \lambda_k)^{-1}I] \begin{bmatrix} Z_1 \\ Z_2 \\ \vdots \\ Z_k \end{bmatrix}$$

When is this realization minimal?

7.13 Find a realization for a rational $n \times n$ matrix function of the form

$$W(\lambda) = D + \sum_{j=1}^{k} Z_j (\lambda - \lambda_j)^{-2}$$

(where $\lambda_1, \ldots, \lambda_k$ are distinct complex numbers). When is the obtained realization minimal?

7.14 Given a realization $W(\lambda) = C(\lambda - A)^{-1}B$, find a realization for the rational matrix function

$$\begin{bmatrix} I & W(\lambda) \\ 0 & I \end{bmatrix}$$

Is it minimal if the realization $W(\lambda) = C(\lambda - A)^{-1}B$ is minimal?

7.15 Given a realization

$$W(\lambda) = D + C(\lambda I - A)^{-1}B \tag{5}$$

of a rational matrix function, find a realization for $W(\alpha\lambda + \beta)$, where $\alpha \neq 0$ and β are fixed complex numbers. Assuming that (5) is minimal, determine whether the obtained realization is minimal as well.

7.16 Given a realization (5), show that $W(\lambda^2)$ has a realization

$$W(\lambda^2) = D + [B \quad 0]\left(\lambda I - \begin{bmatrix} 0 & I \\ A & 0 \end{bmatrix}\right)^{-1} \begin{bmatrix} 0 \\ C \end{bmatrix}$$

If (5) is minimal, is this realization minimal as well?

7.17 Given a realization (5), find a realization for $W(p(\lambda))$, where $p(\lambda)$ is a scalar polynomial of third degree. Is the realization obtained minimal if (5) is minimal?

7.18 Let

$$W(\lambda) = I + C(\lambda I - A)^{-1}B \tag{6}$$

be a minimal realization.

(a) Show that

$$W(\lambda)^2 = I + [C \quad 0]\left(\lambda I - \begin{bmatrix} A & BC \\ 0 & A \end{bmatrix}\right)^{-1}\begin{bmatrix} 2B \\ B \end{bmatrix}$$

is a realization of $W(\lambda)^2$.

(b) Is the realization of $W(\lambda)^2$ minimal?

(c) Is the realization minimal if, in addition, the zeros and poles of $W(\lambda)$ are disjoints?

7.19 For the minimal realization (6), show that

$$W(\lambda)^k = I + [C\, 0 \cdots 0]\left(\lambda I - \begin{bmatrix} A & BC & BC & \cdots & BC \\ 0 & A & BC & \cdots & BC \\ 0 & 0 & A & \cdots & BC \\ \vdots & \vdots & \vdots & & \vdots \\ 0 & 0 & 0 & \cdots & A \end{bmatrix}\right)^{-1}\begin{bmatrix} kB \\ \vdots \\ \vdots \\ 2B \\ B \end{bmatrix}$$

is a realization of $W(\lambda)^k$. Is it minimal? Is it minimal if the zeros and poles of $W(\lambda)$ are disjoint?

7.20 Show that a realization $W(\lambda) = I + C(\lambda I - A)^{-1}B$ is minimal if A and $A - BC$ do not have common eigenvalues. (*Hint*: Use Theorem 7.1.3.)

7.21 Let $W(\lambda)$ be an $n \times n$ rational matrix function with $W(\infty) = I$ and assume that $W(\lambda)$ is hermitian for all real λ that are not poles of $W(\lambda)$. Prove that for every minimal realization

$$W(\lambda) = I + C(\lambda I - A)^{-1}B$$

there exists a unique invertible matrix S such that

$$C = B^*S, \qquad A = S^{-1}A^*S, \qquad B = S^{-1}C^*$$

7.22 Show that the McMillan degree of

$$D + \sum_{j=1}^{k} Z_j(\lambda - \lambda_j)^{-1}$$

where $\lambda_1, \ldots, \lambda_k$ are distinct complex numbers, is equal to the sum of ranks of $Z_1, \ldots, Z_k$.

7.23 Show that for rational $n \times n$ matrix functions $W_1(\lambda)$ and $W_2(\lambda)$ with finite values at infinity the inequalities

$$|\delta(W_1) - \delta(W_2)| \leq \delta(W_1 + W_2) \leq \delta(W_1) + \delta(W_2)$$

$$|\delta(W_1) - \delta(W_2)| \leq \delta(W_1 W_2) \leq \delta(W_1) + \delta(W_2)$$

hold.

7.24 Find the McMillan degree of the circulant rational matrix function

$$W(\lambda) = \begin{bmatrix} a_1(\lambda) & a_2(\lambda) & \cdots & a_n(\lambda) \\ a_n(\lambda) & a_1(\lambda) & \cdots & a_{n-1}(\lambda) \\ \vdots & \vdots & & \vdots \\ a_2(\lambda) & a_3(\lambda) & \cdots & a_1(\lambda) \end{bmatrix}$$

7.25 Find a minimal realization of $W(\lambda)$, and, with respect to this realization, describe all the minimal factorizations $W(\lambda) = W_1(\lambda)W_2(\lambda)$ of $W(\lambda)$ in terms of subspaces $\mathscr{L}$ and $\mathscr{M}$ as in Corollary 7.3.2, for the following scalar rational functions:

(a)
$$\frac{(\lambda - \lambda_1)^2}{(\lambda - \lambda_2)^2}; \quad \lambda_1 \neq \lambda_2$$

(b) $$\frac{(\lambda - \lambda_1)^k}{(\lambda - \lambda_2)^k}, \quad \lambda_1 \neq \lambda_2, \quad \text{where } k \geq 3 \text{ is a fixed integer}$$

(c)
$$\sum_{i=0}^{k} a_j(\lambda - \lambda_0)^{-j}$$

7.26 When is the realization $I_n + I_n(\lambda I_n - A)^{-1}B$, where A is upper triangular with zeros on the main diagonal and B is diagonal with distinct eigenvalues, minimal? Show that in this case $W(\lambda)$ admits a minimal factorization with factors having McMillan degree 1.

7.27 Prove that a circulant rational matrix function (Exercise 7.24) with value I at infinity admits a minimal factorization with factors having McMillan degree 1.

7.28 Let

$$W(\lambda) = I + C(\lambda I - A)^{-1}B$$

be a minimal realization, and assume that $BC = 0$.

(a) Prove that $W(\lambda)^{-1} + W(\lambda) \equiv 2I$.
(b) Prove that $W(\lambda)$ admits a nontrivial minimal factorization if and only if A is not unicellular.

7.29 Let

$$U(\lambda) = \left[\frac{\lambda - \lambda_1}{\lambda - \lambda_2} \right]^2, \quad \lambda_1 \neq \lambda_2$$

be a scalar rational function. Use the recipe of Theorem 7.7.1 to construct all minimal linear fractional decompositions $U(\lambda) = \mathscr{F}_W(V)$, such that $W(\lambda)$ and $V(\lambda)$ take finite values at infinity and $W_{11}(\infty)$, $W_{22}(\infty)$ are invertible. Find all the corresponding reducing pairs of subspaces with respect to a fixed minimal realization of $U(\lambda)$.

7.30 Show that all the following decompositions of a rational matrix function $U(\lambda)$ are particular cases of the linear fractional decomposition:

(a) $U(\lambda) = W_1(\lambda) + W_2(\lambda)$

(b) $U(\lambda) = W_1(\lambda) + W_2(\lambda)W_2(\lambda)$

(c) $U(\lambda) = (W_1(\lambda)^{-1} + W_2(\lambda)^{-1})^{-1}$

7.31 For the rational function $U(\lambda)$ given in Example 7.8.1, find all minimal linear fractional decompositions $U(\lambda) = \mathcal{F}_W(V)$, with $W(\infty) = I$ and $V(\infty) = I$.

Chapter Eight

Linear Systems

In this chapter we show how the concepts and results of previous chapters are applied to the theory of time-invariant linear systems. In fact, this is a short self-contained introduction to linear systems theory. It starts with the analysis of controllability, observability, minimality, and state feedback and continues with a selection of important problems with full solution. These include cascade connections, disturbance decoupling, and output stabilization.

8.1 REDUCTIONS, DILATIONS, AND TRANSFER FUNCTIONS

Consider the system of linear differential equations

$$
\begin{cases}
\dfrac{dx(t)}{dt} = Ax(t) + Bu(t) ; \quad x(0) = x_0 , \quad t \geq 0 \\[2mm]
y(t) = Cx(t) + Du(t)
\end{cases}
\tag{8.1.1}
$$

where $A: \mathbb{C}^m \to \mathbb{C}^m$, $B: \mathbb{C}^n \to \mathbb{C}^m$, $C: \mathbb{C}^m \to \mathbb{C}^r$, and $D: \mathbb{C}^n \to \mathbb{C}^r$ are constant transformations (i.e., independent of t). Here $u(t)$ is an n-dimensional vector function on $t \geq 0$ that is at our disposal and is referred to as the *input* (or *control*) of the linear system [equations (8.1.1)]. The r-dimensional vector function $y(t)$ is the *output* of (8.1.1), and the m-dimensional function $x(t)$ is the *state* of (8.1.1). Usually the state of the system (8.1.1) is unknown to us and must be inferred from the input (which we know) and the output (which we may be able to observe, at least partially).

Let $x(t; x_0, u)$ be the solution of the first equation in (8.1.1) [with the initial value $x(0) = x_0$]. It follows from the basic theory of ordinary differential equations [see Coddington and Levinson (1955), for example] that the solution $x(t; x_0, u)$ is unique and is given by the formula

$$x(t; x_0, u) = e^{tA}x_0 + \int_0^t e^{(t-s)A} Bu(s) \, ds \,, \qquad t \geq 0 \qquad (8.1.2)$$

Substituting into the second equation of (8.1.1), we have

$$y = y(t; x_0, u) = Ce^{tA}x_0 + \int_0^t Ce^{(t-s)A} Bu(s) \, ds + Du(t) \,, \qquad t \geq 0 \qquad (8.1.3)$$

Formula (8.1.3) expresses the output in terms of the input. In other words, the *input–output behaviour* of the system is represented explicitly.

Now we introduce some important operations on linear systems of type (8.1.1). It is convenient to describe (8.1.1) by the quadruple of transformations (A, B, C, D). A linear system (A', B', C', D') with transformations $A'; \mathbb{C}^{m'} \to \mathbb{C}^{m'}$, $B': \mathbb{C}^{n'} \to \mathbb{C}^{m'}$, $C': \mathbb{C}^{m'} \to \mathbb{C}^{r'}$, $D': \mathbb{C}^{n'} \to \mathbb{C}^{r'}$ will be called *similar* to (A, B, C, D) if there exists an invertible transformation $S: \mathbb{C}^{m'} \to \mathbb{C}^{m'}$ such that

$$A' = S^{-1}AS \,, \qquad C' = CS \,, \qquad B' = S^{-1}B \,, \qquad D' = D$$

(In particular, this implies that $m = m'$, $n = n'$, $r = r'$.) We also encounter system (8.1.1) with transformations $A: \mathcal{M} \to \mathcal{M}$, $B: \mathbb{C}^n \to \mathcal{M}$, $C: \mathcal{M} \to \mathbb{C}^r$, and $D: \mathbb{C}^n \to \mathbb{C}^r$, where $\mathcal{M}$ is a subspace of $\mathbb{C}^m$ for some m. The definition of similarity applies equally well to this case. [In particular, similarity with the system (A', B', C', D') described above implies $\dim \mathcal{M} = m'$.]

A system (A', B', C', D') with $A': \mathbb{C}^{m'} \to \mathbb{C}^{m'}$, $B': \mathbb{C}^n \to \mathbb{C}^{m'}$, $C': \mathbb{C}^{m'} \to \mathbb{C}^r$, $D': \mathbb{C}^n \to \mathbb{C}^r$ will be called a *dilation* of (A, B, C, D) if there exists a direct sum decomposition

$$\mathbb{C}^{m'} = \mathcal{L} \dotplus \mathcal{M} \dotplus \mathcal{N} \qquad (8.1.4)$$

with the two following properties: (1) the transformations A', B', C' have the following block forms with respect to this decomposition

$$A' = \begin{bmatrix} * & * & * \\ 0 & \tilde{A} & * \\ 0 & 0 & * \end{bmatrix}, \qquad C' = [\,0 \quad \tilde{C} \quad *\,], \qquad B' = \begin{bmatrix} * \\ \tilde{B} \\ 0 \end{bmatrix} \qquad (8.1.5)$$

where the stars denote entries of no immediate concern (so $\tilde{A}: \mathcal{M} \to \mathcal{M}$, $\tilde{C}: \mathcal{M} \to \mathbb{C}^r$, $\tilde{B}: \mathbb{C}^n \to \mathcal{M}$); (2) the system $(\tilde{A}, \tilde{B}, \tilde{C}, D')$ is similar to (A, B, C, D). In particular, if (A', B', C', D') is a dilation of (A, B, C, D), then $D' = D$. The form (8.1.5) for A' shows that the subspaces $\mathcal{L}$ and $\mathcal{L} \dotplus \mathcal{M}$ are A' invariant; in other words, (8.1.4) is a triinvariant decomposition associated with the A-semiinvariant subspace $\mathcal{M}$. Similarity is actually a particular case of dilation, with $\mathcal{M} = \mathbb{C}^{m'}$ and $\mathcal{L} = \mathcal{N} = \{0\}$.

We say that (A, B, C, D) is a *reduction* of (A', B', C', D') if (A', B', C', D') is a dilation of (A, B, C, D).

The basic property of reductions and dilations is that they have essentially the same input–output behaviour; as follows.

Proposition 8.1.1

Let (A', B', C', D') be a dilation of (A, B, C, D). Then, for $x_0 = 0$, the input–output behaviours of the systems (A', B', C', D') and (A, B, C, D) are the same. In other words, if $u(t)$ is any (say, continuous) n-dimensional vector function, then the output $\tilde{y} = \tilde{y}(t; 0, u)$ of the system (A', B', C', D') and the output $y = y(t; 0, u)$ of the system (A, B, C, D) coincide.

Proof.   Formula (8.1.3) gives

$$\tilde{y}(t; 0, u) = \int_0^t C' e^{(t-s)A'} B' u(s)\, ds + D' u(t), \qquad t \geq 0$$

$$y(t; 0, u) = \int_0^t C e^{(t-s)A} B u(s)\, ds + D u(t), \qquad t \geq 0$$

As $D' = D$, and $e^{(t-s)A}$ (for a fixed t and s) admits a power series representation (see Section 2.6), we have only to show that for $q = 0, 1, \ldots$

$$CA^q B = C' A'^q B' \tag{8.1.6}$$

Using formula (8.1.5), we obtain

$$C'A'^q B' = [0 \quad \tilde{C} \quad *] \begin{bmatrix} * & * & * & * \\ 0 & \tilde{A}^q & * & \tilde{B} \\ 0 & 0 & * & 0 \end{bmatrix} = \tilde{C}\tilde{A}^q\tilde{B}$$

Now $(\tilde{A}, \tilde{B}, \tilde{C}, D')$ and (A, B, C, D) are similar, so there exists an invertible transformation S such that $\tilde{A} = S^{-1}AS$, $\tilde{C} = CS$, and $\tilde{B} = S^{-1}B$. Hence

$$\tilde{C}\tilde{A}^q\tilde{B} = CS(S^{-1}AS)^q S^{-1}B = CA^q B$$

and (8.1.6) follows.   □

In practice one is concerned about the dimension m of the state space of a given system (8.1.1). It is desirable to make this dimension as small as possible without changing the input–output behaviour. We say that the system (8.1.1) is *minimal* if the dimension m of its state space is minimal among all linear systems (A', B', C', D') that exhibit the same input–output behaviour given the initial condition that that state vector is zero [i.e., $x(0) = 0$]. In view of Proposition 8.1.1, the following problem arises: given the linear system (8.1.1), not necessarily minimal, produce a minimal system by reduction of (8.1.1). We see later that this is always possible.

To study this and other problems in linear system theory, it is convenient to introduce the transfer function. Consider the system (8.1.1) with $x(0) = 0$, and apply the Laplace transform. Denote by the capital Roman letter the Laplace transform of the function designated by the corresponding small letter; thus

$$Z(\lambda) = \int_0^\infty e^{-\lambda s} z(s) \, ds$$

[It is assumed here that for $t \geq 0$ $z(t)$ is a continuous function such that $|z(t)| \leq K e^{\mu t}$ for some positive constants K and μ. This ensures that $Z(\lambda)$ is well defined for all complex λ with Re $\lambda > \mu$.] The system (8.1.1) then takes the form

$$\begin{cases} \lambda X(\lambda) = AX(\lambda) + BU(\lambda) \\ Y(\lambda) = CX(\lambda) + DU(\lambda) \end{cases}$$

Solving the first equation for $X(\lambda)$ and substituting in the second equation, we obtain the formula for the input–output behaviour in terms of the Laplace transforms:

$$Y(\lambda) = [D + C(\lambda I - A)^{-1}B]U(\lambda)$$

So the function $W(\lambda) = D + C(\lambda I - A)^{-1}B$ performs the input–outut map of the system (8.1.1), following application of the Laplace transform. This function is called the *transfer function* of the linear system (8.1.1). Observe that the transfer function is a rational matrix function of size $r \times n$ that has finite value $(=D)$ at infinity. Observe also that the transfer functions of two linear systems coincide if and only if the systems have the same input–output behaviour. In particular, systems obtained from each other by reductions and dilations have the same transfer functions.

8.2 MINIMAL LINEAR SYSTEMS: CONTROLLABILITY AND OBSERVABILITY

Consider once more the linear system of the preceding section:

$$\begin{cases} \dfrac{dx(t)}{dt} = Ax(t) + Bu(t) \, ; & t \geq 0 \\ y(t) = Cx(t) + Du(t) \end{cases} \tag{8.2.1}$$

and recall that this system is called *minimal* if the dimension of the state space is minimal. [We omit the initial condition $x(0) = x_0$ from (8.2.1); so (8.2.1) has in general many solutions $x(t)$.]

Applying the results of Section 7.1 to transfer functions, we obtain the following information on minimality of the system (8.2.1).

Theorem 8.2.1

(a) *Any linear system* (8.2.1) *is a dilation of a minimal linear system*; (b) *the linear system* (8.2.1) *is minimal if and only if* (A, B) *is a full-range pair and* (C, A) *is a null kernel pair*:

$$\bigcap_{j=0}^{\infty} \text{Ker } CA^j = \{0\} , \qquad \sum_{j=0}^{\infty} \text{Im}(B^jA) = \mathbb{C}^m \qquad (8.2.2)$$

where m is the dimension of the state space. Moreover, in (8.2.2) *one can replace* $\bigcap_{j=0}^{\infty} \text{Ker } CA^j$ *by* $\bigcap_{j=0}^{p-1} \text{Ker } CA^j$ *and* $\Sigma_{j=0}^{\infty} \text{Im}(B^jA)$ *by* $\Sigma_{j=0}^{p-1} \text{Im}(B^jA)$, *where p is any integer not smaller than the degree of the minimal polynomial of A.*

Indeed, (a) is a restatement of Theorem 7.1.3, and (b) follows from Theorem 7.1.5.

It turns out that the conditions (8.2.2) obtained in Chapter 7 from mathematical considerations have important physical meanings, namely, "controllability" and "observability" of the linear system (8.2.1). Let us introduce these notions.

The system (8.2.1) is called *observable* if for every continuous input $u(t)$ and output $y(t)$ there is at most one solution $x(t)$. In other words, by knowing the input and output one can determine the state (including the initial value) in a unique way.

Theorem 8.2.2

The system (8.2.1) *is observable if and only if* (C, A) *is a null kernel pair*:

$$\bigcap_{j=0}^{\infty} \text{Ker } CA^j = \{0\} \qquad (8.2.3)$$

Proof. Assume that (8.2.1) is observable. With $y(t) \equiv 0$ and $u(t) \equiv 0$, the definition implies that the only solution of the system

$$\frac{dx(t)}{dt} = Ax(t) , \qquad Cx(t) = 0 \qquad (8.2.4)$$

for $t \geq 0$ is $x(t) \equiv 0$. If equality (8.2.3) were not true, there would be a nonzero $x_0 \in \bigcap_{j=0}^{\infty} \text{Ker } CA^j$ and the function $x(t) = e^{tA}x_0$ would be a not identically zero solution of equation (8.2.4). Indeed, for every $t \geq 0$ we have

$$Cx(t) = C \sum_{j=0}^{\infty} \frac{1}{j!} (tA)^j x_0 = 0$$

Thus observability implies the condition stated in equality (8.2.3).

Now assume that (8.2.3) holds but (arguing by contradiction) the system (8.2.1) is not observable. Then there exist continuous vector functions $y(t)$ and $u(t)$ such that for $j = 1, 2$ and all $t \geq 0$, we obtain

$$\frac{dx_j(t)}{dt} = Ax_j(t) + Bu(t) , \qquad y(t) = Cx_j(t) + Du(t) \qquad (8.2.5)$$

for some $x_1(t)$ and $x_2(t)$ that do not coincide everywhere. Subtracting (8.2.5) with $j = 2$ from (8.2.5) with $j = 1$, and denoting $x(t) = x_1(t) - x_2(t) \neq 0$, we have

$$\frac{dx(t)}{dt} = Ax(t) ; \qquad Cx(t) = 0 , \qquad t \geq 0$$

In particular, $C[x^{(k)}(t)]_{t=0} = 0$. Since $x(t) = e^{tA} x_0$ it is found that

$$CA^k x_0 = 0 , \qquad k = 0, 1, \ldots$$

Hence $x_0 = 0$ by (8.2.3); but this contradicts $x(t) \neq 0$. $\square$

The system (8.2.1) is called *controllable* if by a suitable choice of input the state can be driven from any position to any other position in a prescribed period of time. Formally, this means that for every $x_1 \in \mathbb{C}^m$, $x_2 \in \mathbb{C}^m$, and $t_2 > t_1 \geq 0$ there is a continuous function $u(t)$ such that $x(t_1) = x_1$, $x(t_2) = x_2$ for some solution $x(t)$ of

$$\frac{dx(t)}{dt} = Ax(t) + Bu(t) , \qquad t \geq 0 \qquad (8.2.6)$$

Note that in the definition of controllability the second equation $y(t) = Cx(t) + Du(t)$ of equation (8.2.1) is irrelevant. Further, by replacing $x(t)$ by $x(t - t_1)$ we can assume in the definition of controllability that t_1 is always 0.

Theorem 8.2.3

The system (8.2.1) is controllable if and only if (A, B) is a *full-range pair*:

$$\sum_{j=0}^{\infty} \text{Im}(A^j B) = \mathbb{C}^m$$

We need the following lemma for the proof of Theorem 8.2.3.

Lemma 8.2.4

Let $G(t)$, $t \in [0, t_0]$ be an $m \times n$ matrix depending continuously on t. Then

$$\left\{\int_0^{t_0} G(t)u(t)\, dt \mid u(t) \text{ is continuous}\right\} = \text{Im}\left[\int_0^{t_0} G(t)G(t)^*\, dt\right]$$

$$(8.2.7)$$

Proof. Let $W = \int_0^{t_0} G(t)[G(t)]^*\, dt$. Assume $x \in \mathbb{C}^m$ is such that $x = Wy$ for some $y \in \mathbb{C}^n$. Then putting $u(t) = [G(t)]^*y$ we find that x belongs to the left-hand side of (8.2.7).

Conversely, if $x_1 \notin \text{Im } W$, then there exists an $x_2 \in \mathbb{C}^m$ such that $Wx_2 = 0$ and $(x_1, x_2) \neq 0$. [Here we use the property that $W = W^*$ and thus Im $W = (\text{Ker } W)^\perp$.] Arguing by contradiction, assume that there exists a continuous vector function $u(t)$ such that

$$\int_0^{t_0} G(t)u(t) = x_1$$

Then

$$\int_0^{t_0} x_2^* G(t)u(t)\, dt = (x_1, x_2) \neq 0 \qquad (8.2.8)$$

On the other hand

$$0 = x_2^* W x_2 = \int_0^{t_0} x_2^* G(t) G(t)^* x_2\, dt = \int_0^{t_0} \|G(t)^* x_2\|^2\, dt$$

and since the norm is nonnegative and $G(t)^*$ continuous, we obtain $G(t)^* x_2 = 0$, or $x_2^* G(t) = 0$ for all $t \in [0, t_2]$. But this contradicts (8.2.8). $\square$

Proof of Theorem 8.2.3. By formula (8.1.2) for every solution $x(t)$ of (8.2.6) with $x(0) = x_1$ we have

$$x(t) = e^{tA}x_1 + \int_0^t e^{(t-s)A} Bu(s)\, ds, \qquad t \geq 0$$

Hence

$$x(t_2) = e^{t_2 A}x_1 + \int_0^{t_2} e^{(t_2-s)A} Bu(s)\, ds = e^{t_2 A}x_1 + e^{t_2 A}\int_0^{t_2} e^{-sA} Bu(s)\, ds$$

From this equation it is clear that (8.2.1) is controllable if and only if for every $t_2 > 0$ the set of m-dimensional vectors

$$\left\{ \int_0^{t_2} e^{-sA} Bu(s)\, ds \mid u(t) \text{ is continuous} \right\}$$

coincides with the whole space $\mathcal{C}^m$. By Lemma 8.2.4, the controllability of (8.2.1) is equivalent to the condition that Im $W_t = \mathcal{C}^m$ for all $t > 0$, where

$$W_t = \int_0^t e^{-sA} BB^* e^{-sA^*}\, ds: \mathcal{C}^m \to \mathcal{C}^m$$

We prove Theorem 8.2.3 by showing that for all $t > 0$

$$\text{Ker } W_t = \left[\sum_{j=0}^{\infty} \text{Im } BA^j \right]^{\perp} \tag{8.2.9}$$

If $x \in \text{Ker } W_t$, then $x^* W_t x = 0$, that is

$$\int_0^t \| B^* e^{-sA^*} x \|^2\, ds = 0 \tag{8.2.10}$$

So $B^* e^{-sA^*} x = 0$, $0 \le s \le t$. [Otherwise, in view of the continuity of $\| B^* e^{sA^*} x \|^2$ as a function of s, we obtain a contradiction with (8.2.10).] Repeated differentiation with respect to s and putting $s = 0$ gives

$$B^* A^{*i-1} x = 0, \qquad i = 1, 2, \dots, n$$

It follows that

$$x \in \bigcap_{i=1}^{n} \text{Ker}(B^* A^{*i-1}) = \bigcap_{i=1}^{n} [\text{Im}(A^{i-1}B)]^{\perp}$$

$$= \left[\sum_{i=1}^{n} \text{Im}(A^{i-1}B) \right]^{\perp} = \left[\sum_{i=0}^{\infty} \text{Im}(A^i B) \right]^{\perp}$$

Assume now that $x \in [\Sigma_{i=1}^{\infty} \text{Im}(A^i B)]^{\perp}$. Then $B^* A^{*i-1} x = 0$, $i = 1, 2, \dots$. It follows that $B^* e^{sA^*} x = 0$ when $s \ge 0$, and hence $x^* W_t x = 0$ for $t > 0$. But W_t is nonnegative definite, so actually $W_t x = 0$, that is, $x \in \text{Ker } W_t$. $\square$

Combining Theorem 8.2.1 with Theorems 8.2.2 and 8.2.3, we obtain the following important fact.

Corollary 8.2.5

The linear system (8.2.1) is minimal if and only if it is controllable and observable.

This corollary, together with Theorem 7.1.5, shows that the concept of minimality for systems and realizations of rational functions are consistent, in the sense that a system is minimal precisely when it determines a minimal realization for its transfer function.

8.3   CASCADE CONNECTIONS OF LINEAR SYSTEMS

Consider two systems of type (8.1.1) (with initial value zero):

$$\begin{cases} \dfrac{dx_1}{dt} = A_1 x_1(t) + B_1 u_1(t); \quad x_1(0) = 0 \\ y_1(t) = C_1 x_1(t) + D_1 u_1(t) \end{cases} \tag{8.3.1}$$

and

$$\begin{cases} \dfrac{dx_2}{dt} = A_2 x_2(t) + B_2 u_2(t); \quad x_2(0) = 0 \\ y_2(t) = C_2 x_2(t) + D_2 u_2(t) \end{cases} \tag{8.3.2}$$

Suppose also that $u_1(t)$ and $y_2(t)$ are from the same space. The two systems are combined in a "cascade" form when the output y_2 of the second system becomes the input u_1 of the first system. We obtain

$$\frac{dx_1}{dt} = A_1 x_1(t) + B_1 y_2(t) = A_1 x_1(t) + B_1 C_2 x_2(t) + B_1 D_2 u_2(t)$$

and

$$y_1(t) = C_1 x_1(t) + D_1 y_2(t) = C_1 x_1(t) + D_1 C_2 x_2(t) + D_1 D_2 u_2(t)$$

Writing $x(t) = \begin{bmatrix} x_1(t) \\ x_2(t) \end{bmatrix}$, we obtain a new system of the same type:

$$\begin{cases} \dfrac{dx(t)}{dt} = \begin{bmatrix} A_1 & B_1 C_2 \\ 0 & A_2 \end{bmatrix} x(t) + \begin{bmatrix} B_1 D_2 \\ B_2 \end{bmatrix} u_2(t) \\ y_1(t) = [C_1 \quad D_1 C_2] x(t) + D_1 D_2 u_2(t) \end{cases} \tag{8.3.3}$$

The system (8.3.3) is called a *simple cascade* composed of the *first component* (8.3.1) and the *second component* (8.3.2). Note that the dimension of the state space of the simple cascade is the sum of the state space dimensions of its components, and the input of the simple cascade coincides with the input of its second component, whereas the output of the simple cascade coincides with the output of the first component.

Similarly, one can consider the simple cascade of more than two components. Let $(A_1, B_1, C_1, D_1), \ldots, (A_p, B_p, C_p, D_p)$ be linear systems of type (8.1.1). A linear system that is obtained by identifying the output of (A_i, B_i, C_i, D_i) with the input of $(A_{i-1}, B_{i-1}, C_{i-1}, D_{i-1})$, $i = 2, 3, \ldots, p$ will be called the *simple cascade* of the systems $(A_1, B_1, C_1, D_1), \ldots,$ (A_p, B_p, C_p, D_p). By applying formula (8.3.2) $p - 1$ times, we see that such a simple cascade has the form

$$
\left.\begin{aligned}
\frac{dx(t)}{dt} &= \begin{bmatrix} A_1 & B_1C_2 & B_1C_3 & \cdots & B_1C_p \\ 0 & A_2 & B_2C_3 & \cdots & B_2C_p \\ \vdots & \vdots & & \ddots & \vdots \\ 0 & 0 & & \cdots & A_p \end{bmatrix} x(t) + \begin{bmatrix} B_1D_p \\ \vdots \\ B_{p-1}D_p \\ B_p \end{bmatrix} u_p(t) \\[2ex]
y_1(t) &= [C_1, D_1C_2, \ldots, D_1C_p]x(t) + D_1D_2 \cdots D_p u_p(t)
\end{aligned}\right\} \quad (8.3.4)
$$

In the language of transfer functions the simple cascading connection has a very simple interpretation: formula (7.2.9) shows that the transfer function of the simple cascade of two systems is the product of the transfer functions of its first and second components (in this order). More generally, if (A, B, C, D) is the simple cascade of $(A_1, B_1, C_1, D_1), \ldots,$ (A_p, B_p, C_p, D_p), then

$$
D + C(\lambda I - A)^{-1}B = [D_1 + C_1(\lambda I - A_1)^{-1}B_1] \cdots [D_p + C_p(\lambda I - A_p)^{-1}B_p]
$$

The following problem is of considerable interest: describe the representation of a given linear system (A, B, C, D) as a simple cascade of other linear systems. We can assume that (A, B, C, D) is minimal (otherwise replace it by a minimal system with the same input–output behaviour). In order to relate this problem to the factorization problem for rational matrix functions described in Sections 7.3 and 7.5, we shall assume that $D = I$ and that in each component (A_i, B_i, C_i, D_i) of the simple cascade (A, B, C, I) we have $D_i = I$. Equation (8.3.4) shows that if (A, B, C, D) is a simple cascade with components (A_i, B_i, C_i, D_i), $i = 1, \ldots, p$, then the size of A [or, what is the same, the McMillan degree $\delta(W)$ of the transfer function $W(\lambda)$ of (A, B, C, I)] is equal to $m_1 + \cdots + m_p$, where m_i is the size of A_i. Denoting by $W_i(\lambda)$ the transfer function of (A_i, B_i, C_i, D_i), we have $\delta(W_i) \le m_i$. On the other hand, as we have seen in the preceding paragraph

$$
W(\lambda) = W_1(\lambda) \cdots W_p(\lambda) \tag{8.3.5}
$$

which implies

$$
\delta(W) \le \delta(W_1) + \cdots + \delta(W_p) \le m_1 + \cdots + m_p = \delta(W) \tag{8.3.6}
$$

So equality holds throughout (8.3.6), which means that the factorization (8.3.5) is minimal and that each system (A_i, B_i, C_i, D_i), $i = 1, \ldots, p$ is minimal. Now we can use the results of Sections 7.3 and 7.5 concerning minimal factorizations of rational matrix functions to study simple cascading decompositions of minimal linear systems. The following analog of Theorem 7.5.1 is an example.

Theorem 8.3.1

The components of every representation of a minimal system (A, B, C, I) as a simple cascade (with the transfer functions of the components having value I at infinity) are given by

$$(\pi_1 A \pi_1, \pi_1 B, C \pi_1, I), \ldots, (\pi_p A \pi_p, \pi_p B, C \pi_p, I) \qquad (8.3.7)$$

where the projectors $\pi_1, \ldots, \pi_p$ and associated subspaces $\mathscr{L}_1, \ldots, \mathscr{L}_p$ are defined as in Theorem 7.5.1. The transformations $\pi_j A \pi_j$ in (8.3.7) are understood as acting in $\mathscr{L}_j$, and the transformations $C \pi_j$ and $\pi_j B$ are understood as acting from $\mathscr{L}_j$ into $\mathbb{C}^n$, and from $\mathbb{C}^n$ into $\mathscr{L}_j$, respectively, where n is the number of rows in C (which is equal to the number of columns in B).

We now describe a more general way to connect two linear systems. Consider the linear system

$$\frac{dx}{dt} = Ax + Bu , \qquad y = Cx + Du , \qquad x(0) = 0 \qquad (8.3.8)$$

and assume that the input vector $u = u(t)$ and the output vector $y = y(t)$ are divided into two components:

$$y(t) = \begin{bmatrix} y_1(t) \\ y_2(t) \end{bmatrix}, \qquad u(t) = \begin{bmatrix} u_1(t) \\ u_2(t) \end{bmatrix} \qquad (8.3.9)$$

Now let

$$\frac{dw(t)}{dt} = aw(t) + bs(t) , \qquad z(t) = cw(t) + ds(t) , \qquad z(0) = 0 \qquad (8.3.10)$$

be another linear system with the input $s(t)$, output $z(t)$, and the state $w(t)$. (Here a, b, c, and d are constant matrices of appropriate sizes.) We obtain a new system by feeding the first component of the output of (8.3.8) into the input of (8.3.10) and at the same time feeding the output of (8.3.10) into the second component of the input of (8.3.8). [It is assumed, of course, that the vectors $y_1(t)$ and $s(t)$ are in the same space, as well as the vectors $u_2(t)$ and $z(t)$.] This situation is represented diagrammatically by

$$(8.3.11)$$

Here Σ_1 and Σ_2 represent the linear systems described by equations (8.3.8) and (8.3.10), respectively. The new system has $u_1(t)$ as an input and $y_2(t)$ as an output and is called the *cascade* of (8.3.10) by (8.3.8). The "simple cascade" described in the first part of this section is a particular case of a cascade. Indeed, if the first component of the output $y_1(t)$ in the system (8.3.8) depends on $u_1(t)$ only, and $y_2(t) = u_2(t)$, then the cascade described by (8.3.11) is actually a simple cascade.

We turn now to a description of the cascade in terms of transfer functions. First, rewrite (8.3.8) in the form

$$\frac{dx}{dt} = Ax + [B_1 \quad B_2]\begin{bmatrix} u_1 \\ u_2 \end{bmatrix}; \qquad x(0) = 0$$

$$\begin{bmatrix} y_1 \\ y_2 \end{bmatrix} = \begin{bmatrix} C_1 \\ C_2 \end{bmatrix} x + \begin{bmatrix} D_{11} & D_{12} \\ D_{21} & D_{22} \end{bmatrix}\begin{bmatrix} u_1 \\ u_2 \end{bmatrix}$$

where

$$B = [B_1 \quad B_2], \qquad C = \begin{bmatrix} C_1 \\ C_2 \end{bmatrix}, \qquad D = \begin{bmatrix} D_{11} & D_{12} \\ D_{21} & D_{22} \end{bmatrix}$$

are the block matrix representations of B, C, and D conforming with the division [equations (8.3.9)] of $y(t)$ and $u(t)$. The transfer function of this system is

$$W(\lambda) = \begin{bmatrix} D_{11} & D_{12} \\ D_{21} & D_{22} \end{bmatrix} + \begin{bmatrix} C_1 \\ C_2 \end{bmatrix}(\lambda I - A)^{-1}[B_1 \quad B_2] = \begin{bmatrix} W_{11}(\lambda) & W_{12}(\lambda) \\ W_{21}(\lambda) & W_{22}(\lambda) \end{bmatrix}$$

where $W_{ij}(\lambda) = D_{ij} + C_i(\lambda I - A)^{-1}B_j$; $i, j = 1, 2$. So passing to the Laplace transforms, we have

$$\begin{bmatrix} Y_1(\lambda) \\ Y_2(\lambda) \end{bmatrix} = \begin{bmatrix} W_{11}(\lambda) & W_{12}(\lambda) \\ W_{21}(\lambda) & W_{22}(\lambda) \end{bmatrix}\begin{bmatrix} U_1(\lambda) \\ U_2(\lambda) \end{bmatrix} \qquad (8.3.12)$$

where, as usual, the capital Roman letters indicate Laplace transforms of the functions designated by the corresponding lowercase letters. Let $V(\lambda)$ be the transfer function of (8.3.10); then

$$Z(\lambda) = V(\lambda)S(\lambda) \qquad (8.3.13)$$

Now identify

$$S(\lambda) = Y_1(\lambda), \qquad Z(\lambda) = U_2(\lambda) \qquad (8.3.14)$$

Using (8.3.12)–(8.3.14) we have (omitting the variable λ)

$$Y_1 = W_{11}U_1 + W_{12}U_2 = W_{11}U_1 + W_{12}VY_1$$

and hence

$$Y_1 = (I - W_{12}V)^{-1}W_{11}U_1$$

Further

$$Y_2 = W_{21}U_1 + W_{22}U_2 = W_{21}U_1 + W_{22}VY_1 = (W_{21} + W_{22}V(I - W_{12}V)^{-1}W_{11})U_1$$

So the cascade of a linear system with the transfer function $V(\lambda)$ by a linear system with the transfer function

$$W(\lambda) = \begin{bmatrix} W_{11}(\lambda) & W_{12}(\lambda) \\ W_{21}(\lambda) & W_{22}(\lambda) \end{bmatrix}$$

has the transfer function $U(\lambda)$ given by the formula

$$U(\lambda) = W_{21}(\lambda) + W_{22}(\lambda)V(\lambda)(I - W_{12}(\lambda)V(\lambda))^{-1}W_{11}(\lambda)$$

We recognize that $U(\lambda)$ is just a linear fractional transformation, $U(\lambda) = \mathscr{F}_W(V)$, as discussed in Chapter 7. Consequently, the results of Sections 7.6, 7.7, and 7.8 can be interpreted in terms of minimal cascades of linear systems. The cascade of (7.3.10) by (7.3.8) will be called *minimal* if the corresponding linear fractional decomposition $U = \mathscr{F}_W(V)$ is minimal. As an example, let us restate Theorem 7.8.2 in these terms.

Theorem 8.3.2

Any minimal linear system with m-dimensional state space can be represented as a minimal cascade of m linear systems each of which has one-dimensional state space.

8.4   THE DISTURBANCE DECOUPLING PROBLEM

In this and the next section we consider two important problems from linear system theory in which $[A \ B]$-invariant subspaces (as discussed in Chapter 6) appear naturally and play a crucial role.

Consider the linear system

$$\begin{cases} \dfrac{dx(t)}{dt} = Ax(t) + Bu(t) + Eq(t), & t \geq 0 \\[2mm] z(t) = Dx(t) \end{cases} \tag{8.4.1}$$

where $A: \mathbb{C}^n \to \mathbb{C}^n$, $B: \mathbb{C}^m \to \mathbb{C}^n$, $E: \mathbb{C}^p \to \mathbb{C}^n$, and $D: \mathbb{C}^n \to \mathbb{C}^r$ are constant transformations, and $x(t)$, $u(t)$, $q(t)$, and $z(t)$ are vector functions taking values in $\mathbb{C}^n$, $\mathbb{C}^m$, $\mathbb{C}^p$, and $\mathbb{C}^r$, respectively.

As in Section 8.1, $\mathbb{C}^n$ is interpreted as the state space of the underlying dynamical system, and $u(t)$ is the input. The vector function $z(t)$ is interpreted as the *output*. The term $q(t)$ represents a *disturbance* that is supposed to be unknown and unmeasurable. We assume that $q(t)$ is a continuous function of t for $t \geq 0$.

An important transformation of the system (8.4.1) involves "state feedback." This is obtained when the state $x(t)$ is fed through a certain constant linear transformation F into the input, so the input of the new system is actually the sum of the original input $u(t)$ and the feedback. Diagrammatically, we have

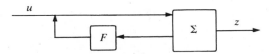

Our problem is to determine (if possible) a state feedback F in such a way that, in the new system, the output is independent of the disturbance $q(t)$.

To express this problem in mathematical terms we introduce the following definition. The system (8.4.1) is called *disturbance decoupled* if for every $x_0 \in \mathbb{C}^n$ the output $z(t)$ of the system (8.4.1) with $x(0) = x_0$ is the same for every continuous function $q(t)$. We have (cf. Section 8.1)

$$x(t) = e^{tA}x(0) + \int_0^t e^{(t-s)A}[Bu(s) + Eq(s)]\, ds\,, \qquad t \geq 0$$

and thus

$$z(t) = De^{tA}x(0) + D\int_0^t e^{(t-s)A} Bu(s)\, ds + D\int_0^t e^{(t-s)A} Eq(s)\, ds\,, \qquad t \geq 0$$

Hence the system (8.4.1) is disturbance decoupled if and only if

$$D\int_0^t e^{(t-s)A} Eq(s)\, ds = 0\,, \qquad t \geq 0$$

for every continuous function $q(t)$.

We need one more notion from linear system theory. Consider the linear system

$$\frac{dx}{dt} = Ax(t) + Bu(t)\,; \qquad t \geq 0\,; \qquad x(0) = 0 \qquad (8.4.2)$$

where $A: \mathbb{C}^n \to \mathbb{C}^n$ and $B: \mathbb{C}^m \to \mathbb{C}^n$ are constant transformations. We say that the state vector $y \in \mathfrak{C}^n$ is *reachable* for the system (8.4.2) if there exist a $t_0 \geq 0$ and a continuous function $u(t)$ such that the solution $x(t)$ of (8.4.2) satisfies $x(t_0) = y$. As

$$x(t) = \int_0^t e^{(t-s)A} Bu(s) \, ds$$

for $t \geq 0$, it follows easily that the set of all reachable state vectors of (8.4.2) is a subspace.

Proposition 8.4.1

The set $\mathfrak{R}$ of reachable states coincides with the minimal A-invariant subspace that contains Im B:

$$\mathfrak{R} = \langle A \mid \text{Im } B \rangle \overset{\text{def}}{=} \text{Im } B + A(\text{Im } B) + \cdots + A^{n-1}(\text{Im } B) \subset \mathfrak{C}^n$$

Proof. By Lemma 8.2.4 we find that $x \in \mathfrak{R}$ if and only if

$$x \in \text{Im}\left[\int_0^{t_0} e^{(t_0-s)A} B[e^{(t_0-s)A} B]^* \, ds\right] = \text{Im} \int_0^{t_0} e^{-sA} BB^* e^{-sA^*} \, ds$$

for some $t_0 > 0$. For any $t_0 > 0$, let

$$W_{t_0} = \int_0^{t_0} e^{-sA} BB^* e^{-sA^*} \, ds$$

By equality (8.2.9)

$$\text{Ker } W_{t_0} = \left[\sum_{j=0}^{\infty} \text{Im}(A^j B)\right]^{\perp}$$

or, taking into account the hermitian property of W_{t_0}

$$\text{Im } W_{t_0} = \sum_{j=0}^{\infty} \text{Im}(A^j B)$$

which coincides with $\langle A \mid \text{Im } B \rangle$ in view of Theorem 2.7.3. $\square$

Using this proposition, we obtain the following characterization of disturbance decoupled systems.

Proposition 8.4.2

The system (8.4.1) is disturbance decoupled if and only if

$$\langle A \mid \operatorname{Im} E \rangle \subset \operatorname{Ker} D$$

Returning to the problem mentioned above, note that state feedback is described by a transformation $F: \mathbb{C}^n \to \mathbb{C}^m$, and substituting $u(t) + Fx(t)$ in place of $u(t)$ in the system (8.4.1), we obtain the system with state feedback:

$$\begin{cases} \dfrac{dx(t)}{dt} = (A + BF)x(t) + Bu(t) + Eq(t), & t \geq 0 \\ z(t) = Dx(t) \end{cases}$$

The new system has the same form as the original system (8.4.1), with A replaced by $A + BF$. Our mathematical problem is: given transformations $A: \mathbb{C}^n \to \mathbb{C}^n$ and $B: \mathbb{C}^m \to \mathbb{C}^n$, and given subspaces $\mathscr{E} \subset \mathbb{C}^n$ (which plays the role of Im E) and $\mathscr{D} \subset \mathbb{C}^n$ (which plays the role of Ker D), find, if possible, a transformation $F: \mathbb{C}^n \to \mathbb{C}^m$ such that the subspace

$$\langle A + BF \mid \mathscr{E} \rangle \overset{\text{def}}{=} \mathscr{E} + (A + BF)\mathscr{E} + \cdots + (A + BF)^{n-1}\mathscr{E}$$

[which is the minimal $(A + BF)$-invariant subspace containing $\mathscr{E}$] is contained in $\mathscr{D}$.

The solution to this problem depends on the notion of $[A\ B]$-invariant subspaces, as developed in Chapter 6.

Theorem 8.4.3

In the preceding notation, there exists a transformation $F: \mathbb{C}^n \to \mathbb{C}^m$ such that

$$\langle A + BF \mid \mathscr{E} \rangle \subset \mathscr{D} \tag{8.4.3}$$

if and only if the $[A\ B]$-invariant subspace $\mathscr{U}$ that is maximal in $\mathscr{D}$ contains $\mathscr{E}$. In this case any transformation $F: \mathbb{C}^n \to \mathbb{C}^m$ with $(A + BF)\mathscr{U} \subset \mathscr{U}$ (which exists by Theorem 6.1.1) has the property (8.4.3).

Proof. Assume that there is an $F: \mathbb{C}^n \to \mathbb{C}^m$ with the property (8.4.3). By Theorem 2.8.4 (applied with $A + BF$ playing the role of A and any transformation whose image is $\mathscr{E}$ playing the role of B) the subspace $\langle A + BF \mid \mathscr{E} \rangle$ is $(A + BF)$ invariant, and thus (Theorem 6.1.1) it is $[A\ B]$ invariant. As $\langle A + BF \mid \mathscr{E} \rangle \supset \mathscr{E}$, and the maximal (in $\mathscr{D}$) $[A\ B]$-invariant subspace $\mathscr{U}$ contains $\langle A + BF \mid \mathscr{E} \rangle$, we obtain $\mathscr{U} \supset \mathscr{E}$.

Conversely, assume $\mathscr{U} \supset \mathscr{E}$. By Theorem 6.1.1 there is a transformation $F: \mathbb{C}^n \to \mathbb{C}^m$ such that $(A + BF)\mathscr{U} \subset \mathscr{U}$. Now

$$\langle A + BF \mid \mathscr{E} \rangle \subset \langle A + BF \mid \mathscr{U} \rangle = \mathscr{U} \subset \mathscr{D}$$

and (8.4.3) follows. $\square$

When applied to the disturbance decoupling problem, Theorem 8.4.3 can be restated in the following form.

Theorem 8.4.4

Given a system (8.4.1), there exists a state feedback $F: \mathbb{C}^n \to \mathbb{C}^m$ such that the system

$$\begin{cases} \dfrac{dx(t)}{dt} = (A + BF)x(t) + Bu(t) + Eq(t) , & t \geq 0 \\ z(t) = Dx(t) \end{cases} \tag{8.4.4}$$

is disturbance decoupled if and only if the $[A \; B]$-invariant subspace $\mathcal{U}$ that is maximal in $\mathrm{Ker}\, D$, contains $\mathrm{Im}\, E$. In this case the system (8.4.4) is disturbance decoupled for every transformation $F: \mathbb{C}^n \to \mathbb{C}^m$ with the property that $\mathcal{U}$ is $(A + BF)$ invariant.

We illustrate Theorem 8.4.4 by a simple example.

EXAMPLE 8.4.1. Let

$$A = \begin{bmatrix} 0 & 1 & 0 \\ 0 & 0 & 1 \\ 0 & 0 & 0 \end{bmatrix} \qquad B = \begin{bmatrix} 0 \\ 0 \\ 1 \end{bmatrix}$$

$$E = \begin{bmatrix} a_1 \\ a_2 \\ a_3 \end{bmatrix} \qquad D = [b_1 \quad b_2 \quad b_3]$$

where a_1, a_2, and a_3, as well as b_1, b_2, and b_3 are complex numbers not all zero. Using Theorem 6.4.1 and its proof, we find that a one-dimensional subspace $\mathcal{M}$ is $[A \; B]$ invariant if and only if

$$\mathcal{M} = \mathrm{Span}\{\langle 1, \lambda, \lambda^2 \rangle\}$$

for some $\lambda \in \mathbb{C}$, and a two-dimensional subspace $\mathcal{M}$ is $[A \; B]$ invariant if and only if either

$$\mathcal{M} = \mathrm{Span}\{\langle 1, \lambda, \lambda^2 \rangle, \langle 1, \mu, \mu^2 \rangle\} ; \qquad \lambda \neq \mu ; \qquad \lambda, \mu \in \mathbb{C}$$

or

$$\mathcal{M} = \mathrm{Span}\{\langle 1, \lambda, \lambda^2 \rangle, \langle 0, 1, 2\lambda \rangle\} ; \qquad \lambda \in \mathbb{C}$$

Consider first the case when $\mathrm{Ker}\, D$ is $[A \; B]$ invariant. This happens if and only if $b_3 \neq 0$. Then obviously $\mathrm{Ker}\, D$ is the maximal $[A \; B]$-invariant subspace in Ker D, and Ker $D \supset \text{Im } E$ if and only if $a_1 b_1 + a_2 b_2 + a_3 b_3 = 0$. So, when $b_3 \neq 0$, there exists a 1×3 matrix $F = [f_1, f_2, f_3]$ such that the system (8.4.4) is disturbance decoupled if and only if $a_1 b_1 + a_2 b_2 + a_3 b_3 = 0$, and in this case one can take f_j in such a way that the polynomial $b_1 + b_2 x + b_3 x^2$ divides $-f_1 - f_2 x - f_3 x^2 + x^3$. Assume now that Ker D is not $[A \ B]$ invariant, that is, $b_3 = 0$. If $b_2 \neq 0$, then the maximal $[A \ B]$-invariant subspace in Ker D is $\text{Span}\{\langle 1, \lambda_0, \lambda_0^2 \rangle\}$, where $\lambda_0 = -b_1/b_2$. In this case we have $\text{Span}\{\langle 1, \lambda_0, \lambda_0^2 \rangle\} \supset \text{Im } E$ if and only if

$$a_1 \neq 0, \qquad a_1 b_1 + a_2 b_2 = 0, \qquad a_3 b_2^2 = a_1 b_1^2 \qquad (8.4.5)$$

So, if $b_3 = 0$ and $b_2 \neq 0$, there exists an $F = [f_1 f_2 f_3]$ as in Theorem 8.4.4 if and only if (8.4.5) holds, in which case one can take f_j in such a way that the polynomial $b_1 + b_2 x$ divides $-f_1 - f_2 x - f_3 x^2 + x^3$. Finally, assume $b_2 = b_3 = 0$. Then the maximal $[A \ B]$-invariant subspace in Ker D is the zero subspace, and there is no F for which the system (8.4.4) is disturbance decoupled. $\square$

8.5   THE OUTPUT STABILIZATION PROBLEM

Consider the system

$$\begin{cases} \dfrac{dx(t)}{dt} = Ax(t) + Bu(t), & t \geq 0 \\[2mm] z(t) = Dx(t) \end{cases} \qquad (8.5.1)$$

where the transformations $A: \mathbb{C}^n \to \mathbb{C}^n$, $B: \mathbb{C}^m \to \mathbb{C}^n$, and $D: \mathbb{C}^n \to \mathbb{C}^r$ are constant. The problem we deal with in this section is that of stabilizing the output $z(t)$ by means of a state feedback while still maintaining the freedom to apply a control function $u(t)$. More exactly, the problem is to find a transformation $F: \mathbb{C}^n \to \mathbb{C}^m$ (which represents the state feedback) such that the solution of the new system

$$\begin{cases} \dfrac{d\bar{x}(t)}{dt} = (A + BF)\bar{x}(t) + Bu(t), & x(0) = x_0 \\[2mm] \bar{z}(t) = D\bar{x}(t) \end{cases}$$

with identically zero input $u(t)$ satisfies $\lim_{t \to \infty} \bar{z}(t) = 0$ for every initial value x_0. As

$$\bar{z}(t) = De^{t(A+BF)} x_0$$

this condition amounts to

$$\lim_{t\to\infty} D e^{t(A+BF)} = 0 \tag{8.5.2}$$

To study the property (8.5.2), we need the following lemma. This is, in fact, a special case of Theorem 5.7.2, but it is convenient to have some of the conclusions recast in the present form.

Lemma 8.5.1

Let $A: \mathfrak{C}^n \to \mathfrak{C}^n$ be a transformation, and let $\mathcal{M}_-$, $\mathcal{M}_0$, $\mathcal{M}_+$ be the sum of root subspaces of A corresponding to the eigenvalues with negative, zero, and positive real parts, respectively. Thus

(a)
$$\lim_{t\to\infty} \| (e^{tA})|_{\mathcal{M}_-} \| = 0$$

(b)
$$\liminf_{t\to\infty} \| (e^{tA})|_{\mathcal{M}_0} \| > 0$$

and for some $x_0 \in \mathcal{M}_0$ we have $\lim_{t\to\infty} \sup \| e^{tA} x_0 \| < \infty;$

(c)
$$\lim_{t\to\infty} \| e^{tA} x \| = \infty \quad \text{for all} \quad x \in \mathcal{M}_+ < \{0\}$$

Note that $\mathcal{M}_-$, $\mathcal{M}_0$, and $\mathcal{M}_+$ are A-invariant subspaces, and therefore these subspaces are also invariant for the transformations e^{tA}, $t \geq 0$.

Given A, B, D as in (8.5.1) and any transformation $F: \mathfrak{C}^n \to \mathfrak{C}^m$, let

$$\mathcal{N}_F = \bigcap_{i=1}^{n} \text{Ker}[D(A + BF)^{i-1}]$$

be the maximal $(A + BF)$-invariant subspace in Ker D. The condition (8.5.2) can be expressed in terms of the root subspaces of $A + BF$ as follows.

Lemma 8.5.2

We have

$$\lim_{t\to\infty} D e^{t(A+BF)} = 0$$

if and only if $\mathcal{R}_{\lambda_0}(A + BF) \subset \mathcal{N}_F$ for every eigenvalue λ_0 of $A + BF$ such that Re $\lambda_0 \geq 0$.

Proof. By Theorem 2.7.4 we have

$$D = [0 \quad D_1], \qquad A + BF = \begin{bmatrix} A_{11} & A_{12} \\ 0 & A_{22} \end{bmatrix}$$

with respect to the direct sum decomposition $\mathbb{C}^n = \mathcal{N}_F \dotplus \mathcal{M}$, where $\mathcal{M}$ is a direct complement to $\mathcal{N}_F$ in $\mathbb{C}^n$. Also, (D_1, A_{22}) is a null kernel pair. Hence

$$De^{t(A+BF)} = D_1 e^{tA_{22}}, \qquad t \geq 0$$

Now clearly $\mathcal{R}_{\lambda_0}(A + BF) = \mathrm{Ker}(A + BF - \lambda_0 I)^n \subset \mathcal{N}_F$ for every $\lambda_0 \in \sigma(A + BF)$ with $\mathcal{R}e\ \lambda_0 \geq 0$ if and only if A_{22} has all its eigenvalues in the open left-half plane. So we have to prove that

$$\lim_{t \to \infty} D_1 e^{tA_{22}} = 0 \tag{8.5.3}$$

if and only if all the eigenvalues of A_{22} are in the open left-half plane.

Let x_0 be an eigenvector of A_{22} corresponding to the eigenvalue λ_0 with $\mathcal{R}e\ \lambda_0 \geq 0$. Then

$$D_1 e^{tA_{22}} x_0 = D_1 e^{t\lambda_0} x_0$$

and by Lemma 8.5.1 (applied to $A = A_{22|\mathrm{Span}\{x_0\}}$)

$$\liminf_{t \to \infty} \| D_1 e^{t\lambda_0} x_0 \| > 0 \tag{8.5.4}$$

unless $x_0 \in \mathrm{Ker}\, D_1$. But if $x_0 \in \mathrm{Ker}\, D_1$, then

$$\mathrm{Span}\{x_0\} + \mathcal{N}_F \subset \bigcap_{i=1}^{n} \mathrm{Ker}[D(A + BF)^{i-1}]$$

which contradicts the definition of $\mathcal{N}_F$. Hence in equality (8.5.4) holds, and thus (8.5.3) does not.

Conversely, if $\sigma(A_{22})$ lies in the open left-half plane, then (8.5.3) holds by Lemma 8.5.1 (where A_{22} plays the role of A). $\quad\square$

Now we can reformulate the problem of stabilizing the output by state feedback as follows: given transformations $A: \mathbb{C}^n \to \mathbb{C}^n$, $B: \mathbb{C}^m \to \mathbb{C}^n$, and a subspace $\mathcal{E} \subset \mathbb{C}^n$ (which plays the role of $\mathrm{Ker}\, D$), find an $F: \mathbb{C}^n \to \mathbb{C}^m$ such that every root subspace of $A + BF$ corresponding to an eigenvalue λ_0 with nonnegative real part is contained in $\mathcal{E}$.

In this formulation there is nothing special about the set of eigenvalues with nonnegative real parts. In general, we can consider any proper subset Ω_b of $\mathbb{C}$ (the "bad" domain) in place of the closed right-half plane.

Now we can prove a general result on solvability of this problem in terms of $[A\ B]$-invariant subspaces.

Theorem 8.5.3

Given transformations $A: \mathbb{C}^n \to \mathbb{C}^n$, $B: \mathbb{C}^m \to \mathbb{C}^n$ and a subspace $\mathcal{E} \subset \mathbb{C}^n$, there exists a transformation $F: \mathbb{C}^n \to \mathbb{C}^m$ such that $\mathcal{R}_{\lambda_0}(A + BF) \subset \mathcal{E}$ for *every eigenvalue* $\lambda_0 \in \Omega_b$ *of* $A + BF$ *if and only if, for every eigenvalue* λ_0 *of* A *in* Ω_b, *we have*

$$\mathcal{R}_{\lambda_0}(A) \subset \langle A \mid \text{Im } B \rangle + \mathcal{U}$$

where $\langle A \mid \text{Im } B \rangle$ *is the minimal* A-*invariant subspace containing* $\text{Im } B$ *and* $\mathcal{U}$ *is the maximal* $[A \; B]$-*invariant subspace in* $\mathcal{E}$.

Proof. For a given transformation $F: \mathbb{C}^n \to \mathbb{C}^m$ let $\mathcal{N}_F$ be the maximal $(A + BF)$-invariant subspace in $\mathcal{E}$. As $\mathcal{N}_F$ is also $[A \; B]$ invariant, we have $\mathcal{N}_F \subset \mathcal{U}$.

Assume now that F is such that

$$\mathcal{R}_{\lambda_0}(A + BF) \subset \mathcal{E}$$

for every eigenvalue λ_0 of $A + BF$ that belongs to Ω_b. Then, by Lemma 8.5.2, for every $\lambda_0 \in \sigma(A + BF) \cap \Omega_b$ we have

$$\mathcal{R}_{\lambda_0}(A + BF) \subset \mathcal{N}_F$$

and hence

$$\mathcal{R}_{\lambda_0}(A + BF) \subset \mathcal{U}$$

Denote by P the canonical linear transformation $\mathbb{C}^n \to \mathbb{C}^n / \langle A \mid \text{Im } B \rangle$ (so $Px = x + \langle A \mid \text{Im } B \rangle$ for every $x \in \mathbb{C}^n$), and for a transformation $X: \mathbb{C}^n \to \mathbb{C}^n$ for which $\langle A \mid \text{Im } B \rangle$ is an invariant subspace, let $\bar{X}$ be the transformation induced by X on $\mathbb{C}^n / \langle A \mid \text{Im } B \rangle$. One easily checks that $\bar{A} = \overline{A + BF}$. Now use Lemma 6.5.4 to obtain

$$P\mathcal{R}_{\lambda_0}(A) = \mathcal{R}_{\lambda_0}(\bar{A}) = \mathcal{R}_{\lambda_0}(\overline{A + BF}) = P\mathcal{R}_{\lambda_0}(A + BF)$$

$$\subset (\langle A \mid \text{Im } B \rangle + \mathcal{U}) / \langle A \mid \text{Im } B \rangle \tag{8.5.5}$$

for every $\lambda_0 \in \sigma(A + BF) \cap \Omega_b$. Similarly

$$P\mathcal{R}_{\lambda_0}(A) \subset (\langle A \mid \text{Im } B \rangle + \mathcal{U}) / \langle A \mid \text{Im } B \rangle$$

for every $\lambda_0 \in \Omega_b$ that is not an eigenvalue of $A + BF$. Consequently

$$\mathcal{R}_{\lambda_0}(A) \subset \langle A \mid \text{Im } B \rangle + \mathcal{U} \tag{8.5.6}$$

for every $\lambda_0 \in \sigma(A) \cap \Omega_b$.

Conversely, assume that (8.5.6) holds for every $\lambda_0 \in \Omega_b$. We have to prove that there exists an F such that $\mathcal{R}_{\lambda_0}(A + BF) \subset \mathcal{E}$ for every eigenvalue of $A + BF$ that belongs to Ω_b. Let $F_0: \mathbb{C}^n \to \mathbb{C}^m$ be a transformation such that $(A + BF_0)\mathcal{U} \subset \mathcal{U}$. It is easily seen that the subspace $\mathcal{M} \stackrel{\text{def}}{=} \langle A \mid \operatorname{Im} B \rangle + \mathcal{U}$ is A invariant. We have $\overline{A + BF_0} = \bar{A}$, where the upper bar denotes the induced transformation on $\mathbb{C}^n/\mathcal{M}$. Denoting by Q the canonical transformation $\mathbb{C}^n \to \mathbb{C}^n/\mathcal{M}$, we see that Lemma 6.5.4 and equality (8.5.6) give for every $\lambda_0 \in \Omega_b$:

$$Q\mathcal{R}_{\lambda_0}(A + BF_0) = \mathcal{R}_{\lambda_0}(\overline{A + BF_0}) = \mathcal{R}_{\lambda_0}(\bar{A}) = Q\mathcal{R}_{\lambda_0}(A) \subset Q\mathcal{M} = \{0\}$$

Hence

$$\mathcal{R}_{\lambda_0}(A + BF_0) \subset \mathcal{M} \tag{8.5.7}$$

Further, the inclusion $(A + BF_0)\mathcal{U} \subset \mathcal{U}$ implies that $\mathcal{U} = \mathcal{N}_{F_0}$. Indeed, we have seen the inclusion $\mathcal{N}_{F_0} \subset \mathcal{U}$ at the beginning of this proof. To prove the opposite inclusion, take $x \in \mathcal{U}$. Then $Dx = 0$. For $i = 2, 3, \ldots, n - 1$ we have $(A + BF_0)^i x \in \mathcal{U}$; hence $D(A + BF_0)^i x = 0$, and the inclusion $\mathcal{U} \subset \mathcal{N}_{F_0}$ follows. Let $P': \mathbb{C}^n \to \mathbb{C}^n/\mathcal{U}$ be the canonical transformation. Denoting by $X': \mathbb{C}^n/\mathcal{U} \to \mathbb{C}^n/\mathcal{U}$ the transformation induced by $X: \mathbb{C}^n \to \mathbb{C}^n$ (it is assumed that $\mathcal{U}$ is X invariant), we have in view of Lemma 6.5.4 and inclusion (8.5.7), for every $\lambda_0 \in \Omega_b$:

$$\begin{aligned}
\mathcal{R}_{\lambda_0}((A + BF_0)') &= P'\mathcal{R}_{\lambda_0}(A + BF_0) \subset P'(\langle A \mid \operatorname{Im} B \rangle + \mathcal{U}) \\
&= P'(\langle A \mid \operatorname{Im} B \rangle) = P'(\langle A + BF_0 \mid \operatorname{Im} B \rangle) \\
&= \langle (A + BF_0)' \mid \operatorname{Im}(P'B) \rangle
\end{aligned}$$

Now Theorem 6.5.3 implies the existence of a transformation $\tilde{F}_1: \mathbb{C}^n/\mathcal{U} \to \mathbb{C}^m$ such that the spectrum of $(A + BF_0)' + P'B\tilde{F}_1$ lies in the complement of Ω_b. Let $F_1 = \tilde{F}_1 P': \mathbb{C}^n \to \mathbb{C}^m$. We have

$$[(A + BF_0)' + P'B\tilde{F}_1]P' = P'[A + B(F_0 + F_1)]$$

which means that

$$(A + BF_0)' + P'B\tilde{F}_1 = (A + B(F_0 + F_1))'$$

By Lemma 6.5.4 again, for every $\lambda_0 \in \Omega_b$

$$P'\mathcal{R}_{\lambda_0}(A + B(F_0 + F_1)) = \mathcal{R}_{\lambda_0}(A + B(F_0 + F_1))' = \{0\}$$

so

$$\mathcal{R}_{\lambda_0}(A + B(F_0 + F_1)) \subset \mathcal{U} \subset \mathcal{E}$$

and $F = F_0 + F_1$ is the desired transformation. $\quad\square$

The proof of Theorem 8.5.3 shows that, assuming $\mathcal{R}_{\lambda_0}(A) \subset \langle A \mid \text{Im } B \rangle + \mathcal{U}$ for every $\lambda_0 \in \Omega_b$, the transformation $F: \mathbb{C}^n \to \mathbb{C}^m$ such that $\mathcal{R}_{\lambda_0}(A + BF) \subset \mathcal{E}$ for every $\lambda_0 \in \Omega_b$ can be constructed as follows: $F = F_0 + \tilde{F}_1 P$, where $F_0: \mathbb{C}^n \to \mathbb{C}^m$ is such that $(A + BF_0)\mathcal{U} \subset \mathcal{U}$; $P: \mathbb{C}^n/\mathcal{U}$ is the canonical transformation; and $\tilde{F}_1: \mathbb{C}^n/\mathcal{U} \to \mathbb{C}^m$ has the property that the spectrum of the transformation on $\mathbb{C}^n/\mathcal{U}$ induced by the transformation $A + B(F_0 + \tilde{F}_1 P)$ on $\mathbb{C}^n$ lies outside Ω_b.

Applying Theorem 8.5.3 to the output stabilization problem, we obtain the following result.

Theorem 8.5.4

Given the linear system

$$\begin{cases} \dfrac{dx(t)}{dt} = Ax(t) + Bu(t), & t \geq 0 \\ z(t) = Dx(t) \end{cases} \qquad (8.5.8)$$

with constant transformations $A: \mathbb{C}^n \to \mathbb{C}^n$, $B: \mathbb{C}^m \to \mathbb{C}^n$, and $D: \mathbb{C}^n \to \mathbb{C}^r$, there exists a transformation (state feedback) $F: \mathbb{C}^n \to \mathbb{C}^m$ such that, for every initial value $x(0)$, the solution of (8.5.8) with $u(t) = Fx(t)$ satisfies $\lim_{t \to \infty} z(t) = 0$ if and only if

$$\mathcal{R}_{\lambda_0}(A) \subset \langle A \mid \text{Im } B \rangle + \mathcal{U}$$

for every eigenvalue λ_0 of A lying in the closed right half plane, and where $\mathcal{U}$ is the maximal $[A \ B]$-invariant subspace in $\text{Ker } D$.

We conclude this section with an example illustrating Theorem 8.5.4.

EXAMPLE 8.5.1.   Let

$$A = \begin{bmatrix} 0 & 1 & 0 \\ 0 & 0 & 0 \\ 0 & 0 & \lambda_0 \end{bmatrix}; \qquad B = \begin{bmatrix} 0 \\ 1 \\ 0 \end{bmatrix}; \qquad \lambda_0 \in \mathbb{C}; \qquad \text{Ker } D = \text{Span}\{\langle a_1, a_2, a_3 \rangle\}$$

where a_1, a_2, a_3 are complex numbers not all zeros. Here $\langle A \mid \text{Im } B \rangle = \text{Span}\{e_1, e_2\}$. If $\mathcal{R}e\, \lambda_0 < 0$, then there is always an $F = [f_1 f_2 f_3]$ with properties as in Theorem 8.5.4 (one can take $f_3 = 0$ and choose f_1 and f_2 so that the equation $\lambda^2 - f_2 \lambda - f_1 = 0$ has its zeros in the open left-half plane). So assume $\mathcal{R}e\, \lambda_0 \geq 0$. Then there exists an F as in Theorem 8.5.4 if and only if

$$\text{Span}\{e_1, e_2\} + \mathcal{U} = \mathbb{C}^3 \qquad (8.5.9)$$

If $a_3 = 0$, then (8.5.9) is always false. If $a_3 \neq 0$, then (8.5.9) happens if and only if the subspace Ker D is $[A\ B]$ invariant, or, equivalently, if Ker D is $(A + BG)$ invariant for some $G = [g_1 g_2 g_3]$. An easy verification shows that this is the case if and only if $a_2 = \lambda_0 a_1$. So there exists an $F = [f_1 f_2 f_3]$ as in Theorem 8.5.4 if and only if $a_3 \neq 0$ and $a_2 = \lambda_0 a_1$. In this case $f_3 = \lambda_0^2 a_1 - f_1 a_1 - \lambda_0 f_2 a_1$ and f_1 and f_2 for which the zeros of $\lambda^2 - f_2 \lambda - f_1 = 0$ are in the left half plane will do. $\square$

8.6   EXERCISES

8.1   For every input $u(t)$ find the output $y(t)$ for the following linear systems:

(a)
$$\begin{cases} \dfrac{dx(t)}{dt} = \begin{bmatrix} -1 & 2 \\ 0 & 1 \end{bmatrix} x(t) + \begin{bmatrix} 3 \\ -1 \end{bmatrix} u(t); & x(0) = 0 \\[2ex] y(t) = \begin{bmatrix} 0 & -2 \\ -2 & 0 \end{bmatrix} x(t) + \begin{bmatrix} 0 \\ 1 \end{bmatrix} u(t) \end{cases}$$

(b)
$$\begin{cases} \dfrac{dx(t)}{dt} = J_n(0)x(t) + \begin{bmatrix} 1 \\ 1 \\ \vdots \\ 1 \end{bmatrix} u(t), & x(0) = 0 \\[3ex] y(t) = [1 \quad \cdots \quad 1]x(t) + u(t) \end{cases}$$

8.2   For every input $u(t)$ find the output $y(t)$ for the following linear systems:

(a)
$$\begin{cases} \dfrac{dx(t)}{dt} = [J_{k_1}(\lambda_0) \oplus J_{k_2}(\lambda_0)]x(t) + Bu(t); & x(0) = x_0 \\[2ex] y(t) = Cx(t) \end{cases}$$

where B is the $(k_1 + k_2) \times 2$ matrix whose first column is e_{k_1} and second column is $e_{k_1 + k_2}$, and C is the $2 \times (k_1 + k_2)$ matrix whose first row is e_1^T and second row is $e_{k_1 + 1}^T$.

(b)
$$\begin{cases} \dfrac{dx(t)}{dt} = Ax(t) + C^T u(t); & x(0) = x_0 \\[2ex] y(t) = Cx(t) \end{cases}$$

where A is an $n \times n$ lower triangular matrix and $C = [0 \cdots 0\ 1\ 0 \cdots 0]$ with 1 in the kth place.

8.3   Consider the linear system

$$\begin{cases} \dfrac{dx(t)}{dt} = \begin{bmatrix} 0 & 1 \\ a_0 & a_1 \end{bmatrix} x(t) + \begin{bmatrix} b_1 \\ b_2 \end{bmatrix} u(t) \\ y(t) = [c_1 \quad c_2] x(t) + u(t) \end{cases}$$

When is this system controllable? observable? minimal?

8.4   Find transfer functions for the linear systems given in Exercises 8.1 and 8.2.

8.5   Build minimal linear systems with the following transfer functions:

(a)
$$\begin{bmatrix} \dfrac{1}{\lambda(\lambda - 1)} & \dfrac{\lambda - 1}{\lambda + 1} \\ 0 & \dfrac{1}{\lambda(\lambda + 1)} \end{bmatrix}$$

(b) $p(\lambda)^{-1}$,   where $p(\lambda) = \sum_{j=0}^{k} \alpha_j \lambda^j$ is a scalar polynomial.

(c) $(L(\lambda))^{-1}$,   where $L(\lambda)$ is a monic $n \times n$ matrix polynomial of degree l.

8.6   Show that the system

$$\begin{cases} \dfrac{dx(t)}{dt} = \begin{bmatrix} 0 & 1 & 0 & \cdots & 0 \\ 0 & 0 & 1 & \cdots & 0 \\ \vdots & \vdots & \vdots & & \vdots \\ & & & & 1 \\ a_0 & a_1 & a_2 & \cdots & a_{n-1} \end{bmatrix} x(t) + \begin{bmatrix} 0 \\ \vdots \\ \vdots \\ 0 \\ 1 \end{bmatrix} u(t) \\ y(t) = [1 \quad 0 \quad \cdots \quad 0] x(t) \end{cases}$$

is controllable and observable.

8.7   For the system in Exercise 8.6, given the n-tuple of complex numbers $\lambda_1, \ldots, \lambda_n$, find a state feedback F such that $A + BF$ has eigenvalues $\lambda_1, \ldots, \lambda_n$. Also, find G such that $A + GC$ has eigenvalues $\lambda_1, \ldots, \lambda_n$.

8.8   Let

$$\frac{dx}{dt} = Ax + Bu ; \qquad y = Cu$$

be a linear system with $n \times n$ circulant matrix A and $n \times 1$ and $1 \times n$ matrices B and C, respectively. When is the system controllable? Observable? Minimal?

8.9   Consider the linear system

$$\begin{cases} \dfrac{dx(t)}{dt} = Jx(t) + Bu(t) \\[2mm] y(t) = Cu(t) + Du(t) \end{cases}$$

where J is a nilpotent $n \times n$ Jordan matrix (i.e., with $J^n = 0$) and B and C are $n \times 1$ and $1 \times n$ matrices, respectively. When is this system controllable? Observable? Minimal?

8.10   Prove or disprove: if the system

$$\frac{dx(t)}{dt} = Ax + Bu ; \qquad y(t) = Cx + Du$$

is minimal, then the system

$$\frac{dx(t)}{dt} = A^2x + Bu ; \qquad y(t) = Cx + Du$$

is minimal as well.

8.11   Let $p(A)$ be a polynomial of the transformation $A: \mathbb{C}^n \to \mathbb{C}^n$. Prove that the minimality of the system

$$\frac{dx(t)}{dt} = p(A)x + Bu , \qquad y(t) = Cx + Du$$

implies the minimality of

$$\frac{dx(t)}{dt} = Ax + Bu , \qquad y(t) = Cx + Du$$

Is the converse true?

8.12   Let

$$\frac{dx(t)}{dt} = A_1x + Bu , \qquad y(t) = Cx + Du$$

and

$$\frac{dx(t)}{dt} = A_2x + Bu , \qquad y(t) = Cx + Du$$

be two systems, and assume that $A_2 = p(A_1)$, where $p(\lambda)$ is a polynomial such that $p(\lambda_1) \neq p(\lambda_2)$ for any pair of different eigenvalues λ_1 and λ_2 of A_1 and $p'(\lambda)|_{\lambda = \lambda_0} \neq 0$ for every eigenvalue λ_0 of A_1 such that $A_1|_{\mathcal{R}_{\lambda_0}(A_1)}$ is not diagonable. Prove that the systems are simultaneously minimal or nonminimal.

8.13 Show that if the system

$$\frac{dx(t)}{dt} = Ax + Bu ; \qquad y(t) = Cx + Du$$

is controllable, then for every $\lambda_0 \in \mathbb{C}$ the system

$$\frac{dx(t)}{dt} = (\lambda_0 I + A)x + Bu ; \qquad y(t) = Cx + Du$$

is controllable as well. Is this property true for the observability of systems?

8.14 For a controllable system

$$\frac{dx(t)}{dt} = J_n(0)x(t) + \begin{bmatrix} b_1 \\ \vdots \\ b_n \end{bmatrix} u(t) \tag{1}$$

where $b_n \neq 0$, find a state feedback F such that the system with feedback

$$\frac{dx(t)}{dt} = \left(J_n(0) + \begin{bmatrix} b_1 \\ \vdots \\ b_n \end{bmatrix} F \right) x(t)$$

is stable, that is, all its solutions $x(t)$ tend to zero as $t \to \infty$.

8.15 For system (1) in Exercise 8.14 and any $k > 0$, find a state feedback F such that all solutions $x(t)$ of the system with feedback satisfy $\|x(t)\| \leq Ke^{-kt}$, where $K > 0$ is constant independent of t.

8.16 Prove that any minimal linear system with n-dimensional state space has a state feedback for which the system with feedback can be represented as a simple cascade of n linear systems with state spaces of dimension 1.

8.17 Prove that controllability is a stable property in the following sense: for every controllable system

$$\frac{dx}{dt} = Ax + Bu$$

there exists an $\epsilon > 0$ such that any linear system

$$\frac{dx}{dt} = A'x + B'u$$

with $\|A' - A\| < \epsilon$, $\|B' - B\| < \epsilon$ is controllable as well.

8.18 Prove that observability and minimality of linear systems are also stable properties. The definition of stability is, in each case, to be similar to that of Exercise 8.17.

8.19 Show that for any system

$$\frac{dx}{dt} = Ax + Bu, \qquad y = Cx + Du$$

there exists a sequence of minimal systems

$$\frac{dx}{dt} = A_p x + B_p u, \qquad y = C_p x + Du$$

where $p = 1, 2, \ldots$ such that $\lim_{p \to \infty} \|A_p - A\| = 0$, $\lim_{p \to \infty} \|B_p - B\| = 0$, $\lim_{p \to \infty} \|C_p - C\| = 0$.

Notes to
Part 1

Chapter 1. The material here is quite elementary and well known, although not everything is readily available in the literature. Part of Section 1.5 is based on the exposition in Chapter S4 of the authors' book (1982). More about angular transformations and matrix quadratic equations can be found in Bart, Gohberg, and Kaashoek (1979). Angular subspaces and operators for the infinite dimensional case were introduced and studied in Krein (1970).

Chapter 2. The proof of the Jordan form presented here is standard and can be found in many books in linear algebra; for example, see Gantmacher (1959) or Lancaster and Tismenetsky (1985). A proof of the Jordan form can be obtained also by analyzing the properties of the set of all invariant subspaces as a lattice. This was done in Soltan (1973a). In this approach, the invariance of the Jordan form follows from the well-known Schmidt–Ore theorem in lattice theory [see, e.g., Kurosh (1965)].

"The A-invariant subspace maximal in $\mathcal{N}$" and "the A-invariant subspace minimal over $\mathcal{N}$" are phrases that are introduced here probably for the first time, although the notions themselves had been developed and are now well known in the context of linear systems theory. In general, the whole material of Sections 2.7 and 2.8 is influenced by linear system theory. However, our presentation here is independent of that theory and leads us to abandon its well-established terminology. In particular, in linear systems theory, "full-range" and "null kernel" pairs are known as "controllable" and "observable" pairs, respectively. Marked invariant subspaces are probably introduced for the first time. The existence of nonmarked invariant subspaces is often overlooked. The description of partial multiplicities and invariant subspaces of functions will hold no surprises for the specialist, but, again, these are results that are not easily found in the standard literature on linear algebra.

Chapter 3. The material of this chapter (except for Theorem 3.3.1) is well known. Theorem 3.3.1 in the infinite dimensional case was proved by Sarason (1965). Here we follow his proof.

Chapter 4. The problem of analysis of partial multiplicities of extensions from an invariant and a coinvariant subspace was stated in Gohberg and Kaashoek (1979). This problem was connected there with the description of partial multiplicities of products of matrix polynomials in terms of partial multiplicities of each factor and reappears in this context in Section 5.2. The first results concerning this description were proved in Sigal (1973). In particular, Theorem 3.3.1 was proved in that paper. Example 4.3.1 and the material in Section 4.4 (except for Proposion 4.4.1) is taken from Rodman and Schaps (1979). For further information and more inequalities concerning the partial multiplicities, see Thijsse (1980, 1984) and Rodman and Schaps (1979).

When this book was finalized, the authors learned about another important line of development concerning the problem of partial multiplicities of products of matrix polynomials. This has been intensively studied (even in a more general setting) by several authors. The reader is referred to recent work of Thompson (1983 and 1985) for details and further references.

Chapter 5. The theory presented in this chapter can be viewed as a generalization of the familiar spectral theory of a matrix A but, in this context, identified with the linear matrix polynomial $\lambda I - A$. This theory of matrix polynomials was developed by the authors and summarized in the book by Gohberg, Lancaster, and Rodman (1982). The material and presentation in this chapter is based on the first four chapters of that book. It also contains further results on matrix polynomials including least common multiples, greatest common divisors, matrix polynomials with hermitian coefficients, nonmonic matrix polynomials, and connections with differential and difference equations. Lists of relevant references and historical comments on this subject are found in the above-mentioned monograph by the authors (1982). In this presentation we focus more closely on decompositions into three or more factors. Theorem 5.2.3 is close to the original theorem of Sigal (1973) concerning matrix-valued functions. See also Thompson (1983 and 1985).

Chapter 6. The main results of this chapter were first obtained in a different form in the theory of linear systems [see, e.g., monographs by Wonham (1974) and Kailath (1980)]. In this chapter the presentation is independent of linear systems theory and is given in a pure linear algebraic form. This approach led us to change the terminology, which is well established in the theory of linear systems, and to make it more suitable for linear algebra.

The ideas of block similarity in Sections 6.2 and 6.6, as well as of $[A \ B]$-invariant and $\begin{bmatrix} A \\ C \end{bmatrix}$-invariant subspaces, are taken from Gohberg,

Kaashoek, and van Schagen (1980). That paper contains a more general theory of invariant subspaces, similarity, canonical forms, and invariants of blocks of matrices in terms of these blocks only. Some applications of these results may be found in Gohberg, Kaashoek, and van Schagen (1981, 1982). Theorem 6.2.5 was proved (by a direct approach, without using the Kronecker canonical form) in Brunovsky (1970). The connection between the Kronecker form for linear polynomials and the state feedback problems is given in Kalman (1971) and Rosenbrock (1970). In Theorem 6.3.2 the equivalence of (a) and (d) is due to Hautus (1969).

The spectral assignment problem is classical, by now, and can be found in many books [see, e.g., Kailath (1980) and Wonham (1974)]. There is a more difficult version of this problem in which the eigenvalues *and their partial multiplicities* are preassigned. This problem is not generally solvable. For further analysis, see Rosenbrock and Hayton (1978) and Djaferis and Mitter (1983).

Chapter 7.   The concept of minimal realization is a well-known and important tool in linear system theory [see, e.g., Wonham (1979) and Kalman (1963)]. See also Bart, Gohberg, and Kaashoek (1979), where the exposition matches the purposes of this chapter. Section 7.1 contains the standard material on realization theory, and Lemma 7.1.1 is a particular case of Theorem 2.2 in Bart, Gohberg, and Kaashoek (1979).

Section 7.2 follows the authors' paper (1983a). Sections 7.3–7.5 are based on Chapters 1 and 4 in Bart, Gohberg, and Kaashoek (1979). Here, we concentrate more on decompositions into three or more factors.

Linear fractional decompositions of rational matrix functions play an important role in network theory; see Helton and Ball (1982). Theorem 7.7.1 is proved in that paper. The exposition in Sections 7.6–7.8 follows that given in Gohberg and Rubinstein (1985).

Chapter 8.   In the last 20 years linear system theory has developed into a major field of research with very important applications. The literature in this field is rich and includes monographs, textbooks, and specialized journals. We mention only the following books where the reader can find further references and historical remarks: Kalman, Falb, and Arbib (1969), Wonham (1974), Kailath (1980), Rosenbrock (1970), and Brockett (1970). This chapter can be viewed as an introduction to some basic concepts of linear systems theory.

The first three sections contain standard material (except for Theorem 8.3.2). In the last two sections we follow the exposition of Wonham (1979).

Part Two

Algebraic Properties of Invariant Subspaces

In Chapters 9–12 we develop material that supplements the theory of Part 1. In particular, we go more deeply into the algebraic structure of invariant subspaces. We include a description of the set of all invariant subspaces for a given transformation and examine to what extent a transformation is defined by its lattice of invariant subspaces. Special attention is paid to invariant subspaces of commuting transformations and of algebras of transformations. In the final chapter the theory of the first two parts (developed for complex linear transformations) is reviewed in the context of real linear transformations.

Chapter Nine

Commuting Matrices and Hyperinvariant Subspaces

In this chapter we study lattices of invariant subspaces that are common to different commuting transformations. The description of all transformations that commute with a given transformation is a necessary part of the investigation of this problem. This description is used later in the chapter to study the hyperinvariant subspaces for a transformation A, that is, those subspaces that are invariant for any transformation commuting with A.

9.1 COMMUTING MATRICES

Matrices A and B (both of the same size $n \times n$) are said to *commute* if $AB = BA$. In this section we describe the set of all matrices which commute with a given matrix A. In other words, we wish to find all the solutions of the equation

$$AX = XA \qquad (9.1.1)$$

where X is an $n \times n$ matrix to be found.

We can restrict ourselves to the case that A is in the Jordan form. Indeed, let $J = S^{-1}AS$ be a Jordan matrix for some nonsingular matrix S. Then X is a solution of equation (9.1.1) if and only if $Z = S^{-1}XS$ is a solution of

$$JZ = ZJ \qquad (9.1.2)$$

So we shall assume that $A = J$ is in the Jordan form. Write

295

$$J = \text{diag}[J_1, \ldots, J_u]$$

where $J_\alpha (\alpha = 1, \ldots, u)$ is a Jordan block of size $m_\alpha \times m_\alpha$, $J_\alpha = \lambda_\alpha I_\alpha + H_\alpha$, where I_α is the unit matrix of size $m_\alpha \times m_\alpha$, and H_α is the $m_\alpha \times m_\alpha$ nilpotent Jordan block:

$$H_\alpha = \begin{bmatrix} 0 & 1 & & & 0 \\ & 0 & \ddots & & \vdots \\ \vdots & \vdots & \ddots & \ddots & \\ & & & & 1 \\ 0 & 0 & & & 0 \end{bmatrix}$$

Let Z be a matrix that satisfies (9.1.2) and write

$$Z = [Z_{\alpha\beta}]_{\alpha,\beta}^u$$

where $Z_{\alpha\beta}$ is a $m_\alpha \times m_\beta$ matrix. Rewrite equality (9.1.2) in the form

$$(\lambda_\alpha - \lambda_\beta)Z_{\alpha\beta} = Z_{\alpha\beta}H_\beta - H_\alpha Z_{\alpha\beta}, \qquad 1 \le \alpha, \beta \le u \qquad (9.1.3)$$

Two cases can occur:

(a) $\lambda_\alpha \ne \lambda_\beta$. We show that in this case $Z_{\alpha\beta} = 0$. Indeed, multiply the left-hand side of equality (9.1.3) by $\lambda_\alpha - \lambda_\beta$ and in each term in the right-hand side replace $(\lambda_\alpha - \lambda_\beta)Z_{\alpha\beta}$ by $Z_{\alpha\beta}H_\beta - H_\alpha Z_{\alpha\beta}$. We obtain

$$(\lambda_\alpha - \lambda_\beta)^2 Z_{\alpha\beta} = Z_{\alpha\beta}H_\beta^2 - 2H_\alpha Z_{\alpha\beta}H_\beta + H_\alpha^2 Z_{\alpha\beta}$$

Repeating this process, we obtain for every $p = 1, 2, \ldots$:

$$(\lambda_\alpha - \lambda_\beta)^p Z_{\alpha\beta} = \sum_{q=0}^{p} (-1)^q \binom{p}{q} H_\alpha^q Z_{\alpha\beta} H_\beta^{p-q} \qquad (9.1.4)$$

Choose p large enough that either $H_\alpha^q = 0$ or $H_\beta^{p-q} = 0$ for every $q = 0, \ldots, p$. Then the right-hand side of equation (9.1.4) is zero, and since $\lambda_\alpha \ne \lambda_\beta$, we find that $Z_{\alpha\beta} = 0$.

(b) $\lambda_\alpha = \lambda_\beta$. Then

$$Z_{\alpha\beta}H_\beta = H_\alpha Z_{\alpha\beta} \qquad (9.1.5)$$

From the structure of H_α and H_β it follows that the product $H_\alpha Z_{\alpha\beta}$ is obtained from $Z_{\alpha\beta}$ by shifting all the rows one place upward and filling the last row with zeros; similarly, $Z_{\alpha\beta}H_\beta$ is obtained from $Z_{\alpha\beta}$

by shifting all the columns one place to the right and filling the first column with zeros. So equation (9.1.5) gives (where ζ_{ik} is the (i, k)th entry in $Z_{\alpha\beta}$, which depends, of course, on α and β):

$$\zeta_{i+i,k} = \zeta_{i,k-1}, \qquad i = 1, \ldots, m_\alpha, \qquad k = 1, \ldots, m_\beta$$

where by definition $\zeta_{i0} = \zeta_{m_\alpha+1,k} = 0$. These equalities mean that the matrix $Z_{\alpha\beta}$ has one of the following structures:

For $m_a = m_\beta$:

$$Z = \begin{bmatrix} c_{\alpha\beta}^{(1)} & c_{\alpha\beta}^{(2)} & \cdots & c_{\alpha\beta}^{(m_\alpha)} \\ 0 & c_{\alpha\beta}^{(1)} & \cdots & c_{\alpha\beta}^{(m_\alpha-1)} \\ 0 & 0 & \cdots & c_{\alpha\beta}^{(1)} \end{bmatrix} \overset{\text{def}}{=} T_{m_\alpha}(c_{\alpha\beta}^{(i)} \in \mathbb{C}) \qquad (9.1.6)$$

For

$$m_\alpha < m_\beta : Z_{\alpha\beta} = [0_{m_\alpha, m_\beta - m_\alpha} T_{m_\alpha}] \qquad (9.1.7)$$

For

$$m_\alpha > m_\beta : Z_{\alpha\beta} = \begin{bmatrix} T_{m_\beta} \\ 0_{m_\alpha - m_\beta, m_\beta} \end{bmatrix} \qquad (9.1.8)$$

where 0_{pq} stands for the zero $p \times q$ matrix. Matrices of types (9.1.6)–(9.1.8) are referred to as *upper triangular Toeplitz matrices*. So we have proved the following result.

Theorem 9.1.1

Let $J = \text{diag}[J_1, \ldots, J_u]$ be an $n \times n$ Jordan matrix with Jordan blocks $J_1, \ldots, J_u$ and eigenvalues $\lambda_1, \ldots, \lambda_u$, respectively. Then an $n \times n$ matrix Z commutes with J if and only if $Z_{\alpha\beta} = 0$ for $\lambda_\alpha \neq \lambda_\beta$ and $Z_{\alpha\beta}$ is an upper triangular Toeplitz matrix for $\lambda_\alpha = \lambda_\beta$, where $Z = [Z_{\alpha\beta}]_{\alpha,\beta=1}^u$ is the partition of Z consistent with the partition of J into Jordan blocks.

We repeat that Theorem 9.1.1 gives, after applying a suitable similarity transformation, a description of all matrices commuting with a fixed matrix A. This theorem has a number of important corollaries.

Corollary 9.1.2

Let A be an $n \times n$ matrix partitioned as follows:

$$A = \begin{bmatrix} A_1 & 0 \\ 0 & A_2 \end{bmatrix} \qquad (9.1.9)$$

where the spectra of the matrices A_1 and A_2 do not intersect. Then any $n \times n$ matrix X that commutes with A has the form

$$X = \begin{bmatrix} X_1 & 0 \\ 0 & X_2 \end{bmatrix}$$

with the same partition as in equality (9.1.9).

    Proof. Let J_1 (resp. J_2) be the Jordan form of A_1 (resp. A_2), so $J_i = S_i^{-1} A_i S_i$ for some nonsingular matrices S_1 and S_2. Then

$$J = \begin{bmatrix} J_1 & 0 \\ 0 & J_2 \end{bmatrix}$$

is the Jordan form of A. By Theorem 9.1.1, and since $\sigma(J_1) \cap \sigma(J_2) = \emptyset$, any matrix Y that commutes with J has the form $Y = Y_1 \oplus Y_2$ with the same partition as in (9.1.9). Now Y commutes with J if and only if $X = SYS^{-1}$ commutes with A, where $S = S_1 \oplus S_2$. So

$$X = \begin{bmatrix} S_1 Y_1 S_1^{-1} & 0 \\ 0 & S_2 Y_2 S_2^{-1} \end{bmatrix}$$

has the desired structure. $\square$

    This corollary, reformulated in terms of transformations, runs as follows: let $A \colon \mathbb{C}^n \to \mathbb{C}^n$ be a transformation, and let $\mathcal{M}_1$ and $\mathcal{M}_2$ be A-invariant subspaces that are complementary to each other and for which the restrictions $A_{|\mathcal{M}_1}$ and $A_{|\mathcal{M}_2}$ have no common eigenvalues. Then $\mathcal{M}_1$ and $\mathcal{M}_2$ are invariant subspaces for every transformation that commutes with A. To prove this, write A in the 2×2 block matrix form with respect to the direct sum decomposition $\mathbb{C}^n = \mathcal{M}_1 \dotplus \mathcal{M}_2$ and use Corollary 9.1.2. The next result is a special case.

Corollary 9.1.3

Every root subspace for a transformation $A \colon \mathbb{C}^n \to \mathbb{C}^n$ is a reducing invariant subspace for any transformation that commutes with A.

    The proof of Theorem 9.1.1 allows us to study the set $\mathcal{C}(A)$ of all matrices (or transformations) that commute with the matrix (or linear transformation) A. First, observe that $\mathcal{C}(A)$ is a linear vector space. Indeed, if $AX_i = X_i A$ for $i = 1$ and 2, then also $A(\alpha X_1 + \beta X_2) = (\alpha X_1 + \beta X_2)A$ for any complex numbers α and β.

    To compute the dimension of $\mathcal{C}(A)$, consider the elementary divisors of A. Thus, for every Jordan block or size $k \times k$ and eigenvalue λ_0 in the Jordan normal form of A we have an *elementary divisor* $(\lambda_0 - \lambda_0)^k$ of A

(which is a polynomial in λ). The greatest common divisor of two elementary divisors $(\lambda - \lambda_1)^{k_1}$ and $(\lambda - \lambda_2)^{k_2}$ of A is $(\lambda - \lambda_1)^{\min(k_1, k_2)}$ if $\lambda_1 = \lambda_2$ and is 1 if $\lambda_1 \neq \lambda_2$. Taking this observation into account, Theorem 9.1.1 shows that the dimension of $\mathscr{C}(A)$ is $\sum_{s,t=1}^{p} \alpha_{st}$, where α_{st} is the degree of the greatest common divisor of $(\lambda - \lambda_s)^{k_s}$ and $(\lambda - \lambda_t)^{k_t}$, and $(\lambda - \lambda_1)^{k_1}, \ldots, (\lambda - \lambda_p)^{k_p}$ are all the elementary divisors of A. In particular

$$\dim \mathscr{C}(A) \geq \sum_{s=1}^{p} \alpha_{ss} = \sum_{s=1}^{p} k_s = n \qquad (9.1.10)$$

where n is the size of A.

We have seen that, quite obviously, any polynomial in A commutes with A, and we now ask about conditions on A such that, conversely, each matrix commuting with A is a polynomial in A.

To this end we need the following notion. An $n \times n$ matrix (or transformation $A: \mathbb{C}^n \to \mathbb{C}^n$) is called *nonderogatory* if there is only one Jordan block in the Jordan form of A associated with each eigenvalue. It turns out that A is nonderogatory if and only if any one of the following four equivalent statements holds: (a) $\dim \operatorname{Ker}(\lambda I - A) \leq 1$ for every $\lambda \in \mathbb{C}$; (b) A is similar to a matrix

$$\begin{bmatrix} 0 & 1 & 0 & \cdots & 0 \\ 0 & 0 & 1 & \cdots & 0 \\ \vdots & \vdots & & \ddots & \vdots \\ 0 & 0 & & \cdots & 1 \\ a_0 & a_1 & & \cdots & a_{n-1} \end{bmatrix} \qquad (9.1.11)$$

for some complex numbers $a_0, \ldots, a_{n-1}$; (c) the minimal polynomial of A coincides with the characteristic polynomial of A; and (d) A is cyclic, that is, there exists an $x \in \mathbb{C}^n$ such that

$$\mathbb{C}^n = \operatorname{Span}\{x, Ax, A^2 x, \ldots\} \qquad (9.1.12)$$

Indeed, by assuming that A is in the Jordan form, condition (a) is clearly equivalent to A having only one Jordan block for each eigenvalue. By Theorem 2.6.1, (d) is equivalent to A being nonderogatory. Further, the minimal polynomial for A is easily seen to be $(\lambda - \lambda_1)^{\alpha_1} \cdots (\lambda - \lambda_p)^{\alpha_p}$, where $\lambda_1, \ldots, \lambda_p$ are all the distinct eigenvalues of A and α_j is the maximal size of the Jordan blocks of A corresponding to λ. From this description it is clear that (c) is equivalent to (a). We have proved, therefore, that (a), (c), and (d) are equivalent to each other and to the condition that A is nonderogatory.

Let A be the matrix (9.1.11). We want to prove that (a) holds. Let

$x = \langle x_1, \ldots, x_n \rangle$ and $y = \langle y_1, \ldots, y_n \rangle$ be eigenvectors of A corresponding to the eigenvalue λ_0. Thus $Ax = \lambda_0 x$, $Ay = \lambda_0 y$, and $x \neq 0$, $y \neq 0$. The structure of A implies that $x_i = \lambda_0^{i-1} x_1$, $y_i = \lambda_0^{i-1} y_1$ for $i = 1, \ldots, n$. But then necessarily $x_1 \neq 0$, $y_1 \neq 0$, and $x = (y_1/x_1)y$, that is, x and y are linearly dependent. Hence (a) holds.

Finally, we show that (d) implies (b). First observe that if (9.1.12) holds, then the vectors $x, Ax, \ldots, A^{n-1}x$ are linearly independent (otherwise $\mathfrak{C}^n$ would be spanned by less than n vectors, which is impossible). In the basis $x, Ax, \ldots, A^{n-1}x$ the matrix A has the form (9.1.11). $\quad\square$

Theorem 9.1.4

Every matrix commuting with A is a polynomial in A if and only if A is nonderogatory.

Proof. First recall that in view of the Cayley–Hamilton theorem the number of linearly independent powers of A does not exceed n. Thus, if $AX = XA$ implies that X is a polynomial of A, then X can be rewritten as $X = p(A)$, where $l = \deg p(\lambda) \leq n$ and all powers $I, A, A^2, \ldots, A^{l-1}$ are linearly independent. So in this case $\dim \mathfrak{C}(A) = l \leq n$. Inequality (9.1.10) then implies that $\dim \mathfrak{C}(A) = n$. This means [again in view of (9.1.10)] that $\alpha_{st} = 0$ for $s \neq t$. So in the Jordan form of A there is only one Jordan block associated with each eigenvalue of A.

Conversely, assume that the Jordan form of A is

$$J = J_{m_1}(\lambda_1) \oplus \cdots \oplus J_{m_s}(\lambda_s)$$

where $\lambda_1, \ldots, \lambda_s$ are different complex numbers. As we have seen, the solution X of $AX = XA$ is then similar to a direct sum of upper triangular Toeplitz matrices

$$Y_1 = \begin{bmatrix} c_i^{(1)} & c_i^{(2)} & \cdots & c_i^{(m_i)} \\ 0 & & & \vdots \\ & \ddots & & \\ & & c_i^{(1)} & c_i^{(2)} \\ 0 & \cdots & 0 & c_i^{(1)} \end{bmatrix} \quad \text{for} \quad i = 1, 2, \ldots, s$$

More exactly, $Y_1 \oplus \cdots \oplus Y_s = S^{-1}XS$, where S is a nonsingular matrix such that $J = S^{-1}AS$. Now a polynomial $p(\lambda)$ satisfying the conditions

$$p(\lambda_i) = c_i^{(1)}, \ldots, \frac{1}{(m_i - 1)!} p^{(m_i-1)}(\lambda_i) = c_i^{(m_i)}$$

for $i = 1, \ldots, s$ gives the desired result:

$$X = S(Y_1 \oplus \cdots \oplus Y_s)S^{-1} = S \operatorname{diag}[p(J_{m_1}(\lambda_1)), \ldots, p(J_{m_s}(\lambda_s))]S^{-1}$$
$$= Sp(J)S^{-1} = P(SJS)^{-1} = p(A)$$

Note that $p(\lambda)$ can be chosen with degree not exceeding $n - 1$. □

We now confine our attention to matrices commuting with a diagonable matrix. Recall that an $n \times n$ matrix A is diagonable if and only if there is a basis in $\mathbb{C}^n$ of eigenvectors of A. The following corollary is obtained from Theorem 9.1.1.

Corollary 9.1.5

If $\lambda_1, \ldots, \lambda_s$ are the distinct eigenvalues of a diagonable matrix A, then

$$\dim \mathscr{C}(A) = \sum_{i=1}^{s} \beta_i^2$$

where

$$\beta_i = \dim \operatorname{Ker}(A - \lambda_i I), \qquad i = 1, \ldots, s$$

For future reference let us also indicate the following fact.

Proposition 9.1.6

An $n \times n$ matrix B commutes with every $n \times n$ matrix A if and only if B is a scalar multiple of I: $B = \lambda I$ for some $\lambda \in \mathbb{C}$.

Proof. The part "if" is obvious. So assume that B commutes with every $n \times n$ matrix A—in particular, taking A to be diagonal with n different eigenvalues with respect to a basis $x_1, \ldots, x_n$ in $\mathbb{C}^n$, Corollary 9.1.2 implies that B is also diagonal in this basis. Therefore, $Bx_1 \in \operatorname{Span}\{x_1\}$. As any nonzero vector x_1 appears in some basis in $\mathbb{C}^n$, we find that $Bx = \lambda x$ for every $x \in \mathbb{C}^n \smallsetminus \{0\}$, where the number can depend on x: $\lambda = \lambda(z)$. However, if $Bx = \lambda(x)x$, $By = \lambda(y)y$ with $\lambda(x) \neq \lambda(y)$, then $B(x + y) \notin \operatorname{Span}\{x + y\}$, a contradiction. Hence λ is independent of x and the proposition is proved. □

9.2 COMMON INVARIANT SUBSPACES FOR COMMUTING MATRICES

In this section we establish a fundamental property of a set of commuting transformations, namely, that there is always a complete chain of subspaces that are invariant for every transformation of the set.

Theorem 9.2.1

Let Ω be a set of commuting transformations from $\mathbb{C}^n$ into $\mathbb{C}^n$ (so $AB = BA$ for any $A, B \in \Omega$). Then there exists a complete chain of subspaces $0 = \mathcal{M}_0 \subset \mathcal{M}_1 \cdots \subset \mathcal{M}_n = \mathbb{C}^n$, $\dim \mathcal{M}_j = j$, such that $\mathcal{M}_0, \mathcal{M}_1, \ldots, \mathcal{M}_n$ are invariant for every transformation from Ω.

Proof. For every nonzero vector $x \in \mathbb{C}^n$ write

$$\mathcal{L}(x) = \text{Span}\{x, A_1 A_2 \cdots A_k x \mid A_1, \ldots, A_k \in \Omega, \quad k = 1, 2, \ldots\}$$

Clearly $\mathcal{L}(x)$ is a nonzero subspace that is invariant for any $A \in \Omega$ (in short, Ω invariant).

Now let $x_1 \in \mathbb{C}^n$ be an eigenvector of some transformation $A_1 \in \Omega$ corresponding to an eigenvalue λ_1; so $A_1 x_1 = \lambda_1 x_1$. Hence for every $B_1, \ldots, B_k \in \Omega$ we have

$$A_1 B_1 \cdots B_k x_1 = B_1 A_1 B_2 \cdots B_k x_1 = \cdots = B_1 B_2 \cdots B_k A_1 x_1$$
$$= \lambda_1 B_1 B_2 \cdots B_k x_1$$

So

$$A_{1|\mathcal{L}(x_1)} = \lambda_1 I$$

Let $x_2 \in \mathcal{L}(x_1)$ be an eigenvector of some $A_2 \in \Omega$: $A_2 x_2 = \lambda_2 x_2$. Then $A_{2|\mathcal{L}(x_2)} = \lambda_2 I$, and $\mathcal{L}(x_2) \subset \mathcal{L}(x_1)$. We continue the construction of nonzero subspaces

$$\mathcal{L}(x_1) \supset \mathcal{L}(x_2) \supset \cdots \supset \mathcal{L}(x_k)$$

where $A_{i|\mathcal{L}(x_i)} = \lambda_i I$, $i = 1, \ldots, k$ for some $A_1, \ldots, A_k \in \Omega$ and complex numbers $\lambda_i, \ldots, \lambda_k$, until we encounter the situation where $\mathcal{L}(y) = \mathcal{L}(x_k)$ for every eigenvector $y \in \mathcal{L}(x_k)$ corresponding to any eigenvalue λ of any transformation $B \in \Omega$. In this case every $B \in \Omega$ has an eigenvalue λ_B with the property that $B_{|\mathcal{L}(x_k)} = \lambda_B I$. Let y_1 be any nonzero vector from $\mathcal{L}(x_k)$. Then the subspace $\mathcal{M}_1 = \text{Span}\{y_1\}$ is Ω invariant.

Let $\mathcal{N}_1$ be a direct complement to $\mathcal{M}_1$ in $\mathbb{C}^n$. With respect to the decomposition $\mathcal{M}_1 \dotplus \mathcal{N}_1 = \mathbb{C}^n$, we have

$$A = \begin{bmatrix} A_1 & A_{12} \\ 0 & A_2 \end{bmatrix} \qquad B = \begin{bmatrix} B_1 & B_{12} \\ 0 & B_2 \end{bmatrix} \quad \text{for any } A, B \in \Omega$$

The condition $AB = BA$ implies that $A_2 B_2 = B_2 A_2$. Repeating the above procedure, we find a common eigenvector $y_2 \in \mathcal{N}_1$ of all linear transformations from Ω. Put $\mathcal{M}_2 = \text{Span}\{y_1, y_2\}$, and so on. Eventually we obtain a complete chain of common Ω-invariant subspaces. $\square$

In terms of bases, Theorem 9.2.1 can be stated as follows.

Theorem 9.2.2

Let Ω be a set of commuting transformations from $\mathbb{C}^n$ into $\mathbb{C}^n$. Then there exists an orthonormal basis $x_1, \ldots, x_n$ in $\mathbb{C}^n$ such that the representation of any $A \in \Omega$ in this basis is an upper triangular matrix.

Proof. Let $\{0\} = \mathcal{M}_0 \subset \mathcal{M}_1 \subset \cdots \subset \mathcal{M}_n = \mathbb{C}^n$ be a complete chain of subspaces as in Theorem 9.2.1. Now construct an orthonormal basis $x_1, \ldots, x_n$ in such a way that $\mathrm{Span}\{x_1, \ldots, x_j\} = \mathcal{M}_j$ for $j = 1, \ldots, n$. $\square$

If every transformation from the set Ω is normal, the upper triangular matrices of Theorem 9.2.2 are actually diagonal (cf. the proof of Theorem 1.9.4). As a result we obtain the "only if" part of the following result.

Theorem 9.2.3

Let Ω be a set of normal transformations $\mathbb{C}^n \to \mathbb{C}^n$. Then $AB = BA$ for any transformations $A, B \in \Omega$ if and only if there is an orthonormal basis consisting of eigenvectors that are common to all transformations in Ω.

The part "if" of this theorem is clear: if $x_1, \ldots, x_n$ is an orthonormal basis in $\mathbb{C}^n$ formed by common eigenvectors of A and B, where $A, B \in \Omega$, then in this basis we have

$$A = \mathrm{diag}[\lambda_1, \lambda_2, \ldots, \lambda_n], \qquad B = \mathrm{diag}[\mu_1, \mu_2, \ldots, \mu_n]$$

9.3 COMMON INVARIANT SUBSPACES FOR MATRICES WITH RANK 1 COMMUTATORS

For $n \times n$ matrices A and B, the *commutator* of A and B is, by definition, the matrix $AB - BA$. So the commutator measures the extent to which A and B fail to commute. We have seen in the preceding section that if A and B commute, that is, if their commutator is zero, then there exists a complete chain of common invariant subspaces of A and B. It turns out that this result is still true if the commutator is small in the sense of rank.

Theorem 9.3.1

Let A and B be $n \times n$ matrices with $\mathrm{rank}(AB - BA) \le 1$. Then there exists a complete chain of subspaces:

$$0 = \mathcal{M}_0 \subset \mathcal{M}_1 \subset \cdots \subset \mathcal{M}_n = \mathbb{C}^n, \qquad \dim \mathcal{M}_j = j$$

such that each $\mathcal{M}_j$ is both A invariant and B invariant.

Proof. We shall assume that $\operatorname{rank}(AB - BA) = 1$. (If $AB - BA = 0$, Theorem 9.3.1 is contained in Theorem 9.2.1.) We can also assume that A is singular. (If necessary, replace A by $A - \lambda_0 I$ for a suitable λ_0, and note that the commutators of A and B and of $A - \lambda_0 I$ and B are the same.) We claim that either $\operatorname{Ker} A$ or $\operatorname{Im} A$ is B invariant. Indeed, if $\operatorname{Ker} A$ is not B invariant, then there exists a nonzero vector $x \in \mathbb{C}^n$ such that $Ax = 0$ and $ABx \neq 0$. Thus

$$(AB - BA)x = ABx$$

span the one-dimensional range of $AB - BA$. Hence for every $y \in \mathbb{C}^n$ there exists a constant $\mu(y)$ such that

$$(AB - BA)y = \mu(y)ABx$$

It follows that

$$BAy = AB(y - \mu(y)x)$$

and hence

$$\operatorname{Im}(BA) \subset \operatorname{Im}(AB) \subset \operatorname{Im} A$$

so $\operatorname{Im} A$ is B invariant. We have shown that there is a nontrivial subspace $\mathcal{N}$ that is invariant for both A and B.

Write A and B as 2×2 block matrices with respect to the decomposition $\mathcal{N} \dotplus \mathcal{N}' = \mathbb{C}^n$, where $\mathcal{N}'$ is some direct complement to $\mathcal{N}$:

$$A = \begin{bmatrix} A_1 & A_{12} \\ 0 & A_2 \end{bmatrix}, \qquad B = \begin{bmatrix} B_1 & B_{12} \\ 0 & B_2 \end{bmatrix}$$

Then $\operatorname{rank}(A_1 B_1 - B_1 A_1) \leq 1$ and $\operatorname{rank}(A_2 B_2 - B_2 A_2) \leq 1$. So we can apply the preceding argument to find a nontrivial common invariant subspace for A_1 and B_1 (if $\dim \mathcal{N} > 1$). Similarly, there exists a nontrivial common invariant subspace for A_2 and B_2 (if $\dim \mathcal{N}' > 1$). Continuing in this way, we ultimately obtain the result of the theorem. $\square$

Theorem 9.3.1 can also be restated in terms of simultaneous triangulizations of A and B, just as Theorem 9.2.1 was recast in the form of Theorem 9.2.2. In contrast with Theorem 9.2.1, the result of Theorem 9.3.1 does not generally hold for sets of more than two matrices.

EXAMPLE 9.3.1. Let

$$A_1 = \begin{bmatrix} 2 & 0 \\ -1 & 1 \end{bmatrix}, \qquad A_2 = \begin{bmatrix} 2 & 1 \\ 0 & 1 \end{bmatrix}, \qquad A_3 = \begin{bmatrix} 1 & 0 \\ 0 & -1 \end{bmatrix}$$

It is easily checked that

$$\text{rank}(A_1A_2 - A_2A_1) = \text{rank}(A_1A_3 - A_3A_1) = \text{rank}(A_2A_3 - A_3A_2) = 1$$

Nevertheless, there is no one-dimensional common invariant subspace for A_1, A_2, and A_3. Indeed, A_3 has exactly two one-dimensional invariant subspaces, $\text{Span}\{e_1\}$ and $\text{Span}\{e_2\}$, and neither of them is invariant for both A_1 and A_2. $\square$

9.4 HYPERINVARIANT SUBSPACES

Let $A: \mathbb{C}^n \to \mathbb{C}^n$ be a transformation. A subspace $\mathcal{M} \subset \mathbb{C}^n$ is called *hyperinvariant* for A (or A *hyperinvariant*) if $\mathcal{M}$ is invariant for any transformation that commutes with A. In particular, an A-hyperinvariant subspace is A invariant. Let us study two simple examples.

EXAMPLE 9.4.1. Let $A = \lambda I$, $\lambda \in \mathbb{C}$. Obviously, any transformation from $\mathbb{C}^n$ to $\mathbb{C}^n$ commutes with A, so the only subspaces which are invariant for every linear transformation that commutes with A are the trivial ones: $\{0\}$ and $\mathbb{C}^n$. Hence A has only two hyperinvariant subspaces: $\{0\}$ and $\mathbb{C}^n$. $\square$

EXAMPLE 9.4.2. Assume that $A: \mathbb{C}^n \to \mathbb{C}^n$ has n distinct eigenvalues $\lambda_1, \ldots, \lambda_n$ with corresponding eigenvectors $x_1, \ldots, x_n$. Then A has exactly 2^n invariant subspaces $\text{Span}\{x_i \mid i \in K\}$, where K is any subset in $\{1, \ldots, n\}$ (see Example 1.1.3). By Theorem 9.1.4, the only transformations that commute with A are the polynomials in A. Since every A-invariant subspace is invariant also for any polynomial of A, we find that every A-invariant subspace is A hyperinvariant. $\square$

More generally, let A be a nonderogatory transformation. Then Theorem 9.1.4 shows that every A-invariant subspace is also A hyperinvariant. This property is characteristic for nonderogatory transformations.

Theorem 9.4.1

For a transformation $A: \mathbb{C}^n \to \mathbb{C}^n$ every A-invariant subspace is A hyperinvariant if and only if A is nonderogatory.

Proof. We have seen already that the part "if" is true. To prove the "only if" part, assume that A is not nonderogatory. We prove that there exists an A-invariant subspace that is not A hyperinvariant. By assumption, $\dim \text{Ker}(A - \lambda_0 I) \geq 2$ for some eigenvalue λ_0 of A. Without loss of generality we can assume that A is a Jordan matrix

$$A = J_{k_1}(\lambda_0) \oplus \cdots \oplus J_{k_m}(\lambda_0) \oplus J_{k_{m+1}}(\lambda_{m+1}) \oplus \cdots \oplus J_{k_p}(\lambda_p)$$

where $m \geq 2$ and the first m Jordan blocks correspond to the eigenvalue λ_0, they are arranged so that $k_1 \le k_2$ and $\lambda_{m+1}, \ldots, \lambda_p$ are different from λ_0. Obviously, Span$\{e_1\}$ is an A-invariant subspace. It turns out that this subspace is not A hyperinvariant. Indeed, by Theorem 9.1.1 the matrix S with 1 in the entries $(k_1 + 1, 1), \ldots, (2k_1, k_1)$ and zero elsewhere, commutes with A. On the other hand, $Se_1 = e_{k_1+1}$, so Span$\{e_1\}$ is not S invariant. $\square$

It is easily seen that all the A-hyperinvariant subspaces form a lattice, that is, the intersection and sum of A-hyperinvariant subspaces are again A hyperinvariant. Denote this lattice by Hinv(A). Now we can state the main result concerning the structure of Hinv(A).

Theorem 9.4.2

The lattice of all A-hyperinvariant subspaces coincides with the smallest lattice $\mathcal{S}_A$ of subspaces in $\mathbb{C}^n$ that contains

$$\text{Im}(A - \lambda I)^k \quad and \quad \text{Ker}(A - \lambda I)^k, \quad \lambda \in \mathbb{C}, \quad k = 1, 2, \ldots$$

Actually, $\mathcal{S}_A$ coincides with the smallest lattice of subspaces in $\mathbb{C}^n$ that contains

$$\text{Ker}(A - \lambda_j I)^k, \quad \text{Im}(A - \lambda_j I)^k, \quad j = 1, \ldots, m; \quad k = 1, \ldots, r_j - 1$$

$$(9.4.1)$$

where $(\lambda - \lambda_1)^{r_1} \cdots (\lambda - \lambda_m)^{r_m}$ is the minimal polynomial of A. Indeed, $\text{Ker}(A - \lambda I)^k = \{0\}$ for $\lambda \notin \{\lambda_1, \ldots, \lambda_m\}$ and $\text{Ker}(A - \lambda I)^k = R_\lambda(A)$ for $\lambda = \lambda_j$ and $k \ge r_j$.

The proof of Theorem 9.4.2 is given in the next section.

The following example shows that, in general, not every A-hyperinvariant subspace is the image or the kernel of a polynomial in A.

EXAMPLE 9.4.3.   Let A be the 6×6 matrix

$$A = \begin{bmatrix} 0 & 1 & 0 & 0 & 0 & 0 \\ 0 & 0 & 1 & 0 & 0 & 0 \\ 0 & 0 & 0 & 1 & 0 & 0 \\ 0 & 0 & 0 & 0 & 0 & 0 \\ 0 & 0 & 0 & 0 & 0 & 1 \\ 0 & 0 & 0 & 0 & 0 & 0 \end{bmatrix}$$

According to Theorem 9.4.2, the subspace $\mathcal{L} = \text{Span}\{e_1, e_2, e_5\} = \text{Ker } A + \text{Im } A^2$ is A hyperinvariant. On the other hand, there is no polynomial $p(\lambda)$ such that $\mathcal{L} = \text{Ker } p(A)$ or $\mathcal{L} = \text{Im } p(A)$. Indeed, for any polynomial $p(\lambda)$ the matrix $p(A)$ has the form (see Section 9.2.10):

$$p(A) = \begin{bmatrix} p_1 & p_2 & p_3 & p_4 & 0 & 0 \\ 0 & p_1 & p_2 & p_3 & 0 & 0 \\ 0 & 0 & p_1 & p_2 & 0 & 0 \\ 0 & 0 & 0 & p_1 & 0 & 0 \\ 0 & 0 & 0 & 0 & p_1 & p_2 \\ 0 & 0 & 0 & 0 & 0 & p_1 \end{bmatrix}$$

for some complex numbers p_1, p_2, p_3, p_4. So Ker $p(A)$ can be only one of the following subspaces: $\{0\}$ (if $p_1 \neq 0$); Span$\{e_1, e_5\}$ (if $p_1 = 0$, $p_2 \neq 0$); Span$\{e_1, e_2, e_5, e_6\}$ (if $p_1 = p_2 = 0$, $p_3 \neq 0$); Span$\{e_1, e_2, e_3, e_5, e_6\}$ (if $p_1 = p_2 = p_3 = 0$, $p_3 \neq 0$); $\math{C}^6$ (if $p_i = 0$, $i = 1, 2, 3, 4$). The subspace Im $p(A)$ can be one of the following: $\math{C}^6$; Span$\{e_1, e_2, e_3, e_5\}$; Span$\{e_1, e_2\}$; Span$\{e_1\}$; $\{0\}$. None of these subspaces coincides with $\mathscr{L}$. $\square$

9.5   PROOF OF THEOREM 9.4.2

The proof of Theorem 9.4.2 requires some preparation. We first prove several auxiliary results that are useful in their own right.

Proposition 9.5.1

For any $\lambda \in \math{C}$ the subspaces

$$\mathrm{Ker}(A - \lambda I)^k, \qquad \mathrm{Im}(A - \lambda I)^k, \qquad k = 1, 2, \ldots$$

are A hyperinvariant.

Proof. Fix $\lambda \in \math{C}$ and a positive integer k, and let x be any vector from Ker$(A - \lambda I)^k$. If B commutes with A, we have

$$(A - \lambda I)^k Bx = B(A - \lambda I)^k x = 0$$

So $Bx \in \mathrm{Ker}(A - \lambda I)^k$, and the subspace $\mathrm{Ker}(A - \lambda I)^k$ is A hyperinvariant. Similarly, let $y \in \mathrm{Im}(A - \lambda I)^k$ and $BA = AB$. Then for any $z \in \math{C}^n$ such that $(A - \lambda I)^k z = y$, we obtain

$$(A - \lambda I)^k Bz = B(A - \lambda I)^k z = By$$

So $By \in \mathrm{Im}(A - \lambda I)^k$; therefore, $\mathrm{Im}(A - \lambda I)^k$ is A hyperinvariant. $\square$

We proceed now with the identification of Hinv(A), assuming that A has only one eigenvalue. Given positive integers $p_1 \geq \cdots \geq p_m$, let $\Lambda(p_1, \ldots, p_m)$ be the set of all m-tuples of integers $(q_1, \ldots, q_m)$ such that $q_1 \geq \cdots \geq q_m \geq 0$ and $p_1 - q_1 \geq p_2 - q_2 \geq \cdots \geq p_m - q_m \geq 0$. For every two sequences $q' = (q'_1, \ldots, q'_m)$ and $q'' = (q''_1, \ldots, q''_m)$ from $\Lambda(p_1, \ldots, p_m)$ put

$$\max(q', q'') = (\max(q'_1, q''_1), \ldots, \max(q'_m, q''_m))$$

It is easily seen that $\max(q', q'') \in \Lambda(p_1, \ldots, p_m)$. Similarly, let

$$\min(q', q''_1) = (\min(q'_1, q''_1), \ldots, \min(q'_m, q''_m))$$

then $\min(q', q'')$ belong to $\Lambda(p_1, \ldots, p_m)$.

Let $B: \mathbb{C}^n \to \mathbb{C}^n$ be a transformation with a single eigenvalue λ_0, and let

$$f_1^{(1)}, \ldots, f_{p_1}^{(1)}; f_1^{(2)}, \ldots, f_{p_2}^{(2)}; \ldots; f_1^{(m)}, \ldots, f_{p_m}^{(m)} \qquad (9.5.1)$$

be a Jordan basis in $\mathbb{C}^n$ for B, where $p_1 \geq p_2 \geq \cdots \geq p_m$. So in this basis B has the form

$$J_{p_1}(\lambda_0) \oplus \cdots \oplus J_{p_m}(\lambda_0)$$

Let

$$\mathcal{H}^i_j = \mathrm{Span}\{f_1^{(i)}, \ldots, f_j^{(i)}\}, \quad i = 1, \ldots, m; \quad j = 1, \ldots, p_i \qquad (9.5.2)$$

Lemma 9.5.2

For every $(q_1, \ldots, q_m) \in \Lambda(p_1, \ldots, p_m)$ the subspace

$$\varphi(q_1, \ldots, q_m) \overset{\text{def}}{=} \mathcal{H}^1_{q_1} \dotplus \cdots \dotplus \mathcal{H}^m_{q_m} \qquad (9.5.3)$$

is B hyperinvariant. Conversely, every B-hyperinvariant subspace $\mathcal{L}$ has the form $\varphi(q_1, \ldots, q_m)$ for some $(q_1, \ldots, q_m) \in \Lambda(p_1, \ldots, p_m)$. Moreover

$$\phi(\max(q', q'')) = \phi(q') + \phi(q'') \qquad (9.5.4)$$

$$\phi(\min(q', q'')) = \phi(q') \cap \phi(q'') \qquad (9.5.5)$$

for every $q', q'' \in \Lambda(p_1, \ldots, p_m)$.

Proof. Let $\mathcal{L}$ be a nonzero B-hyperinvariant subspace, and let $x \in \mathcal{L}$ be an arbitrary nonzero vector. Write x as a linear combination of the basis vectors:

$$x = \sum_{i=1}^{p_1} \xi_i^{(1)} f_i^{(1)} + \cdots + \sum_{i=1}^{p_m} \xi_i^{(m)} f_i^{(m)}$$

Assume that for some j the vector

$$y = \sum_{i=1}^{p_j} \xi_i^{(j)} f_i^{(j)}$$

is nonzero, and let q be the maximal index i $(1 \le i \le p_j)$ such that $\xi_q^{(j)} \ne 0$. We show that the subspace $\mathcal{H}_q^j$ is in $\mathcal{L}$.

Let P_j be the projector on $\mathcal{H}_{p_j}^j$ defined by $P_j f_\alpha^{(i)} = 0$ for $i \ne j$ and $P_j f_\alpha^{(j)} = f_\alpha^{(j)}$ $(\alpha = 1, \ldots, p_j)$. Obviously, $P_j B = B P_j$. Therefore, the subspace $\mathcal{L}$ is P_j invariant. Hence $y = P_j x \in \mathcal{L}$. For every $k = 1, 2, \ldots$ the linear transformation $(B - \lambda_0 I)^k$ commutes with B and hence

$$f_q^{(j)} = \frac{1}{\xi_q^{(j)}} (B - \lambda_0 I)^{q-1} y \in \mathcal{L}$$

Then the vectors

$$f_{q-1}^{(j)} = (B - \lambda_0 I) f_q^{(j)}, \ldots, f_1^{(j)} = (B - \lambda_0 I) f_2^{(j)}$$

also belong to $\mathcal{L}$. Thus $\mathcal{H}_q^j \subset \mathcal{L}$.

Furthermore, we show that if $\mathcal{H}_q^j \subset \mathcal{L}$ $(j \ge 2)$, then also $\mathcal{H}_q^{j-1} \subset \mathcal{L}$. Indeed, let $X: \mathbb{C}^n \to \mathbb{C}^n$ be the linear transformation given in the basis (9.5.1) by the matrix

$$X = [X_{\mu\nu}]_{\mu,\nu=1}^m$$

where $X_{\mu\nu}$ is a $p_\nu \times p_\mu$ matrix, and $X_{\mu\nu} = 0$ for all μ, ν except for $X_{j-1,j}$, which is given as follows:

$$X_{j-1,j} = \begin{bmatrix} I_{p_j} \\ 0 \end{bmatrix}$$

Theorem 9.1.1 shows that X commutes with B. Consequently, $\mathcal{L}$ is X invariant and the vectors

$$f_i^{(j-1)} = X f_i^{(j)} \qquad (i = 1, \ldots, q)$$

belong to $\mathcal{L}$.

We have proved that $\mathcal{L}$ has the form (9.5.3) with $q_1 \ge \cdots \ge q_m$. Let us verify that $p_1 - q_1 \ge \cdots \ge p_m - q_m$. Fix $i_0 < j_0$ and let $C: \mathbb{C}^n \to \mathbb{C}^n$ be defined in the block matrix from $C = [C_{ij}]_{i,j=1}^m$ with respect to the basis (9.5.1) where C_{ij} is the zero $p_i \times p_j$ matrix if $i \ne j_0$ or $j \ne i_0$ and $C_{j_0 i_0}$ is the $p_{i_0} \times p_{j_0}$ matrix $[0 \ I]$. By Theorem 9.1.1, C commutes with A, so $\mathcal{L}$ is C invariant. If $q_{i_0} = 0$ or $p_{i_0} - q_{i_0} \ge p_{j_0}$, then obviously $p_{i_0} - q_{i_0} \ge p_{j_0} - q_{j_0}$. Otherwise

$$C f_{q_{i_0}}^{(i_0)} = f_{p_{j_0} - p_{i_0} + q_{i_0}}^{(j_0)} \in \mathcal{L}$$

which implies $p_{j_0} - p_{i_0} + q_{i_0} \le q_{j_0}$, that is, $p_{i_0} - q_{i_0} \ge p_{j_0} - q_{j_0}$ again.

It remains to show that every subspace

$$\mathscr{L} = \mathscr{K}^1_{q_1} \dotplus \cdots \dotplus \mathscr{K}^m_{q_m}$$

with $(q_1, \ldots, q_m) \in \Lambda(p_1, \ldots, p_m)$ is B hyperinvariant. Let $C \in \mathscr{C}(B)$. We must prove that $\mathscr{L}$ is C invariant. With respect to the basis (9.5.1), write C as the block matrix $C = [C_{ij}]^m_{i,j=1}$, where C_{ij} is a $p_i \times p_j$ matrix of one of the following types (see Theorem 9.1.1):

$$T_{p_i} \quad \text{if} \quad i = j; \qquad [0 \quad T_{p_i}] \quad \text{if} \quad i > j; \qquad \begin{bmatrix} T_{p_j} \\ 0 \end{bmatrix} \quad \text{if} \quad i < j$$

[in the notation of (9.1.6)–(9.1.8)]. From the structure of C it is easily seen that $\mathscr{L}$ is C invariant if and only if the q_jth column in every C_{ij} has all entries zero in the places $q_i + 1, \ldots, p_i$. In case $i > j$ the first nonzero entry in the q_jth column of C_{ij} can be in the $[p_i - (p_j - q_j)]$th place; but $p_i - (p_j - q_j) \leq q_i$ because $(q_1, \ldots, q_m) \in \Lambda(p_1, \ldots, p_m)$. In case $i < j$ the first nonzero entry in the q_jth column of C_{ij} can be in the q_jth place; but $q_j \leq q_i$, so we are done in this case also. Finally, in case $i = j$ obviously the q_jth column of C_{ij} has zeros in places $q_i + 1, \ldots, p_i$. We have verified that $\mathscr{L}$ is indeed C invariant.

Finally, equalities (9.5.4) and (9.5.5) are clear from the definitions of $\min(q', q'')$ and $\max(q', q'')$. $\square$

Now we begin the proof of Theorem 9.4.2 itself. In view of Proposition 9.5.1, every element in the lattice $\mathscr{S}_A$, the smallest lattice containing the subspaces (9.4.1), is A hyperinvariant. Now let $\mathscr{L}$ be an A-hyperinvariant subspace. Then $\mathscr{L}$ is, in particular, A invariant; therefore

$$\mathscr{L} = \mathscr{L} \cap \mathscr{R}_{\lambda_1}(A) \dotplus \cdots \dotplus \mathscr{L} \cap \mathscr{R}_{\lambda_m}(A) \tag{9.5.6}$$

where $\lambda_1, \ldots, \lambda_m$ are all the distinct eigenvalues of A. Now $\mathscr{L} \cap \mathscr{R}_{\lambda_i}(A) = \mathscr{L} \cap \operatorname{Ker}(A - \lambda_i I)^{r_i}$ is also an A-hyperinvariant subspace. [Recall that the integers r_i are defined by the minimal polynomial $(\lambda - \lambda_i)^{r_1} \cdots (\lambda - \lambda_m)^{r_m}$ of A.] Thus, to show that $\mathscr{L} \in \mathscr{S}_A$, we can assume that A has only one eigenvalue λ_0. Letting $p_1 \geq \cdots \geq p_l$ be the partial multiplicities of A, in view of Lemma 9.5.2 it will suffice to verify that

$$\mathscr{K}^1_{q_1} \dotplus \cdots \dotplus \mathscr{K}^l_{q_l} \in \mathscr{S}_A$$

where $(q_1, \ldots, q_l) \in \Lambda(p_1, \ldots, p_l)$ and $\mathscr{K}^i_j$ are defined as in equation (9.5.2) [with respect to a Jordan basis $f^{(i)}_j$ of A]. Actually

$$\mathscr{K}^1_{q_1} \dotplus \cdots \dotplus \mathscr{K}^l_{q_l} = (\operatorname{Ker} N^{q_1} \cap \operatorname{Im} N^{p_1 - q_1}) \dotplus \cdots \dotplus (\operatorname{Ker} N^{q_l} \cap \operatorname{Im} N^{p_l - q_l}) \tag{9.5.7}$$

where $N = A - \lambda_0 I$. Indeed, as $\mathcal{H}_{q_i}^i \subset \operatorname{Ker} N^{q_i} \cap \operatorname{Im} N^{p_i - q_i}$, $i = 1, \ldots, l$, the inclusion $\subset$ in (9.5.7) is obvious. For the opposite inclusion, let $x \in \operatorname{Ker} N^{q_i} \cap \operatorname{Im} N^{p_i - q_i}$ so $x = N^{p_i - q_i} y$ for some y with $N^{p_i} y = 0$. Write $y = y_1 + y_2 + \cdots + y_l$, where $y_j \in \operatorname{Span}\{f_1^{(j)}, \ldots, f_{p_i}^{(j)}\}$. Then $x = \sum_{j=1}^{l} N^{p_i - q_i} y_j$ and

$$N^{p_i} y_j = 0 \quad \text{for} \quad j = 1, \ldots, l \tag{9.5.8}$$

We want to show that $N^{p_i - q_i} y_j \in \mathcal{H}_{q_j}^j$ or, equivalently

$$N^{q_j + p_i - q_i} y_j = 0, \qquad j = 1, \ldots, l \tag{9.5.9}$$

But since $(q_1, \ldots, q_l) \in \Lambda(p_1, \ldots, p_l)$, we have $q_j + p_i - q_i \geq \min(p_i, p_j)$, $1 \leq j \leq l$, and (9.5.9) follows from (9.5.8). Theorem 9.4.2 is proved.

9.6 FURTHER PROPERTIES OF HYPERINVARIANT SUBSPACES

We present here some properties of the lattice Hinv(A) of all A-hyperinvariant subspaces.

Theorem 9.6.1

For any transformation $A: \mathbb{C}^n \to \mathbb{C}^n$ the lattice Hinv(A) *is distributive and self-dual and contains exactly*

$$\prod_{i=1}^{k} \left[\prod_{j=1}^{m_i - 1} (p_j^{(i)} - p_{j+1}^{(i)} + 1) \right] (p_{m_i}^{(i)} + 1) \tag{9.6.1}$$

elements, where $p_1^{(i)} \geq \cdots \geq p_{m_i}^{(i)}$ are the partial multiplicities of A corresponding to the ith eigenvalue, $i = 1, \ldots, k$, and k is the number of different eigenvalues of A (in particular, Hinv(A) *is finite).*

Let us explain the terms that appear in this theorem. By definition, a lattice Λ of subspaces in $\mathbb{C}^n$ is called *distributive* if

$$\mathcal{M} \cap (\mathcal{N}_1 + \mathcal{N}_2) = (\mathcal{M} \cap \mathcal{N}_1) + (\mathcal{M} \cap \mathcal{N}_2)$$

for every $\mathcal{M}, \mathcal{N}_1, \mathcal{N}_2 \in \Lambda$. The lattice Λ is said to be *self-dual* if there exists a bijective map $\psi: \Lambda \to \Lambda$ such that $\psi(\mathcal{M} + \mathcal{N}) = \psi(\mathcal{M}) \cap \psi(\mathcal{N})$, $\psi(\mathcal{M} \cap \mathcal{N}) = \psi(\mathcal{M}) + \psi(\mathcal{N})$ for every $\mathcal{M}, \mathcal{N} \in \Lambda$. [In other words, Λ is isomorphic (as a lattice) to the dual lattice of Λ.]

Proof. Note that every A-hyperinvariant subspace $\mathcal{L}$ admits the representation

$$\mathcal{L} = \mathcal{L} \cap \mathcal{R}_{\lambda_1}(A) \dotplus \cdots \dotplus \mathcal{L} \cap \mathcal{R}_{\lambda_k}(A)$$

where $\lambda_1, \ldots, \lambda_k$ are all the distinct eigenvalues of A. As

$$\mathcal{L}_1 \cap \mathcal{L}_2 = \sum_{i=1}^{k} (\mathcal{L}_1 \cap \mathcal{R}_{\lambda_i}(A)) \cap (\mathcal{L}_2 \cap \mathcal{R}_{\lambda_i}(A))$$

and

$$\mathcal{L}_1 + \mathcal{L}_2 = \sum_{i=1}^{k} [(\mathcal{L}_1 \cap \mathcal{R}_{\lambda_i}(A)) + (\mathcal{L}_2 \cap \mathcal{R}_{\lambda_i}(A))]$$

for any A-hyperinvariant subspaces $\mathcal{L}_1$ and $\mathcal{L}_2$, we assume (without loss of generality) that A has only a single eigenvalue λ_0 [i.e., $\mathcal{R}_{\lambda_0}(A) = \mathbb{C}^n$].

To show that the lattice of A-hyperinvariant subspaces is distributive, first observe the following equality for any real numbers r, s, t:

$$\min(\max(r, s), t) = \max(\min(r, t), \min(s, t)) \qquad (9.6.2)$$

This equality can be easily verified by assuming (without loss of generality) that $r \leq s$, and then by considering three cases separately: (1) $t \leq r \leq s$; (2) $r \leq t \leq s$; (3) $r \leq s \leq t$. Now let $\mathcal{M}_1$, $\mathcal{M}_2$, $\mathcal{M}_3$ be A-hyperinvariant subspaces. According to Lemma 9.5.2, write

$$\mathcal{M}_i = \mathcal{K}^1_{q_1^{(i)}} \dotplus \cdots \dotplus \mathcal{K}^m_{q_m^{(i)}}, \qquad i = 1, 2, 3$$

in the notation of Lemma 9.5.2, where $q^{(i)} = (q_1^{(i)}, \ldots, q_m^{(i)}) \in \Lambda(p_1, \ldots, p_m)$, $i = 1, 2, 3$, and $p_1 \geq \cdots \geq p_m$ are the partial multiplicities of A. Using (9.5.4) and (9.5.5), we have

$$\mathcal{M}_1 \cap (\mathcal{M}_2 + \mathcal{M}_3) = \varphi(\min[\max(q^{(2)}, q^{(3)}), q^{(1)}]) \qquad (9.6.3)$$

and

$$(\mathcal{M}_1 \cap \mathcal{M}_2) + (\mathcal{M}_1 \cap \mathcal{M}_3) = \varphi(\max[\min(q^{(1)}, q^{(2)}), \min(q^{(1)}, q^{(3)})]) \qquad (9.6.4)$$

Using (9.6.2), we obtain equality between (9.6.3) and (9.6.4).

To prove the self-duality of $\text{Hinv}(A)$, observe that, in view of Lemma 9.5.2, the map $\psi: \text{Hinv}(A) \to \text{Hinv}(A)$ defined by

$$\psi(\mathcal{K}^1_{q_1} \dotplus \cdots \dotplus \mathcal{K}^m_{q_m}) = \mathcal{K}^1_{p_1 - q_1} \dotplus \cdots \dotplus \mathcal{K}^m_{p_m - q_m}$$

where $(q_1, \ldots, q_m) \in \Lambda(p_1, \ldots, p_m)$ satisfies the definition of a self-dual lattice. For instance:

$$\psi\left(\sum_{i=1}^{m} \mathscr{K}^i_{q'_i} + \sum_{i=1}^{m} \mathscr{K}^i_{q''_i}\right) = \psi\left(\sum_{i=1}^{m} \mathscr{K}^i_{\max(q'_i,q''_i)}\right)$$

$$= \sum_{i=1}^{m} \mathscr{K}^i_{p_i-\max(q'_i,q''_i)} = \sum_{i=1}^{m} \mathscr{K}^i_{\min(p_i-q'_i,p_i-q''_i)}$$

$$= \left(\sum_{i=1}^{m} \mathscr{K}^i_{p_i-q'_i}\right) \cap \left(\sum_{i=1}^{m} \mathscr{K}^i_{p_i-q''_i}\right)$$

$$= \psi\left(\sum_{i=1}^{m} \mathscr{K}^i_{q'_i}\right) \cap \psi\left(\sum_{i=1}^{m} \mathscr{K}^i_{q''_i}\right)$$

It remains to verify the $\mathrm{Hinv}(A)$ has exactly

$$\left[\prod_{j=1}^{m-1} (p_j - p_{j+1} + 1)\right](p_m + 1) \qquad (9.6.5)$$

elements. Instead of $\mathrm{Hinv}(A)$, we count elements in $\Lambda(p_1, \ldots, p_m)$. Using induction on m [formula (9.6.5) obviously holds for $m = 1$], assume that $\Lambda(p_2, \ldots, p_m)$ has exactly $[\Pi_{j=2}^{m-1}(p_j - p_{j+1} + 1)]$ $(p_m + 1)$ elements. Now observe that $(q_2 + s, q_2, \ldots, q_m)$ belongs to $\Lambda(p_1, \ldots, p_m)$ if and only if $(q_2, \ldots, q_m)$ belongs to $\Lambda(p_2, \ldots, p_m)$ and $0 \le s \le p_1 - p_2$. This completes the induction step. $\square$

We conclude this section by observing that the number of A-hyperinvariant subspaces for $A: \mathbb{C}^n \to \mathbb{C}^n$ lies between 2 and 2^n, and both bounds can be attained. Indeed, the transformation I has only trivial hyperinvariant subspaces, whereas a diagonable transformation with n distinct eigenvalues has 2^n hyperinvariant subspaces (see Examples 9.4.1 and 9.4.2). That the number of A-hyperinvariant subspaces cannot exceed 2^n follows from a general result in lattice theory [see, e.g., Theorem 148 in Donnellan (1968)] using the fact that $\mathrm{Hinv}(A)$ is distributive and each chain in $\mathrm{Hinv}(A)$ contains not more than $n + 1$ different subspaces.

9.7 EXERCISES

9.1 Consider the transformation

$$A = \begin{bmatrix} 2 & 1 & 0 \\ -1 & 0 & 0 \\ 1 & 1 & 2 \end{bmatrix}: \mathbb{C}^3 \to \mathbb{C}^3$$

written as a matrix with respect to the standard basis e_1, e_2, e_3.

(a) Find all transformations that commute with A.
(b) Find all A-hyperinvariant subspaces.

9.2 Show that if a transformation $A: \mathbb{C}^n \to \mathbb{C}^n$ has n distinct eigenvalues, then every transformation commuting with A is diagonable. Conversely, if every transformation commuting with A is diagonable, then A has n distinct eigenvalues.

9.3 Supply a proof for Corollary 9.1.5.

9.4 Show that if $AJ_n(\lambda_0) = J_n(\lambda_0)A$, then A is diagonable if and only if A is a scalar multiple of the identity.

9.5 Prove or disprove each of the following statements for any commuting transformations $A: \mathbb{C}^n \to \mathbb{C}^n$ and $B: \mathbb{C}^n \to \mathbb{C}^n$:

(a)   There exists an orthonormal basis in which A and B have the lower triangular form.

(b)   There exists a basis in which both A and B have Jordan form.

(c)   Both A and B have the same eigenvectors (possibly corresponding to different eigenvalues).

(d)   Both A and B have the same invariant subspaces.

9.6 Show that any matrix commuting with

$$
\begin{bmatrix}
0 & 1 & 0 & \cdots & 0 \\
0 & 0 & 1 & \cdots & 0 \\
\vdots & \vdots & \vdots & & \vdots \\
0 & 0 & 0 & \cdots & 1 \\
1 & 0 & 0 & \cdots & 0
\end{bmatrix}
$$

is a circulant.

9.7 Show that any matrix commuting with

$$
A = \begin{bmatrix}
0 & 1 & 0 & \cdots & 0 \\
0 & 0 & 1 & \cdots & 0 \\
\vdots & \vdots & \vdots & & \vdots \\
0 & 0 & 0 & \cdots & 1 \\
a_0 & a_1 & a_2 & \cdots & a_{n-1}
\end{bmatrix}
$$

where $a_0, a_1, \ldots, a_{n-1}$ are given complex numbers, is a polynomial of A.

9.8 Describe all matrices commuting with

$$
Q = \begin{bmatrix}
0 & 0 & \cdots & 0 & 1 \\
0 & 0 & \cdots & 1 & 0 \\
\vdots & \vdots & & \vdots & \vdots \\
1 & 0 & \cdots & 0 & 0
\end{bmatrix}
$$

Are all of these polynomials of Q? Find all Q-hyperinvariant subspaces.

9.9 Describe all transformations commuting with a transformation $A: \mathbb{C}^n \to \mathbb{C}^n$ of rank 1. Find all A-hyperinvariant subspaces.

9.10 Let $A: \mathbb{C}^n \to \mathbb{C}^n$ be a transformation. Prove that every A-hyperinvariant subspace is the image of some transformation which commutes with A. (*Hint*: Use Lemma 9.5.2.)

9.11 Show that every A-hyperinvariant subspace is the kernel of some transformation which commutes with A.

9.12 Prove that for the matrix A from Exercise 9.7 we have $\text{Hinv}(A) = \text{Inv}(A)$.

9.13 Is $\text{Hinv}(A) = \text{Inv}(A)$ true for any block companion matrix

$$
\begin{bmatrix}
0 & I & 0 & \cdots & 0 \\
0 & 0 & I & \cdots & 0 \\
\vdots & \vdots & \vdots & & \vdots \\
A_0 & A_1 & A_2 & \cdots & A_{n-1}
\end{bmatrix}
$$

where A_j are 2×2 matrices?

9.14 Show that for circulant matrices A in general $\text{Hinv}(A) \neq \text{Inv}(A)$. Find necessary and sufficient conditions on the circulant matrix A in order that $\text{Hinv}(A) = \text{Inv}(A)$.

9.15 Give an example of a transformation A and of an A-hyperinvariant subspace $\mathcal{M}$ that does not belong to the smallest lattice of subspaces containing the images of all polynomials in A.

9.16 Give an example analogous to Exercise 9.15 with "images" replaced by "kernels."

9.17 Give an example of a transformation A such that $\text{Inv}(A)$ is not distributive.

Chapter Ten

Description of Invariant Subspaces and Linear Transformations with the Same Invariant Subspaces

In this chapter we consider two related problems: (a) description of all invariant subspaces of a given transformation and (b) to what extent a transformation is determined by its lattice of all invariant subspaces.

We have seen in Chapter 2 that every invariant subspace of a linear transformation $A: \mathbb{C}^n \to \mathbb{C}^n$ is a direct sum of irreducible A-invariant subspaces, that is, such that the restriction of A to each one of these subspaces has only one Jordan block in its Jordan form. Thus, to solve the first problem mentioned above it will be sufficient to describe all irreducible A-invariant subspaces. This is done in Section 10.1.

The second objective of this chapter is a characterization of transformations having exactly the same set of invariant subspaces. It turns out that, in general, not all such transformations are polynomials of each other. Our characterization (given in Section 10.2) will depend on the description of irreducible invariant subspaces given in Section 10.1.

10.1 DESCRIPTION OF IRREDUCIBLE SUBSPACES

In the description of invariant subspaces upper triangular Toeplitz matrices, and matrices that resemble upper triangular Toeplitz matrices, play an important role, as we see later. We recall first some simple facts about Toeplitz matrices.

A matrix A of size $j \times j$ is called *Toeplitz* if its entries have the following structure

$$A = \begin{bmatrix} a_0 & a_{-1} & \cdots & a_{-j+1} \\ a_1 & a_0 & \cdots & a_{-j+1} \\ \vdots & & & \vdots \\ & & & a_{-1} \\ a_{j-1} & \cdots & a_1 & a_0 \end{bmatrix} = [a_{i-k}]_{i,k=1}^j \qquad (10.1.1)$$

where $a_i \in \mathbb{C}$, $i = -j+1, -j+2, \ldots, j-1$. Denote by T_j the class of all upper triangular Toeplitz matrices of size $j \times j$, that is, such that $a_1 = \cdots = a_{j-1} = 0$ in equation (10.1.1).

Proposition 10.1.1

The class T_j is an algebra, that is, it is closed under the operations of addition, multiplication by scalars, and matrix multiplication. Moreover, if $A \in T_j$ and $\det A \neq 0$, then $A^{-1} \in T_j$.

Proof. All but the last assertions of Proposition 10.1.1 are immediate consequence of the definition of T_j. To prove the last assertion, suppose that

$$\begin{bmatrix} a_0 & a_{-1} & \cdots & a_{-j+1} \\ 0 & a_0 & & \vdots \\ \vdots & \vdots & & a_{-1} \\ 0 & 0 & \cdots & a_0 \end{bmatrix} \begin{bmatrix} b_{11} & \cdots & b_{1j} \\ \vdots & & \vdots \\ \vdots & & \vdots \\ b_{j1} & \cdots & b_{jj} \end{bmatrix} = I$$

One deduces easily that $b_{ik} = 0$ for $i > k$. Further

$$b_{ii} = a_0^{-1} ; \qquad a_0 b_{i-1,i} + a_{-1} b_{ii} = 0$$

and in general

$$\sum_{p=0}^{k} a_{-k+p} b_{i-p,i} = 0 , \qquad k = 0, \ldots, j-1; \; i = 1, \ldots, j \qquad (10.1.2)$$

(It is assumed that $b_{ki} = 0$ whenever $k \leq 0$.) Equations (10.1.2) define $b_{i-k,i}$ recursively:

$$b_{i-k,i} = -a_0^{-1} \left[\sum_{p=0}^{k-1} a_{-k+p} b_{i-p,i} \right] \qquad (10.1.3)$$

Using (10.1.3), we can prove by induction on k (starting with $k = 0$) that $b_{i-k,i}$ does not depend on i. But this means exactly that the matrix $[b_{ik}]_{i,k=1}^j$ is Toeplitz. $\square$

Let $A: \mathbb{C}^n \to \mathbb{C}^n$ be a transformation. It is clear that each A-invariant subspace $\mathcal{M}$ can be represented as a direct sum of nonzero A-invariant subspaces $\mathcal{M}_1, \ldots, \mathcal{M}_k$, each of which is irreducible, that is, not representable as a direct sum of smaller invariant subspaces (indeed, let l be the maximal number of factors in a decomposition

$$\mathcal{M} = \mathcal{M}_1 \dotplus \cdots \dotplus \mathcal{M}_l \tag{10.1.4}$$

into a direct sum of nonzero A-invariant subspaces $\mathcal{M}_i$; then from the choice of l it follows that each $\mathcal{M}_i$ in equality (10.1.4) is irreducible). To describe the A-invariant subspaces, therefore, it is sufficient to describe all the irreducible subspaces.

It follows from Theorem 2.5.1 that an A-invariant subspace $\mathcal{L}$ is irreducible if and only if the Jordan form of $A_{|\mathcal{L}}$ consists of one Jordan block only. In other words, $\mathcal{L}$ is irreducible if and only if there exists a basis $x_1, \ldots, x_p$ in $\mathcal{L}$ and a complex number λ such that

$$(A - \lambda I)x_1 = 0, \quad (A - \lambda I)x_{j+1} = x_j \quad (j = 1, \ldots, p-1) \tag{10.1.5}$$

that is, the system $\{x_i\}_{i=1}^p$ is a Jordan basis in $\mathcal{L}$. Consequently, every irreducible subspace is contained in some root subspace. (One can see this also from Theorem 2.1.5.) Thus it is sufficient to describe all the irreducible subspaces contained in a fixed root subspace corresponding to the eigenvalue λ. Without loss of generality, we assume that $\lambda = 0$. (Otherwise, replace A by $B = A - \lambda I$ and observe that both transformations A and B have the same invariant subspaces.)

The root subspace $\mathcal{R}_0(A)$ is decomposed into a direct sum of Jordan subspaces:

$$\mathcal{R}_0(A) = \mathcal{L}_1 \dotplus \cdots \dotplus \mathcal{L}_m \tag{10.1.6}$$

The description of the Jordan subspaces contained in $\mathcal{R}_0(A)$ is given according to the number m of irreducible subspaces in the decomposition (10.1.6).

If $\mathcal{R}_0(A)$ is an irreducible subspace [i.e., $m = 1$ in (10.1.6)] and the vectors $\{x_i\}_{i=1}^p$ form a Jordan basis in $\mathcal{R}_0(A)$, then $\mathrm{Span}\{x_1, \ldots, x_j\}$, $j = 1, \ldots, p$ are all the A-invariant subspaces in $\mathcal{R}_0(A)$, and all of them are irreducible subspaces.

Consider now the case when $m = 2$ in (10.1.6). We use the following notation: if $\{z_i\}_{i=1}^p$ is a system of vectors $z_i \in \mathbb{C}^n$, denote by $\bar{z}^{(j)}$ the column formed by vectors, as follows:

$$\text{If } j \le p, \text{ then } \bar{z}^{(j)} = \begin{bmatrix} z_j \\ z_{j-1} \\ \vdots \\ z_1 \end{bmatrix}$$

$$\text{If } j > p, \text{ then } \bar{z}^{(j)} = \left.\begin{bmatrix} z_p \\ z_{p-1} \\ \cdots \\ z_1 \\ 0 \\ \cdots \\ 0 \end{bmatrix}\right\} j-p$$

Let $g_1, \ldots, g_p \in \mathscr{L}_1$ and $f_1, \ldots, f_q \in \mathscr{L}_2$ be Jordan bases in $\mathscr{L}_1$ and $\mathscr{L}_2$, respectively. Without loss of generality, suppose that $p \geq q$. It is known that in any irreducible subspace $\mathscr{L}(\neq 0)$ of A there exists only one eigenvector (up to multiplication by a nonzero scalar). We describe first all the irreducible subspaces that contain the eigenvector g_1 [and thus are contained in $\mathscr{R}_0(A)$].

In the following proposition j is a fixed integer, $1 \leq j \leq p$.

Proposition 10.1.2

Let $T^{(v)}$, where $v = \min(j, q)$, be an upper triangular matrix of size $j \times j$, whose diagonal elements are zeros and the block formed by the first v rows and first v columns is a Toeplitz matrix:

$$T^{(v)} = \begin{bmatrix} 0 & \alpha_1 & \alpha_2 & \cdots & \alpha_{v-1} & \beta_{1,v+1} & \cdots & \beta_{1j} \\ 0 & 0 & \alpha_1 & \cdots & \alpha_{v-2} & \beta_{1,v+1} & \cdots & \beta_{2j} \\ 0 & 0 & 0 & \cdots & \alpha_{v-3} & \beta_{3,v+1} & \cdots & \beta_{3j} \\ \cdots & \cdots & \cdots & \cdots & \cdots & \cdots & \cdots & \cdots \\ 0 & 0 & 0 & \cdots & 0 & \beta_{v,v+1} & \cdots & \beta_{vj} \\ \cdots & \cdots & \cdots & \cdots & \cdots & \cdots & \cdots & \cdots \\ 0 & 0 & 0 & \cdots & 0 & 0 & \cdots & 0 \end{bmatrix} \tag{10.1.7}$$

Then the components of the column

$$\bar{z}^{(j)} = \bar{g}^{(j)} + T^{(v)}\bar{f}^{(j)} \tag{10.1.8}$$

form a Jordan basis of some j-dimensional A-invariant irreducible subspace that contains g_1. Conversely, every irreducible subspace of dimension j of A that contains g_1 has a Jordan basis given by the components of (10.1.8), where $T^{(v)}$ is some matrix of type (10.1.7).

The multiplication in $T^{(v)}\bar{f}^{(j)}$ is performed componentwise: for complex numbers x_{rs} and n-dimensional vectors $z_1, \ldots, z_j$ we define

$$\begin{bmatrix} x_{11} & x_{12} & \cdots & x_{1j} \\ x_{21} & x_{22} & \cdots & x_{2j} \\ \vdots & \vdots & & \vdots \\ x_{k1} & x_{k2} & \cdots & x_{kj} \end{bmatrix}\begin{bmatrix} z_j \\ z_{j-1} \\ \vdots \\ z_1 \end{bmatrix} = \begin{bmatrix} x_{11}z_j + x_{12}z_{j-1} + \cdots + x_{1j}z_1 \\ x_{21}z_j + x_{22}z_{j-1} + \cdots + x_{2j}z_1 \\ \vdots \\ x_{k1}z_j + x_{k2}z_{j-1} + \cdots + x_{kj}z_1 \end{bmatrix}$$

Note also that the dimension of every irreducible subspace of A contained in $\mathcal{R}_0(A)$ does not exceed p [recall that $m = 2$ in (10.1.6) and that dim $\mathcal{L}_1 = p \geq \dim \mathcal{L}_2 \geq 1$]; so Proposition 10.1.2 does indeed give the description of all irreducible subspaces that contain g_1.

Proof. First observe that if $\mathcal{L}$ is an irreducible subspace and $g_1 \in \mathcal{L}$, then

$$\mathcal{L} \cap \mathcal{L}_2 = \{0\} \tag{10.1.9}$$

Indeed, if $y \in \mathcal{L} \cap \mathcal{L}_2 \smallsetminus \{0\}$, then for some i $(0 \leq i \leq p - 1)$ and some complex number $\gamma \neq 0$ the equality $A^i y = \gamma f_1$ holds. So $f_1 \in \mathcal{L} \cap \mathcal{L}_2 \subset \mathcal{L}$, and since also $g_1 \in \mathcal{L}$, the irreducible subspace $\mathcal{L}$ contains two linearly independent eigenvectors f_1 and g_1, which is impossible. From (10.1.9) and the inclusion $\mathcal{L} \dotplus \mathcal{L}_2 \subset \mathcal{R}_0(A) = \mathcal{L}_1 \dotplus \mathcal{L}_2$ it follows that dim $\mathcal{L} \leq \dim \mathcal{L}_1 = p$. Now let $\mathcal{L}$ be an irreducible subspace containing g_1 with a Jordan basis $y_1, \ldots, y_j$; so $y_1 = \alpha_0 g_1$ and $A y_{i+1} = y_i (i = 1, \ldots, j - 1)$. We look for the vectors $y_2, \ldots, y_j$ in the form of linear combinations of $g_1, \ldots, g_p$, $f_1, \ldots, f_q$. Two possibilities can occur: (1) $j \leq q$; (2) $q + 1 \leq j \leq p$. Consider first the case when $j \leq q$. Condition $A y_2 = y_1$ implies that

$$y_2 = \alpha_0 g_2 + \alpha_1 g_1 + \beta_1 f_1$$

Condition $A y_3 = y_2$ implies that $y_3 = \alpha_0 g_3 + \alpha_1 g_2 + \alpha_2 g_1 + \beta_1 f_2 + \beta_2 f_1$. Continuing these arguments, we obtain

$$
\begin{aligned}
y_j &= \alpha_0 g_j + \alpha_1 g_{j-1} + \alpha_2 g_{j-2} + \cdots + \alpha_{j-2} g_2 + \alpha_{j-1} g_1 + \beta_1 f_{j-1} + \beta_2 f_{j-2} + \cdots + \beta_{j-2} f_2 + \beta_{j-1} f_1 \\
y_{j-1} &= \alpha_0 g_{j-1} + \alpha_1 g_{j-2} + \alpha_2 g_{j-3} + \cdots + \alpha_{j-2} g_1 + \beta_1 f_{j-2} + \beta_2 f_{j-3} + \cdots + \beta_{j-2} f_1 \\
&\cdots\cdots\cdots\cdots\cdots\cdots\cdots\cdots\cdots \\
y_3 &= \alpha_0 g_3 + \alpha_1 g_2 + \alpha_2 g_1 + \beta_1 f_2 + \beta_2 f_1 \\
y_2 &= \alpha_0 g_2 + \alpha_1 g_1 + \beta_1 f_1 \\
y_1 &= \alpha_0 g_1
\end{aligned}
\tag{10.1.10}
$$

where $\alpha_1, \ldots, \alpha_{j-1}, \beta_1, \ldots, \beta_{j-1}$ are some numbers. In case $q + 1 \leq j \leq p$ one finds analogously

$$
\begin{aligned}
y_j &= \alpha_0 g_j + \alpha_1 g_{j-1} + \alpha_2 g_{j-2} + \cdots + \alpha_{j-2} g_2 + \alpha_{j-1} g_1 + \beta_1 f_q + \beta_2 f_{q-1} + \cdots + \beta_{q-1} f_2 + \beta_q f_1 \\
y_{j-1} &= \alpha_0 g_{j-1} + \alpha_1 g_{j-2} + \alpha_2 g_{j-3} + \cdots + \alpha_{j-2} g_1 + \beta_1 f_{q-1} + \beta_2 f_{q-2} + \cdots + \beta_{q-1} f_1 \\
y_{j-q+1} &= \alpha_0 g_{j-q+1} + \alpha_1 g_{j-q} + \alpha_2 g_{j-q-1} + \cdots + \beta_1 f_1 \\
&\cdots\cdots\cdots\cdots\cdots\cdots\cdots\cdots\cdots \\
y_3 &= \alpha_0 g_3 + \alpha_1 g_2 + \alpha_2 g_1 \\
y_2 &= \alpha_0 g_2 + \alpha_1 g_1 \\
y_1 &= \alpha_0 g_1
\end{aligned}
\tag{10.1.11}
$$

where $\alpha_1, \ldots, \alpha_{j-1}, \beta_1, \ldots, \beta_q$ are some complex numbers.

Formulas (10.1.10) and (10.1.11) can be written in the form

$$\bar{y}^{(j)} = C\bar{g}^{(j)} + S^{(v)}\bar{f}^{(j)}$$

where C and $S^{(v)}$ are $j \times j$ matrices, and C is an upper triangular invertible Toeplitz matrix (invertible because its diagonal element is $\alpha_0 \neq 0$). By Proposition 10.1.1, C^{-1} is also an upper triangular Toeplitz matrix. It is easy to see that the matrix $C^{-1}S^{(v)}$ has the form $T^{(v)}$ [see (10.1.7)]: $T^{(v)} = C^{-1}S^{(v)}$. Put $\bar{z}^{(j)} = C^{-1}\bar{y}^{(j)} = \bar{g}^{(j)} + T^{(v)}\bar{f}^{(j)}$. It is easy to see that $\mathrm{Span}\{y_i\}_1^j = \mathrm{Span}\{z_i\}_1^j$ and the vectors $z_1, \ldots, z_j$ satisfy (10.1.5). So the components of $\bar{z}^{(j)}$ form a Jordan basis in $\mathscr{L}$. $\square$

Now let x_1 ($\neq \alpha g_1$) be an arbitrary eigenvector of A contained in $\mathscr{R}_0(A) = \mathscr{L}_1 \dotplus \mathscr{L}_2$. Evidently, $x_1 = \zeta g_1 + \eta f_1$ ($\zeta \neq 0$). Consider the system of vectors $x_i = \zeta g_i + \eta f_i$, $i = 1, \ldots, q$. Clearly, the vectors $x_1, \ldots, x_q$ satisfy the condition (10.1.5); therefore, they form a Jordan basis of some irreducible subspace $\hat{\mathscr{L}} \subset \mathscr{R}_0(A)$. It can easily be verified that $\mathscr{L}_1 \dotplus \hat{\mathscr{L}} = \mathscr{R}_0(A)$. Hence dim $\hat{\mathscr{L}} = q$. By Proposition 10.1.2, for every irreducible subspace $\mathscr{L}$ containing the vector x_1 (the dimension j of $\mathscr{L}$ is necessarily not larger than q) there exists a matrix $T^{(j)}$ of the form (10.1.7) such that the components of the column $\bar{v}^{(j)} = \bar{x}^{(j)} + T^{(j)}\bar{g}^{(j)}$ form a Jordan basis in $\mathscr{L}$. Conversely, for every matrix $T^{(j)}$ of size $j \times j$ the components of the column $\bar{v}^{(j)}$ form a Jordan basis in some irreducible subspace of A. Thus a complete description of the irreducible subspaces contained in the root subspaces $\mathscr{R}_0(A) = \mathscr{L}_1 \dotplus \mathscr{L}_2$, is obtained.

This description for the case when $m = 2$ in the decomposition (10.1.6) can be generalized for an arbitrary m. This is the content of the following theorem.

Theorem 10.1.3

Let

$$\mathscr{R}_{\lambda_0}(A) = \mathscr{L}_1 \dotplus \cdots \dotplus \mathscr{L}_m$$

be a decomposition of the root subspace $\mathscr{R}_{\lambda_0}(A)$ of the transformation $A: \mathbb{C}^n \to \mathbb{C}^n$ into a direct sum of irreducible subspaces $\mathscr{L}_1, \ldots, \mathscr{L}_m$. Let $g_1, \ldots, g_{p_1} \in \mathscr{L}_1; \ldots; f_1, \ldots, f_{p_r} \in \mathscr{L}_r; \ldots; h_1, \ldots, h_{p_m} \in \mathscr{L}_m$ be Jordan bases in $\mathscr{L}_1, \ldots, \mathscr{L}_r, \ldots, \mathscr{L}_m$, respectively ($p_1 \geq \cdots \geq p_m$). Let j be an integer such that $1 \leq j \leq p_r = \dim \mathscr{L}_r$. For every $i = 1, \ldots, m$ let $v_i = \min(j, p_i)$. Then for every set of matrices $T_1^{(v_1)}, \ldots, T_m^{(v_m)}$ of the form (10.1.7) and of size $j \times j$ the components of the column

$$\bar{z}^{(j)} = T_1^{(v_1)}\bar{g}^{(j)} + \cdots + T_{r-1}^{(v_{r-1})}\bar{u}^{(j)} + \bar{f}^{(j)} + T_{r+1}^{(v_{r+1})}\bar{v}^{(j)} + \cdots + T_m^{(v_m)}\bar{h}^{(j)}$$

$$(10.1.12)$$

form a Jordan basis in some irreducible subspace of A that contains the vector f_1 (here $u_1, \ldots, u_{p_{r-1}} \in \mathcal{L}_{r-1}$ and $v_1, \ldots, v_{p_{r+1}} \in \mathcal{L}_{r+1}$ are Jordan bases in $\mathcal{L}_{r-1}$ and $\mathcal{L}_{r+1}$, respectively). Conversely, for every irreducible subspace $\mathcal{L}$ of dimension j such that $f_1 \in \mathcal{L}$ there exist matrices $T^{(\nu_1)}, \ldots, T^{(\nu_m)}$ such that the components of the column (10.1.12) *form a Jordan basis in $\mathcal{L}$.*

Proof. Use induction on the number m of subspaces in the decomposition (10.1.6). For $m = 2$ this theorem coincides with Proposition 10.1.2. Suppose that the theorem holds for $m \le k - 1$, and assume that $\mathcal{R}_{\lambda_0}(A) = \mathcal{L}_1 + \cdots + \mathcal{L}_k$. If $\mathcal{L}$ is an irreducible subspace such that $f_1 \in \mathcal{L}$, then

$$\mathcal{L} \cap \hat{\mathcal{L}}_r = \{0\} \qquad (10.1.13)$$

where $\hat{\mathcal{L}}_r = \mathcal{L}_1 + \cdots + \mathcal{L}_{r-1} + \mathcal{L}_{r+1} + \mathcal{L}_k$. Indeed, for every $y \in \mathcal{L} \smallsetminus \{0\}$ there exist a nonnegative integer i and a complex number $\gamma \ne 0$ such that $A^i y = \gamma f_1$. If, in addition, $y \in \hat{\mathcal{L}}_r$, then $\gamma f_1 = A^i y \in \hat{\mathcal{L}}_r$, which contradicts the direct decomposition $\mathcal{R}_{\lambda_0}(A) = \mathcal{L}_1 + \cdots + \mathcal{L}_k$. From (10.1.13) and from the inclusion $\mathcal{L} \subset \mathcal{R}_{\lambda_0}(A) = \mathcal{L}_1 + \cdots + \mathcal{L}_k$ we deduce that $\dim \mathcal{L} \le \dim \mathcal{L}_r$. Assume that $r < k$. (The case $r = k$ can be considered in a similar way.) If $\mathcal{L} \subset \mathcal{L}_1 + \cdots + \mathcal{L}_{k-1}$, then by the induction hypothesis the components of a column of the form (10.1.12) form a Jordan basis in $\mathcal{L}$. If $\mathcal{L} \not\subset \mathcal{L}_1 + \cdots + \mathcal{L}_{k-1}$, consider the subspace $\mathcal{L}' = (\mathcal{L} + \mathcal{L}_k) \cap (\mathcal{L}_1 + \cdots + \mathcal{L}_{k-1})$. Since $\mathcal{L} \cap \mathcal{L}_k = \{0\}$, the equality $\dim \mathcal{L}' = \dim(\mathcal{L} + \mathcal{L}_k) + \dim(\mathcal{L}_1 + \cdots + \mathcal{L}_{k-1}) - \dim(\mathcal{L}_1 + \cdots + \mathcal{L}_k) = \dim \mathcal{L}$ holds. Evidently, $\mathcal{L}'$ is A invariant. Let us show that $\mathcal{L}'$ is an irreducible subspace. Suppose the contrary; then there exists an eigenvector $g \in \mathcal{L}'$ of A that is not a scalar multiple of f_1. Since $\mathcal{L}' \subset \mathcal{L}_1 + \mathcal{L}_k$, the vector g is a linear combination of the eigenvectors f_1 and h_1, where $h_1 \in \mathcal{L}_k$. But then $h_1 \in \mathcal{L}' \subset \mathcal{L}_1 + \cdots + \mathcal{L}_{k-1}$, which means that the sum $(\mathcal{L}_1 + \cdots + \mathcal{L}_{k-1}) + \mathcal{L}_k$ is not direct, and this is a contradiction with our assumptions. So $\mathcal{L}'$ is an irreducible subspace. Since $\mathcal{L}' \subset \mathcal{L}_1 + \cdots + \mathcal{L}_{k-1}$, by the assumption of induction the components of the column $\bar{z}^{(j)}$ form a Jordan basis in $\mathcal{L}'$ for some $T_1^{(\nu_1)}, \ldots, T_{k-1}^{(\nu_{k-1})}$. The property that $\mathcal{L}' \subset \mathcal{L} + \mathcal{L}_k$ implies the inclusion $\mathcal{L} \subset \mathcal{L}' + \mathcal{L}_k$. As it has been proved above, there exists a matrix $T_k^{(\nu_k)}$ such that the components of the column $\bar{y}^{(j)} = \bar{z}^{(j)} + T_k^{(\nu_k)} \bar{h}^{(j)} = T_1^{(\nu_1)} \bar{g}^{(j)} + \cdots + \bar{f}^{(j)} + \cdots + T_k^{(\nu_k)} \bar{h}^{(j)}$ form a Jordan basis in $\mathcal{L}$. $\square$

Theorem 10.1.3 also gives a description of all irreducible subspaces of A that contain an arbitrarily given eigenvector of A from the root subspace $\mathcal{R}_{\lambda_0}(A)$.

Indeed, let $x_1 \in \mathcal{R}_{\lambda_0}(A)$ be an eigenvector, and let r be the minimal integer such that $x_1 \in \mathcal{L}_1 + \cdots + \mathcal{L}_r$. Then $x_1 = a_1 g_1 + \cdots + a_r f_1$, where $a_1, \ldots, a_r \in \mathbb{C}$ and $a_r \ne 0$. Consider the system of vectors $x_i = a_1 g_i + \cdots + a_r f_i$, $i = 1, \ldots, p_r$. Evidently, $x_1, \ldots, x_{p_r}$ satisfy the condition (10.1.5).

Therefore, their linear span $\tilde{\mathscr{L}}_r = \text{Span}\{x_1, \ldots, x_{p_r}\}$ is an irreducible subspace. It is easily seen that

$$\tilde{\mathscr{L}}_r \cap (\mathscr{L}_1 + \cdots + \mathscr{L}_{r-1} + \mathscr{L}_{r+1} + \cdots + \mathscr{L}_k) = \{0\}$$

So in the representation (10.1.6) one can replace $\mathscr{L}_r$ by $\tilde{\mathscr{L}}_r$. Then in view of Theorem 10.1.3 the components of the columns of form (10.1.12) describe all the irreducible subspaces of A, which contain the vector x_1 [in (10.1.12) write $\bar{x}^j$ in place of $\bar{f}^{(j)}$].

Observe that every irreducible subspace contains an eigenvector of A. So the description in the preceding paragraph gives all the irreducible subspaces of A (if the vector x_1 is varied).

10.2   TRANSFORMATIONS HAVING THE SAME SET OF INVARIANT SUBSPACES

Consider a transformation $A: \mathbb{C}^n \to \mathbb{C}^n$. In this section we describe the class of all transformations $B: \mathbb{C}^n \to \mathbb{C}^n$ such that $\text{Inv}(A) = \text{Inv}(B)$. A relative simple case of this situation has already been pointed out in Theorem 2.11.3 (when one transformation is a polynomial in the other). Surprisingly enough, it turns out that the set of transformations B such that $\text{Inv}(B) = \text{Inv}(A)$ does not generally consist only of the transformations $f(A)$, where $f(\lambda)$ is a polynomial with the properties indicated in Theorem 2.11.3. It can even happen that noncommuting transformations have the same set of invariant subspaces.

Before we embark on the statement and proof of the main theorem describing the transformations with the same set of invariant subspaces (which is quite complicated), let us study some examples.

EXAMPLE 10.2.1.   Let A be the $n \times n$ Jordan block $J_n(\lambda_0)$. The invariant subspaces of A are $\mathscr{L}_j = \text{Span}\{e_1, \ldots, e_j\}, j = 0, \ldots, n$ (by definition $\mathscr{L}_0 = \{0\}$). Let us find all transformations $B: \mathbb{C}^n \to \mathbb{C}^n$ for which $\text{Inv}(A) = \text{Inv}(B)$. It turns out that $\text{Inv}(A) = \text{Inv}(B)$ if and only if (in the basis $e_1, \ldots, e_n$) B has the form

$$B = \begin{bmatrix} a_{11} & a_{12} & a_{13} & \cdots & a_{1n} \\ 0 & a_{22} & a_{23} & \cdots & a_{2n} \\ 0 & 0 & a_{33} & \cdots & a_{3n} \\ \multicolumn{5}{c}{\dotfill} \\ 0 & 0 & 0 & \cdots & a_{nn} \end{bmatrix} \qquad (10.2.1)$$

where

$$a_{11} = \cdots = a_{nn} \quad \text{and} \quad a_{12}a_{23} \cdots a_{n-1,n} \neq 0 \qquad (10.2.2)$$

Indeed, suppose $\text{Inv}(B) = \text{Inv}(A)$. Then clearly the matrix representing B has the triangular form (10.2.1). Moreover, it is easy to see that $a_{11} = \cdots = a_{nn}$. Indeed, the numbers $a_{11}, \ldots, a_{nn}$ are the eigenvalues of B; if they are not all equal, then there exists a pair of nonzero complemented invariant subspaces of B, namely, the root subspaces corresponding to a pair of complemented nonempty subsets in $\sigma(B)$. But the existence of a pair of nonzero complemented subspaces contradicts the assumption that $\text{Inv}(B) = \text{Inv}(A)$.

Let us show that $a_{12} a_{23} \cdots a_{n-1,n} \neq 0$. Consider the transformation $C = B - a_{11}I$, which has the same invariant subspaces as B. If for some j $(1 \leq j \leq n-1)$ we have $a_{j,j+1} = $, then $C\mathcal{L}_{j+1} \subset \mathcal{L}_{j-1}$. Hence

$$\dim \text{Ker } C \geq 2 \qquad (10.2.3)$$

Since any nonzero vector in $\text{Ker } C$ spans a one-dimensional C-invariant subspace, inequality (10.2.3) contradicts the assumption $\text{Inv}(B) = \text{Inv}(A)$ again.

Conversely, suppose that B satisfies (10.2.1) and (10.2.2). Put $C = B - a_{11}I$. We show that $\text{Ker } C = \mathcal{L}_1$. Let $x = \Sigma_{j=1}^{n} \xi_j e_j \in \text{Ker } C$ and $x \neq 0$. Let p be such that $\xi_{p+1} = \cdots = \xi_n = 0$ and $\xi_p \neq 0$. Then $p = 1$. Indeed, if p were greater than 1, then $Cx = a_{p,p+1}e_{p+1} + \cdots \neq 0$. So $x = \xi_1 e_1$, that is, $\text{Ker } C = \mathcal{L}_1$. This means that any two eigenvectors of B are collinear. Appeal to Theorem 2.5.1 $[(d) \Leftrightarrow (e)]$ and deduce that for any two B-invariant subspaces $\mathcal{M}_1$ and $\mathcal{M}_2$, either $\mathcal{M}_1 \subset \mathcal{M}_2$ or $\mathcal{M}_2 \subset \mathcal{M}_1$. Since $\mathcal{L}_0, \mathcal{L}_1, \ldots, \mathcal{L}_n$ are B invariant and $\dim \mathcal{L}_j = j$ $(j = 0, \ldots, n)$, it follows that any B-invariant subspace coincides with one of $\mathcal{L}_j$. $\square$

Example 10.2.1 provides a situation when $\text{Inv}(A) = \text{Inv}(B)$ but A and B do not commute [take $A = J_n(\lambda_0)$, $n \geq 3$, and B as in (10.2.1) with distinct nonzero numbers $a_{j,j+1}$, $j = 1, \ldots, n-1$].

If A has more than one Jordan block, the situation may be completely different from Example 10.2.1.

EXAMPLE 10.2.2.   Let

$$A = J_3(0) \oplus J_2(0)$$

It turns out that $\text{Inv}(B) = \text{Inv}(A)$ if and only if B is a polynomial in A, $B = p(A)$, such that $p'(\lambda) \neq 0$. In other words, B has the form

$$B = \begin{bmatrix} a & b & c \\ 0 & a & b \\ 0 & 0 & a \end{bmatrix} \oplus \begin{bmatrix} a & b \\ 0 & a \end{bmatrix} \qquad (10.2.4)$$

for some $a, b, c \in \mathbb{C}$ where $b \neq 0$.

As by Theorem 2.11.3 $\text{Inv}(B) = \text{Inv}(A)$ for every B in the form (10.2.4) with $b \neq 0$, we must verify only that every $B: \mathbb{C}^5 \to \mathbb{C}^5$ such that $\text{Inv}(B) = \text{Inv}(A)$ has the form (10.2.4) with $b \neq 0$ (in the basis e_1, e_2, e_3, e_4, e_5).

So assume $\text{Inv}(B) = \text{Inv}(A)$. Then clearly B has upper triangular form, and (see the argument in Example 10.2.1) the elements on the main diagonal of B are all equal. Without loss of generality, we can assume that the main diagonal in B is zero:

$$B = \begin{bmatrix} 0 & a_{12} & a_{13} & a_{14} & a_{15} \\ 0 & 0 & a_{23} & a_{24} & a_{25} \\ 0 & 0 & 0 & a_{34} & a_{35} \\ 0 & 0 & 0 & 0 & a_{45} \\ 0 & 0 & 0 & 0 & 0 \end{bmatrix}$$

As $\text{Span}\{e_4, e_5\}$ is A invariant and hence belongs to $\text{Inv}(B)$, we have $a_{14} = a_{15} = a_{24} = a_{25} = a_{34} = a_{35} = 0$. If one of the numbers a_{12}, a_{23}, or a_{45} were zero, then B would have three one-dimensional invariant subspaces whose sum is direct. This contradicts the assumption $\text{Inv}(B) = \text{Inv}(A)$ (A cannot have more than two one-dimensional invariant subspaces whose sum is direct). Hence a_{12}, a_{23}, and a_{45} are different from zero. It remains to show that $a_{12} = a_{23} = a_{45}$. To this end observe that $\text{Span}\{e_1 + e_4, e_2 + e_5\}$ is A invariant and hence B invariant. So

$$B(e_2 + e_5) = a_{12}e_1 + a_{45}e_4 \in \text{Span}\{e_1 + e_4, e_2 + e_5\}$$

which implies $a_{12} = a_{45}$. A similar analysis of the B-invariant subspace $\text{Span}\{e_1, e_2 + e_4, e_3 + e_5\}$ leads to the conclusion that $a_{23} = a_{45}$. $\square$

Now we state the main theorem, which describes all transformations $B: \mathbb{C}^n \to \mathbb{C}^n$ with $\text{Inv}(B) = \text{Inv}(A)$, where the transformation $A: \mathbb{C}^n \to \mathbb{C}^n$ is given. This description will contain the results of Examples 10.2.1 and 10.2.2 as very special cases. Note that without loss of generality we can assume (and we do) that A is an $n \times n$ matrix in the Jordan form

$$A = \text{diag}[A_1, A_2, \ldots, A_k]$$

where $\sigma(A_j) = \{\lambda_j\}$, $\lambda_1, \ldots, \lambda_k$ are all the different eigenvalues of A, and

$$A_j = \text{diag}[J_{p_1}(\lambda_j), \ldots, J_{p_m}(\lambda_j)]$$

where $p_1 \geq \cdots \geq p_m$. Of course, the number m, as well as $p_1, \ldots, p_m$, depend on j; we suppress this dependence in the notation for the sake of clarity. The notation for upper triangular Toeplitz matrices will be abbreviated to the form

$$T_q(a_0, \ldots, a_{q-1}) = \begin{bmatrix} a_0 & a_1 & a_2 & \cdots & a_{q-2} & a_{q-1} \\ 0 & a_0 & a_1 & \cdots & a_{q-3} & a_{q-2} \\ \vdots & \vdots & \vdots & & \vdots & \vdots \\ 0 & 0 & 0 & & a_0 & a_1 \\ 0 & 0 & 0 & \cdots & 0 & a_0 \end{bmatrix}$$

Finally, we use the notation

$$U_q(a_0, \ldots, a_{p-1}; F) =$$

$$\begin{bmatrix} a_0 & a_1 & a_2 & \cdots & a_{p-1} & f_{11-} & f_{12} & \cdots & f_{1,q-p-1} & f_{1,q-p} \\ 0 & a_0 & a_1 & \cdots a_{p-2} & a_{p-1} & f_{22} & \cdots & f_{2,q-p-1} & f_{2,q-p} \\ 0 & 0 & a_0 & \cdots a_{p-3} & a_{p-2} & a_{p-1} & \cdots & f_{3,q-p-1} & f_{3,q-p} \\ \vdots & \vdots & \vdots & & & & & \vdots & \vdots \\ 0 & 0 & 0 & & & \cdots & & a_{p-1} & f_{q-p,q-p} \\ 0 & 0 & 0 & & & \cdots & & a_{p-2} & a_{p-1} \\ \vdots & \vdots & \vdots & & & & & \vdots & \vdots \\ 0 & 0 & 0 & & & \cdots & & 0 & a_0 \end{bmatrix}$$

where F is the $(q - p) \times (q - p)$ upper triangular matrix whose (i, j) entry is f_{ij} $(i \le j)$. It is assumed, of course, that $p \le q$. In other words, $U_q(a_0, \ldots, a_{p-1}; F)$ is a $q \times q$ matrix whose first p superdiagonals (starting from the main diagonal) have the structure of a Toeplitz matrix, whereas the next $q - p$ superdiagonals contain the upper triangular part of the matrix F, which is not necessarily Toeplitz. If $p = q$, F is empty and $U_p(a_0, \ldots, a_{p-1}; F) = T_p(a_0, \ldots, a_{p-1})$.

Theorem 10.2.1

If $\text{Inv}(B) = \text{Inv}(A)$ *for a transformation* $B: \mathbb{C}^n \to \mathbb{C}^n$, *then*

$$B = \text{diag}[B_1, \ldots, B_k] \tag{10.2.5}$$

(in a chosen Jordan basis for A*), where each block* $B_j = B|_{\mathcal{R}_{\lambda_j}(A)} (j = 1, \ldots, k)$ *has the form*

$$B_j = U_{p_1}(\mu_j, b_2, \ldots, b_{p_2}; F) \oplus T_{p_2}(\mu_j, b_2, \ldots, b_{p_2}) \oplus \cdots$$
$$\oplus T_{p_m}(\mu_j, b_2, \ldots, b_{p_m}) \tag{10.2.6}$$

for some complex numbers $\mu_1, \ldots, \mu_k$, $b_2, \ldots, b_{p_2}$ *with* $\mu_i \ne \mu_j$ $(i \ne j)$, $b_2 \ne 0$ *and an upper triangular matrix* F *of size* $(p_1 - p_2) \times (p_1 - p_2)$; *the numbers* $b_2, \ldots, b_{p_2}$, *as well as the matrix* F, *depend on* j. *Conversely, if* B *has the form (10.2.5), (10.2.6) and* μ_i, b_j *and* F *have the above properties, then* $\text{Inv}(B) = \text{Inv}(A)$.

We relegate the lengthy proof of this theorem to the next section. The proof will be based on the description of irreducible subspaces obtained in Section 10.1.

We conclude this section with two corollaries of Theorem 10.2.1.

Corollary 10.2.2

Suppose that $AB = BA$. Then $\text{Inv}(A) = \text{Inv}(B)$ if and only if $B = f(A)$, where $f(\lambda)$ is a polynomial such that $f(\lambda_i) \neq f(\lambda_j)$ for eigenvalues $\lambda_i \neq \lambda_j$ of A, $f'(\lambda_0) \neq 0$ whenever $\lambda_0 \in \sigma(A)$ and $\text{Ker}(A - \lambda_0 I) \neq \mathcal{R}_{\lambda_0}(A)$.

In other words, the conditions of Theorem 2.11.3 are not only sufficient, but also necessary, provided A and B commute.

Proof. In view of Theorem 2.11.3 it is necessary to prove merely the "only if" statement. So assume $\text{Inv}(A) = \text{Inv}(B)$. Let $\lambda_1, \ldots, \lambda_k$ be the different eigenvalues of A, and let

$$\mathcal{R}_j \overset{\text{def}}{=} \mathcal{R}_{\lambda_j}(A) = \mathcal{L}_{j1} + \cdots + \mathcal{L}_{jm_j}$$

be the decomposition of $\mathcal{R}_{\lambda_j}(A)$ into a direct sum of Jordan subspaces $\mathcal{L}_{j1}, \ldots, \mathcal{L}_{j,m_j}$ such that $\dim \mathcal{L}_{j1} \geq \cdots \geq \dim \mathcal{L}_{j,m_j}$. The restrictions $A|_{\mathcal{L}_{j1}}$ and $B|_{\mathcal{L}_{j1}}$ commute; so in view of Theorem 9.1.1 (observing that $A|_{\mathcal{L}_{j1}}$ has only one Jordan block) there exists a polynomial $p_j(\lambda)$ such that $B|_{\mathcal{L}_{j1}} = p_j(A|_{\mathcal{L}_{j1}})$. It follows now from Theorem 10.2.1 that $B|_{\mathcal{R}_j} = p_j(A|_{\mathcal{R}_j})$. Since the minimal polynomials of $A|_{\mathcal{R}_j}$, $j = 1, \ldots, k$ are relatively prime, there exists a polynomial $p(\lambda)$ such that $B = p(A)$. Indeed, let $p(\lambda)$ be an interpolating polynomial such that $p(\lambda_j) = p_j(\lambda_j)$; $p'(\lambda_j) = p_j'(\lambda_j); \ldots; p^{(k_j-1)}(\lambda_j) = p^{(k_j-1)}(\lambda_j)$, $j = 1, \ldots, k$, where $k_j = \dim \mathcal{L}_{j1}$ and $q^{(\alpha)}(\lambda_0)$ denotes the αth derivative of the polynomial $q(\lambda)$ evaluated at λ_0. (See Gantmacher (1959), Lancaster and Tismenetsky (1985), for example, for information on interpolating polynomials (see also Section 2.10).)

From the definition of a function of the matrix A (see Section 2.10), it follows that $B|_{\mathcal{R}_j} = p(A|_{\mathcal{R}_j})$ for $j = 1, \ldots, k$ and, consequently, $B = p(A)$. Using Theorem 10.2.1 once more, we deduce that $p(\lambda_i) \neq p(\lambda_j)$ for $i \neq j$ and $p'(\lambda_i) \neq 0$ for $i = 1, \ldots, k$. $\square$

Corollary 10.2.3

Let $A: \mathbb{C}^n \to \mathbb{C}^n$ be a transformation. Then every transformation B with $\text{Inv}(B) = \text{Inv}(A)$ commutes with A if and only if the following condition holds: for every eigenvalue λ_0 of A with $\text{Ker}(A - \lambda_0 I) \neq \mathcal{R}_{\lambda_0}(A)$ and $\dim \text{Ker}(A - \lambda_0 I) > 1$ we have

$$\dim \mathcal{R}_{\lambda_0}(A) - \dim \text{Ker}(A - \lambda_0 I)^p \geq 2$$

where $p = p(\lambda_0)$ is the maximal integer such that $\text{Ker}(A - \lambda_0 I)^p \neq \mathcal{R}_{\lambda_0}(A)$. Further, the set of all transformations B with $\text{Inv}(B) = \text{Inv}(A)$ coincides with the set of all transformations commuting with A if and only if $\dim \text{Ker}(A - \lambda_0 I) = 1$ for every eigenvalue λ_0 of A, that is, A is nonderogatory.

The proof is obtained by combining Theorem 10.2.1 with the description of all matrices commuting with A (Theorem 9.1.1).

10.3   PROOF OF THEOREM 10.2.1

We start with three lemmas to be used in the proof of Theorem 10.2.1.

Let $A: \mathbb{C}^n \to \mathbb{C}^n$ be a unicellular transformation. (Recall that A is called *unicellular* if $\mathbb{C}^n$ is its irreducible subspace.) Let $g_1, \ldots, g_n$ be a Jordan basis of A. Let B be a transformation such that its matrix in the basis $g_1, \ldots, g_n$ has the form

$$B = U_n(b_1, \ldots, b_k; F) \qquad (10.3.1)$$

for some $b_i \in \mathbb{C}$ and an $(n - k) \times (n - k)$ upper triangular matrix F.

Lemma 10.3.1

If B has the form (10.3.1) with $b_2 \neq 0$, then in any Jordan basis for B the transformation A has the form

$$A = U_n(a_1, \ldots, a_k; G) \qquad (10.3.2)$$

for some $a_i \in \mathbb{C}$ with $a_2 \neq 0$, and some upper triangular matrix G.

Proof.   Without loss of generality we can assume that $\sigma(A) = \{0\}$ and the Jordan basis $g_1, \ldots, g_n$ coincides with the standard basis: $g_i = e_i$, $i = 1, \ldots, n$. Let $B = B_1 + C$, where

$$B_1 = T_n(b_1, \ldots, b_n), \qquad C = U_n(0, \ldots, 0; F') \qquad (10.3.3)$$

for some $b_{k+1}, \ldots, b_n \in \mathbb{C}$ and upper triangular matrix F' of size $(n - k) \times (n - k)$. Since $b_2 \neq 0$, it follows from Example 10.2.1 that the transformations A, B, and B_1 have the same invariant subspaces. Hence (recalling the equivalence (a) $\Leftrightarrow$ (e) in Theorem 2.5.1) the transformations B and B_1 are also unicellular.

As $AB_1 = B_1 A$ and B_1 is unicellular, it follows from Theorem 9.1.1 that $A = p(B_1)$ for some polynomial $p(\lambda)$.

Let $f_1, \ldots, f_n$ be a Jordan basis for B. We claim that the matrix of C in the basis $f_1, \ldots, f_n$ has the form (10.3.3) again, possibly with another matrix F'. Indeed, the only nonzero B-invariant subspaces are

$\mathrm{Span}\{e_1, e_2, \ldots, e_i\}$, $i = 1, \ldots, n$ (because they are A-invariant subspaces). On the other hand, Example 10.2.1 ensures that the only nonzero B-invariant subspaces are $\mathrm{Span}\{f_1, f_2, \ldots, f_i\}$, $i = 1, \ldots, n$. It follows that $f_i \in \mathrm{Span}\{e_1, e_2, \ldots, e_i\}$, $i = 1, \ldots, n$. Now it is easily seen that the matrix of C in the basis $f_1, \ldots, f_n$ has the form (10.3.3).

Consider the following relations

$$A = p(B_1) = p(B - C) = \sum_{j=0}^{m} \alpha_j (B - C)^j = \sum_{j=0}^{m} \alpha_j B^j + H \quad (10.3.4)$$

where every summand in H contains C as a factor. Consequently, the matrix of H in the basis $f_1, \ldots, f_n$ is upper triangular and the first k diagonals (counting from the main diagonal) are zeros. Now (10.3.2) follows from (10.3.4) and, by Example 10.2.1, $a_2 \neq 0$. $\quad \square$

Let vectors $d_1, \ldots, d_p, f_1, \ldots, f_{p-1}$ be linearly independent in $\mathbb{C}^n$. In the sequel we shall encounter systems of vectors of the form

$$
\begin{aligned}
g_p &= d_p + \alpha f_{p-1} + \beta f_{p-2} + \cdots + \gamma f_2 + \delta f_1 \\
g_{p-1} &= d_{p-1} + \alpha f_{p-2} + \beta f_{p-3} + \cdots + \gamma f_1 \\
&\cdots\cdots\cdots\cdots\cdots\cdots\cdots\cdots\cdots \\
g_3 &= d_3 + \alpha f_2 + \beta f_1 \\
g_2 &= d_2 + \alpha g_1 \\
g_1 &= d_1
\end{aligned}
\quad (10.3.5)
$$

where $\alpha, \beta, \ldots, \gamma, \delta$ are some numbers, and

$$
\begin{aligned}
h_p &= d_p + a_{p,p-1} f_{p-1} + a_{p,p-2} f_{p-2} + \cdots + a_{p2} f_2 + a_{p1} f_1, \\
h_{p-1} &= d_{p-1} + a_{p-1,p-2} f_{p-2} + a_{p-1,p-3} f_{p-3} + \cdots + a_{p-1,1} f_1, \\
&\cdots\cdots\cdots\cdots\cdots\cdots\cdots\cdots\cdots\cdots\cdots\cdots \\
h_3 &= d_3 + a_{32} f_2 + a_{31} f_1, \\
h_2 &= d_2 + a_{21} f_1, \\
h_1 &= d_1
\end{aligned}
\quad (10.3.6)
$$

where a_{ij} are certain numbers.

Lemma 10.3.2

If for every $m = 1, \ldots, p$ the subspaces $\mathrm{Span}\{g_1, \ldots, g_m\}$ and $\mathrm{Span}\{h_1, \ldots, h_m\}$ coincide, then $g_j = h_j$ ($j = 1, \ldots, p$).

Proof. Use induction on p. For $p = 1$ the lemma is evident. Assume the lemma holds true for $p = k$, and $\mathrm{Span}\{g_1, \ldots, g_{k+1}\} =$

Span$\{h_1, \ldots, h_{k+1}\}$. By the induction hypothesis, $g_1 = h_1, \ldots, g_k = h_k$. For every vector $x = \sum_{j=1}^{k+1} \xi_j g_j \in \text{Span}\{g_1, \ldots, g_{k+1}\}$ we have $x = \sum_{j=1}^{k+1} \eta_j h_j$. Rewrite the equation

$$\sum_{j=1}^{k+1} \xi_j g_j = \sum_{j=1}^{k+1} \eta_j h_j$$

in the form

$$\sum_{j=1}^{k} (\xi_j - \eta_j) g_j + \xi_{k+1} g_{k+1} - \eta_{k+1} h_{k+1} = 0 .$$

If $\xi_{k+1} \neq \eta_{k+1}$ or $\xi_{k+1} = \eta_{k+1}$ and $g_{k+1} \neq h_{k+1}$, this will contradict the linear independence of $d_1, \ldots, d_{k+1}, f_1, \ldots, f_k$. So we must have $g_{k+1} = h_{k+1}$. $\square$

Let $\mathcal{J}_A$ be the set of all irreducible subspaces of a transformation $A \colon \mathbb{C}^n \to \mathbb{C}^n$. Clearly, $\mathcal{J}_A \subset \text{Inv}(A)$. Since every invariant subspace for a transformation can be represented as a direct sum of irreducible subspaces, the equality $\text{Inv}(A) = \text{Inv}(B)$ holds if and only if $\mathcal{J}_A = \mathcal{J}_B$. Now consider a special case of this equality.

Lemma 10.3.3

Let $A \colon \mathbb{C}^n \to \mathbb{C}^n$ be such that

$$\mathbb{C}^n = \mathcal{L}_1 \dot{+} \mathcal{L}_2 ,$$

where $\mathcal{L}_i$ $(i = 1, 2)$ are irreducible subspaces of A corresponding to the same eigenvalue. Let $\dim \mathcal{L}_1 = q \geq p = \dim \mathcal{L}_2$, and let $d_1, \ldots, d_q \in \mathcal{L}_1$; $f_1, \ldots, f_p \in \mathcal{L}_2$ be Jordan bases in these subspaces. Then for $B \colon \mathbb{C}^n \to \mathbb{C}^n$ we have $\mathcal{J}_A = \mathcal{J}_B$ if and only if the matrix of B in the basis $d_1, \ldots, d_q$, $f_1, \ldots, f_p$ has the form

$$B = U_q(b_1, \ldots, b_p; F) \oplus T_p(b_1, \ldots, b_p) \qquad (10.3.7)$$

where $b_1, \ldots, b_q$ are complex numbers with $b_2 \neq 0$ and F is an upper triangular matrix of size $(q - p) \times (q - p)$.

Proof. First we prove the necessity, that is, if $\mathcal{J}_A = \mathcal{J}_B$, then B has the form (10.3.7). Consider first the case $p = q$ and prove the necessity by induction on p. For $p = 1$ everything is evident. Suppose that the lemma is true for $p = k$, and let $\mathcal{L}_1, \mathcal{L}_2$ be irreducible subspaces of dimension $k + 1$. Let $\mathcal{L}_1' = \text{Span}\{d_1, \ldots, d_k\}$; $\mathcal{L}_2' = \text{Span}\{f_1, \ldots, f_k\}$. Evidently, $\mathcal{L}_1'$ and $\mathcal{L}_2'$ are irreducible subspaces of A corresponding to the same eigenvalue. Since

$\mathcal{I}_A = \mathcal{I}_B$ by assumption, the subspaces $\mathcal{L}_1$, $\mathcal{L}_1'$, $\mathcal{L}_2$, $\mathcal{L}_2'$ are irreducible subspaces for B. By Example 10.2.1 and the induction hypothesis, the matrix representation of B in the basis $d_1, \ldots, d_{k+1}, f_1, \ldots, f_{k+1}$ has the form

$$
B = \begin{bmatrix}
b_1 & b_2 & b_3 & \cdots & b_k & c_{1,k+1} \\
0 & b_1 & b_2 & \cdots & b_{k-1} & c_{2,k+1} \\
0 & 0 & b_1 & \cdots & b_{k-2} & c_{3,k+1} \\
\multicolumn{6}{c}{\cdots\cdots\cdots\cdots\cdots\cdots\cdots\cdots} \\
0 & 0 & 0 & \cdots & b_1 & c_{k,k+1} \\
0 & 0 & 0 & \cdots & 0 & b_1
\end{bmatrix}
\oplus
\begin{bmatrix}
b_1 & b_2 & b_3 & \cdots & b_k & a_{1,k+1} \\
0 & b_1 & b_2 & \cdots & b_{k-1} & a_{2,k+1} \\
0 & 0 & b_1 & \cdots & b_{k-2} & a_{3,k+1} \\
\multicolumn{6}{c}{\cdots\cdots\cdots\cdots\cdots\cdots\cdots\cdots} \\
0 & 0 & 0 & \cdots & b_1 & a_{k,k+1} \\
0 & 0 & 0 & \cdots & 0 & b_1
\end{bmatrix}
$$

where $b_2, c_{k,k+1}, a_{k,k+1} \neq 0$.

We assume $b_1 \neq 0$; otherwise, consider the linear transformation $B + \lambda_0 I$, where $\lambda_0 \neq -b_1$, in place of B and use the property that $\text{Inv}(B) = \text{Inv}(B + \lambda_0 I)$. This condition means that B is invertible.

Let $\mathcal{L}$ be an irreducible subspace of A such that $\dim \mathcal{L} = k + 1$ and $d_1 \in \mathcal{L}$. By Theorem 10.1.3, there exist numbers $\alpha_1, \ldots, \alpha_k$ such that the vectors

$$
\begin{aligned}
x_{k+1} &= d_{k+1} + \alpha_1 f_k & + \alpha_2 f_{k-1} + \cdots + \alpha_{k-1} f_2 + \alpha_k f_1 \\
x_k &= d_k & + \alpha_1 f_{k-1} + \alpha_2 f_{k-2} + \cdots + \alpha_{k-1} f_1 \\
&\multicolumn{1}{c}{\cdots\cdots\cdots\cdots\cdots\cdots\cdots\cdots} \\
x_3 &= d_3 & + \alpha_1 f_2 + \alpha_2 f_1 \\
x_2 &= d_2 & + \alpha_1 f_1 \\
x_1 &= d_1
\end{aligned}
$$

form a Jordan basis in $\mathcal{L}$. Since $b_1 \neq 0$, it follows that

$$
\text{Span}\{x_1, \ldots, x_k, x_{k+1}\} = \text{Span}\{x_1, \ldots, x_k, Bx_{k+1}\}
$$

It follows from the form of B that

$$
Bx_{k+1} = b_1 d_{k+1} + \sum_{j=1}^{k} c_{j,k+1} d_j + \alpha_1 \left[\sum_{j=1}^{k} b_{k+1-j} f_j \right] + \cdots + \alpha_k b_1 f_1
$$

$$
= b_1 d_{k+1} + \sum_{j=1}^{k} c_{j,k+1} d_j + \alpha_1 b_1 f_k + (\alpha_1 b_2 + \alpha_2 b_1) f_{k-1} + \cdots
$$

$$
+ \left[\sum_{j=1}^{k} \alpha_j b_{k+1-j} \right] f_1
$$

and

$$Bx_{k+1} - \sum_{j=1}^{k} c_{j,k+1}x_j = b_1 d_{k+1} + \alpha_1 b_1 f_k + [\alpha_1(b_2 - c_{k,k+1}) + \alpha_2 b_1]f_{k-1} + \cdots$$

$$+ \left[\sum_{j=1}^{k-1} \alpha_j(b_{k+1-j} - c_{j+1,k+1}) + \alpha_k b_1 \right] f_1$$

Put

$$y = b_1^{-1} \left(Bx_{k+1} - \sum_{j=1}^{k} c_{j,k+1}x_j \right) = d_{k+1} + \alpha_1 f_k + \cdots$$

Evidently, $\text{Span}\{x_1, \ldots, x_k, x_{k+1}\} = \text{Span}\{x_1, \ldots, x_k, y\}$. Then by Lemma 10.3.2 we have

$$\alpha_1 b_1^{-1}(b_2 - c_{k,k+1}) + \alpha_2 = \alpha_2; \ldots; \sum_{j=1}^{k-1} \alpha_j b_1^{-1}(b_{k+1-j} - c_{j+1,k+1}) + \alpha_k = \alpha_k$$

These equalities hold for every $\alpha_1, \ldots, \alpha_k$ (by choosing all possible $\mathcal{L}$; see Theorem 10.1.3). Therefore

$$b_2 = c_{k,k+1}, b_3 = c_{k-1,k+1}, \ldots, b_{k-1} = c_{3,k+1}, b_k = c_{2,k+1}$$

Similarly, considering Jordan bases of the form

$$f_{k+1} + \alpha_1 d_k \quad + \alpha_2 d_{k-1} + \cdots + \alpha_{k-1} d_2 + \alpha_k d_1$$

$$f_k \quad + \alpha_1 d_{k-1} + \alpha_2 d_{k-2} + \cdots + \alpha_{k-1} d_1$$

$$\cdots \cdots \cdots \cdots \cdots \cdots \cdots \cdots \cdots \cdots \cdots \cdots \cdots \cdots$$

$$f_3 \quad + \alpha_1 d_2 \quad + \alpha_2 d_1$$

$$f_2 \quad + \alpha_1 d_1$$

$$f_1$$

we obtain $b_2 = a_{k,k+1}, b_3 = a_{k-1,k+1}, \ldots, b_{k-1} = a_{3,k+1}, b_k = a_{2,k+1}$. Let us show that, in fact, $c_{1,k+1} = a_{1,k+1}$. To this end consider a Jordan basis of A of the form

$$\{z_j = \xi d_j + \eta f_j\}_{j=1}^{k+1}$$

where ξ and η are arbitrary numbers. We have

$$Bz_{k+1} = \sum_{j=2}^{k+1} b_{k+1-j}z_j + \xi c_{1,k+1}d_1 + \eta a_{1,k+1}f_1$$

$$= \sum_{j=2}^{k+1} b_{k+1-j}z_j + c_{1,k+1}z_1 + \eta(a_{1,k+1} - c_{1,k+1})f_1$$

As above, we obtain

$$\mathrm{Span}\{z_1, \ldots, z_k, z_{k+1}\} = \mathrm{Span}\{z_1, \ldots, x_k, Bz_{k+1}\}$$

Further,

$$Bz_{k+1} - \sum_{j=2}^{k} b_{k+1-j}z_j - c_{1,k+1}z_1 = b_1 z_{k+1} + \eta(a_{1,k+1} - c_{1,k+1})f_1$$

Put $y = z_{k+1} + \eta b_1^{-1}(a_{1,k+1} - c_{1,k+1})f_1$. Evidently,

$$\mathrm{Span}\{z_1, \ldots, z_k, z_{k+1}\} = \mathrm{Span}\{z_1, \ldots, z_k, y\}$$

By Lemma 10.3.2, $\eta b_1^{-1}(a_{1,k+1} - c_{1,k+1}) = 0$. Since η can be arbitrary, $a_{1,k+1} = c_{1,k+1}$. Thus the necessity part of Lemma 10.3.3 is proved for the case $p = q$.

Now consider the case $q > p$. Put $h = q - p$ and proceed by induction on h. Assume that the necessity part of Lemma 10.3.3 holds for $h \leq k$, and let $\mathscr{L}_1$, $\mathscr{L}_2$ be irreducible subspaces for A with dim $\mathscr{L}_1 = p + k + 1$ and dim $\mathscr{L}_2 = p$. By Example 10.2.1 and the assumption of induction, the matrix representation of B in the basis $d_1, \ldots, d_{p+k-1}, f_1, \ldots, f_p$ has the form

$$B = \begin{bmatrix} b_1 & b_2 & b_3 & \cdots & b_p & c_{1,p+1} & \cdots & c_{1,p+k} & c_{1,p+k+1} \\ 0 & b_1 & b_2 & \cdots & b_{p-1} & b_p & \cdots & c_{2,p+k} & c_{2,p+k+1} \\ 0 & 0 & b_1 & \cdots & b_{p-2} & b_{p-1} & \cdots & c_{3,p+k} & c_{3,p+k+1} \\ \cdots & \cdots & \cdots & \cdots & \cdots & \cdots & \cdots & \cdots & \cdots \\ 0 & 0 & 0 & \cdots & \cdots & \cdots & \cdots & b_p & c_{k,p+k+1} \\ 0 & 0 & 0 & \cdots & \cdots & \cdots & \cdots & b_{p-1} & c_{k+1,p+k+1} \\ \cdots & \cdots & \cdots & \cdots & \cdots & \cdots & \cdots & \cdots & \cdots \\ 0 & 0 & 0 & \cdots & \cdots & \cdots & \cdots & b_1 & c_{p+k,p+k+1} \\ 0 & 0 & 0 & \cdots & \cdots & \cdots & \cdots & 0 & b_1 \end{bmatrix}$$

$$\oplus \begin{bmatrix} b_1 & b_2 & b_3 & \cdots & b_p \\ 0 & b_1 & b_2 & \cdots & b_{p-1} \\ 0 & 0 & b_1 & \cdots & b_{p-2} \\ \vdots & \vdots & \vdots & & \vdots \\ 0 & 0 & 0 & \cdots & b_1 \end{bmatrix}$$

where $b_2 \neq 0$. Let

$$\begin{aligned} x_{p+k+1} &= d_{p+k+1} + \alpha_1 f_p + \alpha_2 f_{p-1} + \cdots + \alpha_{p-1} f_2 + \alpha_p f_1 \\ x_{p+k} &= d_{p+k} + \alpha_1 f_{p-1} + \alpha_2 f_{p-2} + \cdots + \alpha_{p-1} f_1 \\ &\ \cdots\cdots\cdots\cdots\cdots\cdots\cdots\cdots\cdots\cdots \\ x_{k+2} &= d_{k+2} + \alpha_1 f_1 \\ x_{k+1} &= d_{k+1} \\ &\ \cdots\cdots\cdots \\ x_1 &= d_1 \end{aligned}$$

be a Jordan basis (for A) of an arbitrary irreducible subspace $\mathscr{L}$ of dimension $p + k + 1$ and such that $d_1 \in \mathscr{L}$. As above, we obtain

$$\mathrm{Span}\{x_1, x_2, \ldots, x_{p+k}, x_{p+k+1}\} = \mathrm{Span}\{x_1, \ldots, x_{p+k}, Bx_{p+k+1}\}$$

Now

$$Bx_{p+k+1} = b_1 d_{p+k+1} + \sum_{j=1}^{p+k} c_{j,p+k+1} d_j + \alpha_1 \left[\sum_{j=1}^{p} b_{p+1-j} f_j \right] + \cdots + \alpha_p b_1 f_1$$

$$= b_1 d_{p+k+1} + \sum_{j=1}^{p+k} c_{j,p+k+1} d_j + \alpha_1 b_1 f_p + (\alpha_1 b_2 + \alpha_2 b_1) f_{p-1} + \cdots$$

$$+ \left[\sum_{j=1}^{p} \alpha_j b_{p+1-j} \right] f_1$$

Hence

$$Bx_{p+k+1} - \sum_{j=1}^{p+k} c_{j,p+k+1} x_j = b_1 d_{p+k+1} + \alpha_1 b_1 f_p$$

$$+ [\alpha_1 (b_2 - c_{p+k,p+k+1}) + \alpha_2 b_1] f_{p-1} + \cdots$$

$$+ \left[\sum_{j=1}^{p-1} \alpha_j (b_{p+1-j} - c_{p+k+1-j,p+k+1}) + \alpha_p b_1 \right] f_1$$

Put

$$y = b_1^{-1} \left[Bx_{p+k+1} - \sum_{j=1}^{p+k} c_{j,p+k+1} x_j \right] = d_{p+k+1} + \alpha_1 f_p + \cdots$$

Since $\mathrm{Span}\{x_1, \ldots, x_{p+k}, x_{p+k+1}\} = \mathrm{Span}\{x_1, \ldots, x_{p+k}, y\}$ for every $\alpha_1, \ldots, \alpha_p$, Lemma 10.3.2 implies that

$$b_2 = c_{p+k,p+k+1}, \, b_3 = c_{p+k-1,p+k+1}, \ldots, b_p = c_{k+1,p+k+1}$$

The necessity part of Lemma 10.3.3 is proved.

Let us prove the sufficiency of the conditions of Lemma 10.3.3. Assume that B has the form (10.3.7) in a Jordan basis for A. Let $\mathscr{L}$ be an irreducible subspace for A with $\dim \mathscr{L} = k$ $(1 \le k \le p)$ and $x_1 \in \mathscr{L}$ be an eigenvector. Then $x_1 = \xi d_1 + \eta f_1$ for some numbers ξ and η. Put $x_j = \xi d_j + \eta f_j$ $(j = 2, \ldots, p)$. Suppose that $\eta \ne 0$. In view of Proposition 10.1.2 (see also the remark after its proof), there are some number $\alpha_1, \ldots, \alpha_{k-1}$ for which the vectors

$$v_k = x_k + \alpha_1 d_{k-1} + \cdots + \alpha_{k-2} d_2 + \alpha_{k-1} d_1$$

$$v_{k-1} = x_{k-1} + \alpha_1 d_{k-2} + \cdots + \alpha_{k-2} d_1 \qquad (10.3.8)$$

$$\cdots\cdots\cdots\cdots\cdots\cdots\cdots\cdots\cdots\cdots\cdots\cdots\cdots\cdots$$

$$v_2 = x_2 + \alpha_1 d_1$$

$$v_1 = x_1$$

form a Jordan basis of A in $\mathscr{L}$. [If $\eta = 0$, replace $d_1, \ldots, d_{k-1}$ by $f_1, \ldots, f_{k-1}$ respectively in (10.3.8).] A straightforward computation reveals that $\mathscr{L}$ is B invariant, and in the basis $v_1, \ldots, v_k$ we have:

$$B|_{\mathscr{L}} = T_k(b_1, b_2, \ldots, b_k) \qquad (10.3.9)$$

As in Example 10.2.1, $b_2 \neq 0$ implies that $\mathscr{L}$ is an irreducible subspace for B.

Now let $\mathscr{L}$ be an irreducible subspace for A such that $\dim \mathscr{L} = m$ ($p + 1 \leq m \leq q$). It is easily seen that $d_1 \in \mathscr{L}$ and (by Proposition 10.1.2) there exist numbers $\alpha_1, \ldots, \alpha_p$ such that the vectors

$$u_m = d_m + \alpha_1 f_p + \cdots + \alpha_{p-1} f_2 + \alpha_p f_1$$

$$u_{m-1} = d_{m-1} + \alpha_1 f_{p-1} + \cdots + \alpha_{p-1} f_1$$

$$\cdots\cdots\cdots\cdots\cdots\cdots\cdots\cdots\cdots\cdots\cdots\cdots$$

$$u_{m-p+1} = d_{m-p+1}$$

$$\cdots\cdots\cdots\cdots\cdots$$

$$u_1 = d_1$$

form a Jordan basis of A in $\mathscr{L}$. Again, a straightforward calculation shows that $\mathscr{L}$ is B invariant and in the basis $u_1, \ldots, u_m$

$$B|_{\mathscr{L}} = U_m(b_1, \ldots, b_p; F_0) \qquad (10.3.10)$$

where the (i, j) entry of F_0 is $c_{i,j+p}$ $(i \leq j)$. Since $b_2 \neq 0$, it follows from Example 10.2.1 that the subspace $\mathscr{L}$ is an irreducible subspace of B. So we have proved that $\mathscr{J}_A \subset \mathscr{J}_B$.

Let us prove the opposite inclusion $\mathscr{J}_B \subset \mathscr{J}_A$. Let $g_1, \ldots, g_p$ be a Jordan basis of B in the subspace $\mathscr{L}_2$. Write $g_k = \sum_{i=1}^k \xi_{ki} f_i$ $(k = 1, \ldots, p)$; put $h_k = \sum_{i=1}^k \xi_{ki} d_i$ $(k = 1, \ldots, p)$. Evidently, the vectors $h_1, \ldots, h_p$ form a Jordan basis for B in $\mathscr{L}' = \mathrm{Span}\{d_1, \ldots, d_p\}$.

We show that the sequence $h_1, \ldots, h_p$ can be augmented by vectors $h_{p+1}, \ldots, h_q$ so that $h_1, \ldots, h_q$ is a Jordan basis of B in $\mathscr{L}_1$. (Observe that by Example 10.2.1, $\mathscr{L}_1$ is an irreducible subspace for B.) Assume that the vectors $h_j = \sum_{i=1}^j \xi_{ji} d_i$, $j = p + 1, \ldots, r - 1$ are already constructed. Then for $h_r = \sum_{i=1}^r \xi_{ri} d_i$ the following equation must be satisfied in order that $(B - b_1 I) h_r = h_{r-1}$:

$$Z_r \begin{bmatrix} \xi_{r2} \\ \vdots \\ \xi_{rr} \end{bmatrix} = \begin{bmatrix} \xi_{r-1,1} \\ \vdots \\ \xi_{r-1,r-1} \end{bmatrix} \tag{10.3.11}$$

where Z_r is the $(r-1) \times (r-1)$ submatrix of $B - b_1 I$ formed by the first $r-1$ rows and the columns $2, 3, \ldots, r-1, r$. From (10.3.7) and $b_2 \neq 0$ it follows that Z_r is invertible, so (10.3.11) always has a (unique) solution $\xi_{r2}, \ldots, \xi_{rr}$, and h_r is constructed. By Lemma 10.3.1, A has the following form in the basis $h_1, \ldots, h_q, g_1, \ldots, g_p$:

$$A = U_q(a_1, \ldots, a_p; F) \oplus T_p(a_1, \ldots, a_p)$$

for some F, where $a_2 \neq 0$. The first p diagonals in both blocks are the same in view of the choice of $h_1, \ldots, h_p$.

Now we can repeat the proof of the inclusion $\mathcal{J}_A \subset \mathcal{J}_B$ given above, with A and B interchanged. So $\mathcal{J}_B \subset \mathcal{J}_A$ follows and, therefore, also $\mathcal{J}_B = \mathcal{J}_A$ and $\mathrm{Inv}(B) = \mathrm{Inv}(A)$. $\quad \square$

Now we are prepared to prove Theorem 10.2.1 itself.

Proof of Theorem 10.2.1. As every A-invariant subspace is the sum of its intersections with the root subspaces of A, we may restrict ourselves to the case when $\mathbb{C}^n$ is a root subspace for A. Let

$$\mathbb{C}^n = \mathcal{L}_1 \dotplus \cdots \dotplus \mathcal{L}_m \tag{10.3.12}$$

be the decomposition of $\mathbb{C}^n$ into a direct sum of irreducible subspaces $\mathcal{L}_1, \ldots, \mathcal{L}_m$ of A. Let $d_1^{(1)}, \ldots, d_{p_1}^{(1)} \in \mathcal{L}_1; \ldots; d_1^{(m)}, \ldots, d_{p_m}^{(m)} \in \mathcal{L}_m$ be Jordan bases in $\mathcal{L}_1, \ldots, \mathcal{L}_m$, respectively. Assume (without loss of generality) that $p_1 \geq \cdots \geq p_m$.

Now let $B: \mathbb{C}^n \to \mathbb{C}^n$ be a transformation, and suppose that the invariant subspaces of B and those of A are the same. Applying Lemma 10.3.3 to the restrictions $A|_{\mathcal{L}_1 + \mathcal{L}_i}$, $i = 2, \ldots, m$, we find that B has the form described in Theorem 10.2.1.

Conversely, assume that B has the form

$$B = U_{p_1}(\mu, b_2, \ldots, b_{p_2}; F) \oplus T_{p_2}(\mu, b_2, \ldots, b_{p_2}) \oplus \cdots$$
$$\oplus T_{p_m}(\mu, b_2, \ldots, b_{p_m}) \tag{10.3.13}$$

with $b_2 \neq 0$. We now prove that $\mathrm{Inv}(B) = \mathrm{Inv}(A)$. Suppose for definiteness that $\sigma(A) = \{0\}$. Let us show that every irreducible subspace for A is also an irreducible subspace of B. Let $\mathcal{L}$ be an irreducible subspace for A with $\dim \mathcal{L} = j$, and let $x_1 \in \mathcal{L}$ be an eigenvector of A. Then $x_1 \in \mathrm{Span}\{d_1^{(1)}, \ldots, d_1^{(m)}\} = \mathrm{Ker}\, A$. Write $x = a_1 d_1^{(1)} + \cdots + a_r d_1^{(r)}$ with $a_r \neq 0$, for some r ($1 \leq r \leq m$). Put $x_i = a_1 d_i^{(1)} + \cdots + a_r d_i^{(r)}$, $i = 1, \ldots, p_r$. It is easily seen that $j = \dim \mathscr{L} \leq p_r$. Then the vectors $z_1, \ldots, z_j$ given by (10.1.12) (replacing $f_1, \ldots, f_j$ by $x_1, \ldots, x_j$) form a Jordan basis for A, for some numbers $\alpha_1, \alpha_2, \ldots$. Two possibilities occur for the number j ($=\dim \mathscr{L}$): $j \leq p_2$ or $p_2 + 1 \leq j \leq p_1$. Consider first the case $j \leq p_2$. Taking into account the form of B, it is easy to check that $\mathscr{L}$ is B invariant and the matrix of $B|_{\mathscr{L}}$ in the basis $z_1, \ldots, z_j$ is of the form

$$B|_{\mathscr{L}} = T_j(\mu, b_2, \ldots, b_j)$$

with $b_2 \neq 0$. Then by Example 10.2.1 $\mathscr{L}$ is an irreducible subspace for B.

Now suppose that $p_2 + 1 \leq j \leq p_1$. Since $j \leq p_r$, clearly $r = 1$. This means that the eigenvector $x_1 \in \mathscr{L}$ is collinear with $d_1 \in \mathscr{L}_1$. Taking into account the form of B [given by (10.3.13)] we conclude that $\mathscr{L}$ is B invariant and the matrix of $B|_{\mathscr{L}}$ in the basis $z_1, \ldots, z_j$ is given by (10.3.10) with $b_1 = \mu$ and $b_2 \neq 0$. By Example 10.2.1, $\mathscr{L}$ is an irreducible subspace for B.

We show that every irreducible subspace for B is also an irreducible subspace for A. As we have already proved, the subspaces $\mathscr{L}_1, \ldots, \mathscr{L}_m$ [which appear in (10.3.12)] are also irreducible subspaces for B. Let $\tilde{h}_1, \ldots, \tilde{h}_{p_m}$ be a Jordan basis of B in $\mathscr{L}_m$; then

$$\tilde{h}_k = \sum_{i=1}^{k} \xi_{ki} h_i \qquad (k = 1, \ldots, p_m)$$

where $h_1, \ldots, h_{p_m}$ is the Jordan basis of A in $\mathscr{L}_m$. Let $\varphi_1, \ldots, \varphi_{p_{m-1}}$ be a Jordan basis of A in $\mathscr{L}_{m-1}$. Construct the vectors $\tilde{\varphi}_1, \ldots, \tilde{\varphi}_{p_m}$ as follows (recall that $p_m \leq p_{m-1}$):

$$\tilde{\varphi}_k = \sum_{i=1}^{k} \xi_{ki} \varphi_i$$

Since the vectors $\tilde{h}_1, \ldots, \tilde{h}_{p_m}$ form a Jordan basis in $\mathscr{L}_m$, the vectors $\tilde{\varphi}_1, \ldots, \tilde{\varphi}_{p_m}$ satisfy the equalities $(B - \mu I)\tilde{\varphi}_1 = 0$; $(B - \mu I)\tilde{\varphi}_{j+1} = \tilde{\varphi}_j$ for $j = 2, \ldots, p_m$. Because $\mathscr{L}_{m-1}$ is an irreducible subspace for B, there exists vectors $\tilde{\varphi}_{p_m+1}, \ldots, \tilde{\varphi}_{p_{m-1}}$ such that the system $\tilde{\varphi}_1, \ldots, \tilde{\varphi}_{p_{m-1}}$ forms a Jordan basis for B in $\mathscr{L}_{m-1}$. (See the last paragraph in the proof of Lemma 10.3.3.) Express $\tilde{\varphi}_{p_m+1}, \ldots, \tilde{\varphi}_{p_{m-1}}$ by means of the Jordan basis for A in $\mathscr{L}_{m-1}$:

$$\tilde{\varphi}_k = \sum_{i=1}^{k} \xi_{ki} \varphi_i, \qquad k = p_m + 1, \ldots, p_{m-1}$$

Continuing these constructions, we obtain Jordan bases for B in each of the subspaces $\mathscr{L}_m, \mathscr{L}_{m-1}, \ldots, \mathscr{L}_1$. From the choice of these bases and Lemma 10.3.1 it follows that the matrix of A in the union of these bases has the form

$$A = U_{p_1}(\lambda, a_2, \ldots, a_{p_2}, F') \oplus T_{p_2}(\lambda, a_2, \ldots, a_{p_2}) \oplus \cdots$$
$$\oplus T_{p_m}(\lambda, a_2, \ldots, a_{p_m})$$

where λ is the eigenvalue of A and $a_2 \neq 0$. As it was proved above, every irreducible subspace for B is also irreducible for A. Thus the equality $\mathrm{Inv}(A) = \mathrm{Inv}(B)$ holds. $\square$

10.4 EXERCISES

10.1 Let

$$A = J_2(0) \oplus J_3(0)$$

(a) Describe all irreducible A-invariant subspaces that contain e_1.
(b) Describe all irreducible A-invariant subspaces that contain e_3.

10.2 Let $A = (J_n(0))^2$. Describe all irreducible A-invariant subspaces that contain e_1.

10.3 Prove or disprove the following statement: if $A, B: \mathbb{C}^n \to \mathbb{C}^n$ are transformations with $\sigma(A) = \sigma(B) = \{\lambda_0\}$, $\lambda_0 \in \mathbb{C}$ and with $\mathrm{Inv}(A) = \mathrm{Inv}(B)$, then A and B are similar.

10.4 Show that if $A, B: \mathbb{C}^n \to \mathbb{C}^n$ have the same set of hyperinvariant subspaces and if $\sigma(A) = \sigma(B) = \{\lambda_0\}$, $\lambda_0 \in \mathbb{C}$, then A and B are similar.

10.5 Show that two lower triangular Toeplitz matrices have the same invariant subspaces if and only if each matrix is a polynomial in the other.

10.6 Show that two circulants have the same invariant subspaces if and only if each circulant is a polynomial in the other.

10.7 Is the property expressed in Exercise 10.6 true for two block circulants of type

$$\begin{bmatrix} A_1 & A_2 & \cdots & A_n \\ A_n & A_1 & \cdots & A_{n-1} \\ \vdots & \vdots & & \vdots \\ A_2 & A_3 & \cdots & A_1 \end{bmatrix}$$

where A_j are 2×2 matrices? What happens if A_j are 3×3 matrices?

10.8 Show that two companion matrices have the same invariant subspaces if and only if each is a polynomial in the other. Is this property true for block companion matrices

$$\begin{bmatrix} 0 & I & 0 & \cdots & 0 \\ 0 & 0 & I & \cdots & 0 \\ \vdots & \vdots & \vdots & & \vdots \\ 0 & 0 & 0 & \cdots & I \\ A_0 & A_1 & A_2 & \cdots & A_{n-1} \end{bmatrix}$$

with 2×2 blocks A_j? For block companion matrices with 3×3 blocks A_j?

Chapter Eleven

Algebras of Matrices and Invariant Subspaces

In this chapter we consider subspaces that are invariant for every transformation from a given algebra of transformations. In fact, this framework includes general finite-dimensional algebras over $\mathbb{C}$. The key result, that every algebra of $n \times n$ matrices that is not the algebra of *all* $n \times n$ matrices has a nontrivial invariant subspace, is developed with a complete proof. Some results concerning characterization of lattices of subspaces that are invariant for every transformation from an algebra are presented. Finally, in the last section we study algebras of transformations for which the orthogonal complement of an invariant subspace is again invariant.

11.1 FINITE-DIMENSIONAL ALGEBRAS

A linear space V (over the field of complex numbers $\mathbb{C}$) is called an *algebra* if an operation (usually called *multiplication*) is defined in V, which associates an element in V (denoted xy or $x \cdot y$) with every (ordered) pair of elements x, y from V with the following properties: (a) $\alpha(xy) = (\alpha x)y = x(\alpha y)$ for every $\alpha \in \mathbb{C}$ and every $x, y \in V$; (b) $(xy)z = x(yz)$ for every $x, y, z \in V$ (associativity of multiplication); (c) $(x + y)z = xz + yz$, $x(y + z) = xy + xz$ for every $x, y, z \in V$ (distributivity of multiplication with respect to addition).

Note that generally speaking $xy \neq yx$ in the algebra V. The algebra V may or may not have an identity, that is, an element $e \in V$ such that $ae = ea = a$ for every $a \in V$.

We consider only finite-dimensional algebras, that is, those that are finite-dimensional linear spaces. The basic example of an algebra is $M_{n,n}$, the algebra of all $n \times n$ matrices with complex entries, with the usual multiplication operation. Another important example is the algebra of upper triangular $n \times n$ (complex) matrices.

The following theorem shows that actually every (finite-dimensional) algebra is an algebra of (not necessarily all) matrices. This is the basic simple result concerning representations of finite-dimensional algebras.

Theorem 11.1.1

Let V be an algebra of dimension n (as a linear space). If V has identity, then V can be identified with an algebra of $n \times n$ matrices. If V does not have identity, it can be identified with an algebra of $(n + 1) \times (n + 1)$ matrices.

Proof. Assume first that V has the identity e. Let $x_1, \ldots, x_n$ be a basis in V. For every $a \in V$ the mapping $\hat{a}: V \to V$ defined by $\hat{a}(x) = ax$, $x \in V$ is a linear transformation. Denote by $M(a)$ the $n \times n$ matrix that represents the linear transformation $\hat{a}$ in the fixed basis $x_1, \ldots, x_n$. It is easy to check that the mapping $M: V \to M_{n,n}$ defined above is an algebraic homomorphism:

$$M(a + b) = M(a) + M(b)$$

$$M(ab) = M(a)M(b)$$

$$M(\alpha a) = \alpha M(a)$$

for any elements $a, b \in V$ and any $\alpha \in \mathbb{C}$. Further, the only element $a \in V$ for which $M(a) = 0$ is $a = 0$. Indeed, if $M(a) = 0$, then $ax = 0$ for every $x \in V$. Taking $x = e$, we obtain $a = 0$. Hence we can identify V with the algebra $\{M(a) \mid a \in V\}$, which is simply an algebra of $n \times n$ matrices.

Assume now that V does not have identity. Define a new algebra $\tilde{V}$ as all ordered pairs (x, α) with $x \in V$, $\alpha \in \mathbb{C}$ and with the following operations:

$$(x, \alpha) + (y, \beta) = (x + y, \alpha + \beta)$$

$$(x, \alpha) \cdot (y, \beta) = (xy + \alpha y + \beta x, \alpha \beta)$$

$$\gamma(x, \alpha) = (\gamma x, \gamma \alpha)$$

for any $x, y \in V$ and any $\alpha, \beta, \gamma \in \mathbb{C}$. Obviously, the algebra $\tilde{V}$ has the identity $(0, 1)$ and dimension $n + 1$. According to the part of Theorem 11.1.1 already proved, we can identify $\tilde{V}$ with an algebra of $(n + 1) \times (n + 1)$ matrices (clearly, $\dim \tilde{V} = n + 1$). As V can be identified in turn with the subalgebra $\{(x, 0) \mid x \in V\}$ of $\tilde{V}$, the conclusion of Theorem 11.1.1 follows. $\square$

In view of Theorem 11.1.1 we consider only algebras of matrices in the sequel.

11.2   CHAINS OF INVARIANT SUBSPACES

Let V be an algebra of (not necessarily all) $n \times n$ matrices. A subspace $\mathcal{M} \subset \mathbb{C}^n$ is called V *invariant* if $\mathcal{M}$ is invariant for any matrix from V. The following basic fact (known as Burnside's theorem) establishes the existence of nontrivial invariant subspaces for algebras of matrices.

Theorem 11.2.1

Let V be an algebra of $n \times n$ (complex) matrices with $V \neq M_{n,n}$ and $n \geq 2$. Then there exists a nontrivial V-invariant subspace.

We exclude the case $n = 1$, when every subspace in $\mathbb{C}^n$ is trivial (in this case the theorem fails for $V = \{0\}$). The proof of Theorem 11.2.1 is lengthy and based on a series of auxiliary results; it is given in the next section.

Taking a maximal chain of V-invariant subspaces and using Burnside's theorem we arrive at the following conclusion.

Theorem 11.2.2

For any algebra V of $n \times n$ matrices, there is a chain of V-invariant subspaces

$$\{0\} = \mathcal{M}_1 \subset \mathcal{M}_2 \subset \cdots \subset \mathcal{M}_{k+1} = \mathbb{C}^n \qquad (11.2.1)$$

such that, with respect to a direct sum decomposition

$$\mathbb{C}^n = \mathcal{N}_1 \dotplus \cdots \dotplus \mathcal{N}_k \qquad (11.2.2)$$

where $\mathcal{N}_p$ is a direct complement to $\mathcal{M}_p$ in $\mathcal{M}_{p+1}$ ($p = 1, \ldots, k$), every transformation $A \in V$ has a block triangular form

$$A = [A_{pq}]_{p,q=1}^k \qquad with \qquad A_{pq} = 0 \qquad for \qquad p > q$$

and the set $\{A_{pp} \mid A \in V\}$, coincides with the algebra of all transformations from $\mathcal{N}_p$ into $\mathcal{N}_p$, for $p = 1, \ldots, k$. The chain (11.2.1) is maximal, and every maximal chain of V-invariant subspaces has the property stated above.

The case when V is the algebra of all block upper triangular matrices with respect to the decomposition (11.2.2) is of special interest. Then $M_{n,n}$ is a direct sum of two subspaces: V and W, where W is the algebra of all lower block triangular matrices with zeros on the main block diagonal:

$$W = \left\{ X \in M_{n,n} \mid X = \begin{bmatrix} 0 & 0 & 0 & \cdots & 0 \\ X_{21} & 0 & 0 & \cdots & 0 \\ X_{31} & X_{32} & 0 & \cdots & 0 \\ & \vdots & & & \\ X_{k1} & X_{k2} & X_{k3} & \cdots & 0 \end{bmatrix}, \; X_{ij}: \mathcal{N}_j \to \mathcal{N}_i \right\}$$

The subspaces

$$\mathcal{L}_1 = \{0\}, \qquad \mathcal{L}_2 = \mathcal{N}_k, \ldots, \qquad \mathcal{L}_k = \mathcal{N}_2 \dotplus \cdots \dotplus \mathcal{N}_k, \qquad \mathcal{L}_{k+1} = \mathbb{C}^n$$

are all the invariant subspaces for W. In particular, we have the following direct sum decompositions:

$$\mathbb{C}^n = \mathcal{M}_1 \dotplus \mathcal{L}_k = \mathcal{M}_2 \dotplus \mathcal{L}_{k-1} = \cdots = \mathcal{M}_k \dotplus \mathcal{L}_1$$

This motivates the following conjecture.

Conjecture 11.2.3

Let V_1 and V_2 be nonzero subalgebras in $M_{n,n}$ such that $V_1 \cap V_2 = \{0\}$, $V_1 + V_2 = M_{n,n}$. Then there exist nonzero invariant subspaces $\mathcal{M}_1$ and $\mathcal{M}_2$ for V_1 and V_2, respectively, which are direct complements of each other in $\mathbb{C}^n$.

We are able to prove a partial result in the direction of this conjecture. Namely, if V_1 and V_2 are subalgebras in $\mathcal{M}_{n,n}$ such that $V_1 + V_2 = M_{n,n}$, then for every V_1-invariant subspace $\mathcal{M}_1$ and every V_2-invariant subspace $\mathcal{M}_2$ either $\mathcal{M}_1 \cap \mathcal{M}_2 = \{0\}$ or $\mathcal{M}_1 + \mathcal{M}_2 = \mathbb{C}^n$ (or both) holds. Indeed, assuming the contrary, let $\mathcal{M}'_i$ be a direct complement to $\mathcal{M}_1 \cap \mathcal{M}_2$ in $\mathcal{M}_i$, $i = 1, 2$, and let $\mathcal{N}$ be a direct complement to $\mathcal{M}_1 + \mathcal{M}_2$ in $\mathbb{C}^n$. Then we have a direct sum decomposition

$$\mathbb{C}^n = \mathcal{M}'_1 \dotplus (\mathcal{M}_1 \cap \mathcal{M}_2) \dotplus \mathcal{M}'_2 \dotplus \mathcal{N}$$

With respect to this decomposition, every $X \in V_1$ has a block matrix representation of type

$$\begin{bmatrix} * & * & * & * \\ * & * & * & * \\ 0 & 0 & * & * \\ 0 & 0 & * & * \end{bmatrix}$$

[the zeros appear because of the V_1 invariance of $\mathcal{M}_1 = \mathcal{M}'_1 \dotplus (\mathcal{M}_1 \cap \mathcal{M}_2)$], whereas every $Y \in V_2$ has a block matrix representation of type

$$\begin{bmatrix} * & 0 & 0 & * \\ * & * & * & * \\ * & * & * & * \\ * & 0 & 0 & * \end{bmatrix}$$

[the zeros appear because of the V_2 invariance of $\mathcal{M}_2 = (\mathcal{M}_1 \cap \mathcal{M}_2) \dotplus \mathcal{M}'_2$]. So every matrix in $V_1 + V_2$ has a zero in the $(4, 2)$ block entry, which contradicts the assumption that $V_1 + V_2 = M_{n,n}$.

11.3 PROOF OF THEOREM 11.2.1

We start with auxiliary results. A subset Q of an algebra U of $n \times n$ matrices is called an *ideal* if Q is a subalgebra; that is, $A_1, A_2 \in Q$ implies $A_1 + A_2 \in Q$, $A_1 A_2 \in Q$, and $\alpha A_1 \in Q$ for every complex number α; and, in addition, AB and BA belong to Q as long as $A \in U$ and $B \in Q$. Trivial examples of ideals are $Q = \{0\}$ and $Q = M_{n,n}$.

Lemma 11.3.1

The algebra $M_{n,n}$ has no nontrivial ideals.

Proof. Let Q be a nonzero ideal in $M_{n,n}$, and let $A \in Q$, $A \neq 0$. It is easily seen that for every pair of indices (i, j) $(1 \leq i, j \leq n)$ there are matrices G_{ij}, H_{ij} such that $G_{ij} A H_{ij}$ has a one in the (i, j) entry and zeros elsewhere. Now any $n \times n$ matrix $B = [b_{ij}]_{i,j=1}^n$ can be written

$$B = \sum_{i,j=1}^n b_{ij} G_{ij} A H_{ij}$$

and thus belongs to Q. Hence $Q = M_{n,n}$. □

Now let U be an algebra of $n \times n$ matrices $(n \geq 2)$ that has no nontrivial invariant subspaces. We prove that $U = M_{n,n}$, thereby proving Theorem 11.2.1. The first observation is that without loss of generality we can assume $I \in U$. Indeed, consider the algebra $\tilde{U} = \{A + \alpha I \mid A \in U, \alpha \in \mathbb{C}\}$. Obviously, $\tilde{U}$ has no nontrivial invariant subspaces as well. Also, U is an ideal in $\tilde{U}$. Hence, if we know already that $\tilde{U} = M_{n,n}$, then Lemma 11.3.1 implies that either $U = M_{n,n}$ or $U = \{0\}$. But the latter case is excluded by the definition of $\tilde{U}$ and the condition $n \geq 2$. So it is assumed that $I \in U$. U.

Lemma 11.3.2   For every nonzero vector x in $\mathbb{C}^n$ and every $y \in \mathbb{C}^n$ there exists a matrix $A \in U$ such that $Ax = y$.

Proof. The set $\mathcal{M} = \{Ax \mid A \in U\}$ is an invariant subspace for U. This subspace is nonzero because $x = I \cdot x$ is a nonzero vector in $\mathcal{M}$ (recall that $I \in U$). By our assumption on U the subspace $\mathcal{M}$ coincides with $\mathbb{C}^n$. Hence for every $y \in \mathbb{C}^n$ there exists an $A \in U$ such that $y = Ax$. □

Lemma 11.3.3

The only matrices that commute with every matrix in U are the scalar multiples of I.

Proof. Let $S \in M_{n,n}$ be such that $SA = AS$ for every $A \in U$. Let λ_0 be an eigenvalue of S with corresponding eigenvector x_0. Then for every $A \in U$ we have

$$SAx_0 = ASx_0 = \lambda_0 Ax_0 \qquad (11.3.1)$$

By Lemma 11.3.2, for every $y \in \mathbb{C}^n$ there is an A in U with $Ax_0 = y$. So equations (11.3.1) mean that $S = \lambda_0 I$. $\quad\square$

Lemma 11.3.4

If x_1 and x_2 are linearly independent vectors in $\mathbb{C}^n$, then for every pair of vectors $y_1, y_2 \in \mathbb{C}^n$ there exists a matrix A from U such that $Ax_1 = y_1$, and $Ax_2 = y_2$.

Proof. It is sufficient to show that there exist $A_1, A_2 \in U$ such that $A_1 x_1 \neq 0$, $A_1 x_2 = 0$ and $A_2 x_1 = 0$, $A_2 x_2 \neq 0$. Indeed, we may then use Lemma 11.3.2 to find $B_1, B_2 \in U$ with $B_1 A_1 x_1 = y_1$, $B_2 A_2 x_2 = y_2$. Hence

$$(B_1 A_1 + B_2 A_2)x_i = y_i, \qquad i = 1, 2$$

We now prove the existence of A_1. (The existence of A_2 is proved similarly.) Arguing by contradiction, assume that $Ax_2 = 0$ implies $Ax_1 = 0$ for every $A \in U$. Then one can define a transformation $T: \mathbb{C}^n \to \mathbb{C}^n$ by the requirement that $TAx_2 = Ax_1$ for all $A \in U$. Indeed, if $Ax_2 = Bx_2$ for some A and B in U, then $(A - B)x_2 = 0$ and thus also $(A - B)x_1 = 0$, which means $Ax_1 = Bx_1$. So T is correctly defined. Further, $\{Ax_2 \mid A \in U\} = \mathbb{C}^n$ by Lemma 11.3.2; hence T is defined on the whole of $\mathbb{C}^n$. Now for any A and B in U we have

$$TABx_2 = ABx_1 = ATBx_2$$

and since $\{Bx_2 \mid B \in U\} = \mathbb{C}^n$, we find that $TA = AT$ for all $A \in U$. By Lemma 11.3.3, $T = \alpha I$ for some $\alpha \in \mathbb{C}$. Therefore, $A(x_1 - \alpha x_2) = 0$ for all $A \in U$. But this contradicts Lemma 11.3.2. $\quad\square$

We say that an algebra V of $n \times n$ matrices is k *transitive* if for every set of k linearly independent vectors $x_1, \ldots, x_k$ in $\mathbb{C}^n$ and every set of k vectors $y_1, \ldots, y_k$ in $\mathbb{C}^n$ there exists a matrix $A \in V$ such that $Ax_i = y_i$, $i = 1, \ldots, k$. Evidently, every k-transitive algebra is p transitive for $p < k$. Lemma 11.3.4 says that the algebra U is 2 transitive.

Proof of Theorem 11.2.1 In view of Lemma 11.3.4 it is sufficient to prove that every 2-transitive algebra V of $n \times n$ matrices is n transitive. Assume by induction that V is k transitive, and we will prove that V is $(k + 1)$ transitive (here $2 \leq k \leq n - 1$).

So let $x_1, \ldots, x_{k+1}$ be linearly independent vectors in $\mathbb{C}^n$. It will suffice to verify that for every i ($1 \leq i \leq k + 1$) there exists a matrix $A_i \in V$ such that $A_i x_i \neq 0$, $A_i x_j = 0$, $j \neq i$ (indeed, for given $y_1, \ldots, y_{k+1} \in \mathbb{C}^n$ the 1 transitivity of V implies the existence of $B_i \in V$ such that $B_i A_i x_i = y_i$; then for $A = \Sigma_{i=1}^{k+1} B_i A_i$ we have $A x_i = y_i$, $i = 1, \ldots, k + 1$).

We will prove the existence of A_{k+1} (for A_i, $1 \le i \le k$ one has simply to permute the indices). Suppose that no such A_{k+1} exists; that is, $A x_1 = \cdots = A x_k = 0$, $A \in V$ implies that $A x_{k+1} = 0$. Consider the algebra

$$V^{(2)} = \left\{ \begin{bmatrix} A & 0 \\ 0 & A \end{bmatrix} \middle| A \in V \right\}$$

of $2n \times 2n$ matrices. It turns out (because of the 2 transitivity of V) that any $V^{(2)}$-invariant subspace is one of the subspaces $\{0\}$, $\mathbb{C}^{2n}$, $\{0\} \oplus \mathbb{C}^n$, $\left\{ \begin{bmatrix} x \\ \lambda x \end{bmatrix} \middle| x \in \mathbb{C}^n \right\}$ for some $\lambda \in \mathbb{C}$. Indeed, the $V^{(2)}$-invariant subspace $\mathcal{M}$ (which we can assume to be nonzero) is a sum of cyclic $V^{(2)}$-invariant subspaces: $\mathcal{M} = \Sigma_{i=1}^p \mathcal{M}_i$, where

$$\mathcal{M}_i = \left\{ \begin{bmatrix} A & 0 \\ 0 & A \end{bmatrix} \begin{bmatrix} x_{i1} \\ x_{i2} \end{bmatrix} \middle| A \in V \right\}, \qquad \begin{bmatrix} x_{i1} \\ x_{i2} \end{bmatrix} \ne 0$$

Fix an index i. For any $n \times n$ matrix B, assuming x_{i1}, x_{i2} are linearly independent, and by the assumption of 2 transitivity of V, we have $B x_{i1} = A x_{i1}$, $B x_{i2} = A x_{i2}$ for some $A \in V$; hence $\mathcal{M}_i$ is $M_{n,n}^{(2)}$ invariant, where

$$M_{n,n}^{(2)} = \left\{ \begin{bmatrix} B & 0 \\ 0 & B \end{bmatrix} \begin{bmatrix} x_{i1} \\ x_{i2} \end{bmatrix} \middle| B \in M_{n,n} \right\}$$

Now because of the obvious 2 transitivity of $M_{n,n}$, we find that $\mathcal{M}_i = \mathbb{C}^n \oplus \mathbb{C}^n$. Assume now that x_{i1} and x_{i2} are linearly dependent. Then 1 transitivity of V implies again that $\mathcal{M}_i$ is $M_{n,n}^{(2)}$ invariant. If $x_{i1} = 0$, we get $\mathcal{M}_i = \{0\} \oplus \mathbb{C}^n$, and if $x_{i2} = \lambda x_{i1}$ for some $\lambda \in \mathbb{C}$, we get

$$\mathcal{M}_i = \left\{ \begin{bmatrix} x \\ \lambda x \end{bmatrix} \middle| x \in \mathbb{C}^n \right\}$$

Consequently, $\mathcal{M} = \Sigma_{i=1}^p \mathcal{M}_i$ is equal to $\mathbb{C}^{2n}$ except for the two cases: (1) $x_{i1} = 0$ for all $i = 1, \ldots, p$; (2) $x_{i2} = \lambda x_{i1}$, $i = 1, \ldots, p$ for the same $\lambda \in \mathbb{C}$. In the first case $\mathcal{M} = \{0\} \oplus \mathbb{C}^n$, and in the second case $\mathcal{M} = \left\{ \begin{bmatrix} x \\ \lambda x \end{bmatrix} \middle| x \in \mathbb{C}^n \right\}$.

Now we return to the proof of the existence of A_{k+1}. By the induction hypothesis, for each j ($1 \le j \le k$) there is some $C_j \in V$ with $C_j x_j \ne 0$ and $C_j x_i = 0$ for $i \ne j$, $1 \le i \le k$. The subspace

$$\mathcal{M}_j = \left\{ \begin{bmatrix} AC_j & 0 \\ 0 & AC_j \end{bmatrix} \begin{bmatrix} x_j \\ x_{k+1} \end{bmatrix} \middle| A \in V, \qquad j = 1, \ldots, k \right\}$$

is $V^{(2)}$ invariant; therefore (according to the fact proved in the preceding paragraph), there exists a complex number α such that $AC_jx_{k+1} = \alpha_j AC_jx_j$ for all $A \in V$. The induction hypothesis implies that

$$\mathbb{C}^n \oplus \cdots \oplus \mathbb{C}^n_{\underset{k \text{ times}}{}} = \left\{ \begin{bmatrix} A & & & 0 \\ & A & & \\ & & \ddots & \\ 0 & & & A \end{bmatrix} \begin{bmatrix} x_1 \\ x_2 \\ \vdots \\ x_k \end{bmatrix} \middle| A \in V \right\}$$

and the assumption that $Ax_{k+1} = 0$ whenever $Ax_j = 0$ for $j = 1, \ldots, k$ shows that a mapping $T: \mathbb{C}^{nk} \to \mathbb{C}^n$ is unambiguously defined by

$$T(Ax_1 \oplus \cdots \oplus Ax_k) = Ax_{k+1}, \qquad A \in V$$

Obviously, T is linear. Further, for $A \in V$ and $j = 1, \ldots, k$ we have (where the term AC_jx_j appears in the jth place)

$$T(0 \oplus \cdots \oplus AC_jx_j \oplus 0 \oplus \cdots \oplus 0) = T \begin{bmatrix} A & & & 0 \\ & A & & \\ & & \ddots & \\ 0 & & & A \end{bmatrix} \begin{bmatrix} C_j & & & 0 \\ & C_j & & \\ & & \ddots & \\ 0 & & & C_j \end{bmatrix} \begin{bmatrix} x_1 \\ x_2 \\ \vdots \\ x_k \end{bmatrix}$$

$$= AC_jx_{k+1} = \alpha_j AC_jx_j$$

Since $C_jx_j \neq 0$, the subspace $\{AC_jx_j \mid A \in V\}$ coincides with $\mathbb{C}^n$ by the 1 transitivity of V. So the linearity of T gives

$$T(y_1 \oplus \cdots \oplus y_k) = \sum_{j=1}^{k} \alpha_j y_j, \qquad y_j \in \mathbb{C}^n$$

Then, for $A \in V$

$$A\left(x_{k+1} - \sum_{j=1}^{k} \alpha_j x_j\right) = Ax_{k+1} - T(Ax_1 \oplus \cdots \oplus Ax_k) = Ax_{k+1} - Ax_{k+1} = 0$$

Hence $\{x \mid Ax = 0 \text{ for all } A \in V\}$ is a nontrivial V-invariant subspace. This contradicts the 1 transitivity of V. $\square$

11.4   REFLEXIVE LATTICES

Let Λ be a lattice of subspaces in $\mathbb{C}^n$. The set of all $n \times n$ matrices A such that $A\mathcal{L} \subset \mathcal{L}$ for every $\mathcal{L} \in \Lambda$, denoted $\text{Alg}(\Lambda)$, is an algebra. Indeed, if $A, B \in \text{Alg}(\Lambda)$, then

$$(A + B)\mathcal{L} \subset A\mathcal{L} + B\mathcal{L} \subset \mathcal{L}$$

$$(AB)\mathcal{L} = A(B\mathcal{L}) \subset A\mathcal{L} \subset \mathcal{L}$$

$$(\alpha A)\mathcal{L} = \alpha(A\mathcal{L}) \subset \alpha\mathcal{L} \subset \mathcal{L}, \qquad (\alpha \in \mathbb{C})$$

for every subspace $\mathscr{L} \in \Lambda$. On the other hand, for an algebra V of $n \times n$ matrices the set $\mathrm{Inv}(V)$ of all V-invariant subspaces in $\mathbb{C}^n$ is easily seen to be a lattice of subspaces [i.e., $\mathscr{L}, \mathscr{M} \in \mathrm{Inv}(V)$ implies $\mathscr{L} + \mathscr{M} \in \mathrm{Inv}(V)$ and $\mathscr{L} \cap \mathscr{M} \in \mathrm{Inv}(V)$]. The following properties of $\mathrm{Alg}(\Lambda)$ and $\mathrm{Inv}(V)$ are immediate consequences of the definitions.

Proposition 11.4.1

(a) If Λ_1 and Λ_2 are two lattices of subspaces in $\mathbb{C}^n$, and $\Lambda_1 \subset \Lambda_2$, then $\mathrm{Alg}(\Lambda_1) \supset \mathrm{Alg}(\Lambda_2)$. (b) If V_1 and V_2 are algebras of $n \times n$ matrices and $V_1 \supset V_2$, then $\mathrm{Inv}(V_1) \subset \mathrm{Inv}(V_2)$; (c) $\mathrm{Inv}(\mathrm{Alg}(\Lambda)) \supset \Lambda$; (d) $\mathrm{Alg}(\mathrm{Inv}(V)) \supset V$.

Let us check property (c), for example. Assume $\mathscr{L} \in \Lambda$; then $A\mathscr{L} \subset \mathscr{L}$ for every $A \in \mathrm{Alg}(\Lambda)$. Hence $\mathscr{L}$ is $\mathrm{Alg}(\Lambda)$ invariant; that is, $\mathscr{L} \in \mathrm{Inv}(\mathrm{Alg}(\Lambda))$.

EXAMPLE 11.4.1. Let Λ be the chain

$$0 \subset \mathrm{Span}\{e_1\} \subset \mathrm{Span}\{e_1, e_2\} \subset \cdots \subset \mathrm{Span}\{e_1, \ldots, e_{n-1}\} \subset \mathbb{C}^n$$

Then $\mathrm{Alg}(\Lambda)$ is the algebra of all upper triangular matrices. $\square$

EXAMPLE 11.4.2. Let Λ be the set of subspaces $\mathrm{Span}\{e_i \mid i \in K\}$, where K runs over all subsets of $\{1, \ldots, n\}$. Clearly Λ is a lattice. The algebra $\mathrm{Alg}(\Lambda)$ is easily seen to be the algebra of all diagonal matrices. $\square$

EXAMPLE 11.4.3. For a fixed subspace $\mathscr{M} \subset \mathbb{C}^n$, let Λ be the lattice of all subspaces that are contained in $\mathscr{M}$. Then $\mathrm{Alg}(\Lambda)$ is the algebra of all transformations A having the form

$$A = \begin{bmatrix} \alpha I & * \\ 0 & * \end{bmatrix}, \qquad \alpha \in \mathbb{C}$$

with respect to the direct sum decomposition $\mathbb{C}^n = \mathscr{M} \dotplus \mathscr{N}$ (for a fixed direct complement $\mathscr{N}$ to $\mathscr{M}$). $\square$

EXAMPLE 11.4.4. Let V be the algebra of polynomials $\sum_{j=0}^{l} \alpha_j A^j$, $\alpha_j \in \mathbb{C}$, where $A: \mathbb{C}^n \to \mathbb{C}^n$ is a fixed linear transformation. Then $\mathrm{Inv}(V)$ is the lattice of all A-invariant subspaces. $\square$

EXAMPLE 11.4.5. Let $A: \mathbb{C}^n \to \mathbb{C}^n$ be a fixed transformation, and let V be the algebra of all transformations that commute with A. Then $\mathrm{Inv}(V)$ is the lattice of all A-hyperinvariant subspaces. $\square$

Note that

$$\mathrm{Alg}(\mathrm{Inv}(\mathrm{Alg}(\Lambda))) = \mathrm{Alg}(\Lambda) \tag{11.4.1}$$

for every lattice Λ of the subspaces in $\mathbb{C}^n$. Indeed, the inclusion $\subset$ in equation (11.4.1) follows from (c) and (a). To prove the opposite inclusion, let $A \in \mathrm{Alg}(\Lambda)$. Then any subspace $\mathcal{M}$ belonging to $\mathrm{Inv}(\mathrm{Alg}(\Lambda))$ is invariant for every transformation in $\mathrm{Alg}(\Lambda)$; in particular, $\mathcal{M}$ is A invariant. This shows that $A \in \mathrm{Alg}(\mathrm{Inv}(\mathrm{Alg}(\Lambda)))$. Similarly, one proves that

$$\mathrm{Inv}(\mathrm{Alg}(\mathrm{Inv}(V))) = \mathrm{Inv}(V) \tag{11.4.2}$$

for every algebra V of transformations $\mathbb{C}^n \to \mathbb{C}^n$.

A lattice Λ of subspaces in $\mathbb{C}^n$ is called *reflexive* if $\mathrm{Inv}(\mathrm{Alg}(\Lambda)) = \Lambda$. Equality (11.4.2) shows, for example, that any lattice of the form $\mathrm{Inv}(V)$ for some algebra V is reflexive. Let us give an example of a nonreflexive lattice.

EXAMPLE 11.4.6. Let Λ be the following lattice of subspaces in $\mathbb{C}^2$: $\{0\}$, $\mathcal{L} = \mathrm{Span}\{e_2\}$, $\mathcal{M} = \mathrm{Span}\{e_1\}$, $\mathcal{N} = \mathrm{Span}\{e_1 + e_2\}$, $\mathbb{C}^2$. Let us find the algebra $\mathrm{Alg}(\Lambda)$. The 2×2 matrix $A = \begin{bmatrix} a & b \\ c & d \end{bmatrix}$ has invariant subspaces $\mathcal{L}$ and $\mathcal{M}$ if and only if $b = c = 0$. Further, $\mathcal{N}$ is A invariant if and only if $a + b = c + d$. So

$$\mathrm{Alg}(\Lambda) = \left\{ \begin{bmatrix} a & 0 \\ 0 & a \end{bmatrix} \,\middle|\, a \in \mathbb{C} \right\} = \{aI \mid a \in \mathbb{C}\}$$

and $\mathrm{Lat}(\mathrm{Alg}(\Lambda))$ consists of all subspaces in $\mathbb{C}^2$. □

Many results are known about sufficient conditions for reflexivity of a lattice of subspaces. Often the key ingredient in such conditions is distributivity. Recall the definition of a distributive lattice of subspaces given in Section 9.6.

Theorem 11.4.2

A distributive lattice of subspaces in $\mathbb{C}^n$ is reflexive. Conversely, every finite reflexive lattice of subspaces is distributive.

The proof of Theorem 11.4.2 is beyond the scope of this book, and we refer the reader to the original papers by Johnson (1964) and Harrison (1974) for the full story. Here, we shall only prove two particular cases in the form of Theorems 11.4.3 and 11.4.4.

Theorem 11.4.3

A complete chain of subspaces

$$\{0\} \subset \mathcal{M}_1 \subset \mathcal{M}_2 \subset \cdots \subset \mathcal{M}_{n-1} \subset \mathbb{C}^n, \ \dim \mathcal{M}_i = i\,; \qquad i = 1, \ldots, n-1$$

is reflexive.

Proof. Let $f_1, \ldots, f_n$ be a basis in $\mathbb{C}^n$ such that $\text{Span}\{f_1, \ldots, f_i\} = \mathcal{M}_i$, $i = 1, \ldots, n-1$, and write linear transformations as matrices with respect to this basis. Example 11.4.1 shows that $\text{Alg}(\Lambda)$ consists of all upper triangular matrices. As the linear transformation

$$K = \begin{bmatrix} 0 & 1 & 0 & \cdots & 0 \\ 0 & 0 & 1 & \cdots & 0 \\ \vdots & \vdots & & \ddots & \vdots \\ & & & & 1 \\ 0 & 0 & & \cdots & 0 \end{bmatrix}$$

obviously belongs to $\text{Alg}\,\Lambda$, and its only invariant subspaces are $\{0\}$, $\mathcal{M}_i$, $i = 1, \ldots, n-1$, and $\mathbb{C}^n$, we have $\text{Inv}(\text{Alg}(\Lambda)) \subset \Lambda$. Since the reverse inclusion is clear, the conclusion of Theorem 11.4.3 follows. $\square$

The next theorem deals with lattices that are as unlike chains as possible.

A lattice Λ of subspaces in $\mathbb{C}^n$ is called a *Boolean algebra* if it is distributive and for every $\mathcal{M} \in \Lambda$ there is a *unique* complement $\mathcal{M}'$ (i.e., $\mathcal{M} + \mathcal{M}' = \mathbb{C}^n$, $\mathcal{M} \cap \mathcal{M}' = \{0\}$) that belongs to Λ. We say that a nonzero subspace $\mathcal{H} \in \Lambda$ is an *atom* if there are no subspaces for the lattice Λ strictly between $\mathcal{H}$ and $\{0\}$. The Boolean algebra Λ is called *atomic* if any $\mathcal{M} \in \Lambda$ is a sum of all atoms $\mathcal{H}$ contained in $\mathcal{M}$. A typical example of an atomic Boolean algebra of subspaces is $\Lambda = \{\text{Span}\{x_i \mid i \in E\}$, where E is any subset in $\{1, 2, \ldots, n\}\}$, and $x_1, \ldots, x_n$ is a fixed basis in $\mathbb{C}^n$.

Theorem 11.4.4

Every atomic Boolean algebra Λ of subspaces of $\mathbb{C}^n$ is reflexive.

Proof. Let K be the set of all atoms in Λ, and for every $\mathcal{H} \subset K$ let $P_{\mathcal{H}}$ be the projector on $\mathcal{H}$ along the complement $\mathcal{H}'$ of $\mathcal{H}$ in the lattice Λ. We shall show that $\Lambda = \text{Inv}(V)$, where V is the algebra generated by the transformations of type $P_{\mathcal{H}} A P_{\mathcal{H}}$, where $A: \mathbb{C}^n \to \mathbb{C}^n$ and $\mathcal{H} \in K$. In other words, V consists of all linear combinations of transformations of type

$$P_{\mathcal{H}_1} A_1 P_{\mathcal{H}_1} P_{\mathcal{H}_2} A_2 P_{\mathcal{H}_2} \cdots P_{\mathcal{H}_m} A_m P_{\mathcal{H}_m}$$

where $A_i: \mathbb{C}^n \to \mathbb{C}^n$, and $\mathcal{H}_1, \ldots, \mathcal{H}_m$ are atoms in Λ.

Let $\mathcal{L}$ be an atom in Λ. For any atom $\mathcal{H}$, we have either $\mathcal{L} = \mathcal{H}$ or $\mathcal{L} \subset \mathcal{H}'$. (This follows from the distributivity of Λ:

$$\mathcal{L} = \mathcal{L} \cap (\mathcal{H} \cup \mathcal{H}') = (\mathcal{L} \cap \mathcal{H}) \cup (\mathcal{L} \cap \mathcal{H}');$$

as $\mathcal{L}$ is an atom, either $\mathcal{L} \cap \mathcal{H} = \mathcal{L}$ or $\mathcal{L} \cap \mathcal{H}' = \mathcal{L}$ holds.) In the former case $\text{Im}\, P_{\mathcal{H}} A P_{\mathcal{H}} \subset \mathcal{L}$ for every transformation $A: \mathbb{C}^n \to \mathbb{C}^n$, and in the latter case $\mathscr{L} \subset \mathrm{Ker}\, P_{\mathscr{H}} A P_{\mathscr{H}}$. In either case $\mathscr{L}$ is $P_{\mathscr{H}} A P_{\mathscr{H}}$ invariant. Hence $\mathscr{L} \in \mathrm{Inv}(V)$. Now every $\mathscr{M} \in \Lambda$ is a sum of the (finitely many) atoms contained in $\mathscr{M}$. Hence $\mathscr{M} \in \mathrm{Inv}(V)$. In other words, $\Lambda \subset \mathrm{Inv}(V)$.

To prove the reverse inclusion, it is convenient to use the following fact: if $\mathscr{H}$ is an atom in Λ and $\mathscr{M} \in \mathrm{Inv}(V)$, then either $\mathscr{H} \subset \mathscr{M}$ or $\mathscr{M} \subset \mathscr{H}'$. Indeed, suppose that $\mathscr{M}$ is not contained in $\mathscr{H}'$, so there exists a vector $f \in \mathbb{C}^n$ such that $f \in \mathscr{M} \setminus \mathscr{H}'$. Since $P_{\mathscr{H}} f \neq 0$, it follows that every vector x in $\mathscr{H}$ has the form $A P_{\mathscr{H}} f$ for some transformation $A : \mathbb{C}^n \to \mathbb{C}^n$. Then also $x = P_{\mathscr{H}} A P_{\mathscr{H}} f$. As $f \in \mathscr{M}$ and $\mathscr{M} \in \mathrm{Inv}(V)$, we have $x \in \mathscr{M}$, that is, $\mathscr{H} \subset \mathscr{M}$.

Return to the proof of the inclusion $\mathrm{Inv}(V) \subset \Lambda$. Let $\mathscr{M} \in \mathrm{Inv}(V)$, and let $\mathscr{M}_0 \in \Lambda$ be the sum of all the atoms in Λ that are contained in $\mathscr{M}$. Also, let $\mathscr{M}_1 \in \Lambda$ be the intersection of all the complements of atoms in Λ such that these complements contain $\mathscr{M}$. Obviously

$$\mathscr{M}_0 \subset \mathscr{M} \subset \mathscr{M}_1 \tag{11.4.3}$$

Since Λ is atomic, the complement $\mathscr{M}_0'$ of $\mathscr{M}_0$ is the sum of all atoms that are not contained in $\mathscr{M}$. (Indeed, if an atom $\mathscr{H}$ is contained in $\mathscr{M}_0'$, then $\mathscr{H}$ is not contained in $\mathscr{M}_0$ and thus by the definition of $\mathscr{M}_0$, $\mathscr{H}$ is not contained in $\mathscr{M}$. Conversely, if an atom $\mathscr{H}$ is not contained in $\mathscr{M}$, then obviously $\mathscr{H}$ is not contained in $\mathscr{M}_0$, and since $\mathscr{H}$ is an atom, it must be contained in $\mathscr{M}_0'$.) The fact proved in the preceding paragraph shows that $\mathscr{M}_0'$ is the sum of all the atoms $\mathscr{H}$ with the property that $\mathscr{H}' \supset \mathscr{M}$. For any finite set $\mathscr{H}_1, \ldots, \mathscr{H}_p$ of atoms with $\mathscr{H}_i' \supset \mathscr{M}$, $i = 1, \ldots, p$, we have (using the distributivity of Λ):

$$(\mathscr{H}_1 + \cdots + \mathscr{H}_p) + (\mathscr{H}_1' \cap \cdots \cap \mathscr{H}_p')$$
$$= (\mathscr{H}_1 + \cdots + \mathscr{H}_p + \mathscr{H}_1') \cap \cdots \cap (\mathscr{H}_1 + \cdots + \mathscr{H}_p + \mathscr{H}_p') = \mathbb{C}^n$$

and

$$(\mathscr{H}_1 + \cdots + \mathscr{H}_p) \cap (\mathscr{H}_1' \cap \cdots \cap \mathscr{H}_p')$$
$$= (\mathscr{H}_1 \cap \mathscr{H}_1' \cap \cdots \cap \mathscr{H}_p') + \cdots + (\mathscr{H}_p \cap \mathscr{H}_1' + \cdots + \mathscr{H}_p') = \{0\}$$

so actually

$$\mathscr{H}_1 + \cdots + \mathscr{H}_p = (\mathscr{H}_1' \cap \cdots \cap \mathscr{H}_p')'$$

This shows that $\mathscr{M}_0' = \mathscr{M}_1'$; hence $\mathscr{M}_0 = \mathscr{M}_1$. Combining this with (11.4.3), we see that $\mathscr{M} = \mathscr{M}_0 = \mathscr{M}_1$ and thus $\mathscr{M} \in \Lambda$. $\square$

11.5 REDUCTIVE AND SELF-ADJOINT ALGEBRAS

We have seen in Corollary 3.4.4 that the set $\mathrm{Inv}(A)$ of invariant subspaces of a transformation $A : \mathbb{C}^n \to \mathbb{C}^n$ has the property that $\mathscr{M} \in \mathrm{Inv}(A)$ exactly when $\mathscr{M}^\perp \in \mathrm{Inv}(A)$ if and only if A is normal. This property makes it natural to introduce the following definition: an algebra V of $n \times n$ matrices is called *reductive* if it contains I and for every subspace belonging to $\text{Inv}(V)$ its orthogonal complement belongs to $\text{Inv}(V)$ as well. Thus the algebra $P(A)$ of all polynomials $\sum_{i=0}^{m} \alpha_i A^i$, where A is a normal transformation, is reductive. This algebra $P(A)$ has the property that $X \in P(A)$ implies $X^* \in P(A)$. Indeed, we have only to show that, for the normal transformation A, the adjoint A^* is a polynomial in A. Passing, if necessary, to the orthonormal basis of eigenvectors of A, we can assume that A is diagonal: $A = \text{diag}[\lambda_1, \lambda_2, \ldots, \lambda_n]$. Now let $f(\lambda)$ be a scalar polynomial satisfying the conditions $f(\lambda_i) = \bar{\lambda}_i$, $i = 1, \ldots, n$. Then clearly $A^* = f(A)$.

The next theorem shows that this property of the reductive algebra $P(A)$ is a particular case of a much more general fact.

Theorem 11.5.1

An algebra V of $n \times n$ matrices with $I \in V$ is reductive if and only if V is self-adjoint, that is, $X \in V$ implies $X^* \in V$.

As a subspace $\mathcal{M}$ is A invariant if and only if $\mathcal{M}^\perp$ is A^* invariant, it follows immediately that every self-adjoint algebra with identity is reductive. To prove the converse, we need the following basic property of invariant subspaces of reductive algebras.

Lemma 11.5.2

Let V be a reductive algebra of $n \times n$ matrices, and let $\mathcal{M}_1, \ldots, \mathcal{M}_m$ be a set of mutually orthogonal V-invariant subspaces such that

$$\mathbb{C}^n = \mathcal{M}_1 \oplus \cdots \oplus \mathcal{M}_m$$

and for every i the set of restrictions $\{A|_{\mathcal{M}_i} \mid A \in V\}$ coincides with the algebra $M(\mathcal{M}_i)$ of all transformations from $\mathcal{M}_i$ into $\mathcal{M}_i$. Then V is self-adjoint.

Proof. We proceed by induction on m. For $m = 1$, that is, $\mathcal{M}_1 = \mathbb{C}^n$ and $V = M(\mathbb{C}^n)$, the lemma is obvious. So assume that the lemma is proved already for $m - 1$ subspaces, and we prove the lemma for m subspaces.

It is convenient to distinguish two cases. In case 1, there exist distinct integers j and k between 1 and m and an algebraic isomorphism $\varphi: M(\mathcal{M}_j) \to M(\mathcal{M}_k)$ such that $A|_{\mathcal{M}_k} = \varphi(A|_{\mathcal{M}_j})$ for every $A \in V$. This means that φ is a one-to-one and onto map with the following properties:

(a) $\varphi(\alpha A|_{\mathcal{M}_j} + \beta B|_{\mathcal{M}_j}) = \alpha A|_{\mathcal{M}_k} + \beta A|_{\mathcal{M}_k}$ for every $A, B \in V$ and $\alpha, \beta \in \mathbb{C}$

(b) $\quad\quad \varphi(A|_{\mathcal{M}_j} \cdot B|_{\mathcal{M}_j}) = A|_{\mathcal{M}_k} \cdot B|_{\mathcal{M}_k}$ for every $A, B \in V$

(c) $\quad\quad\quad\quad \varphi(I|_{\mathcal{M}_j}) = I|_{\mathcal{M}_k}$

As dim $M(\mathcal{M}_j) = (\dim \mathcal{M}_j)^2$ is equal to dim $M(\mathcal{M}_k) = (\dim \mathcal{M}_k)^2$, we have dim $\mathcal{M}_j = \dim \mathcal{M}_k$.

We show first that there exists an invertible transformation $S: \mathcal{M}_j \to \mathcal{M}_k$ such that $\varphi(X) = SXS^{-1}$ for every $X \in M(\mathcal{M}_j)$. Note that φ takes rank 1 projectors into rank 1 projectors. Indeed, if $P^2 = P \in M(\mathcal{M}_j)$ and rank $P = 1$, then $(\varphi(P))^2 = \varphi(P)$, so $\varphi(P)$ is a projector. Moreover, the one-dimensional subspace

$$\{ PXP \mid X \in M(\mathcal{M}_j) \} \subset M(\mathcal{M}_j)$$

is mapped by φ into the subspace

$$\{ \varphi(P)Y\varphi(P) \mid Y \in M(\mathcal{M}_k) \} \subset M(\mathcal{M}_k)$$

so the subspace $\{ \varphi(P)Y\varphi(P) \mid Y \in M(\mathcal{M}_k) \}$ is also one-dimensional; hence rank $\varphi(P) = 1$. Now fix any nonzero vector $f \in \mathcal{M}_j$, and let $A_0: \mathcal{M}_j \to \mathcal{M}_j$ be the orthogonal projector on Span$\{f\}$. As $\varphi(A_0): \mathcal{M}_k \to \mathcal{M}_k$ is also a one-dimensional projector, there exists an invertible transformation $S_0: \mathcal{M}_j \to \mathcal{M}_k$ such that $\varphi(A_0) = S_0 A_0 S_0^{-1}$. (This follows from the fact that the Jordan form of any one-dimensional projector in $\mathbb{C}^n$ is the same: diag$[1, 0, \ldots, 0]$.) Define $S: \mathcal{M}_j \to \mathcal{M}_k$ by

$$S(Af) = \varphi(A)S_0 f , \qquad A \in M(\mathcal{M}_j)$$

Let us show that this definition is correct. Indeed, if $A_1 f = A_2 f$, then $(A_1 - A_2)A_0 = 0$. Consequently, $(\varphi(A_1) - \varphi(A_2))\varphi(A_0) = 0$, and since $\varphi(A_0)$ is a projector onto Span$\{S_0 f\}$, we obtain $(\varphi(A_1) - \varphi(A_2))S_0 f = 0$. In other words, $A_1 f = A_2 f$ happens only if $\varphi(A_1)S_0 f = \varphi(A_2)S_0$. Hence S is correctly defined. Clearly, S is linear and onto. If $\varphi(A)S_0 f = 0$, then $\varphi(A)\varphi(A_0) = 0$, which implies $AA_0 = 0$ and $Af = 0$. This shows that Ker $S = \{0\}$. Hence S is invertible. Finally, for every $A, B \in M(\mathcal{M}_j)$ we have

$$S(AB)f = \varphi(A)\varphi(B)S_0 f = \varphi(A)SBf ,$$

and thus $SAg = \varphi(A)Sg$ for every $g \in \mathcal{M}_j$. Thus $\varphi(A) = SAS^{-1}$ for all $A \in M(\mathcal{M}_j)$.

Next, we show that S can be taken to be unitary, that is, $S^{-1} = S^*$. Let $\mathcal{M}$ be a subspace in $\mathbb{C}^n$ consisting of all vectors of the form $x_1 + \cdots + x_m$, where $x_1 \in \mathcal{M}_1, \ldots, x_m \in \mathcal{M}_m$ and $x_k = Sx_j$. As

$$A|_{\mathcal{M}_k} = SA|_{\mathcal{M}_j} S^{-1}$$

for every $A \in V$, it follows that $\mathcal{M}$ is V invariant. Since V is reductive, $\mathcal{M}^\perp$ is V invariant as well. A computation shows that

$$\mathcal{M}^\perp = \{x_j + x_k \mid x_j \in \mathcal{M}_j, x_k \in \mathcal{M}_k, x_j = -S^*x_k\}$$

The fact that $A\mathcal{M}^\perp \subset \mathcal{M}^\perp$ for all $A \in V$ implies that if $x_j = -S^*x_k$ for $x_j \in \mathcal{M}_j$ and $x_k \in \mathcal{M}_k$, then

$$Ax_j = -S^*Ax_k = -S^*A|_{\mathcal{M}_k}x_k = -S^*SA|_{\mathcal{M}}S^{-1}x_k = S^*SA|_{\mathcal{M}}S^{-1}S^{*-1}x_j$$

As $\{A|_{\mathcal{M}_j} \mid A \in V\}$ coincides with $M(\mathcal{M}_j)$ and in the preceding equality x_j can be an arbitrary vector from $\mathcal{M}_j$, we obtain $B = S^*SBS^{-1}S^{*-1}$ for all $B \in M(\mathcal{M}_j)$. By Proposition 9.1.6, $S^*S = \lambda I$ for some number λ that must be positive because S^*S is positive definite. Letting $U = \lambda^{-1/2}S$, we obtain a unitary transformation U such that $\varphi(B) = UBU^{-1}$ for all $B \in M(\mathcal{M}_j)$.

We next show that $V|_{\mathcal{M}_k^\perp}$ is reductive. Indeed, let $\mathcal{N} \subset \mathcal{M}_k^\perp = \mathcal{M}_1 + \cdots + \mathcal{M}_{k-1} + \mathcal{M}_{k+1} + \cdots + \mathcal{M}_m$ be $V|_{\mathcal{M}_k^\perp}$ invariant. Then clearly $\mathcal{N}$ is V invariant, and by the reductive property of V, so is $\mathcal{N}^\perp$, and hence also $\mathcal{N}^\perp \cap \mathcal{M}_k^\perp$. It remains to notice that $\mathcal{N}^\perp \cap \mathcal{M}_k^\perp$ coincides with the orthogonal complement to $\mathcal{N}^\perp$ in $\mathcal{M}_k^\perp$.

By the induction hypothesis, $V|_{\mathcal{M}_k}$ is self-adjoint. Therefore, for every matrix

$$A = A_1 \oplus \cdots \oplus A_{k-1} \oplus A_k \oplus \cdots \oplus A_m \in V$$

the transformation

$$A_1^* \oplus \cdots \oplus A_{k-1}^* \oplus A_{k+1}^* \oplus \cdots \oplus A_m^* : \mathcal{M}_k^\perp \to \mathcal{M}_k^\perp$$

belongs to $V|_{\mathcal{M}_k^\perp}$. As for every $B = B_1 \oplus \cdots \oplus B_k \oplus \cdots \oplus B_m \in V$ we have $B_k = UB_jU^{-1}$, it follows that

$$A_1^* \oplus \cdots \oplus A_{k-1}^* \oplus UA_j^*U^{-1} \oplus A_{k+1}^* \oplus \cdots \oplus A_m^* \in V \quad (11.5.1)$$

But $A_k = UA_jU^{-1}$ and U is unitary. So the transformation (11.5.1) is just A^*. We have proved that V is self-adjoint (in case 1).

Consider now case 2. For any pair of distinct integers j and k between 1 and m, there is no algebraic isomorphism φ as in case 1. If for fixed $j \neq k$, $A|_{\mathcal{M}_j} = 0$ implies $A|_{\mathcal{M}_k} = 0$ for any $A \in V$ and vice versa, then we can correctly define an algebraic isomorphism $\varphi : M(\mathcal{M}_j) \to M(\mathcal{M}_k)$ by putting $\varphi(A|_{\mathcal{M}_j}) = A|_{\mathcal{M}_k}$ for all $A \in V$ (recall that $V|_{\mathcal{M}_j} = M(\mathcal{M}_j)$). Thus our assumption in case 2 implies the following. For each pair j, k of distinct integers between 1 and m there exists a matrix $A \in V$ such that exactly one of the transformations $A|_{\mathcal{M}_j}$ and $A|_{\mathcal{M}_k}$ is zero.

We now prove that there exists a matrix $A \in V$ such that $A|_{\mathcal{M}_j}$ is different from zero for exactly one index j. Choose $A \in V$ different from zero so that the number p of indices j ($1 \leq j \leq m$) with $A|_{\mathcal{M}_j} \neq 0$ is minimal. Permuting

$\mathcal{M}_1, \ldots, \mathcal{M}_m$ if necessary, we can assume that $A|_{\mathcal{M}_1} \neq 0, \ldots, A|_{\mathcal{M}_p} \neq 0$, $A|_{\mathcal{M}_j} = 0$ for $j > p$. We must show that $p = 1$. Assume the contrary, that is, $p > 1$. Interchanging $\mathcal{M}_1$ and $\mathcal{M}_p$ if necessary, we can assume that $C|_{\mathcal{M}_1} \neq 0$, $C|_{\mathcal{M}_p} = 0$ for some matrix $C \in V$. Let $\mathcal{J}_1$ denote the set of all transformations $B: \mathcal{M}_1 \to \mathcal{M}_1$ such that $B = \hat{B}|_{\mathcal{M}_1}$ for some $\hat{B} \in V$ with $\hat{B}|_{\mathcal{M}_p} = 0$. The fact that $V|_{\mathcal{M}_1} = M(\mathcal{M}_1)$ implies that $\mathcal{J}_1$ is an ideal in $M(\mathcal{M}_1)$. Since $C \in \mathcal{J}_1$ and $C \neq 0$, Lemma 11.3.1 shows that actually $\mathcal{J}_1 = M(\mathcal{M}_1)$. Similarly, the set $\mathcal{J}_2$ of all transformations $B: \mathcal{M}_1 \to \mathcal{M}_1$ such that $B = \hat{B}|_{\mathcal{M}_1}$ for some $\hat{B} \in V$ with $\hat{B}|_{\mathcal{M}_{p+1}} = 0, \ldots, \hat{B}|_{\mathcal{M}_m} = 0$, is a nonzero ideal in $M(\mathcal{M}_1)$ and thus $\mathcal{J}_2 = M(\mathcal{M}_1)$. Now the identity transformation $I: \mathcal{M}_1 \to \mathcal{M}_1$ belongs to both $\mathcal{J}_1$ and $\mathcal{J}_2$. Therefore, there exist transformations $B_j: \mathcal{M}_j \to \mathcal{M}_j$ ($j \neq 1$, $j \neq p$) and $C_j: \mathcal{M}_j \to \mathcal{M}_j$ ($j = 2, 3, \ldots, p$) such that

$$\tilde{B} \stackrel{\text{def}}{=} I \oplus B_2 \oplus \cdots \oplus B_{p-1} \oplus 0 \oplus B_{p+1} \oplus \cdots \oplus B_m$$

and

$$\tilde{C} \stackrel{\text{def}}{=} I \oplus C_2 \oplus \cdots \oplus C_{p-1} \oplus C_p \oplus 0 \oplus \cdots \oplus 0$$

belongs to V. Then also $\tilde{B}\tilde{C}$ belongs to V, and $(\tilde{B}\tilde{C})|_{\mathcal{M}_i} = 0$ for $i \geq p$. However, this contradicts the choice of p. So, indeed, $p = 1$.

As the ideal $\mathcal{J}_2$ constructed above coincides with $M(\mathcal{M}_1)$, it follows that every matrix B from V is a sum of two transformations $B|_{\mathcal{M}_1}$ and $B|_{\mathcal{M}_1^{\perp}}$. Since $V|_{\mathcal{M}_1} = M(\mathcal{M}_1)$ we find that V is self-adjoint provided $V|_{\mathcal{M}_1^{\perp}}$ is. But the algebra $V|_{\mathcal{M}_1^{\perp}}$ is easily seen to be reductive because V is. Now the self-adjointness of $V|_{\mathcal{M}_1}$ follows from the induction hypothesis. Lemma 11.5.2 is proved completely. $\square$

Now we are ready to prove the converse statement of Theorem 11.5.1. If V has no nontrivial invariant subspaces, then by Theorem 11.2.1 $V = M_{n,n}$, and obviously V is self-adjoint. If V has nontrivial invariant subspaces, then it has a minimal one, say, $\mathcal{M}_1$. As V is reductive, $\mathcal{M}_1^{\perp}$ is also V invariant, and the restriction $V|_{\mathcal{M}_1^{\perp}}$ is reductive. If $V|_{\mathcal{M}_1^{\perp}}$ is not the algebra of all transformations $\mathcal{M}_1^{\perp} \to \mathcal{M}_1^{\perp}$, then there exist a minimal nontrivial V-invariant subspace $\mathcal{M}_2 \subset \mathcal{M}_1^{\perp}$. Proceeding in this manner, we obtain a sequence of mutually orthogonal V-invariant subspaces $\mathcal{M}_1, \ldots, \mathcal{M}_m$ such that

$$\mathbb{C}^n = \mathcal{M}_1 + \cdots + \mathcal{M}_m$$

and for each j there are no nontrivial V-invariant subspaces in $\mathcal{M}_j$. By Theorem 11.2.1 the restriction $V|_{\mathcal{M}_j}$ ($j = 1, \ldots, m$) coincides with the algebra of all transformations $\mathcal{M}_j \to \mathcal{M}_j$. It remains to apply Lemma 11.5.2.

11.6 EXERCISES

11.1 Prove or disprove that the following sets of $n \times n$ matrices are algebras:

(a) Upper triangular Toeplitz matrices:

$$\begin{bmatrix} a_1 & a_2 & a_3 & \cdots & a_n \\ 0 & a_1 & a_2 & \cdots & a_{n-1} \\ \vdots & \vdots & \vdots & & \vdots \\ 0 & 0 & 0 & \cdots & a_1 \end{bmatrix}, \qquad a_j \in \mathbb{C} \qquad (1)$$

(b) Toeplitz matrices:

$$\begin{bmatrix} a_0 & a_1 & a_2 & \cdots & a_{n-1} \\ a_{-1} & a_0 & a_1 & \cdots & a_{n-1} \\ \vdots & \vdots & \vdots & & \vdots \\ a_{-n+1} & a_{-n+2} & a_{-n+3} & \cdots & a_0 \end{bmatrix}, \qquad a_j \in \mathbb{C} \quad (2)$$

(c) Circulant matrices:

$$\begin{bmatrix} a_1 & a_2 & \cdots & a_n \\ a_n & a_1 & \cdots & a_{n-1} \\ \vdots & \vdots & & \vdots \\ a_2 & a_3 & \cdots & a_1 \end{bmatrix}, \qquad a_j \in \mathbb{C} \qquad (3)$$

(d) Companion matrices:

$$\begin{bmatrix} 0 & 1 & 0 & \cdots & 0 \\ 0 & 0 & 1 & \cdots & 0 \\ \vdots & \vdots & \vdots & & \vdots \\ 0 & 0 & 0 & \cdots & 1 \\ a_0 & a_1 & a_2 & \cdots & a_{n-1} \end{bmatrix}, \qquad a_j \in \mathbb{C} \qquad (4)$$

(e) Upper triangular matrices $[a_{ij}]_{i,j=1}^n$ where $a_{ij} = 0$ if $i > j$.

11.2 Prove or disprove that the following sets of $nk \times nk$ matrices are algebras:

(a) Block upper triangular Toeplitz matrices (1), where a_j are $k \times k$ matrices, $j = 1, \ldots, n$.

(b) Block Toeplitz matrices (2), where a_j are $k \times k$ matrices, $j = -n + 1, \ldots, n - 1$.

(c) Block circulant matrices (3), where a_j are $k \times k$ matrices, $j = 1, \ldots, n$.

(d)   Block upper triangular matrices $[a_{ij}]_{i,j=1}^{n}$, where a_{ij} are $k \times k$ matrices and $a_{ij} = 0$ if $i > j$.

(e)   Matrices of type

$$
\begin{bmatrix}
0 & a_{12} & 0 & \cdots & 0 \\
0 & 0 & a_{13} & \cdots & 0 \\
\vdots & \vdots & \vdots & & \vdots \\
0 & 0 & 0 & \cdots & a_{n-1,n} \\
a_{n1} & a_{n2} & a_{n3} & \cdots & a_{nn}
\end{bmatrix}
$$

where a_{ij} are $k \times k$ matrices.

11.3   Show that the set of all $n \times n$ matrices of type

$$
\begin{bmatrix}
a_1 & 0 & 0 & \cdots & 0 & b_n \\
0 & a_2 & 0 & \cdots & b_{n-1} & 0 \\
0 & 0 & a_3 & \cdots & 0 & 0 \\
\vdots & \vdots & & & \vdots & \vdots \\
0 & b_2 & 0 & \cdots & a_{n-1} & 0 \\
b_1 & 0 & 0 & \cdots & 0 & a_n
\end{bmatrix}
$$

is an algebra. Find all invariant subspaces of this algebra.

11.4   Let A be an $n \times n$ matrix.

(a)   Show that the set

$$Q = \{X \in M_{n,n} \mid \mathrm{Inv}(A) = \mathrm{Inv}(X)\}$$

is not necessarily an algebra.

(b)   Prove that the closure of Q, that is, the set of all $n \times n$ matrices X for which there exists a sequence $\{X_m\}_{m=1}^{\infty}$ with $X_m \in Q$ for $m = 1, 2, \ldots$ and $\lim_{m \to \infty} X_m = X$, is an algebra with identity.

(c)   Describe all invariant subspaces of the closure of Q.

11.5   Show that the algebra of all $n \times n$ upper triangular Toeplitz matrices and the algebra of all $n \times n$ upper triangular matrices have exactly the same lattice of invariant subspaces.

11.6   Show that the algebra of all upper triangular $n \times n$ matrices contains any algebra A for which

$$\mathrm{Inv}(A) = \{\{0\}, \mathrm{Span}\{e_1\}, \ldots, \mathrm{Span}\{e_1, \ldots, e_{n-1}\}, \mathbb{C}^n\}$$

11.7   Show that there is no algebra A with identity strictly contained in the algebra $UT(n)$ of upper triangular Toeplitz matrices for which

$$\mathrm{Inv}(A) = \{\{0\}, \mathrm{Span}\{e_1\}, \ldots, \mathrm{Span}\{e_1, \ldots, e_{n-1}\}, \mathbb{C}^n\}$$

11.8 Prove that the algebra $U(n)$ of $n \times n$ upper triangular matrices is the unique reflexive algebra for which the lattice of all invariant subspaces is the chain

$$\{0\} \subset \text{Span}\{e_1\} \subset \text{Span}\{e_1, e_2\} \subset \cdots \subset \text{Span}\{e_1, \ldots, e_{n-1}\} \subset \mathbb{C}^n$$
(5)

11.9 Show that there exist n different algebras $V_1, \ldots, V_n$ whose set of invariant subspaces coincides with (5) and for which

$$UT(n) = V_1 \subset V_2 \subset \cdots \subset V_n = U(n)$$

11.10 Find all invariant subspaces of the algebra of all $2n \times 2n$ matrices of type

$$\begin{bmatrix} A & B \\ C & D \end{bmatrix}$$

where A, B, C, and D are upper triangular matrices.

11.11 As Exercise 11.10 but now, in addition, B and C have zeros along the main diagonal.

11.12 Find all invariant subspaces of the algebra of all $2n \times 2n$ matrices $\begin{bmatrix} A & B \\ C & D \end{bmatrix}$, where A, B, C, and D are $n \times n$ circulant matrices.

11.13 Let A be an $n \times n$ matrix that is not a scalar multiple of the identity. Find a nontrivial invariant subspace for the algebra of all matrices that commute with A. Does there exist such a subspace of dimension 1?

11.14 Let A be an $n \times n$ matrix and

$$V = \left\{ \sum_{j=0}^{n-1} \alpha_j A^j \ \middle|\ \alpha_0, \ldots, \alpha_{n-1} \in \mathbb{C} \right\}$$

be the algebra of polynomials in A. Give necessary and sufficient conditions for reflexivity of V in terms of the structure of the Jordan form of A.

11.15 Indicate which of the following algebras are reflexive:

(a) $n \times n$ upper triangular Toeplitz matrices.
(b) $n \times n$ upper triangular matrices.
(c) $n \times n$ circulant matrices.
(d) $nk \times nk$ block circulant matrices (with $k \times k$ blocks).
(e) $nk \times nk$ block upper triangular matrices (with $k \times k$ blocks).
(f) $nk \times nk$ block upper triangular Toeplitz matrices (with $k \times k$ blocks).
(g) the algebra from Exercise 11.3.

11.16 Let Q be as in Exercise 11.4. When is the closure of Q a reflexive algebra?

11.17 Given a chain of subspaces

$$\{0\} \subset \mathcal{M}_1 \subset \cdots \subset \mathcal{M}_k \subset \mathbb{C}^n \tag{6}$$

construct reflexive and nonreflexive algebras whose set of invariant subspaces coincides with (6).

11.18 Let $x_1, \ldots, x_n$ be a basis in $\mathbb{C}^n$, and let Λ be the minimal lattice of subspaces that contains $\text{Span}\{x_1\}, \ldots, \text{Span}\{x_n\}$. Prove that there exists a unique algebra V for which $\Lambda = \text{Inv}(V)$. Is V reflexive?

11.19 Let V be an algebra of $n \times n$ matrices without identity and such that $A^n = 0$ for every $A \in V$. Prove that $A_1 A_2 \cdots A_n = 0$ for every n-tuple of matrices $A_1, \ldots, A_n$ from V. (*Hint*: Use Theorem 11.2.2.)

Chapter Twelve

Real Linear Transformations

In this chapter we review the basic facts concerning invariant subspaces for transformations $A: \mathbb{R}^n \to \mathbb{R}^n$, focusing mainly on those results that are different (or their proofs are different) in the real case, or cannot be obtained as immediate corollaries, from the corresponding results for transformations from $\mathbb{C}^n$ into $\mathbb{C}^n$.

We note here that the applications presented in Chapters 5, 7, and 8 also hold in the real case. That is, applications to matrix polynomials $\sum_{j=0}^{l} \lambda^j A_j$ with real $n \times n$ matrices A_j and to rational matrix functions $W(\lambda)$ whose values are real $n \times n$ matrices for the real values of λ that are not poles of $W(\lambda)$. In fact, the description of multiplication and divisibility of matrix polynomials and rational matrix functions in terms of invariant subspaces (as developed in Chapters 5 and 7) holds for matrices over any field. This remark applies for the linear fractional decompositions of rational matrix functions as well. In contrast, the Brunovsky canonical form (Section 6.2) is not available in the framework of real matrices, so all the results of Chapter 6 that are based on the Brunovsky canonical form fail, in general, in this context. Also, the results of Chapter 11 do not generally hold in the context of finite-dimensional algebras over the field of real numbers.

12.1 DEFINITION, EXAMPLES, AND FIRST PROPERTIES OF INVARIANT SUBSPACES

Let $A: \mathbb{R}^n \to \mathbb{R}^n$ be a linear transformation. As in the case of linear transformations on a complex space, we say that a subspace $\mathcal{M} \subset \mathbb{R}^n$ is invariant for A (or A invariant) if $Ax \in \mathcal{M}$ for every $x \in \mathcal{M}$. The whole of $\mathbb{R}^n$ and the zero subspace are trivially A invariant, and the same applies to Im A and Ker A. As in the complex case, one checks that all the nonzero invariant subspaces of the $n \times n$ Jordan block with real eigenvalue (considered as a transformation from $\mathbb{R}^n$ into $\mathbb{R}^n$ written as a matrix in the standard orthonormal basis $e_1, \ldots, e_n$) are $\mathrm{Span}\{e_1, \ldots, e_k\}$, $k = 1, \ldots, n$. Also, for the diagonal matrix $A = \mathrm{diag}[\lambda_1, \ldots, \lambda_n]$, where $\lambda_1, \ldots, \lambda_n$ are distinct real numbers, all the invariant subspaces are of the form $\mathrm{Span}\{e_i \mid i \in K\}$ with $K \subset \{1, \ldots, n\}$ ($\mathrm{Span}\{e_i \mid i \in \emptyset\}$ is interpreted as the zero subspace).

In addition to these examples, the following example is basic and specially significant for real transformations.

EXAMPLE 12.1.1.   Let

$$
A = \begin{bmatrix}
\sigma & \tau & 1 & 0 & \cdots & 0 & 0 \\
-\tau & \sigma & 0 & 1 & \cdots & 0 & 0 \\
0 & 0 & \sigma & \tau & \cdots & 0 & 0 \\
0 & 0 & -\tau & \sigma & \cdots & 0 & 0 \\
\vdots & \vdots & \vdots & \vdots & & 1 & 0 \\
 & & & & & 0 & 1 \\
0 & 0 & 0 & 0 & \cdots & \sigma & \tau \\
0 & 0 & 0 & 0 & \cdots & \tau & \sigma
\end{bmatrix} : \mathbb{R}^n \to \mathbb{R}^n
$$

where σ and τ are real numbers and $\tau \neq 0$. The size n of the matrix A is obviously an even number. It is easily seen that $\mathrm{Span}\{e_1, \ldots, e_{2k}\}$, $k = 1, \ldots, n/2$ are A-invariant subspaces. It turns out that A has no other nontrivial invariant subspaces. Indeed, replacing A by $A - \sigma I$, we can assume without loss of generality that $\sigma = 0$. We prove that if $\mathcal{M}$ is an A-invariant subspace and $x = \Sigma_{j=1}^{2k} \alpha_j e_j \in \mathcal{M}$ with at least one of the real numbers α_{2k-1} and α_{2k} different from zero, then $\mathcal{M} \supset \mathrm{Span}\{e_1, \ldots, e_{2k}\}$, and proceed by induction on k.

In the case $k = 1$ we have $\alpha_1 e_1 + \alpha_2 e_2 \in \mathcal{M}$ and $A(\alpha_1 e_1 + \alpha_2 e_2) = \tau \alpha_2 e_1 - \tau \alpha_1 e_2 \in \mathcal{M}$. The conditions $\tau \neq 0$ and $\alpha_1^2 + \alpha_2^2 \neq 0$ ensure that both vectors e_1 and e_2 are linear combinations of $\alpha_1 e_1 + \alpha_2 e_2$ and $\tau \alpha_2 e_1 - \tau \alpha_1 e_2$, and the assertion is proved for $k = 1$. Assuming that the assertion is proved for $k - 1$, let $x = \Sigma_{j=1}^{2k} \alpha_j e_j \in \mathcal{M}$ with $\alpha_{2k-1}^2 + \alpha_{2k}^2 \neq 0$. A computation shows that the vector $y = (A^2 + \tau^2)x$ belongs to $\mathrm{Span}\{e_1, \ldots, e_{2k-2}\}$ and in the linear combination $y = \Sigma_{j=1}^{2k-2} \beta_j e_j$ at least one of the numbers β_{2k-3}, β_{2k-2} is different from zero. Obviously, $y \in \mathcal{M}$, so the induction assumption implies $\mathcal{M} \supset \mathrm{Span}\{e_1, \ldots, e_{2k-2}\}$. Hence $\alpha_{2k-1} e_{2k-1} + \alpha_{2k} e_{2k} \in \mathcal{M}$; as the difference $Ax - (\tau \alpha_{2k} e_{2k-1} - \tau \alpha_{2k-1} e_{2k})$ belongs to $\mathrm{Span}\{e_1, \ldots, e_{2k-2}\}$, also $\tau \alpha_{2k} e_{2k-1} - \tau \alpha_{2k-1} e_{2k} \in \mathcal{M}$. Consequently, the vectors e_{2k-1} and e_{2k} belong to $\mathcal{M}$, and $\mathcal{M} \supset \mathrm{Span}\{e_1, \ldots, e_{2k}\}$. In particular, A has no odd-dimensional invariant subspaces. $\square$

We say that a complex number λ_0 is an *eigenvalue* of A if $\det(\lambda_0 I - A) = 0$. Note that we admit nonreal numbers as eigenvalues of the real transformation A. As before, the set of all eigenvalues of A will be called the *spectrum* of A and denoted by $\sigma(A)$. Since the polynomial $\det(\lambda I - A)$ has real coefficients (as one can see by writing A in matrix form in some basis in $\mathbb{R}^n$), it follows that the spectrum of A is symmetrical with respect to the real axis: if λ_0 is an eigenvalue of A, so is $\bar{\lambda}_0$, and the multiplicity of λ_0 as a zero of $\det(\lambda I - A)$ is equal to that of $\bar{\lambda}_0$.

Not every transformation $A: \mathbb{R}^n \to \mathbb{R}^n$ has real eigenvalues. For instance, in Example 12.1.1 the eigenvalues of A are $\sigma + i\tau$ and $\sigma - i\tau$. However, if n is odd, then A must have at least one real eigenvalue. Indeed, $\det(\lambda I - A)$ is a monic polynomial of degree n with real coefficients; hence for n odd $\det(\lambda I - A)$ has real zeros. This implies the following fact (which has already been observed in the case of Example 12.1.1).

Proposition 12.1.1

If the transformation $A: \mathbb{R}^n \to \mathbb{R}^n$ has no real eigenvalues, then A has no odd-dimensional invariant subspaces.

Proof. If $\mathcal{M} \subset \mathbb{R}^n$ were an odd-dimensional A-invariant subspace, the restriction $A|_{\mathcal{M}}$ would have a real eigenvalue, which contradicts the fact that A has no real eigenvalues. (As in the complex case, the eigenvalues of any restriction $A|_N$ to an A-invariant subspace are necessarily eigenvalues of A.) $\square$

The Jordan chains for real transformations are defined in the same way as for complex transformations: vectors $x_0, \ldots, x_k \in \mathbb{R}^n$ form a *Jordan chain* of the transformation $A: \mathbb{R}^n \to \mathbb{R}^n$ corresponding to the eigenvalue λ_0 of A if $x_0 \neq 0$ and $Ax_0 = \lambda_0 x_0$; $Ax_j - \lambda_0 x_j = x_{j-1}$, $j = 1, \ldots, k$. The vector x_0 is called an *eigenvector*. The eigenvalue λ_0 for which a Jordan chain exists must obviously be real. Since not every real transformation has real eigenvalues, it follows that there exist transformations $A: \mathbb{R}^n \to \mathbb{R}^n$ without Jordan chains (and in particular without eigenvectors). On the other hand, for every real eigenvalue λ_0 of $A: \mathbb{R}^n \to \mathbb{R}^n$ there exists an eigenvector (which is any nonzero vector from $\text{Ker}(\lambda_0 I - A) \subset \mathbb{R}^n$). In particular, A has eigenvectors provided n is odd.

As we have seen (e.g., in Example 12.1.1), not every real transformation has one-dimensional invariant subspaces. In contrast, two-dimensional invariant subspaces always exist, as shown in the following proposition.

Proposition 12.1.2

Any transformation $A: \mathbb{R}^n \to \mathbb{R}^n$ with $n \geq 2$ has at least one two-dimensional invariant subspace.

Proof. Assume first that A has a pair of nonreal eigenvalues $\sigma + i\tau$, $\sigma - i\tau$ (σ, τ are real, $\tau \neq 0$). Then

$$0 = \det((\sigma + i\tau)I - A) \det((\sigma - i\tau)I - A) = \det((\sigma^2 + \tau^2)I - 2\sigma A + A^2)$$

Let $x \in \mathbb{R}^n \smallsetminus \{0\}$ be such that

$$[(\sigma^2 + \tau^2)I - 2\sigma A + A^2]x = 0 \tag{12.1.1}$$

Then clearly the subspace $\mathcal{M} = \text{Span}\{x, Ax\}$ is A-invariant. Further, $\mathcal{M}$ cannot be one-dimensional because otherwise $Ax = \mu x$ for some $\mu \in \mathbb{R}$, which in view of equality (12.1.1) would imply $\mu^2 - 2\mu\sigma + (\sigma^2 + \tau^2) = 0$, or $(\mu - \sigma)^2 + \tau^2 = 0$, which is impossible since $\tau \neq 0$.

If A has no nonreal eigenvalues, then (leaving aside the trivial case when A is a scalar multiple of I) the subspace $\text{Span}\{x, y\}$, where x and y are eigenvectors of A corresponding to different eigenvalues, is two-dimensional and A invariant. $\square$

It is clear now that Theorem 1.9.1 is generally false for real transformations. The next result is the real analog of that theorem.

Theorem 12.1.3

Let $A: \mathbb{R}^n \to \mathbb{R}^n$ be a transformation and assume that $\det(\lambda I - A)$ has exactly s real zeros (counting multiplicities). Then there exists an orthonormal basis $x_1, \ldots, x_n$ in $\mathbb{R}^n$ such that, with respect to this basis, the transformation A has the form $[a_{ij}]_{i,j=1}^n$ where all the entries a_{ij} with $i > j$ are zeros except for $a_{s+2,s+1}, a_{s+4,s+3}, \ldots, a_{n,n-1}$.

So, the matrix $[a_{ij}]_{i,j=1}^n$ is "almost" upper triangular.

Proof. Apply induction on n. If A has a real eigenvalue, then use the proof of Theorem 1.9.1. If A has no real eigenvalues, then pick a two-dimensional A-invariant subspace (which exists by Proposition 12.1.2) with an orthonormal basis x, y. Write A as the 2×2 block matrix with respect to the orthogonal decomposition $\mathbb{C}^n = \mathcal{M} \dotplus \mathcal{M}^\perp$:

$$A = \begin{bmatrix} A_{11} & A_{12} \\ 0 & A_{22} \end{bmatrix}$$

and apply the induction hypothesis to the transformation $A_{22}: \mathcal{M}^\perp \to \mathcal{M}^\perp$. $\square$

It follows from Theorem 12.1.3 that a transformation $A: \mathbb{R}^n \to \mathbb{R}^n$ with $\det(\lambda I - A)$ having s real zeros has a chain of $p + 1 \stackrel{\text{def}}{=} \frac{1}{2}(n + s) + 1$ invariant subspaces:

$$\{0\} = \mathcal{M}_0 \subset \mathcal{M}_1 \subset \cdots \subset \mathcal{M}_p = \mathbb{R}^n$$

(Observe that $n - s$ is the number of nonreal zeros of $\det(\lambda I - A)$. So $n - s$ and $n + s$ are even numbers.) We leave it to the reader to verify that $\frac{1}{2}(n + s) + 1$ is the maximal number of elements in a chain of A-invariant subspaces.

We say that a transformation $A: \mathbb{R}^n \to \mathbb{R}^n$ is *self-adjoint* if $(Ax, y) = (x, Ay)$ for every $x, y \in \mathbb{R}^n$, [As usual, $(\cdot, \cdot)$ stands for the standard scalar product in $\mathbb{R}^n$.] In other words, A is self-adjoint if $A = A^*$. Also, a transformation A is called *unitary* if $A^* = A^{-1}$ and *normal* if $AA^* = A^*A$. Note that in an orthonormal basis a self-adjoint transformation is represented by a symmetric matrix, and a unitary transformation is represented by an orthogonal matrix. (Recall that a real matrix U is called *orthogonal* if $UU^T = U^TU = I$.)

For normal transformations the "almost" triangular form of Theorem 12.1.3 is actually "almost" diagonal:

Theorem 12.1.4

Let A be as in Theorem 12.1.3 and assume, in addition, that A is normal. Then there exists an orthonormal basis in $\mathbb{R}^n$ with respect to which A has the matrix form $[a_{ij}]_{i,j=1}^n$, where $a_{ij} = 0$ for $i \neq j$ except for $a_{s+2,s+1}$, $a_{s+1,s+2}, \ldots, a_{n,n-1}, a_{n-1,n}$.

Proof. Use an orthonormal basis in $\mathbb{R}^n$ with the properties described in Theorem 12.1.3, and observe that the equality $A^*A = AA^*$ implies that actually $a_{ij} = 0$ for $i > j$ except $a_{s+1,s+2}, \ldots, a_{n-1,n}$. $\square$

12.2   ROOT SUBSPACES AND THE REAL JORDAN FORM

Let $A: \mathbb{R}^n \to \mathbb{R}^n$ be a transformation. The root subspace $\mathcal{R}_{\lambda_0}(A)$ corresponding to the real eigenvalue λ_0 of A is defined to be $\text{Ker}(\lambda_0 I - A)^n$, as in the complex case. Then $\mathcal{R}_{\lambda_0}(A)$ is spanned by the members of all Jordan chains of A corresponding to λ_0. For a pair of nonreal eigenvalues $\sigma + i\tau, \sigma - i\tau$ of A (here σ, τ are real and $\tau \neq 0$) the root subspace is defined by

$$\mathcal{R}_{\sigma \pm i\tau}(A) = \text{Ker}[(\sigma^2 + \tau^2)I - 2\sigma A + A^2]^p$$

where p is a positive integer such that

$$\text{Ker}[(\sigma^2 + \tau^2)I - 2\sigma A + A^2]^k \subset \text{Ker}[(\sigma^2 + \tau^2)I - 2\sigma A + A^2]^p$$

for every positive integer k.

Note that, if $\lambda_1, \ldots, \lambda_r$ are the distinct real eigenvalues of A (if any) and

$\sigma_1 + i\tau_1, \ldots, \sigma_s + i\tau_s$ are the district eigenvalues of A in the open upper half of the complex plane (if any), then

$$\det(\lambda I - A) = \prod_{j=1}^{r} (\lambda - \lambda_j)^{\alpha_j} \prod_{k=1}^{s} [(\sigma_k^2 + \tau_k^2) - 2\sigma_k\lambda + \lambda^2]^{\beta_k}$$

for some positive integers $\alpha_1, \ldots, \alpha_r, \beta_1, \ldots, \beta_s$. Using this observation, it can be proved that there is a direct sum decomposition

$$\mathcal{R}^n = \mathcal{R}_{\lambda_1}(A) \dotplus \cdots \dotplus \mathcal{R}_{\lambda_r}(A) \dotplus \mathcal{R}_{\sigma_1 \pm i\tau_1}(A) \dotplus \cdots \dotplus \mathcal{R}_{\sigma_s \pm i\tau_s}(A)$$

(see the remark following the proof of Theorem 2.1.2). Moreover, we have:

Theorem 12.2.1

For every A-invariant subspace $\mathcal{M}$ the direct sum decomposition

$$\mathcal{M} = (\mathcal{M} \cap \mathcal{R}_{\lambda_1}(A)) \dotplus \cdots \dotplus (\mathcal{M} \cap \mathcal{R}_{\lambda_r}(A)) \dotplus (\mathcal{M} \cap \mathcal{R}_{\sigma_1 \pm i\tau_1}(A)) \dotplus \cdots$$
$$\dotplus (\mathcal{M} \cap \mathcal{R}_{\sigma_s \pm i\tau_s}(A))$$

holds.

For the deeper study of properties of invariant subspaces, the real Jordan form of a real transformation, to be described in the following theorem, is most useful. As usual, $J_k(\lambda)$ denotes the $k \times k$ Jordan block with eigenvalue λ. Also, we introduce the $2l \times 2l$ matrix

$$J_l(\mu, w) = \begin{bmatrix} K & I_2 & 0 & \cdots & 0 \\ 0 & K & I_2 & \cdots & 0 \\ \vdots & \vdots & \vdots & & \vdots \\ 0 & 0 & 0 & \cdots & I_2 \\ 0 & 0 & 0 & \cdots & K \end{bmatrix}$$

where $K = \begin{bmatrix} \mu & w \\ -w & \mu \end{bmatrix}$ and μ, w are real numbers with $w \neq 0$ and I_2 represents the 2×2 identity matrix.

Theorem 12.2.2

For every transformation $A \colon \mathcal{R}^n \to \mathcal{R}^n$ there exists a basis in $\mathcal{R}^n$ in which A has the following matrix form:

$$A = J_{k_1}(\lambda_1) \oplus \cdots \oplus J_{k_p}(\lambda_p) \oplus J_{l_1}(\mu_1, w_1) \oplus \cdots \oplus J_{l_q}(\mu_q, w_q) \qquad (12.2.1)$$

where $\lambda_1, \ldots, \lambda_p; \mu_1, \ldots, \mu_q; w_1, \ldots, w_q$ are real numbers (not necessarily *distinct) and* $w_1, \ldots, w_q$ *are positive. In the representation* (12.2.1) *the blocks* $J_{k_i}(\lambda_i)$ *and* $J_{l_j}(\mu_j, w_j)$ *are uniquely determined by* A *up to permutation.*

The proof of Theorem 12.2.2 will be relegated to the next section.

The right-hand side of equality (12.2.1) is called a *real Jordan form* of A. Clearly, $\lambda_1, \ldots, \lambda_p$ are the real eigenvalues of A, and $\mu_1 \pm iw_1, \ldots, \mu_q \pm iw_q$ are the nonreal eigenvalues of A. Given $\lambda_0 \in \sigma(A)$, λ_0 real, the partial multiplicities and the algebraic and geometric multiplicity of A corresponding to λ_0 are defined as in the complex case. For a nonreal eigenvalue $\mu + iw$ of A, the *partial multiplicities* of A corresponding to $\mu + iw$ are, by definition, the half-sizes l_j of the blocks $J_{l_j}(\mu_j, w_j)$ with $\mu_j = \mu$ and $w_j = \pm w$. The number of partial multiplicities of A corresponding to $\mu + iw$ is the *geometric multiplicity* of $\mu + iw$, and the sum of partial multiplicities is the *algebraic multiplicity* of $\mu + iw$.

By use of the real Jordan form, it is not difficult to prove the following fact, which we need later.

Proposition 12.2.3

If n is odd, then every transformation $A: \mathbb{R}^n \to \mathbb{R}^n$ has an invariant subspace of any dimension k with $0 \leq k \leq n$.

Proof. Without loss of generality we can assume that A is given by an $n \times n$ matrix in the real Jordan form. As n is odd, A has a real eigenvalue, so that blocks $J_{k_i}(\lambda_i)$ in the real Jordan form (12.2.1) of A are present. Since the subspaces $\mathrm{Span}\{e_1, \ldots, e_j\}$, $j = 1, \ldots, k_i$ are $J_{k_i}(\lambda_i)$ invariant, and the subspaces $\mathrm{Span}\{e_1, \ldots, e_{2j}\}$, $j = 1, \ldots, l_j$ are $J_{l_j}(\mu_j, w_j)$ invariant, we obtain the existence of A-invariant subspaces of any dimension k, $0 \leq k \leq n$. $\square$

Analogs of the results on spectral and irreducible invariant subspaces proved in Chapter 2 can be stated and proved for transformations from $\mathbb{R}^n$ to $\mathbb{R}^n$. (As in the complex case we say that an A-invariant subspace $\mathcal{M}$ is irreducible if $\mathcal{M}$ cannot be represented as a direct sum of two A-invariant subspaces.) For example, see Theorem 12.2.4.

Theorem 12.2.4 *Let $A: \mathbb{R}^n \to \mathbb{R}^n$ be a transformation. The following statements are equivalent for an A-invariant subspace $\mathcal{M}$:*

(a) $\mathcal{M}$ *is irreducible.*
(b)   *Each A-invariant subspace contained in $\mathcal{M}$ is irreducible.*
(c)   *The Jordan form of the restriction $A|_{\mathcal{M}}$ is either $J_n(\lambda)$, $\lambda \in \mathbb{R}$, or (in case n is even) $J_{n/2}(\mu, w)$, $\mu, w \in \mathbb{R}$, $w \neq 0$.*

(d)   *There is either a unique eigenvector (up to multiplication by a nonzero real number) of A in $\mathcal{M}$ or (in case $A|_{\mathcal{M}}$ has no eigenvectors) a unique two-dimensional A-invariant subspace in $\mathcal{M}$.*

(e)   *The lattice of A-invariant subspaces is a chain.*

(f)   *The spectrum of $A|_{\mathcal{M}}$ is either a singleton $\{\lambda_0\}$, $\lambda_0 \in \mathbb{R}$, or a pair of nonreal eigenvalues $\{\mu + iw, \mu - iw\}$, and*

$$\operatorname{rank}[(A|_{\mathcal{M}} - \lambda_0 I)^i] = \max\{0, \dim \mathcal{M} - i\}, \qquad i = 0, 1, \ldots$$

in the former case and

$$\operatorname{rank}[[(\mu^2 + w^2)I - 2\mu A + A^2]|_{\mathcal{M}}]^i = \max\{0, \dim \mathcal{M} - 2i\}, \qquad i = 0, 1, \ldots$$

in the latter case.

The real Jordan form can be used instead of the (complex) Jordan form to produce results for real transformations analogous to those presented in Chapters 3 and 4 (with the exception of Proposition 3.1.4). For this purpose we say that a transformation $A: \mathbb{R}^n \to \mathbb{R}^n$ is *diagonable* if its real Jordan form has only 1×1 blocks $J_1(\lambda_j)$, $\lambda_1, \ldots, \lambda_p \in \mathbb{R}$ or 2×2 blocks $J_1(\mu_j, w_j)$, $j = 1, \ldots, q$. Also, we use the fact that the Jordan form of the transformation $A: \mathbb{R}^n \to \mathbb{R}^n$ with the real Jordan form (12.2.1) is

$$J_{k_1}(\lambda_1) \oplus \cdots \oplus J_{k_p}(\lambda_p) \oplus J_{l_1}(\mu_1 + iw_1) \oplus J_{l_1}(\mu_1 - iw_1) \oplus \cdots$$
$$\oplus J_{l_q}(\mu_q + iw_q) \oplus J_{l_q}(\mu_q - iw_q)$$

12.3   COMPLEXIFICATION AND PROOF OF THE REAL JORDAN FORM

We describe here a standard method for constructing a transformation $\mathbb{C}^n \to \mathbb{C}^n$ from a given transformation $\mathbb{R}^n \to \mathbb{R}^n$ with similar spectral properties. In many cases this method allows us to obtain results on real transformations from the corresponding results on complex transformations. In particular, it is used in the proof of Theorem 12.2.2.

Let $A: \mathbb{R}^n \to \mathbb{R}^n$ be a transformation. Define the *complexification* $A^c: \mathbb{C}^n \to \mathbb{C}^n$ of A as follows: $A^c(x + iy) = Ax + iAy$, where $x, y \in \mathbb{R}^n$. Obviously, A^c is a linear transformation. If A is given by an $n \times n$ matrix in some basis in $\mathbb{R}^n$, then this same basis may be considered as a basis in $\mathbb{C}^n$ and A^c is given by the same matrix. It is clear from this observation that the eigenvalues and the corresponding partial multiplicities of A and of A^c are the same.

Let $\mathcal{M}$ be a subspace in $\mathbb{R}^n$. Then $\mathcal{M} + i\mathcal{M} \overset{\text{def}}{=} \{x + iy \mid x, y \in \mathcal{M}\}$ is a subspace in $\mathbb{C}^n$. Moreover, if $\mathcal{M}$ is A invariant, then $\mathcal{M} + i\mathcal{M}$ is easily seen to be A^c invariant.

We need the following basic connection between the invariant subspaces of a real transformation and the invariant subspaces of its complexification.

Theorem 12.3.1

Assume that the transformation $A: \mathbb{R}^n \to \mathbb{R}^n$ does not have real eigenvalues. Let $\mathcal{R}_+ \subset \mathbb{C}^n$ be the spectral subspace of A^c corresponding to the eigenvalues in the open upper half plane. Then for every A-invariant subspace $\mathcal{L}(\subset \mathbb{R}^n)$ the subspace $(\mathcal{L} + i\mathcal{L}) \cap \mathcal{R}_+$ is A^c invariant and contained in $\mathcal{R}_+$. Conversely, for every A^c-invariant subspace $\mathcal{M} \subset \mathcal{R}_+$ there exists a unique A-invariant subspace $\mathcal{L}$ such that $(\mathcal{L} + i\mathcal{L}) \cap \mathcal{R}_+ = \mathcal{M}$.

Proof. The direct statement of Theorem 12.3.1 has already been observed. To prove the converse statement, let $\mathcal{M} \subset \mathcal{R}_+$ be an A^c-invariant subspace. Fix a basis $z_1, \ldots, z_k$ in $\mathcal{M}$, and write $z_j = x_j + iy_j, j = 1, \ldots, k$, where $x_j, y_j \in \mathbb{R}^n$. Put $\mathcal{L} = \mathrm{Span}\{x_1, \ldots, x_k, y_1, \ldots, y_k\} \subset \mathbb{R}^n$. Let us check that $\mathcal{L}$ is A invariant. Indeed, for each j, $A^c z_j$ is a linear combination (with complex coefficients) of $z_1, \ldots, z_k$, say

$$A^c z_j = \sum_{p=1}^{k} \alpha_p^{(j)} z_p \qquad (12.3.1)$$

Letting $\alpha_p^{(j)} = \beta_p^{(j)} + i\gamma_p^{(j)}$, where $\beta_p^{(j)}$ and $\gamma_p^{(j)}$ are real, use the definition of A^c to rewrite (12.3.1) in the form

$$Ax_j + iAy_j = \sum_{p=1}^{k} (\beta_p^{(j)} + i\gamma_p^{(j)})(x_p + iy_p), \qquad j = 1, \ldots, k$$

After separation of real and imaginary parts, these equations clearly imply that $\mathcal{L}$ is A invariant. Further, it is easily seen that

$$\mathcal{L} + i\mathcal{L} = \mathrm{Span}\{z_1, \ldots, z_k, \bar{z}_1, \ldots, \bar{z}_k\} \subset \mathbb{C}^n$$

where $\bar{z}_j = x_j - iy_j, j = 1, \ldots, k$. Equality (12.3.1) implies that the subspace $\mathcal{M} \overset{\mathrm{def}}{=} \mathrm{Span}\{\bar{z}_1, \ldots, \bar{z}_k\}$ is A^c invariant and

$$\sigma(A^c|_{\bar{\mathcal{M}}}) = \overline{\sigma(A^c|_{\mathcal{M}})}$$

This statement is easily verified; by letting $z_1, \ldots, z_k$ be a Jordan basis for $A^c|_{\mathcal{M}}$, for example. As $\mathcal{M} \subset \mathcal{R}_+$, we have $\bar{\mathcal{M}} \subset \mathcal{R}_-$, where $\mathcal{R}_-$ is the spectral subspace of A^c corresponding to the eigenvalues in the open lower half plane. Now

$$\mathscr{L} + i\mathscr{L} = [(\mathscr{L} + i\mathscr{L}) \cap \mathscr{R}_+] \dotplus [(\mathscr{L} + i\mathscr{L}) \cap \mathscr{R}_-]$$
$$\supset \mathrm{Span}\{z_1, \ldots, z_k\} \dotplus \mathrm{Span}\{\bar{z}_1, \ldots, \bar{z}_k\} = \mathscr{L} + i\mathscr{L}$$

Hence

$$(\mathscr{L} + i\mathscr{L}) \cap \mathscr{R}_+ = \mathrm{Span}\{z_1, \ldots, z_k\} = \mathscr{M}$$

It remains to prove the uniqueness of $\mathscr{L}$. Let $\mathscr{L}'$ be another A-invariant subspace such that

$$(\mathscr{L}' + i\mathscr{L}') \cap \mathscr{R}_+ = \mathscr{M} \qquad (12.3.2)$$

For a given subspace $\mathscr{N} \subset \mathbb{C}^n$, define its complex conjugate:

$$\bar{\mathscr{N}} = \{\langle \bar{z}_1, \ldots, \bar{z}_n \rangle \mid \langle z_1, \ldots, z_n \rangle \in \mathscr{N}, z_j \in \mathbb{C}\}$$

Obviously, $\bar{\mathscr{N}}$ is also a subspace in $\mathbb{C}^n$. We have $\overline{\mathscr{L}' + i\mathscr{L}'} = \mathscr{L}' + i\mathscr{L}'$. Also, it is easy to check (e.g., by taking complex conjugates of a Jordan basis in $\mathscr{R}_+$ for $A^c|_{\mathscr{R}_+}$) that $\overline{\mathscr{R}_+} = \mathscr{R}_-$. Taking complex conjugates in (12.3.2), we have

$$(\mathscr{L}' + i\mathscr{L}') \cap \mathscr{R}_- = \bar{\mathscr{M}}$$

and

$$\mathscr{L}' + i\mathscr{L}' = [(\mathscr{L}' + i\mathscr{L}') \cap \mathscr{R}_+] \dotplus [(\mathscr{L}' + i\mathscr{L}') \cap \mathscr{R}_-]$$
$$= \mathscr{M} \dotplus \bar{\mathscr{M}} = [(\mathscr{L} + i\mathscr{L}) \cap \mathscr{R}_+] \dotplus [(\mathscr{L} + i\mathscr{L}) \cap \mathscr{R}_-] = \mathscr{L} + i\mathscr{L}$$

As $\mathscr{L} + i\mathscr{L} = \{x + iy \mid x, y \in \mathscr{L}\}$, and similarly for $\mathscr{L}' + i\mathscr{L}'$, the equality of $\mathscr{L}'$ and $\mathscr{L}$ follows. $\square$

The proof shows that Theorem 12.3.1 remains valid if the subspace $\mathscr{R}_+$ is replaced by the spectral subspace of A^c corresponding to any set S of eigenvalues of A^c such that $\lambda_0 \in S$ implies $\bar{\lambda}_0 \notin S$ and S is maximal with respect to this property.

We pass now to the proof of Theorem 12.2.2. First, let us observe that in terms of matrices Theorem 12.2.2 can be restated as follows.

Theorem 12.3.2

Given an $n \times n$ matrix A whose entries are real numbers, there exists an invertible $n \times n$ matrix with real entries S such that

$$SAS^{-1} = J_{k_1}(\lambda_1) \oplus \cdots \oplus J_{k_p}(\lambda_p) \oplus J_{l_1}(\mu_1, w_1) \oplus \cdots \oplus J_{l_q}(\mu_q, w_q)$$

$$(12.3.3)$$

where λ_j, μ_j, and w_j are as in Theorem 12.2.2. The right-hand side of (12.3.3) is uniquely determined by A up to permutations of blocks $J_{k_i}(\lambda_i)$ and $J_{l_j}(\mu_j, w_j)$.

We now prove the result in the latter form. The Jordan form for transformations from $\mathbb{C}^n$ into $\mathbb{C}^n$ is used in the proof.

Proof. Let A^c be the complexification of A. Let $\mathcal{R}_{\lambda_0}(A^c) \subset \mathbb{C}^n$ be the root subspace of A^c corresponding to a *real* eigenvalue λ_0. As the matrices $(A^c - \lambda_0 I)^i$, $i = 0, 1, 2, \ldots$ have real entries, there exists a basis in each subspace $\text{Ker}(A^c - \lambda_0 I)^i \subset C^n$ that consists of n-dimensional vectors with real coordinates. (Here, we use the fact that vectors $x_1, \ldots, x_k \in \mathbb{R}^n$ are linearly independent over $\mathbb{R}$ if and only if they are linearly independent over $\mathbb{C}$.) Further, if m is such that $\text{Ker}(A^c - \lambda_0 I)^m = \mathcal{R}_{\lambda_0}(A^c)$ but $\text{Ker}(A^c - \lambda_0 I)^{m-1} \neq \mathcal{R}_{\lambda_0}(A^c)$, then, by using the same fact, we see that there is a basis in $\mathcal{R}_{\lambda_0}(A^c)$ modulo $\text{Ker}(A^c - \lambda_0 I)^{m-1}$ consisting of real vectors. We can now repeat the arguments from the proof of the Jordan form (Section 12.2.3) to show that there exists a basis in $\mathcal{R}_{\lambda_0}(A^c)$ consisting of Jordan chains of A^c (in short, a *Jordan basis*) with real coordinates.

Further, let $x_{i1}, \ldots, x_{i,m_i}$; $i = 1, \ldots, p$ be a Jordan basis in $\mathcal{R}_{\lambda_0}(A^c)$ where λ_0 is a nonreal eigenvalue of A^c (so for each i the vectors $x_{i1}, \ldots, x_{i,m_i}$ form a Jordan chain of A^c corresponding to λ_0). By taking complex conjugates in the equalities

$$(A^c - \lambda_0 I)x_{ij} = x_{i,j-1}; \qquad j = 1, \ldots, m_i; \qquad i = 1, \ldots, p$$

(by definition, $x_{i0} = 0$) and using the fact that A^c is given by a real matrix in the standard basis, we see that

$$\bar{x}_{i1}, \ldots, \bar{x}_{i,m_i}, \qquad i = 1, \ldots, p \qquad (12.3.4)$$

are the Jordan chains of A^c corresponding to $\bar{\lambda}_0$. The vectors (12.3.4) inherit linear independence from the vectors x_{ij}. Further, $\dim \mathcal{R}_{\lambda_0}(A^c) = \dim \mathcal{R}_{\bar{\lambda}_0}(A^c)$ (because the algebraic multiplicities of A^c at λ_0 and at $\bar{\lambda}_0$ are the same); hence the vectors (12.3.4) form a basis in $\mathcal{R}_{\bar{\lambda}_0}(A^c)$.

Putting together Jordan bases for each $\mathcal{R}_{\lambda_0}(A^c)$, where $\lambda_0 \in \mathbb{R} \cap \sigma(A^c)$, which consist of vectors with real coordinates, and Jordan bases for each pair of subspaces $\mathcal{R}_{\lambda_0}(A^c)$ and $\mathcal{R}_{\bar{\lambda}_0}(A^c)$ (where λ_0 is a nonreal eigenvalue of A^c) that are obtained from each other by complex conjugation, we obtain the following equality:

$$AR = R\{J_{m_1}(\lambda_1) \oplus \cdots \oplus J_{m_p}(\lambda_p) \oplus [J_{l_1}(\lambda_{p+1}) \oplus J_{l_1}(\bar{\lambda}_{p+1})] \oplus \cdots$$
$$\oplus [J_{l_q}(\lambda_{p+q}) \oplus J_{l_q}(\bar{\lambda}_{p+q})]\} \qquad (12.3.5)$$

Here $\lambda_1, \ldots, \lambda_p$ are real numbers, $\lambda_{p+1}, \ldots, \lambda_{p+q}$ are nonreal numbers (which can be assumed to have positive imaginary parts), and R is an invertible $n \times n$ matrix that, when partitioned according to the sizes of Jordan blocks in the right-hand side of (12.3.5), say

$$R = [R_1 \cdots R_p R_{p+1} R_{p+2} \cdots R_{p+2q-1} R_{p+2q}]$$

has the property that R_i $(i = 1, \ldots, p)$ are real and $R_{p+2j-1} = \bar{R}_{p+2j}$, $j = 1, \ldots, q$.

Fix j $(1 \le j \le q)$, and consider the $2l_j \times 2l_j$ matrix

$$U_j = \frac{1}{\sqrt{2}} \begin{bmatrix} 1 & -i & 0 & 0 & \cdots & 0 & 0 \\ 0 & 0 & 1 & -i & \cdots & 0 & 0 \\ \vdots & \vdots & \vdots & \vdots & & \vdots & \vdots \\ 0 & 0 & 0 & 0 & \cdots & 1 & -i \\ 1 & i & 0 & 0 & \cdots & 0 & 0 \\ 0 & 0 & 1 & i & \cdots & 0 & 0 \\ \vdots & \vdots & \vdots & \vdots & & \vdots & \vdots \\ 0 & 0 & 0 & 0 & \cdots & 1 & i \end{bmatrix}$$

One checks easily that U_j is unitary, that is, $U_j U_j^* = I$, and that

$$U_j^* [J_{l_j}(\lambda_{p+j}) \oplus J_{l_j}(\bar{\lambda}_{p+j})] U_j = J_{l_j}(\mu_j, w_j)$$

and μ_j and w_j are the real and imaginary parts of λ_{p+j}, repectively (see the paragraph preceding Theorem 12.2.2 for the definition of $J_{l_j}(\mu_j, w_j)$). Also, it is easily seen that the matrix

$$[R_{p+2j-1}, R_{p+2j}] U = [R_{p+2j-1}, \bar{R}_{p+2j-1}] U$$

has real entries. Multiplying (12.3.5) from the right by

$$U \overset{\text{def}}{=} \text{diag}[I_{m_1}, \ldots, I_{m_p}, U_{l_1}, \ldots, U_{l_q}]$$

and denoting the real invertible matrix RU by Q, we have

$$\begin{aligned} AQ &= RUU^* \{ J_{m_1}(\lambda_1) \oplus \cdots \oplus J_{m_p}(\lambda_p) \oplus [J_{l_1}(\lambda_{p+1}) \oplus J_{l_1}(\bar{\lambda}_{p+1})] \\ &\qquad \oplus [J_{l_q}(\lambda_{p+q}) \oplus J_{l_q}(\bar{\lambda}_{p+q})] \} U \\ &= Q \{ J_{m_1}(\lambda_1) \oplus \cdots \oplus J_{m_p}(\lambda_p) \oplus J_{l_1}(\mu_1, w_1) \oplus \cdots \oplus J_{l_q}(\mu_q, w_q) \} \end{aligned}$$

and formula (12.3.3) follows.

The uniqueness of the right-hand side of (12.3.3) follows from the uniqueness of the Jordan form of A^c. [Indeed, the right-hand side of (12.3.3) is uniquely determined by the eigenvalues and partial multiplicities of A^c.]

$\square$

12.4 COMMUTING MATRICES

Let A be an $n \times n$ matrix with real entries. In this section we study the general form of real matrices that commute with A. This result is applied in the next section to characterize the lattice of hyperinvariant subspaces of a real transformation.

In view of Theorem 12.2.2, we can assume that

$$A = \text{diag}[J_1, \ldots, J_u] \tag{12.4.1}$$

where each J_α is either a Jordan block of size $m_\alpha \times m_\alpha$ with real eigenvalue λ_α, or $J = J_{m_\alpha/2}(\mu_\alpha, w_\alpha)$ (in the notation introduced before Theorem 12.2.2). Let Z be a real matrix such that $AZ = ZA$. Partition Z according to (12.4.1): $Z = [Z_{\alpha\beta}]_{\alpha,\beta=1}^u$, where $Z_{\alpha\beta}$ is an $m_\alpha \times m_\beta$ real matrix. Then we have

$$J_\alpha Z_{\alpha\beta} = Z_{\alpha\beta} J_\beta ; \qquad \alpha, \beta = 1, \ldots, u \tag{12.4.2}$$

If $\sigma(J_\alpha) \cap \sigma(J_\beta) = \emptyset$, then equation (12.4.2) has only the trivial solution $Z_{\alpha\beta} = 0$ (Corollary 9.1.2). Assume $\sigma(J_\alpha) = \sigma(J_\beta) = \{\lambda_0\}$, where λ_0 is real. Then, as in the proof of Theorem 9.1.1, $Z_{\alpha\beta}$ is an upper triangular Toeplitz matrix.

To study the case $\sigma(J_\alpha) = \sigma(J_\beta) = \{\mu_0 + iw_0, \mu_0 - iw_0\}$, it is convenient to first verify the following lemma.

Lemma 12.4.1

Let $K = \begin{bmatrix} \mu & w \\ -w & \mu \end{bmatrix}$ be a 2×2 matrix with real μ, w such that $w \neq 0$. Then the system of equations

$$KA + C = AK ; \qquad KC = CK$$

for unknown 2×2 matrices A and C implies $C = 0$.

The lemma is verified by a direct computation after writing out the entries in A and C explicitly.

Now return to the case $\sigma(J_\alpha) = \sigma(J_\beta) = \{\mu_0 + iw_0, \mu_0 - iw_0\}$, $\mu_0, w_0 \in$ $\mathbb{R}$, $w_0 > 0$ in equations (12.4.2). Letting $K = \begin{bmatrix} \mu_0 & w_0 \\ -w_0 & \mu_0 \end{bmatrix}$ and writing $Z_{\alpha\beta}$

as a $(m_\alpha/2) \times (m_\beta/2)$ block matrix $[U_{ij}]_{i,j=1}^{m_\alpha/2, m_\beta/2}$ with 2×2 blocks U_{ij}, we have

$$
\begin{bmatrix}
K & I & 0 & \cdots & 0 \\
0 & K & I & \cdots & 0 \\
\vdots & \vdots & \vdots & & \vdots \\
0 & 0 & 0 & \cdots & I \\
0 & 0 & 0 & \cdots & K
\end{bmatrix}
\begin{bmatrix}
U_{11} & U_{12} & \cdots & U_{1,m_\beta/2} \\
U_{21} & U_{22} & \cdots & U_{2,m_\beta/2} \\
\vdots & & \vdots & & \vdots \\
U_{m_\alpha/2,1} & U_{m_\alpha/2,2} & \cdots & U_{m_\alpha/2,m_\beta/2}
\end{bmatrix}
$$

$$
=
\begin{bmatrix}
U_{11} & U_{12} & \cdots & U_{1,m_\beta/2} \\
U_{21} & U_{22} & \cdots & U_{2,m_\beta/2} \\
\vdots & \vdots & & \vdots \\
U_{m_\alpha/2,1} & U_{m_\alpha/2,2} & \cdots & U_{m_\alpha/2,m_\beta/2}
\end{bmatrix}
\begin{bmatrix}
K & I & 0 & \cdots & 0 \\
0 & K & I & \cdots & 0 \\
\vdots & \vdots & \vdots & & \vdots \\
0 & 0 & 0 & \cdots & I \\
0 & 0 & 0 & \cdots & K
\end{bmatrix}
\tag{12.4.3}
$$

Comparing the block entries $(m_\alpha/2, 1)$ and then $(m_\alpha/2 - 1, 1)$ in this equation, we obtain

$$
KU_{m_\alpha/2,1} = U_{m_\alpha/2,1}K , \qquad KU_{m_\alpha/2-1,1} + U_{m_\alpha/2,1} = U_{m_\alpha/2-1,1}K
$$

By Lemma 12.4.1, $U_{m_\alpha/2,1} = 0$. Now compare the block entries in positions $(m_\alpha/2 - 1, 1)$ and $(m_\alpha/2 - 2, 1)$, and reapplying Lemma 12.4.1, it follows that $U_{m_\alpha/2-1,1} = 0$. Continue in this way, and it is found that

$$
Z_{\alpha\beta} = [0, \tilde{Z}_{\alpha\beta}] \qquad (\text{if } m_\alpha \le m_\beta)
$$

$$
Z_{\alpha\beta} = \begin{bmatrix} \tilde{Z}_{\alpha\beta} \\ 0 \end{bmatrix} \qquad (\text{if } m_\alpha \ge m_\beta)
$$

where $\tilde{Z}_{\alpha\beta} = [\tilde{U}_{ij}]_{i,j=1}^{p}$ is a square $p \times p$ matrix, $p = \min(m_\alpha, m_\beta)$ with $\tilde{U}_{ij} = 0$ for $i > j$. So

$$
\begin{bmatrix}
K & I & 0 & \cdots & 0 \\
0 & K & I & \cdots & 0 \\
\vdots & \vdots & \vdots & & \vdots \\
0 & 0 & 0 & \cdots & I \\
0 & 0 & 0 & \cdots & K
\end{bmatrix}
\begin{bmatrix}
\tilde{U}_{11} & \tilde{U}_{12} & \cdots & \tilde{U}_{1,p/2} \\
0 & \tilde{U}_{22} & \cdots & \tilde{U}_{2,p/2} \\
\vdots & \vdots & & \vdots \\
0 & 0 & \cdots & \tilde{U}_{p/2,p/2}
\end{bmatrix}
$$

$$
=
\begin{bmatrix}
\tilde{U}_{11} & \tilde{U}_{12} & \cdots & \tilde{U}_{1,p/2} \\
0 & \tilde{U}_{22} & \cdots & \tilde{U}_{2,p/2} \\
\vdots & \vdots & & \vdots \\
0 & 0 & \cdots & \tilde{U}_{p/2,p/2}
\end{bmatrix}
\begin{bmatrix}
K & I & 0 & \cdots & 0 \\
0 & K & I & \cdots & 0 \\
\vdots & \vdots & \vdots & & \vdots \\
0 & 0 & 0 & \cdots & I \\
0 & 0 & 0 & \cdots & K
\end{bmatrix}
\tag{12.4.4}
$$

Equality (12.4.4) implies that for $j = 1, \ldots, p/2$, $K\tilde{U}_{jj} = \tilde{U}_{jj}K$ and for $j = 2, \ldots, p/2$

$$K\tilde{U}_{j-1,j} + \tilde{U}_{jj} = \tilde{U}_{j-1,j-1} + \tilde{U}_{j-1,j}K$$

In view of Lemma 12.4.1, $\tilde{U}_{11} = \tilde{U}_{22} = \cdots = \tilde{U}_{p/2,p/2}$; hence $\tilde{U}_{j-1,j}$ commutes with K for $j = 1, \ldots, p/2$. Further, $K\tilde{U}_{j-2,j} + \tilde{U}_{j-1,j} = \tilde{U}_{j-2,j-1} + \tilde{U}_{j-2,j}K$ for $j = 3, \ldots, p/2$. Using Lemma 12.4.1 again, $\tilde{U}_{j-1,j} = \tilde{U}_{j-2,j-1}$ and $K\tilde{U}_{j-2,j} = \tilde{U}_{j-2,j}K$. Continuing in this way, we find that $\tilde{U}_{ij}$ $(i \leq j)$ depends only on the difference between j and i and commutes with K. Because of the latter property $\tilde{U}_{ij}$ must have the form $\begin{bmatrix} a & b \\ -b & a \end{bmatrix}$ for some real numbers a and b (which depend, of course, on i and j).

Putting all the above information together, we arrive at the following description of all real matrices that commute with a given real $n \times n$ matrix A.

Theorem 12.4.2

Let A be an $n \times n$ matrix with the real Jordan form $\mathrm{diag}[J_1, \ldots, J_\mu]$, so

$$A = S^{-1}[\mathrm{diag}\, J_1, \ldots, J_\mu]S$$

for some invertible real $n \times n$ matrix S, where each J_α is either a Jordan block of size $m_\alpha \times m_\alpha$ with real eigenvalue or a matrix of type

$$\begin{bmatrix}
\mu_\alpha & w_\alpha & 1 & 0 & 0 & 0 & \cdots & 0 & 0 \\
-w_\alpha & \mu_\alpha & 0 & 1 & 0 & 0 & \cdots & 0 & 0 \\
0 & 0 & \mu_\alpha & w_\alpha & 1 & 0 & \cdots & 0 & 0 \\
0 & 0 & -w_\alpha & \mu_\alpha & 0 & 1 & \cdots & 0 & 0 \\
& & & & & & & \vdots & \vdots \\
\vdots & \vdots & \vdots & \vdots & \vdots & \vdots & & 1 & 0 \\
& & & & & & & 0 & 1 \\
0 & 0 & 0 & 0 & 0 & 0 & \cdots & \mu_\alpha & w_\alpha \\
0 & 0 & 0 & 0 & 0 & 0 & \cdots & -w_\alpha & \mu_\alpha
\end{bmatrix}$$

with real μ_α, w_α and $w_\alpha > 0$. Then every real $n \times n$ matrix X that commutes with A has the form $X = S^{-1}ZS$, where the matrix $Z = [Z_{\alpha\beta}]_{\alpha,\beta=1}^\mu$ partitioned conformally with the Jordan form $\mathrm{diag}[J_1, \ldots, J_\mu]$ has the following structure: If $\sigma(J_\alpha) \cap \sigma(J_\beta) = \emptyset$, then $Z_{\alpha\beta} = 0$. If $\sigma(J_\alpha) = (J_\beta) = \{\lambda_0\}$, λ_0 real, then

$$Z = [0 \quad T_{\alpha\beta}] \quad \text{in case} \quad m_\alpha \leq m_\beta$$

or

$$Z = \begin{bmatrix} T_{\alpha\beta} \\ 0 \end{bmatrix} \quad \text{in case} \quad m_\alpha \geq m_\beta$$

where

$$T_{\alpha\beta} = \begin{bmatrix} x_{\alpha\beta}^{(1)} & x_{\alpha\beta}^{(2)} & \cdots & x_{\alpha\beta}^{(p)} \\ 0 & x_{\alpha\beta}^{(1)} & \cdots & x_{\alpha\beta}^{(p-1)} \\ \vdots & \vdots & & \vdots \\ 0 & 0 & \cdots & x_{\alpha\beta}^{(1)} \end{bmatrix}, \qquad p = \min(m_\alpha, m_\beta)$$

is a real $p \times p$ upper triangular Toeplitz matrix. If $\sigma(J_\alpha) = \sigma(J_\beta) = \{\mu + iw, \mu - iw\}$, where μ and $w > 0$ are real, then again

$$Z_{\alpha\beta} = [0 \quad T_{\alpha\beta}] \quad \text{in case} \quad m_\alpha \leq m_\beta$$

or

$$Z_{\alpha\beta} = \begin{bmatrix} T_{\alpha\beta} \\ 0 \end{bmatrix} \quad \text{in case} \quad m_\alpha \geq m_\beta$$

and in this case

$$T_{\alpha\beta} = \begin{bmatrix} X_{\alpha\beta}^{(1)} & X_{\alpha\beta}^{(2)} & \cdots & X_{\alpha\beta}^{(q)} \\ 0 & X_{\alpha\beta}^{(1)} & \cdots & X_{\alpha\beta}^{(q-1)} \\ \vdots & \vdots & & \vdots \\ 0 & 0 & \cdots & X_{\alpha\beta}^{(1)} \end{bmatrix}, \qquad q = \tfrac{1}{2} \min(m_\alpha, m_\beta)$$

where the 2×2 blocks $X_{\alpha\beta}^{(j)}$ have the form

$$X_{\alpha\beta}^{(j)} = \begin{bmatrix} u_{\alpha\beta}^{(j)} & v_{\alpha\beta}^{(j)} \\ -v_{\alpha\beta}^{(j)} & u_{\alpha\beta}^{(j)} \end{bmatrix}$$

for some real numbers $u_{\alpha\beta}^{(j)}$ and $v_{\alpha\beta}^{(j)}$.

12.5   HYPERINVARIANT SUBSPACES

Let $A: \mathbb{R}^n \to \mathbb{R}^n$ be a transformation. A subspace $\mathcal{M} \subset \mathbb{R}^n$ is called A *hyperinvariant* if $\mathcal{M}$ is invariant for every transformation $X: \mathbb{R}^n \to \mathbb{R}^n$ that commutes with A. It is easily seen that the set of all A-hyperinvariant subspaces is a lattice. In this section we obtain another characterization of this lattice, one that is analogous to Theorem 9.4.2. The description of commuting matrices obtained in Theorem 12.4.2 is used in the proof.

Theorem 12.5.1

Let a transformation $A: \mathbb{R}^n \to \mathbb{R}^n$ have the minimal polynominal

$$f(\lambda) = \prod_{i=1}^{k} (\lambda - \lambda_i)^{r_i} \prod_{j=1}^{m} [(\lambda - \mu_j)^2 + w_j^2]^{s_j}$$

where λ_i, μ_j, and w_j are real and $w_j > 0$, $\lambda_1, \ldots, \lambda_k$ are distinct, and so are $\mu_1 + iw_1, \ldots, \mu_s + i\omega_s$. Then the lattice of all A hyperinvariant subspaces coincides with the smallest lattice $\mathscr{S}_A$ of subspaces in $\mathbf{C}^n$ that contains $\mathrm{Ker}(A - \lambda_j I)^k$, $\mathrm{Im}(A - \lambda_j I)^k$ for $k = 1, \ldots, r_j$; $j = 1, \ldots, k$, and $\mathrm{Ker}[(A - \mu_j I)^2 + w_j^2 I]^k$, $\mathrm{Im}[(A - \mu_j I)^2 + w_j^2 I]^k$ for $k = 1, \ldots, s_j$; $j = 1, \ldots, m$.

We consider first a particular case of Theorem 12.5.1 when the spectrum of A consists only of one pair of nonreal eigenvalues $\mu + iw$, $\mu - iw$ ($\mu, w \in \mathbf{R}$, $w \neq 0$). Let

$$f_1^{(1)}, \ldots, f_{2p_1}^{(2)}; f_1^{(2)}, \ldots, f_{2p_2}^{(2)}; \ldots; f_1^{(m)}, \ldots, f_{2p_m}^{(m)} \qquad (12.5.1)$$

be a Jordan basis in $\mathbf{R}^n$, where $p_1 \geq \cdots \geq p_m$ so that, in this basis, A is represented by the matrix

$$J_{p_1}(\mu, w) \oplus \cdots \oplus J_{p_m}(\mu, w)$$

Let

$$\mathscr{K}_j^i = \mathrm{Span}\{f_1^{(i)}, \ldots, f_{2j}^{(i)}\}, \qquad j = 1, \ldots, p_i; \qquad i = 1, \ldots, m$$

The following lemma is an analog of Lemma 9.5.2.

Lemma 12.5.2

Every A-hyperinvariant subspace is of the form

$$\mathscr{K}_{q_1}^1 \dotplus \cdots \dotplus \mathscr{K}_{q_m}^m \qquad (12.5.2)$$

where $q_1, \ldots, q_m$ is a non-decreasing sequence of nonnegative integers such that $p_1 - q_1 \geq \cdots \geq p_m - q_m \geq 0$.

If $q_i = 0$ for some i, then, of course, $\mathscr{K}_{q_i}^i$ is interpreted as the zero subspace. We see later that conversely, every subspace of the form (12.5.2) is A hyperinvariant.

Proof. Let $\mathscr{L}$ be a nonzero A-hyperinvariant subspace, and let $x \in \mathscr{L}$. Write x as a linear combination of the vectors (12.5.1):

$$x = \sum_{i=1}^{2p_1} \xi_i^{(1)} f_i^{(1)} + \cdots + \sum_{i=1}^{2p_m} \xi_i^{(m)} f_i^{(m)}$$

We claim that each vector $y_r = \sum_{i=1}^{2p_r} \xi_i^{(r)} f_i^{(r)}$ belongs to $\mathscr{L}$. Indeed, let P_r be the projector on $\mathscr{K}_{p_r}^r$ defined by $P_r f_i^{(s)} = 0$ for $s \neq r$ and $P_r f_i^{(r)} = f_i^{(r)}$ for $i = 1, \ldots, 2p_r$. It follows from Theorem 12.4.2 that P_r commutes with A. Hence $\mathscr{L}$ is P_r invariant, and $y_r = P_r x \in \mathscr{L}$.

Fix an integer r between 1 and m and denote by α the maximal index i of a nonzero coefficient $\xi_i^{(r)}$ $(i = 1, \ldots, 2p_r)$. Without loss of generality, we can assume that $\alpha = 2\beta$ is even (otherwise consider Ax in place of x). Let us show that all the vectors $f_1^{(r)}, \ldots, f_\alpha^{(r)}$ belong to $\mathscr{L}$. Indeed, the vectors

$$z_1 = [(A - \mu I)^2 + w^2 I]^{\beta - 1} y_r = \xi_{\alpha-1}^{(r)} f_1^{(r)} + \xi_\alpha^{(r)} f_2^{(r)}$$

and $z_2 = Az_1$ belong to $\mathscr{L}$ and also to $\text{Span}\{f_1^{(r)}, f_2^{(r)}\}$. Now z_1 and z_2 are not collinear; otherwise, A would have a real eigenvalue, and this is impossible. It follows that $\text{Span}\{z_1, z_2\} = \text{Span}\{f_1^{(r)}, f_2^{(r)}\}$, and hence $f_1^{(r)}, f_2^{(r)} \in \mathscr{L}$. If we already know that $f_1^{(r)}, \ldots, f_{2i-2}^{(r)} \in \mathscr{L}$ for some $i \geq 2$, then by a similar argument using the vectors

$$z_{2i-1} = [(A - \mu I)^2 + w^2 I]^{\beta - i} y_r \in \mathscr{L}$$

and $z_{2i} = Az_{2i-1} \in \mathscr{L}$, we find that $f_{2i-1}^{(r)}, f_{2i}^{(r)} \in \mathscr{L}$. For $i = \beta$ we have $f_1^{(r)}, \ldots, f_\alpha^{(r)} \in \mathscr{L}$.

As the vector $x \in \mathscr{L}$ was arbitrary, it follows that $\mathscr{L} = \mathscr{K}_{q_1}^1 \dotplus \cdots \dotplus \mathscr{K}_{q_m}^m$ for some integers q_i such that $0 \leq q_i \leq p_i$, $i = 1, \ldots, m$. To prove that $q_1 \geq \cdots \geq q_r$, we must show that $\mathscr{K}_\alpha^r \subset \mathscr{L}$ implies $\mathscr{K}_\alpha^{r-1} \subset \mathscr{L}$. Consider the transformation $B: \mathbb{R}^n \to \mathbb{R}^n$ that, in the basis (12.5.1), has the block matrix form $B = [X_{ij}]_{i,j=1}^m$ where X_{ij} is the $2p_i \times 2p_j$ zero matrix $(i, j = 1, \ldots, m)$, except for

$$X_{r-1,r} = \begin{bmatrix} I_{2p_{r-1}} \\ 0 \end{bmatrix}$$

Theorem 12.4.2 ensures that B commutes with A. Hence $\mathscr{L}$ is B invariant, and $f_i^{(r-1)} = Bf_i^{(r)} \in \mathscr{L}$, $i = 1, \ldots, 2\alpha$. In other words, $\mathscr{K}_\alpha^{r-1} \subset \mathscr{L}$.

Further, consider the transformation $C: \mathbb{R}^n \to \mathbb{R}^n$ that, in the basis (12.5.1), has the block matrix form $C = [Y_{ij}]_{i,j=1}^m$, where Y_{ij} is the $2p_i \times 2p_j$ zero matrix except for

$$Y_{r+1,r} = [0 \quad I_{2p_{r+1}}]$$

Then by Theorem 12.4.2, C commutes with A, and assuming $2q_r > 2(p_r - p_{r+1})$, we have

$$Cf_{2q_r}^{(r)} = f_{2q_r - 2(p_r - p_{r+1})}^{(r+1)} \in \mathscr{L}$$

This implies $2q_r - 2(p_r - p_{r+1}) \leq 2q_{r+1}$, or $p_r - q_r \geq p_{r+1} - q_{r+1}$. If $q_r \leq p_r - p_{r+1}$, then the inequality $p_r - q_r \geq p_{r+1} - q_{r+1}$ is obvious. $\square$

We are now in a position to prove Theorem 12.5.1 for the case $\sigma(A) = \{\mu + iw, \mu - iw\}$. As in the proof of Theorem 9.4.2, one shows that every subspace of the form $\mathrm{Ker}[(A - \mu I)^2 + w^2 I]^k$, or $\mathrm{Im}[(A - \mu I)^2 + w^2 I]^k$ is A hyperinvariant. So we have only to show that every A-hyperinvariant subspace $\mathscr{L}$ belongs to the lattice $\mathscr{S}_A$. By Lemma 12.5.2

$$\mathscr{L} = \mathscr{K}^1_{q_1} \dotplus \cdots \dotplus \mathscr{K}^m_{q_m} \tag{12.5.3}$$

for some sequence of integers $q_1, \ldots, q_m$ such that $q_1 \geq \cdots \geq q_m \geq 0$ and $p_1 - q_1 \geq p_2 - q_2 \geq \cdots \geq p_m - q_m \geq 0$. We prove that $\mathscr{L} \in \mathscr{S}_A$ by induction on q_1. Assume first $q_1 = 1$. Then $\mathscr{L} = \mathscr{K}^1_1 \dotplus \cdots \dotplus \mathscr{K}^t_1$ for some $t \leq m$. As $p_t > p_{t+1}$, we have

$$\mathscr{L} = \mathrm{Ker}[(A - \mu I)^2 + w^2 I] \cap \mathrm{Im}[(A - \mu I)^2 + w^2 I]^{p_t - 1} \in \mathscr{S}_A$$

Now assume that the inclusion $\mathscr{L} \in \mathscr{S}_A$ is proved for $q_1 = \nu - 1$, and let $\mathscr{L}$ be a subspace of the form (12.5.3) with $q_1 = \nu$. Let r, α be the maximal integers for which $q_1 = \cdots = q_r$ and $p_\alpha - p_r + \nu > 0$. Consider the subspace

$$\mathscr{M} = \mathscr{K}^1_\nu \dotplus \cdots \dotplus \mathscr{K}^r_\nu \dotplus \mathscr{K}^{r+1}_{p_{r+1} - p_r + \nu} \dotplus \cdots \dotplus \mathscr{K}^\alpha_{p_\alpha - p_r + \nu}$$

It is easily seen that

$$\mathscr{M} = \mathrm{Ker}[(A - \mu I)^2 + w^2 I]^\nu \cap \mathrm{Im}[(A - \mu I)^2 + w^2 I]^{p_r + \nu} \in \mathscr{S}_A$$

The inequalities $p_i - q_i \geq p_{i+1} - q_{i+1}$ imply that $\mathscr{M} \subset \mathscr{L}$. Further, the subspace

$$\mathscr{N} = \mathscr{L} \cap \mathrm{Ker}[(A - \mu I)^2 + w^2 I]^\nu$$

is A hyperinvariant, and since

$$\mathscr{N} = \mathscr{K}^1_{\nu - 1} \dotplus \cdots \dotplus \mathscr{K}^r_{\nu - 1} \dotplus \mathscr{K}^{r+1}_{q_{r+1}} \dotplus \cdots \dotplus \mathscr{K}^m_{q_m}$$

the induction hypothesis ensures that $\mathscr{N} \in \mathscr{S}_A$. Finally, $\mathscr{L} = \mathscr{M} + \mathscr{N}$ belongs to $\mathscr{S}_A$ as well.

We have proved Theorem 12.5.1 for the case when the spectrum of A consists of exactly one pair of nonreal eigenvalues. As the proof shows, the converse statement of Lemma 12.5.2 is also true: every subspace of the form (12.5.2) is A hyperinvariant.

Proof of Theorem 5.1 (the general case). Again, it is easily seen that each subspace $\mathrm{Ker}(A - \lambda_j I)^k$, $\mathrm{Im}(A - \lambda_j I)^k$, $\mathrm{Ker}[(A - \lambda_j I)^2 + w_j^2 I]^k$, $\mathrm{Im}[(A - \lambda_j I)^2 + w_j^2 I]^k$ is A hyperinvariant. So we must show that each A-hyperinvariant subspace belongs to $\mathscr{S}_A$. Let $\mathscr{M}$ be an A-hyperinvariant subspace. By Theorem 12.2.1 we have

$$\mathcal{M} = (\mathcal{M} \cap \mathcal{R}_{\lambda_1}(A)) \dotplus \cdots \dotplus (\mathcal{M} \cap \mathcal{R}_{\lambda_k}(A)) \dotplus (\mathcal{M} \cap \mathcal{R}_{\mu_1 \pm iw_1}(A)) \dotplus \cdots$$
$$\dotplus (\mathcal{M} \cap \mathcal{R}_{\mu_s \pm iw_s}(A))$$

Write A in the real Jordan form (as in Theorem 12.2.2) and use Theorem 12.4.2 to deduce that each intersection $\mathcal{M} \cap \mathcal{R}_j(A)$ is $A|_{\mathcal{R}_{\lambda_j}(A)}$ hyperinvariant and $\mathcal{M} \cap \mathcal{R}_{\mu_p \pm iw_p}$ is $A|_{\mathcal{R}_{\mu_p \pm iw_p}(A)}$ hyperinvariant ($p = 1, \ldots, s$). With the use of Theorem 9.4.2, it follows that $\mathcal{M} \cap \mathcal{R}_{\lambda_j}(A)$ belongs to the smallest lattice that contains the subspaces

$$\operatorname{Ker}(A|_{\mathcal{R}_{\lambda_j}(A)} - \lambda_j I)^k = \operatorname{Ker}(A - \lambda_j I)^k, \qquad k = 1, \ldots, r_j$$

and

$$\operatorname{Im}(A|_{\mathcal{R}_{\lambda_j}(A)} - \lambda_j I)^k = \operatorname{Im}(A - \lambda_j I)^k \cap \operatorname{Ker}(A - \lambda_j I)^{r_j}, \qquad k = 1, \ldots, r_j$$

Similarly, by the part of the theorem already proved, we find that $\mathcal{M} \cap \mathcal{R}_{\mu_p \pm iw_p}(A)$ belongs to the smallest lattice that contains the subspaces

$$\operatorname{Ker}[(A|_{\mathcal{R}_{\mu_p \pm iw_p}(A)} - \mu_p I)^2 + w_p^2 I]^k = \operatorname{Ker}[(A - \mu_p I)^2 + w_p^2 I]^k$$

for $k = 1, 2, \ldots, s_p$, and

$$\operatorname{Im}[(A|_{\mathcal{R}_{\mu_p \pm iw_p}(A)} - \mu_p I)^2 + w_p^2 I]^k$$
$$= \operatorname{Im}[(A - \mu_p I)^2 + w_p^2 I]^k \cap \operatorname{Ker}[(A - \mu_p I)^2 + w_p^2 I]^{s_p}$$

It follows that $\mathcal{M} \in \mathscr{S}_A$, and Theorem 12.5.1 is proved completely. $\square$

12.6 REAL TRANSFORMATIONS WITH THE SAME INVARIANT SUBSPACES

In this section we describe transformations $B: \mathbb{R}^n \to \mathbb{R}^n$, which have the same invariant subspaces as a given transformation $A: \mathbb{R}^n \to \mathbb{R}^n$. This description is a real analog of Theorem 10.2.1.

By Theorem 12.2.2, we can assume that, in a certain basis in $\mathbb{R}^n$, A has the matrix form

$$A = \operatorname{diag}[J_1, \ldots, J_p, K_1, \ldots, K_q] \tag{12.6.1}$$

where

$$J_i = \operatorname{diag}[J_{k_{i1}}(\lambda_i), \ldots, J_{k_{im_i}}(\lambda_i)], \qquad i = 1, \ldots, p$$

with different real numbers $\lambda_1, \ldots, \lambda_p$; and

$$K_i = \text{diag}[J_{l_{i1}}(\mu_i, w_i), \ldots, J_{l_{ir_i}}(\mu_i, w_i)], \qquad i = 1, \ldots, q$$

with different complex numbers $\mu_1 + iw_1, \ldots, \mu_q + iw_q$ in the open upper half plane. We use the notation introduced in Section 12.2, and also assume that $k_{i1} \geq \cdots \geq k_{im_i}$, $l_{i1} \geq \cdots \geq l_{ir_i}$.

Now introduce the following notation (partly used in Section 10.2): given real numbers $a_0, \ldots, a_{s-1}$, denote by $T_s(a_0, \ldots, a_{s-1})$ the $s \times s$ upper triangular Toeplitz matrix

$$\begin{bmatrix} a_0 & a_1 & \cdots & a_{s-1} \\ 0 & a_0 & \cdots & a_{s-2} \\ \vdots & \vdots & & \vdots \\ 0 & 0 & \cdots & a_0 \end{bmatrix} \qquad (12.6.2)$$

Further, for positive integers $s \leq t$ let

$$U_t(a_0, \ldots, a_{s-1}, F) = \begin{bmatrix} a_0 & a_1 & a_2 & \cdots & a_{s-1} & f_{11} & f_{12} & \cdots & f_{1,t-s} \\ 0 & a_0 & a_1 & \cdots & a_{s-2} & a_{s-1} & f_{22} & \cdots & f_{2,t-2} \\ \vdots & \vdots & & & & & & & \vdots \\ 0 & 0 & & & \cdots & & & a_{s-1} & f_{t-s,t-s} \\ 0 & 0 & & & \cdots & & & a_{s-2} & a_{s-1} \\ \vdots & \vdots & & & & & & & \vdots \\ 0 & 0 & & & \cdots & & & 0 & a_0 \end{bmatrix}$$

$$(12.6.3)$$

where F is a real $(t-s) \times (t-s)$ upper triangular matrix

$$\begin{bmatrix} f_{11} & f_{12} & \cdots & f_{1,t-s} \\ 0 & f_{22} & \cdots & f_{2,t-s} \\ \vdots & \vdots & & \vdots \\ 0 & 0 & \cdots & f_{t-s,t-s} \end{bmatrix} \qquad (12.6.4)$$

Similarly, if $a_j = \begin{bmatrix} b_j & c_j \\ -c_j & b_j \end{bmatrix}$, $j = 1, \ldots, s-1$ are 2×2 real matrices, we define the $2s \times 2s$ upper triangular Toeplitz matrix $T_s^{2 \times 2}(a_0, \ldots, a_{s-1})$ by the same formula (12.6.2). If, in addition, the real 2×2 matrices f_{jk} $(1 \leq j \leq k \leq t-s)$ are given, denote by $U_t^{2 \times 2}(a_0, \ldots, a_{s-1}; F)$ the $2t \times 2t$ matrix given by (12.6.3) with F given by (12.6.4). By definition, for $s = t$ we have

$$U_s(a_0, \ldots, a_{s-1}; F) = T_s(a_0, \ldots, a_{s-1})$$

and

$$U_s^{2\times2}(a_0, \ldots, a_{s-1}; F) = T_s^{2\times2}(a_0, \ldots, a_{s-1})$$

We can now give a description of all transformations $B: \mathbb{R}^n \to \mathbb{R}^n$ with the same invariant subspaces as A.

Theorem 12.6.1

Let the transformation $A: \mathbb{R}^n \to \mathbb{R}^n$ be given by (12.6.1), in some basis in $\mathbb{R}^n$. Then a transformation $B: \mathbb{R}^n \to \mathbb{R}^n$ has the same invariant subspaces as A if and only if B has the following matrix form (in the same basis):

$$B = \text{diag}[B_1, \ldots, B_p, C_1, \ldots, C_q]$$

where

$$B_i = U_{k_{i1}}(b_1^{(i)}, \ldots, b_{k_{i2}}^{(i)}; F^{(i)}) \oplus T_{k_{i2}}(b_1^{(i)}, \ldots, b_{k_{i2}}^{(i)}) \oplus \cdots$$
$$\oplus T_{k_{im_i}}(b_1^{(i)}, \ldots, b_{k_{im_i}}^{(i)})$$

for some real numbers $b_1^{(i)}, \ldots, b_{k_{i2}}^{(i)}$ with $b_2^{(i)} \neq 0$ and some $(k_{i1} - k_{i2}) \times (k_{i1} - k_{i2})$ matrix $F^{(i)}$;

$$C_j = U_{l_{j1}}^{2\times2}(c_1^{(j)}, \ldots, c_{l_{j2}}^{(j)}; G^{(j)}) \oplus T_{l_{j2}}^{2\times2}(c_1^{(j)}, \ldots, c_{l_{j2}}^{(j)}) \oplus \cdots$$
$$\oplus T_{l_{jr_j}}^{2\times2}(c_1^{(j)}, \ldots, c_{l_{jr_j}}^{(j)})$$

for some 2×2 real blocks

$$c_s^{(j)} = \begin{bmatrix} d_s^{(j)} & f_s^{(j)} \\ -f_s^{(j)} & d_s^{(j)} \end{bmatrix}, \qquad s = 1, \ldots, l_{j2}$$

with $f_1^{(j)} \neq 0$ and $\det c_2^{(j)} \neq 0$ and some $2(l_{j1} - l_{j2}) \times 2(l_{j1} - l_{j2})$ real matrix $G^{(j)}$. Moreover, the real numbers $b_1^{(1)}, \ldots, b_1^{(p)}$ are different and the complex numbers $d_1^{(1)} + i|f_1^{(1)}|, \ldots, d_1^{(q)} + i|f_1^{(q)}|$ are different as well.

For the proof of Theorem 12.6.1, we refer the reader to Soltan (1974).

12.7   EXERCISES

12.1   Prove that the transformation of rotation through an angle φ:

$$\begin{bmatrix} \cos \varphi & \sin \varphi \\ \sin \varphi & \cos \varphi \end{bmatrix} : \mathbb{R}^2 \to \mathbb{R}^2$$

has no nontrivial invariant subspaces except when φ is an integer multiple of π.

12.2 Given an example of a transformation $A: R^{2n} \rightarrow R^{2n}$ such that A has no eigenvectors but A^2 has a basis of eigenvectors in R^{2n}.

12.3 Show that if $A: R^n \rightarrow R^n$ is such that A^2 has an eigenvector corresponding to a nonnegative eigenvalue λ_0, then A has an eigenvector as well.

12.4 Show that if $A: R^{2n} \rightarrow R^{2n}$ is a transformation with det $A < 0$, then A has at least two distinct real eigenvalues.

12.5 Find the real Jordan form of the $n \times n$ matrix

$$
\begin{bmatrix}
0 & 1 & 0 & \cdots & 0 \\
0 & 0 & 1 & \cdots & 0 \\
\vdots & \vdots & \vdots & & \vdots \\
0 & 0 & 0 & \cdots & 1 \\
1 & 0 & 0 & \cdots & 0
\end{bmatrix}
$$

Find all the invariant subspaces in R^n of this matrix.

12.6 Describe the real Jordan form and all invariant subspaces in R^3 of the 3×3 real circulant matrix

$$
\begin{bmatrix}
a & b & c \\
c & a & b \\
b & c & a
\end{bmatrix}, \quad a, b, c \in R
$$

12.7 Find the real Jordan form of an $n \times n$ real circulant matrix

$$
\begin{bmatrix}
a_1 & a_2 & \cdots & a_n \\
a_n & a_1 & \cdots & a_{n-1} \\
\vdots & \vdots & & \vdots \\
a_2 & a_3 & \cdots & a_1
\end{bmatrix}
$$

12.8 Find the real Jordan form and all invariant subspaces in R^n of the real companion matrix

$$
\begin{bmatrix}
0 & 1 & 0 & \cdots & 0 \\
0 & 0 & 1 & \cdots & 0 \\
\vdots & \vdots & \vdots & & \vdots \\
0 & 0 & 0 & \cdots & 1 \\
a_0 & a_1 & a_2 & \cdots & a_{n-1}
\end{bmatrix}
$$

assuming that the polynomial $\lambda^n - a_{n-1}\lambda^{n-1} - \cdots - a_1\lambda - a_0$ has n distinct complex zeros.

12.9 What is the real Jordan form of real $n \times n$ companion matrix?

12.10   Find the real Jordan form and all invariant subspaces in $\mathbb{R}^n$ of the matrix

$$\begin{bmatrix} 0 & 0 & \cdots & 0 & \alpha_1 \\ 0 & 0 & \cdots & \alpha_2 & 0 \\ \vdots & \vdots & & \vdots & \vdots \\ 0 & \alpha_{n-1} & \cdots & 0 & 0 \\ \alpha_n & 0 & \cdots & 0 & 0 \end{bmatrix}$$

where $\alpha_1, \ldots, \alpha_n \in \mathbb{R}$.

12.11   Two linear matrix polynomials $\lambda A_1 + B_1$ and $\lambda A_2 + B_2$ with real matrices A_1, B_1, A_2, and B_2 are called *strictly equivalent* (*over* $\mathbb{R}$) if there exist invertible real matrices P and Q such that $P(\lambda A_1 + B_1)Q = \lambda A_2 + B_2$. Prove the following result on the canonical form for the strict equivalence (over $\mathbb{R}$) (the real analog of Theorem A.7.3). A real linear matrix polynomial $\lambda A + B$ is strictly equivalent (over $\mathbb{R}$) to a real linear polynomial of the type

$$0_{pq} \oplus L_{k_1} \oplus \cdots \oplus L_{k_r} \oplus M_{l_1} \oplus \cdots \oplus M_{l_s} \oplus (I_{m_1} + \lambda J_{m_1}(0)) \oplus \cdots$$
$$\oplus (I_{m_t} + \lambda J_{m_t}(0)) \oplus (\lambda I_{n_1} + J_{n_1}(\lambda_1)) \oplus \cdots \oplus (\lambda I_{n_u} + J_{n_u}(\lambda_u))$$
$$\oplus (\lambda I_{h_1} + J_{h_1}(\mu_1, \omega_1)) \oplus \cdots \oplus (\lambda I_{h_v} + J_{h_v}(\mu_v, \omega_v)) \tag{1}$$

where 0_{pq} is the $p \times q$ zero matrix; L_ϵ is the $\epsilon \times (\epsilon + 1)$ matrix

$$\begin{bmatrix} \lambda & 1 & 0 & \cdots & 0 \\ 0 & \lambda & 1 & \cdots & 0 \\ \vdots & \vdots & \vdots & & \vdots \\ 0 & 0 & 0 & \cdots & \lambda\ 1 \end{bmatrix}$$

M_ϵ is the transpose of L_ϵ; $\lambda_1, \ldots, \lambda_u$ are real numbers;

$$J_l(\mu, \omega) = \begin{bmatrix} K & I_2 & 0 & \cdots & 0 \\ 0 & K & I_2 & \cdots & 0 \\ \vdots & \vdots & \vdots & & \vdots \\ 0 & 0 & 0 & \cdots & K \end{bmatrix}$$

with $K = \begin{bmatrix} \mu & \omega \\ -\omega & \mu \end{bmatrix}$; and μ_j, ω_j are real numbers with $\omega_j > 0$ for $j = 1, \ldots, v$. Moreover, the form (1) is uniquely determined by $\lambda A + B$ up to permutations of blocks. (*Hint*: In the proof of Theorem A.7.3 use the real Jordan form in place of the complex Jordan form.)

12.12 Prove the following analog of the Brunovsky canonical form for real transformations. Two transformations $[A_1 \ \ B_1]: R^n \dotplus R^m \to R^n$ and $[A_2 \ \ B_2]: R^n \oplus R^m \to R^n$ are called *block similar* if there exist invertible transformations $M: R^m \to R^m$ and $N: R^n \to R^n$ and a transformation $F: R^n \to R^m$ such that

$$N^{-1}[A_1 \ \ B_1]\begin{bmatrix} N & 0 \\ F & M \end{bmatrix} = [A_2 \ \ B_2]$$

Prove that every transformation $[A \ B]: R^n \dotplus R^m \to R^n$ is block similar to a transformation $[A_0 \ \ B_0]$ of the following form (written as matrices with respect to certain bases in R^m and R^n):

$$A_0 = J_{k_1}(0) \oplus \cdots \oplus J_{k_r}(0) \oplus J$$

where J is a matrix in the real Jordan form; B_0 has all zero entries except for the entries $(k_1, 1), (k_1 + k_2, 2), \ldots, (k_1 + \cdots + k_r, r)$, and these exceptional entries are equal to 1. (*Hint*: Use Exercise 12.11 in the proof of the Brunovsky canonical form.)

12.13 Let $A: R^n \to R^n$ and $B: R^m \to R^n$ be a full-range pair of transformations. Prove that given a sequence $S = \{\lambda_1, \ldots, \lambda_n\}$ of n (not necessarily distinct) complex numbers such that $\lambda_0 \in S$ implies $\bar{\lambda}_0 \in S$ and $\bar{\lambda}_0$ appears in S exactly as many times as λ_0, there exists a transformation $F: R^n \to R^m$ such that $\lambda_1, \ldots, \lambda_n$ are the eigenvalues of $A + BF$ (counted with multiplicities). (*Hint*: Use Exercise 12.12.)

Notes to Part 2

Chapter 9. The first two sections contain standard material in linear algebra [see, e.g., Gantmacher (1959)]. Theorem 9.3.1 is due to Laffey (1978) and Guralnick (1979). The proof presented here follows Choi, Laurie, and Radjavi (1981). Theorem 9.4.2 appears in Soltan (1976) and Fillmore, Herrero, and Longstaff (1977). Our expositions of Theorem 9.4.2 and Section 9.6 follow the latter paper.

Chapter 10. The results and proofs of this chapter are from Soltan (1973b).

Chapter 11. Theorem 11.2.1 is a well-known result (Burnside's theorem). It may be found in books on general algebra [see, e.g., Jacobson (1953)] but generally not in books on linear algebra. In the proof of Theorem 11.2.1 we follow the exposition from Chapter 8 in Radjavi and Rosenthal (1973). Other proofs are also available [see Jacobson (1953); Halperin and Rosenthal (1980); E. Rosenthal (1984)]. Example 11.4.6 and Theorem 11.4.4 are from Halmos (1971). In the proof of Theorem 5.1 we are following Radjavi and Rosenthal (1973).

Chapter 12. The real Jordan form is a standard result, although not so frequently included in books on linear algebra as the (complex) Jordan form. The real Jordan form can be found in Lancaster and Tismenetsky (1985), for instance. The proof of Theorem 5.1 is taken from Soltan (1981).

Part Three

Topological Properties of Invariant Subspaces and Stability

There are a number of practical problems in which it is necessary to obtain an invariant subspace of a transformation or a matrix by numerical methods. In practice, numerical computation can be performed with only a finite degree of precision and, in addition, the data for a problem will generally be imprecise. In this situation, the best that we can hope to do is to obtain an invariant subspace of a transformation that is close to the one we really have in mind. However, simple examples show that although two transformations may be close (in any reasonable sense), their invariant subspaces can be completely different. This leads us to the problem of identifying all invariant subspaces of a given transformation that are "stable" under small perturbations of the transformation—that is, to identify those invariant subspaces for which the perturbed transformation will have a "close" or "neighbouring" invariant subspace, in an appropriate sense.

To develop these ideas, we must introduce a measure of distance between subspaces and to analyze further the structure of the invariant subspaces of a given transformation. This is done in Part 3, together with descriptions of stable invariant subspaces, using different notions of stability.

This machinery is then applied to the study of stability of divisors of polynomial and rational matrix functions and other problems. The reader whose interest is confined to the applications of Chapter 17 needs only to study the material presented in Chapter 13, Section 14.3, and Chapter 15.

385

Chapter Thirteen

The Metric Space
of Subspaces

This chapter is of an auxiliary character. We set forth the basic facts about the topological properties of the set of subspaces in $\mathbb{C}^n$. Observe that all the results and proofs of this chapter hold for the set of subspaces in $\mathbb{R}^n$ as well.

13.1 THE GAP BETWEEN SUBSPACES

We consider $\mathbb{C}^n$ endowed with the standard scalar product. If $x = \langle x_1, \ldots, x_n \rangle$, $y = \langle y_1, \ldots, y_n \rangle \in \mathbb{C}^n$, then $(x, y) = \sum_{i=1}^{n} x_i \bar{y}_i$, and the corresponding norm is

$$\|x\| = \left(\sum_{i=1}^{n} |x_i|^2 \right)^{1/2}$$

The norm of an $n \times n$ matrix A (or a transformation $A: \mathbb{C}^n \to \mathbb{C}^n$) is defined accordingly:

$$\|A\| = \max_{x \in \mathbb{C}^n \smallsetminus \{0\}} \|Ax\| / \|x\|$$

Now we introduce a concept that serves as a measure of distance between subspaces. The *gap* between subspaces $\mathscr{L}$ and $\mathscr{M}$ (in $\mathbb{C}^n$) is defined as

$$\theta(\mathscr{L}, \mathscr{M}) = \|P_{\mathscr{M}} - P_{\mathscr{L}}\| \tag{13.1.1}$$

where $P_{\mathscr{L}}$ and $P_{\mathscr{M}}$ are the orthogonal projectors on $\mathscr{L}$ and $\mathscr{M}$, respectively. It is clear from the definition that $\theta(\mathscr{L}, \mathscr{M})$ is a *metric* in the set of all subspaces in $\mathbb{C}^n$; that is, $\theta(\mathscr{L}, \mathscr{M})$ enjoys the following properties: (a) $\theta(\mathscr{L}, \mathscr{M}) > 0$ if $\mathscr{L} \neq \mathscr{M}$, $\theta(\mathscr{L}, \mathscr{L}) = 0$; (b) $\theta(\mathscr{L}, \mathscr{M}) = \theta(\mathscr{M}, \mathscr{L})$; (c) $\theta(\mathscr{L}, \mathscr{M}) \leq \theta(\mathscr{L}, \mathscr{N}) + \theta(\mathscr{N}, \mathscr{M})$ (the triangle inequality).

Note also that $\theta(\mathscr{L}, \mathscr{M}) \leq 1$. [This property follows immediately from the characterization given in condition (13.1.3).] It follows from (13.1.1) that

$$\theta(\mathscr{L}, \mathscr{M}) = \theta(\mathscr{L}^{\perp}, \mathscr{M}^{\perp}) \tag{13.1.2}$$

where $\mathscr{L}^{\perp}$ and $\mathscr{M}^{\perp}$ denote orthogonal complements. Indeed, $P_{\mathscr{L}^{\perp}} = I - P_{\mathscr{L}}$, so $\|P_{\mathscr{M}} - P_{\mathscr{L}}\| = \|P_{\mathscr{M}^{\perp}} - P_{\mathscr{L}^{\perp}}\|$.

In the following paragraphs denote by $S_{\mathscr{L}}$ the unit sphere in a subspace $\mathscr{L} \subset \mathbb{C}^n$, that is, $S_{\mathscr{L}} = \{x \in \mathscr{L} \mid \|x\| = 1\}$. We also need the concept of the distance of $d(x, Z)$ from $x \in \mathbb{C}^n$ to a set $Z \subset \mathbb{C}^n$. This is defined by $d(x, Z) = \inf_{t \in Z} \|x - t\|$.

Theorem 13.1.1

Let $\mathscr{M}$, $\mathscr{L}$ be subspaces in $\mathbb{C}^n$. Then

$$\theta(\mathscr{L}, \mathscr{M}) = \max\{\sup_{x \in S_{\mathscr{M}}} d(x, \mathscr{L}), \sup_{x \in S_{\mathscr{L}}} d(x, \mathscr{M})\} \tag{13.1.3}$$

If exactly one of the subspaces $\mathscr{L}$ and $\mathscr{M}$ is the zero subspace, then the right-hand side of (13.1.3) is interpreted as 1; if $\mathscr{L} = \mathscr{M} = \{0\}$, then the right-hand side of (13.1.3) is interpreted as 0. If P_1 and P_2 are projectors with $\operatorname{Im} P_2 = \mathscr{L}$ and $\operatorname{Im} P_2 = \mathscr{M}$, not necessarily orthogonal, then

$$\theta(\mathscr{L}, \mathscr{M}) \leq \|P_1 - P_2\| \tag{13.1.4}$$

Proof. For every $x \in S_{\mathscr{L}}$ we have

$$\|x - P_2 x\| = \|(P_1 - P_2)x\| \leq \|P_1 - P_2\|$$

Therefore

$$\sup_{x \in S_{\mathscr{L}}} d(x, \mathscr{M}) \leq \|P_1 - P_2\|$$

Similarly, $\sup_{x \in S_{\mathscr{M}}} d(x, \mathscr{L}) \leq \|P_1 - P_2\|$; so

$$\max\{p_{\mathscr{L}}, p_{\mathscr{M}}\} \leq \|P_1 - P_2\| \tag{13.1.5}$$

where $p_{\mathscr{L}} = \sup_{x \in S_{\mathscr{L}}} d(x, \mathscr{M})$, $p_{\mathscr{M}} = \sup_{x \in S_{\mathscr{M}}} d(x, \mathscr{L})$.

Observe that $p_{\mathscr{L}} = \sup_{x \in S_{\mathscr{L}}} \|(I - P_{\mathscr{M}})x\|$, $p_{\mathscr{M}} = \sup_{x \in S_{\mathscr{M}}} \|(I - P_{\mathscr{L}})x\|$. Consequently, for every $x \in \mathbb{C}^n$ we have

$$\|(I - P_{\mathscr{L}})P_{\mathscr{M}}x\| \leq p_{\mathscr{M}}\|P_{\mathscr{M}}x\|, \qquad \|(I - P_{\mathscr{M}})P_{\mathscr{L}}x\| \leq p_{\mathscr{L}}\|P_{\mathscr{L}}x\| \tag{13.1.6}$$

Now

$$\|P_{\mathcal{M}}(I - P_{\mathcal{L}})x\|^2 = ((I - P_{\mathcal{L}})P_{\mathcal{M}}(I - P_{\mathcal{L}})x, (I - P_{\mathcal{L}})x)$$
$$\leq \|(I - P_{\mathcal{L}})P_{\mathcal{M}}(I - P_{\mathcal{L}})x\| \cdot \|(I - P_{\mathcal{L}})x\|$$

Hence by (13.1.6)

$$\|P_{\mathcal{M}}(I - P_{\mathcal{L}})x\|^2 \leq p_{\mathcal{M}}\|P_{\mathcal{M}}(I - P_{\mathcal{L}})x\| \cdot \|(I - P_{\mathcal{L}})x\|$$

$$\|P_{\mathcal{M}}(I - P_{\mathcal{L}})X\| \leq p_{\mathcal{M}}\|(I - P_{\mathcal{L}})x\| \tag{13.1.7}$$

On the other hand, using the relation

$$P_{\mathcal{M}} - P_{\mathcal{L}} = P_{\mathcal{M}}(I - P_{\mathcal{L}}) - (I - P_{\mathcal{M}})P_{\mathcal{L}}$$

and the orthogonality of $P_{\mathcal{M}}$, we obtain

$$\|(P_{\mathcal{M}} - P_{\mathcal{L}})x\|^2 = \|P_{\mathcal{M}}(I - P_{\mathcal{L}})x\|^2 + \|(I - P_{\mathcal{M}})P_{\mathcal{L}}x\|^2$$

Taking advantage of (13.1.6) and (13.1.7) we obtain

$$\|(P_{\mathcal{M}} - P_{\mathcal{L}})x\|^2 \leq p_{\mathcal{M}}^2\|(I - P_{\mathcal{L}})x\|^2 + p_{\mathcal{L}}^2\|P_{\mathcal{L}}x\|^2 \leq \max\{p_{\mathcal{M}}^2, p_{\mathcal{L}}^2\}\|x\|^2$$

So

$$\|P_{\mathcal{M}} - P_{\mathcal{L}}\| \leq \max\{p_{\mathcal{L}}, p_{\mathcal{M}}\}$$

Using (13.1.5) (with $P_1 = P_{\mathcal{L}}$, $P_2 = P_{\mathcal{M}}$), we obtain (13.1.3). The inequality (13.1.4) follows now from (13.1.5). $\square$

It is an important property of the metric $\theta(\mathcal{L}, \mathcal{M})$ that, in a neighbourhood of every subspace $\mathcal{L} \in \mathbb{C}^n$, all the subspaces have the same dimension (equal to dim $\mathcal{L}$). This is a consequence of the following theorem.

Theorem 13.1.2

If $\theta(\mathcal{L}, \mathcal{M}) < 1$, then dim $\mathcal{L}$ = dim $\mathcal{M}$.

Proof. The condition $\theta(\mathcal{L}, \mathcal{M}) < 1$ implies that $\mathcal{L} \cap \mathcal{M}^{\perp} = \{0\}$ and $\mathcal{L}^{\perp} \cap \mathcal{M} = \{0\}$. Indeed, suppose the contrary, and assume, for instance, that $\mathcal{L} \cap \mathcal{M}^{\perp} \neq \{0\}$. Let $x \in S_{\mathcal{L}} \cap \mathcal{M}^{\perp}$. Then $d(x, \mathcal{M}) = 1$, and by (13.1.3) $\theta(\mathcal{L}, \mathcal{M}) \geq 1$, a contradiction. Now $\mathcal{L} \cap \mathcal{M}^{\perp} = \{0\}$ implies that dim $\mathcal{L} \leq$ dim $\mathcal{M}$, and $\mathcal{L}^{\perp} \cap \mathcal{M} = \{0\}$ implies that dim $\mathcal{L} \geq$ dim $\mathcal{M}$. $\square$

It also follows directly from this proof that the hypothesis $\theta(\mathcal{L}, \mathcal{M}) < 1$ implies $\mathbb{C}^n = \mathcal{L} \dotplus \mathcal{M}^{\perp} = \mathcal{L}^{\perp} \dotplus \mathcal{M}$. In addition, we have

$$P_{\mathcal{M}}(\mathcal{L}) = \mathcal{M}, \qquad P_{\mathcal{L}}(\mathcal{M}) = \mathcal{L}$$

For example, to see the first of these observe that for any $x \in \mathcal{M}$ there is the unique decomposition $x = y + z$, $y \in \mathcal{L}$, $z \in \mathcal{M}^\perp$. Hence $x = P_\mathcal{M} x = P_\mathcal{M} y$ so that $\mathcal{M} \subset P_\mathcal{M}(\mathcal{L})$. But the reverse inclusion is obvious, and so we must have equality.

The following result makes precise the idea that direct sum decompositions of $\mathbb{C}^n$ are stable under small perturbations of the subspaces, as measured in the gap metric.

Theorem 13.1.3

Let $\mathcal{M}, \mathcal{M}_1 \subset \mathbb{C}^n$ be subspaces such that

$$\mathcal{M} \dotplus \mathcal{M}_1 = \mathbb{C}^n$$

If $\mathcal{N}$ is a subspace in $\mathbb{C}^n$ such that $\theta(\mathcal{M}, \mathcal{N})$ is sufficiently small, then

$$\mathcal{N} \dotplus \mathcal{M}_1 = \mathbb{C}^n \tag{13.1.8}$$

and

$$\theta(\mathcal{M}, \mathcal{N}) \le \|\tilde{P}_\mathcal{M} - \tilde{P}_\mathcal{N}\| \le C\theta(\mathcal{M}, \mathcal{N}) \tag{13.1.9}$$

where $\tilde{P}_\mathcal{M}(\tilde{P}_\mathcal{N})$ projects $\mathbb{C}^n$ onto $\mathcal{M}$ (onto $\mathcal{N}$) along $\mathcal{M}_1$ and C is a constant depending on $\mathcal{M}$ and $\mathcal{M}_1$ but not on $\mathcal{N}$. In fact

$$C = 2\|\tilde{P}_\mathcal{M}\| \max_{x \in \mathcal{M}_1, \|x\|=1} \{d(x, \mathcal{M})^{-1}\}$$

Proof. Let us prove first that the sum $\mathcal{N} \dotplus \mathcal{M}_1$ is indeed direct. The condition that $\mathcal{M} \dotplus \mathcal{M}_1 = \mathbb{C}^n$ is a direct sum implies that $\|x - y\| \ge \delta > 0$ for every $x \in S_{\mathcal{M}_1}$ and every $y \in \mathcal{M}$. Here δ is a fixed positive constant. Take $\mathcal{N}$ so close to $\mathcal{M}$ that $\theta(\mathcal{M}, \mathcal{N}) \le \delta/2$. Then $\|z - y\| \le \delta/2$ for every $z \in S_\mathcal{N}$, where $y = y(z)$ is the orthogonal projection of z on $\mathcal{M}$. Thus for $x \in S_{\mathcal{M}_1}$ and $z \in S_\mathcal{N}$ we have

$$\|x - z\| \ge \|x - y\| - \|z - y\| \ge \frac{\delta}{2}$$

so $\mathcal{N} \cap \mathcal{M}_1 = \{0\}$. By Theorem 13.1.2 dim $\mathcal{N} =$ dim $\mathcal{M}$ if $\theta(\mathcal{M}, \mathcal{N}) < 1$, so dimensional considerations tell us that $\mathcal{N} + \mathcal{M}_1 = \mathbb{C}^n$ for $\theta(\mathcal{M}, \mathcal{N}) < 1$, and equation (13.1.8) follows.

To establish the right-hand inequality in (13.1.9) two preliminary remarks are needed. First note that for any $x \in \mathcal{M}$, and $y \in \mathcal{M}_1$ we have $x = \tilde{P}_\mathcal{M}(x + y)$ so that

$$\|x + y\| \ge \|\tilde{P}_\mathcal{M}\|^{-1}\|x\| \tag{13.1.10}$$

It is claimed that, for $\theta(\mathcal{M}, \mathcal{N})$ small enough

$$\|z + y\| \geq \tfrac{1}{2}\|\tilde{P}_{\mathcal{M}}\|^{-1}\|z\| \tag{13.1.11}$$

for all $z \in \mathcal{N}$ and $y \in \mathcal{M}_1$.

Without loss of generality, assume $\|z\| = 1$. Suppose that $\theta(\mathcal{M}, \mathcal{N}) < \delta$ and let $x \in \mathcal{M}$. Then, using (13.1.10), we obtain

$$\|z + y\| \geq \|x + y\| - \|z - x\| \geq \|\tilde{P}_{\mathcal{M}}\|^{-1}\|x\| - \delta$$

But then $x = (x - z) + z$ implies $\|x\| \geq 1 - \delta$, and so

$$\|z + y\| \geq \|\tilde{P}_{\mathcal{M}}\|^{-1}(1 - \delta) - \delta$$

and, for δ small enough, (13.1.11) is established.

The second remark is that, for any $x \in \mathbb{C}^n$

$$\|x - \tilde{P}_{\mathcal{M}}x\| \leq C_0 d(x, \mathcal{M}) \tag{13.1.12}$$

for some constant C_0. To establish (13.1.12), it is sufficient to consider the case that $x \in \mathcal{M}_1$ and $\|x\| = 1$. But then, obviously, we can take

$$C_0 = \max_{x \in \mathcal{M}_1, \|x\|=1} \{d(x, \mathcal{M})^{-1}\}$$

Now for any $x \in S_{\mathcal{N}}$, by use of (13.1.12) and (13.1.3), we obtain

$$\|(\tilde{P}_{\mathcal{M}} - \tilde{P}_{\mathcal{N}})x\| = \|x - \tilde{P}_{\mathcal{M}}x\| \leq C_0 d(x, \mathcal{M}) \leq C_0 \theta(\mathcal{M}, \mathcal{N})$$

Then, if $w \in \mathbb{C}^n$, $\|w\| = 1$, and $w = y + z$, $y \in \mathcal{N}$, $z \in \mathcal{M}_1$, it follows that

$$\|(\tilde{P}_{\mathcal{M}} - \tilde{P}_{\mathcal{N}})w\| = \|(\tilde{P}_{\mathcal{M}} - \tilde{P}_{\mathcal{N}})y\| \leq \|y\| C_0 \theta(\mathcal{M}, \mathcal{N}) \leq 2C_0 \|\tilde{P}_{\mathcal{M}}\| \theta(\mathcal{M}, \mathcal{N})$$

and the last inequality follows from (13.1.11). This completes the proof of the theorem. $\square$

We remark that the definition and analysis of the gap between subspaces presented in this section extends verbatim to a finite-dimensional vector space V over $\mathbb{C}$ (or over $\mathbb{R}$) on which a *scalar product* is defined. Namely, there exists a complex-valued (or real-valued) function defined on all the ordered pairs, x, y, where $x, y \in V$, denoted by (x, y), which satisfies the following properties: (a) $(\alpha x + \beta y, z) = \alpha(x, z) + \beta(y, z)$ for every $x, y, z \in V$ and every $\alpha, \beta \in \mathbb{C}$ (or $\alpha, \beta \in \mathbb{R}$); (b) $(x, y) = \overline{(y, x)}$, $x, y \in V$; (c) $(x, x) \geq 0$ for all $x \in V$; and $(x, x) = 0$ if and only if $x = 0$.

13.2 *THE MINIMAL ANGLE AND THE SPHERICAL GAP*

There are notions of the "minimal angle" and the "spherical gap" between two subspaces that are closely related to the gap between the subspaces. The basic facts about these notions are exposed in this and the next sections. It should be noted, however, that these notions and their properties are used (apart from Sections 13.2 and 13.3) only in Section 13.8 and in the proof of Theorem 15.2.1.

Given two subspaces $\mathcal{L}, \mathcal{M} \subset \mathbb{C}^n$, the *minimal angle* $\varphi_{\min}(\mathcal{L}, \mathcal{M})$ ($0 \leq \varphi_{\min}(\mathcal{L}, \mathcal{M}) \leq \pi/2$) between $\mathcal{L}$ and $\mathcal{M}$ is determined by

$$\sin \varphi_{\min}(\mathcal{L}, \mathcal{M}) = \inf\{\|x + y\| \mid x \in \mathcal{L}, \ y \in \mathcal{M}, \ \max\{\|x\|, \|y\|\} = 1\} \tag{13.2.1}$$

The minimal angle can also be defined by the equality

$$\cos \varphi_{\min}(\mathcal{L}, \mathcal{M}) = \sup_{x \in S_{\mathcal{L}}, S_{\mathcal{M}}} |(x, y)| \tag{13.2.2}$$

Indeed, writing

$$b_{x,y} = \inf_{\max\{|\alpha|, |\beta|\} = 1} \|\alpha x + \beta y\|$$

for any $x, y \in \mathbb{C}^n$, we have

$$b_{x,y}^2 = \min\{ \inf_{|\beta| \leq 1} \|x + \beta y\|^2, \ \inf_{|\alpha| \leq 1} \|\alpha x + y\|^2\}$$

Now for $\|x\| = \|y\| = 1$

$$\|x + \beta y\|^2 = (x, x) + \beta(x, y) + \bar{\beta}(y, x) + |\beta|^2(y, y)$$
$$= 1 + \beta(x, y) + \bar{\beta}(y, x) + |\beta|^2$$

and writing $\beta = u + iv$, where u and v are real, we see easily that the function $f(u, v) = 1 + \beta(x, y) + \bar{\beta}(y, x) + |\beta|^2$ of two real variables u and v has its minimum for $u = -\frac{1}{2}((x, y) + (y, x))$ and $v = \frac{1}{2}(i(y, x) - i(x, y))$, that is, when $\beta = -(y, x)$. Thus

$$\inf_{|\beta| \leq 1} \|x + \beta y\|^2 = 1 - |(x, y)|^2 \tag{13.2.3}$$

Similarly, if $\|x\| = \|y\| = 1$

$$\inf_{|\alpha| \leq 1} \|\alpha x + y\|^2 = 1 - |(x, y)|^2$$

Denote by a and b the right-hand sides of equations (13.2.2) and (13.2.1), respectively. Then

$$1 - a^2 = \inf_{x \in S_{\mathscr{L}}, y \in S_{\mathscr{M}}} (1 - |(x, y)|^2) \, ; \qquad b^2 = \inf_{x \in S_{\mathscr{L}}, y \in S_{\mathscr{M}}} b_{x, y}^2$$

In view of (13.2.3) the equality $1 - a^2 = b^2$ follows, and this means that, indeed, formulas (13.2.1) and (13.2.2) define the same angle $\varphi_{\min}(\mathscr{L}, \mathscr{M})$ with $0 \le \varphi_{\min}(\mathscr{L}, \mathscr{M}) \le \pi/2$.

Proposition 13.2.1

For two nontrivial subspaces $\mathscr{L}$ and $\mathscr{M}$ of $\mathbb{C}^n$, $\mathscr{L} \cap \mathscr{M} = \{0\}$ if and only if

$$\sin \varphi_{\min}(\mathscr{L}, \mathscr{M}) > 0$$

Proof. Obviously, if $x \in \mathscr{L} \cap \mathscr{M}$ is a vector of norm 1, then

$$\sin \varphi_{\min}(\mathscr{L}, \mathscr{M}) \le \|x + (-x)\| = 0$$

so $\varphi_{\min}(\mathscr{L}, \mathscr{M}) = 0$. Conversely, assume $\varphi_{\min}(\mathscr{L}, \mathscr{M}) = 0$. As the set $\Phi \stackrel{\text{def}}{=} \{(x, y) \in \mathbb{C}^n \times \mathbb{C}^n \mid \max\{\|x\|, \|y\|\} = 1\}$ is closed and bounded, the continuous function $\|x + y\|$ has a minimum in the set Φ, which in our case is zero. In other words, $\|x_0 + y_0\| = 0$ for some $x_0 \in \mathscr{L}$, $y_0 \in \mathscr{M}$, where at least one of $\|x_0\|$ and $\|y_0\|$ is equal to 1. But then, clearly, $x_0 \in \mathscr{L} \cap \mathscr{M} \setminus \{0\}$. $\square$

We also need the notion of the "spherical gap" between subspaces. For nonzero subspaces $\mathscr{L}$, $\mathscr{M}$ in $\mathbb{C}^n$ the spherical gap $\tilde{\theta}(\mathscr{L}, \mathscr{M})$ is defined by

$$\tilde{\theta}(\mathscr{L}, \mathscr{M}) = \max\{\sup_{x \in S_{\mathscr{M}}} d(x, S_{\mathscr{L}}), \sup_{x \in S_{\mathscr{L}}} d(x, S_{\mathscr{M}})\}$$

We also put $\tilde{\theta}(\{0\}, \mathscr{L}) = \tilde{\theta}(\mathscr{L}, \{0\}) = 1$ for every nonzero subspace $\mathscr{L}$ in $\mathbb{C}^n$ and $\tilde{\theta}(\{0\}, \{0\}) = 0$. The spherical gap is also a metric in the set of all subspaces in $\mathbb{C}^n$. Indeed, the only nontrivial statement that we have to verify for this purpose is the triangle inequality:

$$\tilde{\theta}(\mathscr{L}, \mathscr{M}) + \tilde{\theta}(\mathscr{M}, \mathscr{N}) \ge \tilde{\theta}(\mathscr{L}, \mathscr{N}) \tag{13.2.4}$$

for all subspaces $\mathscr{L}$, $\mathscr{M}$, and $\mathscr{N}$ in $\mathbb{C}^n$. If at least one of $\mathscr{L}$, $\mathscr{M}$, and $\mathscr{N}$ is the zero subspace, (13.2.4) is evident (observe that $\tilde{\theta}(\mathscr{L}, \mathscr{M}) \le 2$ for all subspaces $\mathscr{L}$, $\mathscr{M} \subset \mathbb{C}^n$). So we can assume that $\mathscr{L}$, $\mathscr{M}$ and $\mathscr{N}$ are nonzero. Given $x \in S_{\mathscr{N}}$, let $z_x \in S_{\mathscr{M}}$ be such that $\|x - z_x\| = d(x, S_{\mathscr{M}})$. Then for every $y \in S_{\mathscr{L}}$ we have

$$\|x - y\| \le \|x - z_x\| + \|z_x - y\| = d(x, S_\mathcal{M}) + \|z_x - y\|$$

and taking the infimum with respect to y, it follows that

$$d(x, S_\mathcal{L}) \le d(x, S_\mathcal{M}) + d(z_x, S_\mathcal{L}) \le \bar\theta(\mathcal{N}, \mathcal{M}) + \bar\theta(\mathcal{M}, \mathcal{L})$$

It remains to take the supremum over $x \in S_\mathcal{N}$ and repeat the argument with the roles of $\mathcal{N}$ and $\mathcal{L}$ interchanged, in order to verify (13.2.4).

In fact, the spherical gap $\bar\theta$ is not far away from the gap θ in the following sense:

$$\theta(\mathcal{L}, \mathcal{M}) \le \bar\theta(\mathcal{L}, \mathcal{M}) \le \sqrt{2}\theta(\mathcal{L}, \mathcal{M}) \tag{13.2.5}$$

The left inequality here follows from (13.1.3). To prove the right inequality in (13.2.5) it is sufficient to check that for every $x \in \mathbb{C}^n$ with $\|x\| = 1$ we have

$$d(x, S_\mathcal{L}) \le \sqrt{2}d(x, \mathcal{L}) \tag{13.2.6}$$

where $\mathcal{L} \subset \mathbb{C}^n$ is a subspace. Let $y = P_\mathcal{L}x$, where $P_\mathcal{L}$ is the orthogonal projector on $\mathcal{L}$. If $y = 0$, then $x \perp \mathcal{L}$, and for every $z \in S_\mathcal{L}$ we have

$$\|x - z\|^2 = \|x\|^2 + \|z\|^2 = 2 = 2[d(x, \mathcal{L})]^2$$

So (13.2.6) follows. If $y \ne 0$, then, in the two-dimensional real subspace spanned by x and y, there is an acute angle between the vectors x and y. Consider the isosceles triangle with sides x, $y/\|y\|$ and enclosing this acute angle. In this triangle the angle between the sides $y/\|y\|$ and $x - y/\|y\|$ is greater than $\pi/4$. Consequently

$$\left\|x - \frac{y}{\|y\|}\right\| < \sqrt{2}\|x - y\| = \sqrt{2}d(x, \mathcal{L})$$

and (13.2.6) follows again.

Proposition 13.2.2

For any three subspaces $\mathcal{L}, \mathcal{M}, \mathcal{N} \subset \mathbb{C}^n$,

$$\sin\varphi_{\min}(\mathcal{L}, \mathcal{N}) \ge \sin\varphi_{\min}(\mathcal{L}, \mathcal{M}) - \bar\theta(\mathcal{M}, \mathcal{N}) \tag{13.2.7}$$

Proof. Let $y_1 \in \mathcal{L}$ and $y_3 \in \mathcal{N}$ be arbitrary vectors satisfying $\max\{\|y_1\|, \|y_3\|\} = 1$. Letting ϵ be any fixed positive number, choose $y_2 \in \mathcal{M}$ such that $\|y_2\| = \|y_3\|$ and

$$\|y_3 - y_2\| \le (\bar\theta(\mathcal{M}, \mathcal{N}) + \epsilon)\|y_3\| \le \bar\theta(\mathcal{M}, \mathcal{N}) + \epsilon$$

Indeed, if $y_3 = 0$, choose $y_2 = 0$; if $y_3 \neq 0$, then the definition of $\theta(\mathcal{M}, \mathcal{N})$ allows us to choose a suitable y_2. Now

$$\| y_1 + y_3 \| \geq \| y_1 + y_2 \| - \| y_3 - y_2 \| \geq \sin \varphi_{min}(\mathcal{L}, \mathcal{M}) - (\tilde{\theta}(\mathcal{M}, \mathcal{N}) + \epsilon)$$

As $\epsilon > 0$ was arbitrary, the inequality (13.2.7) follows. $\square$

The angle between subspaces allows us to give a qualitative description of the result of Theorem 13.1.3.

Theorem 13.2.3

Let $\mathcal{M}, \mathcal{N}$ be subspaces in $\mathbb{C}^n$ such that $\mathcal{M} \cap \mathcal{N} = \{0\}$. Then for every pair of subspaces $\mathcal{M}_1, \mathcal{N}_1 \subset \mathbb{C}^n$ such that

$$\tilde{\theta}(\mathcal{M}, \mathcal{M}_1) + \tilde{\theta}(\mathcal{N}, \mathcal{N}_1) \leq \sin \varphi_{min}(\mathcal{M}, \mathcal{N}) \qquad (13.2.8)$$

we have $\mathcal{M}_1 \cap \mathcal{N}_1 = \{0\}$. If, in addition, $\mathcal{M} + \mathcal{N} = \mathbb{C}^n$, then every pair of subspaces $\mathcal{M}_1, \mathcal{N}_1$ satisfying (13.2.8) has the additional property that $\mathcal{M}_1 + \mathcal{N}_1 = \mathbb{C}^n$.

Proof.   In view of Proposition 13.2.2 we have

$$\sin \varphi_{min}(\mathcal{M}_1, \mathcal{N}_1) \geq \sin \varphi_{min}(\mathcal{M}_1, \mathcal{N}) - \tilde{\theta}(\mathcal{N}, \mathcal{N}_1)$$

and

$$\sin \varphi_{min}(\mathcal{M}_1, \mathcal{N}) \geq \sin \varphi_{min}(\mathcal{N}, \mathcal{M}) - \tilde{\theta}(\mathcal{M}, \mathcal{M}_1)$$

Adding these inequalities, and using (13.2.8) and Proposition 13.2.1, we find that $\mathcal{M}_1 \cap \mathcal{N}_1 = \{0\}$.

Assume now that, in addition, $\mathcal{M} + \mathcal{N} = \mathbb{C}^n$. Suppose first that $\mathcal{M} = \mathcal{M}_1$. Let $\epsilon > 0$ be so small that

$$\frac{\tilde{\theta}(\mathcal{N}, \mathcal{N}_1) + \epsilon}{\sin \varphi_{min}(\mathcal{M}, \mathcal{N})} = \delta < 1$$

If $\mathcal{M} + \mathcal{N}_1 \neq \mathbb{C}^n$, then there exists a vector $x \in \mathbb{C}^n$ with $\| x \| = 1$ and $\| x - y \| > \delta$ for all $y \in \mathcal{M} + \mathcal{N}_1$ [e.g., one can take $x \in (\mathcal{M} + \mathcal{N}_1)^\perp$]. We can represent the vector x as $x = y + z$, $y \in \mathcal{M}$, $z \in \mathcal{N}$. It follows from the definition of $\sin \varphi_{min}(\mathcal{M}, \mathcal{N})$ that

$$\| z \| \leq (\sin \varphi_{min}(\mathcal{M}, \mathcal{N}))^{-1}$$

Indeed, denoting $u = \max\{\| y \|, \| z \|\}$, we have

$$\sin \varphi_{\min}(\mathcal{L}, \mathcal{M}) = \inf\{\|x_1 + x_2\| \mid x_1 \in \mathcal{L}, x_2 \in \mathcal{M}, \max\{\|x_1\|, \|x_2\|\} = 1\}$$

$$\leq \left\|\frac{y}{u} + \frac{z}{u}\right\| = \frac{1}{u} \leq \frac{1}{\|z\|}$$

By the definition of $\bar{\theta}(\mathcal{L}, \mathcal{M})$ we can find a vector z_1 from $\mathcal{N}_1$ with

$$\|z_1\| = \|z\|, \qquad \|z - z_1\| < [\bar{\theta}(\mathcal{N}, \mathcal{N}_1) + \epsilon]\|z\| \leq \delta$$

The last inequality contradicts the choice of x, because $z - z_1 = x - t$, where $t = y + z_1 \in \mathcal{M} \dot{+} \mathcal{N}_1$, and $\|x - t\| < \delta$.

Now consider the general case. Inequality (13.2.8) implies $\bar{\theta}(\mathcal{N}, \mathcal{N}_1) < \sin \varphi_{\min}(\mathcal{M}, \mathcal{N})$ and, in view of Proposition 13.2.2, $\bar{\theta}(\mathcal{M}, \mathcal{M}_1) < \sin \varphi_{\min}(\mathcal{M}, \mathcal{N}_1)$. Applying the part of Theorem 13.2.3 already proved, we obtain $\mathcal{M} + \mathcal{N}_1 = \mathbb{C}^n$ and then $\mathcal{M}_1 + \mathcal{N}_1 = \mathbb{C}^n$. $\square$

13.3 MINIMAL OPENING AND ANGULAR LINEAR TRANSFORMATION

In this section we study the properties of angular transformations in terms of the minimal angle between subspaces.

Let $\mathcal{M}_1, \mathcal{M}_2$ be subspaces in $\mathbb{C}^n$. The number

$$\eta(\mathcal{M}_1, \mathcal{M}_2) = \inf\{\|x + y\| \mid x \in \mathcal{M}_1, y \in \mathcal{M}_2, \max(\|x\|, \|y\|) = 1\}$$

is called the *minimal opening* between $\mathcal{M}_1$ and $\mathcal{M}_2$. So

$$\eta(\mathcal{M}_1, \mathcal{M}_2) = \sin \varphi_{\min}(\mathcal{M}_1, \mathcal{M}_2)$$

where $\varphi_{\min}(\mathcal{M}_1, \mathcal{M}_2)$ is the minimal angle between $\mathcal{M}_1$ and $\mathcal{M}_2$. By convention, $\eta(\{0\}, \{0\}) = \infty$. If Π is any projector defined on $\mathbb{C}^n$, then

$$\max\{\|\Pi\|, \|I - \Pi\|\} \leq (\eta(\text{Im}\,\Pi, \text{Ker}\,\Pi))^{-1} \qquad (13.3.1)$$

To see this, note that for each $z \in \mathbb{C}^n$

$$\|z\| = \|\Pi z + (I - \Pi)z\| \geq \eta(\text{Im}\,\Pi, \text{Ker}\,\Pi) \cdot \max(\|\Pi z\|, \|(I - \Pi)z\|)$$

We would like to mention also the following properties of the minimal opening. If Q_1 and Q_2 are nontrivial (i.e., different from 0 and I) orthogonal projectors of $\mathbb{C}^n$ onto the subspaces $\mathcal{M}_1$ and $\mathcal{M}_2$, respectively, then

$$\eta(\mathcal{M}_1, \mathcal{M}_2) = \inf_{0 \neq x \in \mathcal{M}_1} \frac{\|x - Q_2 x\|}{\|x\|} = \inf_{0 \neq y \in \mathcal{M}_2} \frac{\|y - Q_1 y\|}{\|y\|}$$

and

$$1 - \eta(\mathcal{M}_1, \mathcal{M}_2)^2 = \sup_{0 \neq x \in \mathcal{M}_1} \frac{\|Q_2 x\|^2}{\|x\|^2} = \sup_{0 \neq y \in \mathcal{M}_2} \frac{\|Q_1 y\|^2}{\|y\|^2} \qquad (13.3.2)$$

Indeed, these formulas follow from the equality

$$\inf\{\|x + y\| \mid y \in \mathcal{M}_2\} = \|x - Q_2 x\|$$

for every $x \in \mathcal{M}_1$, and from

$$\inf_{0 \neq x \in \mathcal{M}_1} \left[\frac{\|x - Q_2 x\|}{\|x\|} \right]^2 = \inf_{0 \neq x \in \mathcal{M}_1} \frac{\|x\|^2 - \|Q_2 x\|^2}{\|x\|^2}$$

$$= 1 - \sup_{0 \neq x \in \mathcal{M}_1} \frac{\|Q_2 x\|^2}{\|x\|^2} = 1 - \sup_{\substack{x \in \mathcal{M}_1 \\ x \neq 0}} \sup_{\substack{y \in \mathcal{M}_2 \\ y \neq 0}} \frac{|(x, y)|^2}{\|x\|^2 \|y\|^2}$$

$$= 1 - \sup_{\substack{y \in \mathcal{M}_2 \\ y \neq 0}} \sup_{\substack{x \in \mathcal{M}_1 \\ x \neq 0}} \frac{|(x, y)|^2}{\|x\|^2 \|y\|^2} = 1 - \sup_{0 \neq y \in \mathcal{M}_2} \frac{\|Q_1 y\|^2}{\|y\|^2}$$

$$= \inf_{0 \neq y \in \mathcal{M}_2} \left[\frac{\|y - Q_1 y\|}{\|y\|} \right]^2$$

As a consequence of (13.3.2) we obtain the following connection between the minimal opening and the distance from one subspace to another. For two subspaces $\mathcal{M}_1$ and $\mathcal{M}_2$ in $\mathbb{C}^n$, put

$$\rho(\mathcal{M}_1, \mathcal{M}_2) = \sup_{x \in S_{\mathcal{M}_1}} d(x, \mathcal{M}_2)$$

[if $\mathcal{M}_1 = \{0\}$, then define $\rho(\mathcal{M}_1, \mathcal{M}_2) = 0$]. Then we have

$$\rho(\mathcal{M}_2, \mathcal{M}_1^{\perp}) = (1 - \eta(\mathcal{M}_1, \mathcal{M}_2)^2)^{1/2} = \cos \varphi_{\min}(\mathcal{M}_1, \mathcal{M}_2) \qquad (13.3.3)$$

whenever $\mathcal{M}_1 \neq \{0\}$. To see this, note that for $\mathcal{M}_2 \neq \{0\}$

$$\rho(\mathcal{M}_2, \mathcal{M}_1^{\perp}) = \sup_{0 \neq y \in \mathcal{M}_2} \frac{\|y - (I - Q_1)y\|}{\|y\|} = \sup_{0 \neq y \in \mathcal{M}_2} \frac{\|Q_1 y\|}{\|y\|}$$

where Q_1 is the orthogonal projector onto $\mathcal{M}_1$. But then we can use (13.3.2) to obtain formula (13.3.3). If $\mathcal{M}_2 = \{0\}$, then (13.3.3) holds trivially.

We use the notion of the minimal opening between two subspaces to describe the behaviour of angular transformations when the corresponding projectors are allowed to change.

Lemma 13.3.1

Let Π_0 be a projector defined on $\mathbb{C}^n$, and let Π be another projector on $\mathbb{C}^n$ such that $\operatorname{Ker} \Pi_0 = \operatorname{Ker} \Pi$. Then, provided

$$\rho(\operatorname{Im} \Pi, \operatorname{Im} \Pi_0) < \eta(\operatorname{Ker} \Pi_0, \operatorname{Im} \Pi_0)$$

we have the following estimate for the norm of the angular transformation R of $\operatorname{Im} \Pi$ with respect to Π_0:

$$\|R\| \leq \rho(\operatorname{Im} \Pi, \operatorname{Im} \Pi_0)(\eta(\operatorname{Ker} \Pi_0, \operatorname{Im} \Pi_0) - \rho(\operatorname{Im} \Pi, \operatorname{Im} \Pi_0))^{-1} \quad (13.3.4)$$

Proof. Put $\rho_0 = \rho(\operatorname{Im} \Pi, \operatorname{Im} \Pi_0)$ and $\eta_0 = \eta(\operatorname{Ker} \Pi_0, \operatorname{Im} \Pi_0)$. Recall that

$$R = (\Pi - \Pi_0)|_{\operatorname{Im} \Pi_0} \quad (13.3.5)$$

For $x \in \operatorname{Im} \Pi$ and $z \in \operatorname{Im} \Pi_0$ we have

$$\|(\Pi - \Pi_0)x\| = \|(I - \Pi_0)x\| = \|(I - \Pi_0)(x - z)\| \leq \|I - \Pi_0\| \, \|x - z\|$$

Taking the infimum over all $z \in \operatorname{Im} \Pi_0$ and using inequality (3.1), one sees that

$$\|(\Pi - \Pi_0)x\| \leq \rho_0 \eta_0^{-1}\|x\|, \qquad x \in \operatorname{Im} \Pi \quad (13.3.6)$$

Now recall that $Ry + y \in \operatorname{Im} \Pi$ for each $y \in \operatorname{Im} \Pi_0$. As $Ry \in \operatorname{Ker} \Pi_0 = \operatorname{Ker} \Pi$, we see from (13.3.5) that

$$(\Pi - \Pi_0)(Ry + y) = Ry$$

So, using (13.3.6), we obtain

$$\|Ry\| \leq \rho_0 \eta_0^{-1}\|Ry + y\|, \qquad y \in \operatorname{Im} \Pi_0 \quad (13.3.7)$$

It follows from (13.3.7) that $(1 - \rho_0 \eta_0^{-1})\|Ry\| \leq \rho_0 \eta_0^{-1}\|y\|$ for each $y \in \operatorname{Im} \Pi_0$, which proves the inequality (13.3.4). $\square$

The following lemma will be useful.

Lemma 13.3.2

Let P and $P^\times$ be projectors defined on $\mathbb{C}^n$ such that $\mathbb{C}^n = \operatorname{Im} P \dotplus \operatorname{Im} P^\times$. Then for any pair of projectors Q and $Q^\times$ defined on $\mathbb{C}^n$ with $\|P - Q\| + \|P^\times - Q^\times\|$ sufficiently small, we have $\mathbb{C}^n = \operatorname{Im} Q \dotplus \operatorname{Im} Q^\times$, and there exists an invertible transformation $S: \mathbb{C}^n \to \mathbb{C}^n$ which maps $\operatorname{Im} Q$ on $\operatorname{Im} P$, $\operatorname{Im} Q^\times$ on $\operatorname{Im} P^\times$, and

$$\max\{\|S - I\|, \|S^{-1} - I\|\} \le \beta(\|P - Q\| + \|P^\times - Q^\times\|) \quad (13.3.8)$$

where the positive constant β depends on P and $P^\times$ only.

Proof. Let $\alpha = \frac{1}{6}\eta(\operatorname{Im} P, \operatorname{Im} P^\times)(\|P^\times\| + 1)^{-1}$, and assume that the projectors Q and $Q^\times$ satisfy

$$\|P - Q\| + \|P^\times - Q^\times\| < \alpha \quad (13.3.9)$$

As $\theta(\operatorname{Im} P, \operatorname{Im} Q) \le \|P - Q\|$ and $\theta(\operatorname{Im} P^\times, \operatorname{Im} Q^\times) \le \|P^\times - Q^\times\|$, condition (13.3.9) implies that

$$\sqrt{2}\theta(\operatorname{Im} P, \operatorname{Im} Q) + \sqrt{2}\theta(\operatorname{Im} P^\times, \operatorname{Im} Q^\times) < \eta(\operatorname{Im} P, \operatorname{Im} P^\times)$$

But then we may apply Theorem 13.2.3 combined with (13.2.5) to show that $\mathbb{C}^n = \operatorname{Im} Q + \operatorname{Im} Q^\times$.

Note that (13.3.9) implies that $\|P - Q\| < \frac{1}{4}$. Hence $S_1 = I + P - Q$ is invertible, and we can write $S_1^{-1} = I + V$ with $\|V\| \le \frac{4}{3}\|P - Q\| < \frac{1}{3}$. As $I - P + Q$ is invertible also, we have

$$\operatorname{Im} P = P(I - P + Q) = PQ = (I + P - Q)Q = S_1(\operatorname{Im} Q) \quad (13.3.10)$$

Further

$$\begin{aligned}
S_1 Q^\times S_1^{-1} - P^\times &= (I + P - Q)Q^\times(I + V) - P^\times \\
&= Q^\times + (P - Q)Q^\times + Q^\times V + (P - Q)Q^\times V - P^\times \\
&= Q^\times - P^\times + (P - Q)(Q^\times - P^\times) + (P - Q)P^\times \\
&\quad + (Q^\times - P^\times)V + P^\times V + (P - Q)(Q^\times - P^\times)V \\
&\quad + (P - Q)P^\times V
\end{aligned}$$

So $\|S_1 Q^\times S_1^{-1} - P^\times\| \le 3\|Q^\times - P^\times\| + 3\|P - Q\| \cdot \|P^\times\|$. But then

$$\begin{aligned}
\rho(\operatorname{Im} S_1 Q^\times S_1^{-1}, \operatorname{Im} P^\times) &\le \|S_1 Q^\times S_1^{-1} - P^\times\| \\
&\le 3(\|P - Q\| + \|P^\times - Q^\times\|)(\|P^\times\| + 1) \\
&< \tfrac{1}{2}\eta(\operatorname{Im} P, \operatorname{Im} P^\times)
\end{aligned}$$

Let $\Pi_0(\Pi)$ be the projector of $\mathbb{C}^n$ along $\operatorname{Im} P$ ($\operatorname{Im} Q$) onto $\operatorname{Im} P^\times$ ($\operatorname{Im} Q^\times$) and put $\tilde{\Pi} = S_1 \Pi S_1^{-1}$. Then $\tilde{\Pi}$ is again a projector, and by (13.3.10) we have $\operatorname{Ker} \tilde{\Pi} = \operatorname{Ker} \Pi_0$. Further, $\operatorname{Im} \tilde{\Pi} = \operatorname{Im} S_1 Q^\times S_1^{-1}$, and so we have

$$\rho(\operatorname{Im} \tilde{\Pi}, \operatorname{Im} \Pi_0) \le \tfrac{1}{2}\eta(\operatorname{Ker} \Pi_0, \operatorname{Im} \Pi_0)$$

Hence, if R denotes the angular transformation of $\operatorname{Im} \tilde{\Pi}$ with respect to Π_0, then because of equation (13.3.4) of Lemma 13.3.1, we obtain

$$\|R\| \le 2\rho(\operatorname{Im} \tilde{\Pi}, \operatorname{Im} \Pi_0)[\eta(\operatorname{Ker} \Pi_0, \operatorname{Im} \Pi_0)]^{-1}$$

As $\rho(\operatorname{Im} \tilde{\Pi}, \operatorname{Im} \Pi_0) \le 3(\|P - Q\| + \|P^\times - Q^\times\|)(\|P^\times\| + 1)$, this implies that

$$\|R\| \le \frac{1}{\alpha}(\|P - Q\| + \|P^\times - Q^\times\|) \tag{13.3.11}$$

Next, put $S_2 = I - R\Pi_0$, and take $S = S_2 S_1$. Clearly, S_2 is invertible; in fact, $S_2^{-1} = I + R\Pi_0$. It follows that S is invertible also. From the properties of the angular transformation one easily sees that $S(\operatorname{Im} Q) = \operatorname{Im} P$, $S(\operatorname{Im} Q^\times) = \operatorname{Im} P^\times$.

To prove (13.3.8), we simplify our notation. Put $d = \|P - Q\| + \|P^\times - Q^\times\|$, and let $\eta = \eta(\operatorname{Im} P, \operatorname{Im} P^\times)$. From $S = (I - R\Pi_0)(I + P - Q)$ and the fact that $\|P - Q\| < \frac{1}{4}$, one deduces that $\|S - I\| \le \|P - Q\| + \frac{5}{4}\|R\| \cdot \|\Pi_0\|$. For $\|R\|$ an upper bound is given by (13.3.11), and from (13.3.1) we know that $\|\Pi_0\| \le \eta^{-1}$. It follows that

$$\|S - I\| \le d + \tfrac{5}{4}d(\alpha\eta)^{-1} \tag{13.3.12}$$

Finally, we consider S^{-1}. Recall that $S_1^{-1} = I + V$ with $\|V\| \le \frac{4}{3}\|P - Q\| < \frac{1}{3}$. Hence

$$\|S^{-1} - I\| \le \|V\| + \|V\| \cdot \|\Pi_0\| \cdot \|R\| + \|R\| \cdot \|\Pi_0\|$$
$$\le \tfrac{4}{3}\|P - Q\| + \tfrac{4}{3}\|R\| \cdot \|\Pi_0\| \le \tfrac{4}{3}d + \tfrac{4}{3}d(\alpha\eta)^{-1}$$

and (13.3.8) follows in view of (13.3.12). $\square$

13.4   THE METRIC SPACE OF SUBSPACES

We have already seen in Section 13.1 that the set $G(\mathbb{C}^n)$ of all subspaces in $\mathbb{C}^n$ is a metric space with respect to the gap $\theta(\mathcal{L}, \mathcal{M})$. In this section we investigate some topological properties of $G(\mathbb{C}^n)$, that is, those properties that depend on convergence (or divergence) in the sense of the gap metric.

Theorem 13.4.1

The metric space $G(\mathbb{C}^n)$ is compact, and, therefore, complete (as a metric space).

Recall that compactness of $G(\mathbb{C}^n)$ means that for every sequence $\mathcal{L}_1, \mathcal{L}_2, \ldots$ of subspaces in $G(\mathbb{C}^n)$ there exists a converging subsequence $\mathcal{L}_{i_1}, \mathcal{L}_{i_2}, \ldots$, that is, such that

$$\lim_{k \to \infty} \theta(\mathcal{L}_{i_k}, \mathcal{L}_0) = 0$$

for some $\mathcal{L}_0 \in \mathcal{G}(\mathbb{C}^n)$. Completeness of $\mathcal{G}(\mathbb{C}^n)$ means that every sequence of subspaces $\mathcal{L}_i$, $i = 1, 2, \ldots$, for which $\lim_{i,j \to \infty} \theta(\mathcal{L}_i, \mathcal{L}_j) = 0$ is convergent.

Proof. In view of Theorem 13.1.2, the metric space $\mathcal{G}(\mathbb{C}^n)$ is decomposed into components $\mathcal{G}_m$, $m = 0, \ldots, n$, where $\mathcal{G}_m$ is a closed and open set in $\mathcal{G}(\mathbb{C}^n)$ consisting of all m-dimensional subspaces in $\mathbb{C}^n$.

Obviously, it is sufficient to prove the compactness of each $\mathcal{G}_m$. To this end consider the set $\$_m$ of all orthonormal systems $u = \{u_k\}_{k=1}^m$ consisting of m vectors $u_1, \ldots, u_m$ in $\mathbb{C}^n$.

For $u = \{u_k\}_{k=1}^m \in \$_m$, $v = \{v_k\}_{k=1}^m \in \$_m$ define

$$\delta(u, v) = \left[\sum_{k=1}^m \| u_k - v_k \|^2 \right]^{1/2}$$

It is easily seen that $\delta(u, v)$ is a metric in $\$_m$, thus turning $\$_m$ into a metric space. For each $u = \{u_k\}_{k=1}^m \in \$_m$ define $A_m u = \text{Span}\{u_1, \ldots, u_m\} \in \mathcal{G}_m$. In this way we obtain a map $A_m : \$_m \to \mathcal{G}_m$ of metric spaces $\$_m$ and $\mathcal{G}_m$.

We prove that the map A_m is continuous. Indeed, let $\mathcal{L} \in \mathcal{G}_m$ and let $v_1, \ldots, v_m$ be an orthonormal basis in $\mathcal{L}$. Pick some $u = \{u_k\}_{k=1}^m \in \$_m$ (which is supposed to be in a neighbourhood of $v = \{v_k\}_{k=1}^m \in \$_m$). For v_i, $i = 1, \ldots, m$, we have (where $\mathcal{M} = A_m u$ and $P_{\mathcal{N}}$ stands for the orthogonal projector on the subspace $\mathcal{N}$):

$$\|(P_{\mathcal{M}} - P_{\mathcal{L}}) v_i\| = \| P_{\mathcal{M}} v_i - v_i \| \leq \| P_{\mathcal{M}} (v_i - u_i) \| + \| u_i - v_i \|$$
$$\leq \| P_{\mathcal{M}} \| \, \| v_i - u_i \| + \| u_i - v_i \| \leq 2\delta(u, v)$$

and thus for $x = \sum_{i=1}^m \alpha_i v_i \in S_{\mathcal{L}}$

$$\|(P_{\mathcal{M}} - P_{\mathcal{L}}) x\| \leq 2 \sum_{i=1}^m |\alpha_i| \delta(u, v)$$

Now, since $\|x\| = \sum_{i=1}^m |\alpha_i|^2 = 1$, we find that $|\alpha_i| \leq 1$ and $\sum_{i=1}^m |\alpha_i| \leq m$, and so

$$\|(P_{\mathcal{M}} - P_{\mathcal{L}})|_{\mathcal{L}}\| \leq 2m\delta(u, v) \tag{13.4.1}$$

Fix some $y \in S_{\mathcal{L}^\perp}$. We wish to evaluate $P_{\mathcal{M}} y$. For every $x \in \mathcal{L}$, write

$$(x, P_{\mathcal{M}} y) = (P_{\mathcal{M}} x, y) = ((P_{\mathcal{M}} - P_{\mathcal{L}}) x, y) + (x, y) = ((P_{\mathcal{M}} - P_{\mathcal{L}}) x, y)$$

and

$$|(x, P_{\mathcal{M}} y)| \leq 2m \|x\| \delta(u, v) \tag{13.4.2}$$

by (13.4.1). On the other hand, write

$$P_{\mathcal{M}} y = \sum_{i=1}^{m} \alpha_i u_i$$

then for every $z \in \mathcal{L}^{\perp}$

$$(z, P_{\mathcal{M}} y) = \left(z, \sum_{i=1}^{m} \alpha_i(u_i - v_i)\right) + \left(z, \sum_{i=1}^{m} \alpha_i v_i\right) = \left(z, \sum_{i=1}^{m} \alpha_i(u_i - v_i)\right),$$

and

$$|(z, P_{\mathcal{M}} y)| \leq \|z\| \left\| \sum_{i=1}^{m} \alpha_i(u_i - v_i) \right\| \leq \|z\| m \max_{1 \leq i \leq m} |\alpha_i| \, \|u_i - v_i\|$$

But $\|y\| = 1$ implies that $\sum_{i=1}^{m} |\alpha_i|^2 \leq 1$, so $\max\{|\alpha_1|, \ldots, |\alpha_m|\} \leq 1$. Hence

$$|(z, P_{\mathcal{M}} y)| \leq \|z\| m \delta(u, v) \qquad (13.4.3)$$

Combining (13.4.2) and (13.4.3), we find that $|(t, P_{\mathcal{M}} y)| \leq 3m\delta(u, v)$ for every $t \in \mathbb{C}^n$ with $\|t\| = 1$. Thus

$$\|P_{\mathcal{M}} y\| \leq 3m\delta(u, v) \qquad (13.4.4)$$

Now we can easily prove the continuity of A_m. Pick an $x \in \mathbb{C}^n$ with $\|x\| = 1$. Thus, using (13.4.1) and (13.4.4) we have

$$\|(P_{\mathcal{M}} - P_{\mathcal{L}})x\| \leq \|(P_{\mathcal{M}} - P_{\mathcal{L}})P_{\mathcal{L}} x\| + \|P_{\mathcal{M}}(x - P_{\mathcal{L}} x)\| \leq 5m \cdot \delta(u, v)$$

so

$$\theta(\mathcal{M}, \mathcal{L}) = \|P_{\mathcal{M}} - P_{\mathcal{L}}\| \leq 5m\delta(u, v)$$

which obviously implies the continuity of A_m.

It is easily seen that $\$_m$ is compact. Indeed, this follows from the compactness of the unit sphere $\{x \in \mathbb{C}^n \mid \|x\| = 1\}$ in $\mathbb{C}^n$. Since $A_m: \$_m \to \mathbb{G}_m$ is a continuous map onto $\mathbb{G}_m$, the metric space $\mathbb{G}_m$ is compact as well.

Finally, let us prove the completeness of $\mathbb{G}_m$. Let $\mathcal{L}_1, \mathcal{L}_2, \ldots$ be a Cauchy sequence in $\mathbb{G}_m$, that is, $\theta(\mathcal{L}_i, \mathcal{L}_j) \to 0$ as $i, j \to \infty$. By compactness, there exists a subsequence $\mathcal{L}_{i_k}$ such that $\lim_{k \to \infty} \theta(\mathcal{L}_{i_k}, \mathcal{L}) = 0$ for some $\mathcal{L} \in \mathbb{G}_m$. But then it is easily seen that in fact $\mathcal{L} = \lim_{i \to \infty} \mathcal{L}_i$. $\square$

Next we develop a useful characterization of limits in $\mathbb{G}(\mathbb{C}^n)$.

Theorem 13.4.2

Let $\mathcal{M}_1, \mathcal{M}_2, \ldots$ be a sequence of m-dimensional subspaces in $\mathbb{G}(\mathbb{C}^n)$, such that $\theta(\mathcal{M}_p, \mathcal{M}) \to 0$ as $p \to \infty$ for some subspace $\mathcal{M} \subset \mathbb{C}^n$. Then $\mathcal{M}$ consists of exactly those vectors $x \in \mathbb{C}^n$ for which there exists a sequence of vectors $x_p \in \mathbb{C}^n$, $p = 1, 2, \ldots$ such that $x_p \in \mathcal{M}_p$, $p = 1, 2, \ldots$ and $x = \lim_{p \to \infty} x_p$.

Proof. Denoting by $P_{\mathcal{N}}$ the orthogonal projector on the subspace $\mathcal{N} \subset \mathbb{C}^n$, for every $x \in \mathcal{M}$ we have:

$$\|P_{\mathcal{M}_p} x - x\| = \|(P_{\mathcal{M}_p} - P_{\mathcal{M}})x\| \leq \|P_{\mathcal{M}_p} - P_{\mathcal{M}}\| \cdot \|x\|$$
$$\leq \theta(\mathcal{M}_p, \mathcal{M})\|x\| \to 0 \text{ as } p \to \infty$$

So $x_p = P_{\mathcal{M}_p} x$ has the properties that $x_p \in \mathcal{M}_p$ and $\lim_{p \to \infty} x_p = x$.

Conversely, let $x_p \in \mathcal{M}_p$, $p = 1, 2, \ldots$ be such that $\lim_{p \to \infty} x_p = x$. Then

$$\|P_{\mathcal{M}} x - x\| \leq \|P_{\mathcal{M}} x - P_{\mathcal{M}} x_p\| + \|P_{\mathcal{M}} x_p - P_{\mathcal{M}_p} x_p\| + \|x_p - x\|$$
$$\leq \theta(\mathcal{M}, \mathcal{M}_p)\|x\| + \|P_{\mathcal{M}_p}\| \cdot \|x - x_p\| + \|x_p - x\|$$
$$\leq \theta(\mathcal{M}, \mathcal{M}_p)\|x\| + 2\|x - x_p\| \to 0 \text{ as } p \to \infty$$

(in the last inequality we have used the fact that the norm of an orthogonal projector is 1); so $P_{\mathcal{M}} x = x$, and $x \in \mathcal{M}$. $\square$

Using Theorems 13.4.2 and 13.1.3, one obtains the following fact.

Theorem 13.4.3

Let $\mathcal{L}$ and $\mathcal{M}$ be direct complements to each other in $\mathbb{C}^n$, and let $\{\mathcal{L}_m\}_{m=1}^\infty$, $\{\mathcal{M}_m\}_{m=1}^\infty$ be sequences of subspaces such that

$$\lim_{m \to \infty} \theta(\mathcal{L}_m, \mathcal{L}) = \lim_{m \to \infty} \theta(\mathcal{M}_m, \mathcal{M}) = 0$$

Then, denoting by P (resp. P_m) the projector on $\mathcal{L}$ along $\mathcal{M}$ (resp. on $\mathcal{L}_m$ along $\mathcal{M}_m$) we have

$$\lim_{m \to \infty} \|P_m - P\| = 0$$

Moreover, there exists a constant $K > 0$ depending on $\mathcal{L}$ and $\mathcal{M}$ only such that

$$\|P_m - P\| \leq K\{\theta(\mathcal{L}_m, \mathcal{L}) + \theta(\mathcal{M}_m, \mathcal{M})\} \tag{13.4.5}$$

for all sufficiently large m.

Observe that, in view of Theorem 13.1.3, the subspaces $\mathcal{L}_m$ and $\mathcal{M}_m$ are direct complements to each other for sufficiently large m.

Proof. Let $P_{m,\mathcal{L}}$ be the projector on $\mathcal{M}_m$ along $\mathcal{L}$ and $P_{\mathcal{M},m}$ be the projector on $\mathcal{M}$ along $\mathcal{L}_m$ (for sufficiently large m). By Theorem 13.1.3 we have

$$\|P_{m,\mathscr{L}} - P\| \le C_1 \theta(\mathscr{M}_m, \mathscr{M}) \tag{13.4.6}$$

$$\|P_{\mathscr{M},m} - P\| \le C_2 \theta(\mathscr{L}_m, \mathscr{L})$$

where

$$C_1 = 2\|P\| \max_{x \in \mathscr{L}, \|x\|=1} \{d(x, \mathscr{M})^{-1}\}, \ C_2 = 2\|P\| \max_{x \in \mathscr{M}, \|x\|=1} \{d(x, \mathscr{L})^{-1}\}$$

As usual, $d(x, N) = \inf\{\|x - y\| \mid y \in \mathscr{N}\}$ is the distance between $x \in \mathbb{C}^n$ and a subset $\mathscr{N} \subset \mathbb{C}^n$. In particular, for m large enough we find that

$$\|P_{m,\mathscr{L}}\| \le \|P\| + 1$$

When Theorem 13.1.3 is applied again, it follows that

$$\|P_m - P_{m,\mathscr{L}}\| \le 2\|P_{m,\mathscr{L}}\| \max_{x \in \mathscr{M}_m, \|x\|=1} \{d(x, \mathscr{L})^{-1}\} \theta(\mathscr{L}_m, \mathscr{L})$$

Now use (13.4.6) and deduce (for sufficiently large m):

$$\|P_m - P\| \le \|P_m - P_{m,\mathscr{L}}\| + \|P_{m,\mathscr{L}} - P\|$$
$$\le 2(\|P\| + 1) \max_{x \in \mathscr{M}_m, \|x\|=1} \{d(x, \mathscr{L})^{-1}\} \theta(\mathscr{L}_m, \mathscr{L}) + C_1 \theta(\mathscr{M}_m, \mathscr{M})$$

We finish the proof by showing that

$$\max_{x \in \mathscr{M}_m, \|x\|=1} \{d(x, \mathscr{L})^{-1}\} \le 2 \max_{x \in \mathscr{M}, \|x\|=1} \{d(x, \mathscr{L})^{-1}\} \tag{13.4.7}$$

for m sufficiently large.

Arguing by contradiction, assume that (13.4.7) does not hold. Then there exists a subsequence $\{\mathscr{M}_{m_k}\}_{k=1}^{\infty}$ and vectors $x_{m_k} \in \mathscr{M}_{m_k}$ with norm 1 such that

$$d(x_{m_k}, \mathscr{L})^{-1} > 2 \max_{x \in \mathscr{M}, \|x\|=1} \{d(x, \mathscr{L})^{-1}\} \tag{13.4.8}$$

As the sequence $\{x_{m_k}\}_{k=1}^{\infty}$ of n-dimensional vectors is bounded, it has a converging subsequence. So we can assume that $x_{m_k} \to x_0$ as $k \to \infty$. Clearly, $\|x_0\| = 1$, and by Theorem 13.4.2, $x_0 \in \mathscr{M}$. In view of (13.4.8) for each $k = 1, 2, \ldots$, there is a vector $y_k \in \mathscr{L}$ such that

$$\|x_k - y_k\| < (2 \max_{x \in \mathscr{M}, \|x\|=1} \{d(x, \mathscr{L})^{-1}\})^{-1} \tag{13.4.9}$$

In particular, the sequence $\{y_k\}_{k=1}^{\infty}$ is bounded, and we can assume that $y_k \to y_0$ as $k \to \infty$, for some $y_0 \in \mathscr{L}$. Passing to the limit in (13.4.9) when k tends to infinity, we obtain the inequality

$$2 \max_{x \in \mathcal{M}, \|x\|=1} \{d(x, \mathcal{L})^{-1}\} \leq \|x_0 - y_0\|^{-1} \leq \max_{x \in \mathcal{M}, \|x\|=1} \{d(x, \mathcal{L})^{-1}\}$$

which is contradictory. $\square$

The proof of Theorem 13.4.3 shows that actually equation (13.4.5) holds with

$$K = 4(\|P\| + 1) \max_{x \in \mathcal{M}, \|x\|=1} \{d(x, \mathcal{L})^{-1}\}$$

We conclude this section with the following simple observation.

Proposition 13.4.4

The set $\mathbb{G}_m(\mathbb{C}^n)$ of all m-dimensional subspaces in $\mathbb{C}^n$ is connected. That is, for every $\mathcal{M}, \mathcal{N} \in \mathbb{G}_m(\mathbb{C}^n)$ there exists a continuous function $f: [0, 1] \to \mathbb{G}_m(\mathbb{C})$ such that $f(0) = \mathcal{M}$, $f(1) = \mathcal{N}$ (and the continuity of f is understood in the gap metric).

Proof. Using the proof of Theorem 13.4.1 and the notation introduced there, we must show that the set $\mathcal{S}_m$ is connected. As any orthonormal system $u_1, \ldots, u_m$ in $\mathbb{C}^n$ can be completed to an orthonormal basis in $\mathbb{C}^n$, the connectedness of $\mathcal{S}_m$ would follow from the connectedness of the group $U(\mathbb{C}^n)$ of all $n \times n$ unitary matrices. To show that $U(\mathbb{C}^n)$ is connected, observe that any $X \in U(\mathbb{C}^n)$ has the form

$$X = S \operatorname{diag}[e^{i\theta_1}, \ldots, e^{i\theta_n}]S^{-1},$$

where S is unitary and $\theta_1, \ldots, \theta_n$ are real numbers (see Section 1.9). So

$$f(t) = S \operatorname{diag}[e^{it\theta_1}, \ldots, e^{it\theta_n}]S^{-1}, \qquad t \in [0, 1]$$

is a continuous $U(\mathbb{C}^n)$-valued function that connects I and X. $\square$

Similarly, one can prove that the set $\mathbb{G}_m(\mathbb{R}^n)$ of all m-dimensional subspaces in $\mathbb{R}^n$ is connected. To this end use the facts that any orthonormal systems $u_1, \ldots, u_m$ in $\mathbb{R}^n$ ($m < n$) can be completed to an orthonormal basis $u_1, \ldots, u_n$ with $\det[u_1, u_2, \ldots, u_n] = 1$ and that the set $U_+(\mathbb{R}^n)$ of all orthogonal $n \times n$ matrices with determinant 1 is connected. Recall that a real $n \times n$ matrix U is called orthogonal if $U^T U = UU^T = I$.

For completeness, let us prove the connectedness of $U_+(\mathbb{R}^n)$. It follows from Theorem 12.1.4 that any $x \in U_+(\mathbb{R}^n)$ admits the representation

$$X = S^{-1} \operatorname{diag}[K_1, K_2, \ldots, K_p]S$$

where S is orthogonal and each K_j is either the scalar ± 1 or the 2×2 matrix

$\Psi_\theta = \begin{bmatrix} \cos\theta & \sin\theta \\ -\sin\theta & \cos\theta \end{bmatrix}$ for some θ, $0 \le \theta \le 2\pi$, which depends on j. As det $X = 1$, also $\det[K_1, K_2, \ldots, K_p] = 1$, which means that the number of indices j such that $K_j = -1$ is even. Since $\begin{bmatrix} -1 & 0 \\ 0 & -1 \end{bmatrix} = \Psi_\pi$, we can assume that each K_j is either 1 or Ψ_θ, $\theta = \theta(j)$. Putting

$$X(t) = S^{-1} \operatorname{diag}[K_1(t), K_2(t), \ldots, K_p(t)] S, \qquad 0 \le t \le 1$$

where $K_j(t) \equiv K_j$ if $K_j = 1$ and $K_j(t) = \Psi_{t\theta}$ if $K_j = \Psi_\theta$, we obtain a $U_+(\mathbb{R}^n)$-valued continuous function that connects I and X.

13.5   KERNELS AND IMAGES OF LINEAR TRANSFORMATIONS

Important examples of subspaces in $\mathbb{C}^n$ are images of transformations into $\mathbb{C}^n$ and kernels of transformations from $\mathbb{C}^n$. We study here the behaviour of these subspaces when the transformation is allowed to change. The main result in this direction is the following theorem.

Theorem 13.5.1

Let X: $\mathbb{C}^n \to \mathbb{C}^m$ be a transformation, and let P_X be a projector on Ker X. *Then there exists a constant $K > 0$, depending only on X and P_X, with the following property: for every transformation Y: $\mathbb{C}^n \to \mathbb{C}^m$ with* dim Ker $Y =$ dim Ker X *there exists a projector P_Y on* Ker Y *such that*

$$\|P_Y - P_X\| \le K\|X - Y\|. \tag{13.5.1}$$

In particular

$$\theta(\operatorname{Ker} X, \operatorname{Ker} Y) \le K\|X - Y\|. \tag{13.5.2}$$

   Proof.   It will suffice to prove (13.5.1) for all those Y with dim Ker $Y =$ dim Ker X that are sufficiently close to X, that is, $\|X - Y\| \le \epsilon$, where $\epsilon > 0$ depends on X and P_X only. Indeed, for Y with dim Ker $Y =$ dim Ker X and $\|X - Y\| > \epsilon$, use the orthogonal projector P_Y on Y and the fact that $\|P_Y - P_X\| \le \|P_Y\| + \|P_X\| = 1 + \|P_X\|$ to obtain (13.5.1) (maybe with a bigger constant K).
   Consider first the case when X is right invertible. There exists a right inverse X^I of X such that Im $X^I = \operatorname{Im}(I - P_X)$ (cf. Theorem 1.5.5), and then $X^I X = I - P_X$ (indeed, both sides are projectors with the same kernel and the same image). It is easy to verify that any transformation Y: $\mathbb{C}^n \to \mathbb{C}^m$ with the property

$$\|Y - X\| \le \tfrac{1}{2}\|X^I\|^{-1} \tag{13.5.3}$$

is also right invertible and one of the right inverses Y^I is given by the formula $Y^I = ZX^I$ where

$$Z = \sum_{n=0}^{\infty} (-1)^n (X^I(Y - X))^n$$

Indeed, we have

$$Y = X + (Y - X) = X(I + X^I(Y - X))$$

and hence

$$YZX^I = \lim_{k \to \infty} X(I + X^I(Y - X)) \left(\sum_{n=0}^{k} (-1)^n (X^I(Y - X))^n \right) X^I$$

$$= \lim_{k \to \infty} X(I + (-1)^k (X^I(Y - X))^k) X^I = XX^I = I$$

where the penultimate equality follows from (13.5.3), because

$$\| (X^I(Y - X))^k \| \le 2^{-k}$$

A similar argument shows that Z is invertible and $\| Z \| \le 2$, $\| I - Z \| \le 2 \| X^I \| \, \| X - Y \|$. Now put $P_Y = I - Y^I Y$. We have

$$\| P_X - P_Y \| = \| X^I X - Y^I Y \| = \| X^I X - X^I ZY \|$$

$$\le \| X^I \| \cdot \| X - ZY \| \le \| X^I \| \cdot \{ \| I - Z \| \cdot \| X \| + \| Z \| \cdot \| X - Y \| \}$$

$$\le \| X^I \| \{ 2 \| X^I \| \cdot \| X - Y \| \cdot \| X \| + 2 \| X - Y \| \}$$

So (13.5.1) holds for every Y satisfying $\| Y - X \| \le \frac{1}{2} \| X^I \|^{-1}$, with $K = 2 \| X^I \|^2 \| X \| + 2 \| X^I \|$.

Now consider the case when X is not right invertible, and let r be the dimension of a complementary subspace $\mathcal{N}$ to Im X in $\mathbb{C}^m$. Consider the transformation

$$\tilde{X} \colon \mathbb{C}^n \oplus \mathbb{C}^r \to \mathbb{C}^m$$

defined by $\tilde{X}(x + y) = Xx + Ly$; $x \in \mathbb{C}^n$, $y \in \mathbb{C}^r$, where $L \colon \mathbb{C}^r \to \mathcal{N}$ is some invertible transformation. As the image of $\tilde{X}$ is the whole space $\mathbb{C}^m$ the transformation $\tilde{X}$ is right invertible. Also Ker $\tilde{X} = $ Ker X. Let $P_{\tilde{X}}$ be a projector on Ker $\tilde{X}$ defined by $P_{\tilde{X}}(x + y) = P_X x$; $x \in \mathbb{C}^n$, $y \in \mathbb{C}^r$. Applying the part of Theorem 13.5.1 already proved to $\tilde{X}$, we find positive constants ϵ and K such that, for every transformation $\tilde{Y} \colon \mathbb{C}^n \oplus \mathbb{C}^r \to \mathbb{C}^m$ with $\| \tilde{X} - \tilde{Y} \| \le \epsilon$, there exists a projector $P_{\tilde{Y}}$ on Ker $\tilde{Y}$ such that

$$\|P_{\tilde{X}} - P_{\tilde{Y}}\| \le K \|\tilde{X} - \tilde{Y}\| \tag{13.5.4}$$

Note that the equality $\dim \operatorname{Ker} \tilde{Y} = \dim \operatorname{Ker} \tilde{X}$ holds automatically for ϵ small enough because then such $\tilde{Y}$ will also be right invertible (see the first part of this proof). Apply (13.5.4) for $\tilde{Y}$ of the form $\tilde{Y}(x + y) = Yx + Ly$; $x \in \mathbb{C}^n$, $y \in \mathbb{C}^r$, where $Y: \mathbb{C}^n \to \mathbb{C}^m$ is a transformation such that $\|X - Y\| \le \epsilon$ and $\dim \operatorname{Ker} Y = \dim \operatorname{Ker} X$. Let us check that $\operatorname{Ker} \tilde{Y} \subset \mathbb{C}^n$. Indeed

$$\dim \operatorname{Ker} \tilde{Y} = \dim \operatorname{Ker} \tilde{X} = \dim \operatorname{Ker} X = \dim \operatorname{Ker} Y$$

and since $\operatorname{Ker} Y \subset \operatorname{Ker} \tilde{Y}$, we have in fact $\operatorname{Ker} \tilde{Y} = \operatorname{Ker} Y$ and thus $\operatorname{Ker} \tilde{Y} \subset \mathbb{C}^n$. Now put $P_X = P_{\tilde{X}|_{\mathbb{C}^n}}$, $P_Y = P_{\tilde{Y}|_{\mathbb{C}^n}}$ to satisfy (13.5.1), for transformations $Y: \mathbb{C}^n \to \mathbb{C}^m$ such that $\|X - Y\| \le \epsilon$.

Finally, observe that (13.5.2) follows from (13.5.1) in view of Theorem 13.1.3. $\square$

The condition $\dim \operatorname{Ker} Y = \dim \operatorname{Ker} X$ is clearly necessary for the inequality (13.5.1), since otherwise we obtain a contradiction with Theorem 13.1.2 on taking a $Y: \mathbb{C}^n \to \mathbb{C}^n$ such that $\|X - Y\| < K^{-1}$.

A result analogous to Theorem 13.5.1 also holds for the images of linear transformations. The statement of this result is obtained from Theorem 13.5.1 by replacing $\operatorname{Ker} X$ and $\operatorname{Ker} Y$ by $\operatorname{Im} X$ and $\operatorname{Im} Y$, respectively, and its proof is reduced to Theorem 13.5.1 by observing that $\operatorname{Im} A = (\operatorname{Ker} A^*)^{\perp}$ for a linear transformation A and that $\theta(\mathcal{M}, \mathcal{N}) = \theta(\mathcal{M}^{\perp}, \mathcal{N}^{\perp})$ for any subspaces $\mathcal{M}, \mathcal{N} \subset \mathbb{C}^n$.

13.6   CONTINUOUS FAMILIES OF SUBSPACES

As before, we denote by $\mathcal{G}(\mathbb{C}^n)$ the set of all subspaces in $\mathbb{C}^n$ seen as a metric space in the gap metric.

In this section we consider subspace-valued families $\mathcal{L}(t)$ defined on some fixed compact set $K \subset \mathbb{R}^m$, that is, for each $t \in K$, $\mathcal{L}(t)$ is a subspace in $\mathbb{C}^n$. The family $\mathcal{L}(t)$ will be called *continuous* (on K) if for every $t_0 \in K$ and every $\epsilon > 0$ there is $\delta > 0$ such that $\|t - t_0\| < \delta$, $t \in K$ implies $\theta(\mathcal{L}(t), \mathcal{L}(t_0)) < \epsilon$ (the norm $\|t - t_0\|$ is understood as the Euclidean norm, that is, generated by the standard scalar product $(x, y) = \sum_{i=1}^{m} x_i y_i$ for $x = \langle x_1, \ldots, x_m \rangle$, $y = \langle y_1, \ldots, y_m \rangle \in \mathbb{R}^m$). In other words, the continuity is understood in the sense of the gap metric.

Examples of continuous families of subspaces are provided by the following proposition.

Proposition 13.6.1

Let $B(t)$ be a continuous $m \times n$ complex matrix function on K such that rank $B(t) \overset{\text{def}}{=} p$ *is independent of t on K. Then $\operatorname{Ker} B(t)$ and $\operatorname{Im} B(t)$ are continuous families of subspaces on K.*

Proof. Take $t_0 \in K$. There exists a nonzero minor of size $p \times p$ of $B(t_0)$. For simplicity of notation assume that this minor is in the upper left corner of $B(t_0)$. By continuity the $p \times p$ minor in the upper left corner of $B(t)$ is also nonzero as long as t belongs to some neighbourhood U_0 of t_0. So [here we use the assumption that rank $B(t)$ is independent of t] for $t \in U_0$

$$\text{Im } B(t) = \text{Span}\{b_1(t), \ldots, b_p(t)\} \tag{13.6.1}$$

where $b_i(t)$ is the ith column of $B(t)$. Let $b_{ij}(t)$ be the (i, j)th entry in $B(t)$; and let $D(t) = [b_{ij}(t)]_{i=p+1; j=1}^{i=m; j=p}$; $C(t) = [b_{ij}(t)]_{i,j=1}^{p}$. Then the matrix

$$P(t) = \begin{bmatrix} I_p & 0 \\ D(t)C(t) & 0 \end{bmatrix}, \quad t \in U_0$$

is a continuous projector with Im $P(t) = \text{Im } B(t)$. Hence $P(t)$ is uniformly continuous on U_1, where U_1 is a neighbourhood of t_0 in K such that $\bar{U}_1 \subset U_0$. By Theorem 13.1.1 [inequality (13.1.4)] the orthogonal projector on Im $B(t)$ is also uniformly continuous on U_1.

The statement concerning Ker $B(t)$ can be reduced to that already considered because Ker $B(t)$ is the orthogonal complement to $\text{Im}(B(t))^*$ (note that $B(t)^*$ is continuous in t if $B(t)$ is). $\square$

In particular, we obtain an important case.

Corollary 13.6.2

Let $P(t)$ be a continuous projector-valued function on K. Then Im $P(t)$ and Ker $P(t)$ are continuous families of subspaces on K.

We have to show that rank $P(t)$ is constant if the projector function $P(t)$ is continuous. But this follows from inequality (13.1.4) and the fact that the set of subspaces of fixed dimension is open in the set of all subspaces in $\mathbb{C}^n$ (Theorem 13.1.2).

The following characterization of continuous families of subspaces is very useful.

Theorem 13.6.3

Let $\mathcal{L}(t)$ be a family of subspaces (of $\mathbb{C}^n$) on a connected compact subset K of $\mathbb{R}^m$. Then the following properties are equivalent: (a) $\mathcal{L}(t)$ is continuous; (b) for each $t \in K$ there exists an invertible transformation $S(t): \mathbb{C}^n \to \mathbb{C}^n$ which depends continuously on t for $t \in K$, and there exists a subspace $\mathcal{M} \subset \mathbb{C}^n$ such that $\mathcal{L}(t) = S(t)\mathcal{M}$ for all $t \in K$; (c) for each $t_0 \in K$ there exist a neighbourhood U_{t_0} of t_0 in K, an invertible transformation $S_{t_0}(t): \mathbb{C}^n \to \mathbb{C}^n$ that depends continuously on t in U_{t_0}, and a subspace $\mathcal{M}_{t_0} \subset \mathbb{C}^n$ such that $\mathcal{L}(t) = S_{t_0}(t)\mathcal{M}_{t_0}$, $t \in U_{t_0}$.

We prove Theorem 13.6.3 only for the case $K = [0, 1]$ (of course, the case when $K \subset \dot{R}$ is easily reduced to this one). The proof when K is a connected compact set of $\dot{R}^m$ requires mathematical tools that are beyond the scope of this book [see Gohberg and Leiterer (1972) for the complete proof].

Proof. Assume that $\mathcal{L}(t)$ is continuous on $K = [0, 1]$. Let $0 = t_0 < t_1 < t_2 < \cdots < t_{p-1} < t_p = 1$ be points with the property that

$$\|P_{\mathcal{M}(t_i)} - P_{\mathcal{M}(\eta)}\| < 1 \quad \text{for} \quad t_i \le \eta \le t_{i+1}, \qquad i = 0, \ldots, p - 1$$

Here $P_{\mathcal{N}}$ is the orthogonal projector on the subspace $\mathcal{N} \subset \mathbb{C}^n$. For each $i = 0, \ldots, p - 1$, the transformation $S_i(\eta)$, $t_i \le \eta \le t_{i+1}$, defined by $S_i(\eta) = I - (P_{\mathcal{M}(t_i)} - P_{\mathcal{M}(\eta)})$ maps $\mathcal{M}(t_i)$ on $\mathcal{M}(\eta)$, is invertible and $S_i(t_i) = I$. Now put

$$S(t) = S_i(t) \cdots S_1(t_2) S_0(t_1) \quad \text{for} \quad t_i \le t \le t_{i+1}; \qquad \mathcal{M} = \mathcal{M}(0)$$

to satisfy (b).

Obviously, (b) implies (c). Finally, let us prove that (c) implies (a). Given S_{t_0} and $\mathcal{M}_{t_0}$ as in (c), let P_0 be the orthogonal projector on $\mathcal{M}_{t_0}$. Then $S_{t_0}(t) P_0 (S_{t_0}(t))^{-1}$ is a projector on $\mathcal{L}(t)$; therefore, for $t \in U_{t_0}$ we have

$$\theta(\mathcal{L}(t), \mathcal{L}(t_0)) \le \|S_{t_0}(t) P_0 S_{t_0}(t)^{-1} - S_{t_0}(t_0) P_0 S_{t_0}(t_0)^{-1}\|$$

$$\le \|S_{t_0}(t) P_0 S_{t_0}(t)^{-1} - S_{t_0}(t_0) P_0 S_{t_0}(t)^{-1}\|$$

$$+ \|S_{t_0}(t_0) P_0 S_{t_0}(t)^{-1} - S_{t_0}(t_0) P_0 S_{t_0}(t_0)^{-1}\|$$

$$\le \|S_{t_0}(t) - S_{t_0}(t_0)\| \cdot \|P_0\| \cdot \|S_{t_0}(t)^{-1}\|$$

$$+ \|S_{t_0}(t_0)\| \cdot \|P_0\| \cdot \|S_{t_0}(t)^{-1} - S_{t_0}(t_0)^{-1}\|.$$

As $S_{t_0}(t)$ is continuous and invertible in U_{t_0}, its inverse is continuous as well, and the continuity of $\mathcal{L}(t)$ follows from the preceding inequality. $\square$

Corollary 13.6.4

Let $\mathcal{L}(t)$ be a continuous family of subspaces (of $\mathbb{C}^n$) on K, where $K \subset \dot{R}^m$ is a connected compact set. Then there exists a continuous basis $x_1(t), \ldots, x_p(t)$ in $\mathcal{L}(t)$, where $p = \dim \mathcal{L}(t)$. (Note that because of the connectedness of K the dimension of $\mathcal{L}(t)$ is independent of t on K.)

Indeed, use Theorem 13.6.3, (b) and put $x_j(t) = S(t) x_j$, $j = 1, \ldots, p$, where $x_1, \ldots, x_p$ is a basis in $\mathcal{M}$.

Corollary 13.6.5

Let $B(t)$ be a continuous $m \times n$ matrix function on a connected compact set $K \subset \dot{R}^q$, such that $\operatorname{rank} B(t) \overset{\text{def}}{=} p$ is independent of t. Then there exists a *continuous basis* $x_1(t), \ldots, x_{n-p}(t)$ *in* $\operatorname{Ker} B(t)$ *and a continuous basis* $y_1(t), \ldots, y_p(t)$ *in* $\operatorname{Im} B(t)$.

This corollary follows from Corollary 13.6.4, taking into account Proposition 13.6.1.

13.7 APPLICATIONS TO GENERALIZED INVERSES

In this section we apply results of the preceding sections to study the behaviour of a generalized inverse of a transformation when this transformation is allowed to change. Recall that a transformation $B: \mathbb{C}^m \to \mathbb{C}$ is called a *generalized inverse* of a transformation $A: \mathbb{C}^n \to \mathbb{C}^m$ if the equalities $BAB = B$, $ABA = A$ hold (see Section 1.5).

As an application of Theorem 13.5.1, we have the following result concerning close generalized inverses for close linear transformations.

Theorem 13.7.1

Let $X: \mathbb{C}^n \to \mathbb{C}^m$ *be a transformation with a generalized inverse* $X^I: \mathbb{C}^m \to \mathbb{C}^n$. *Then there exist constants* $K > 0$ *and* $\epsilon > 0$ *with the property that every transformation* $Y: \mathbb{C}^n \to \mathbb{C}^m$ *with* $\|Y - X\| < \epsilon$ *and* $\dim \operatorname{Ker} Y = \dim \operatorname{Ker} X$ *has a generalized inverse* Y^I *satisfying*

$$\|Y^I - X^I\| \le K\|Y - X\| \tag{13.7.1}$$

Proof. By Theorem 1.5.5, the generalized inverse X^I is determined by a direct complement $\mathcal{N}$ to $\operatorname{Ker} X$ in $\mathbb{C}^n$ and by a direct complement $\mathcal{M}$ to $\operatorname{Im} X$ in $\mathbb{C}^m$, as follows:

$$X^I y = X_1^{-1}(P_X y), \qquad y \in \mathbb{C}^m$$

where P_X is the projector on $\operatorname{Im} X$ along $\mathcal{M}$, and $X_1: \mathcal{N} \to \operatorname{Im} X$ is the invertible transformation defined by $X_1 x = Xx$, $X \in \mathcal{N}$. Denote by $\mathcal{K}(X)$ the set of all transformations $Y: \mathbb{C}^n \to \mathbb{C}^m$ such that $\dim \operatorname{Ker} Y = \dim \operatorname{Ker} X$. Using Theorem 13.1.3 and inequality (13.5.2), choose $\epsilon_1 > 0$ in such a way that $\mathcal{N}$ is a direct complement to $\operatorname{Ker} Y$ for every $Y \in \mathcal{K}(X)$ with $\|X - Y\| < \epsilon_1$. Using the analog of Theorem 13.5.1 for images of linear transformations, we find a projector P_Y on $\operatorname{Im} Y$ such that

$$\|P_X - P_Y\| \le K_1\|X - Y\| \tag{13.7.2}$$

for every $Y \in \mathcal{K}(X)$. Here the constant K_1 depends on X and P_X only.

Our next observation is that, by Lemma 13.3.2 and (13.7.2), there exists a positive number $\epsilon_2 \leq \epsilon_1$ such that for any $Y \in \mathcal{H}(X)$ with $\|X - Y\| < \epsilon_2$ we can find an invertible transformation $S_Y \colon \mathbb{C}^m \to \mathbb{C}^m$ with $S_Y(\operatorname{Im} Y) = \operatorname{Im} X$ and

$$\max(\|S_Y - I\|, \|S_Y^{-1} - I\|) \leq K_2 \|X - Y\|$$

where the positive constant K_2 depends on X and P_X only. Let $\tilde{Y} = S_Y Y$, and note that for every generalized inverse $\tilde{Y}^I$ of $\tilde{Y}$ the transformation $\tilde{Y}^I S_Y$ is a generalized inverse for Y. Now for $Y \in \mathcal{H}(X)$ with $\|X - Y\| < \epsilon_2$ we have

$$\|\tilde{Y}^I S_Y - X^I\| \leq \|\tilde{Y}^I - X^I\| + \|\tilde{Y}^I S_Y - \tilde{Y}^I\| \leq \|\tilde{Y}^I - X^I\| + \|Y^I\| K_2 \|X - Y\|$$

so it is sufficient to prove Theorem 13.7.1 for $\tilde{Y}$ in place of Y. In other words, we can (and will) assume that the transformation Y from Theorem 13.7.1 satisfies the additional property that $\operatorname{Im} Y = \operatorname{Im} X$.

Now we verify (13.7.1) for the generalized inverse $Y^I = Y_1^{-1} P_Y$, where $Y_1 \colon \mathcal{N} \to \operatorname{Im} Y = \operatorname{Im} X$ is defined by $Y_1 x = Yx$, $x \in \mathcal{N}$. Indeed

$$\|Y_1^{-1} P_Y - X_1^{-1} P_X\| \leq \|Y_1^{-1}\| \cdot \|P_Y - P_X\| + \|Y_1^{-1} - X_1^{-1}\| \cdot \|P_X\|$$

and

$$\|Y_1^{-1} - X_1^{-1}\| = \|Y_1^{-1}(X_1 - Y_1)X_1^{-1}\| \leq \|Y_1^{-1}\| \, \|X_1 - Y_1\| \, \|X_1^{-1}\|$$
$$\leq \|Y_1^{-1}\| \, \|X - Y\| \, \|X_1^{-1}\|$$

But the norms $\|Y_1^{-1}\|$ are bounded provided the transformation $Y \in \mathcal{H}(X)$ with $\operatorname{Im} Y = \operatorname{Im} X$ is such that $\|X - Y\| \leq \frac{1}{2}\|X_1^{-1}\|$. Theorem 13.7.1 is proved. $\square$

Observe that the complete analog of Theorem 13.5.1 does not hold for the case of generalized inverses. Namely, given X and X^I as in Theorem 13.7.1, in general there is no positive constant K such that any transformation $Y \colon \mathbb{C}^n \to \mathbb{C}^m$ with $\dim \operatorname{Ker} Y = \dim \operatorname{Ker} X$ has a generalized inverse Y^I satisfying (13.7.1). To produce an example of such a situation, take $n = m$ and let $X \colon \mathbb{C}^n \to \mathbb{C}^n$ be invertible. Then there is only one generalized inverse of X, namely, its inverse X^{-1}. Further, let $Y = \alpha X$, where $\alpha \neq 0$. If (13.7.1) were true, we would have for some $K > 0$ and all α:

$$|\alpha^{-1} - 1| \cdot \|X^{-1}\| \leq K \|\alpha - 1\| \cdot \|X^{-1}\|$$

which is contradictory for α close to zero.

Now we consider continuous families of transformations and their generalized inverses. It is convenient to use the language of matrices with the usual understanding that $n \times m$ matrices represent transformations from $\mathbb{C}^m$ into $\mathbb{C}^n$ in fixed bases in $\mathbb{C}^m$ and $\mathbb{C}^n$.

Theorem 13.7.2

Let $B(t)$ be a continuous $m \times n$ matrix function on a connected compact set $K \subset \mathbb{R}^q$ such that rank $B(t) \overset{\text{def}}{=} p$ is independent of t. Then there exists a continuous $n \times m$ matrix function $X(t)$ on K such that, for every $t \in K$, $X(t)$ is a generalized inverse of $B(t)$.

Proof. In view of Corollary 13.6.5 there exists a continuous basis $x_1(t), \ldots, x_{n-p}(t)$ in Ker $B(t)$, as well as a continuous basis $y_1(t), \ldots, y_p(t)$ in Im $B(t)$. By the same corollary there exist a continuous basis $x_{n-p+1}(t), \ldots, x_n(t)$ in Im $B(t)^*$ and a continuous basis $y_{p+1}(t), \ldots, y_m(t)$ in Ker $B(t)^*$. As Im $B(t)^* = (\text{Ker } B(t))^\perp$, it follows that $x_1(t), \ldots, x_n(t)$ is a basis in $\mathbb{C}^m$ for all $t \in K$. Also, $y_1(t), \ldots, y_m(t)$ is a basis in $\mathbb{C}^m$ for all $t \in K$. Define a transformation $X(t): \mathbb{C}^m \to \mathbb{C}^n$ as follows: $X(t)y_j(t) = 0$, $j = p + 1, \ldots, m$; and for $j = 1, \ldots, p$ $X(t)y_j(t)$ is the unique vector in Im $B(t)^*$ such that $B(t)X(t)y_j(t) = y_j(t)$. Theorem 1.5.5 shows that $X(t)$ is indeed a generalized inverse of $B(t)$ for all $t \in K$. It remains to show that $X(t)$ is continuous.

For a fixed vector $z \in \mathbb{C}^m$ and any $t \in K$, write $z = \sum_{i=1}^m z_i(t)y_i(t)$, for some complex numbers $z_i(t)$ that depend on t. These numbers $z_i(t)$ turn out to be continuous, because

$$\begin{bmatrix} z_1(t) \\ \vdots \\ z_m(t) \end{bmatrix} = [y_1(t) \cdots y_m(t)]^{-1} z$$

Further, the transformation

$$B|_{\text{Im } B(t)^*}: \text{Im } B(t)^* \to \text{Im } B(t)$$

is invertible, so

$$B(t)^{-1} y_j(t) = \sum_{i=n-p+1}^n \alpha_{ji}(t) x_i(t), \qquad j = 1, \ldots, p$$

for some complex numbers $\alpha_{ji}(t)$ that also depend on t. Again, $\alpha_{ji}(t)$ are continuous on K. Indeed, $\alpha_{ji}(t)$ is the unique solution of the linear system of equations

$$y_j(t) = \sum_{i=n-p+1}^n \alpha_{ji}(t) B(t) x_i(t), \qquad j = 1, \ldots, p \qquad (13.7.3)$$

Writing $y_j(t)$, $j = 1, \ldots, p$ in terms of linear combinations of the standard basis vectors $e_1, \ldots, e_m$, and writing $x_i(t)$, $i = n - p + 1, \ldots, n$ in terms of linear combinations of $e_1, \ldots, e_n$ we can represent the system (13.7.3) in the form

$$A(t)\alpha(t) = C(t), \qquad t \in K \tag{13.7.4}$$

where $\alpha(t)$ is the p^2-dimensional vector formed by $\alpha_{ji}(t)$, $j = 1, \ldots, p$; $i = n - p + 1, \ldots, n$, and $A(t)$ and $C(t)$ are suitable matrix and vector functions, respectively, which are continuous in t. As the solution of (13.7.4) exists and is unique for every $t \in K$, it follows that the columns of $A(t)$ are linearly independent for every $t \in K$. Now fix $t_0 \in K$, and assume for simplicity of notation that the upper p^2 rows of $A(t_0)$ are linearly independent. Partition

$$A(t) = \begin{bmatrix} A_0(t) \\ A_1(t) \end{bmatrix}; \qquad C(t) = \begin{bmatrix} C_0(t) \\ C_1(t) \end{bmatrix}$$

where $A_0(t)$ and $C_0(t)$ are the top p^2 rows of $A(t)$ and $C(t)$, respectively. Then $A_0(t_0)$ is nonsingular; as $A(t)$ is continuous in t, the matrix $A_0(t)$ is nonsingular for every t from some neighbourhood U_{t_0} of t_0 in K. It follows that

$$\alpha(t) = (A_0(t))^{-1} C_0(t)$$

is continuous in t for $t \in U_{t_0}$. As $t_0 \in K$ was arbitrary, the functions $\alpha_{ji}(t)$ are continuous on K.

Returning to our generalized inverse $X(t)$, we have for every

$$z = \sum_{i=1}^{m} z_i(t) y_i(t) \in \mathbb{C}^m$$

the following equalities:

$$X(t)z = \sum_{i=1}^{p} z_i(t) X(t) y_i(t) = \sum_{i=1}^{p} z_i(t) B(t)^{-1} y_i(t)$$

$$= \sum_{i=1}^{p} z_i(t) \sum_{j=n-p+1}^{p} \alpha_{ij}(t) x_j(t)$$

and so $X(t)$ is continuous on K. $\square$

A particular case of Theorem 13.7.2 deserves to be mentioned explicitly.

Corollary 13.7.3.

Let $B(t)$ be a continuous $m \times n$ matrix function on a connected compact set $K \subset \mathbb{R}^q$ such that, for every $t \in K$, the matrix $B(t)$ is left invertible (resp. right invertible). Then there exists a left inverse (resp. a right inverse) $X(t)$ of $B(t)$ such that $X(t)$ is a continuous function of t on K.

13.8 SUBSPACES OF NORMED SPACES

Until now we have studied the notions of gaps, minimal angle, minimal opening, and so on for subspaces of $\mathbb{C}^n$ where the norm of a vector $x = \langle x_1, \ldots, x_n \rangle$ is Euclidean: $\|x\| = (\sum_{i=1}^{n} |x_i|^2)^{1/2}$. Here we show how these notions can be extended to the framework of a finite-dimensional linear space with a norm that is not necessarily generated by a scalar product.

Let V be a finite-dimensional linear space over $\mathbb{C}$ or over $\mathbb{R}$. A real-valued function defined for all elements $x \in V$, denoted by $\|x\|$, is called a *norm* if the following properties are satisfied: (a) $\|x\| \geq 0$ for all $x \in V$; $\|x\| = 0$ if and only if $x = 0$; (b) $\|\lambda x\| = |\lambda| \, \|x\|$ for every $x \in V$ and every scalar λ (so $\lambda \in \mathbb{C}$ or $\lambda \in \mathbb{R}$ according as V is over $\mathbb{C}$ or over $\mathbb{R}$); (c) $\|x + y\| \leq \|x\| + \|y\|$, for all $x, y \in V$ (the triangle inequality).

EXAMPLE 13.8.1. Let $f_1, \ldots, f_n$ be a basis in V, and fix a number $p \geq 1$. For every $x = \sum_{i=1}^{n} \alpha_i f_i \in V$, put

$$\|x\|_p = \left(\sum_{i=1}^{n} |\alpha_i|^p \right)^{1/p}$$

Also, define $\|x\|_\infty = \max(|\alpha_1|, \ldots, |\alpha_n|)$. We leave it to the reader to verify that $\|\cdot\|_p \, (p \geq 1)$ and $\|\cdot\|_\infty$ are norms (one should use the Minkowski inequality for this purpose): for any complex numbers $x_1, \ldots, x_n$, $y_1, \ldots, y_n$ and any $p \geq 1$ we have

$$\left(\sum_{j=1}^{n} |x_j + y_j|^p \right)^{1/p} \leq \left(\sum_{j=1}^{n} |x_j|^p \right)^{1/p} + \left(\sum_{j=1}^{n} |y_j|^p \right)^{1/p} \quad \square$$

EXAMPLE 13.8.2. For $V = \mathbb{C}^n$ (or $V = \mathbb{R}^n$) let

$$\|x\| = \left(\sum_{i=1}^{n} |x_i|^2 \right)^{1/2}$$

where $x = \langle x_1, \ldots, x_n \rangle$ belongs to $\mathbb{C}^n$ (or to $\mathbb{R}^n$). We have used this norm throughout the book. Actually, this is a particular case of Example 13.8.1 (with the basis $f_i = e_i$, $i = 1, \ldots, n$ in $\mathbb{C}^n$ (or $\mathbb{R}^n$) and $p = 2$). $\square$

Any norm on V is continuous, as proved in the following proposition.

Proposition 13.8.1

Let $f_1, \ldots, f_n$ be a basis in V, and let $\|\cdot\|$ be a norm in V. Then, given $\epsilon > 0$ there exists a $\delta > 0$ such that the inequality

$$| \, \|x\| - \|y\| \, | < \epsilon$$

holds provided $|x_j - y_j| < \delta$ *for* $j = 1, \ldots, n$, *where* $x = \Sigma_{j=1}^n x_j f_j$ *and* $y = \Sigma_{j=1}^n y_j f_j$.

Proof. Letting $M = \max_{1 \le j \le n} \|f_j\|$, choose $\delta = \epsilon M^{-1} n^{-1}$. Then for every $x = \Sigma_{j=1}^n x_j f_j$, $y = \Sigma_{j=1}^n y_j f_j$ with $|x_j - y_j| < \delta$, $j = 1, \ldots, n$, we have

$$\|x - y\| \le \sum_{j=1}^n |x_j - y_j| \, \|f_j\| \le M \sum_{j=1}^n |x_j - y_j| < Mn\delta = \epsilon$$

It remains to use the inequality

$$| \, \|x\| + \|y\| \, | \le \|x - y\|$$

which follows easily from the axioms of a norm. $\square$

It is important to recognize that different norms on a given finite-dimensional vector space are equivalent in the following sense.

Theorem 13.8.2

Let $\| \cdot \|'$ *and* $\| \cdot \|''$ *be two norms in* V. *Then there exists a constant* $K \ge 1$ *such that*

$$K^{-1} \|x\|' \le \|x\|'' \le K \|x\|' \tag{13.8.1}$$

for every $x \in V$.

We stress the fact that K depends on $\| \cdot \|'$, $\| \cdot \|''$ only (and of course on the underlying linear space V).

Proof. Let $f_1, \ldots, f_n$ be a basis in V. It is sufficient to prove the theorem for the case when

$$\left\| \sum_{i=1}^n \alpha_i f_i \right\|' = \left(\sum_{i=1}^n |\alpha_i|^2 \right)^{1/2}$$

Consider the real-valued continuous function g defined on $\mathbb{C}^n$ by

$$g(\alpha_1, \ldots, \alpha_n) = \left\| \sum_{i=1}^n \alpha_i f_i \right\|'', \qquad \alpha_j \in \mathbb{C} \quad \text{for} \quad j = 1, \ldots, n$$

As the set $\{(\alpha_1, \ldots, \alpha_n) \in \mathbb{C}^n \mid \Sigma_{i=1}^n |\alpha_i|^2 = 1\}$ is closed and bounded, the function g attains its maximum and minimum on this bounded set. So there exist $x_1, x_2 \in V$ such that $\|x_1\|' = \|x_2\|' = 1$ and

$$\|x_1\|'' \le \|v\|'' \le \|x_2\|''$$

for every $v \in V$ with $\|v\|' = 1$. Now for $x \in V$, $x \ne 0$ we have $\|x/\|x\|'\|' = 1$ and hence

$$\|x_1\|'' \le \frac{\|x\|''}{\|x\|'} \le \|x_2\|''.$$

Thus inequality (13.8.1) holds with $K = \max(\|x_2\|'', 1/\|x_1\|'')$. $\quad\Box$

In the rest of this section we assume that an arbitrary norm $\|\cdot\|$ is given in the finite-dimensional linear space V.

For any subspace $\mathcal{M} \subset V$, let

$$S(\mathcal{M}) = \{x \in \mathcal{M} \mid \|x\| = 1\}$$

be the unit sphere of $\mathcal{M}$. Now the *gap* $\theta(\mathcal{L}, \mathcal{M})$ between the subspaces $\mathcal{L}$ and $\mathcal{M}$ in V is defined by formula (13.1.3):

$$\theta(\mathcal{L}, \mathcal{M}) = \max\{\sup_{x \in S(\mathcal{M})} d(x, \mathcal{L}), \sup_{x \in S(\mathcal{L})} d(x, \mathcal{M})\}$$

where $d(x, Z) = \inf_{t \in Z} \|x - t\|$ for a set $Z \subset V$.

The gap has two properties of a metric: (a) $\theta(\mathcal{L}, \mathcal{M}) = \theta(\mathcal{M}, \mathcal{L})$ for all subspaces $\mathcal{M}, \mathcal{L} \subset V$; (b) $\theta(\mathcal{L}, \mathcal{M}) > 0$ if $\mathcal{L} \ne \mathcal{M}$; $\theta(\mathcal{L}, \mathcal{L}) = 0$. However, the triangle inequality

$$\theta(\mathcal{L}, \mathcal{M}) \le \theta(\mathcal{L}, \mathcal{N}) + \theta(\mathcal{N}, \mathcal{M}) \tag{13.8.2}$$

for all subspaces $\mathcal{L}, \mathcal{M}, \mathcal{N}$ in V fails in general, although it is true when the norm is defined by means of a scalar product (x, y), as Theorem 13.1.1 shows. The following example illustrates this fact.

EXAMPLE 13.8.3. Let $\mathbb{R}^2$ be the normed space with the norm

$$\|\langle x_1, x_2 \rangle\|_1 = |x_1| + |x_2|, \qquad \langle x_1, x_2 \rangle \in \mathbb{R}^2$$

Consider a family of one-dimensional subspaces

$$\mathcal{L}(\alpha) = \operatorname{Span}\{e_1 + \alpha e_2\}, \qquad \alpha \in \mathbb{R}$$

We compute $\theta(\mathcal{L}(\alpha), \mathcal{L}(\beta))$. Take $x \in S(\mathcal{L}(\beta))$, so that $x = \langle \gamma, \gamma\beta \rangle$, where $|\gamma| = (1 + |\beta|)^{-1}$. Now

$$\inf_{y \in \mathscr{L}(\alpha)} \|x - y\|_1 = \inf_{\mu \in \mathscr{R}} \left\| \begin{bmatrix} \gamma \\ \gamma\beta \end{bmatrix} - \begin{bmatrix} \mu \\ \mu\alpha \end{bmatrix} \right\|_1 = \inf_{\mu \in \mathscr{R}} \{|\gamma - \mu| + |\gamma\beta - \mu\alpha|\}$$

As the function $f(\mu) = |\gamma - \mu| + |\gamma\beta - \mu\alpha|$ is piecewise linear, we have

$$\inf_{\mu \in \mathscr{R}} \{|\gamma - \mu| + |\gamma\beta - \mu\alpha|\} = \min\left\{|\gamma\beta - \gamma\alpha|, \left|\gamma - \frac{\gamma\beta}{\alpha}\right|\right\}$$

$$= \frac{|\beta - \alpha|}{1 + |\beta|} \min(1, |\alpha|^{-1})$$

So

$$\theta(\mathscr{L}(\alpha), \mathscr{L}(\beta))$$
$$= |\beta - \alpha| \max\{(1 + |\beta|)^{-1} \min(1, |\alpha|^{-1}), (1 + |\alpha|)^{-1} \min(1, |\beta|^{-1})\}$$

Let $\alpha < \beta < \gamma$ be positive numbers such that $\beta < 1 < \gamma$ and $\beta\gamma < 1$. We compute

$$\theta(\mathscr{L}(\alpha), \mathscr{L}(\beta)) + \theta(\mathscr{L}(\beta), \mathscr{L}(\gamma)) = \frac{\beta - \alpha}{1 + \alpha} + \frac{\gamma - \beta}{\gamma + \beta\gamma}$$

and

$$\theta(\mathscr{L}(\alpha), \mathscr{L}(\gamma)) = \frac{\gamma - \alpha}{\gamma + \alpha\gamma}$$

However, clearly

$$\beta + \frac{\gamma - \beta}{\gamma + \beta\gamma} < 1$$

so the inequality

$$\frac{\beta - \alpha}{1 + \alpha} + \frac{\gamma - \beta}{\gamma + \beta\gamma} < \frac{\gamma - \alpha}{\gamma + \alpha\gamma}$$

holds for sufficiently small positive α, and the triangle inequality for the gap fails in this particular case. $\square$

In contrast, the spherical gap

$$\bar{\theta}(\mathscr{L}, \mathscr{M}) = \max\{\sup_{x \in S(\mathscr{M})} d(x, S(\mathscr{L})), \sup_{x \in S(\mathscr{L})} d(x, S(\mathscr{M}))\}$$

is a metric. (The verification of this fact is exactly the same as that given in Section 13.2.) Instead of inequality (13.2.5), we have in the case of a general normed space the weaker inequality

$$\theta(\mathscr{L}, \mathscr{M}) \le \bar{\theta}(\mathscr{L}, \mathscr{M}) \le 2\theta(\mathscr{L}, \mathscr{M}) \tag{13.8.3}$$

for any subspaces $\mathscr{L}, \mathscr{M} \subset V$. Indeed, the left-hand inequality of (13.8.3) is evident from the definitions of $\theta(\mathscr{L}, \mathscr{M})$ and $\bar{\theta}(\mathscr{L}, \mathscr{M})$. To prove the right-hand inequality in (13.8.3), it is sufficient to verify that for every vector $v \in V$ with $\|v\| = 1$ and every subspace $\mathscr{M} \subset V$ we have

$$d(u, S(\mathscr{N})) \le 2d(u, \mathscr{N}) \tag{13.8.4}$$

For a given $\epsilon > 0$ there exists a $v \in \mathscr{N}$ such that

$$\|u - v\| < d(u, \mathscr{N}) + \epsilon \tag{13.8.5}$$

and we can assume that $v \ne 0$. [Otherwise, replace v by a nonzero vector sufficiently close to zero so that (13.8.5) still holds.] Then $v_0 \stackrel{\text{def}}{=} v/\|v\| \in S(\mathscr{N})$ and hence

$$d(u, S(\mathscr{N})) \le \|u - v_0\| \le \|u - v\| + \|v - v_0\|$$

But

$$\|v - v_0\| = |\,\|v\| - 1\,| = |\,\|v\| - \|u\|\,| \le \|v - u\|$$

and we have

$$d(u, S(\mathscr{N})) \le 2\|v - u\| < 2d(u, \mathscr{N}) + 2\epsilon$$

As $\epsilon > 0$ is arbitrary, the desired inequality (13.8.4) follows.

The minimal angle between two subspaces is defined in a normed space by the formula (13.2.1). With this definition, Proposition 13.2.2 and Theorem 13.2.3 are valid in this case. Without going into details, we remark that Lemmas 13.3.1 and 13.3.2 also can be extended to the normed space context.

Concerning the metric space properties of the set of all subspaces in the spherical gap metric (such as compactness, completeness), it follows from inequality (13.8.3) and the following result that these do not depend on the particular choice of the norm.

Theorem 13.8.3

Let $\|\cdot\|'$ and $\|\cdot\|''$ be two norms in V, with the corresponding gaps $\theta'(\mathscr{M}, \mathscr{N})$ and $\theta''(\mathscr{M}, \mathscr{N})$ between subspaces $\mathscr{M}$ and $\mathscr{N}$ in V. Then there exists a constant $L \ge 1$ such that

$$L^{-1}\theta'(\mathscr{M}, \mathscr{N}) \le \theta''(\mathscr{M}, \mathscr{N}) \le L\theta'(\mathscr{M}, \mathscr{N}) \tag{13.8.6}$$

for all subspaces $\mathscr{M}$ and $\mathscr{N}$.

Again, the constant L depends on the norm $\| \cdot \|'$ and $\| \cdot \|''$ only.

Proof. By Theorem 13.8.2 we have for any $x \in V$

$$K^{-1}\|x\|' \le \|x\|'' \le K\|x\|'' ,$$

where the constant $K \ge 1$ is independent of x. Hence

$$\sup_{\substack{x \in M \\ \|x\|'=1}} \inf_{t \in L} \|x - t\|' = \sup_{\substack{x \in M \\ \|x\|' \le 1}} \inf_{t \in L} \|x - t\|'$$

$$\le K \sup_{\substack{x \in M \\ \|x\|'' \le K}} \inf_{t \in L} \|x - t\|'' = K^2 \sup_{\substack{x \in M \\ \|x\|'' \le 1}} \inf_{t \in L} \|x - t\|''$$

$$= K^2 \sup_{\substack{x \in M \\ \|x\|'' = 1}} \inf_{t \in L} \|x - t\|'' .$$

In view of the definition of $\theta(L, M)$ we obtain the left-hand inequality in (13.8.6) with $L = K^2$. The right-hand inequality in (13.8.6) follows similarly. $\square$

13.9   EXERCISES

13.1   Compute the gap $\theta(M, N)$, where

$$M = \mathrm{Span}\begin{bmatrix} 1 \\ x \end{bmatrix}, \qquad N = \mathrm{Span}\begin{bmatrix} 1 \\ y \end{bmatrix} \subset \mathbb{C}^2$$

and x and y are complex numbers such that $|x| = |y|$.

13.2   Compute the gap $\theta(M, N)$, spherical gap $\bar{\theta}(M, N)$, minimal opening $\eta(M, N)$ and minimal angle $\varphi_{\min}(M, N)$, where

$$M = \mathrm{Span}\begin{bmatrix} 1 \\ x \end{bmatrix}, \qquad N = \mathrm{Span}\begin{bmatrix} 1 \\ y \end{bmatrix} \subset \mathbb{R}^2$$

and x and y are real numbers such that $|x| = |y|$.

13.3   Compute $\bar{\theta}(M, N)$, $\eta(M, N)$, and $\varphi_{\min}(M, N)$ for any two one-dimensional subspaces M and N in $\mathbb{R}^n$.

13.4   Let $U: \mathbb{C}^n \to \mathbb{C}^n$ be a unitary transformation. Prove that

$$\theta(M, N) = \theta(UM, UN) ; \qquad \bar{\theta}(M, N) = \bar{\theta}(UM, UN)$$

$$\eta(M, N) = \eta(UM, UN)$$

for any pair of subspaces $M, N \subset \mathbb{C}^n$.

13.5   Prove that for subspaces $\mathcal{L}, \mathcal{M}$ in $\mathbb{C}^n$

$$\theta(\mathcal{L}^\perp, \mathcal{M}^\perp) = \theta(\mathcal{L}, \mathcal{M})$$

13.6   Show that the equality $\theta(\mathcal{L}, \mathcal{M}) = 1$ holds if and only if either $\mathcal{L}^\perp \cap \mathcal{M} \neq \{0\}$ or $\mathcal{L} \cap \mathcal{M}^\perp \neq \{0\}$ (or both).

13.7   Let $\mathcal{M}_1, \mathcal{N}_1$ be subspaces in $\mathbb{C}^n$ and $\mathcal{M}_2, \mathcal{N}_2$ be subspaces in $\mathbb{C}^m$. Prove that

$$\theta(\mathcal{M}_1 \oplus \mathcal{M}_2, \mathcal{N}_1 \oplus \mathcal{N}_2) = \max\{\theta(\mathcal{M}_1, \mathcal{N}_1), \theta(\mathcal{M}_2, \mathcal{N}_2)\}$$

where $\mathcal{M}_1 \oplus \mathcal{M}_2, \mathcal{N}_1 \oplus \mathcal{N}_2 \subset \mathbb{C}^n \oplus \mathbb{C}^m$.

13.8   Find the gaps $\theta(\text{Ker } A, \text{Ker } B)$ and $\theta(\text{Im } A, \text{Im } B)$ for the following pairs of transformations $A, B\colon \mathbb{C}^n \to \mathbb{C}^n$:

(a) A and B are diagonal in the same orthonormal basis.
(b) A and B are commuting normal transformations.
(c) A and B are circulant matrices in the same orthonormal basis. (*Hint*: A and B can be simultaneously diagonalized by a unitary matrix.)

(d) $A = \begin{bmatrix} 0 & & \cdots & 0 & \alpha_n \\ \vdots & & & 0 & \vdots \\ & & \ddots & & \\ 0 & \alpha_2 & & & 0 \\ \alpha_1 & 0 & \cdots & & 0 \end{bmatrix}$ $B = \begin{bmatrix} 0 & & \cdots & 0 & \beta_n \\ \vdots & & & 0 & \vdots \\ & & \ddots & & \\ 0 & \beta_2 & & & 0 \\ \beta_1 & 0 & \cdots & & 0 \end{bmatrix}$

in the same orthonormal basis, where α_j and β_j are complex numbers.

13.9   For each of cases (a)–(d) in Exercise 13.8, find

$$\tilde{\theta}(\text{Ker } A, \text{Ker } B), \qquad \tilde{\theta}(\text{Im } A, \text{Im } B), \qquad \eta(\text{Ker } A, \text{Ker } B),$$

$$\eta(\text{Im } A, \text{Im } B), \qquad \varphi_{\min}(\text{Ker } A, \text{Ker } B), \quad \text{and} \quad \varphi_{\min}(\text{Im } A, \text{Im } B)$$

13.10   Let $A\colon \mathbb{C}^n \to \mathbb{C}^n$ be a transformation. Then $\theta(\mathcal{M}, \mathcal{N}) = 1$ for any distinct A-invariant subspaces $\mathcal{M}$ and $\mathcal{N}$ if and only if A is normal with n distinct eigenvalues.

13.11   Show that if $A(t)$, $t \in [0, 1]$ is a continuous family of $n \times n$ circulant matrices and $\dim \text{Ker } A(t)$ is constant (i.e., independent of t), then the subspaces $\text{Ker } A(t)$ and $\text{Im } A(t)$ are constant.

13.12   Prove or disprove the following:

(a)   If $A(t)$ is a continuous family of upper triangular Toeplitz $n \times n$ matrices for $t \in [0, 1]$, then $\dim \text{Ker } A(t)$ is constant if and only if $\text{Ker } A(t)$ and $\text{Im } A(t)$ are constant.

(b)   Same as (a) for

$$A(t) = \sum_{j=0}^{n-1} \alpha_j(t) A^j$$

where $\alpha_j(t)$ are continuous scalar functions of $t \in [0, 1]$ and A is a fixed $n \times n$ matrix.

13.13 Show that a circulant matrix has a generalized inverse that is also a circulant.

13.14 Let $A(t)$ be a continuous family of circulant matrices with dim Ker $A(t)$ constant for $t \in [0, 1]$. Show that there exists a continuous family $B(t)$ of generalized inverses of $A(t)$ on $[0, 1]$ that also consists of circulant matrices.

13.15 Solve Exercises 13.13 and 13.14 with "circulant" replaced by "upper triangular Toeplitz."

13.16 Assume the hypotheses of Lemma 13.3.1 and, in addition, assume that the projector Π_0 is orthogonal. Prove that $\|R\| = \cotan \varphi_{\min}$, where $\varphi_{\min}$ is the minimal angle between Ker Π_0 and Im Π.

13.17 Find the minimal angle between any two one-dimensional subspaces in the normed space $\mathbb{R}^2$ with the following norms:

(a) $\|\langle x, y \rangle\|_1 = |x| + |y|$.
(b) $\|\langle x, y \rangle\|_\infty = \max(|x|, |y|)$.

Chapter Fourteen

The Metric Spaces of Invariant Subspaces

We study the structure of the set Inv(A) of all invariant subspaces of a transformation $A: \mathbb{C}^n \to \mathbb{C}^n$ in the context of the metric space $\mathcal{G}(\mathbb{C}^n)$ of all subspaces in $\mathbb{C}^n$. Throughout this chapter $\mathbb{C}^n$ is considered with the standard scalar product and the gap metric determined by this scalar product on $\mathcal{G}(\mathbb{C}^n)$, as studied in the preceding chapter. With the exception of Section 14.3, the results of this chapter are not used subsequently in this book.

14.1 CONNECTED COMPONENTS: THE CASE OF ONE EIGENVALUE

Let $\mathcal{A} \subset \mathcal{B}$ be two sets of subspaces of $\mathbb{C}^n$. We say that $\mathcal{A}$ is *connected in* $\mathcal{B}$ if for any subspaces $\mathcal{L}, \mathcal{M} \subset \mathcal{A}$ there is a continuous function $f: [0, 1] \to \mathcal{B}$ such that $f(0) = \mathcal{L}$, $f(1) = \mathcal{M}$. [The continuity of f is understood in the gap metric. Thus, for every $t_0 \in [0, 1]$ and every $\epsilon > 0$ there is a $\delta > 0$ such that $|t - t_0| < \delta$ and $t \in [0, 1]$ imply $\theta(f(t), f(t_0)) < \epsilon$.] The set $\mathcal{A}$ is called *connected* if $\mathcal{A}$ is connected in $\mathcal{A}$.

We start the study of connectedness of the set Inv(A) with the case when $A = J$, a Jordan matrix with $\sigma(J) = \{0\}$. Let r be the geometric multiplicity of the eigenvalue 0 of J, and let $k_1 \geq \cdots \geq k_r$ be the sizes of the Jordan blocks in J. Also, denote the set of all p-dimensional J-invariant subspaces by Inv$_p$.

Let $l = (l_1, \ldots, l_r)$ be an ordered r-tuple of integers such that $0 \leq l_i \leq k_i$, $\sum_{i=1}^r l_i = p$, and let Φ_p be the set of all such r-tuples. We associate every $l = (l_1, \ldots, l_r) \in \Phi_p$ with the subspace $\mathcal{G}(l) \in$ Inv$_p$, spanned by vectors $u_j^{(i)}$; $j = 0, \ldots, l_i - 1$; $i = 1, \ldots, r$, where $u_j^{(i)}$ are unit coordinate vectors in $\mathbb{C}^n$ and the sole nonzero coordinate of $u_j^{(i)}$ is equal to one and is in the place $k_1 + \cdots + k_{i-1} + j + 1$ (we assume $k_0 = 0$) for $j = 0, \ldots, k_i - 1$ and $i = 1, \ldots, r$. There is a one-to-one correspondence between elements of Φ_p and subspaces from Inv_p spanned by unit coordinate vectors. So we can assume that $\Phi_p \subset \text{Inv}_p$.

Lemma 14.1.1

Φ_p *is connected in* Inv_p.

Proof. Let $l = (l_1, \ldots, l_r)$ and $\bar{l} = (\bar{l}_1, \ldots, \bar{l}_r)$ be r-tuples from Φ_p, and suppose, for example, that $l_1 > \bar{l}_1$ and $l_2 < \bar{l}_2$. Let $\mathcal{F}(\epsilon) \in \text{Inv}_p$ be the subspace spanned by vectors $u_{l_1-1}^{(1)} + \epsilon u_{l_2}^{(2)}$, $u_0^{(1)}, \ldots, u_{l_1-2}^{(1)}$, $u_j^{(i)}$ for $j = 0, \ldots, l_i - 1$ and $i = 2, \ldots, r$, where ϵ is a complex number. Then $\mathcal{F}(0) = \mathcal{G}(l)$ (the subspace corresponding to the r-tuple l) and

$$\mathcal{F}(\infty) = \mathcal{G}(l_1 - 1, l_2 + 1, l_3, \ldots, l_r)$$

So $l = (l_1, \ldots, l_r)$ and $(l_1 - 1, l_2 + 1, l_3, \ldots, l_r)$ are connected in Inv_p. Applying this procedure several times, we obtain a connection between l and $\bar{l}$. $\square$

Lemma 14.1.2

Let $\mathcal{F}_1 \in \text{Inv}_p$. *Then* $\mathcal{F}_1$ *is connected in* Inv_p *with some* $\mathcal{F}_2 \in \Phi_p$.

Proof. For $i = 0, 1, 2, \ldots$, let

$$\mathcal{R}_i = \{x \in \mathbb{C}^n \mid J^i x = 0\}$$

Then $0 = \mathcal{R}_0 \subset \mathcal{R}_1 \subset \cdots \subset \mathcal{R}_s = \mathbb{C}^n$ for some integer s (s is the minimal integer such that $J^s = 0$). We construct the basic set of vectors in $\mathcal{F}_1$ in the following way (see the proof of the Jordan form in Section 2.3). Let i_0 be the greatest index that satisfies $(\mathcal{R}_{i_0} \smallsetminus \mathcal{R}_{i_0-1}) \cap \mathcal{F}_1 \neq \emptyset$. Take a basis $v_{i_0 1}, \ldots, v_{i_0 q_{i_0}}$ in $\mathcal{F}_1 \cap \mathcal{R}_{i_0}$ modulo $\mathcal{R}_{i_0-1}$. Then the vectors $J^i v_{i_0 1}, \ldots, J^i v_{i_0 q_{i_0}}$ are linearly independent in $\mathcal{F}_1 \cap \mathcal{R}_{i_0-i}$ modulo $\mathcal{R}_{i_0-i-1}$; $i = 1, \ldots, i_0 - 1$. We complete the set $J v_{i_0 1}, \ldots, J v_{i_0 q_{i_0}}$ by additional vectors $v_{i_0-1,1}, \ldots, v_{i_0-1,q_{i_0-1}}$ to form a basis in $\mathcal{F}_1 \cap \mathcal{R}_{i_0-1}$ modulo $\mathcal{R}_{i_0-2}$. Then the vectors

$$J^{i-1} v_{i_0-1,1}, \ldots, J^{i-1} v_{i_0-1,q_{i_0-1}}, \qquad J^i v_{i_0 1}, \ldots, J^i v_{i_0 q_{i_0}}$$

are linearly independent in $\mathcal{F}_1 \cap \mathcal{R}_{i_0-i}$ modulo $\mathcal{R}_{i_0-i-1}$ for $i = 2, \ldots, i_0 - 1$. Complete the set

$$J v_{i_0-1,1}, \ldots, J v_{i_0-1,q_{i_0-1}}; \qquad J^2 v_{i_0 1}, \ldots, J^2 v_{i_0 q_{i_0}}$$

by additional vectors $v_{i_0-2,1}, \ldots, v_{i_0-2,q_{i_0-2}}$ to a basic set of vectors in

$\mathscr{F}_1 \cap \mathscr{R}_{i_0-2}$ modulo $\mathscr{R}_{i_0-3}$, and so on. So we obtain the basic set of vectors in $\mathscr{F}_1$:

$$\{J^t v_{i1}, \ldots, J^t v_{iq_i}; \quad i = 1, \ldots, i_0; \quad t = 0, \ldots, i-1\}$$

To connect $\mathscr{F}_1$ with some subspace $\mathscr{F}_2 \in \Phi_p$, we use the following procedure. Take a set of q_{i_0} coordinate unit vectors $y_{i1}, \ldots, y_{i_0 q_{i_0}}$ in $\mathscr{R}_{i_0}$ that are independent modulo $\mathscr{R}_{i_0-1}$. For $j = 1, 2, \ldots, q_{i_0}$, put

$$v_{i_0 j}(\lambda) = \lambda v_{i_0 j} + (1 - \lambda) y_{i_0 j}$$

where λ is a complex parameter. Then the $v_{i_0 j}(\lambda)$ are linearly independent modulo $\mathscr{R}_{i_0-1}$ for every $\lambda \in \mathbb{C}$ except possibly for a finite set S_1. Indeed, let $x_1, \ldots, x_k$ be a basis in $\mathscr{R}_{i_0-1}$, and put

$$B(\lambda) = [v_{i_0 1}(\lambda), \ldots, v_{i_0 q_{i_0}}(\lambda), x_1, \ldots, x_k]$$

Then $v_{i_0 j}(\lambda)$, $j = 1, \ldots, q_{i_0}$ are linearly independent modulo $\mathscr{R}_{i_0-1}$ if and only if the columns of $B(\lambda)$ are linearly independent. Let $b(\lambda)$ be a minor of $B(\lambda)$ of order $i_0 + k$ such that $b(0) \neq 0$ (such a minor exists because $y_{i_0 j}$, $j = 1, \ldots, q_{i_0}$ are linearly independent modulo $\mathscr{R}_{i_0-1}$). So $b(\lambda)$ is a polynomial that is not identically zero. Clearly, for every λ that does not belong to the finite set S_1 of zeros of $b(\lambda)$, the vectors $v_{i_0 j}(\lambda)$, $j = 1, \ldots, q_{i_0}$ are linearly independent modulo $\mathscr{R}_{i_0-1}$. Observe that S_1 does not contain 0 and 1.

Further, take a set of q_{i_0-1} coordinate unit vectors $y_{i_0-1,1}, \ldots, y_{i_0-1,q_{i_0-1}}$ in $\mathscr{R}_{i_0-1}$ such that the vectors

$$y_{i_0-1,1}, \ldots, y_{i_0-1,q_{i_0-1}}, \quad J y_{i_0 1}, \ldots, J y_{i_0 q_{i_0}}$$

are independent modulo $\mathscr{R}_{i_0-2}$. Putting

$$v_{i_0-1,j}(\lambda) = \lambda v_{i_0-1,j} + (1 - \lambda) y_{i_0-1,j}$$

for $j = 1, \ldots, q_{i_0-1}$, we see similarly that the vectors

$$v_{i_0-1,j}(\lambda), \quad j = 1, \ldots, q_{i_0-1}; \quad J v_{i_0 j}(\lambda), \quad j = 1, \ldots, q_{i_0}$$

are independent modulo $\mathscr{R}_{i_0-2}$ for $\lambda \in \mathbb{C} \smallsetminus S_2$, where $S_2 \supset S_1$ is a finite set of complex numbers (not including 0 and 1). We continue this procedure and obtain vectors

$$v_{ij}(\lambda); \quad j = 1, \ldots, q_i; \quad i = i_0, i_0 - 1, \ldots, 1$$

such that

$$\{J^t v_{i+t,j}(\lambda) \, ; \quad j = 1, \ldots, q_i \, ; \quad t = 0, 1, \ldots, i_0 - i \, ,$$
$$i = i_0, i_0 - 1, \ldots, 1\} \tag{14.1.1}$$

are linearly independent for $\lambda \in \mathbb{C}^n \smallsetminus S$, where S is finite set of complex numbers not including 0 and 1 and $v_{ij}(1)$ are coordinate unit vectors. From this procedure it follows also that $v_{ij}(\lambda) \in \mathcal{R}_i$ for $\lambda \in \mathbb{C} \smallsetminus S$. Therefore, the subspace $\mathcal{F}(\lambda)$ in $\mathbb{C}^n$ spanned by vectors (14.1.1) for $\lambda \in \mathbb{C} \smallsetminus S$ is a J-invariant subspace with dimension not depending on λ. Since S is finite we can connect between 0 and 1 by a continuous curve Γ such that $\Gamma \cap S = \emptyset$. Then $\mathcal{F}(\lambda)$, $\lambda \in \Gamma$ carries out the connection between $\mathcal{F}_1 = \mathcal{F}(0)$ and $\mathcal{F}_2 = (1)$, where $\mathcal{F}_2 \in \Phi_p$. $\quad \square$

We say that a set $\mathcal{A} \subset \mathbb{G}(\mathbb{C}^n)$ has connected components $\mathcal{A}_1, \ldots, \mathcal{A}_m$ if each $\mathcal{A}_i$, $i = 1, \ldots, m$ is a nonempty connected set, but there is no continuous function $f: [0, 1] \to \mathbb{G}(\mathbb{C}^n)$ such that $f(0) \in \mathcal{A}_i$, $f(1) \in \mathcal{A}_j$, and $i \neq j$. (In other words, each $\mathcal{A}_i$ is a maximal connected set in $\mathcal{A}$.)

Lemmas 14.1.1 and 14.1.2 allow us to settle the question of connected components of the set $\text{Inv}(A)$ when the transformation A has only one eigenvalue.

Theorem 14.1.3

Assume that the transformation $A: \mathbb{C}^n \to \mathbb{C}^n$ has only one eigenvalue λ_0. Then $\text{Inv}(A)$ has exactly $n + 1$ connected components, and each connected component consists of all A-invariant subspaces of fixed dimension.

Proof.   Without loss of generality we can assume $\lambda_0 = 0$. Let J be the Jordan form of A, and $A = S^{-1}JS$ for some invertible transformation S. Obviously, $\text{Inv}(A) = S^{-1}(\text{Inv}(J))$ and $\text{Inv}_p(A) = S^{-1}(\text{Inv}_p(J))$, where $\text{Inv}_p(A)$ is the set of all A-invariant subspaces of dimension p. Lemmas 14.1.1 and 14.1.2 show that $\text{Inv}_p(J)$ and, therefore, $\text{Inv}_p(A)$ are connected. On the other hand, if $\mathcal{L} \in \text{Inv}_p(A)$ and $\mathcal{M} \in \text{Inv}_q(A)$ with $p \neq q$, then there is no continuous function $f: [0, 1] \to \mathbb{G}(\mathbb{C}^n)$ with $f(0) = \mathcal{L}$ and $f(1) = \mathcal{M}$. Indeed, if there were such a function f, then $\dim f(t)$ would not be constant in a neighbourhood of some point $t_0 \in [0, 1]$. This contradicts the continuity of f in view of Theorem 13.1.2. $\quad \square$

14.2   CONNECTED COMPONENTS: THE GENERAL CASE

The description of connected components in $\text{Inv}(A)$ for a general transformation $A: \mathbb{C}^n \to \mathbb{C}^n$ is given in the following theorem.

Theorem 14.2.1

Let $\lambda_1, \ldots, \lambda_c$ be all the different eigenvalues of A, and let $\psi_1, \ldots, \psi_c$ be their respective algebraic multiplicities. Then for every integer p, $0 \leq p \leq n$, *and for every ordered c-tuple of integers* $(\chi_1, \ldots, \chi_c)$ *such that* $0 \le \chi_i \le \psi_i$, $i = 1, \ldots, c$ *and* $\Sigma_{i=1}^c \chi_i = p$

$\{\mathcal{L} \in \text{Inv } A \mid \dim \mathcal{L} = p$ and the algebraic multiplicity of

$$A|_{\mathcal{L}} \text{ corresponding to } \lambda_i \text{ is } \chi_i \text{ for } i = 1, \ldots, c\} \qquad (14.2.1)$$

is a connected component of Inv(A), *and each connected component of* Inv(A) *has the form* (14.2.1) *for a suitable p and suitable c-tuple* $(\chi_1, \ldots, \chi_c)$.

Proof. In the proof we use the following well-known properties of the trace of a transformation $A: \mathbb{C}^n \to \mathbb{C}^n$, denoted by $\text{tr}(A)$ [e.g., see Section 3.5 in Hoffman and Kunze (1967)]. We may define $\text{tr}(A)$ to be the sum of eigenvalues of A. If A is written as an $n \times n$ matrix in any basis in $\mathbb{C}^n$, then $\text{tr}(A)$ is also the sum of diagonal elements of A. We have $\text{tr}(AB) = \text{tr}(BA)$ for any transformations $A, B: \mathbb{C}^n \to \mathbb{C}^n$; in particular, $\text{tr}(S^{-1}AS) = \text{tr}(A)$ for any invertible S. The trace (considered as a map from the set of all transformations $\mathbb{C}^n \to \mathbb{C}^n$ onto $\mathbb{C}$) is a continuous function.

Returning to the proof of Theorem 14.2.1, let Γ_i be a small circle around λ_i with no other eigenvalue of A inside or on Γ_i. Let $\mathcal{N}$ be an A-invariant subspace, and let $\chi_i(\mathcal{N})$ be the geometric multiplicity of λ_i for the transformation $A|_{\mathcal{N}}$. Using the Jordan form of $A|_{\mathcal{N}}$, for instance, it is easily seen that

$$\chi_i(\mathcal{N}) = \text{tr}\left(\frac{1}{2\pi i} \int_{\Gamma_i} (\lambda I - A|_{\mathcal{N}})^{-1} \, d\lambda\right)$$

Let $a_1, \ldots, a_p$ be an orthonormal basis in $\mathcal{N}$. Then in some neighbourhood $V(\mathcal{N})$ of $\mathcal{N}$, $P_{\mathcal{N}'}a_1, \ldots, P_{\mathcal{N}'}a_p$ will be a basis in the subspace $\mathcal{N}' \in V(\mathcal{N})$, where $P_{\mathcal{N}'}$ is the orthogonal projector on $\mathcal{N}'$. We have

$$\theta(\mathcal{N}, \mathcal{N}') = \|P_{\mathcal{N}} - P_{\mathcal{N}'}\| \ge \|P_{\mathcal{N}'}a_i - a_i\| \qquad (14.2.2)$$

Write $A|_{\mathcal{N}}$ as a matrix in the basis $a_1, \ldots, a_p$, and for every A-invariant subspace $\mathcal{N}'$ that belongs to $V(\mathcal{N})$, write $A|_{\mathcal{N}'}$ as a matrix in the basis $P_{\mathcal{N}'}a_1, \ldots, P_{\mathcal{N}'}a_p$. Using formula (14.2.2) and the continuity of the trace, we see that there exists a $\delta > 0$ such that, if $\theta(\mathcal{N}, \mathcal{N}') < \delta$ and $\mathcal{N}'$ is A invariant, then

$$|\chi_i(\mathcal{N}) - \chi_i(\mathcal{N}')| < 1$$

Since $\chi_i(\mathcal{N}')$ assumes only integer values, it follows that $\chi_i(\mathcal{N}')$ is constant in some neighbourhood of $\mathcal{N}$ in Inv(A) and, therefore, constant in the connected component of Inv(A) that contains $\mathcal{N}$.

We show now that if $\mathcal{N}$ and $\mathcal{N}'$ are p-dimensional A-invariant subspaces such that $\chi_i(\mathcal{N}) = \chi_i(\mathcal{N}')$ for $i = 1, \ldots, c$, then $\mathcal{N}$ and $\mathcal{N}'$ are connected in Inv(A). Indeed, applying Theorem 14.1.3 to each restriction $A|_{\mathscr{R}_{\lambda_i}}(A)$ for $i = 1, \ldots, c$, we find that $\mathcal{N} \cap \mathscr{R}_{\lambda_i}(A)$ is connected with $\mathcal{N}' \cap \mathscr{R}_{\lambda_i}(A)$ in the set of all A-invariant subspaces of dimension $\chi_i(\mathcal{N})$ in $\mathscr{R}_{\lambda_i}(A)$. Since

$$\mathcal{N} = (\mathcal{N} \cap \mathscr{R}_{\lambda_1}(A)) \dotplus (\mathcal{N} \cap \mathscr{R}_{\lambda_2}(A)) \dotplus \cdots \dotplus (\mathcal{N} \cap \mathscr{R}_{\lambda_c}(A))$$

and similarly for $\mathcal{N}'$, it follows that $\mathcal{N}$ and $\mathcal{N}'$ are connected in Inv(A).

It remains to show that, given integers $\chi_1, \ldots, \chi_c$ such that $0 \le \chi_i \le \psi_i$ and $\Sigma_{i=1}^c \chi_i = p$, there exists a subspace $\mathcal{N} \in \text{Inv}(A)$ with $\chi_i(\mathcal{N}) = \chi_i$, for $i = 1, \ldots, c$. But assuming that A is in Jordan form, we can always choose an $\mathcal{N}$ spanned by appropriate coordinate unit vectors. $\square$

Corollary 14.2.2

The set Inv(A) *has exactly* $\prod_{i=1}^c (\psi_i + 1)$ *connected components, where* $\psi_1, \ldots, \psi_c$ *are the algebraic multiplicities of the different eigenvalues* $\lambda_1, \ldots, \lambda_c$ *of* A, *respectively.*

The proof of Theorems 14.1.3 and 14.2.1 shows in more detail how the subspaces in Inv A belonging to the same connected component are connected. We say that a vector function $x(t)$ defined for $t \in [0, 1]$ and with values in $\mathbb{C}^n$ is *piecewise linear continuous* if there exist m points $0 < t_1 < \cdots < t_m < 1$ and vectors $y_1, \ldots, y_{m+1}$ and $z_1, \ldots, z_{m+1}$ such that, for $i = 1, \ldots, m + 1$

$$x(t) = y_i + tz_i, \qquad t_{i-1} \le t \le t_i$$

(by definition, $t_0 = 0$, $t_{m+1} = 1$), and for $i = 1, \ldots, m$, we obtain

$$y_i + t_i z_i = y_{i+1} + t_i z_{i+1}$$

Corollary 14.2.3

Let $\mathcal{M}$ *and* $\mathcal{N}$ *be p-dimensional A-invariant subspaces that belong to the same connected component in* Inv A. *Then there exist piecewise linear continuous vector functions* $v_1(t), \ldots, v_p(t)$ *such that, for all* $t \in [0, 1]$, *the subspace* $\text{Span}\{v_1(t), \ldots, v_p(t)\}$ *is p-dimensional, A invariant, and*

$$\mathcal{M} = \text{Span}\{v_1(0), \ldots, v_p(0)\}, \qquad \mathcal{N} = \text{Span}\{v_1(1), \ldots, v_p(1)\}$$

14.3 ISOLATED INVARIANT SUBSPACES

Let $A: \mathbb{C}^n \to \mathbb{C}^n$ be a transformation. An A-invariant subspace $\mathcal{M}$ is called *isolated* if there is an $\epsilon > 0$ such that the only A-invariant subspace $\mathcal{N}$ satisfying $\theta(\mathcal{M}, \mathcal{N}) < \epsilon$ is $\mathcal{M}$ itself.

Theorem 14.3.1

An A-invariant subspace $\mathcal{M}$ is isolated if and only if, for every eigenvalue λ_0 of A with $\dim \mathrm{Ker}(A - \lambda_0 I) \geq 2$, either $\mathcal{M} \supset \mathcal{R}_{\lambda_0}(A)$ or $\mathcal{M} \cap \mathcal{R}_{\lambda_0}(A) = \{0\}$.

To prove Theorem 14.3.1, we use a lemma that allows us to reduce the problem to the case when A has only one eigenvalue.

Lemma 14.3.2

An A-invariant subspace $\mathcal{M}$ is isolated if and only if for every eigenvalue λ_0 of A the subspace $\mathcal{M} \cap \mathcal{R}_{\lambda_0}(A)$ is isolated as an $A|_{\mathcal{R}_{\lambda_0}(A)}$-invariant subspace.

Proof. We have

$$\mathcal{M} = \mathcal{M} \cap \mathcal{R}_{\lambda_1}(A) \dotplus \mathcal{M} \cap \mathcal{R}_{\lambda_2}(A) \dotplus \cdots \dotplus \mathcal{M} \cap \mathcal{R}_{\lambda_r}(A)$$

where $\lambda_1, \ldots, \lambda_r$ are all the different eigenvalues of A.

Assume that $\mathcal{M}$ is isolated. If for some λ_i the subspace $\mathcal{M} \cap \mathcal{R}_{\lambda_i}(A)$ is not isolated [as an $A|_{\mathcal{R}_{\lambda_i}(A)}$-invariant subspace], then there exists a sequence of A-invariant subspaces $\mathcal{M}_m \subset \mathcal{R}_{\lambda_i}(A)$, $m = 1, 2, \ldots$, such that $\mathcal{M}_m \neq \mathcal{M} \cap \mathcal{R}_{\lambda_i}(A)$ and $\theta(\mathcal{M}_m, \mathcal{M} \cap \mathcal{R}_{\lambda_i}(A)) \to 0$. For $m = 1, 2, \ldots$, let

$$\mathcal{N}_m = \mathcal{M} \cap \mathcal{R}_{\lambda_1}(A) \dotplus \cdots \dotplus \mathcal{M} \cap \mathcal{R}_{\lambda_{i-1}}(A) \dotplus \mathcal{M}_m \dotplus \mathcal{M} \cap \mathcal{R}_{\lambda_{i+1}}(A) \dotplus \cdots$$
$$\dotplus \mathcal{M} \cap \mathcal{R}_{\lambda_r}(A)$$

Obviously, $\mathcal{N}_m$ is A invariant. Let $\mathcal{L}_j$ be a direct complement to $\mathcal{M} \cap \mathcal{R}_{\lambda_j}(A)$ in $\mathcal{R}_{\lambda_j}(A)$ for $j = 1, \ldots, r$, and put $\mathcal{L} = \mathcal{L}_1 \dotplus \mathcal{L}_2 \dotplus \cdots \dotplus \mathcal{L}_r$. Then $\mathcal{L}$ is a direct complement to $\mathcal{M}$ in $\mathbb{C}^n$. Theorem 13.1.3 shows that for m sufficiently large, $\mathcal{L}_i$ is a direct complement to $\mathcal{M}_m$ in $\mathcal{R}_{\lambda_i}(A)$, and therefore $\mathcal{L}$ is a direct complement to $\mathcal{N}_m$ in $\mathbb{C}^n$. Letting Q (resp. Q_m) be the projector on $\mathcal{M}$ (resp. $\mathcal{N}_m$) along $\mathcal{L}$, we have (cf. (13.1.4))

$$\theta(\mathcal{M}, \mathcal{N}_m) \leq \|Q - Q_m\| \leq \|\hat{Q} - \hat{Q}_m\| \cdot \|P_i\| \tag{14.3.1}$$

where P_i is the projector on $\mathcal{R}_{\lambda_i}(A)$ along

$$\mathcal{R}_{\lambda_1}(A) \dotplus \cdots \dotplus \mathcal{R}_{\lambda_{i-1}}(A) \dotplus \mathcal{R}_{\lambda_{i+1}}(A) \dotplus \cdots \dotplus \mathcal{R}_{\lambda_r}(A)$$

and $\hat{Q}: \mathcal{R}_{\lambda_i}(A) \to \mathcal{R}_{\lambda_i}(A)$ [resp. $\hat{Q}_m: \mathcal{R}_{\lambda_i}(A) \to \mathcal{R}_{\lambda_i}(A)$] is the projector on $\mathcal{M} \cap \mathcal{R}_{\lambda_i}(A)$ (resp. on $\mathcal{M}_m$) along $\mathcal{L}_i$. Theorem 13.1.3 shows that for large m

$$\|\hat{Q} - \hat{Q}_m\| \leq C\theta(\mathcal{M}_m, \mathcal{M} \cap \mathcal{R}_{\lambda_i}(A))$$

where the constant $C > 0$ is independent of m. Comparing with (14.3.1), we obtain $\theta(\mathcal{M}, \mathcal{N}_m) \to 0$ as $m \to \infty$, a contradiction with the fact that $\mathcal{M}$ is isolated.

Assume now that, for $i = 1, \ldots, r$, $\mathcal{M} \cap \mathcal{R}_{\lambda_i}(A)$ is isolated as an $A|_{\mathcal{R}_{\lambda_i}(A)}$-invariant subspace. So there exists an $\epsilon_i > 0$ such that the only A-invariant subspace $\mathcal{N}_i \subset \mathcal{R}_{\lambda_i}(A)$ satisfying

$$\theta(\mathcal{M} \cap \mathcal{R}_{\lambda_i}(A), \mathcal{N}_i) < \epsilon_i$$

is $\mathcal{M} \cap \mathcal{R}_{\lambda_i}(A)$ itself.

We show now that, for every $\epsilon > 0$, there exists a $\delta > 0$ such that, for any A-invariant subspace $\mathcal{N}$ with $\theta(\mathcal{M}, \mathcal{N}) < \delta$, the inequalities $\theta(\mathcal{M} \cap \mathcal{R}_{\lambda_i}(A), \mathcal{N} \cap \mathcal{R}_{\lambda_i}(A)) < \epsilon$ hold for $i = 1, \ldots, r$. Indeed, arguing by contradiction, assume that for some $\epsilon > 0$ and some i there exists a sequence $\{\mathcal{N}_m\}_{m=1}^{\infty}$ of A-invariant subspaces such that $\theta(\mathcal{M}, \mathcal{N}_m) \to 0$ as $m \to \infty$ but

$$\theta(\mathcal{M} \cap \mathcal{R}_{\lambda_i}(A), \mathcal{N}_m \cap \mathcal{R}_{\lambda_i}(A)) \geq \epsilon \qquad (14.3.2)$$

Let $y \in \mathcal{M} \cap \mathcal{R}_{\lambda_i}(A)$. Then, in particular, $y \in \mathcal{M}$ and by Theorem 13.4.2 there exists a sequence $\{x_m\}_{m=1}^{\infty}$ such that $x_m \in \mathcal{N}_m$ for $m = 1, 2, \ldots$ and

$$y = \lim_{m \to \infty} x_m \qquad (14.3.3)$$

Write $x_m = x_{m1} + \cdots + x_{mr}$, where $x_{mj} \in \mathcal{N}_m \cap \mathcal{R}_{\lambda_j}(A)$, $j = 1, \ldots, r$. Apply the projector on $\mathcal{R}_{\lambda_i}(A)$ along the sum of all other root subspaces of A to both sides of (14.3.3). We see that $y = \lim_{m \to \infty} x_{mi}$. Conversely, if $y = \lim_{m \to \infty} x_{mi}$ for some $x_{mi} \in \mathcal{N}_m \cap \mathcal{R}_{\lambda_i}(A)$, then obviously $y \in \mathcal{R}_{\lambda_i}(A)$, and, by Theorem 13.4.2, we also have $y \in \mathcal{M}$. Now by the same Theorem 13.4.2 any limit point of the sequence $\mathcal{N}_m \cap \mathcal{R}_{\lambda_i}(A)$, $m = 1, 2, \ldots$ coincides with $\mathcal{M} \cap \mathcal{R}_{\lambda_i}(A)$. Since Theorem 13.4.1 ensures that the limit points of $\{\mathcal{N}_m \cap \mathcal{R}_{\lambda_j}(A)\}_{m=1}^{\infty}$ exist, we obtain a contradiction with (14.3.2).

Now take $\epsilon = \min(\epsilon_1, \ldots, \epsilon_r)$. Then for $\delta > 0$ with the property described in the preceding paragraph, we find that for every A-invariant subspace $\mathcal{N}$ with $\theta(\mathcal{M}, \mathcal{N}) < \delta$ the equalities $\mathcal{N} \cap \mathcal{R}_{\lambda_j}(A) = \mathcal{M} \cap \mathcal{R}_{\lambda_j}(A)$ hold for $j = 1, \ldots, r$. But these equalities imply $\mathcal{N} = \mathcal{M}$, that is, $\mathcal{M}$ is isolated. $\square$

Proof of Theorem 14.3.1   In view of Lemma 14.3.2, we can assume that $\sigma(A) = \{\lambda_0\}$. If $\dim \mathrm{Ker}(A - \lambda_0 I) = 1$, then A is unicellular and has a unique complete chain of invariant subspaces. Obviously, every A-invariant subspace is isolated. Now assume that $\dim \mathrm{Ker}(A - \lambda_0 I) \geq 2$. In view of Theorem 14.1.3, the set $\mathrm{Inv}_p(A)$ of all A-invariant subspaces of fixed dimension p is connected. So to prove that the only isolated A-invariant subspaces are $\{0\}$ and $\mathbb{C}^n$, we must show that $\mathrm{Inv}_p(A)$ has at least two members for $0 \neq p \neq n$. However, for every p with $0 < p < n$, and in a fixed Jordan basis for A, the transformation A has at least two invariant subspaces of the same dimension p spanned by some vectors from this basis. $\square$

An A-invariant subspace $\mathcal{M}$ is called *inaccessible* if the only continuous mapping of the interval $[0, 1]$ into the lattice $\text{Inv}(A)$ of A-invariant subspaces with $\varphi(0) = \mathcal{M}$ is the constant map $\varphi(t) \equiv \mathcal{M}$. Clearly, every isolated invariant subspace is inaccessible. The converse is also true, as follows.

Proposition 14.3.3

Every inaccessible A-invariant subspace is isolated.

Indeed, if A has only one eigenvalue λ_0 and $\dim \text{Ker}(A - \lambda_0 I) = 1$, then any A-invariant subspace is obviously inaccessible and isolated. It can be proved by using the arcwise connectedness of $\text{Inv}_p(A)$ for $0 < p < n$ that, if $\sigma(A) = \{\lambda_0\}$ and $\dim \text{Ker}(A - \lambda_0 I) > 1$, then any nontrivial A-invariant subspace is not inaccessible (Corollary 14.2.3). The reduction of the general case to this special case is achieved with the following lemma.

Lemma 14.3.4

An A-invariant subspace $\mathcal{M}$ is inaccessible if and only if, for every eigenvalue λ_0 of A, the subspace $\mathcal{M} \cap \mathcal{R}_{\lambda_0}(A)$ is inaccessible as an $A|_{\mathcal{R}_{\lambda_0}(A)}$-invariant subspace.

The proof of Lemma 14.3.4 is left to the reader. (It can be obtained along the same lines as the proof of Lemma 14.3.2.)

Theorem 14.3.5

Every inaccessible (equivalently, isolated) A-invariant subspace is A hyperinvariant.

Proof. Let $\lambda_1, \ldots, \lambda_s$ be the distinct eigenvalues of A (if any) with $\dim \text{Ker}(A - \lambda_i I) = 1$ for $i = 1, \ldots, s$, and let $\lambda_{s+1}, \lambda_{s+2}, \ldots, \lambda_r$ be other distinct eigenvalues of A (if any). For a given isolated A-invariant subspace $\mathcal{M}$ we have, by Theorem 14.3.1

$$\mathcal{M} \cap \mathcal{R}_{\lambda_i}(A) = \{0\}, \quad \text{for} \quad i = s+1, s+2, \ldots, t$$
$$\mathcal{M} \supset \mathcal{R}_{\lambda_i}, \quad \text{for} \quad i = t+1, \ldots, r$$

and some t with $s + 1 \le t \le r$. Letting α_i be the dimension of $\mathcal{M} \cap \mathcal{R}_{\lambda_i}$, we have $\mathcal{M} = \text{Ker } p(A)$, where $p(\lambda) = \Pi_{i=1}^{s}(\lambda - \lambda_i)^{\alpha_i} \cdot \Pi_{i=t+1}^{r}(\lambda - \lambda_i)^{\alpha_i}$. As every transformation that commutes with A also commutes with $p(A)$, the subspace $\mathcal{M}$ is A hyperinvariant. $\square$

The converse of Theorem 14.3.5 does not hold in general, as the next example shows.

EXAMPLE 14.3.1.   Let

$$T = \begin{bmatrix} 0 & 0 & 0 \\ 0 & 0 & 1 \\ 0 & 0 & 0 \end{bmatrix}$$

The subspace $\mathcal{M} = \mathrm{Span}\{e_1, e_2\}$ is the kernel of T and is thus T hyperinvariant. For any complex number α, the subspace $\mathcal{M}(\alpha) = \mathrm{Span}\{e_1 + \alpha e_3, e_2\}$ is easily seen to be T invariant. We have

$$P_{\mathcal{M}} = \begin{bmatrix} 1 & 0 & 0 \\ 0 & 1 & 0 \\ 0 & 0 & 0 \end{bmatrix}; \qquad P_{\mathcal{M}(\alpha)} = \begin{bmatrix} \dfrac{1}{\sqrt{1+|\alpha|^2}} & 0 & \dfrac{\bar{\alpha}}{\sqrt{1+|\alpha|^2}} \\ 0 & 0 & 0 \\ \dfrac{\alpha}{\sqrt{1+|\alpha|^2}} & & \dfrac{|\alpha|^2}{\sqrt{1+|\alpha|^2}} \end{bmatrix}$$

so

$$\theta(\mathcal{M}(\beta), \mathcal{M}(\alpha)) = \| P_{\mathcal{M}(\beta)} - P_{\mathcal{M}(\alpha)} \|$$

and as the norm of a hermitian matrix is equal to the maximal absolute value of its eigenvalues, a computation shows that

$$\theta(\mathcal{M}(\beta), \mathcal{M}(\alpha)) = \tfrac{1}{2} \max\{|p + q + \sqrt{(p-q)^2 + 4r}|,$$
$$|p + q - \sqrt{(p-q)^2 + 4r}|\}$$

where

$$p = \frac{1}{\sqrt{1+|\beta|^2}} - \frac{1}{\sqrt{1+|\alpha|^2}}, \qquad q = \frac{|\beta|^2}{\sqrt{1+|\beta|^2}} - \frac{|\alpha|^2}{\sqrt{1+|\alpha|^2}}$$

$$r = \left| \frac{\beta}{\sqrt{1+|\beta|^2}} - \frac{\alpha}{\sqrt{1+|\alpha|^2}} \right|^2$$

So the subspace valued function F defined on $\mathbb{C}$ by $F(\alpha) = \mathcal{M}(\alpha)$ is continuous and nonconstant and takes T-invariant values. As $F(0) = \mathcal{M}$, the T-invariant subspace $\mathcal{M}$ is not inaccessible. $\square$

14.4   REDUCING INVARIANT SUBSPACES

Recall that an invariant subspace $\mathcal{M}$ of a transformation $A: \mathbb{C}^n \to \mathbb{C}^n$ is called *reducing* if there exists an A-invariant subspace $\mathcal{N}$ that is a direct complement to $\mathcal{M}$ in $\mathbb{C}^n$.

The question of existence and openness of the set of reducing A-invariant subspaces of fixed dimension p is settled by the following theorem.

Theorem 14.4.1

Let $A: \mathbb{C}^n \to \mathbb{C}^n$ be a transformation with partial multiplicities $m_1, \ldots, m_k$ (so $m_1 + \cdots + m_k = n$). Then there exists a reducing A-invariant subspace of dimension $p \neq 0$ if and only if p is admissible, that is, is the sum of some partial multiplicities $m_{i_s}, \ldots, m_{i_q}$. In this case the set of all reducing A-invariant subspaces of dimension p is open in the set of all A-invariant subspaces.

Proof. If p is admissible, then obviously a reducing A-invariant subspace of dimension p exists. Conversely, assume that $\mathcal{M}$ is a reducing A-invariant subspace of dimension p with an A-invariant complement $\mathcal{N}$. Write

$$A = \begin{bmatrix} A_1 & 0 \\ 0 & A_2 \end{bmatrix}$$

with respect to the direct sum decomposition $\mathcal{M} \dotplus \mathcal{N} = \mathbb{C}^n$. Taking Jordan forms of A_1 and A_2, we see that p is admissible.

For an admissible p, let $\mathrm{Rinv}_p(A)$ be the set of all p-dimensional reducing A-invariant subspaces. For a subspace $\mathcal{M} \in \mathrm{Rinv}_p(A)$, let $\mathcal{N}$ be a direct complement to $\mathcal{M}$ that is A invariant. Theorem 13.1.3 shows that there exists an $\epsilon < 0$ such that $\mathcal{N}$ is a direct complement for any A-invariant subspace $\mathcal{M}_1$ with $\theta(\mathcal{M}, \mathcal{M}_1) < \epsilon$. Hence $\mathrm{Rinv}_p(A)$ is open in the set $\mathrm{Inv}_p(A)$ of all p-dimensional A-invariant subspaces. $\square$

Now consider the question of whether (for admissible p) the set $\mathrm{Rinv}_p(A)$ of all p-dimensional reducing subspaces for A is dense in the set $\mathrm{Inv}_p(A)$ of all p-dimensional A-invariant subspaces. We see later that the answer is, in general, no. So a problem arises as to how one can describe the situations when $\mathrm{Rinv}_p(A)$ is dense in $\mathrm{Inv}_p(A)$ in terms of the Jordan structure of A.

We need some preparation to state the results. Let $A: \mathbb{C}^n \to \mathbb{C}^n$ be a transformation with single eigenvalue λ_0 and partial multiplicities $m_1 \geq \cdots \geq m_r$. It follows from Section 4.1 that the partial multiplicities $p_1 \geq \cdots \geq p_l$ of the restriction $A|_{\mathcal{M}}$ to an A-invariant subspace $\mathcal{M}$ satisfy the inequalities

$$l \leq r, \qquad p_j \leq m_j, \qquad j = 1, \ldots, l \qquad (14.4.1)$$

Given an integer p with $1 \leq p \leq n$, let $p_1 \geq \cdots \geq p_l$ be a sequence of positive integers such that (14.4.1) holds and $p_1 + \cdots + p_l = p$; a sequence with these properties is called *p admissible*. For a p admissible sequence $p_1 \geq \cdots \geq p_l$ denote by $\mathrm{Inv}_p(A; p_1, \ldots, p_l)$ the (nonempty) set of all A-invariant subspaces $\mathcal{M}$ such that the restriction $A|_{\mathcal{M}}$ has the partial multiplicities $p_1, \ldots, p_l$, Clearly, dim $\mathcal{M} = p$ for every $\mathcal{M} \in \mathrm{Inv}_p(A; p_1, \ldots, p_l)$. Moreover

$$\mathrm{Inv}_p(A) = \cup \; \mathrm{Inv}_p(A; p_1, \ldots, p_l)$$

where the union is taken over the finite set of all p-admissible sequences $p_1 \geq \cdots \geq p_l$. For each p-admissible sequence $p_1 \geq \cdots \geq p_l$ let

$$\Xi(A; p_1, \ldots, p_l) = \sum_{i=1}^{p_1} q_i(c_i - q_i) \qquad (14.4.2)$$

where $c_i = \{ j \mid 1 \leq j \leq r, \; m_j \geq i \}^{\#}$, $q_i = \{ j \mid 1 \leq j \leq l, \; p_j \geq i \}^{\#}$, and $K^{\#}$ indicates the number of elements in the finite set K. In connection with the definition of $\Xi(A; p_1, \ldots, p_l)$, observe that $c_j \geq q_j$ for $j = 1, 2, \ldots$ (so each summand on the right-hand side of (14.4.2) is a nonnegative integer), and p_1 is the maximal index with $q_{p_1} > 0$.

We now give a necessary and sufficient condition for the denseness of $\mathrm{Rinv}_p(A)$ in $\mathrm{Inv}_p(A)$, for a transformation $A: \mathbb{C}^n \to \mathbb{C}^n$ with single eigenvalue and partial multiplicities $m_1 \geq \cdots \geq m_r$.

Theorem 14.4.2

For a fixed admissible integer p, the set $\mathrm{Rinv}_p(A)$ is dense in $\mathrm{Inv}_p(A)$ if and only if the following condition holds: any p-admissible sequence $p_1 \geq \cdots \geq p_l$ for which the number $\Xi(A; p_1, \ldots, p_l)$ attains its maximal value among all p-admissible sequences has the form $p_1 = m_{i_1}, \ldots, p_l = m_{i_l}$ for some indices $1 \leq i_1 < i_2 < \cdots < i_l \leq r$. In particular, $\mathrm{Rinv}_p(A)$ is dense in $\mathrm{Inv}_p(A)$ provided there is only one p-admissible sequence $p_1 \geq \cdots \geq p_l$ for which $\Xi(A; p_1, \ldots, p_l)$ is maximal.

In the proof of Theorem 14.4.2 we apply a result proved in Shayman (1982) concerning a representation of $\mathrm{Inv}_p(A)$ as a union of complex (analytic) manifolds. In this proof (and only in this proof) we assume some familiarity with the definition and simple properties of complex manifolds that can be found, for instance, in Wells (1980).

Theorem 14.4.3

For every p-admissible sequence $p_1 \geq \cdots \geq p_l$ the set $\mathrm{Inv}_p(A; p_1, \ldots, p_l)$ is, in the topology induced by the gap metric, a connected complex manifold whose (complex) dimension is equal to $\Xi(A; p_1, \ldots, p_l)$.

For the proof of Theorem 14.4.3 we refer the reader to Shayman (1982).

Proof of Theorem 14.4.2.   Assume that the condition fails, that is, there exists a p-admissible sequence $p_1 \geq \cdots \geq p_l$ with maximal $\Xi(A; p_1, \ldots, p_l)$

that is not of the form $p_1 = m_{i_1}, \ldots, p_l = m_{i_l}$, $1 \le i_1 < i_2 < \cdots < i_l \le r$. By Theorem 14.4.3 the complex manifold $\text{Inv}_p(A; p_1, \ldots, p_l)$ has maximal dimension among all the complex manifolds whose union is $\text{Inv}_p(A)$. On the other hand, it is easily seen that $\text{Inv}_p(A; p_1, \ldots, p_l)$ does not contain any reducing subspace for A (cf. the proof of Theorem 14.4.1). So $\text{Rinv}_p(A)$ is not dense in $\text{Inv}_p(A)$.

Assume now that the condition holds. Then every complex manifold $\text{Inv}_p(A; p_1, \ldots, p_l)$ with maximal $\Xi(A; p_1, \ldots, p_l)$ will contain a reducing subspace $\mathcal{M}(p_1, \ldots, p_l)$ for A. Fix such a p-admissible sequence $p_1 \ge \cdots \ge p_l$, and let $\mathcal{N}$ be an A-invariant direct complement to $\mathcal{M}(p_1, \ldots, p_l)$ in $\mathbb{C}^n$. It follows from Theorem 7 in Shayman (1982) that the complex manifold $\text{Inv}_p(A; p_1, \ldots, p_l)$ can be covered by a finite number of analytic charts and that each chart is of the form $\varphi: \mathbb{C}^q \to \text{Inv}_p(A; p_1, \ldots, p_l)$ $(q = \Xi(A; p_1, \ldots, p_l))$ with $\varphi(z) = \text{Span}\{x_1(z), \ldots, x_p(z)\}$, where $x_1(z), \ldots, x_p(z)$ are analytic vector functions in $\mathbb{C}^q$. Now it is easily seen that the set of all subspaces $\mathcal{M} \in \text{Inv}_p(A; p_1, \ldots, p_l)$ that are not direct complements to $\mathcal{N}$ is an analytic set (i.e., the union of the sets of zeros of a finite number of analytic functions that are not identically zero) in each of the charts mentioned above. Denoting by K the union of all $\text{Inv}_p(A; p_1, \ldots, p_l)$ for which $\Xi(A; p_1, \ldots, p_l)$ is maximal, it follows that $\text{Rinv}_p(A) \cap K$ is dense in K. As $\text{Inv}_p(A)$ is connected (Theorem 14.1.3), it follows from Theorem 14.4.3 that the closure of K coincides with $\text{Inv}_p(A)$; hence $\text{Rinv}_p(A)$ is dense in $\text{Inv}_p(A)$.

Finally, suppose that there exists only one p-admissible sequence $p_1' \ge \cdots \ge p_{l'}'$, for which $\Xi(A; p_1', \ldots, p_{l'}')$ is maximal. As the set $\text{Inv}_p(A)$ is connected, and Theorem 14.4.3 implies that $\text{Inv}_p(A)$ is the closure of $\text{Inv}_p(A; p_1', \ldots, p_{l'}')$. Since p is admissible, there exists a p-dimensional A-invariant subspace $\mathcal{M}_0$ such that $\mathcal{M}_0 \dotplus \mathcal{N}_0 = \mathbb{C}^n$ for some A-invariant subspace $\mathcal{N}_0$. So there exists a subspace $\mathcal{M}$ in $\text{Inv}_p(A; p_1', \ldots, p_{l'}')$ (sufficiently close to $\mathcal{M}_0$) for which $\mathcal{N}_0$ is a direct complement. Now we can repeat the arguments in the preceding paragraph to show that $\text{Rinv}_p(A)$ is dense in $\text{Inv}_p(A)$. $\square$

Let us give an example showing that, for an admissible p, $\text{Rinv}_p(A)$ is not generally dense in $\text{Inv}_p(A)$.

EXAMPLE 14.4.1. Let

$$A = J_5(0) \oplus J_3(0) \oplus J_1(0)$$

where $J_m(0)$ is the Jordan block of size m with eigenvalue 0. Clearly, $p = 5$ is admissible. However, $\text{Rinv}_5(A)$ is not dense in $\text{Inv}_5(A)$. According to Theorem 14.4.3, the connected set $\text{Inv}_5(A)$ is the disjoint union of five analytic manifolds S_1, S_2, S_3, S_4, S_5 described as follows: let $\gamma_1 = \{5\}$; $\gamma_2 = \{4, 1\}$; $\gamma_3 = \{3, 2\}$; $\gamma_4 = \{3, 1, 1\}$; $\gamma_5 = \{2, 2, 1\}$. Then for $j = 1, \ldots, 5$,

S_j consists of all five dimensional A-invariant subspaces $\mathcal{M}$ such that the restriction $A|_{\mathcal{M}}$ has partial multiplicities given by γ_j. Further, the (complex) dimensions of S_1, S_2, S_3, S_4, S_5 are 4, 4, 3, 2, 0, respectively. It is easily seen that there is no reducing subspace for A in S_2. Indeed, the sum of a subspace from S_2 and any four-dimensional A-invariant subspace fails to contain the vector $e_5 \in \mathbb{C}^9$. Since the dimension of S_2 is maximal among the dimensions of S_j, $j = 1, \ldots, 5$, it follows that $\mathrm{Rinv}_5(A)$ is not dense in $\mathrm{Inv}_5(A)$. $\square$

In the next example $\mathrm{Rinv}_p(A)$ is dense in $\mathrm{Inv}_p(A)$, for all admissible p.

EXAMPLE 14.4.2.   Let

$$A = \begin{bmatrix} 0 & 1 & 0 \\ 0 & 0 & 0 \\ 0 & 0 & 0 \end{bmatrix}$$

Obviously, all $p = 0$, 1, 2, 3 are admissible. Among the one-dimensional A-invariant subspaces $\mathrm{Span}\{\alpha e_1 + (1 - \alpha)e_3\}$ (where $\alpha \in \mathbb{C}$), all are reducing with the exception of $\mathrm{Span}\{e_1\}$ (i.e., when $\alpha = 1$). Indeed

$$\mathrm{Span}\{e_1, e_2\} + \mathrm{Span}\{\alpha e_1 + (1 - \alpha)e_3\} = \mathbb{C}^3$$

for $\alpha \neq 1$. So $\mathrm{Rinv}_1(A)$ is dense in $\mathrm{Inv}_1(A)$. Further, in the set

$$\mathrm{Span}\{e_1, e_3\} \cup \left(\bigcup_{\alpha \in \mathbb{C}} \mathrm{Span}\{e_1, e_2 + \alpha e_3\} \right)$$

of two-dimensional A-invariant subspaces the reducing ones are $\mathrm{Span}\{e_1, e_2 + \alpha e_3\}$, $\alpha \in \mathbb{C}$, that is, again a dense set. $\square$

We note the following corollary from Theorem 14.4.2.

Corollary 14.4.4

If the transformation $A: \mathbb{C}^n \rightarrow \mathbb{C}^n$ has only one eigenvalue λ_0 and $\dim \mathrm{Ker}(\lambda_0 I - A) = 2$, then $\mathrm{Rinv}_p(A)$ is dense in $\mathrm{Inv}_p(A)$ for every p such that $\mathrm{Rinv}_p(A)$ is not empty.

Proof. Indeed, let $m_1 \geq m_2$ be the partial multiplicities of A. A simple calculation shows that for every p-admissible sequence $p_1 \geq p_2$ we have $\Xi(A; p_1, p_2) = m_2 - p_2$, and for the p-admissible sequence consisting of one integer p_1 only, we have $\Xi(A; p_1) = m_2$. Hence there exists only one p-admissible sequence $p_1 \geq \cdots \geq p_l$ for which $\Xi(A; p_1, \ldots, p_l)$ is maximal, and the second part of Theorem 14.4.2 applies. $\square$

14.5 COVARIANT AND SEMIINVARIANT SUBSPACES

In this section we study topological properties of the sets of coinvariant and semiinvariant subspaces for a transformation $A: \mathbb{C}^n \to \mathbb{C}^n$. As usual, the topology on these sets is the metric topology induced by the gap metric.

For the coinvariant subspaces we have the following basic result.

Theorem 14.5.1

The set $\mathrm{Coinv}(A)$ *of all coinvariant subspaces for a transformation* $A: \mathbb{C}^n \to \mathbb{C}^n$ *is open and dense in the set* $\mathcal{G}(\mathbb{C}^n)$ *of all subspaces in* $\mathbb{C}^n$. *Furthermore, the set* $\mathrm{Coinv}_p(A)$ *of all A-coinvariant subspaces of a fixed dimension p is connected.*

Proof. Let $\mathcal{M}$ be A coinvariant, so there is an A-invariant subspace $\mathcal{N}$ that is a direct complement to $\mathcal{M}$ in $\mathbb{C}^n$. By Theorem13.1.3 there exists an $\epsilon > 0$ such that $\mathcal{N}$ is a direct complement to any subspace $\mathcal{L} \in \mathcal{G}(\mathbb{C}^n)$ with $\theta(\mathcal{M}, \mathcal{L}) < \epsilon$. Hence $\mathrm{Coinv}(A)$ is open.

We now prove that $\mathrm{Coinv}(A)$ is dense. Let $\mathcal{M} = \mathrm{Span}\{v_1, \ldots, v_p\}$ be a p-dimensional subspace in $\mathbb{C}^n$. There exists an $(n - p)$-dimensional A-invariant subspace $\mathcal{N}$. Let $u_1, \ldots, u_{n-p}$ be a basis for $\mathcal{N}$. Denoting by $w_1, \ldots, w_p$ a basis for some direct complement to $\mathcal{N}$, put

$$\mathcal{M}(\eta) = \mathrm{Span}\{v_1 + \eta w_1, \ldots, v_p + \eta w_p\}$$

where $\eta \neq 0$ is a complex number. As $v_1, \ldots, v_p$ are linearly independent, for η close enough to zero the vectors $v_1 + \eta w_1, \ldots, v_1 + \eta w_p$ are linearly independent as well. Hence $\dim \mathcal{M}(\eta) = p$ for η close enough to zero. Further, the determinant of the $n \times n$ matrix $[u_1 \cdots u_{n-p} w_1 \cdots w_p]$ is nonzero. If $\xi \in \mathbb{C}$ the determinant of $[u_1 \cdots u_{n-p}, \xi v_1 + w_1, \ldots, \xi v_p + w_p]$ is a polynomial in ξ that is not identically zero, and it follows that

$$\det[u_1, \ldots, u_{n-p}, \xi v_1 + w_1, \ldots, \xi v_p + w_p] \neq 0$$

for all ξ such that $|\xi|$ is large enough. For such ξ, the subspace $\mathrm{Span}\{\xi v_1 + w_1, \ldots, \xi v_p + w_p\}$ is a direct complement to $\mathcal{N}$. As $\mathcal{M}(\eta) = \mathrm{Span}\{(1/\eta)v_1 + w_1, \ldots, (1/\eta)v_p + w_p\}$ it follows that for $\eta \neq 0$ and close enough to zero $\mathcal{M}(\eta) \dotplus \mathcal{N} = \mathbb{C}^n$. To show that $\mathcal{M}$ belongs to the closure of the set of all A-coinvariant subspaces, it remains to prove that

$$\lim_{\eta \to 0} (\mathcal{M}, \mathcal{M}(\eta)) = 0 \tag{14.5.1}$$

To prove this, assume for simplicity of notation that the upper p rows in $[v_1 \cdots v_p]$ are linearly independent. Then the same will be true for the upper p rows of $[v_1 + \eta w_1, \ldots, v_p + \eta w_p]$ (for η close enough to zero). Write

$$[v_1 + \eta w_1, \ldots, v_p + \eta w_p] = \begin{bmatrix} B(\eta) \\ C(\eta) \end{bmatrix}$$

where $B(\eta)$ is a nonsingular $p \times p$ matrix and $C(\eta)$ is an $(n - p) \times p$ matrix. Then the matrix

$$P(\eta) = \begin{bmatrix} L(\eta) & L(\eta)X(\eta)^* \\ X(\eta)L(\eta) & X(\eta)L(\eta)X(\eta)^* \end{bmatrix}$$

where $X(\eta) = C(\eta)B(\eta)^{-1}$ and $L(\eta) = (I + X(\eta)^*X(\eta))^{-1}$ is the orthogonal projector on $\mathcal{M}(\eta)$. As the entries of $P(\eta)$ are continuous functions of η, equality (14.5.1) follows.

Finally, let us verify the connectedness of $\text{Coinv}_p(A)$. Let $\mathcal{M}_1$, $\mathcal{M}_2 \in \text{Coinv}_p(A)$. So $\mathcal{M}_1 \dotplus \mathcal{L}_1 = \mathcal{M}_2 \dotplus \mathcal{L}_2 = \mathbb{C}^n$ for some $(n - p)$-dimensional A-invariant subspaces $\mathcal{L}_1$ and $\mathcal{L}_2$. Let $v_1, \ldots, v_p$ and $u_1, \ldots, u_p$ be bases in $\mathcal{M}_1$ and $\mathcal{M}_2$, respectively, and consider the subspaces $\mathcal{M}(\eta) = \text{Span}\{v_1 + \eta u_1, \ldots, v_p + \eta u_p\}$ where $\eta \in \mathbb{C}$. As in the preceding proof of the denseness of $\text{Coinv}(A)$, one verifies that for all η with the possible exception of a finite set Φ (= the set of zeros of a certain polynomial), $\mathcal{M}(\eta)$ is a direct complement to the least one of the subspaces $\mathcal{L}_1$ and $\mathcal{L}_2$. Pick a continuous curve $\Gamma(t)$ in $\mathbb{C} \cup \{\infty\}$ where $t \in [0, 1]$ and that does not intersect Φ and such that $\Gamma(0) = 0$, $\Gamma(1) = \infty$. Then $\mathcal{M}(\Gamma(t))$ for $t \in [0, 1]$ is the desired connection between $\mathcal{M}_1$ and $\mathcal{M}_2$ in the set $\text{Coinv}_p(A)$. $\square$

Now we consider the semiinvariant subspaces. As any A-coinvariant subspace is also A semiinvariant, Theorem 14.5.1 implies that the set $\text{Sinv}(A)$ of all A-semiinvariant subspaces is dense in $\mathbb{G}(\mathbb{C}^n)$. However, $\text{Sinv}(A)$ is not necessarily open, as the following example shows.

EXAMPLE 14.5.1. Let $A = J_4(0)$: $\mathbb{C}^4 \to \mathbb{C}^4$. The two-dimensional subspace $\text{Span}\{e_2, e_3\}$ is obviously A semiinvariant, and

$$\lim_{\eta \to 0} \theta(\text{Span}\{e_2, e_3\}, \text{Span}\{e_2, e_3 + \eta e_4\}) = 0$$

(see the proof of Theorem 14.5.1). But the subspace $\text{Span}\{e_2, e_3 + \eta e_4\}$ is not A semiinvariant for $\eta \neq 0$. Indeed, suppose that

$$\text{Span}\{e_2, e_3 + \eta e_4\} \dotplus \mathcal{N} = \mathcal{M} \qquad (14.5.2)$$

where $\mathcal{N}$ and $\mathcal{M}$ are A invariant. As the only nonzero A-invariant subspaces are $\text{Span}\{e_i \mid 1 \leq i \leq j\}$ for $j = 1, 2, 3, 4$, and (14.5.2) implies $e_3 + \eta e_4 \in \mathcal{M}$, it follows that $\mathcal{M} = \mathbb{C}^4$. Then dim $\mathcal{N} = 2$. Hence $\mathcal{N}$ must be $\text{Span}\{e_1, e_2\}$, which contradicts (14.5.2). $\square$

Theorem 14.5.2

For any transformation $A: \math{C}^n \to \math{C}^n$ the set $\mathrm{Sinv}_p(A)$ of all A-semiinvariant subspaces of a fixed dimension p is connected.

Proof. Given an A-invariant subspace $\mathcal{N}$ with dimension not less than p, denote by $S_p(\mathcal{N})$ the set of all A-semiinvariant subspaces $\mathcal{L}$ of dimension p such that $\mathcal{L} \dot{+} \mathcal{M} = \mathcal{N}$ for some A-invariant subspace $\mathcal{M}$ (in other words, $\mathcal{L}$ is $A|_{\mathcal{N}}$ coinvariant). It will suffice to show that for any $\mathcal{N}$ and any $\mathcal{L}_1 \in S_p(\mathcal{N})$, $\mathcal{L}_2 \in S_p(\math[C]^n)$ there exists a continuous function $f: [0, 1] \to \mathrm{Sinv}_p(A)$ such that $f(0) = \mathcal{L}_2$, $f(1) = \mathcal{L}_1$. Let $\mathcal{L}_2 \dot{+} \mathcal{M}_2 = \math{C}^n$, where $\mathcal{M}_2$ is A invariant, and let $f_1, \ldots, f_p$ and $g_1, \ldots, g_p$ be bases in $\mathcal{L}_2$ and $\mathcal{L}_1$, respectively. Denote by S the finite set of all $\eta \in \math{C}$ for which $\mathrm{Span}\{f_1 + \eta g_1, \ldots, f_p + \eta g_p\}$ is not a direct complement to $\mathcal{M}_2$ in $\math{C}^n$. Then put $f(t) = \mathrm{Span}\{f_1 + \Gamma(t)g_1, \ldots, f_p + \Gamma(t)g_p\}$ for $0 \le t \le 1$ and $f(1) = \mathcal{L}_1$, where $\Gamma: [0, 1] \to (\math{C} \cup \{\infty\}) \smallsetminus S$ is any continuous function with $\Gamma(0) = 0$, $\Gamma(1) = \infty$. $\square$

14.6 THE REAL CASE

Consider now a transformation $A: \math{R}^n \to \math{R}^n$. We study here the connected components and isolated subspaces in the set $\mathrm{Inv}^R(A)$ of all A-invariant subspaces in $\math{R}^n$.

Theorem 14.6.1

If A has only one eigenvalue, and this eigenvalue is real, then the set $\mathrm{Inv}_p^R(A)$ of all A-invariant subspaces of fixed dimension p is connected.

The proof of Theorem 14.6.1 will be modeled after the proof of Theorem 14.3.1, taking into account the fact that in some basis in $\math{R}^n$ the transformation A has the real Jordan form (see Section 12.2). We apply the following fact.

Lemma 14.6.2

The set $GL_r(n)$ of all real invertible $n \times n$ matrices has two connected components; one contains the matrices with positive determinant, the other contains those with negative determinant.

Proof. Let T be a real matrix with $\det T > 0$ and let J be a real Jordan form for T. We first show that J can be connected in $GL_r(n)$ to a diagonal matrix K with diagonal entries ± 1. Indeed, J may have blocks J_p of two types: first

$$J_p = \begin{bmatrix} \lambda_p & 1 & 0 & \cdots & 0 \\ 0 & \lambda_p & 1 & \cdots & 0 \\ \vdots & \vdots & & \ddots & \vdots \\ & & & & 1 \\ 0 & 0 & & \cdots & \lambda_p \end{bmatrix}, \qquad \lambda_p \in \mathbb{R}, \qquad \lambda_p \neq 0$$

in this case we define

$$J_p(t) = \begin{bmatrix} \lambda_p(t) & 1-t & \cdots & 0 \\ 0 & \lambda_p(t) & \ddots & \vdots \\ \vdots & & \ddots & 1-t \\ 0 & & \cdots & 0 & \lambda_p(t) \end{bmatrix}$$

for any $t \in [0, 1]$, where $\lambda_p(t)$ is a continuous path of nonzero real numbers such that $\lambda_p(0) = \lambda_p$, and $\lambda_p(1) = 1$ or -1 according as $\lambda_p > 0$ or $\lambda_p < 0$.
Second, a Jordan block J_p may have the form

$$J_p = \begin{bmatrix} K_p & I & 0 & \cdots & 0 \\ 0 & K_p & I & \cdots & 0 \\ \vdots & \vdots & & \ddots & \vdots \\ & & & & I \\ 0 & 0 & & \cdots & K_p \end{bmatrix}$$

where $I = \begin{bmatrix} 1 & 0 \\ 0 & 1 \end{bmatrix}$, $K_p = \begin{bmatrix} \sigma & \tau \\ -\tau & \sigma \end{bmatrix}$ for real σ and τ with $\tau \neq 0$. Then $J_p(t)$ is defined to have the same zero blocks as J_p, whereas the diagonal and superdiagonal blocks are replaced by

$$\begin{bmatrix} (1-t)\sigma + t & (1-t)\tau \\ -(1-t)\tau & (1-t)\sigma + t \end{bmatrix}, \qquad \begin{bmatrix} 1-t & 0 \\ 0 & 1-t \end{bmatrix}$$

respectively, for $t \in [0, 1]$. Then $J_p(t)$ determines a continuous path of real invertible matrices such that $J_p(0) = J_p$ and $J_p(1)$ is an identity matrix.

Applying the above procedures to every diagonal block in J, we see that J is connected to K by a path in $GL_r(n)$. Now observe that the path in $GL_r(2)$ defined for $t \in [0, 2]$ by

$$\begin{bmatrix} -(1-t) & t \\ -t & -(1-t) \end{bmatrix} \quad \text{when } t \in [0, 1]$$

$$\begin{bmatrix} t-1 & t-2 \\ -(2-t) & t-1 \end{bmatrix} \quad \text{when } t \in [1, 2]$$

connects $\begin{bmatrix} -1 & 0 \\ 0 & -1 \end{bmatrix}$ to $\begin{bmatrix} 1 & 0 \\ 0 & 1 \end{bmatrix}$. Consequently K, and hence J, is connected in $GL_r(n)$ with either I or diag$[-1, 1, 1, \ldots, 1]$. But det $T > 0$ implies det $J > 0$, and so the latter case is excluded. Since $T = S^{-1}JS$ for some invertible real S, we can hold S fixed and observe that the path in $GL_r(n)$ connecting J and I will also connect T and I.

Now assume $T \in GL_r(n)$ and det $T < 0$. Then det $T' > 0$, where $T' = T$ diag$[-1, 1, \ldots, 1]$. Using the argument above, we find that T' is connected with I in $GL_r(n)$. Hence T' is connected with diag$[-1, 1, \ldots, 1]$ in $GL_r(n)$. $\square$

Proof of Theorem 14.6.1. Without loss of generality we can assume that $A = J_n(0)$. Let $k_1 \geq \cdots \geq k_r$ be the sizes of Jordan blocks in A. Let Φ_p the set of all ordered r-tuples of nonnegative integers $l_1, \ldots, l_r$ such that $0 \leq l_i \leq k_i$, $\Sigma_{i=1}^r l_i = p$. As in Section 14.1, each $(l_1, \ldots, l_r) \in \Phi_p$ is identified with a certain p-dimensional A-invariant subspace; so Φ_p can be supposed to be contained in $\text{Inv}_p^R(A)$. The proof of Lemma 14.1.1 shows that Φ_p is connected in $\text{Inv}_p^R(A)$. Further, we apply the proof of Lemma 14.1.2 to show that any $\mathcal{F}_1 \in \text{Inv}_p^R(A)$ is connected in $\text{Inv}_p^R(A)$ with some $\mathcal{F}_2 \in \Phi_p$. Take vectors $v_{ij} \in R^n$, $j = 1, \ldots, q_i$; $i = i_0, i_0 - 1, \ldots, 1$ as in the proof of Lemma 14.1.2. Let $p_i = \dim \mathcal{R}_i - \dim \mathcal{R}_{i-1}$ for $i = i_0, i_0 - 1, \ldots, 1$. As the vectors $v_{i_0 1}, \ldots, v_{i_0 q_{i_0}}$ are linearly independent modulo $\mathcal{R}_{i_0-1}$, the $p_{i_0} \times q_{i_0}$ matrix Q_{i_0} formed by the rows $i_0, k_1 + i_0, \ldots, k_1 + \cdots + k_{p_{i_0}-1} + i_0$ of the $n \times q_{i_0}$ matrix $[v_{i_0 1}, \ldots, v_{i_0 q_{i_0}}]$ has linearly independent columns. For simplicity of notation assume that the top $q_{i_0} \times q_{i_0}$ submatrix $\hat{Q}_{i_0}$ of Q_{i_0} is nonsingular. Now Lemma 14.6.2 allows us to connect the vectors $v_{i_0 1}, \ldots, v_{i_0 q_{i_0}}$ with $\pm e_{i_0}, e_{k_1+i_0}, \ldots, e_{k_1+\cdots+k_{q_{i_0}-1}+i_0}$, respectively (the sign $+$ or $-$ coincides with the sign of the nonzero real number det $\hat{Q}_{i_0}$) in the set of all q_{i_0}-tuples of vectors in $\mathcal{R}_{i_0}$ that are linearly independent modulo $\mathcal{R}_{i_0-1}$. Put $y_{i_0 1} = \pm e_{i_0}$, $y_{i_0 j} = e_{k_1+\cdots+k_{j-1}+i_0}$ for $j = 2, \ldots, q_{i_0}$ in the proof of Lemma 14.1.2. Using an analogous rule for the choice of y_{ij} at each step of the procedure described in the proof of Lemma 14.1.2, we finish the proof of Theorem 14.6.1. $\square$

Theorem 14.6.3

If the transformation $A: R^n \to R^n$ has the only eigenvalues $\alpha \pm i\beta$, where α and β are real and $\beta \neq 0$, then again the set $\text{Inv}_p^R(A)$ of all A-invariant subspaces of fixed dimension p is connected.

Note that under the condition of Theorem 14.6.3, A does not have odd-dimensional invariant subspaces (in particular, n is even), so we can assume that p is even (see Proposition 12.1.1).

Proof. Consider A as the $n \times n$ real matrix that represents the transformation A in the basis $e_1, \ldots, e_n$ in R^n, and let A^c be the complexification of A; so $A^c: \mathbb{C}^n \to \mathbb{C}^n$. By Theorem 12.3.1, there exists a one-to-one correspondence between the A^c-invariant ($p/2$)-dimensional subspaces $\mathcal{M}$ in $\mathcal{R}_{\alpha+i\beta}(A^c)$ and the A-invariant p-dimensional subspaces $\mathcal{L}$, which is given by the formula

$$\mathcal{M} = (\mathcal{L} + i\mathcal{L}) \cap \mathcal{R}_{\alpha+i\beta}(A^c) \stackrel{\text{def}}{=} \varphi(\mathcal{L}) \tag{14.6.1}$$

It is easily seen from the proof of Theorem 12.3.1 that this correspondence is actually a homeomorphism $\varphi: \text{Inv}_p^R(A) \to \text{Inv}_{p/2}(A^c|_{\mathcal{R}_{\alpha+i\beta}(A^c)})$.

Now the connectedness of $\text{Inv}_p^R(A)$ follows from the connectedness of $\text{Inv}_{p/2}(A^c|_{\mathcal{R}_{\alpha+i\beta}(A^c)})$ (see Theorem 14.1.3). $\square$

Recall that as shown in Chapter 12, any A-invariant subspace $\mathcal{L}$ admits the decomposition

$$\mathcal{L} = (\mathcal{L} \cap \mathcal{R}_{\lambda_1}(A)) \dotplus \cdots \dotplus (\mathcal{L} \cap \mathcal{R}_{\lambda_s}(A)) \dotplus (\mathcal{L} \cap \mathcal{R}_{\alpha_1 \pm i\beta_1}(A)) \dotplus \cdots$$
$$\dotplus (\mathcal{L} \cap \mathcal{R}_{\alpha_t \pm i\beta_t}(A)),$$

where $\lambda_1, \ldots, \lambda_s$ are all the distinct real eigenvalues of A (if any) and $\alpha_1 + i\beta_1, \ldots, \alpha_t + i\beta_t$ are all the distinct eigenvalues of A in the open upper half plane. Using this observation, the proof of Theorem 14.2.1 yields the following description of the connected components in the metric space $\text{Inv}^R(A)$ of all A-invariant subspaces in $\mathbb{R}^n$ for the general transformation $A: \mathbb{R}^n \to \mathbb{R}^n$.

Theorem 14.6.4

Let $\lambda_1, \ldots, \lambda_s$ be all the different real eigenvalues of A, let their algebraic multiplicities be $\psi_1, \ldots, \psi_s$, respectively, and let $\alpha_1 + i\beta_1, \ldots, \alpha_t + i\beta_t$ be all the distinct eigenvalues of A in the open upper half plane with the algebraic multiplicities $\varphi_1, \ldots, \varphi_t$, respectively. Then for every $(s + t)$-tuple of integers $\chi = (\chi_1, \ldots, \chi_{s+t})$ such that $0 \le \chi_i \le \psi_i$, $i = 1, \ldots, s$; $0 \le \chi_{s+i} \le \varphi_i$, $i = 1, \ldots, t$ the set $\{\mathcal{L} \in \text{Inv}^R(A) \mid \dim \mathcal{L} = p; \chi_i$ is the algebraic multiplicity of $A|_{\mathcal{L}}$ corresponding to λ_i for $i = 1, \ldots, s; \chi_{s+j}$ is that corresponding to $\alpha_j + i\beta_j$ for $j = 1, \ldots, t\}$, where $p = \chi_1 + \cdots + \chi_s + 2(\chi_{s+1} + \cdots + \chi_{s+t})$ is a connected component of $\text{Inv}^R(A)$ and every connected component of $\text{Inv}^R(A)$ has this form. In particular, $\text{Inv}^R(A)$ has exactly $\Pi_{i=1}^s (\psi_i + 1) \cdot \Pi_{j=1}^t (\varphi_j + 1)$ connected components.

Finally, consider the isolated subspaces in $\text{Inv}^R(A)$.

Theorem 14.6.5

Let $A: \mathbb{R}^n \to \mathbb{R}^n$ be a transformation. Then an A-invariant subspace $\mathcal{M}$ is isolated in $\text{Inv}^R(A)$ if and only if either $\mathcal{M} \cap \mathcal{R}_{\lambda_0}(A) = \{0\}$ or $\mathcal{M} \supset \mathcal{R}_{\lambda_0}(A)$

for every real eigenvalue λ_0 of A with $\dim \operatorname{Ker}(\lambda_0 I - A) \geq 2$, *and either* $\mathcal{M} \cap \mathcal{R}_{\alpha \pm i\beta}(A) = \{0\}$ *or* $\mathcal{M} \supset \mathcal{R}_{\alpha \pm i\beta}(A)$ *for any nonreal eigenvalue $\alpha + i\beta$ of A with geometric multiplicity greater than 1.*

Proof. Using the real analog of Lemma 14.3.2 (its proof is similar to that of Lemma 14.3.2), we can assume that one of two cases holds: (a) $\sigma(A) = \{\lambda_0\}$, $\lambda_0 \in \mathcal{R}$; (b) $\sigma(A) = \{\alpha + i\beta, \alpha - i\beta\}$, $\alpha, \beta \in \mathcal{R}$, $\beta \neq 0$. In the first case Theorem 14.6.5 is proved in the same way as Theorem 14.3.1. In the second case use Theorem 14.3.1 and the homeomorphism between $\operatorname{Inv}_p^{\mathcal{R}}(A)$ and $\operatorname{Inv}_{p/2}(A^c|_{\mathcal{R}_{\alpha+i\beta}(A^c)})$ given by formula (14.6.1).   □

14.7   EXERCISES

14.1   Supply the details for the proof of Lemma 14.3.4.

14.2   Prove that for a transformation A the sets of A-hyperinvariant subspaces and isolated A-invariant subspaces coincide if and only if A is diagonable. In this case an A-invariant subspace is isolated if and only if it is a root subspace.

14.3   What is the number of isolated invariant subspaces of the companion matrix

$$\begin{bmatrix} 0 & 1 & 0 & \cdots & 0 \\ 0 & 0 & 1 & \cdots & 0 \\ \vdots & \vdots & \vdots & & \vdots \\ 0 & 0 & 0 & \cdots & 1 \\ a_0 & a_1 & a_2 & \cdots & a_{n-1} \end{bmatrix}, \quad a_j \in \mathcal{C}?$$

14.4   Let $A = \operatorname{diag}[J_2(0), J_2(0), J_2(0)]: \mathcal{C}^6 \to \mathcal{C}^6$. Is the set of all reducing A-invariant subspaces dense in $\operatorname{Inv}(A)$?

14.5   Show that there exists a converging sequence of semiinvariant subspaces for the matrix $J_3(0)$ whose limit is not $J_3(0)$-semiinvariant.

Chapter Fifteen

Continuity and Stability of Invariant Subspaces

It has already been mentioned that computational problems for invariant subspaces naturally lead to the problem of describing a class of invariant subspaces that are stable after small perturbations. Only such subspaces can be amenable to numerical computations. The analysis of stability of invariant subspaces is the main topic of this chapter. We also include related material on stability of other classes of subspaces (notably, $[A \ B]$-invariant subspaces), and on stability of lattices of invariant subspaces. Different types of stability are analyzed.

15.1 SEQUENCES OF INVARIANT SUBSPACES

In this section we consider the continuity of invariant subspaces for transformations from $\mathbb{C}^n$ into $\mathbb{C}^n$. We start with the following simple fact.

Theorem 15.1.1

Let $\{A_m\}_{m=1}^{\infty}$ be a sequence of transformations from $\mathbb{C}^n$ into $\mathbb{C}^n$ that converges to a linear transformation $A: \mathbb{C}^n \to \mathbb{C}^n$. If $\mathcal{M}_m$ is an A_m-invariant subspace for $m = 1, 2, \ldots$ such that $\mathcal{M}_m \to \mathcal{M}$ for some subspace $\mathcal{M} \subset \mathbb{C}^n$, then $\mathcal{M}$ is A invariant.

Proof. Let $x \in \mathcal{M}$. Then, by Theorem 13.4.2, there exists a sequence $\{x_m\}_{m=1}^{\infty}$ such that $x_m \in \mathcal{M}_m$ for each m and $\lim_{m \to \infty} \|x_m - x\| = 0$. Now

$$\|Ax - A_m x_m\| \leq \|Ax - A_m x\| + \|A_m x - A_m x_m\|$$

$$\leq \|A - A_m\| \cdot \|x\| + \|A_m\| \cdot \|x - x_m\|$$

As $A_m \to A$, the norms $\|A_m\|$ are bounded; $\|A_m\| \le K$ for some positive constant K independent of m. So as $m \to \infty$,

$$\|Ax - A_m x_m\| \le \|A - A_m\| \cdot \|x\| + K \cdot \|x - x_m\| \to 0$$

As $\mathcal{M}_m$ is A_m invariant, we have $A_m x_m \in \mathcal{M}_m$ for each m, and Theorem 13.4.2 can be applied to conclude that $Ax \in \mathcal{M}$. $\square$

The continuity property of invariant subspaces expressed in Theorem 15.1.1 does not hold for the classes of coinvariant and semiinvariant subspaces.

EXAMPLE 15.1.1. For $m = 1, 2, \ldots$, let

$$A_m = \begin{bmatrix} 0 & 1 \\ 0 & \dfrac{1}{m} \end{bmatrix}$$

The subspace $\mathrm{Span}\{e_1\}$ is A_m coinvariant for every m. (Indeed, $\mathrm{Span}\begin{bmatrix} m \\ 1 \end{bmatrix}$ is a direct complement to $\mathrm{Span}\{e_1\}$, which is A_m invariant.) However, $\mathrm{Span}\{e_1\}$ is not A coinvariant, where $A = \begin{bmatrix} 0 & 1 \\ 0 & 0 \end{bmatrix}$ is the limit of A_m. The same subspace $\mathrm{Span}\{e_1\}$ is also A_m reducing, but not A reducing. $\square$

EXAMPLE 15.1.2. For $m = 1, 2, \ldots$, let

$$A_m = \begin{bmatrix} 0 & 1 & 0 \\ 0 & \dfrac{1}{m} & 1 \\ 0 & 0 & -\dfrac{1}{m} \end{bmatrix}$$

The eigenvectors of A_m are (up to multiplication by a scalar) e_1, $me_1 + me_2$, $m^2 e_1 - me_2 + 2e_3$. Consequently, the subspace $\mathrm{Span}\{e_1, e_3\}$ is A_m semi-invariant for all m (because $\mathrm{Span}\{me_1 + e_2\}$ is a direct complement to $\mathrm{Span}\{e_1, e_3\}$, which is an A_m-invariant subspace). However, $\mathrm{Span}\{e_1, e_3\}$ is not A semiinvariant, where

$$A = \begin{bmatrix} 0 & 1 & 0 \\ 0 & 0 & 1 \\ 0 & 0 & 0 \end{bmatrix}$$

is the limit of A_m if $m \to \infty$. $\square$

Corollary 15.1.2

The set of A-invariant subspaces is closed; that is, if $\{\mathcal{M}_m\}_{m=1}^{\infty}$ is a sequence of A-invariant subspaces with limit $\mathcal{M} = \lim_{m\to\infty} \mathcal{M}$, then $\mathcal{M}$ is also A invariant.

Simple examples show that the A-invariant subspaces Ker A and Im A are not generally continuous in the sense of Theorem 15.1.1. Thus it may happen that $\{\text{Ker } A_m\}_{m=1}^{\infty}$ does not converge to Ker A and $\{\text{Im } A_m\}_{m=1}^{\infty}$ does not converge to Im A as $A_m \to A$. The following result shows that the only obstruction to convergence of Ker A_m and Im A_m is the dimension.

Theorem 15.1.3

Let $\{A_m\}_{m=1}^{\infty}$ be a sequence of transformations on $\mathbb{C}^n$ that converges to a transformation A on $\mathbb{C}^n$. Then Ker A contains the limit of every convergent subsequence of the sequence $\{\text{Ker } A_m\}_{m=1}^{\infty}$. In particular, if $\dim \text{Ker } A_m = \dim \text{Ker } A$ for every $m = 1, 2, \ldots$ then Ker A_m and Im A_m converge, and

$$\text{Ker } A = \lim_{m\to\infty} \text{Ker } A_m, \qquad \text{Im } A = \lim_{m\to\infty} \text{Im } A_m$$

Proof. For $k = 1, 2, \ldots$, let Ker A_{m_k} converge to some $\mathcal{M} \subset \mathbb{C}^n$. Then for every $x \in \mathcal{M}$ there exists a sequence $x_{m_k} \in \text{Ker } A_{m_k}$, such that $x_{m_k} \to x$. As $A_{m_k} x_{m_k} = 0$, we have also $Ax = 0$, that is, $x \in \text{Ker } A$.

Now let Im A_{m_k} be a sequence converging to some $\mathcal{N} \subset \mathbb{C}^n$. Then [see formula (13.1.1)]

$$\text{Ker } A_{m_k}^* = (\text{Im } A_{m_k})^{\perp} \to \mathcal{N}^{\perp}$$

Since $A_m^* \to A^*$, by the part of the theorem already proved, $\mathcal{N}^{\perp} \subset \text{Ker } A^* = (\text{Im } A)^{\perp}$ and so $\mathcal{N} \supset \text{Im } A$.

Assume in addition that $\dim \text{Ker } A_m = \dim \text{Ker } A$ for all $m = 1, 2, \ldots$. If $\mathcal{L}$ is a limit of a converging subsequence from the sequence $\{\text{Ker } A_m\}_{m=1}^{\infty}$, then (see Theorem 13.1.2) $\dim \mathcal{L} = \dim \text{Ker } A$. From the first part of the theorem we know that $\mathcal{L} \subset \text{Ker } A$. So actually $\mathcal{L} = \text{Ker } A$. Hence Ker A is a limit of every converging subsequence of $\{\text{Ker } A_m\}_{m=1}^{\infty}$. It follows [using the compactness of $\mathbb{G}(\mathbb{C}^n)$] that Ker A_m converges to Ker A. Further, we also have $\dim \text{Im } A_m = \dim \text{Im } A$ for each m. A similar argument shows that Im A_m converges to Im A. $\square$

Let $\mathcal{M}$ be an A-invariant subspace and Ω be an open set in $\mathbb{C}$. We conclude this section by showing that the inclusion $\sigma(A|_{\mathcal{M}}) \subset \Omega$ is preserved under small perturbations. Recall that θ denotes the "gap" metric introduced in Chapter 13.

Theorem 15.1.4

Let $\mathcal{M}$ be an invariant subspace for the transformation $A: \mathbb{C}^n \to \mathbb{C}^n$, and let $\Omega \subset \mathbb{C}$ be an open set such that all eigenvalues of $A|_{\mathcal{M}}$ are inside Ω. Then for transformations B on $\mathbb{C}^n$ and B-invariant subspaces $\mathcal{N}$, $\sigma(B|_{\mathcal{N}}) \subset \Omega$ as long as $\|B - A\| + \theta(\mathcal{M}, \mathcal{N})$ is sufficiently small.

Proof. Arguing by contradiction, suppose that there exists a sequence of transformations $\{B_m\}_{m=1}^{\infty}$ on $\mathbb{C}^n$ and a sequence of subspaces $\{\mathcal{N}_m\}_{m=1}^{\infty}$ such that $\mathcal{N}_m$ is B_m invariant,

$$\|B_m - A\| + \theta(\mathcal{M}, \mathcal{N}_m) < \frac{1}{m}, \qquad m = 1, 2, \ldots$$

and $\sigma(B_m|_{\mathcal{N}_m}) \not\subset \Omega$. For each m, let λ_m be an eigenvalue of $B_m|_{\mathcal{N}_m}$ outside Ω:

$$B_m x_m = \lambda_m x_m, \qquad \|x_m\| = 1, \qquad x_m \in \mathcal{N}_m \qquad (15.1.1)$$

Since $\|B_m - A\| \to 0$ as $m \to \infty$, the norms $\{\|B_m\|\}_{m=1}^{\infty}$ are bounded; hence the sequence $\{\lambda_m\}_{m=1}^{\infty}$ is bounded as well. Passing to subsequences in formula (15.1.1), if necessary, we can assume that $\lambda_m \to \lambda_0$ and $x_m \to x_0$ (as $m \to \infty$), for some $\lambda_0 \in \mathbb{C}$ and $x_0 \in \mathbb{C}^n$. By Theorem 13.4.2, $x_0 \in \mathcal{M}$, and clearly $x_0 \neq 0$. As $Ax_0 = \lambda_0 x_0$, λ_0 is an eigenvalue of $A|_{\mathcal{M}}$, which, by hypothesis, belongs to Ω. But this contradicts $\lambda_m \not\subset \Omega$ for $m = 1, 2, \ldots$. $\quad\square$

15.2   STABLE INVARIANT SUBSPACES: THE MAIN RESULT

Let $A: \mathbb{C}^n \to \mathbb{C}^n$ be a transformation. An A-invariant subspace $\mathcal{N}$ is called *stable* if, given $\epsilon > 0$, there exists a $\delta > 0$ such that $\|B - A\| < \delta$ for a transformation $B: \mathbb{C}^n \to \mathbb{C}^n$ implies that B has an invariant subspace $\mathcal{M}$ with $\theta(\mathcal{M}, \mathcal{N}) < \epsilon$. The same definition applies for matrices.

This concept is particularly important from the point of view of numerical computation. It is generally true that the process of finding a matrix representation for a linear transformation and then finding invariant subspaces can be performed only approximately. Consequently, the stable invariant subspaces will generally be the only ones amenable to numerical computation.

Suppose that $\mathcal{N}$ is a direct sum of root subspaces of A. The $\mathcal{N}$ is a stable invariant subspace for A. This follows from the fact that $\mathcal{N}$ appears as the image of a *Riesz projector*

$$R_A = \frac{1}{2\pi i} \int_{\Gamma} (I\lambda - A)^{-1} \, d\lambda \qquad (15.2.1)$$

where Γ is a suitable closed rectifiable contour in $\mathbb{C}$ such that the eigenvalue

λ_0 of A is inside Γ if $\mathcal{R}_{\lambda_0}(A) \subset \mathcal{N}$ and outside Γ if $\mathcal{R}_{\lambda_0}(A) \cap \mathcal{N} = \{0\}$ (see Proposition 2.4.3). Further, the function $F(\lambda) = (I\lambda - A)^{-1}$ is a continuous function of λ on Γ. This follows from the formula

$$(I\lambda - A)^{-1} = [\det(I\lambda - A)]^{-1}\mathrm{Adj}(I\lambda - A)$$

where $\mathrm{Adj}(I\lambda - A)$ is the matrix of algebraic adjoints of $I\lambda - A$, and from the continuity of $\det(I\lambda - A)$ and $\mathrm{Adj}(I\lambda - A)$ as functions of λ. Since Γ is compact, the number $K_A \overset{\text{def}}{=} \max_{\lambda \in \Gamma} \|(I\lambda - A)^{-1}\|$ is well defined. Now any transformation $B: \mathbb{C}^n \to \mathbb{C}^n$ with $\|B - A\| < K_A^{-1}$ has the property that $I\lambda - B$ is invertible for all $\lambda \in \Gamma$. [Indeed, for $\lambda \in \Gamma$ we have

$$I\lambda - B = (I\lambda - A) + (A - B) = (I\lambda - A)[I + (I\lambda - A)^{-1}(A - B)]$$

and since $\|(I\lambda - A)^{-1}(A - B)\| < 1$, the invertibility of $I\lambda - B$ follows.] Moreover

$$\|(I\lambda - A)^{-1} - (I\lambda - B)^{-1}\| \le K_A K_B \|A - B\|$$

which implies that $\|R_B - R_A\|$ is arbitrarily small if $\|A - B\|$ is small enough.

Theorem 13.1.1 shows that

$$\theta(\mathcal{N}, \mathcal{M}) \le \|R_B - R_A\| \tag{15.2.2}$$

so $\theta(\mathcal{N}, \mathcal{M})$ is small together with $\|R_B - R_A\|$.

However, it will turn out that not every stable invariant subspace is spectral. On the other hand, if $\dim \mathrm{Ker}(\lambda_j I - A) > 1$ and $\mathcal{N}$ is a one-dimensional subspace of $\mathrm{Ker}(\lambda_j I - A)$, it is intuitively clear that a small perturbation of A can result in a large change in the gap between invariant subspaces. The following simple example provides such a situation. Let A be the 2×2 zero matrix, and let $\mathcal{N} = \mathrm{Span}\left\{\begin{bmatrix} 1 \\ 1 \end{bmatrix}\right\} \subset \mathbb{C}^2$. Clearly, $\mathcal{N}$ is A invariant, but $\mathcal{N}$ is unstable. Indeed, let $B = \mathrm{diag}[0, \epsilon]$, where $\epsilon \ne 0$ is close enough to zero. The only one-dimensional B-invariant subspaces are $\mathcal{M}_1 = \mathrm{Span}\left\{\begin{bmatrix} 1 \\ 0 \end{bmatrix}\right\}$ and $\mathcal{M}_2 = \mathrm{Span}\left\{\begin{bmatrix} 0 \\ 1 \end{bmatrix}\right\}$, and both are far from $\mathcal{N}$: computation shows that

$$\theta(\mathcal{N}, \mathcal{M}) = 1/\sqrt{2}, \qquad i = 1, 2$$

The following theorem gives the description of all stable invariant subspaces.

Theorem 15.2.1

Let $\lambda_1, \ldots, \lambda_r$ be the different eigenvalues of the transformation A. A subspace $\mathcal{N}$ of $\mathbb{C}^n$ is A invariant and stable if and only if $\mathcal{N} = \mathcal{N}_1 \dotplus \cdots \dotplus \mathcal{N}_r$, *where for each j the space $\mathcal{N}_j$ is an arbitrary A-invariant subspace of $\mathcal{R}_{\lambda_j}(A)$ if $\dim \mathrm{Ker}(\lambda_j I - A) = 1$; if $\dim \mathrm{Ker}(\lambda_j I - A) \neq 1$ then either $\mathcal{N}_j = \{0\}$ or $\mathcal{N}_j = \mathcal{R}_{\lambda_j}(A)$.*

Comparing this theorem with Theorem 14.3.1, we obtain the following important fact: *an A-invariant subspace $\mathcal{N}$ is stable if and only if $\mathcal{N}$ is isolated in the metric space* $\mathrm{Inv}(A)$ *of all A-invariant subspaces.*

An interesting corollary is easily detained from Theorem 15.2.1.

Corollary 15.2.2

All invariant subspaces of a transformation $A: \mathbb{C}^n \to \mathbb{C}^n$ are stable if and only if A is nonderogatory [i.e., $\dim \mathrm{Ker}(A - \lambda_0 I) = 1$ for every eigenvalue λ_0 of A].

The proof of Theorem 15.2.1 will be based on a series of lemmas and an auxiliary theorem that is of some interest in itself. We will also take advantage of an observation that follows immediately from the definition of a stable subspace: the A-invariant subspace $\mathcal{N}$ is stable if and only if the SAS^{-1}-invariant subspace $S\mathcal{N}$ is stable. Here $S: \mathbb{C}^n \to \mathbb{C}^n$ is an arbitrary invertible transformation.

First we present results leading to the proof of Theorem 15.2.1 for the case when A has only one eigenvalue. To state the next theorem we need the following notion: a chain $\mathcal{M}_1 \subset \mathcal{M}_2 \subset \cdots \subset \mathcal{M}_{n-1}$ of A-invariant subspaces is said to be *complete* if $\dim \mathcal{M}_j = j$ for $j = 1, \ldots, n-1$.

Theorem 15.2.3

Given $\epsilon > 0$, there exists a $\delta > 0$ such that the following holds true: if B is a transformation with $\|B - A\| < \delta$ and $\{\mathcal{M}_j\}$ is a complete chain of B-invariant subspaces, then there exists a complete chain $\{\mathcal{N}_j\}$ of A-invariant subspaces such that $\theta(\mathcal{N}_j, \mathcal{M}_j) < \epsilon$ for $j = 1, \ldots, n-1$.

In general, the chain $\{\mathcal{M}_j\}$ for A will depend on the choice of B. To see this, consider

$$A = \begin{bmatrix} 0 & 0 \\ 0 & 0 \end{bmatrix}, \qquad B_v = \begin{bmatrix} 0 & 0 \\ v & 0 \end{bmatrix}, \qquad B'_v = \begin{bmatrix} 0 & v \\ 0 & 0 \end{bmatrix}$$

where $v \in \mathbb{C}$. Observe that for $v \neq 0$ the only one-dimensional invariant subspace of B_v is $\mathrm{Span}\{e_2\}$, and for B'_v, $v \neq 0$, the only one-dimensional invariant subspace is $\mathrm{Span}\{e_1\}$.

Proof. Assume that the conclusion of the theorem is not correct. Then there exists an $\epsilon > 0$ with the property that for every positive integer m there exists a transformation B_m satisfying $\|B_m - A\| < 1/m$ and a complete chain $\{\mathcal{M}_{mj}\}$ of B_m-invariant subspaces such that for every complete chain $\{\mathcal{N}_j\}$ of A-invariant subspaces

$$\max_{1 \le j \le k-1} \theta(\mathcal{N}_j, \mathcal{M}_{mj}) \ge \epsilon \qquad m = 1, 2, \dots \qquad (15.2.3)$$

Denote by P_{mj} the orthogonal projector on $\mathcal{M}_{mj}$.

Since $\| P_{mj} \| = 1$, there exists a subsequence $\{m_i\}$ of the sequence of positive integers and transformations $P_1, \dots, P_{n-1}$ on $\mathbb{C}^n$, such that

$$\lim_{i \to \infty} P_{m_i, j} = P_j, \qquad j = 1, \dots, n-1 \qquad (15.2.4)$$

Observe that $P_1, \dots, P_{n-1}$ are orthogonal projectors. Indeed, passing to the limit in the equalities $P_{m_i, j} = (P_{m_i, j})^2$, we find that $P_j = P_j^2$. Further, equation (15.2.4) combined with $P_{m_i, j}^* = P_{m_i, j}$ implies that $P_j^* = P_j$; so P_j is an orthogonal projector (see Section 1.5).

Further, the subspace $\mathcal{N}_j = \mathrm{Im}\, P_j$ has dimension j, $j = 1, \dots, n-1$. This is a consequence of Theorem 13.1.2.

By passing to the limits it follows from $B_m P_{mj} = P_{mj} B_m P_{mj}$ that $AP_j = P_j AP_j$. Hence $\mathcal{N}_j$ is A invariant. Since $P_{mj} = P_{m,j+1} P_{mj}$ we have $P_j = P_{j+1} P_j$, and thus $\mathcal{N}_j \subset \mathcal{N}_{j+1}$. It follows that $\mathcal{N}_j$ is a complete chain of A-invariant subspaces. Finally, $\theta(\mathcal{N}_j, \mathcal{M}_{m_i, j}) = \| P_j - P_{mj} \| \to 0$. But this contradicts (15.2.3), and the proof is complete. $\square$

Corollary 15.2.4

If A has only one eigenvalue, λ_0, say, and if $\dim \mathrm{Ker}(\lambda_0 I - A) = 1$, then each invariant subspace of A is stable.

Proof. The conditions on A are equivalent to the requirement that for each $1 \le j \le n-1$ the operator A has only one j-dimensional invariant subspace and the nontrivial invariant subspaces form a complete chain (see Section 2.5). So we may apply the previous theorem to obtain the desired results. $\square$

Lemma 15.2.5

If A has only one eigenvalue, λ_0 say, and if $\dim \mathrm{Ker}(\lambda_0 I - A) \ge 2$, then the only stable A-invariant subspaces are $\{0\}$ and $\mathbb{C}^n$.

Proof. Let $J = \mathrm{diag}[J_{k_1}(\lambda_0), \dots, J_{k_s}(\lambda_0)]$ be the Jordan form for A. As $\dim \mathrm{Ker}(\lambda_0 I - A) \ge 2$, we have $s \ge 2$. By similarity, it suffices to prove that J has no nontrivial stable invariant subspace.

For $\epsilon \in \mathbb{C}$, define the transformation T_ϵ on $\mathbb{C}^n$ by setting

$$T_\epsilon e_i = \begin{cases} \epsilon e_{i-1} & \text{if } i = k_1 + \cdots + k_j + 1, \qquad j = 1, \dots, s-1 \\ 0 & \text{otherwise} \end{cases}$$

and put $B_\epsilon = J + T_\epsilon$. Then $\| B_\epsilon - J \|$ tends to 0 as $\epsilon \to 0$. For $\epsilon \ne 0$ the linear transformation B_ϵ has exactly one j-dimensional invariant subspace, namely,

$\mathcal{N}_j = \text{Span}\{e_1, \ldots, e_j\}$. Here $1 \le j \le k-1$. It follows that $\mathcal{N}_j$ is the only candidate for a stable J-invariant subspace of dimension j.

Now consider $\tilde{J} = \text{diag}[J_{k_s}(\lambda_0), \ldots, J_{k_2}(\lambda_0), J_{k_1}(\lambda_0)]$. Repeating the argument of the previous paragraph for $\tilde{J}$ instead of J, we see that $\mathcal{N}_j$ is the only candidate for a stable $\tilde{J}$-invariant subspace of dimension j. But $J = S\tilde{J}S^{-1}$, where S is the similarity transformation that reverses the order of the blocks in J. It follows that $S\mathcal{N}_j$ is the only candidate for a stable J-invariant subspace of dimension j. As $s \ge 2$, however, we have $S\mathcal{N}_j \ne \mathcal{N}_j$ for $1 \le j \le k-1$, and the proof is complete. $\square$

Corollary 15.2.4 and Lemma 15.2.5 together prove Theorem 15.2.1 for the case when A has one eigenvalue only.

15.3 PROOF OF THEOREM 15.2.1 IN THE GENERAL CASE

The proof of Theorem 15.2.1 in the general case is reduced to the case of one eigenvalue considered in the preceding section. Recall the notion of the minimal opening

$$\eta(\mathcal{M}, \mathcal{N}) = \inf\{\|x + y\| \mid x \in \mathcal{M}, \ y \in \mathcal{N}, \ \max(\|x\|, \|y\|) = 1\}$$

between subspaces $\mathcal{M}$ and $\mathcal{N}$ (Section 13.3). Always $0 \le \eta(\mathcal{M}, \mathcal{N}) \le 1$, except when both $\mathcal{M}$ and $\mathcal{N}$ are the zero subspace, in which case $\eta(\mathcal{M}, \mathcal{N}) = \infty$. Note that $\eta(\mathcal{M}, \mathcal{N}) > 0$ if and only if $\mathcal{M} \cap \mathcal{N} = \{0\}$ (Proposition 13.2.1). We need to apply the following fact.

Proposition 15.3.1

Let $\{\mathcal{M}_m\}_{m=1}^{\infty}$ be a sequence of subspaces in $\mathbb{C}^n$. If $\lim_{m \to \infty} \theta(\mathcal{M}_m, \mathcal{L}) = 0$ for some subspace $\mathcal{L}$, then

$$\eta(\mathcal{M}_m, \mathcal{N}) \to \eta(\mathcal{L}, \mathcal{N}) \tag{15.3.1}$$

for every subspace $\mathcal{N}$.

Indeed, if both $\mathcal{L}$ and $\mathcal{N}$ are nonzero, then also $\mathcal{M}_m$ are nonzero (at least for m large enough; see Theorem 13.1.2). Then (15.3.1) follows from formula (13.3.2). If at least one of $\mathcal{L}$ and $\mathcal{N}$ is the zero subspace, then (15.3.1) is trivial.

Let us introduce some terminology and notation that will be used in the next two lemmas and their proofs. We use the shorthand $A_m \to A$ for $\lim_{m \to \infty} \|A_m - A\| = 0$, where A_m, $m = 1, 2, \ldots$, and A are transformations on $\mathbb{C}^n$. Note that $A_m \to A$ if and only if the entries of the matrix representations of A_m (in some fixed basis) converge to the corresponding entries of

A (represented as a matrix in the same basis). We say that a simple rectifiable contour Γ *splits the spectrum* of a transformation T if $\sigma(T) \cap \Phi = \emptyset$. In that case we can associate with T and Γ the Riesz projector

$$P(T; \Gamma) = \frac{1}{2\pi i} \int_{\Gamma} (I\lambda - T)^{-1} \, d\lambda$$

The following observation is used subsequently. If T is a transformation for which Γ splits the spectrum, then Γ splits the spectrum for every transformation S that is sufficiently close to T (i.e., $\|S - T\|$ is close enough to zero). Indeed, this follows from the continuity of eigenvalues of a linear transformation as functions of this transformation.

Lemma 15.3.2

Let Γ be a simple rectifiable contour that splits the spectrum of T, let T_0 be the restriction of T to $\operatorname{Im} P(T; \Gamma)$, and let $\mathcal{N}$ be a subspace of $\operatorname{Im} P(T; \Gamma)$. Then $\mathcal{N}$ is a stable invariant subspace for T if and only if $\mathcal{N}$ is a stable invariant subspace for T_0.

Proof. Suppose that $\mathcal{N}$ is a stable invariant subspace for T_0, but not for T. Then one can find an $\epsilon > 0$ such that for every positive integer m there exists a transformation S_m such that

$$\|S_m - T\| < \frac{1}{m} \tag{15.3.2}$$

and

$$\theta(\mathcal{N}, \mathcal{M}) \geq \epsilon, \qquad \mathcal{M} \in \operatorname{Inv}(S_m) \tag{15.3.3}$$

From (15.3.2) it is clear that $S_m \to T$. By assumption, Γ splits the spectrum of T. Thus, for m sufficiently large, the contour Γ will split the spectrum of S_m. Moreover, $P(S_m; \Gamma) \to P(T; \Gamma)$, and hence $\operatorname{Im} P(S_m; \Gamma)$ tends to $\operatorname{Im} P(T; \Gamma)$ in the gap topology. But then, for m sufficiently large,

$$\operatorname{Ker} P(T; \Gamma) \dotplus \operatorname{Im} P(S_m; \Gamma) = \mathbb{C}^n$$

(cf. Theorem 13.1.3).

Let R_m be the angular transformation of $\operatorname{Im} P(S_m; T)$ with respect to $P(T; \Gamma)$. Here, as in what follows, m is supposed to be sufficiently large. As $P(S_m; \Gamma) \to P(T; \Gamma)$, we have $R_m \to 0$. Put

$$E_m = \begin{bmatrix} I & R_m \\ 0 & I \end{bmatrix}$$

where the matrix representation corresponds to the decomposition

$$\mathbb{C}^n = \text{Ker } P(T; \Gamma) \dot{+} \text{Im } P(T; \Gamma) \tag{15.3.4}$$

Then E_m is invertible with inverse

$$E_m^{-1} = \begin{bmatrix} 1 & -R_m \\ 0 & 1 \end{bmatrix}$$

Also, $E_m \text{ Im } P(T; \Gamma) = \text{Im } P(S_m; \Gamma)$, and $E_m \to I$.

Put $T_m = E_m^{-1} S_m E_m$. Then $T_m \text{ Im } P(T; \Gamma) \subset \text{Im } P(T; \Gamma)$ and $T_m \to T$. Let T_{m_0} be the restriction of T_m to $\text{Im } P(T; \Gamma)$. Then $T_{m_0} \to T_0$. As $\mathcal{N}$ is a stable invariant subspace for T_0, there exists a sequence $\{\mathcal{N}_m\}$ of subspaces of $\text{Im } P(T; \Gamma)$ such that $\mathcal{N}_m$ is T_{m_0} invariant and $\theta(\mathcal{N}_m, \mathcal{N}) \to 0$. Note that $\mathcal{N}_m$ is also T_m invariant.

Now put $\mathcal{M}_m = E_m \mathcal{N}_m$. Then $\mathcal{M}_m$ is an invariant subspace for S_m. From $E_m \to I$ one can easily deduce that $\theta(\mathcal{M}_m, \mathcal{N}_m) \to 0$. Together with $\theta(\mathcal{M}_m, \mathcal{N}) \to 0$, this gives $\theta(\mathcal{M}_m, \mathcal{N}) \to 0$, which contradicts (15.3.3).

Next assume that $\mathcal{N} \subset \text{Im } P(T; \Gamma)$ is a stable invariant subspace for T, but not for T_0. Then one can find an $\epsilon > 0$ such that, for every positive integer m, there exists a transformation S_{m_0} on $\text{Im } P(T; \Gamma)$ satisfying

$$\|S_{m_0} - T_0\| < \frac{1}{m} \tag{15.3.5}$$

and

$$\theta(\mathcal{N}, \mathcal{M}) \geq \epsilon, \qquad \mathcal{N} \in \text{Inv}(S_{m_0}) \tag{15.3.6}$$

Let T_1 be the restriction of T to $\text{Ker } P(T; \Gamma)$ and write

$$S_m = \begin{bmatrix} T_1 & 0 \\ 0 & S_{m_0} \end{bmatrix}$$

where the matrix representation corresponds to the decomposition (15.3.4). From (15.3.5) it is clear that $S_m \to T$. Hence, as $\mathcal{N}$ is a stable invariant subspace for T, there exists a sequence $\{\mathcal{N}_m\}$ of subspaces of $\mathbb{C}^n$ such that $\mathcal{N}_m$ is S_m invariant and $\theta(\mathcal{N}_m, \mathcal{M}) \to 0$. Put $\mathcal{M}_m = P(T; \Gamma)\mathcal{N}_m$. Since $P(T; \Gamma)$ commutes with S_m, then $\mathcal{M}_m$ is an invariant subspace for S_{m_0}. We now prove that $\theta(\mathcal{M}_m, \mathcal{N}) \to 0$, thus obtaining a contradiction with (15.3.6).

Take $y \in \mathcal{M}_m$ with $\|y\| \leq 1$, and let $x \in \mathcal{N}_m$ be such that $y = P(T; \Gamma)x$. Then

$$\|y\| = \|P(T; \Gamma)x\| \geq \inf\{\|x - u\| \mid u \in \text{Ker } P(T; \Gamma)\}$$
$$\geq \eta(\mathcal{N}_m, \text{Ker } P(T; \Gamma)) \cdot \|x\| \tag{15.3.7}$$

By Proposition 15.3.1, $\theta(\mathcal{N}_m, \mathcal{N}) \to 0$ implies that $\eta(\mathcal{N}_m, \text{Ker } P(T; \Gamma)) \to \eta_0$, where $\eta_0 = \eta(\mathcal{N}, \operatorname{Ker} P(T; \Gamma))$. So, for m sufficiently large, $\eta(\mathcal{N}_m, \operatorname{Ker} P(T; \Gamma)) \geq \frac{1}{2}\eta_0$. Together with (15.3.7), this gives

$$\| y \| \geq \tfrac{1}{2}\eta_0 \| x \|$$

for m sufficiently large. Using this inequality, we obtain

$$\sup_{\substack{y \in \mathcal{M}_m \\ \|y\|=1}} \inf_{z \in \mathcal{N}} \| y - z \| \leq \sup_{\substack{x \in \mathcal{N} \\ \|x\| \leq 2/\eta_0}} \inf_{z \in \mathcal{N}} \| P(T; \Gamma)x - z \|$$

$$= \sup_{\substack{z \in \mathcal{N}_m \\ \|x\| \leq 2/\eta_0}} \inf_{z \in \mathcal{N}} \| P(T; \Gamma)x - P(T; \Gamma)z \|$$

$$\geq \| P(T; \Gamma) \| \left(\frac{2}{\eta_0} \right) \theta(\mathcal{N}_m, \mathcal{N})$$

and

$$\sup_{\substack{z \in \mathcal{N} \\ \|z\|=1}} \inf_{y \in \mathcal{M}_m} \| z - y \| \leq \sup_{z \in \mathcal{N}} \inf_{x \in \mathcal{N}_m} \| P(T; \Gamma)z - P(T; \Gamma)x \|$$

$$\leq \| P(T; \Gamma) \| \theta(\mathcal{N}_m, \mathcal{N})$$

So

$$\theta(\mathcal{M}_m, \mathcal{N}) \leq \left(1 + \frac{2}{\eta_0} \right) \| P(T; \Gamma) \| \theta(\mathcal{N}_m, \mathcal{N})$$

for m sufficiently large. We conclude that $\theta(\mathcal{M}_m, \mathcal{N}) \to 0$, and the proof is complete. □

Lemma 15.3.3

Let $\mathcal{N}$ be an invariant subspace for T, and assume that the contour Γ splits the spectrum of T. If $\mathcal{N}$ is stable for T, then $P(T; \Gamma)\mathcal{N}$ is a stable invariant subspace for the restriction T_0 of T to $\operatorname{Im} P(T; \Gamma)$.

Proof. It is clear that $\mathcal{M} = P(T; \Gamma)\mathcal{N}$ is T_0 invariant.

Assume that $\mathcal{M}$ is not stable for T_0. Then $\mathcal{M}$ is not stable for T, either, by Lemma 15.3.2. Hence there exist $\epsilon > 0$ and a sequence $\{S_m\}$ such that $S_m \to T$ and

$$\theta(\mathcal{L}, \mathcal{M}) \geq \epsilon, \qquad \mathcal{L} \in \operatorname{Inv}(S_m), \qquad m = 1, 2, \ldots \qquad (15.3.8)$$

As $\mathcal{N}$ is stable for T, one can find a sequence of subspaces $\{\mathcal{N}_m\}$ such that $S_m \mathcal{N}_m \subset \mathcal{N}_m$ and $\theta(\mathcal{N}_m, \mathcal{N}) \to 0$. Further, since Γ splits the spectrum of T and $S_m \to T$, the contour Γ will split the spectrum of S_m for m sufficiently large. But then, without loss of generality, we may assume that Γ splits the spectrum of each S_m. Again using $S_m \to T$, it follows that $P(S_m; \Gamma) \to P(T; \Gamma)$.

Let $\mathscr{L}$ be a direct complement of $\mathscr{N}$ in $\mathbb{C}^n$. As $\theta(\mathscr{N}_m, \mathscr{N}) \to 0$, we have $\mathbb{C}^n = \mathscr{L} + \mathscr{N}_m$ for m sufficiently large (Theorem 13.1.3). So, without loss of generality, we may assume that $\mathbb{C}^n = \mathscr{L} + \mathscr{N}_m$ for each m. Let R_m be the angular transformation of $\mathscr{N}_m$ with respect to the projector of $\mathbb{C}^n$ along $\mathscr{L}$ onto $\mathscr{N}$, and put

$$E_m = \begin{bmatrix} I & R_m \\ 0 & I \end{bmatrix}$$

where the matrix corresponds to the decomposition $\mathbb{C}^n = \mathscr{L} \dotplus \mathscr{N}$. Note that $T_m = E_m^{-1} S_m E_m$ leaves $\mathscr{N}$ invariant. Because $R_m \to 0$, we have $E_m \to I$, and so $T_m \to T$.

Clearly, Γ splits the spectrum of $T|_{\mathscr{N}}$. As $T_m \to T$ and $\mathscr{N}$ is invariant for T_m, the contour Γ will split the spectrum of $T_m|_{\mathscr{N}}$ too, provided m is sufficiently large. But then we may assume that this happens for all m. Also, we have

$$\lim_{m \to \infty} P(T_m|_{\mathscr{N}}; \Gamma) \to P(T|_{\mathscr{N}}; \Gamma)$$

Hence $\mathscr{M}_m = \operatorname{Im} P(T_m|_{\mathscr{N}}; \Gamma) \to \operatorname{Im} P(T|_{\mathscr{N}}; \Gamma) = \mathscr{M}$ in the gap topology.

Now consider $\mathscr{L}_m = E_m \mathscr{M}_m$. Then $\mathscr{L}_m$ is an S_m-invariant subspace. From $E_m \to I$ it follows that $\theta(\mathscr{L}_m, \mathscr{M}_m) \to 0$. This, together with $\theta(\mathscr{M}_m, \mathscr{M}) \to 0$, gives $\theta(\mathscr{L}_m, \mathscr{M}) \to 0$. So we arrive at a contradiction to (15.3.8) and the proof is complete. $\square$

After this long preparation we are now able to give a short proof of Theorem 15.2.1.

Proof of Theorem 15.2.1. Suppose that $\mathscr{N}$ is a stable invariant subspace for A. Put $\mathscr{N}_j = \mathscr{N} \cap \mathscr{R}_{\lambda_j}(A)$. Then $\mathscr{N} = \mathscr{N}_1 \dotplus \cdots \dotplus \mathscr{N}_r$. By Lemma 15.3.3, the space $\mathscr{N}_j$ is a stable invariant subspace for the restriction A_j of A to $\mathscr{R}_{\lambda_j}(A)$. But, by Lemma 2.1.3, A_j has one eigenvalue only, namely, λ_j. So we may apply Lemma 15.2.5 to prove that $\mathscr{N}_j$ has the desired form.

Conversely, assume that each $\mathscr{N}_j$ has the desired form, and let us prove that $\mathscr{N} = \mathscr{N}_1 \dotplus \cdots \dotplus \mathscr{N}_r$ is a stable invariant subspace for A. By Corollary 15.2.4, the space $\mathscr{N}_j$ is a stable invariant subspace for the restriction A_j of A to $\mathscr{R}_{\lambda_j}(A)$. Hence we may apply Lemma 15.3.2 to show that each $\mathscr{N}_j$ is a stable invariant subspace for A. But then the same is true for the direct sum $\mathscr{N} = \mathscr{N}_1 \dotplus \cdots \dotplus \mathscr{N}_r$. $\square$

15.4 PERTURBED STABLE INVARIANT SUBSPACES

In this section we show that the stability of an A-invariant subspace $\mathscr{M}$ is preserved under small perturbations of $\mathscr{M}$ and A. This is true also when we restrict our attention to the intersection of $\mathscr{M}$ and a fixed spectral subspace of A. To state this result precisely, denote by $\mathcal{R}_\Omega(A)$ the spectral subspace of A (the sum of root subspaces for A) corresponding to those eigenvalues of A that lie in an open set Ω.

Theorem 15.4.1

Let $A: \mathbb{C}^n \to \mathbb{C}^n$ be a transformation, and let $\Omega \subset \mathbb{C}$ be an open set whose boundary does not intersect $\sigma(A)$. Assume that $\mathcal{M}$ is an A-invariant subspace for which the intersection $\mathcal{M} \cap \mathcal{R}_\Omega(A)$ is stable (with respect to A). Then any B-invariant subspace $\mathcal{N}$ has the property that $\mathcal{N} \cap \mathcal{R}_\Omega(B)$ is stable (with respect to B) provided $\|B - A\|$ and $\theta(\mathcal{M}, \mathcal{N})$ are small enough.

The particular case of Theorem 15.4.1 when $\Omega = \mathbb{C}$ is especially important.

Corollary 15.4.2

Let $\mathcal{M}$ be a stable A-invariant subspace. Then there exists an $\epsilon > 0$ such that any B-invariant subspace $\mathcal{N}$ is stable provided

$$\|B - A\| + \theta(\mathcal{M}, \mathcal{N}) < \epsilon$$

We need the following lemma for the proof of Theorem 15.4.1.

Lemma 15.4.3

Let A and Ω be as in Theorem 15.4.1, and let $\mathcal{M}$ be an A-invariant subspace. Then for every $\epsilon > 0$ there exists a $\delta > 0$ such that every B-invariant subspace $\mathcal{N}$ with $\|B - A\| + \theta(\mathcal{M}, \mathcal{N}) < \delta$ satisfies the inequality $\theta(\mathcal{M} \cap \mathcal{R}_\Omega(A), \mathcal{N} \cap \mathcal{R}_\Omega(B)) < \epsilon$.

Proof. Arguing by contradiction, assume that there is a sequence of transformations $\{B_m\}_{m=1}^\infty$ and a sequence of subspaces $\{\mathcal{N}_m\}_{m=1}^\infty$ such that $\lim_{m\to\infty} \|B_m - A\| = 0$, $\lim_{m\to\infty} \theta(\mathcal{M}, \mathcal{N}_m) = 0$, $\mathcal{N}_m$ is B_m invariant for each m, but

$$\theta(\mathcal{M} \cap \mathcal{R}_\Omega(A), \mathcal{N}_m \cap \mathcal{R}_\Omega(B_m)) \geq \epsilon > 0 \qquad (15.4.1)$$

where ϵ does not depend on m.

Denote by $P_\Omega(B_m)$ [resp. $P_\Omega(A)$] the Riesz projector onto $\mathcal{R}_\Omega(B_m)$ [resp. onto $\mathcal{R}_\Omega(A)$]. By Lemma 13.3.2, for m large enough there exists an invertible transformation $S_m: \mathbb{C}^n \to \mathbb{C}^n$ such that

$$S_m(\mathcal{R}_\Omega(A)) = \mathcal{R}_\Omega(B_m), \qquad S_m(\mathrm{Ker}\, P_\Omega(A)) = \mathrm{Ker}\, P_\Omega(B_m)$$

and, moreover,

$$\max\{\|S_m - I\|, \|S_m^{-1} - I\|\} \leq C_1 \|A - B_m\|$$

Here $C_1, C_2, \ldots$ are positive constants that depend on A only. Actually, one can take S_m defined as follows:

$$S_m x = (I - P_\Omega(B_m) + P_\Omega(A))x , \qquad x \in \operatorname{Ker} P_\Omega(A)$$
$$S_m x = (I + P_\Omega(B_m) - P_\Omega(A))x , \qquad x \in \mathcal{R}_\Omega(A)$$

Put $\tilde{B}_m = S_m^{-1} B_m S_m$ and $\tilde{\mathcal{N}}_m = S_m^{-1} \mathcal{N}_m$ (so that $\tilde{\mathcal{N}}_m$ is $\tilde{B}_m$ invariant). Let $P_{\mathcal{M}}$ (resp. $P_{\mathcal{N}_m}$) be the orthogonal projector onto $\mathcal{M}$ (resp. $\mathcal{N}_m$). As $S_m^{-1} P_{\mathcal{N}_m} S_m$ is a projector onto $\tilde{\mathcal{N}}_m$ (not necessarily orthogonal), we have

$$\theta(\mathcal{M}, \tilde{\mathcal{N}}_m) \le \|S_m^{-1} P_{\mathcal{N}_m} S_m - P_{\mathcal{M}}\| \le C_2 \theta(\mathcal{M}, \mathcal{N}_m) \tag{15.4.2}$$

where the first inequality follows from (13.1.4). Hence

$$\theta(\mathcal{M}, \tilde{\mathcal{N}}_m) \to 0 \quad \text{as} \quad m \to \infty \tag{15.4.3}$$

It is easily seen that $\mathcal{R}_\Omega(\tilde{B}_m) = \mathcal{R}_\Omega(A)$ and $\operatorname{Ker} P_\Omega(\tilde{B}_m) = \operatorname{Ker} P_\Omega(A)$ (for m large enough). Consequently

$$\tilde{\mathcal{N}}_m = (\tilde{\mathcal{N}}_m \cap \mathcal{R}_\Omega(A)) \dotplus (\tilde{\mathcal{N}}_m \cap \operatorname{Ker} P_\Omega(A))$$

Since also

$$\mathcal{M} = (\mathcal{M} \cap \mathcal{R}_\Omega(A)) \dotplus (\mathcal{M} \cap \operatorname{Ker} P_\Omega(A))$$

Theorem 13.4.2, together with (15.4.2), implies that

$$\theta(\mathcal{M} \cap \mathcal{R}_\Omega(A), \tilde{\mathcal{N}}_m \cap \mathcal{R}_\Omega(A)) \to 0 \quad \text{as} \quad m \to \infty \tag{15.4.4}$$

(cf. the proof of Lemma 14.3.2). Now, as in (15.4.2), we have

$$\theta(\mathcal{M} \cap \mathcal{R}_\Omega(A), \mathcal{N}_m \cap \mathcal{R}_\Omega(B_m)) \le C_3 \theta(\mathcal{M} \cap \mathcal{R}_\Omega(A), \tilde{\mathcal{N}}_m \cap \mathcal{R}_\Omega(A))$$

which contradicts (15.4.1) in view of (15.4.4). $\quad\square$

Proof of Theorem 15.4.1. Consider first the case $\Omega = \mathbb{C}$ (i.e., $\mathcal{R}_\Omega(A) = \mathbb{C}^n$, where n is the size of A). Arguing by contradiction, assume that the statement of the theorem is not true (for $\Omega = \mathbb{C}$). Then there exist an $\epsilon > 0$ and a sequence $\{B_m\}_{m=1}^\infty$ of transformations on $\mathbb{C}^n$ converging to A such that $\theta(\mathcal{M}, \mathcal{N}) \ge \epsilon$ for every stable B_m-invariant subspace $\mathcal{N}$, $m = 1, 2, \ldots$ Since $\mathcal{M}$ is stable and $B_m \to A$, there exists a sequence $\{\mathcal{M}_m\}_{m=1}^\infty$ of subspaces in $\mathbb{C}^n$ with $B_m \mathcal{M}_m \subset \mathcal{M}_m$ for each m and $\theta(\mathcal{M}_m, \mathcal{M}) \to 0$. For m sufficiently large we have $\theta(\mathcal{M}_m, \mathcal{M}) < \epsilon$, and hence the B_m-invariant subspace $\mathcal{M}_m$ is not stable.

Let $\mathscr{L}$ be a direct complement of $\mathcal{M}$ in $\mathbb{C}^n$. We may assume that $\mathscr{L}$ is also a direct complement to each $\mathcal{M}_m$ (Theorem 13.1.3). Let R_m be the angular transformation of $\mathcal{M}_m$ with respect to the projector onto $\mathcal{M}$ along $\mathscr{L}$. Then $R_m \to 0$. Write

$$E_m = \begin{bmatrix} I & R_m \\ 0 & I \end{bmatrix}$$

where the matrix representation is taken with respect to the decomposition $\mathbb{C}^n = \mathscr{L} \dotplus \mathcal{M}$. Then E_m is invertible, $E_m \mathcal{M} = \mathcal{M}_m$, and $E_m \to I$. Put $A_m = E_m^{-1} B_m E_m$. Obviously, $A_m \to A$ and $A_m \mathcal{M} \subset \mathcal{M}$. Note that $\mathcal{M}$ is not stable for A_m.

With respect to the decomposition $\mathbb{C}^n = \mathcal{M} \dotplus \mathscr{L}$, we write

$$A = \begin{bmatrix} U & V \\ 0 & W \end{bmatrix}, \qquad A_m = \begin{bmatrix} U_m & V_m \\ 0 & W_m \end{bmatrix}$$

Then $U_m \to U$ and $W_m \to W$. Since $\mathcal{M}$ is not stable for A_m, Theorem 15.2.1 ensures the existence of a common eigenvalue λ_m of U_m and W_m such that

$$\dim \operatorname{Ker}(\lambda_m I - A_m) \geq 2, \qquad m = 1, 2, \ldots \qquad (15.4.5)$$

Now $|\lambda_m| \leq \|U_m\|$ and $\{U_m\}$ converges to U. Hence the sequence $\{\lambda_m\}$ is bounded. Passing, if necessary, to a subsequence, we may assume that $\lambda_m \to \lambda_0$ for some $\lambda_0 \in \mathbb{C}$. But then $\lambda_m I - U_m \to \lambda_0 I - U$ and $\lambda_m I - W_m \to \lambda_0 I - W$. It follows that λ_0 is a common eigenvalue of U and W. Again applying Theorem 15.2.1, we see that λ_0 is an eigenvalue of geometric multiplicity one: $\dim \operatorname{Ker}(\lambda_0 I - A) = 1$. So there exists a nonzero $(n-1) \times (n-1)$ minor in $\lambda_0 I - A$. Then, for m large enough, the corresponding minor in $\lambda_m I - A_m$ is also nonzero, a contradiction with (15.4.5).

Now consider the general case of Theorem 15.4.1. It is seen from the proof of Lemma 15.4.3 that we can assume that B satisfies $\mathscr{R}_\Omega(B) = \mathscr{R}_\Omega(A)$. But then we can apply the part of Theorem 15.4.1 already proved with $\mathbb{C}^n$, A and B replaced by $\mathscr{R}_\Omega(A)$, $A|_{\mathscr{R}_\Omega(A)}$ and $B|_{\mathscr{R}_\Omega(B)}$, respectively. $\square$

Now let us focus attention on the spectral A-invariant subspaces, that is, sums of root subspaces for A (the zero subspace will also be called spectral). Theorem 15.2.1 shows that each spectral invariant subspace is stable. The converse is not true in general: every invariant subspace of a unicellular transformation is stable, but the only spectral subspaces in this case are the trivial ones.

For the spectral subspaces, an analog of Theorem 15.4.1 holds.

Theorem 15.4.4

Let A and Ω be as in Theorem 15.4.1. Assume that $\mathcal{M}$ is an A-invariant subspace for which $\mathcal{M} \cap \mathscr{R}_\Omega(A)$ is a spectral invariant subspace for A. Then *any B-invariant subspace $\mathcal{N}$ has the property that $\mathcal{N} \cap \mathcal{R}_\Omega(B)$ is spectral (as a B-invariant subspace) provided $\|B - A\| + \theta(\mathcal{M}, \mathcal{N})$ is small enough.*

Proof. As in the proof of Theorem 15.4.1, the general case can be reduced to the case $\Omega = \mathbb{C}$. So assume $\Omega = \mathbb{C}$.

Since every invariant subspace is the sum of its intersections with the root subspaces, it follows that an A-invariant subspace $\mathcal{L}$ is spectral if and only if there is an A-invariant direct complement $\mathcal{L}'$ to $\mathcal{L}$ such that $\sigma(A|_\mathcal{L}) \cap \sigma(A|_{\mathcal{L}'}) = \emptyset$. Let Δ be an open set containing $\sigma(A|_\mathcal{M})$, and let Δ' be an open set disjoint with Δ that contains all other eigenvalues of A (if any). Then $\sigma(A|_{\mathcal{M}'}) \subset \Delta'$ for an A-invariant direct complement $\mathcal{M}'$ to $\mathcal{M}$ (actually, $\mathcal{M}'$ is the spectral A-invariant subspace corresponding to the eigenvalues in Δ'). By Theorem 15.1.4, any B-invariant subspace $\mathcal{N}$ satisfies $\sigma(B|_\mathcal{N}) \subset \Delta$ provided $\|B - A\| + \theta(\mathcal{M}, \mathcal{N})$ is small enough. On the other hand, by Theorems 15.2.1 and 15.1.4 there exists a B-invariant subspace $\mathcal{N}'$ such that $\sigma(B|_{\mathcal{N}'}) \subset \Delta'$ and $\theta(\mathcal{M}', \mathcal{N}')$ is as small as we wish provided $\|B - A\|$ is small enough. As $\mathcal{N}'$ is a direct complement to $\mathcal{N}$ (Theorem 13.1.3) and $\sigma(B|_\mathcal{M}) \cap \sigma(B|_{\mathcal{N}'}) = \emptyset$, it follows that $\mathcal{N}$ is spectral. $\square$

The proof of Theorem 15.4.4 shows that if $\mathcal{M}$ is a spectral A-invariant subspace with $\sigma(A|_\mathcal{M}) \subset \Omega$, where $\Omega \subset \mathbb{C}$ is an open set, then for any B-invariant subspace $\mathcal{N}$ such that $\|B - A\| + \theta(\mathcal{M}, \mathcal{N})$ is small enough, we also have $\sigma(B|_\mathcal{N}) \subset \Omega$.

15.5 LIPSCHITZ STABLE INVARIANT SUBSPACES

In this section we study a stronger version of stability for invariant subspaces. A subspace $\mathcal{M} \subset \mathbb{C}^n$ that is invariant for a transformation $A: \mathbb{C}^n \to \mathbb{C}^n$ is said to be *Lipschitz stable* (with respect to A) if there exist positive constants K and ϵ such that every transformation $B: \mathbb{C}^n \to \mathbb{C}^n$ with $\|B - A\| < \epsilon$ has an invariant subspace $\mathcal{N}$ with $\theta(\mathcal{M}, \mathcal{N}) \leq K\|B - A\|$. Clearly, every Lipschitz stable subspace is stable; the converse is not true in general.

The following theorem decribes Lipschitz stability.

Theorem 15.5.1

For a transformation A and an A-invariant subspace $\mathcal{M}$ the following statements are equivalent: (a) *$\mathcal{M}$ is Lipschitz stable;* (b) *$\mathcal{M} = \{0\}$ or else $\mathcal{M} = \mathcal{R}_{\lambda_1}(A) \dotplus \cdots \dotplus \mathcal{R}_{\lambda_r}(A)$ for some different eigenvalues $\lambda_1, \ldots, \lambda_r$ of A; in other words, $\mathcal{M}$ is a spectral A-invariant subspace;* (c) *for every sufficiently small $\epsilon > 0$ there exists a $\delta > 0$ such that any transformation B with $\|A - B\| < \delta$ has a unique invariant subspace $\mathcal{N}$ for which $\theta(\mathcal{M}, \mathcal{N}) < \epsilon$.*

The emphasis in (c) is on the uniqueness of $\mathcal{N}$; if the word "unique" is omitted in (c), we obtain the definition of stability of $\mathcal{M}$.

Proof. First, arguing as in the proof of Lemma 15.3.2, one shows that $\mathcal{M}$ is a Lipschitz stable A-invariant subspace if and only if each intersection $\mathcal{M} \cap \mathcal{R}_{\mu_j}(A)$ is Lipschitz stable (with respect to the restriction $A|_{\mathcal{R}_{\mu_j}(A)}$) for $j = 1, \ldots, s$, where $\mu_1, \ldots, \mu_s$ are all the distinct eigenvalues of A.

Assume that (c) holds but (b) does not. Then $\mathcal{M}$ is a stable subspace, and Theorem 15.2.1 ensures that for some eigenvalue λ_0 of A with $\dim \mathrm{Ker}(\lambda_0 I - A) = 1$ we have $\{0\} \neq \mathcal{M} \cap \mathcal{R}_{\lambda_0}(A) \neq \mathcal{R}_{\lambda_0}(A)$. Let

$$A|_{\mathcal{R}_{\lambda_0}(A)} = \begin{bmatrix} \lambda_0 & 1 & 0 & \cdots & 0 \\ 0 & \lambda_0 & 1 & & \vdots \\ \vdots & \vdots & & \ddots & \\ & & & & 1 \\ 0 & 0 & & \cdots & \lambda_0 \end{bmatrix}$$

in a Jordan basis for A in $\mathcal{R}_{\lambda_0}(A)$, and define the transformation $B(\alpha)$, where $0 < \alpha < 1$, as follows:

$$B(\alpha)|_{\mathcal{R}_{\lambda_0}(A)} = \begin{bmatrix} \lambda_0 & 1 & & 0 \\ 0 & \lambda_0 & 1 & \\ \vdots & & & \ddots \\ 0 & & & 1 \\ \alpha & 0 & \cdots & 0 \lambda_0 \end{bmatrix} \tag{15.5.1}$$

$B(\alpha) = A$ on all root subspaces of A other than $\mathcal{R}_{\lambda_0}(A)$. Then $B(\alpha) \to A$ as $\alpha \to 0$. Let $p = \dim \mathcal{R}_{\lambda_0}(A)$; $q = \dim \mathrm{Ker}\, \mathcal{M} \cap \mathcal{R}_{\lambda_0}(A)$; so $0 < q < p$. For brevity, denote the right-hand side of (15.5.1) by $K(\alpha)$. To obtain a contradiction, it is sufficient to show that for α small enough the number of q-dimensional $K(\alpha)$-invariant subspaces $\mathcal{N}$ such that $\theta(\mathcal{M} \cap \mathcal{R}_{\lambda_0}(A), \mathcal{N}) \leq C_1 \alpha^{1/p}$ is exactly $\binom{p}{q} > 1$ (we denote by $C_1, C_2, \ldots$ positive constants that depend on p and q only).

Let us prove this assertion. The matrix $K(\alpha)$ has p different eigenvalues $\epsilon_1, \ldots, \epsilon_p$, which are the p different roots of the equation $x^p = \alpha$. The corresponding eigenvectors are $y_i = \langle 1, \epsilon_i, \ldots, \epsilon_i^{p-1} \rangle$, $i = 1, \ldots, p$. The only q-dimensional $K(\alpha)$-invariant subspaces are those spanned by any q vectors among $y_1, \ldots, y_p$. Take such a subspace $\mathcal{N}$ and suppose for notational convenience that $\mathcal{N} = \mathrm{Span}\{y_1, \ldots, y_q\}$. The projector $Q_{\mathcal{N}}$ onto $\mathcal{N}$ along the subspace spanned by $e_{q+1}, \ldots, e_p$ is given by the formula

$$Q_{\mathcal{N}} \begin{bmatrix} I_q & 0 \\ \tilde{Y}_{p-q} Y_q^{-1} & 0 \end{bmatrix}$$

where Y_q (resp. $\bar{Y}_{p-q}$) is the $q \times q$ [resp. $(p-q) \times q$] matrix formed by the first q (resp. last $p-q$) rows of the matrix $[y_1 \ y_2 \ \cdots \ y_q]$. As Y_q is a Vandermonde matrix, $\det Y_q = \Pi_{1 \le i < j \le q} (\epsilon_j - \epsilon_i) \ne 0$ (cf. Example 2.6.4). Let $Z_q = \operatorname{Adj} Y_q$ be the matrix of algebraic adjoints to the elements of Y_q, so that $Y_q^{-1} = 1/(\det Y_q) Z_q$. From the form of Y_q it is easily seen that $\|Z_q\| \le C_2 \alpha^{r/p}$, where $r = 1 + 2 + \cdots + (q-2) = \frac{1}{2}(q-1)(q-2)$. Further, $|\det Y_q| = C_3 \alpha^{s/p}$, where $s = \frac{1}{2} q(q-1)$ is the number of all pairs of integers (i, j) such that $1 \le i < j \le q$. As $\|\bar{Y}_{p-q}\| < C_4 \alpha^{q/p}$, it follows that $\|\bar{Y}_{p-q} Y_q^{-1}\| \le C_5 \alpha^{(r+s+q)/p} = C_5 \alpha^{1/p}$. Consequently,

$$\|Q - Q_{\mathcal{N}}\| \le C_6 \cdot \alpha^{1/p} \tag{15.5.2}$$

where $Q = \begin{bmatrix} I_q & 0 \\ 0 & 0 \end{bmatrix}$. As Q (resp. $Q_{\mathcal{N}}$) is a projector onto $\mathcal{M} \cap \mathcal{R}_{\lambda_0}(A)$ (resp. onto $\mathcal{N}$) we have

$$\theta(\mathcal{M} \cap \mathcal{R}_{\lambda_0}(A), \mathcal{N}) \le \|Q - Q_{\mathcal{N}}\|$$

[see (13.1.4)]. Combining this inequality with (15.5.2), we find that $\theta(\mathcal{M} \cap \mathcal{R}_{\lambda_0}(A), \mathcal{N}) \le C_6 \cdot \alpha^{1/p}$ for $\alpha > 0$ small enough. Since the number of q-dimensional $K(\alpha)$-invariant subspaces N is exactly $\binom{p}{q}$, the required assertion is proved.

Conversely, assume that (b) holds but (c) does not. Since $\mathcal{M}$ is a stable subspace (by Theorem 15.2.1), this implies the existence of a sequence $\{B_m\}_{m=1}^{\infty}$ and the existence of two different B_m-invariant subspaces $\mathcal{N}_{1m}$ and $\mathcal{N}_{2m}$ such that $\|B_m - A\| < (1/m)$ and

$$\theta(\mathcal{M}, \mathcal{N}_{im}) < \frac{1}{m} \tag{15.5.3}$$

for $i = 1$ and 2. Let Γ (resp. Δ) be a closed simple rectifiable contour such that $\sigma(A) \cap \Gamma = \emptyset$ [resp. $\sigma(A) \cap \Delta = \emptyset$] and $\lambda_1, \ldots, \lambda_r$ are the only eigenvalues of A inside Γ (resp. outside Δ). Letting $\mathcal{R}_\Gamma(C)$ be the image of the Riesz projector $(2\pi i)^{-1} \int_\Gamma (\lambda I - C)^{-1} d\lambda$, where the matrix C has no eigenvalues on Γ, we have $\mathcal{M} = \mathcal{R}_\Gamma(A)$. Since $\theta(\mathcal{R}_\Gamma(B_m), \mathcal{R}_\Gamma(A)) \to 0$ as $m \to \infty$, we find in view of (15.5.3) that

$$\theta(\mathcal{R}_\Gamma(B_m), \mathcal{N}_{im}) \to 0 \quad \text{as } m \to \infty, \quad i = 1, 2 \tag{15.5.4}$$

Now $\mathcal{N}_{im} = (\mathcal{N}_{im} \cap \mathcal{R}_\Gamma(B_m)) \dotplus (\mathcal{N}_{im} \cap \mathcal{R}_\Delta(B_m))$; combining this with (15.5.4), it is easily seen that $\mathcal{N}_{im} \cap \mathcal{R}_\Delta(B_m) = \{0\}$, at least for large m. (Indeed, argue by contradiction and use the properties that the set of all subspaces in $\mathbb{C}^n$ is compact and that the limit of a converging sequence of nonzero subspaces is again nonzero.) So $\mathcal{N}_{im} \subset \mathcal{R}_\Gamma(B_m)$. But (15.5.3) implies that (for large m) $\dim \mathcal{N}_{im} = \dim \mathcal{M} = \dim \mathcal{R}_\Gamma(B_m)$. Hence $\mathcal{N}_{im} = \mathcal{R}_\Gamma(B_m)$,

$i = 1, 2$ (for large m), contradicting the assumption that $\mathcal{N}_{1m}$ and $\mathcal{N}_{2m}$ are different.

Now we prove the equivalence of (a) and (b). In view of Theorem 15.2.1, we have to check that the only Lipschitz stable invariant subspaces of the Jordan block

$$J = \begin{bmatrix} 0 & 1 & 0 & & & 0 \\ 0 & 0 & 1 & & & \vdots \\ \vdots & \vdots & & \ddots & & \\ & & & & & 1 \\ 0 & 0 & 0 & \cdots & & 0 \end{bmatrix} : \mathbb{C}^n \to \mathbb{C}^n$$

are the trivial spaces $\{0\}$ and $\mathbb{C}^n$. For $\alpha > 0$, let

$$J_\alpha = \begin{bmatrix} 0 & 1 & 0 & \cdots & 0 \\ 0 & 0 & 1 & & \vdots \\ \vdots & \vdots & \vdots & \ddots & \\ & & & & 1 \\ \alpha & 0 & 0 & \cdots & 0 \end{bmatrix}$$

For $k = 1, \ldots, n - 1$, the only k-dimensional J-invariant subspace $\mathcal{N}_k$ is spanned by the first k unit coordinate vectors. Denote by P_k the orthogonal projector onto $\mathcal{N}_k$, and let $P_{k,\alpha}$ denote the orthogonal projector onto a k-dimensional J_α-invariant subspace $\mathcal{N}_{k,\alpha}$ $(1 \le k \le n - 1)$. We have $y \overset{\text{def}}{=} \langle 1, \epsilon, \ldots, \epsilon^{n-1} \rangle \in \mathcal{N}_{k,\alpha}$, where $\epsilon = \alpha^{1/n}$. So

$$\theta(\mathcal{N}_k, \mathcal{N}_{k,\alpha}) = \| P_k - P_{k,\alpha} \| \ge \frac{1}{\|y\|} \| P_k y - P_{k,\alpha} y \|$$

$$= \left\{ \frac{\sum_{j=k}^{n-1} |\epsilon^j|^2}{\sum_{j=0}^{n-1} |\epsilon^j|^2} \right\}^{1/2}$$

Now use $|\epsilon| = \sqrt[n]{\alpha}$. One finds that for α sufficiently small

$$\theta(\mathcal{N}_k, \mathcal{N}_{k,\alpha}) \ge \tfrac{1}{2} \alpha^{k/n}$$

On the other hand, $\| J - J_\alpha \| = \alpha$. But then it is clear that for $1 \le k \le n - 1$ the space $\mathcal{N}_k$ is not a Lipschitz stable invariant subspace of J, and thus J has no nontrivial Lipschitz stable invariant subspace. $\quad\square$

The property of being a Lipschitz stable subspace is stable in the following sense: let $\mathcal{M}$ be an A-invariant Lipschitz stable subspace. Then any B-invariant subspace $\mathcal{N}$ is Lipschitz stable (with respect to B) provided $\| B - A \|$ and $\theta(\mathcal{M}, \mathcal{N})$ are small enough. In view of Theorem 15.5.1, this is simply a reformulation of Theorem 15.4.4.

It follows from Theorem 15.5.1 that a transformation $A : \mathbb{C}^n \to \mathbb{C}^n$ has exactly 2^r different Lipschitz stable invariant subspaces, where r is the number of distinct eigenvalues of A.

15.6   STABILITY OF LATTICES OF INVARIANT SUBSPACES

In this section we extend the notion of stable invariant subspaces to the lattices of invariant subspaces.

Recall that a set Λ of subspaces in $\mathbb{C}^n$ is called a *lattice* if $\mathcal{M}, \mathcal{N} \in \Lambda$ implies $\mathcal{M} + \mathcal{N} \in \Lambda$ and $\mathcal{M} \cap \mathcal{N} \in \Lambda$. Two lattices Λ and Λ' of subspaces in $\mathbb{C}^n$ are *isomorphic* if there exists a bijective map $S: \Lambda \to \Lambda'$ such that $S(\mathcal{M} \cap \mathcal{N}) = S\mathcal{M} \cap S\mathcal{N}$ and $S(\mathcal{M} + \mathcal{N}) = S\mathcal{M} + S\mathcal{N}$ for any two members $\mathcal{M}$ and $\mathcal{N}$ of Λ. In this case S is called an *isomorphism* of Λ onto Λ'.

Let Λ be a lattice of (not necessarily all) invariant subspaces of a transformation $A: \mathbb{C}^n \to \mathbb{C}^n$. The lattice Λ is called *stable* if for every $\epsilon > 0$ there exists a $\delta > 0$ such that, for any transformation $B: \mathbb{C}^n \to \mathbb{C}^n$ with $\|A - B\| < \delta$, there exists a lattice Λ' of (not necessarily all) B-invariant subspaces that is isomorphic to Λ and satisfies $\sup_{\mathscr{L} \in \Lambda} \theta(\mathscr{L}, S(\mathscr{L})) < \epsilon$ for some isomorphism $S: \Lambda \to \Lambda'$. If Λ consists of just one subspace, we obtain the definition of a stable invariant subspace.

Theorem 15.6.1

A lattice Λ of A-invariant subspaces is stable if and only if it consists of stable A-invariant subspaces.

Proof.   Without loss of generality we can assume that $\{0\}$ and $\mathbb{C}^n$ belong to Λ.

Suppose first that Λ contains an A-invariant subspace $\mathcal{M}$ that is not stable. Then there exist an $\epsilon > 0$ and a sequence of transformations $\{B_m\}_{m=1}^{\infty}$ tending to A such that $\theta(\mathcal{M}, \mathcal{N}) \geq \epsilon_0$ for any B_m-invariant subspace $\mathcal{N}$ and any m. Obviously, Λ cannot be stable.

Assume now that every member of Λ is a stable A-invariant subspace. As the number of stable A-invariant subspaces is finite (by Theorem 15.2.1), the lattice Λ is finite. Let $\mathcal{M}_1, \ldots, \mathcal{M}_p$ be all the elements in Λ. Denote by $\lambda_1, \ldots, \lambda_r$ the different eigenvalues of A ordered so that

$$\dim \operatorname{Ker}(A - \lambda_i I) = 1 \qquad \text{for } i = 1, \ldots, s$$

$$\dim \operatorname{Ker}(A - \lambda_i I) > 1 \qquad \text{for } i = s+1, \ldots, r$$

Then

$$\mathcal{M}_i = \mathcal{N}_{i1} \dotplus \mathcal{N}_{i2} \dotplus \cdots \dotplus \mathcal{N}_{ir}, \qquad i = 1, \ldots, p$$

where $\mathcal{N}_{ij} = \mathcal{M}_i \cap \mathcal{R}_{\lambda_j}(A)$, and $\mathcal{N}_{ij}$ is equal either to $\{0\}$ or to $\mathcal{R}_{\lambda_j}(A)$ for

$j = s + 1, \ldots, r$. Let Γ_j $(j = 1, \ldots, r)$ be a small circle around λ_j such that λ_j is the only eigenvalue of A inside or on Γ_j. There exists a $\delta_0 > 0$ such that all transformations $B: \mathbb{C}^n \to \mathbb{C}^n$ with $\|B - A\| < \delta_0$ have all their eigenvalues inside the circles Γ_j; for such a B denote by $\mathcal{R}_j(B)$ the sum of root subspaces of B corresponding to the eigenvalues inside Γ_j. Now put

$$\mathcal{M}'_i = \mathcal{N}'_{i1} \dotplus \cdots \dotplus \mathcal{N}'_{ir}, \qquad i = 1, \ldots, p$$

where for $j = s + 1, \ldots, r$, $\mathcal{N}'_{ij} = \{0\}$ if $\mathcal{N}_{ij} = \{0\}$ and $\mathcal{N}'_{ij} = \mathcal{R}_j(B)$ if $\mathcal{N}_{ij} = \mathcal{R}_{\lambda_j}(A)$; for $j = 1, \ldots, s$ we take $\mathcal{N}'_{ij}$ as follows. Let $\{0\} = \mathcal{L}_0 \subset \mathcal{L}_1 \subset \cdots \subset \mathcal{L}_m = \mathcal{R}_j(B)$ be a complete chain of B-invariant subspaces in $\mathcal{R}_j(B)$; then $\mathcal{N}'_{ij}$ is equal to that subspace $\mathcal{L}_k$ whose dimension coincides with the dimension of $\mathcal{N}_{ij}$. Clearly, $\mathcal{M}'_i$ is B invariant. Further, it is clear from the construction that $\mathcal{M}'_i \subset \mathcal{M}'_k$ if and only if $\mathcal{M}_i \subset \mathcal{M}_k$. Using Theorem 15.2.3, it is not difficult to see that, given $\epsilon < 0$, there exists a positive $\delta \leq \delta_0$ such that $\max_{1 \leq i \leq p} \theta(\mathcal{M}_i, \mathcal{M}'_i) < \epsilon$ for any transformation $B: \mathbb{C}^n \to \mathbb{C}^n$ with $\|B - A\| < \delta$. Putting $\Lambda' = \{\mathcal{M}'_1, \ldots, \mathcal{M}'_p\}$, we find that Λ is stable. $\square$

The case when the lattice Λ is a chain is of special interest for us.

We say that a chain $\mathcal{L}_1 \subset \cdots \subset \mathcal{L}_r$ of A-invariant subspaces is *stable* if for every $\epsilon > 0$ there exists a $\delta > 0$ such that any transformation $B: \mathbb{C}^n \to \mathbb{C}^n$ with $\|B - A\| < \delta$ has a chain $\mathcal{L}'_1 \subset \cdots \subset \mathcal{L}'_r$ of invariant subspaces such that $\theta(\mathcal{L}'_i, \mathcal{L}_i) < \epsilon$ for $i = 1, \ldots, r$. It follows from Theorem 15.6.1 that a chain of A-invariant subspaces is stable if and only if each member of this chain is a stable A-invariant subspace.

The notion of Lipschitz stability of a lattice of invariant subspaces is introduced naturally: a lattice Λ of (not necessarily all) A-invariant subspaces is called *Lipschitz stable* if there exist positive constants ϵ and K such that every transformation B with $\|B - A\| < \epsilon$ has a lattice Λ' of invariant subspaces that is isomorphic to Λ and satisfies

$$\inf_S \sup_{\mathcal{L} \in \Lambda} \theta(\mathcal{L}, S(\mathcal{L})) \leq K\|B - A\|$$

where S runs through the set of all isomorphisms of Λ onto Λ'. Obviously, every Lipschitz stable lattice of invariant subspaces is stable. We leave the proof of the following result to the readers.

Theorem 15.6.2

A lattice Λ of A-invariant subspaces is Lipschitz stable if and only if Λ consists only of a spectral subspaces for A.

15.7   STABILITY IN METRIC OF THE LATTICES OF INVARIANT SUBSPACES

If the lattice Λ consists of all A-invariant subspaces, then a different notion of stability (based on the distance between sets) is also of interest. To introduce this notion, we start with some terminology.

Given two sets X and Y of subspaces in $\mathbb{C}^n$, the distance between X and Y is introduced naturally:

$$\text{dist}(X, Y) = \max\{\sup_{\mathcal{M} \in X} \inf_{\mathcal{N} \in Y} \theta(\mathcal{M}, \mathcal{N}), \sup_{\mathcal{M} \in Y} \inf_{\mathcal{N} \in X} \theta(\mathcal{M}, \mathcal{N})\}$$

Borrowing notation from set theory, denote by 2^Z the set of all subsets in a set Z. Then $\text{dist}(X, Y)$ is a metric in $2^{\mathcal{G}(\mathbb{C}^n)}$ [as before, $\mathcal{G}(\mathbb{C}^n)$ represents the set of all subspaces in $\mathbb{C}^n$]. Indeed, the only nontrivial property that we have to check is the triangle inequality:

$$\text{dist}(X, Y) \le \text{dist}(X, Z) + \text{dist}(Z, Y)$$

for any subsets X, Y, Z in $\mathcal{G}(\mathbb{C}^n)$. For $\mathcal{M} \in X$, $\mathcal{N} \in Y$, $\mathcal{L} \in Z$ we have

$$\theta(\mathcal{M}, \mathcal{N}) \le \theta(\mathcal{M}, \mathcal{L}) + \theta(\mathcal{L}, \mathcal{N}) \tag{15.7.1}$$

Fix $\mathcal{M}$ and $\epsilon > 0$ and take $\mathcal{L}$ in such a way that $\theta(\mathcal{M}, \mathcal{L}) \le \inf_{\mathcal{F} \in Z} \theta(\mathcal{M}, \mathcal{F}) + \epsilon$. Taking the infimum in (15.7.1) with respect to $\mathcal{N}$, we obtain

$$\inf_{\mathcal{N} \in Y} \theta(\mathcal{M}, \mathcal{N}) \le \inf_{\mathcal{F} \in Z} \theta(\mathcal{M}, \mathcal{F}) + \inf_{\mathcal{N} \in Y} \theta(\mathcal{L}, \mathcal{N}) + \epsilon$$
$$\le \text{dist}(X, Z) + \text{dist}(Z, Y) + \epsilon$$

Now take the supremum with respect to $\mathcal{M}$, and, from the resulting inequality with the roles of $\mathcal{M}$ and $\mathcal{N}$ interchanged, it follows that

$$\text{dist}(X, Y) \le \text{dist}(X, Z) + \text{dist}(Z, Y) + \epsilon$$

As $\epsilon > 0$ was arbitrary, the triangle inequality follows.

Note also that $\text{dist}(X, Y) \le 1$ for any $X, Y \in Z^{\mathcal{G}(\mathbb{C}^n)}$.

The lattice $\text{Inv}(A)$ of all invariant subspaces of a transformation $A: \mathbb{C}^n \to \mathbb{C}^n$ is called *stable in metric* if for every $\epsilon > 0$ there exists a $\delta > 0$ such that the lattice $\text{Inv}(B)$ of any transformation $B: \mathbb{C}^n \to \mathbb{C}^n$ with $\|B - A\| < \delta$ satisfies $\text{dist}(\text{Inv}(B), \text{Inv}(A)) < \epsilon$. The following theorem describes all transformations with stable lattices of invariant subspaces.

Theorem 15.7.1

$\text{Inv}(A)$ *is stable in metric if and only if A is nonderogatory, that is,* $\dim \text{Ker}(A - \lambda_0 I) = 1$ *for every eigenvalue λ_0 of A.*

Proof. Assume that A is derogatory. Then obviously $\text{Inv}(A)$ is an infinite set. Without loss of generality we can assume that A is a matrix in the Jordan form:

$$A = J_{k_1}(\lambda_1) \oplus \cdots \oplus J_{k_r}(\lambda_r)$$

Here $\lambda_1, \ldots, \lambda_r$ are (not necessarily distinct) eigenvalues of A. For $i = 1, \ldots, r$ let $\{\epsilon_i(m)\}_{m=1}^{\infty}$ be a sequence of numbers such that $\lim_{m \to \infty} \epsilon_i(m) = 0$ and

$$\lambda_i + \epsilon_i(m) \neq \lambda_j + \epsilon_j(m)$$

for any $i \neq j$ and any positive integer m. (Such sequences can obviously be arranged.) Letting

$$A_m = J_{k_1}(\lambda_1 + \epsilon_1(m)) \oplus \cdots \oplus J_{k_r}(\lambda_r + \epsilon_r(m))$$

we obtain $\|A_m - A\| \to 0$ as $m \to \infty$. Moreover, the number of A_m-invariant subspaces is exactly $(k_1 + 1) \cdots (k_r + 1)$, and the lattice of A_m-invariant subspaces is independent of m. As $\mathrm{Inv}(A)$ is infinite, clearly, $\mathrm{dist}(\mathrm{Inv}(A), \mathrm{Inv}(A_m)) \geq \epsilon > 0$, where ϵ does not depend on m. Hence $\mathrm{Inv}(A)$ is not stable.

Assume now that A is nonderogatory. Then the lattice $\mathrm{Inv}(A)$ is finite. Let $\mathcal{M}_1, \ldots, \mathcal{M}_p$ be all the A-invariant subspaces. Theorem 15.2.1 shows that every $\mathcal{M}_i$ is stable. That is, given $\epsilon > 0$, there exists a $\delta_i > 0$ such that any transformation $B: \mathbb{C}^n \to \mathbb{C}^n$ with $\|B - A\| < \delta_i$ has an invariant subspace $\mathcal{N}_i$ such that $\theta(\mathcal{M}_i, \mathcal{N}_i) < \epsilon$. Taking $\delta' = \min(\delta_1, \ldots, \delta_p)$, we have

$$\max_{\mathcal{M} \in \mathrm{Inv}(A)} \inf_{\mathcal{N} \in \mathrm{Inv}(B)} \theta(\mathcal{M}, \mathcal{N}) < \epsilon$$

for every transformation with $\|B - A\| < \delta'$. We prove now that given $\epsilon > 0$ there exists a $\delta'' > 0$ such that

$$\sup_{\mathcal{N} \in \mathrm{Inv}(B)} \inf_{\mathcal{M} \in \mathrm{Inv}(A)} \theta(\mathcal{M}, \mathcal{N}) < \epsilon$$

for every transformation B with $\|B - A\| < \delta''$. Suppose not. Then there is a sequence of transformations on $\mathbb{C}^n$, $\{B_m\}_{m=1}^{\infty}$, such that $B_m \to A$ as $m \to \infty$ and for every m there exists a B_m-invariant subspace $\mathcal{N}_m$ with

$$\inf_{\mathcal{M} \in \mathrm{Inv}(A)} \theta(\mathcal{M}, \mathcal{N}_m) \geq \epsilon_0 > 0 \tag{15.7.2}$$

where ϵ_0 is independent of m. Using the compactness of the set of all subspaces in $\mathbb{C}^n$, we can assume that $\lim_{m \to \infty} \theta(\mathcal{N}_m, \mathcal{N}) = 0$ for certain subspace $\mathcal{N}$ in $\mathbb{C}^n$. Then (15.7.2) gives

$$\inf_{\mathcal{M} \in \mathrm{Inv}(A)} \theta(\mathcal{M}, \mathcal{N}) \geq \epsilon_0 \tag{15.7.3}$$

However, by Theorem 15.1.1, $\mathcal{N} \in \mathrm{Inv}(A)$, which contradicts (15.7.3).

Now given $\epsilon > 0$, let $\delta = \min(\delta', \delta'')$ to see that $\text{dist}(\text{Inv}(B), \text{Inv}(A)) < \epsilon$ for every transformation B with $\|B - A\| < \delta$. $\square$

It follows from Theorems 15.6.1 and 15.7.1 that $\text{Inv}(A)$ is stable if and only if it is stable in metric.

Also, let us introduce the notion of Lipschitz stability in metric. We say that the lattice $\text{Inv}(A)$ of all invariant subspaces of a transformation $A: \mathbb{C}^n \to \mathbb{C}^n$ is *Lipschitz stable in metric* if there exist positive constants K and ϵ such that for any transformation $B: \mathbb{C}^n \to \mathbb{C}^n$ with $\|B - A\| < \epsilon$ the inequality $\text{dist}(\text{Inv}(B), \text{Inv}(A)) \le K\|B - A\|$ holds.

Theorem 15.7.2

The lattice $\text{Inv}(A)$ *for a transformation* $A: \mathbb{C}^n \to \mathbb{C}^n$ *is Lipschitz stable in metric if and only if* A *has* n *distinct eigenvalues.*

Proof. Assume that A has n distinct eigenvalues. Then every A-invariant subspace is spectral and by Theorem 15.5.1 every A-invariant subspace is Lipschitz stable. let $\mathcal{M}_1, \ldots, \mathcal{M}_p$ be all A-invariant subspaces (their number is finite). So there exist positive constants K_i, ϵ_i such that any transformation B with $\|B - A\| < \epsilon_i$ has a invariant subspace $\mathcal{N}_i$ with $\theta(\mathcal{M}_i, \mathcal{N}_i) \le K_i\|B - A\|$. Letting $K = \max(K_1, \ldots, K_p)$, $\epsilon = \min(\epsilon_1, \ldots, \epsilon_p)$, we find that

$$\sup_{\mathcal{M} \in \text{Inv}(A)} \inf_{\mathcal{N} \in \text{Inv}(B)} \theta(\mathcal{M}, \mathcal{N}) \le K\|B - A\| \tag{15.7.4}$$

provided $\|B - A\| < \epsilon$. Now consider the invariant subspaces of B. As A has n distinct eigenvalues, the same is true for any transformation B sufficiently close to A. So every B-invariant subspace $\mathcal{N}$ is spectral:

$$\mathcal{N} = \text{Im}\left[\frac{1}{2\pi i} \int_\Gamma (\lambda I - B)^{-1} \, d\lambda \right]$$

for a suitable contour Γ. We can assume that $\Gamma \cap \sigma(A) = \emptyset$. Then, letting

$$\mathcal{M} = \text{Im}\left[\frac{1}{2\pi i} \int_\Gamma (\lambda I - A)^{-1} \, d\lambda \right]$$

we find that

$$\theta(\mathcal{M}, \mathcal{N}) \le \left\| \frac{1}{2\pi i} \int_\Gamma (\lambda I - A)^{-1} \, d\lambda - \frac{1}{2\pi i} \int_\Gamma (\lambda I - B)^{-1} \, d\lambda \right\|$$
$$\le K_1 \|B - A\|$$

for every transformation B sufficiently close to A (cf. the verification of stability of a direct sum of root subspaces in the beginning of Section 15.2). Hence

$$\sup_{\mathcal{M} \in \mathrm{Inv}(B)} \inf_{\mathcal{N} \in \mathrm{Inv}(A)} \theta(\mathcal{M}, \mathcal{N}) \le K_1 \|B - A\| \qquad (15.7.5)$$

for all such B. In view of (15.7.4) and (15.7.5), $\mathrm{Inv}(A)$ is Lipschitz stable in metric.

Conversely, if A has less than n distinct eigenvalues, then by Theorem 15.5.1 there exists an A-invariant subspace that is not Lipschitz stable. Then clearly $\mathrm{Inv}(A)$ cannot be Lipschitz stable in metric. $\square$

15.8 STABILITY OF [A B]-INVARIANT SUBSPACES

In this section we treat the stability of $[A \ B]$-invariant subspaces. In view of the important part they play in our applications (see Section 17.7), the reader can anticipate subsequent applications of this material.

Let $A: \mathbb{C}^n \to \mathbb{C}^n$ and $B: \mathbb{C}^n \to \mathbb{C}^n$ be linear transformations. Recall from Chapter 6 that a subspace $\mathcal{M} \subset \mathbb{C}^n$ is called $[A \ B]$ invariant if there exists a transformation $F: \mathbb{C}^n \to \mathbb{C}^m$ such that $(A + BF)\mathcal{M} \subset \mathcal{M}$ (actually, this is a property that is equivalent to the definition of an $[A \ B]$-invariant subspace, as proved in Theorem 6.1.1). *We restrict our attention to the case most important in applications, when the pair (A, B) is a full-range pair*; thus

$$\sum_{j=0}^{\infty} \mathrm{Im}(A^j B) = \mathbb{C}^n$$

It turns out that, in contrast with the case of invariant subspaces for a transformation, every $[A \ B]$-invariant subspace is stable and, moreover, the stability is understood in the Lipschitz sense. More exactly, we have the following theorem.

Theorem 15.8.1

Let $A: \mathbb{C}^n \to \mathbb{C}^n$ and $B: \mathbb{C}^m \to \mathbb{C}^n$ be a full-range pair of transformations. Then for every $[A \ B]$-invariant subspace $\mathcal{M}$ there exist positive constants ϵ and K such that, for every pair of transformations $A': \mathbb{C}^n \to \mathbb{C}^n$ and $B': \mathbb{C}^m \to \mathbb{C}^n$, with

$$\|A - A'\| + \|B - B'\| < \epsilon$$

there exists an $[A' \ B']$-invariant subspace $\mathcal{M}'$ satisfying

$$\theta(\mathcal{M}', \mathcal{M}) \le K(\|A - A'\| + \|B - B'\|) \qquad (15.8.1)$$

Proof. Let $F: \mathbb{C}^n \to \mathbb{C}^m$ be such that $(A + BF)\mathcal{M} \subset \mathcal{M}$. Write $A + BF$

and B as block matrices with respect to the decomposition $\mathbb{C}^n = \mathcal{M} \dotplus \mathcal{N}$, where $\mathcal{N}$ is some direct complement to $\mathcal{M}$:

$$A + BF = \begin{bmatrix} A_{11} & A_{12} \\ 0 & A_{22} \end{bmatrix}, \quad B = \begin{bmatrix} B_1 \\ B_2 \end{bmatrix}$$

We claim that (A_{22}, B_2) is a full-range pair. Indeed, since (A, B) is a full-range pair, so is $(A + BF, B)$ (Lemma 6.3.1). Now for every $x = x_\mathcal{M} + x_\mathcal{N} \in \mathbb{C}^n$ with $x_\mathcal{M} \in \mathcal{M}$, $x_\mathcal{N} \in \mathcal{N}$ we have

$$(A + BF)^i Bx \in A_{22}^i B_2 x_\mathcal{N} + \mathcal{M}$$

Hence in view of the full-range property of $(A + BF, B)$ we find that

$$\text{Span}\{A_{22}^i B_2 x_\mathcal{N} \mid x_\mathcal{N} \in \mathcal{N} ; \quad i = 0, 1, \ldots\} = \mathcal{N}$$

This implies the full-range property of (A_{22}, B_2).

We appeal to the spectral assignment theorem (Theorem 6.5.1). According to this theorem, there exists a transformation $G: \mathcal{N} \to \mathbb{C}^m$ such that

$$\sigma(A_{22} + B_2 G) \cap \sigma(A_{11}) = \emptyset \qquad (15.8.2)$$

Put $F_0 = F + \tilde{G}$, where the transformation $\tilde{G}: \mathbb{C}^n \to \mathbb{C}^m$ is defined by the properties that $\tilde{G}x = 0$ for all $x \in \mathcal{M}$ and $\tilde{G}x = Gx$ for all $x \in \mathcal{N}$. Clearly

$$A + BF_0 = \begin{bmatrix} A_{11} & A_{12} + B_1 G \\ 0 & A_{22} + B_2 G \end{bmatrix}$$

Condition (15.8.2) ensures that $\mathcal{M}$ is a spectral invariant subspace for $A + BF_0$. By Theorem 15.5.1, $\mathcal{M}$ is Lipschitz stable [as an $(A + BF_0)$-invariant subspace]. So there exist constants $\epsilon', K' > 0$ such that every transformation $H: \mathbb{C}^n \to \mathbb{C}^n$ with $\|A + BF_0 - H\| < \epsilon'$ has an invariant subspace $\mathcal{M}'$ such that

$$\theta(\mathcal{M}, \mathcal{M}') \le K' \|A + BF_0 - H\|$$

It remains to choose ϵ in such a way that

$$\|A - A'\| + \|B - B'\| < \epsilon \quad \text{implies} \quad \|(A + BF_0) - (A' + B'F_0)\| < \epsilon'$$

and put $K = K' \max(1, \|F_0\|)$ to ensure (15.8.1). $\square$

We emphasize that the full-range property of (A, B) is crucial in Theorem 15.8.1. Indeed, in the extreme case when $B = 0$ the $[A \; B]$-invariant subspaces coincide with A-invariant subspaces and, in general, not every A-invariant subspace is Lipschitz stable.

The proof of Theorem 15.8.1 reveals some additional information about the stability of $[A \; B]$-invariant subspaces:

Corollary 15.8.2

Let $A: \mathbb{C}^n \to \mathbb{C}^n$ and $B: \mathbb{C}^m \to \mathbb{C}^n$ be a full-range pair of transformations, and let $\mathcal{M}$ be an $[A \; B]$-invariant subspace. Then for every transformation $F: \mathbb{C}^n \to \mathbb{C}^m$ such that $(A + BF)\mathcal{M} \subset \mathcal{M}$ and every direct complement $\mathcal{N}$ to $\mathcal{M}$ in $\mathbb{C}^n$ there exists positive constants K and ϵ with the property that to any pair of transformations $A': \mathbb{C}^n \to \mathbb{C}^n$, $B': \mathbb{C}^m \to \mathbb{C}^n$ with $\|A - A'\| + \|B - B'\| < \epsilon$ there corresponds a transformation $F': \mathbb{C}^n \to \mathbb{C}^m$ with $\mathrm{Ker}\, F' \supset \mathcal{M}$, and a subspace $\mathcal{M}' \subset C^n$, such that $(A' + B'(F + F'))\mathcal{M}' \subset \mathcal{M}'$ and

$$\theta(\mathcal{M}, \mathcal{M}') \le K(\|A - A'\| + \|B - B'\|)$$

A dual version of Theorem 15.8.1 also holds. Namely, given a null kernel pair of transformations $G: \mathbb{C}^n \to \mathbb{C}^m$ and $A: \mathbb{C}^n \to \mathbb{C}^n$, every $\begin{bmatrix} A \\ G \end{bmatrix}$-invariant subspace is Lipschitz stable in the above sense. The proof can be obtained by using Theorem 15.8.1 and the fact that a subspace $\mathcal{M}$ is $\begin{bmatrix} A \\ G \end{bmatrix}$ invariant if and only if its orthogonal complement is $[A^* \; G^*]$ invariant. We leave it to the reader to state and prove this dual version of Corollary 15.8.2.

15.9 STABLE INVARIANT SUBSPACES FOR REAL TRANSFORMATIONS

Let $A: \mathbb{R}^n \to \mathbb{R}^n$ be a transformation. The definition of stable invariant subspaces of A is analogous to that for transformations from $\mathbb{C}^n$ to $\mathbb{C}^n$. Namely, an A-invariant subspace $\mathcal{M} \subset \mathbb{R}^n$ is called *stable* if for every $\epsilon > 0$ there exists a $\delta > 0$ such that any transformation $B: \mathbb{R}^n \to \mathbb{R}^n$ with $\|B - A\| < \delta$ has an invariant subspace $\mathcal{N}$ with $\theta(\mathcal{M}, \mathcal{N}) < \epsilon$. However, it turns out that, in contrast with the complex case, the classes of stable and of isolated invariant subspaces no longer coincide. More exactly, every stable invariant subspace is isolated, but, in general, not every isolated invariant subspace is stable.

To describe the stable invariant subspaces of real transformations, we start with several basic particular cases.

Lemma 15.9.1

Let $A: \mathbb{R}^n \to \mathbb{R}^n$ be a transformation such that $\sigma(A)$ consists of either exactly one real eigenvalue or exactly one pair of nonreal eigenvalues. Let the geometric multiplicity (multiplicities) be greater than one in either case. Then there is no nontrivial stable A-invariant subspaces.

The proof of this lemma is similar to the proof of Lemma 15.2.5.

Lemma 15.9.2

Assume that n is odd and the transformation $A: \mathbb{R}^n \to \mathbb{R}^n$ has exactly one eigenvalue (which is real) and the geometric multiplicity of this eigenvalue is one. Then each A-invariant subspace is stable.

Proof. As n is odd, every transformation $X: \mathbb{R}^n \to \mathbb{R}^n$ has an invariant subspace of every dimension k for $1 \le k \le n - 1$ (this follows from the real Jordan form for X, because X must have a real eigenvalue). Arguing as in the proof of Theorem 15.2.3, one proves that for every $\epsilon > 0$ there exists a $\delta > 0$ such that, if B is a transformation with $\|B - A\| < \delta$ and $\mathcal{M}$ is a k-dimensional B-invariant subspace, there exists a k-dimensional A-invariant subspace $\mathcal{N}$ with $\theta(\mathcal{M}, \mathcal{N}) < \epsilon$. Since A is unicellular, this subspace $\mathcal{N}$ is unique, and its stability follows. $\square$

Lemma 15.9.3

Let n be even, and $A: \mathbb{R}^n \to \mathbb{R}^n$ have exactly one real eigenvalue. Let its geometric multiplicity be one. Then the even dimensional A-invariant subspaces are stable and the odd dimensional A-invariant subspaces are not stable.

Proof. If k is even, then the stability of the k-dimensional A-invariant subspace (which is unique) is proved in the same way as Lemma 15.9.2, using the existence of a k-dimensional invariant subspace for every transformation $X: \mathbb{R}^n \to \mathbb{R}^n$.

Now let $\mathcal{M}$ be a k-dimensional A-invariant subspace where k is odd. Without loss of generality we can assume $A = J_n(0)$. For every positive ϵ, the transformation $A(\epsilon) = S(\epsilon) + A$, where $S(\epsilon)$ has ϵ in the entries $(2, 1)$, $(4, 3), \ldots, (n - 2, n - 3)$, $(n, n - 1)$, and zeros in all other entries, has no real eigenvalues. Hence $A(\epsilon)$ has no k-dimensional invariant subspaces, so $\theta(\mathcal{M}, \mathcal{N}) \ge 1$ for every $A(\epsilon)$-invariant subspace $\mathcal{N}$ (Theorem 13.1.2). Therefore, $\mathcal{M}$ is not stable. $\square$

Lemma 15.9.4

Assume that $A: \mathbb{R}^n \to \mathbb{R}^n$ has exactly one pair of nonreal eigenvalues $\alpha \pm i\beta$, and their geometric multiplicity is one. Then every A-invariant subspace is stable.

Proof. Using the real Jordan form of A, we can assume that

$$
A = \begin{bmatrix} K & I & 0 & \cdots & 0 & 0 \\ 0 & K & I & \cdots & 0 & 0 \\ \vdots & \vdots & \vdots & & \vdots & \vdots \\ 0 & 0 & 0 & \cdots & K & I \\ 0 & 0 & 0 & \cdots & 0 & K \end{bmatrix}, \qquad K = \begin{bmatrix} \alpha & \beta \\ -\beta & \alpha \end{bmatrix}
$$

(In particular, n is even.) Theorem 12.2.4 shows that the lattice of A-invariant subspaces is a chain; so for every even integer k with $0 \le k \le n$, there exists exactly one A-invariant subspaces of dimension k. Also, there exists an $\epsilon > 0$ such that any transformation B with $\|B - A\| < \epsilon$ has no real eigenvalues. [Indeed, for a suitable ϵ all the eigenvalues of B will be in the union of two discs $\{\lambda \in \mathbb{C} \mid |\lambda - (\alpha \pm i\beta)| < (\beta/2)\}$ that do not intersect the real axis.] Now one can use the proof of Lemma 15.9.2. $\square$

Now we are prepared to handle the general case of a transformation $A: \mathbb{R}^n \to \mathbb{R}^n$. Let $\lambda_1, \dots, \lambda_r$ be all the distinct real eigenvalues of A, and let $\alpha_1 + i\beta_1, \dots, \alpha_s + i\beta_s$ be all the distinct eigenvalues of A in the open upper half plane (so α_i are real and β_i are positive). We have

$$\mathbb{R}^n = \mathcal{R}_{\lambda_1}(A) \dotplus \cdots \dotplus \mathcal{R}_{\lambda_r}(A) \dotplus \mathcal{R}_{\alpha_1 \pm i\beta_1}(A) \dotplus \cdots \dotplus \mathcal{R}_{\alpha_s \pm i\beta_s}(A)$$

For every A-invariant subspace $\mathcal{N}$ we also have

$$\mathcal{N} = (\mathcal{N} \cap \mathcal{R}_{\lambda_1}(A)) \dotplus \cdots \dotplus (\mathcal{N} \cap \mathcal{R}_{\lambda_r}(A)) \dotplus (\mathcal{N} \cap \mathcal{R}_{\alpha_1 \pm i\beta_1}(A)) \dotplus \cdots$$
$$\dotplus (\mathcal{N} \cap \mathcal{R}_{\alpha_s \pm i\beta_s}(A))$$

(see Theorem 12.2.1). In this notation we have the following general result that describes all stable A-invariant subspaces.

Theorem 15.9.5

Let A be a transformation on $\mathbb{R}^n$. The A-invariant subspace $\mathcal{N}$ is stable if and only if all the following properties hold: (a) $\mathcal{N} \cap \mathcal{R}_{\lambda_j}(A)$ *is an arbitrary even dimensional A-invariant subspace of $\mathcal{R}_{\lambda_j}(A)$ whenever the algebraic multiplicity of λ_j is even and the geometric multiplicity of λ_j is 1;* (b) $\mathcal{N} \cap \mathcal{R}_{\lambda_j}(A)$ *is an arbitrary A-invariant subspace of $\mathcal{R}_{\lambda_j}(A)$ whenever the algebraic multiplicity of λ_j is odd and the geometric multiplicity of λ_j is 1;* (c) $\mathcal{N} \supset \mathcal{R}_{\lambda_j}(A)$, or $\mathcal{N} \cap \mathcal{R}_{\lambda_j}(A) = \{0\}$ *whenever λ_j has geometric multiplicity at least 2;* (d) $\mathcal{N} \cap \mathcal{R}_{\alpha_j \pm i\beta_j}(A)$ *is an arbitrary A-invariant subspace of $\mathcal{R}_{\alpha_j \pm i\beta_j}(A)$ whenever the geometric multiplicity of $\alpha_j + i\beta_j$ is 1;* (e) $\mathcal{N} \supset \mathcal{R}_{\alpha_j \pm i\beta_j}(A)$ *or $\mathcal{N} \cap \mathcal{R}_{\alpha_j \pm \beta_j(A)} = \{0\}$ whenever $\alpha_j + i\beta_j$ has geometric multiplicity of at least 2.*

Proof. As in Lemma 15.3.2 one proves that $\mathcal{N}$ is stable if and only if each intersection $\mathcal{N} \cap \mathcal{R}_{\lambda_j}(A)$ is stable as an $A|_{\mathcal{R}_{\lambda_j}(A)}$-invariant subspace, and each intersection $\mathcal{N} \cap \mathcal{R}_{\alpha_j \pm i\beta_j}(A)$ is stable as an $A|_{\mathcal{R}_{\alpha_j \pm i\beta_j(A)}}$-invariant subspace. Now apply Lemmas 15.9.1–15.9.4. $\square$

Comparing Theorem 15.9.5 with Theorem 14.6.5, we obtain the following corollary.

Corollary 15.9.6

For a transformation $A: \mathbb{R}^n \to \mathbb{R}^n$, every stable A-invariant subspace is iso-lated. Conversely, every isolated A-invariant subspace is stable if and only if A has no real eigenvalues with even algebraic multiplicity and geometric multiplicity 1.

We pass now to Lipschitz stable invariant subspaces for real transfor-mations. The definition of Lipschitz stability is the same as for transfor-mations on $\mathbb{C}^n$. Clearly, every Lipschitz stable invariant subspace is stable. Also, for a transformation $A: \mathbb{R}^n \to \mathbb{R}^n$ every root subspace $\mathcal{R}_\lambda(A)$ corres-ponding to a real eigenvalue λ of A, as well as every root subspace $\mathcal{R}_{\alpha \pm i\beta}(A)$ corresponding to a pair $\alpha \pm i\beta$ of nonreal eigenvalues of A, is a Lipschitz stable A-invariant subspace. Moreover, every spectral subspace for A (i.e., a sum of root subspaces) is also a Lipschitz stable A-invariant subspace. As in the complex case, these are all Lipschitz stable subspaces:

Theorem 15.9.7

For a transformation $A: \mathbb{R}^n \to \mathbb{R}^n$ and an A-invariant subspace $\mathcal{N} \subset \mathbb{R}^n$, the following statements are equivalent: (a) $\mathcal{M}$ is Lipschitz stable; (b) $\mathcal{M} = \mathcal{R}_{\lambda_1}(A) \dotplus \cdots \dotplus \mathcal{R}_{\lambda_r}(A) \dotplus \mathcal{R}_{\alpha_1 \pm i\beta_1}(A) \dotplus \cdots \dotplus \mathcal{R}_{\alpha_s \pm i\beta_s}(A)$ for some distinct real eigenvalues $\lambda_1, \ldots, \lambda_r$ of A and some distinct eigenvalues $\alpha_1 + i\beta_1, \ldots, \alpha_s + i\beta_s$ in the open upper half plane (here terms $\mathcal{R}_{\lambda_j}(A)$ or terms $\mathcal{R}_{\alpha_j \pm i\beta_j}(A)$, or even both (in which case $\mathcal{M}$ is interpreted as the zero subspace) may be absent); (c) for every $\epsilon > 0$ small enough there exists a $\delta > 0$ such that every transformation $B: \mathbb{R}^n \to \mathbb{R}^n$ for which $\|B - A\| < \delta$ has a unique invariant subspace $\mathcal{N}$ for which $\theta(\mathcal{M}, \mathcal{N}) < \epsilon$.

Proof. As in Lemma 15.3.2, one proves that $\mathcal{M}$ is Lipschitz stable if and only if for every real eigenvalue λ of A the intersection $\mathcal{M} \cap \mathcal{R}_\lambda(A)$ is Lipschitz stable as an $A|_{\mathcal{R}_\lambda(A)}$-invariant subspace and for every nonreal eigenvalue $\alpha + i\beta$ of A $\mathcal{M} \cap \mathcal{R}_{\alpha \pm i\beta}(A)$ is Lipschitz stable as an $A|_{\mathcal{R}_{\alpha \pm i\beta}(A)}$-invariant subspace.

Let us prove the equivalence (a)$\Leftrightarrow$(b). In view of the above remark, we can assume that A has either exactly one real eigenvalue or exactly one pair of nonreal eigenvalues. By Theorem 15.9.6 we have only to prove that the transformations represented by the matrices

$$
A_1 = \begin{bmatrix} \lambda & 1 & 0 & \cdots & 0 \\ 0 & \lambda & 1 & \cdots & 0 \\ \vdots & \vdots & & \ddots & \vdots \\ & & & & 1 \\ 0 & 0 & & \cdots & \lambda \end{bmatrix}, \qquad \lambda \in \mathbb{R}
$$

and

$$A_2 = \begin{bmatrix} K & I_2 & 0 & \cdots & 0 \\ 0 & K & I_2 & \cdots & 0 \\ \vdots & & & \ddots & \vdots \\ & & & & I_2 \\ 0 & 0 & & \cdots & K \end{bmatrix}, \qquad K = \begin{bmatrix} \sigma & \tau \\ -\tau & \sigma \end{bmatrix}, \qquad \delta, \tau \in \mathbb{R}, \qquad \tau \neq 0$$

have no nontrivial Lipschitz stable invariant subspaces. For A_1 one shows this as in the proof of Theorem 15.5.1. Consider now A_2. By a direct computation one shows that

$$A_2 = S[J_{n/2}(\sigma - i\tau) \oplus J_{n/2}(\sigma + i\tau)]S^{-1}$$

where n is the size of A_2 and

$$S = \begin{bmatrix} 1 & 0 & \cdots & 0 & 1 & 0 & \cdots & 0 \\ -i & 0 & \cdots & 0 & i & 0 & \cdots & 0 \\ 0 & 1 & \cdots & 0 & 0 & 1 & \cdots & 0 \\ 0 & -i & \cdots & 0 & 0 & i & \cdots & 0 \\ \vdots & \vdots & & \vdots & \vdots & \vdots & & \vdots \\ 0 & 0 & \cdots & 1 & 0 & 0 & \cdots & 1 \\ 0 & 0 & \cdots & -i & 0 & 0 & \cdots & i \end{bmatrix}$$

(For convenience, note that

$$S^{-1} = \frac{1}{2i} \begin{bmatrix} i & -1 & 0 & 0 & \cdots & 0 & 0 \\ 0 & 0 & i & -1 & \cdots & 0 & 0 \\ \hdotsfor{7} \\ 0 & 0 & 0 & 0 & & i & -1 \\ i & 1 & 0 & 0 & \cdots & 0 & 0 \\ 0 & 0 & i & 1 & \cdots & 0 & 0 \\ \hdotsfor{7} \\ 0 & 0 & 0 & 0 & \cdots & i & 1 \end{bmatrix})$$

Moreover, denoting by T the $n \times n$ matrix that has 1 in the entries $(n/2, 1)$ and $(n, n/2 + 1)$ and zeros elsewhere, we have (for $\alpha \in \mathbb{R}$)

$$A_2(\alpha) \overset{\text{def}}{=} \begin{bmatrix} K & I & 0 & \cdots & 0 \\ 0 & K & I & \cdots & 0 \\ \vdots & \vdots & \vdots & & \vdots \\ 0 & 0 & 0 & \cdots & I \\ \alpha I & 0 & 0 & \cdots & K \end{bmatrix} = S^{-1}[(J_{n/2}(\sigma - i\tau) \oplus J_{n/2}(\sigma + i\tau)) + \alpha T]S$$

$$(15.9.1)$$

Now the proof of Theorem 15.5.1 shows that the only candidates for nontrivial Lipschitz invariant subspaces for A_2 are $S^{-1}(\text{Span}\{e_1, \ldots, e_{n/2}\})$ and $S^{-1}(\text{Span}\{e_{n/2+1}, \ldots, e_n\})$. But since these subspaces are not real (i.e., cannot be obtained from subspaces in $\mathbb{R}^n$ by complexification), A_2 has no nontrivial Lipschitz invariant subspaces.

The implication (b) $\Rightarrow$ (c) is proved as in the proof of Theorem 15.5.1. To prove the converse implication, observe that, as we have seen in the proof of Theorem 15.5.1, it is sufficient to show that for any A_2-invariant subspace $\mathcal{M}(\subset \mathbb{R}^n)$ of dimension q $(0 < q < n)$ the number of q-dimensional invariant subspaces $\mathcal{N}$ of $A_2(\alpha)$ such that

$$\theta(\mathcal{M}, \mathcal{N}) \le C\alpha^{2/n} \tag{15.9.2}$$

is at least $\binom{n/2}{q/2}$ (Here α is positive and sufficiently close to zero, and C is a positive constant depending on q and n only.) Observe that q, as well as n, must be even. Using formula (15.9.1) and arguing as in the proof of Theorem 15.5.1, we find that for any choice of different complex numbers $\epsilon_1, \ldots, \epsilon_{q/2}$ with $\epsilon_i^{n/2} = 1$, $i = 1, \ldots, q/2$, the subspace $\mathcal{N}$ spanned by the columns of the real matrix

$$S\begin{bmatrix} V_1 & \cdots & V_{q/2} \\ \bar{V}_1 & \cdots & \bar{V}_{q/2} \end{bmatrix}, \qquad V_i = \langle 1, \epsilon_i, \ldots, \epsilon_i^{n/2-1} \rangle, \qquad i = 1, \ldots, q/2$$

satisfies (15.9.2). □

15.10 PARTIAL MULTIPLICITIES OF CLOSE LINEAR TRANSFORMATIONS

In this chapter we have studied up to now the behaviour of invariant subspaces under perturbations of the given linear transformation. We have found that certain information about the transformation (e.g., its spectral invariant subspaces) remains stable under small changes in the transformation. Here we study the corresponding problem of stability of the partial multiplicities of transformations.

Given a transformation $A: \mathbb{C}^n \to \mathbb{C}^n$, denote by $k_1(\lambda, A), \ldots, k_p(\lambda, A)$ the partial multiplicities of A corresponding to its eigenvalue λ, and put $k_r(\lambda, A) = 0$ for $r > p$ (here p is the geometric multiplicity of the eigenvalue λ). For a closed contour Γ in the complex plane that does not intersect the spectrum of A, let

$$k_j(\Gamma, A) = \sum_{k=1}^{r} k_j(\lambda_k, A), \qquad j = 1, 2, \ldots$$

where $\lambda_1, \ldots, \lambda_r$ are all the distinct eigenvalues of A inside Γ. If there are no eigenvalues of A inside Γ, put formally $k_j(\Gamma, A) = 0$ for $j = 1, 2, \ldots$.

Theorem 15.10.1

Given a transformation $A: \mathbb{C}^n \to \mathbb{C}^n$ and a closed contour Γ with $\Gamma \cap \sigma(A) = \emptyset$, there exists an $\epsilon > 0$ such that any transformation $B: \mathbb{C}^n \to \mathbb{C}^n$ with $\|B - A\| < \epsilon$ has no eigenvalues on Γ and satisfies the inequalities

$$\sum_{j=s}^{\infty} k_j(\Gamma, B) \leq \sum_{j=s}^{\infty} k_j(\Gamma, A); \qquad s = 2, 3, \ldots \qquad (15.10.1)$$

and the equality

$$\sum_{j=1}^{\infty} k_j(\Gamma, B) = \sum_{j=1}^{\infty} k_j(\Gamma, A) \qquad (15.10.2)$$

Proof. Let $n(\Gamma, f)$ be the number of zeros (counting multiplicities) of a scalar polynomial f inside Γ. (It is assumed that f does not have zeros on Γ.) For $s = 1, 2, \ldots, n$, we have the relations

$$\sum_{j=1}^{s} k_{n+1-j}(\Gamma; A) = n(\Gamma; f_s) \qquad (15.10.3)$$

where $f_s(\lambda)$ is the greatest common divisor of all determinants of $s \times s$ submatrices in $\lambda I - A$. (Here and in the sequel all transformations on $\mathbb{C}^n$ are regarded as $n \times n$ matrices in a fixed basis in $\mathbb{C}^n$.) Indeed, (15.10.3) follows from Theorem A.4.3 (in the appendix).

Consider the Smith form of $\lambda I - A$ (see the appendix):

$$\lambda I - A = F(\lambda) \operatorname{diag}[a_1(\lambda), a_2(\lambda), \ldots, a_n(\lambda)] G(\lambda)$$

where $F(\lambda)$ and $G(\lambda)$ are $n \times n$ matrix polynomials with constant nonzero determinant, and $a_1(\lambda), \ldots, a_n(\lambda)$ are scalar polynomials such that $a_i(\lambda)$ is divisible by $a_{i-1}(\lambda)$ for $i = 2, \ldots, n$. By the Binet–Cauchy formula (Theorem A.2.1) $f_s(\lambda)$ coincides with the greatest common divisor of all determinants of $s \times s$ submatrices in $\operatorname{diag}[a_1(\lambda), \ldots, a_n(\lambda)]$, and this is equal to the product $a_1(\lambda) \cdots a_s(\lambda)$ in view of the properties of $a_1(\lambda), \ldots, a_n(\lambda)$. So for $s = 1, 2, \ldots, n$

$$\sum_{j=n+1-s}^{\infty} k_j(\Gamma; A) = \sum_{i=1}^{s} k_{n+1-i}(\Gamma; A) = n(\Gamma; f_s) = n(\Gamma; a_1(\lambda) \cdots a_s(\lambda))$$

$$(15.10.4)$$

Now let $\epsilon > 0$ be so small that if $\|B - A\| < \epsilon$, the determinant of the top $s \times s$ submatrix in $F(\lambda)^{-1}(\lambda I - B)G(\lambda)^{-1}$ has exactly $n(\Gamma; a_1(\lambda) \cdots a_s(\lambda))$

zeros in Γ. [Such an ϵ exists by Rouché's theorem in the theory of functions of a complex variable; e.g., see Marsden (1973).] Denote by $h_s(\lambda)$ the greatest common divisor of determinants of all $s \times s$ submatrices of $F(\lambda)^{-1}$ $(\lambda I - B)G(\lambda)^{-1}$. Then $h_s(\lambda)$ coincides (again by the Binet–Cauchy formula) with the greatest common divisor of determinants of all $s \times s$ submatrices in $\lambda I - B$. When $\|B - A\| < \epsilon$ we obviously have $n(\Gamma; a_1(\lambda) \cdots a_s(\lambda)) \geq n(\Gamma; h_s)$. Combining this inequality with (15.10.4) and using (15.10.3) with A replaced by B, we find that, for $s = 1, \ldots, n$

$$\sum_{j=n+1-s}^{\infty} k_j(\Gamma; A) \geq \sum_{j=n+1-s}^{\infty} k_j(\Gamma; B)$$

As the inequalities (15.10.1) with $s > n$ are trivial, (15.10.1) is proved. Further, $\sum_{j=1}^{\infty} k_j(\Gamma; A)$ coincides with the number of zeros of $\det(\lambda I - A)$ inside Γ, counting multiplicities. This number does not change after sufficiently small perturbations of A, again by the Rouché theorem. $\square$

The following question arises in connection with Theorem 15.10.1: Are the restrictions (15.10.1) and (15.10.2) imposed on the transformation B sufficient for existence of such a B arbitrarily close to A? Before we answer this question (it turns out that the answer is yes), let us introduce a convenient notation for the partial multiplicities of a transformation.

Given a transformation $A: \math01{C}^n \to \math01{C}^n$, let

$$\{s; r_1, r_2, \ldots, r_s; m_{11}, \ldots, m_{1r_1}; m_{21}, \ldots, m_{2r_2}; \ldots; m_{s1}, \ldots, m_{sr_s}\}$$
$$(15.10.5)$$

be an ordered sequence, where s is the number of distinct eigenvalues of A, and the ith eigenvalue has geometric multiplicity r_i and partial multiplicities $m_{1i}, \ldots, m_{ir_i}$. So $\sum_{i=1}^{s} \sum_{j=1}^{r_i} m_{ij} = n$. The order in (15.10.5) is determined by the following properties: (a) $r_1 \geq r_2 \cdots \geq r_s$; (b) if $r_i = r_{i+1}$, then

$$\sum_{j=1}^{r_i} m_{ij} \geq \sum_{j=1}^{r_i} m_{i+1,j} \tag{15.10.6}$$

(c) if $r_i = r_{i+1}$ and equality holds in (15.10.6), then

$$\sum_{j=1}^{k} m_{ij} \geq \sum_{j=1}^{k} m_{i+1,j}, \qquad k = 1, 2, \ldots, r_i - 1$$

We say that (15.10.5) is the *Jordan structure sequence* of A. Denote by Φ the finite set of all ordered sequences of positive integers (15.10.5) such that properties (a)–(c) hold and $\sum_{i=1}^{s} \sum_{j=1}^{r_i} m_{ij} = n$ (here n is fixed).

Given the sequence $\Omega \in \Phi$ as in (15.10.5), for every nonempty subset $\Delta \subset \{1, \ldots, s\}$ define

$$k_j(\Omega; \Delta) = \sum_{p \in \Delta} m_{pj}, \qquad j = 1, 2, \ldots$$

(m_{pj} is interpreted as zero for $j > r_p$). Now we have the following.

Theorem 15.10.2

Let $A: \mathbb{C}^n \to \mathbb{C}^n$ be a transformation with s distinct eigenvalues and Jordan structure sequence Ω. Then, given a sequence

$$\Omega' = \{s'; r'_1, \ldots, r'_{s'}; m'_{11}, \ldots, m'_{1r'_1}; \ldots; m'_{s'1}, \ldots, m'_{s'r'_{s'}}\} \in \Phi$$

there exists a sequence of transformations on $\mathbb{C}^n$, say, $\{B_m\}_{m=1}^\infty$ that converges to A and has a common Jordan structure sequence Ω' if and only if there is a partition $\{1, 2, \ldots, s'\}$ into s disjoint nonempty sets $\Delta_1, \ldots, \Delta_s$ such that the following inqualities hold:

$$\sum_{j=1}^t k_j(\Omega; \{p\}) \le \sum_{j=1}^t k_j(\Omega'; \Delta_p); \qquad t = 1, 2, \ldots; \qquad p = 1, \ldots, s$$

$$(15.10.7)$$

$$\sum_{j=1}^\infty k_j(\Omega; \{p\}) = \sum_{j=1}^\infty k_j(\Omega'; \Delta_p); \qquad p = 1, \ldots, s \qquad (15.10.8)$$

Informally, if $\lambda_1, \ldots, \lambda_s$ are the distinct eigenvalues of A ordered as in Ω, and if $\lambda_{1m}, \ldots, \lambda_{s'm}$ are the distinct eigenvalues of B_m ordered as in Ω', then the eigenvalues $\{\lambda_{jm}\}_{j \in \Delta_p}$ cluster around λ_p, for $p = 1, \ldots, s$.

Proof of Theorem 15.10.2. The necessity of conditions (15.10.7) and (15.10.8) follows from Theorem 15.10.1.

To prove sufficiency, we can restrict our attention to the case $s = 1$. Let λ_0 be the eigenvalue of A, and write $\tilde{m}_j = \sum_{p=1}^{s'} m'_{pj}$ (recall that $r'_1 = \max\{r'_1, \ldots, r'_s\}$ and that m'_{pj} is zero by definition if $j > r'_p$). We then have the inequalities

$$\sum_{j=1}^t m_{ij} \le \sum_{j=1}^t \tilde{m}_j, \qquad t = 1, 2, \ldots$$

and the equality

$$\sum_{j=1}^\infty m_{1j} = \sum_{j=1}^\infty \tilde{m}_j$$

Now we construct a sequence $\{\tilde{B}_q\}_{q=1}^\infty$ converging to A such that λ_0 is the only eigenvalue of $\tilde{B}_q$ and, for each q, the Jordan structure sequence of $\tilde{B}_q$ is $\tilde{\Omega} = \{1; r'_1; \tilde{m}_1, \ldots, \tilde{m}_{r_1}\}$. Using induction on the number $\sum_{t=1}^\infty (\sum_{j=1}^t \tilde{m}_j -$

$\Sigma^t_{j=1} m_{1j}$), it is sufficient to consider only the case when, for some indices $l < q$, we have $\tilde{m}_l = m_{1l} + 1$, $\tilde{m}_q = m_{1q} - 1$, whereas $\tilde{m}_j = m_{1j}$ for $j \neq l, j \neq q$. Write

$$A = J_{m_1}(\lambda_0) \oplus \cdots \oplus J_{m_{1r_1}}(\lambda_0)$$

as a matrix in some Jordan basis for A. Let

$$\tilde{B}_q = A + \frac{1}{q} Q$$

where the matrix Q has all zero entries except for the entry in position $(m_{11} + \cdots + m_{1l}, m_{11} + \cdots + m_{1q})$ that is equal to 1. One verifies without difficulty that the partial multiplicities of $\tilde{B}_q$ are $\tilde{m}_l = m_{1l} + q$, $\tilde{m}_q = m_{1q} - 1$, and $\tilde{m}_j = m_{1j}$ for $j \neq l, q$.

Given a sequence $\{\tilde{B}_q\}^\infty_{q=1}$ converging to A such that $\sigma(\tilde{B}_q) = \{\lambda_0\}$ and the Jordan structure sequence of $\tilde{B}_q$ is $\tilde{\Omega}$ (for each q). For a fixed q, let

$$x_{11}, \ldots, x_{1,\tilde{m}_1}; x_{21}, \ldots, x_{2,\tilde{m}_2}; \ldots; x_{r_1,1}, \ldots, x_{r_1,\tilde{m}_{r_1}} \quad (15.10.9)$$

be a Jordan basis for $\tilde{B}_q$; in other words, $x_{j1}, \ldots, x_{j,\tilde{m}_j}$ is a Jordan chain for $\tilde{B}_q$ for $j = 1, \ldots, r_1'$. Let $\mu_1, \ldots, \mu_{s'}$ be distinct complex numbers; define the transformation $B_q(\mu_1, \ldots, \mu_{s'})$ by the requirement that in the basis (15.10.9) it has the matrix form

$$\tilde{B}_q + \mathrm{diag}[\mu_1 I_{m'_{11}}, \mu_2 I_{m'_{21}}, \ldots, \mu_{s'} I_{m'_{s'1}}, \mu_1 I_{m'_{12}}, \mu_2 I_{m'_{22}}, \ldots,$$

$$\mu_{s'} I_{m'_{s'2}}, \ldots, \mu_1 I_{m'_{1r'_1}}, \mu_2 I_{m'_{2r'_1}}, \ldots, \mu_{s'} I_{m'_{s'r'_1}}] \quad (15.10.10)$$

where I_i is the $i \times i$ unit matrix, and $\mu_i I_{m'_{ik}}$ does not appear in (15.10.10) if $k > r'_i$. Clearly, $B_q(\mu_1, \ldots, \mu_{s'})$ has the Jordan structure sequence Ω', and by suitable choice of μ_i values one can ensure that

$$\|B_q(\mu_1, \ldots, \mu_{s'}) - \tilde{B}_q\| \leq \frac{1}{q}$$

With this choice of μ_i values (which depend on q), put $B_m = B_m(\mu_1, \ldots, \mu_{s'})$ to satisfy the requirements of Theorem 15.10.2. □

15.11 EXERCISES

15.1 When are all invariant subspaces of the following transformations $A: \mathbb{C}^n \to \mathbb{C}^n$ (written as matrices in the standard orthonormal basis) stable?

(a) A is an upper triangular Toeplitz matrix.
(b) A is a circulant matrix.
(c) A is a companion matrix.

15.2   Describe all stable invariant subspaces for the classes (a), (b), and (c) in Exercise 15.1.

15.3   Describe all stable invariant subspaces of a block circulant matrix with blocks of size 2×2.

15.4   Show that any transformation $A: \mathbb{C}^n \to \mathbb{C}^n$ with rank $A \leq n - 2$ has a nonstable invariant subspace and identify it.

15.5   Prove that for every transformation A there exists a transformation B such that every invariant subspace of $A + \epsilon B$ is stable. Show that one can always ensure, in addition, that rank $B = n - 1$.

15.6   Give an example of a transformation $A: \mathbb{C}^n \to \mathbb{C}^n$ such that there is no transformation $B: \mathbb{C}^n \to \mathbb{C}^n$ with rank $B \leq n - 2$ such that, for some $\epsilon \in \mathbb{C}$, all invariant subspaces of $A + \epsilon B$ are stable.

15.7   Given transformations $A: \mathbb{C}^n \to \mathbb{C}^n$ and $B: \mathbb{C}^n \to \mathbb{C}^n$, an A-invariant subspace $\mathcal{L}$ will be called B *stable* if for every $\epsilon_0 > 0$ there exists $\delta_0 > 0$ such that each transformation $A + \delta B$, with $|\delta| < \delta_0$ has an invariant subspace $\mathcal{M}$ such that $\theta(\mathcal{L}, \mathcal{M}) < \epsilon_0$. Clearly, every stable A-invariant subspace is B stable for every B. Give an example of a B-stable A-invariant subspace that is not stable.

15.8   Show that if A and B commute, then there is a complete chain of B-stable A-invariant subspaces.

15.9   Give an example of transformations A and B with the property that an A-invariant subspace is stable if and only if it is B stable.

15.10   Show that an A-invariant subspace is stable if and only if it is B stable for every B.

15.11   Show that the set of all stable invariant subspace of a transformation $A: \mathbb{C}^n \to \mathbb{C}^n$ is a lattice. When is this lattice trivial, that is, when does it consist of $\{0\}$ and $\mathbb{C}^n$ only? When does this lattice coincide with $\text{Inv}(A)$?

15.12   Show that every stable invariant subspace is hyperinvariant. Is the converse true?

15.13   Prove that the transformation $A: \mathbb{C}^n \to \mathbb{C}^n$ has the following property if and only if A is nonderogatory: for every orthonormal basis $x_1, \ldots, x_n$ in which A has an upper triangular form and any $\epsilon > 0$ there exists a $\delta > 0$ such that any transformation $B: \mathbb{C}^n \to \mathbb{C}^n$ with $\|B - A\| < \delta$ has an upper triangular form in some orthonormal basis $y_1, \ldots, y_n$ that satisfies

$$\sum_{i=1}^{n} \|y_i - x_i\| < \epsilon$$

15.14 Let $A: \mathbb{R}^n \to \mathbb{R}^r$ and $B: \mathbb{R}^m \to \mathbb{R}^n$ be a full-range pair of real transformations. Show that every $[A \ B]$-invariant subspace is stable (in the class of real transformations and real subspaces). [*Hint*: Use the spectral assignment theorem for real transformations (Exercise 12.13).]

15.15 Let A be an upper triangular Toeplitz matrix. Find all possible partial multiplicities for upper triangular Toeplitz matrices that are arbitrarily close to A.

15.16 Let A and B be circulant matrices. Compute $\text{dist}(\text{Inv}(A), \text{Inv}(B))$.

Chapter Sixteen

Perturbations of Lattices of Invariant Subspaces with Restrictions on the Jordan Structure

In this chapter we study the behaviour of the lattice Inv(X) of all invariant subspaces of a matrix X, when X is perturbed within the class of matrices with fixed Jordan structure (i.e., with isomorphic lattices of invariant subspaces). A larger class of matrices with fixed Jordan structure corresponding to the eigenvalues of geometric multiplicity greater than 1 is also studied. For transformations A and B on $\mathbb{C}^n$, our main concern is the relationship of the distance between the lattices of invariant subspaces for A and B to $\|A - B\|$.

16.1 PRESERVATION OF JORDAN STRUCTURE AND ISOMORPHISM OF LATTICES

We start with a definition. Transformations $A, B: \mathbb{C}^n \to \mathbb{C}^n$ are said to have the *same Jordan structure* if they have the same number of distinct eigenvalues [so that we may write $\sigma(A) = \{\lambda_1, \ldots, \lambda_s\}$ and $\sigma(B) = \{\mu_1, \ldots, \mu_s\}$], and the eigenvalues can be ordered in such a way that the partial multiplicities of λ_i as an eigenvalue of A coincide with the partial multiplicities of μ_i as an eigenvalue of B, $i = 1, \ldots, s$.

Given a transformation A, denote by $J(A)$ the set of all transformations with the same Jordan structure as A. This structure is determined by the sequence of positive integers (which was also a useful tool in Section 15.10):

$$\{s; r_1, r_2, \ldots, r_s; m_{11}, \ldots, m_{1r_1}; m_{21}, \ldots, m_{2r_2}; \ldots; m_{s1}, \ldots, m_{sr_s}\}$$

$$(16.1.1)$$

where s is the number of distinct eigenvalues of A, and the ith eigenvalue has geometric multiplicity r_i and partial multiplicities $m_{i1}, \ldots, m_{ir_i}$. Thus $\Sigma_{i=1}^{s} \Sigma_{j=1}^{r_i} m_{ij} = n$. The parameters of this sequence are ordered in such a way that $r_1 \geq r_2 \geq \cdots \geq r_s$, and if $r_i = r_{i+1}$, then

$$\sum_{j=1}^{r_i} m_{ij} \geq \sum_{j=1}^{r_i} m_{i+1,j} \qquad (16.1.2)$$

and, furthermore, if $r_i = r_{i+1}$ and equality holds in inequality (16.1.2), the integers m_{ij} and $m_{i+1,j}$ are ordered in such a way that

$$\sum_{j=1}^{k} m_{ij} \geq \sum_{j=1}^{k} m_{i+1,j}, \qquad k = 1, 2, \ldots, r_i - 1$$

Clearly, the property of having the same Jordan structure induces an equivalence relation on the set of all transformations on $\mathbb{C}^n$. The number of equivalence classes under the relation is finite and is equal to the number of all different sequences of type (16.1.1) with the order properties described.

It is shown in the first theorem that transformations have the same Jordan structure if and only if they have isomorphic (or linearly isomorphic) lattices of invariant subspaces.

Let us define the notion of isomorphism of lattices. First, let $\mathscr{S}_1$ and $\mathscr{S}_2$ be two lattices of subspaces in $\mathbb{C}^n$. A map $\varphi: \mathscr{S}_1 \to \mathscr{S}_2$ is called a *lattice homomorphism* if $\varphi(\{0\}) = \{0\}$, $\varphi(\mathbb{C}^n) = \mathbb{C}^n$, and $\varphi(\mathcal{M} + \mathcal{N}) = \varphi(\mathcal{M}) + \varphi(\mathcal{N})$, $\varphi(\mathcal{M} \cap \mathcal{N}) = \varphi(\mathcal{M}) \cap \varphi(\mathcal{N})$ for every two subspaces $\mathcal{M}, \mathcal{N} \in \mathscr{S}_1$. Then a lattice homomorphism φ is called a *lattice isomorphism* if φ is one-to-one and onto; in this case the lattices $\mathscr{S}_1$ and $\mathscr{S}_2$ are said to be *isomorphic*. An example of a lattice isomorphism is provided by the following proposition.

Proposition 16.1.1

If $S: \mathbb{C}^n \to \mathbb{C}^n$ is an invertible transformation and $\mathscr{S}$ is a lattice of subspaces in $\mathbb{C}^n$, then $S\mathscr{S} = \{S\mathcal{M} \mid \mathcal{M} \in \mathscr{S}\}$ is also a lattice of subspaces and the correspondence $\varphi(\mathcal{M}) = S\mathcal{M}$ is a lattice isomorphism of $\mathscr{S}$ onto $S\mathscr{S}$.

Proof. The definition of φ ensures that φ is onto, and invertibility of S ensures that φ is one-to-one. Furthermore,

$$S(\mathcal{M} \cap \mathcal{N}) = S\mathcal{M} \cap S\mathcal{N} \qquad (16.1.3)$$

for any subspaces $\mathcal{M}$ and $\mathcal{N}$ in $\mathbb{C}^n$. Indeed, the inclusion $\subset$ in equation

(16.1.3) is evident. To prove the opposite inclusion, take $x \in S\mathcal{M} \cap S\mathcal{N}$, so $x = Sm = Sn$ for some $m \in \mathcal{M}$, $n \in \mathcal{N}$. As S is invertible, actually $m = n$ and $x \in S(\mathcal{M} \cap \mathcal{N})$. Finally, the equality $S(\mathcal{M} + \mathcal{N}) = S\mathcal{M} + S\mathcal{N}$ is evident. □

The lattice isomorphisms described in Proposition 16.1.1 are called *linear*. So the lattices $\mathcal{S}_1$ and $\mathcal{S}_2$ are called *linearly isomorphic* if there exists a transformation $S \colon \mathbb{C}^n \to \mathbb{C}^n$ (necessarily invertible) such that $\mathcal{L} \in \mathcal{S}_1$ if and only if $S\mathcal{L} \in \mathcal{S}_2$.

It is easy to provide examples of lattices of subspaces that are isomorphic but not linearly isomorphic. For instance, two chains of subspaces

$$\{0\} = \mathcal{M}_0 \subset \mathcal{M}_1 \subset \mathcal{M}_2 \subset \cdots \subset \mathcal{M}_{k-1} \subset \mathcal{M}_k = \mathbb{C}^n$$

and

$$\{0\} = \mathcal{L}_0 \subset \mathcal{L}_1 \subset \mathcal{L}_2 \subset \cdots \subset \mathcal{L}_{l-1} \subset \mathcal{L}_l = \mathbb{C}^n$$

are lattice isomorphic if and only if $k = l$ (it is assumed that $\mathcal{M}_i \neq \mathcal{M}_j$ for $i \neq j$ and $\mathcal{L}_i \neq \mathcal{L}_j$ for $i \neq j$). However, there exists an invertible matrix S such that $S\mathcal{M}_i = \mathcal{L}_i$ for $i = 1, \ldots, k$ if and only if $\dim \mathcal{M}_i = \dim \mathcal{L}_i$ for each i.

The following theorem shows, in particular, that for the lattices of all invariant subspaces isomorphism and linear isomorphism are the same.

Theorem 16.1.2

Let a transformation $A \colon \mathbb{C}^n \to \mathbb{C}^n$ be given. The following statements are equivalent for a transformation $B \colon \mathbb{C}^n \to \mathbb{C}^n$: (a) B has the same Jordan structure as A; (b) the lattices $\mathrm{Inv}(B)$ and $\mathrm{Inv}(A)$ are isomorphic; (c) the lattices $\mathrm{Inv}(B)$ and $\mathrm{Inv}(A)$ are linearly isomorphic.

Proof. Assume $B \in J(A)$. Let $\lambda_1, \ldots, \lambda_p$ and $\mu_1, \ldots, \mu_p$ be all the distinct eigenvalues of A and B, respectively, and let them be numbered so that the partial multiplicities of A and λ_j coincide with the partial multiplicities of B at μ_j for $j = 1, \ldots, p$. For a fixed j, let

$$x_{11}, \ldots, x_{1k_1}; x_{21}, \ldots, x_{2k_2}; \ldots; x_{q1}, \ldots, x_{qk_q}$$

be a Jordan basis in $\mathcal{R}_{\lambda_j}(A)$, and let

$$y_{11}, \ldots, y_{1,k_1}; y_{21}, \ldots, y_{2k_2}; \ldots; y_{q1}, \ldots, y_{q,k_q}$$

be a Jordan basis in $\mathcal{R}_{\mu_j}(B)$ (so $k_1, k_2, \ldots, k_q$ are the partial multiplicities of A at λ_j and of B at μ_j). Given an A-invariant subspace $\mathcal{M} \subset \mathcal{R}_{\lambda_j}(A)$ spanned by the vectors

$$v_1 = \sum_{r=1}^{q} \sum_{s=1}^{k_r} \alpha_{rs}^{(1)} x_{rs}, \ldots, v_l = \sum_{r=1}^{q} \sum_{s=1}^{k_r} \alpha_{rs}^{(l)} x_{rs}$$

[here $\alpha_{rs}^{(t)}$ are complex numbers], put

$$\psi_j(\mathcal{M}) = \text{Span}\{\mu_1, \ldots, \mu_l\}$$

where

$$u_t = \sum_{r=1}^{q} \sum_{s=1}^{k_r} \alpha_{rs}^{(t)} y_{rs}, \qquad t = 1, \ldots, l$$

Clearly, $\psi_j(\mathcal{M})$ is a B-invariant subspace that belongs to $\mathcal{R}_{\mu_j}(B)$. Now for any A-invariant subspace $\mathcal{M}$ put

$$\psi(\mathcal{M}) = \psi_1(\mathcal{M} \cap \mathcal{R}_{\lambda_1}(A)) \dotplus \cdots \dotplus \psi_p(\mathcal{M} \cap \mathcal{R}_{\lambda_p}(A))$$

It is easily seen that ψ is a desired isomorphism between $\text{Inv}(A)$ and $\text{Inv}(B)$; moreover, $\psi(\mathcal{M}) = S\mathcal{M}$, where S is the invertible transformation defined by $Sx_{rs} = y_{rs}$; $s = 1, \ldots, k_r$; $r = 1, \ldots, q$.

Conversely, suppose that $\psi: \text{Inv}(A) \to \text{Inv}(B)$ is an isomorphism of lattices. Let $\lambda_1, \ldots, \lambda_p$ be all the distinct eigenvalues of A, and let $\mathcal{N}_i = \psi(\mathcal{R}_{\lambda_i}(A))$, $i = 1, \ldots, p$. Then $\mathbb{C}^n$ is a direct sum of the B-invariant subspaces $\mathcal{N}_1, \ldots, \mathcal{N}_p$. We claim that $\sigma(B|_{\mathcal{N}_i}) \cap \sigma(B|_{\mathcal{N}_j}) = \emptyset$ for $i \neq j$. Indeed, assume the contrary, that is, $\mu_0 \in \sigma(B|_{\mathcal{N}_i}) \cap \sigma(B|_{\mathcal{N}_j})$ for some $\mathcal{N}_i$ and $\mathcal{N}_j$ with $i \neq j$. Let $\mathcal{N} = \text{Span}\{y_1 + y_2\}$, where y_1 (resp. y_2) is some eigenvector of $B|_{\mathcal{N}_i}$ (resp. of $B|_{\mathcal{N}_j}$) corresponding to the eigenvalue μ_0. Then $\mathcal{N}$ is B invariant. Let $\mathcal{M}$ be the A-invariant subspace such that $\psi(\mathcal{M}) = \mathcal{N}$. Since $\mathcal{M}$ must contain a one-dimensional A-invariant subspace, and since ψ is a lattice isomorphism, the subspace $\mathcal{M}$ is one-dimensional. Therefore, $\mathcal{M} \subset \mathcal{R}_{\lambda_k}(A)$ for some k. This implies $\mathcal{N} = \psi(\mathcal{M}) \subset \psi(\mathcal{R}_{\lambda_k}(A)) = \mathcal{N}_k$, a contradiction with the choice of $\mathcal{N}$.

Further, the spectrum of each restriction $B|_{\mathcal{N}_i}$ is a singleton. To verify this, assume the contrary. Then for some i the subspace $\mathcal{N}_i$ is a sum of at least two root subspaces for B:

$$\mathcal{N}_i = \mathcal{R}_{\mu_i}(B) \dotplus \cdots \dotplus \mathcal{R}_{\mu_k}(B), \qquad k \geq 2$$

Letting $\mathcal{M}_j$ be the A-invariant subspace such that $\psi(\mathcal{M}_j) = \mathcal{R}_{\mu_j}(B)$, $j = 1, \ldots, k$, we have

$$\mathcal{R}_{\lambda_i}(A) = \mathcal{M}_1 \dotplus \cdots \dotplus \mathcal{M}_k$$

If x_1 and x_2 are eigenvectors of $A|_{\mathcal{M}_1}$ and of $A|_{\mathcal{M}_2}$, respectively, then

Span$\{x_1 + x_2\}$ is A-invariant and does not belong to any subspace $\mathcal{M}_j$. Hence $\psi(\text{Span}\{x_1 + x_2\})$ is B invariant, belongs to $\mathcal{N}_i$ but does not belong to any subspace $\mathcal{R}_{\mu_j}(B)$. This is impossible because $\psi(\text{Span}\{x_1 + x_2\})$ is one-dimensional.

We have proved, therefore, that $\mathcal{N}_i = \mathcal{R}_{\mu_i}(B)$, $i = 1, \ldots, p$, where $\mu_1, \ldots, \mu_p$ are all the distinct eigenvalues of B.

For a fixed i, the number of partial multiplicities of A corresponding to λ_i that are greater than or equal to a fixed integer q, coincides with the maximal number of summands in a direct sum $\mathcal{L}_1 \dotplus \cdots \dotplus \mathcal{L}_s$, where, for $j = 1, \ldots, s$, $\mathcal{L}_j \subset \mathcal{R}_{\lambda_i}(A)$ are irreducible subspaces with dimension not less than q. As ψ induces an isomorphism between $\text{Inv}(A|_{\mathcal{R}_{\lambda_i}(A)})$ and $\text{Inv}(B|_{\mathcal{R}_{\mu_i}(B)})$, it follows that the number of partial multiplicities of A corresponding to λ_i that are greater than or equal to q coincides with the number of partial multiplicities of B corresponding to μ_i that are not less than q. Hence A and B have the same Jordan structure. $\square$

Corollary 16.1.3

Assume that A and B are transformations on $\mathbb{C}^n$ with one and only one eigenvalue λ_0: $\sigma(A) = \sigma(B) = \{\lambda_0\}$. Then the lattices $\text{Inv}(A)$ and $\text{Inv}(B)$ are isomorphic if and only if A and B are similar.

16.2 PROPERTIES OF LINEAR ISOMORPHISMS OF LATTICES: THE CASE OF SIMILAR TRANSFORMATIONS

In view of Theorem 16.1.2, for transformations A and B with the same Jordan structure, the set $\mathcal{S}(A, B)$ of all invertible transformations S such that $\mathcal{L} \in \text{Inv}(A)$ if and only if $S\mathcal{L} \in \text{Inv}(B)$, is not empty. Denote

$$\Omega(A, B) = \inf\{\|I - S\| \mid S \in \mathcal{S}(A, B)\}$$

Note that the set $\mathcal{S}(A, B)$ contains transformations arbitrarily close to zero [Indeed, take a fixed $S \in \mathcal{S}(A, B)$ and consider αS with $\alpha \to 0$, $\alpha \neq 0$.] Hence $\Omega(A, B) \leq 1$ for any A and B with the same Jordan structure. This observation will be used frequently in the sequel.

The following example shows that the equality $\Omega(A, B) = 1$ is possible.

EXAMPLE 16.2.1. Let

$$A = \begin{bmatrix} 0 & 0 \\ 1 & 0 \end{bmatrix}, \qquad B = \begin{bmatrix} 0 & 1 \\ 0 & 0 \end{bmatrix}$$

Then

$$\mathcal{S}(A, B) = \left\{ \begin{bmatrix} a & b \\ c & 0 \end{bmatrix} \middle| b, c \neq 0 \right\}$$

and

$$\Omega(A, B) = \inf_{a,b,c \in \mathbb{C}} \left\| \begin{bmatrix} 1-a & -b \\ -c & 1 \end{bmatrix} \right\|$$

However, it is easily seen that the norm of $\begin{bmatrix} 1-a & -b \\ -c & 1 \end{bmatrix}$ is at least 1, for any choice of a, b, and c, and can be arbitrarily close to 1. Hence $\Omega(A, B) = 1$. $\square$

The number $\Omega(A, B)$ is closely related to the distance between $\mathrm{Inv}(A)$ and $\mathrm{Inv}(B)$, as we shall see in the next theorem. Recall that

$$\mathrm{dist}(\mathrm{Inv}(A), \mathrm{Inv}(B)) = \max\{ \sup_{\mathcal{L} \in \mathrm{Inv}(A)} \inf_{\mathcal{M} \in \mathrm{Inv}(B)} \theta(A, B),$$

$$\sup_{\mathcal{L} \in \mathrm{Inv}(B)} \inf_{\mathcal{M} \in \mathrm{Inv}(B)} \theta(A, B)\}$$

Theorem 16.2.1

If A and B have the same Jordan structure and $\Omega(A, B) < 1$, then

$$\mathrm{dist}(\mathrm{Inv}(A), \mathrm{Inv}(B)) \leq 2\Omega(A, B)(1 - \Omega(A, B))^{-1}$$

Proof. For positive $\epsilon < 1 - \Omega(A, B)$ let $S \in \mathcal{S}(A, B)$ be such that

$$\|I - S\| \leq q \overset{\mathrm{def}}{=} \Omega(A, B) + \epsilon$$

For every nonzero $x \in \mathbb{C}^n$, denoting $y = S^{-1}x$, we have

$$\frac{\|y\|}{\|x\|} \leq \frac{\|y - Sy\| + \|Sy\|}{\|x\|} \leq \frac{q\|y\|}{\|x\|} + 1$$

Hence

$$\frac{\|y\|}{\|x\|} \leq \frac{1}{1-q}$$

so $\|S^{-1}\| \leq (1-q)^{-1}$ and $\|I - S^{-1}\| \leq \|S^{-1}\| \, \|I - S\| \leq q(1-q)^{-1}$. Now for any subspace $\mathcal{M} \subset \mathbb{C}^n$ the transformation $SP_{\mathcal{M}}S^{-1}$ is a projector on $S\mathcal{M}$ (we denote by $P_{\mathcal{M}}$ the orthogonal projector on $\mathcal{M}$). So

$$\theta(\mathcal{M}, S\mathcal{M}) \leq \|P_{\mathcal{M}} - SP_{\mathcal{M}}S^{-1}\| \leq \|P_{\mathcal{M}} - SP_{\mathcal{M}}\| + \|SP_{\mathcal{M}} - SP_{\mathcal{M}}S^{-1}\|$$

$$\leq \|I - S\| \cdot \|P_{\mathcal{M}}\| + \|S\| \cdot \|P_{\mathcal{M}}\| \cdot \|I - S^{-1}\|$$

$$\leq \|I - S\| + \|I - S^{-1}\| + \|I - S\| \cdot \|I - S^{-1}\| \leq 2q(1 - q)^{-1}$$

Consequently

$$\text{dist}(\text{Inv}(A), \text{Inv}(B)) \leq 2q(1 - q)^{-1}$$

and since $\epsilon < 0$ was arbitrary, Theorem 16.2.1 follows. □

Now consider the case when A and B are similar. Then, evidently, A and B have the same Jordan structure. Clearly, in this case $\mathcal{S}(A, B)$ contains all the similarity transformations between A and B:

$$\mathcal{S}(A, B) \supset \{S: \mathbb{C}^n \to \mathbb{C}^n \mid S \text{ is invertible and } A = S^{-1}BS\}$$

We remark that this inclusion can be proper. Indeed, in Example 16.2.1 above, the similarity transformations between A and B have the form

$$\left\{ \begin{bmatrix} a & b \\ b & 0 \end{bmatrix} \middle| b \neq 0 \right\}$$

which is a proper subset of $\mathcal{S}(A, B)$.

Theorem 16.2.2

For every transformation $A: \mathbb{C}^n \to \mathbb{C}^n$ we have

$$\sup \frac{\Omega(B, A)}{\|B - A\|} < \infty \quad and \quad \sup \frac{\text{dist}(\text{Inv}(B), \text{Inv}(A))}{\|B - A\|} < \infty \quad (16.2.1)$$

where the suprema are taken over all transformations B that are similar to A.

In other words, the first inequality in (16.2.1) means that there exists a positive constant $K > 0$ (depending on A) such that for every B that is similar to A we have

$$\|I - T\| \leq K\|B - A\|$$

for some invertible transformation T satisfying $A = TBT^{-1}$.

In the next section the result of Theorem 16.2.2 is generalized to include all the transformations B with the same Jordan structure as A.

Proof of Theorem 16.2.2. As $\theta(\mathcal{M}, \mathcal{N}) \leq 1$ for any subspaces $\mathcal{M}, \mathcal{N}$ in $\mathbb{C}^n$ [this follows, for instance, from formula (13.1.3)], we have

$$\text{dist}(\text{Inv}(X), \text{Inv}(Y)) \leq 1 \qquad (16.2.2)$$

for any transformations $X, Y: \mathbb{C}^n \to \mathbb{C}^n$. So, by Theorem 16.2.1, the second inequality in (16.2.1) follows from the first one.

To prove the first inequality in (16.2.1), consider the linear space $L(\mathbb{C}^n)$ of all transformations $A: \mathbb{C}^n \to \mathbb{C}^n$, with the scalar product $(X, Y) = \text{tr}(XY^*)$ for $X, Y \in L(\mathbb{C}^n)$ (where Y^* denotes the adjoint of Y defined by the standard scalar product on $\mathbb{C}^n$) and the corresponding norm $\|X\|_t = \sqrt{(X, X)}$ for all $X \in L(\mathbb{C}^n)$. For every $B \in L(\mathbb{C}^n)$ consider the linear transformation

$$W_B(X) = AX - XB, \quad X \in L(\mathbb{C}^n)$$

so that, in particular, $W_A(X) = AX - XA$. If B is similar to A, then $\dim \text{Ker } W_B = \dim \text{Ker } W_A$ (indeed, $\text{Ker } W_B = \{XS \mid X \in \text{Ker } W_A\}$, where S is a fixed invertible transformation such that $B = S^{-1}AS$). Let P_A be a fixed projector on $\text{Ker } W_A$. [Thus $P_A: L(\mathbb{C}^n) \to L(\mathbb{C}^n)$.] By Theorem 13.5.1 there exists a positive constant K_1 such that, if B is similar to A, then $\|P_A - P_B\|_t \le K_1 \|W_A - W_B\|_t$ for some projector P_B on $\text{Ker } W_B$. Here

$$\|P_A - P_B\|_t = \max_{X \in L(\mathbb{C}^n), \|X\|_t = 1} \|(P_A - P_B)X\|_t$$

is the norm induced by $\| \cdot \|_t$, and similarly for $\|W_A - W_B\|_t$.

Observe that the norm $\| \cdot \|_t$ is multiplicative: $\|XY\|_t \le \|X\|_t \cdot \|Y\|_t$ for all transformations $X, Y: \mathbb{C}^n \to \mathbb{C}^n$. Indeed, if $\| \cdot \|$ is the norm induced on transformations by the standard norm on $\mathbb{C}^n$, then it is easily verified that $\|Y\|^2 I - YY^*$ is positive semidefinite and hence that $\|Y\|^2 XX^* - XYY^*X^*$ is positive semidefinite. Thus

$$\|XY\|_t^2 = \text{tr}(XYY^*X^*) \le \text{tr}(\|Y\|^2 XX^*) = \|Y\|^2 \cdot \|X\|_t^2 \quad (16.2.3)$$

Further, denoting by $\lambda_1 \ge \cdots \ge \lambda_n$ (≥ 0) the eigenvalues of the positive semidefinite transformation Y^*Y, we have every $f \in \mathbb{C}^n$ with $\|f\| = 1$:

$$\|Yf\|^2 = (Yf, Yf) = (Y^*Yf, f) \le \lambda_1 \le \lambda_1 + \cdots + \lambda_n$$
$$= \text{tr}(Y^*Y) = \text{tr}(YY^*) = \|Y\|_t^2$$

so $\|Y\|^2 \le \|Y\|_t^2$. Substitution in (16.2.3) yields the desired inequality $\|XY\|_t \le \|X\|_t \|Y\|_t$.

Note that $(W_A - W_B)(X) = X(B - A)$; so the multiplicative property of $\| \cdot \|_t$ implies

$$\|W_A - W_B\|_t \le \|B - A\|_t$$

Now $\|P_A - P_B\|_t \le K_1 \|B - A\|_t$ for every transformation B similar to A. The identity transformation I belongs to $\text{Ker } W_A$; so $P_A(I) = I$ and

$$\|I - P_B(I)\|_t \le K_1 \|B - A\|_t \|I\|_t = \sqrt{n} K_1 \|B - A\|_t$$

If, in addition, $\|B - A\|_t < (\sqrt{n} K_1)^{-1}$, then $P_B(I)$ is an invertible transformation. In this case $P_B(I) \in \mathcal{S}(B, A)$; hence

$$\Omega(B, A) \le K_2 \|B - A\| \qquad (16.2.4)$$

for every $B \in L(\mathbb{C}^n)$ that is similar to A and such that $\|B - A\|_t \le (\sqrt{n} K_1)^{-1}$, where the constant $K_2 > 0$ depends on A only.

Taking into account the fact that $\Omega(B, A) \le 1$ for all B similar to A, we find that (16.2.1) follows from (16.2.4). $\square$

Results analogous to Theorem 16.2.2 hold also for other classes of subspaces connected with a linear transformation. For example, the lattice of all invariant subspaces in Theorem 16.2.2 can be replaced by any one of the following sets: coinvariant subspaces, semiinvariant subspaces, hyperinvariant subspaces, reducing invariant subspaces, and root subspaces. The proof remains the same in all these cases.

Theorem 16.2.2 fails in general if we drop the requirement that B is similar to A. The next example illustrates this fact.

EXAMPLE 16.2.2.   Let

$$A = \begin{bmatrix} 0 & 0 \\ 0 & 0 \end{bmatrix} : \mathbb{C}^2 \to \mathbb{C}^2 ; \qquad B(\delta) = \begin{bmatrix} 0 & \delta \\ 0 & 0 \end{bmatrix}, \delta \in \mathbb{C}$$

Let us compute $\mathrm{dist}(\mathrm{Inv}(A), \mathrm{Inv}(B(\delta)))$ for $\delta \ne 0$. We have for a complex number α

$$\theta\left(\mathrm{Span}\{e_1\}, \mathrm{Span}\left\{ \begin{bmatrix} \alpha \\ 1 \end{bmatrix} \right\} \right) = \left\| \begin{bmatrix} 1 & 0 \\ 0 & 0 \end{bmatrix} - (|\alpha|^2 + 1)^{-1} \begin{bmatrix} |\alpha|^2 & \alpha \\ \bar{\alpha} & 1 \end{bmatrix} \right\|$$

$$= (|\alpha|^2 + 1)^{-1} \left\| \begin{bmatrix} 1 & -\alpha \\ -\bar{\alpha} & 1 \end{bmatrix} \right\|$$

$$= (|\alpha| + 1)(|\alpha|^2 + 1)^{-1}$$

$$\theta\left(\{0\}, \mathrm{Span}\left\{ \begin{bmatrix} \alpha \\ 1 \end{bmatrix} \right\} \right) = \theta\left(\mathbb{C}^2, \mathrm{Span}\left\{ \begin{bmatrix} \alpha \\ 1 \end{bmatrix} \right\} \right) = 1$$

So

$$\inf_{\mathscr{L} \in \mathrm{Inv}\, B(\delta)} \theta(\mathcal{M}, \mathscr{L}) = \begin{cases} 0, \text{ if } \mathcal{M} = \{0\},\ \mathcal{M} = \mathrm{Span}\{e_1\},\ \mathcal{M} = \mathbb{C}^2 \\ \min(1, (|\alpha| + 1)(|\alpha|^2 + 1)^{-1}), \text{ if } \mathcal{M} = \mathrm{Span}\begin{bmatrix} \alpha \\ 1 \end{bmatrix} \end{cases}$$

Hence

$$\sup_{\mathcal{M} \in \mathrm{Inv}(A)} \inf_{\mathcal{L} \in \mathrm{Inv}\, B(\delta)} \theta(\mathcal{M}, \mathcal{L}) = 1$$

As any subspace in $\mathbb{C}^2$ is A invariant, we obviously have

$$\inf_{\mathcal{M} \in \mathrm{Inv}(A)} \theta(\mathcal{L}, \mathcal{M}) = 0$$

for every $\mathcal{L} \in \mathrm{Inv}\, B(\delta)$. Thus $\mathrm{dist}(\mathrm{Inv}(A), \mathrm{Inv}(B(\delta))) = 1$. In the limit as $\delta \to 0$, we see that the conclusion of Theorem 16.2.2 fails for this particular A if we drop the condition that B be similar to A. $\square$

We conclude this section with a simple example in which $\Omega(B, A)$ and $\mathrm{dist}(\mathrm{Inv}(B), \mathrm{Inv}(A))$ can be calculated explicitly.

EXAMPLE 16.2.3.   Let

$$A = \begin{bmatrix} 0 & 0 \\ 0 & 1 \end{bmatrix}, \qquad B = \begin{bmatrix} 0 & x \\ 0 & 1 \end{bmatrix} (x \in \mathbb{C})$$

Then

$$\mathscr{S}(A, B) = \left\{ \begin{bmatrix} a & xd \\ 0 & d \end{bmatrix} \,\middle|\, a, d \neq 0 \right\} \cup \left\{ \begin{bmatrix} xd & a \\ d & 0 \end{bmatrix} \,\middle|\, a, d \neq 0 \right\}$$

and

$$\Omega(A, B)^2 = \min \left\{ \inf_{a,d \in \mathbb{C}} \; \max_{\substack{u,v \in \mathbb{C} \\ |u|^2 + |v|^2 = 1}} \{|(1-a)u - xdv|^2 + |(1-d)v|^2\} \right.$$

$$\left. \inf_{a,d \in \mathbb{C}} \; \max_{\substack{u,v \in \mathbb{C} \\ |u|^2 + |v|^2 = 1}} \{|(1-xd)u - av|^2 + |-du + v|^2\} \right\}$$

Taking $a = 1$, it follows that

$$\Omega(A, B)^2 \leq \inf_{d \in \mathbb{C}} \{|xd|^2 + |1 - d|^2\}$$

On the other hand, taking $u = 0$, we have

$$\Omega(A, B)^2 \geq \min \left\{ \inf_{a,d \in \mathbb{C}} \max_{|v|^2 = 1} \{|xdv|^2 + |(1-d)v|^2\} \right\}$$

$$\inf_{a,d \in \mathbb{C}} \max_{|v|^2 = 1} \{|av|^2 + |v|^2\} = \min \{ \inf_{d \in \mathbb{C}} \{|xd|^2 + |1 - d|^2\}, 1 \}$$

So

$$\Omega(A, B)^2 = \inf_{d \in \mathbb{C}} \{|xd|^2 + |1 - d|^2\}$$

An elementary calculation (using the stationary points of $|xd|^2 + |1 - d|^2$ considered as a function of two real variables $\Re\, d$ and $\Im\, d$) yields

$$\Omega(A, B) = |x|(|x|^2 + 1)^{-1/2}$$

To calculate the distance between $\mathrm{Inv}(A)$ and $\mathrm{Inv}(B)$, note that the unique different invariant subspaces of A and B (if $x \neq 0$) are $\mathrm{Span}\begin{bmatrix} 0 \\ 1 \end{bmatrix}$ and $\mathrm{Span}\begin{bmatrix} x \\ 1 \end{bmatrix}$, respectively, with corresponding orthogonal projectors

$$P_1 = \begin{bmatrix} 0 & 0 \\ 0 & 1 \end{bmatrix} \quad \text{and} \quad P_2 = (|x|^2 + 1)^{-1}\begin{bmatrix} |x|^2 & x \\ \bar{x} & 1 \end{bmatrix}$$

Observe that

$$\|P_1 - P_2\| = |x|(|x|^2 + 1)^{-1/2}$$

and, letting $P_3 = I - P_1$, we find that

$$\|P_1 - P_3\| = 1, \qquad \|P_3 - P_2\| = (|x|^2 + 1)^{-1/2}$$

These inequalities, together with the fact that $\theta(\mathcal{M}, \mathcal{N}) = 1$ if $\dim \mathcal{M} \neq \dim \mathcal{N}$ (see Theorem 13.1.2), allow us to verify that

$$\mathrm{dis}(\mathrm{Inv}(A), \mathrm{Inv}(B)) = |x|(|x|^2 + 1)^{-1/2}$$

It is curious that $\Omega(A, B) = \mathrm{dist}(\mathrm{Inv}(A), \mathrm{Inv}(B))$ in this example. □

16.3 DISTANCE BETWEEN INVARIANT SUBSPACES FOR TRANSFORMATIONS WITH THE SAME JORDAN STRUCTURE

We state the main result of this chapter.

Theorem 16.3.1

Given a transformation A on $\mathbb{C}^n$, we have

$$\sup \frac{\Omega(A, B)}{\|B - A\|} < \infty \tag{16.3.1}$$

and

$$\sup \frac{\mathrm{dist}(\mathrm{Inv}(A), \mathrm{Inv}(B))}{\|B - A\|} < \infty \tag{16.3.2}$$

where the suprema are taken over the set $J(A)$ of all transformations $B: \mathbb{C}^n \to \mathbb{C}^n$ which have the same Jordan structure as A.

Before we proceed with the proof of Theorem 16.4.1 (which is quite long), let us mention the following result on Lipschitz continuity of dist(Inv(A), Inv(B)), whose proof is facilitated by the use of Theorem 16.3.1.

Theorem 16.3.2

Let J be a class of all linear transformations having the same Jordan structure. Then the real function defined on J by

$$\varphi(A, B) = \text{dist}(\text{Inv}(A), \text{Inv}(B))$$

for all $A, B \in J$ is Lipschitz continuous at every pair $A_0, B_0 \in J$, that is

$$|\varphi(A, B) - \varphi(A_0, B_0)| \leq K(\|A - A_0\| + \|B - B_0\|)$$

for every $A, B \in J$, where the constant $K > 0$ depends on A_0 and B_0 only.

Proof. We need the following observation (proved in Section 15.6):

$$\text{dist}(\text{Inv}(A), \text{Inv}(B)) \leq \text{dist}(\text{Inv}(A), \text{Inv}(C)) + \text{dist}(\text{Inv}(C), \text{Inv}(B))$$

$$(16.3.3)$$

for any transformations $A, B, C: \math4{C}^n \to \math4{C}^n$. Using (16.3.3) and (16.3.2), we obtain for a fixed $A_0, B_0 \in J$:

$$|\varphi(A, B) - \varphi(A_0, B_0)| \leq |\varphi(A, B) - \varphi(A_0, B)| + |\varphi(A_0, B) - \varphi(A_0, B_0)|$$
$$\leq \varphi(A, A_0) + \varphi(B, B_0) \leq K(\|A - A_0\| + \|B - B_0\|) \quad \square$$

Proof of Theorem 16.3.1. Since Theorem 16.2.1, together with (16.3.1), implies (16.3.2), we have only to prove (16.3.1). The main idea of the proof is to reduce it to Theorem 16.2.2. For the reader's convenience the proof is divided into three parts.

(a) Let $\lambda_1, \ldots, \lambda_p$ be all the distinct eigenvalues of A, and let Γ_i be a circle around λ_i, $i = 1, \ldots, p$ chosen so small that $\Gamma_i \cap \Gamma_j = \emptyset$ for $i \neq j$ and λ_i is the unique eigenvalue of A inside Γ_i. For every Γ_i and every transformation $B: \math4{C}^n \to \math4{C}^n$ that has no eigenvalues on Γ_i, define

$$k_j(\Gamma_i, B) = \sum_{l=1}^{q} k_j(\mu_l, B), \qquad j = 1, 2, \ldots$$

where $\mu_1, \ldots, \mu_q$ are all the eigenvalues of B inside Γ_i and

$k_1(\mu_m, B) \ge k_2(\mu_m, B) \ge \cdots$ are the partial multiplicities of B at μ_m (we put $k_r(\mu_m, B) = 0$ for r greater than the geometric multiplicity of μ_m as an eigenvalue of B). By Theorem 13.5.1 there exists an $\epsilon_1 > 0$ such that any transformation B with $\|B - A\| < \epsilon_1$ has all its eigenvalues in the union of the interiors of $\Gamma_1, \ldots, \Gamma_p$; and, moreover, the sum of algebraic multiplicities of the eigenvalues of B inside a fixed circle Γ_i is equal to the algebraic multiplicity of the eigenvalue λ_i of A, for $i = 1, \ldots, p$; further

$$\sum_{j=s}^{\infty} k_j(\Gamma_i, B) \le \sum_{j=s}^{\infty} k_j(\Gamma_i, A), \qquad j = 1, 2, \ldots; \qquad i = 1, \ldots, p$$

(16.3.4)

provided $\|B - A\| < \epsilon_1$.

(b) Assume now that $\|B - A\| < \epsilon_1$ and $B \in J(A)$. As the numbers of different eigenvalues of B and of A coincide, there is exactly one eigenvalue of B, denoted μ_i, inside each circle Γ_i. We claim that for every $i = 1, \ldots, p$ the eigenvalue λ_i of A and the eigenvalue μ_i of B have the same partial multiplicities. Indeed, assuming the contrary, it follows from (16.3.4) that

$$\sum_{j=s_0}^{\infty} k_j(\Gamma_{i_0}, B) < \sum_{j=s_0}^{\infty} k_j(\Gamma_{i_0}, A)$$

(16.3.5)

for some i_0 $(1 \le i_0 \le p)$ and some s_0 (note that the equality

$$\sum_{j=1}^{\infty} k_j(\Gamma_i, B) = \sum_{j=1}^{\infty} k_j(\Gamma_i, A)$$

holds for $i = 1, \ldots, p$). For notational simplicity assume that $i_0 = 1$, and that $\lambda_1, \lambda_2, \ldots, \lambda_{p_0}$ are exactly those eigenvalues of A whose algebraic multiplicities are equal to the algebraic multiplicity of λ_1. As $B \in J(A)$, there is a permutation π of $\{1, 2, \ldots, p_0\}$ such that

$$k_j(\Gamma_i, A) = k_j(\Gamma_{\pi(i)}, B), \qquad i = 1, \ldots, p_0; \qquad j = 1, 2, \ldots$$

Consequently,

$$\sum_{i=1}^{p_0} \sum_{j=s_0}^{\infty} k_j(\Gamma_i, B) = \sum_{i=1}^{p_0} \sum_{j=s_0}^{\infty} k_j(\Gamma_i, A)$$

However, (16.3.4) and (16.3.5) imply

$$\sum_{i=1}^{p_0} \sum_{j=s_0}^{\infty} k_j(\Gamma_i, B) < \sum_{i=1}^{p_0} \sum_{j=s_0}^{\infty} k_j(\Gamma_i, A)$$

which is a contradiction.

(c)   Observe that a transformation $F: \mathbb{C}^n \to \mathbb{C}^n$ with $\|F - A\| \le \epsilon_1/2$ has no eigenvalue on $\Gamma_1 \cup \Gamma_2 \cup \cdots \cup \Gamma_p$. So the number

$$M = \max_{\substack{F: \mathbb{C}^n \to \mathbb{C}^n \\ \|F-A\| \le \epsilon_1/2}} \max_{\lambda \in \Gamma_1 \cup \cdots \cup \Gamma_p} \|(\lambda I - F)^{-1}\|$$

is well defined. For the transformation $B \in J(A)$ with $\|B - A\| \le \epsilon_1/2$ we have [using (13.1.4)]

$$\theta(\mathscr{R}_{\lambda_i}(A), \mathscr{R}_{\mu_i}(B))$$

$$\le \left\| \frac{1}{2\pi i} \int_{\Gamma_i} (\lambda I - A)^{-1} \, d\lambda - \frac{1}{2\pi i} \int_{\Gamma_i} (\lambda I - B)^{-1} \, d\lambda \right\|$$

$$\le \frac{1}{2\pi} \int_{\Gamma_i} \|(\lambda I - A)^{-1} - (\lambda I - B)^{-1}\| \, |d\lambda|$$

$$\le \frac{1}{2\pi} \int_{\Gamma_i} \|(\lambda I - A)^{-1}\| \cdot \|A - B\| \cdot \|(\lambda I - B)^{-1}\| \cdot |d\lambda|$$

$$\le \frac{M^2 \Delta_i}{2\pi} \|A - B\|$$

where Δ_i is the length of Γ_i. Let

$$S_i = I - \frac{1}{2\pi i} \int_{\Gamma_i} [(\lambda I - A)^{-1} - (\lambda I - B)^{-1}] \, d\lambda , \qquad i = 1, \ldots, p$$

Then $\|I - S_i\| \le (M^2 \Delta_i/2\pi)\|A - B\|$ and (provided, in addition, $(M^2 \Delta_i/2\pi)\|A - B\| < 1$) $S_i(\mathscr{R}_{\lambda_i}(A)) = \mathscr{R}_{\mu_i}(B)$, $i = 1, \ldots, p$.

Put

$$\epsilon_2 = \min\left\{ \frac{\epsilon_1}{2}, \frac{1}{2}\left[\frac{M^2 \Delta_1}{2\pi} \right]^{-1}, \ldots, \frac{1}{2}\left[\frac{M^2 \Delta_p}{2\pi} \right]^{-1} \right\}$$

and for fixed i $(1 \le i \le p)$ let S_i be the transformation constructed above for the transformation $B \in J(A)$ with $\|B - A\| < \epsilon_2$. Define the transformation $\tilde{B}_i: \mathscr{R}_{\lambda_i}(A) \to \mathscr{R}_{\lambda_i}(A)$ by

$$\tilde{B}_i = \tilde{S}_i^{-1} B|_{\mathscr{R}_{\mu_i}(B)} \tilde{S}_i$$

where $\tilde{S}_i = S_{i|\mathscr{R}_{\lambda_i}(A)}$. Obviously, μ_i is the only eigenvalue of $\tilde{B}_i$. Further, for the transformation $\tilde{A}_i = A|_{\mathscr{R}_{\lambda_i}(A)}$ we have (here $x \in \mathscr{R}_{\lambda_i}(A)$):

$$\|(\tilde{A}_i - \tilde{B}_i)x\| = \|Ax - S_i^{-1}BS_i x\|$$

$$\leq \|Ax - S_i^{-1}Bx\| + \|S_i^{-1}B(I - S_i)x\|$$

$$\leq \|Ax - Bx\| + \|(I - S_i^{-1})Bx\| + \|S_i^{-1}B(I - S_i)x\| \qquad (16.3.6)$$

Now

$$\|B\| \leq \|A\| + \|A - B\| \leq \|A\| + 1 \qquad (16.3.7)$$

and $\|S_i^{-1}\| \leq (1 - q_i)^{-1}$, $\|I - S_i^{-1}\| \leq q_i(1 - q_i)^{-1}$ where $q_i = \|I - S_i\|$ (cf. the proof of Theorem 16.2.1). Since $(1 - q_i)^{-1} \leq 2$, (16.3.6) gives

$$\|\tilde{A}_i - \tilde{B}_i\| \leq K_i \|A - B\| \qquad (16.3.8)$$

where $K_i = 1 + (4M^2\Delta_i/\pi)(\|A\| + 1) + 4(\|A\| + 1)(M^2\Delta_i/\pi)$. Now we have we

$$\det(\lambda I - \tilde{A}_i) = (\lambda - \lambda_i)^{k_i}, \qquad \det(\lambda I - \tilde{B}_i) = (\lambda - \mu_i)^{k_i},$$

$$\operatorname{tr} \tilde{A}_i = k_i \lambda_i, \qquad \operatorname{tr} \tilde{B}_i = k_i \lambda_i \qquad (16.3.9)$$

On the other hand, for any orthonormal basis $f_i, \ldots, f_{k_i}$ in $\mathcal{R}_{\lambda_i}(A)$ the inequality (16.3.8) gives

$$|\operatorname{tr} \tilde{A}_i - \operatorname{tr} \tilde{B}_i| = \left| \sum_{j=1}^{k_i} (\tilde{A}_i f_j, f_j) - (\tilde{B}_i f_j, f_j) \right|$$

$$\leq \sum_{j=1}^{k_i} |(\tilde{A}_i f_j, f_j) - (\tilde{B}_i f_j, f_j)| \leq k_i \|\tilde{A}_i - \tilde{B}_i\| \leq k_i K_i \|A - B\|$$

Taking into account (16.3.9), we obtain

$$|\mu_i - \lambda_i| \leq k_i^2 K_i \|A - B\| \qquad (16.3.10)$$

Now define the transformation $B'\colon \mathbb{C}^n \to \mathbb{C}^n$ by

$$B'x = (B - \mu_i I + \lambda_i I)x, \qquad x \in \mathcal{R}_{\mu_i}(B)$$

Then B' is clearly similar to A. As every invariant subspace of a transformation is the direct sum of its intersections with all the root subspaces of this transformation, it follows that $\operatorname{Inv}(B) = \operatorname{Inv}(B')$. Moreover, inequality (16.3.10) shows that for all $x_i \in \mathcal{R}_{\mu_i}(B)$,

$$\|(B' - A)x_i\| \leq \|(B - A)x_i\| + \|(\mu_i - \lambda_i)x_i\| \leq (1 + k_i^2 K_i)\|A - B\| \cdot \|x_i\|$$

$$(16.3.11)$$

For every $x \in \mathbb{C}^n$ write $x = x_1 + \cdots + x_p$, where $x_j = P_j(B)x$, and $P_j(B)$ is the projector on $\mathcal{R}_{\mu_j}(B)$ along $\Sigma_{l \neq j} \mathcal{R}_{\mu_l}(B)$. As $P_j(B) = (1/2\pi i) \int_{\Gamma_j} (\lambda I - B)^{-1} d\lambda$, we have

$$\| P_j(B) - P_j(A) \| \leq \frac{M^2 \Delta_j}{2\pi} \| A - B \|$$

where $P_j(A)$ is the projector on $\mathcal{R}_{\lambda_j}(A)$ along $\Sigma_{l \neq j} \mathcal{R}_{\lambda_l}(A)$. Denoting

$$Q_1 = \max_{1 \leq j \leq p} \left(\frac{M^2 \Delta_j}{2\pi} \| A - B \| + \| P_j(A) \| \right) \qquad (16.3.12)$$

we see that $\| P_j(B) \| \leq Q_1$, $j = 1, \ldots, p$. Now using (16.3.11) with these inequalities we obtain

$$\| (B' - A)x \| \leq \sum_{j=1}^{p} \| (B' - A)x_j \| \leq \left(\sum_{j=1}^{p} (1 + k_j^2 K_j) \| x_j \| \right) \| A - B \|$$

$$\leq \left\{ \sum_{j=1}^{p} (1 + k_j^2 K_j) Q_1 \right\} \| A - B \| \cdot \| x \|$$

and thus

$$\| B' - A \| \leq Q_2 \| A - B \|$$

where $Q_2 = Q_1 \Sigma_{j=1}^{p} (1 + k_j^2 K_j)$. By Theorem 16.2.2 there exists $Q_3 > 0$ such that for any transformation X that is similar to A there exists an invertible S with $A = S^{-1} X S$ and $\| I - S \| \leq Q_3 \| X - A \|$. Applying this result for $X = B'$ and bearing in mind that $\text{Inv}(B') = \text{Inv}(B)$, we obtain

$$\Omega(A, B) \leq Q_2 Q_3 \| B - A \| \qquad (16.3.13)$$

for any $B \in J(A)$ with $\| B - A \| < \epsilon_3$.

As $\Omega(A, B) \leq 1$ for any $B \in J(A)$, (16.3.1) follows from (16.3.13). $\square$

16.4 TRANSFORMATIONS WITH THE SAME DEROGATORY JORDAN STRUCTURE

The result on continuity of $\text{Inv}(A)$ that is contained in Theorem 16.4.1 can be extended to admit pairs of transformations that are close to one another and have *different* Jordan structures, provided the variations in this structure are confined to those eigenvalues with geometric multiplicity 1. To make this idea precise, we introduce the following definition. We say that trans-formations $A: \mathbb{C}^n \to \mathbb{C}^n$ and $B: \mathbb{C}^n \to \mathbb{C}^n$ have the *same derogatory Jordan structure* if $A|_{\mathscr{C}(A)}$ and $B|_{\mathscr{C}(B)}$ have the same Jordan structure, where $\mathscr{C}(A)$ is the sum of the root subspaces of A corresponding to eigenvalues λ_0 with $\dim \text{Ker}(\lambda_0 I - A) > 1$. By definition, $\mathscr{C}(A) = 0$ if $\dim \text{Ker}(\lambda_0 I - A) = 1$ for every eigenvalue λ_0 of A.

Denote by $DJ(A)$ the set of all transformations that have the same derogatory Jordan structure as A.

We need one more definition to state the next theorem. For a transformation A, the *height* of A is the maximal partial multiplicity of A corresponding to the eigenvalues λ_0 with $\dim \text{Ker}(\lambda_0 I - A) = 1$. If A has no such eigenvalues, its height is defined to be 1.

Theorem 16.4.1

Let $A: \math{C}^n \to \math{C}^n$ be a transformation with height α. Then

$$\sup \frac{\text{dist}(\text{Inv}(A), \text{Inv}(B))}{\|B - A\|^{1/\alpha}} < \infty$$

where the supremum is taken over all $B \in DJ(A)$.

The inequality in Theorem 16.4.1 is exact in the sense that in general α cannot be replaced by a smaller number. Namely, given a transformation A with height α, there exists a sequence $\{B_m\}_{m=1}^\infty$ of transformations converging to A with $B_m \in DJ(A)$ such that

$$\liminf_{m \to \infty} \frac{\text{dist}(\text{Inv}(A), \text{Inv}(B_m))}{\|B_m - A\|^{1/\alpha}} > 0 \tag{16.4.1}$$

Indeed, it is sufficient to consider the case when $A = J_n(0)$ is a Jordan block. Then the sequence

$$B_m = \begin{bmatrix} 0 & 1 & 0 & \cdots & 0 \\ 0 & 0 & 1 & \cdots & 0 \\ \vdots & & & \ddots & \vdots \\ & & & & 1 \\ \dfrac{1}{m} & 0 & 0 & \cdots & 0 \end{bmatrix}$$

satisfies (16.4.1). This is not difficult to verify using the fact that B_m has n distinct eigenvalues $\epsilon m^{-1/n}$ with corresponding eigenvectors

$$\text{Span}\langle 1, \epsilon m^{-1/n}, \ldots, \epsilon^{n-1} m^{-(n-1)/n} \rangle$$

where ϵ is an nth root of unity. Indeed, writing $\zeta = \epsilon m^{-1/n}$, we see that the orthogonal projector on $\text{Span}\langle 1, \zeta, \ldots, \zeta^{n-1} \rangle$ is

$$(1 + |\zeta|^2 + \cdots + |\zeta|^{2(n-1)})^{-1} \begin{bmatrix} 1 & \bar{\zeta} & \cdots & \bar{\zeta}^{n-1} \\ \zeta & |\zeta|^2 & \cdots & \zeta\bar{\zeta}^{n-1} \\ \zeta^2 & \zeta^2\bar{\zeta} & \cdots & \zeta^2\bar{\zeta}^{n-1} \\ \vdots & \vdots & & \vdots \\ \zeta^{n-1} & \zeta^{n-1}\bar{\zeta} & \cdots & |\zeta|^{2(n-1)} \end{bmatrix}$$

so

$$\theta(\text{Span } e_1, \text{Span}\langle 1, \zeta, \ldots, \zeta^{n-1}\rangle) \geq C|\zeta| = Cm^{-1/n}$$

where the positive constant C is independent of m. Hence for m large enough (such that $C|\zeta| < 1$) we have

$$\text{dist}(\text{Inv}(A), \text{Inv}(B_m)) \geq Cm^{-1/n}$$

and (16.4.1) follows.

The proof of Theorem 16.4.1 is given in the next section. For the time being, note the following important special case.

Corollary 16.4.2

Let $A: \mathbb{C}^n \to \mathbb{C}^n$ be a nonderogatory transformation with height α. Then there exists a neighbourhood $\mathcal{U}$ of A in the set of all transformations on $\mathbb{C}^n$ such that

$$\sup_{B \in \mathcal{U}} \frac{\text{dist}(\text{Inv}(A), \text{Inv}(B))}{\|B - A\|^{1/\alpha}} < \infty$$

Recall that a transformation A is called *nonderogatory* if $\dim \text{Ker}(\lambda I - A) = 1$ for every eigenvalue λ of A, and note that the set of all nonderogatory transformations is open. Indeed, if $A: \mathbb{C}^n \to \mathbb{C}^n$ is nonderogatory, then $\text{rank}(A - \lambda_0 I) = n - 1$ for every eigenvalue λ_0 of A. Write A as an $n \times n$ matrix in some basis in $\mathbb{C}^n$, and let A_0 be an $(n - 1) \times (n - 1)$ nonsingular submatrix of $A - \lambda_0 I$. Then, for B sufficiently close to A and λ sufficiently close to λ_0, the corresponding $(n - 1) \times (n - 1)$ submatrix B_0 of $B - \lambda I$ will also be nonsingular. Consequently

$$\text{rank}(B - \lambda I) \geq n - 1 \tag{16.4.2}$$

for all such B and λ. Now the eigenvalues of a transformation depend continuously on that transformation. So the set of λ values for which (16.4.2) holds will contain all eigenvalues of B (if B is close enough to A), which means that B is nonderogatory.

Using the openness of the set of all nonderogatory linear transformations, we see that Corollary 16.4.2 follows immediately from Theorem 16.4.1.

The following result on continuity of dist(Inv(A), Inv(B)) can be obtained from Theorem 16.4.1 in the same way that Theorem 16.3.2 was obtained from Theorem 16.3.1.

Theorem 16.4.3

Let DJ be a class of all transformations having the same derogatory Jordan structure. Then the real function defined on DJ by

$$\varphi(A, B) = \text{dist}(\text{Inv}(A), \text{Inv}(B))$$

for every A, $B \in DJ$ is continuous. Moreover, for every pair A_0, $B_0 \in J$ there exists a constant $K > 0$ such that

$$|\varphi(A, B) - \varphi(A_0, B_0)| \leq K(\|A - A_0\|^{1/\alpha} + \|B - B_0\|^{1/\beta})$$

for every A, $B \in DJ$ that is sufficiently close to A_0, B_0, and where α, β are the heights of A_0 and B_0, respectively.

Now we consider stable invariant subspaces. Recall from Section 15.2 that an A-invariant subspace $\mathcal{M}$ is called *stable* if for every $\epsilon > 0$ there exists $\delta > 0$ such that any transformation B with $\|B - A\| < \delta$ has an invariant subspace $\mathcal{N}$ with the property that $\theta(\mathcal{M}, \mathcal{N}) < \epsilon$. Using Theorem 16.4.1 and its proof, we can prove a stronger property of stable invariant subspaces:

Theorem 16.4.4

Let $A: \mathbb{C}^n \to \mathbb{C}^n$ be a transformation with height α, and let $\mathcal{M}$ be a stable A-invariant subspace. Then

$$\sup \frac{\displaystyle\inf_{\mathcal{N} \in \text{Inv}(B)} \theta(\mathcal{M}, \mathcal{N})}{\|B - A\|^{1/\alpha}} < \infty$$

where the supremum is taken over all transformations $B: \mathbb{C}^n \to \mathbb{C}^n$.

It will be convenient to prove Theorem 16.4.4 in the next section, following the proof of Theorem 16.4.1.

16.5 PROOFS OF THEOREMS 16.4.1 AND 16.4.4

We start with a preliminary result.

Lemma 16.5.1

Let $A: \mathbb{C}^n \to \mathbb{C}^n$ be a transformation with $\sigma(A) = \{0\}$ and $\dim \text{Ker } A = 1$. Then, given a constant $M > 0$, there exists a $K > 0$ such that

$$|\lambda_0| \leq K\|B - A\|^{1/n} \tag{16.5.1}$$

for every eigenvalue λ_0 of every transformation $B: \mathbb{C}^n \to \mathbb{C}^n$ satisfying $\|B - A\| \leq M$.

Proof.   Let $B: \mathbb{C}^n \to \mathbb{C}^n$ be such that $\|B - A\| \leq M$. We have $A^n = 0$ and thus

$$\|B^n\| = \|B^n - A^n\| = \|B^{n-1}(B - A) + B^{n-2}(B - A)A + \cdots$$
$$+ B(B - A)A^{n-2} + (B - A)A^{n-1}\|$$

$$\leq \|B - A\| \sum_{j=0}^{n-1} \|B\|^{n-1-j} \|A\|^j$$

$$\leq \|B - A\| \sum_{j=0}^{n-1} (M + \|A\|)^{n-1-j} \|A\|^j$$

On the other hand, if λ_0 is an eigenvalue of B, then λ_0^n is an eigenvalue of B^n (as one can easily see by passing to the Jordan form of B). Hence $|\lambda_0|^n = |\lambda_0^n| \leq \|B^n\|$. If this inequality is combined with the preceding one and nth roots of both sides are taken, the lemma follows. $\square$

Now we prove Theorem 16.4.1 for the case when $A: \mathbb{C}^n \to \mathbb{C}^n$ is nonderogatory and has only one eigenvalue.

Lemma 16.5.2

Let $\sigma(A) = \{\lambda_0\}$ and $\dim \operatorname{Ker}(\lambda_0 I - A) = 1$. *Then there exists a constant* $K > 0$ *such that the inequality*

$$\operatorname{dist}(\operatorname{Inv}(A), \operatorname{Inv}(B)) \leq K \|B - A\|^{1/n} \tag{16.5.2}$$

holds for every transformation $B: \mathbb{C}^n \to \mathbb{C}^n$.

Proof.   It will suffice to prove (16.5.2) for all B belonging to some neighbourhood of A. We can assume $\lambda_0 = 0$. By Lemma 16.5.1 there exist $K_1 > 0$ and $\epsilon_1 > 0$ such that any eigenvalue λ_0 of a B with $\|B - A\| < \epsilon_1$ satisfies $|\lambda_0| \leq K_1 \|B - A\|^{1/n}$. As the set of nonderogatory transformations is open, we can assume also that every B with $\|B - A\| < \epsilon_1$ is nonderogatory. Now for such a B and its eigenvalue λ_0 let x_0 be the corresponding eigenvector: $(B - \lambda_0 I)x_0 = 0$, $x_0 \neq 0$. Then $\dim \operatorname{Ker}(B - \lambda_0 I) = \dim \operatorname{Ker} A = 1$, and using Theorem 13.5.1, we find that

$$\theta(\operatorname{Ker} A, \operatorname{Ker}(B - \lambda_0 I)) \leq K_2 \|A - B\|^{1/n} \tag{16.5.3}$$

for any eigenvalue λ_0 of any B satisfying $\|B - A\| < \epsilon_2$, where the positive constants K_2 and $\epsilon_2 \leq \epsilon_1$ depend on A only.

It is convenient to assume that A is the Jordan block with respect to the standard orthonormal basis in $\mathbb{C}^n$: $A = J_n(0)$. For any B sufficiently close to A write $B - A = [b_{ij}]_{i,j=1}^n$. Inequality (16.5.3) shows that there is an eigenvector x of B corresponding to an eigenvalue λ_0 of the form $x = \langle 1, x_2, x_3, \ldots, x_n \rangle$, where $x_2, \ldots, x_n \in \mathbb{C}$. The equation $(B - \lambda_0 I)x = 0$ has the form

$$
\begin{bmatrix}
b_{11} - \lambda_0 & 1 + b_{12} & b_{13} & \cdots & & b_{1n} \\
b_{21} & b_{22} - \lambda_0 & 1 + b_{23} & \cdots & & b_{2n} \\
b_{31} & b_{32} & b_{33} - \lambda_0 & \cdots & & b_{3n} \\
\vdots & \vdots & \vdots & & & \vdots \\
b_{n-1,1} & b_{n-1,2} & b_{n-1,3} & \cdots & b_{n-1,n-1} - \lambda_0 & 1 + b_{n-1,n} \\
b_{n1} & b_{n2} & b_{n3} & \cdots & b_{n,n-1} & b_{n,n} - \lambda_0
\end{bmatrix}
\begin{bmatrix}
1 \\ x_2 \\ x_3 \\ \vdots \\ \\ x_n
\end{bmatrix}
=
\begin{bmatrix}
1 \\ 0 \\ 0 \\ \vdots \\ \\ 0
\end{bmatrix}
$$

Rewrite the first $n - 1$ equations in the form

$$
\begin{bmatrix}
1 + b_{12} & b_{13} & \cdots & b_{1n} \\
b_{22} - \lambda_0 & 1 + b_{23} & \cdots & b_{2n} \\
\vdots & \vdots & & \vdots \\
b_{n-1,2} & b_{n-1,3} & \cdots & 1 + b_{n-1,n}
\end{bmatrix}
\begin{bmatrix}
x_2 \\ x_3 \\ \vdots \\ x_n
\end{bmatrix}
=
\begin{bmatrix}
-(b_{11} - \lambda_0) \\ -b_{21} \\ \vdots \\ -b_{n-1,1}
\end{bmatrix}
$$

Using $|\lambda_0| \le K_1 \|B - A\|^{1/n}$ and Cramer's rule, we see that for $j = 2, 3, \ldots, n$, x_j has the following structure:

$$
x_j = \lambda_0^{j-1}(1 + f_{j,j-1}(b_{pq})) + \lambda_0^{j-2} f_{j,j-2}(b_{pq}) + \cdots + \lambda_0 f_{j1}(b_{pq}) + f_{j0}(b_{pq})
$$

$$(16.5.4)$$

where $f_{jk}(b_{pq})$ are scalar functions of n^2 variables $\{b_{pq}\}_{p,q=1}^n$ such that

$$
|f_{jk}(b_{pq})| \le L_0 \|A - B\|
$$

where B satisfies $\|B - A\| < \epsilon_2$. Here and in the sequel $L_0, L_1, \ldots,$ denote positive constants that depend on A only.

Now let $x^{(1)}, \ldots, x^{(k)}$ be k eigenvectors of B corresponding to k different eigenvalues $\lambda_1, \ldots, \lambda_k$. Construct new vectors using divided differences:

$$
u^{(12)} = \frac{x^{(1)} - x^{(2)}}{\lambda_1 - \lambda_2}, \, u^{(23)} = \frac{x^{(2)} - x^{(3)}}{\lambda_2 - \lambda_3}, \ldots,
$$

$$
u^{(k-1,k)} = \frac{x^{(k-1)} - x^{(k)}}{\lambda_{k-1} - \lambda_k}
$$

$$
u^{(13)} = \frac{u^{(12)} - u^{(23)}}{\lambda_1 - \lambda_3}, \, u^{(24)} = \frac{u^{(23)} - u^{(34)}}{\lambda_2 - \lambda_4}, \ldots,
$$

$$
u^{(k-2,k)} = \frac{u^{(k-2,k-1)} - u^{(k-1,k)}}{\lambda_{k-2} - \lambda_k}; \ldots;
$$

$$
u^{(1,k)} = \frac{u^{(1,k-1)} - u^{(2,k)}}{\lambda_1 - \lambda_k}
$$

Let

$$p_k(y_1, \ldots, y_l) = \sum_{\substack{\alpha_i \geq 0 \\ \alpha_1 + \cdots + \alpha_l = k}} y_1^{\alpha_1} y_2^{\alpha_2} \cdots y_l^{\alpha_l}$$

be the homogeneous polynomial of degree k in variables $y_1, \ldots, y_l$. A simple induction argument [using (16.5.4)] shows that $u^{(jk)}$ has the following form (where $s = k - j$ and the first s coordinates in $u^{(jk)}$ are zeros):

$$u^{(jk)} = \begin{bmatrix} 0 \\ \vdots \\ 0 \\ 1 + f_{s+1,s} \\ p_1(\lambda_j, \lambda_{j+1}, \ldots, \lambda_k)(1 + f_{s+2,s+1}) + f_{s+2,s} \\ p_2(\lambda_j, \lambda_{j+1}, \ldots, \lambda_k)(1 + f_{s+3,s+2}) + p_1(\lambda_j, \ldots, \lambda_k)f_{s+3,s+1} + f_{s+3,s} \\ \vdots \\ p_{n-1-s}(\lambda_j, \lambda_{j+1}, \ldots, \lambda_k)(1 + f_{n,n-1}) + p_{n-s-2}(\lambda_j, \ldots, \lambda_k)f_{n,n-2} \\ \quad + \cdots + p_1(\lambda_j, \ldots, \lambda_k)f_{n,s+1e} + f_{ns} \end{bmatrix}$$

$$(16.5.5)$$

Here $f_{uw} = f_{uw}(b_{pq})$. The induction argument is based on the following equality (where we put formally $p_0 \equiv 1$):

$$\frac{p_u(\lambda_j, \ldots, \lambda_q) - p_u(\lambda_{j+1}, \ldots, \lambda_{q+1})}{(\lambda_j - \lambda_{q+1})}$$

$$= \frac{[\sum_{w=0}^u \lambda_j^w p_{u-w}(\lambda_{j+1}, \ldots, \lambda_q) - \sum_{w=0}^u \lambda_{q+1}^u p_{u-w}(\lambda_{j+1}, \ldots, \lambda_q)]}{\lambda_j - \lambda_{q+1}}$$

$$= \sum_{w=1}^u p_{w-1}(\lambda_j, \lambda_{q+1})p_{u-w}(\lambda_{j+1}, \ldots, \lambda_q) = p_{u-1}(\lambda_j, \ldots, \lambda_{q+1})$$

Now consider the subspace

$$\mathcal{L} = \text{Span}\{x^{(1)}, u^{(12)}, u^{(13)}, \ldots, u^{(1k)}\}.$$

Obviously

$$\mathcal{L} = \text{Span}\{x^{(1)}, x^{(2)}, \ldots, x^{(k)}\}$$

On the other hand, the matrix

$$Q = \begin{bmatrix} I_k & 0 \\ Y_{n-k}Y_k^{-1} & 0 \end{bmatrix}$$

is a projector on $\mathscr{L}$, where Y_k (resp. Y_{n-k}) is the $k \times k$ [resp. $(n-k) \times k$] matrix formed by the upper k (resp. lower $n-k$) rows of the $n \times k$ matrix

$$[x^{(1)} u^{(12)} u^{(13)} \cdots u^{(1k)}]$$

Using formulas (16.5.5), we see that

$$\det Y_k = (1 + f_{21}) \cdots (1 + f_{k,k-1})$$

and thus, Y_k is invertible (for B sufficiently close to A). Using the estimates $|f_{uv}| \le L_0 \|A - B\|$, $|\lambda_j| \le K_1 \|B - A\|^{1/n}$, we easily find from (16.5.5) that $\|Y_k^{-1}\| \le L_1$.

Further, $\|Y_{n-k}\| \le L_2 \|A - B\|^{1/n}$. Hence

$$\left\| Q - \begin{bmatrix} 1 & 0 \\ 0 & 0 \end{bmatrix} \right\| \le L_3 \|A - B\|^{1/n}$$

So

$$\theta(\mathscr{L}, \operatorname{Span}\{e_1, \ldots, e_k\}) \le L_3 \|A - B\|^{1/n} \tag{16.5.6}$$

Consequently

$$\operatorname{dist}(\operatorname{Inv}(A), \operatorname{Inv}(B)) \le L_4 \|A - B\|^{1/n}$$

for every transformation B such that $\|B - A\| < \epsilon_2$ and every B-invariant subspace is spanned by its eigenvectors. As B must be nonderogatory, the last condition means that B has n distinct eigenvalues.

Assume now that B is such that $\|B - A\| < \epsilon_2$, but B does not have n distinct eigenvalues. In particular, B is nonderogatory. Let $\{B_m\}_{m=1}^{\infty}$ be a sequence of transformations such that $\|B_m - A\| < \epsilon_2$ for all m, $B_m \to B$ as $m \to \infty$, and B_m has n distinct eigenvalues for each m. Let $\mathscr{M}$ be a k-dimensional B-invariant subspace. As $\mathscr{M}$ is a stable subspace (see Theorem 15.2.1), there exists a sequence $\{\mathscr{M}_m\}_{m=1}^{\infty}$, where $\mathscr{M}_m$ is a k-dimensional B_m-invariant subspace such that $\theta(\mathscr{M}_m, \mathscr{M}) \to 0$ as $m \to \infty$. By (16.5.6)

$$\theta(\mathscr{M}, \operatorname{Span}\{e_1, \ldots, e_k\}) \le \theta(\mathscr{M}, \mathscr{M}_m) + \theta(\mathscr{M}_m, \operatorname{Span}\{e_1, \ldots, e_k\})$$
$$\le \theta(\mathscr{M}, \mathscr{M}_m) + L_3 \|A - B_m\|^{1/n}$$

Passing to the limit in this inequality as $m \to \infty$, we obtain

$$\theta(\mathscr{M}, \operatorname{Span}\{e_1, \ldots, e_k\} \le L_3 \|A - B\|^{1/n}$$

hence

$$\text{dist}(\text{Inv}(A),\ \text{Inv}(B)) \le L_4 \|A - B\|^{1/n}$$

for all B with $\|B - A\| < \epsilon_2$. $\square$

Proof of Theorem 16.4.1. We now start to prove Theorem 16.4.1 in full generality. Let Γ_1 and Γ_2 be two closed contours in the complex plane such that $\Gamma_1 \cap \Gamma_2 = \emptyset$ and the eigenvalues λ_0 of A lying inside Γ_1 (resp. Γ_2) are exactly those for which $\dim \text{Ker}(\lambda_0 I - A) = 1$ (resp. $\dim \text{Ker}(\lambda_0 I - A) > 1$).

Let $\delta_1 > 0$ be chosen so that any transformation $B: \mathfrak{C}^n \to \mathfrak{C}^n$ with $\|B - A\| \le \delta_1$ has no eigenvalues on $\Gamma_1 \cup \Gamma_2$. For such a B, let

$$S_j = I - \frac{1}{2\pi i} \int_{\Gamma_j} [(\lambda I - A)^{-1} - (\lambda I - B)^{-1}]\, d\lambda, \qquad j = 1, 2$$

and define the transformation $S: \mathfrak{C}^n \to \mathfrak{C}^n$ by $Sx = S_j x$ for $x \in \mathscr{R}_j(A)$, the spectral subspace associated with the eigenvalues of A inside Γ_j. Denote by P_1 the projector on $\mathscr{R}_1(A)$ along $\mathscr{R}_2(A)$; then for any $x \in \mathfrak{C}^n$ with $\|x\| = 1$ we have

$$\|(I - S)x\| = \|(P_1 - S_1 P_1)x + ((I - P_1) - S_2(I - P_1))x\|$$
$$\le \|I - S_1\| \cdot \|P_1\| + \|I - S_2\| \cdot \|I - P_1\|$$
$$\le \frac{M^2 \Delta_1}{2\pi} \|A - B\| \cdot \|P_1\| + \frac{M^2 \Delta_2}{2\pi} \|A - B\| \cdot \|I - P_1\|$$

where Δ_j is the length of Γ_j and

$$M = \max_{\substack{F:\, \mathfrak{C}^n \to \mathfrak{C}^n \\ \|F - A\| \le \delta_1}} \max_{\lambda \in \Gamma_1 \cup \Gamma_2} \|(\lambda I - F)^{-1}\|$$

(cf. the proof of Theorem 16.3.1). Letting $N = (2\pi)^{-1} M^2 (\Delta_1 \|P_1\| + \Delta_2 \|I - P_1\|)$, we have $\|I - S\| \le N\|A - B\|$. Hence for $\|A - B\| \le \min(\delta_1, (2N)^{-1})$ the transformation S is invertible and $S\mathscr{R}_j(A) = \mathscr{R}_j(B)$, $j = 1, 2$. Now put $\tilde{B} = S^{-1} B S$. Then (cf. the proof of Theorem 16.2.1)

$$\text{dist}(\text{Inv}(B),\ \text{Inv}(\tilde{B})) \le 2N\|A - B\|(1 - 2N\|A - B\|)^{-1}$$

As

$$\text{dist}(\text{Inv}(A),\ \text{Inv}(B)) \le \text{dist}(\text{Inv}(A),\ \text{Inv}(\tilde{B})) + \text{dist}(\text{Inv}(B),\ \text{Inv}(\tilde{B}))$$

$$(16.5.7)$$

it is sufficient to prove Theorem 16.4.1 only for those $B: \mathfrak{C}^n \to \mathfrak{C}^n$ that are close enough to A and satisfy $\mathscr{R}_j(B) = \mathscr{R}_j(A)$, $j = 1, 2$.

Note that for any transformation B sufficiently close to A with $\mathcal{R}_j(B) = \mathcal{R}_j(A)$, every B-invariant subspace $\mathcal{N}$ is of the form $\mathcal{N} = \mathcal{N}_1 + \mathcal{N}_2$, where $\mathcal{N}_j = \mathcal{N} \cap \mathcal{R}_j(A)$. Let $\mathcal{M}$ be an A-invariant subspace, and let $\mathcal{M} = \mathcal{M}_1 + \mathcal{M}_2$, where $\mathcal{M}_j = \mathcal{M} \cap \mathcal{R}_j(A)$. Then, denoting by $P_{\mathcal{L}_1}$ (resp. $Q_{\mathcal{L}_2}$) the orthogonal projector on the subspace $\mathcal{L}_1$ (resp. $\mathcal{L}_2$) in $\mathcal{R}_1(A)$ [resp. $\mathcal{R}_2(A)$], we have

$$\theta(\mathcal{M}, \mathcal{N}) \le \|(P_{\mathcal{M}_1} + Q_{\mathcal{M}_2}) - (P_{\mathcal{N}_1} + Q_{\mathcal{N}_2})\|$$

$$\le \|P_{\mathcal{M}_1} - P_{\mathcal{N}_1}\| + \|Q_{\mathcal{M}_2} - Q_{\mathcal{N}_2}\| = \theta(\mathcal{M}_1, \mathcal{N}_1) + \theta(\mathcal{M}_2, \mathcal{N}_2)$$

Hence

$$\text{dist}(\text{Inv}(A), \text{Inv}(B)) \le \text{dist}(\text{Inv}(A|_{\mathcal{R}_1(A)}), \text{Inv}(B|_{\mathcal{R}_1(A)}))$$

$$+ \text{dist}(\text{Inv}(A|_{\mathcal{R}_2(A)}), \text{Inv}(B|_{\mathcal{R}_2(A)})) \quad (16.5.8)$$

Further, we remark that if B is sufficiently close to A, and $\mathcal{R}_j(B) = \mathcal{R}_j(A)$ for $j = 1, 2$, then $B|_{\mathcal{R}_1(A)}$ is nonderogatory, that is, $\dim \text{Ker}(\lambda_0 I - B|_{\mathcal{R}_1(A)}) = 1$ for every eigenvalue λ_0 of $B|_{\mathcal{R}_1(A)}$. Indeed, this follows from the choice of $\mathcal{R}_1(A)$, which ensures that $A|_{\mathcal{R}_1(A)}$ is nonderogatory and from the openness of the set of nonderogatory transformations. If, in addition, $B \in DJ(A)$, it now follows that $A|_{\mathcal{R}_2(A)}$ and $B|_{\mathcal{R}_2(A)}$ have the same Jordan structure. Hence in view of (16.5.8) and Theorem 16.3.1, we only need to prove the inequality

$$\text{dist}(\text{Inv}(A|_{\mathcal{R}_1(A)}), \text{Inv}(B|_{\mathcal{R}_1(A)})) \le K\|B - A\|^{1/\alpha}$$

In other words, we can assume that A is nonderogatory. Moreover, using the arguments similar to those employed above, we can assume in addition that A has only one eigenvalue, and this case is covered already in Lemma 16.5.2.

Theorem 16.4.1 is proved completely. $\square$

Proof of Theorem 16.4.4. It is sufficient to prove that there exist positive constants ϵ and K such that the inequality

$$\inf_{\mathcal{N} \in \text{Inv}(B)} \theta(\mathcal{M}, \mathcal{N}) \le K\|B - A\|^{1/\alpha} \quad (16.5.9)$$

holds for every transformation B satisfying $\|B - A\| \le \epsilon$.

Observe that for any transformations $B, \bar{B}: \mathfrak{C}^n \to \mathfrak{C}^n$ the inequality

$$\inf_{\mathcal{N} \in \text{Inv}(B)} \theta(\mathcal{M}, \mathcal{N}) \le \inf_{\bar{\mathcal{N}} \in \text{Inv}(\bar{B})} \theta(\mathcal{M}, \bar{\mathcal{N}}) + \text{dist}(\text{Inv}(B), \text{Inv}(\bar{B})) \quad (16.5.10)$$

holds. Indeed, for every $\mathcal{N} \in \text{Inv}(B)$ and $\bar{\mathcal{N}} \in \text{Inv}(\bar{B})$ we have

$$\theta(\mathcal{M}, \mathcal{N}) \leq \theta(\mathcal{M}, \tilde{\mathcal{N}}) + \theta(\tilde{\mathcal{N}}, \mathcal{N})$$

Taking the infimum over all $\mathcal{N} \in \text{Inv}(B)$ it follows that

$$\inf_{\mathcal{N} \in \text{Inv}(B)} \theta(\mathcal{M}, \mathcal{N}) \leq \theta(\mathcal{M}, \tilde{\mathcal{N}}) + \inf_{\mathcal{N} \in \text{Inv}(B)} \theta(\tilde{\mathcal{N}}, \mathcal{N})$$

$$\leq \theta(\mathcal{M}, \tilde{\mathcal{N}}) + \text{dist}(\text{Inv}(B), \text{Inv}(\tilde{B}))$$

It remains to take the infimum over all $\tilde{\mathcal{N}} \in \text{Inv}(\tilde{B})$ to obtain (16.5.10).

Using the arguments from the proof of Theorem 16.4.1 [when (16.5.10) is used instead of (16.5.7)], we reduce the proof of (16.5.9) to the case when B has the property that every root subspace $\mathcal{R}_{\lambda_j}(A)$ of A is a spectral subspace for B and, moreover, the spectra of $B|_{\mathcal{R}_{\lambda_1}(A)}$ and $B|_{\mathcal{R}_{\lambda_2}(A)}$ do not intersect if $\lambda_1 \neq \lambda_2$. Let $\lambda_1, \ldots, \lambda_r$ be all the distinct eigenvalues of A; then

$$\mathcal{M} = (\mathcal{M} \cap \mathcal{R}_{\lambda_1}(A)) \dotplus \cdots \dotplus (\mathcal{M} \cap \mathcal{R}_{\lambda_r}(A))$$

Also, for every B-invariant subspace $\mathcal{N}$ we have

$$\mathcal{N} = (\mathcal{N} \cap \mathcal{R}_{\lambda_1}(A)) \dotplus \cdots \dotplus (\mathcal{N} \cap \mathcal{R}_{\lambda_r}(A))$$

Arguing as in the proof of Theorem 16.4.1, we obtain

$$\theta(\mathcal{M}, \mathcal{N}) \leq \sum_{i=1}^{r} \theta(\mathcal{M} \cap \mathcal{R}_{\lambda_i}(A), \mathcal{N} \cap \mathcal{R}_{\lambda_i}(A))$$

So in order to prove (16.5.9), we can assume without loss of generality that A has only one eigenvalue, say λ_1. If $\dim \text{Ker}(\lambda_1 I - A) > 1$, then by Theorem 15.2.1 (here we use the assumption that $\mathcal{M}$ is stable) $\mathcal{M} = \{0\}$ or $\mathcal{M} = \mathcal{C}^n$, in which case (16.5.9) is trivial. If $\dim \text{Ker}(\lambda_1 I - A) = 1$, then (16.5.9) follows from Theorem 16.4.1. [Note that in this case $B \in DJ(A)$ for all B sufficiently close to A.] $\quad\square$

16.6 DISTANCE BETWEEN INVARIANT SUBSPACES FOR TRANSFORMATIONS WITH DIFFERENT JORDAN STRUCTURES

In this section we investigate the behaviour of $\text{dist}(\text{Inv}(A), \text{Inv}(B))$ when A and B have different Jordan structures or different derogatory Jordan structures. The basic result in this direction is as follows.

Theorem 16.6.1

We have

$$\inf \text{dist}(\text{Inv}(A), \text{Inv}(B)) > 0 \tag{16.6.1}$$

where the infimum is taken over all pairs of transformations A, B: $\mathbb{C}^n \to \mathbb{C}^n$ such that A is derogatory and B is nonderogatory. [The infimum in (16.6.1) depends on n.]

Proof. Recall that B is nonderogatory if and only if the set of its invariant subspaces is finite.

By assumption, dim $\mathrm{Ker}(\lambda_0 I - A) > 1$ from some eigenvalue λ_0 of A. Let x and y be orthonormal vectors belonging to $\mathrm{Ker}(\lambda_0 I - A)$, and put

$$\mathcal{M}(t) = \mathrm{Span}\{x + ty\}, \qquad 0 \le t \le 1$$

Clearly, the subspaces $\mathcal{M}(t)$ are A invariant.

On the other hand, for every nonderogatory B: $\mathbb{C}^n \to \mathbb{C}^n$ it is easily seen that the number of B-invariant subspaces does not exceed

$$\max \prod_{i=1}^{s} (p_i + 1) = 2^n$$

where the maximum is taken over all sequences $p_1, \ldots, p_s$ of positive integers with $p_1 + \cdots + p_s = n$.

Now for any set of 2^n subspaces $\mathcal{L}_1, \ldots, \mathcal{L}_{2^n}$ in $\mathbb{C}^n$ put

$$F(\mathcal{L}_1, \ldots, \mathcal{L}_{2^n}) = \max_{0 \le t \le 1} \min_{1 \le i \le 2^n} \theta(\mathcal{M}(t), \mathcal{L}_i)$$

As $\theta(\mathcal{M}(t), \mathcal{L})$ is a continuous function of t on $[0, 1]$, so is $\min_{1 \le i \le 2^n} \theta(\mathcal{M}(t), \mathcal{L}_i)$, hence $F(\mathcal{L}_1, \ldots, \mathcal{L}_{2^n})$ is well defined. Let us show that $F(\mathcal{L}_1, \ldots, \mathcal{L}_{2^n})$ is a continuous function of $\mathcal{L}_1, \ldots, \mathcal{L}_{2^n}$. For some $\delta > 0$, let $\mathcal{N}_i, i = 1, \ldots, 2^n$ be subspaces in $\mathbb{C}^n$ such that $\theta(\mathcal{N}_i, \mathcal{L}_i) < \delta$ for each i. Then for $i = 1, \ldots, 2^n$ and $t \in [0, 1]$, we obtain

$$\theta(\mathcal{M}(t), \mathcal{N}_i) \le \theta(\mathcal{M}(t), \mathcal{L}_i) + \delta$$

First take the minimum with respect to i on the left-hand side and then on the right-hand side. We obtain

$$\min_{1 \le i \le 2^n} \theta(\mathcal{M}(t), \mathcal{N}_i) \le \min_{1 \le i \le 2^n} \theta(\mathcal{M}(t), \mathcal{L}_i) + \delta$$

for all $t \in [0, 1]$. Taking the maximum with respect to t on the right-hand side first, and then on the left-hand side, we obtain

$$F(\mathcal{N}_1, \ldots, \mathcal{N}_{2^n}) \le F(\mathcal{L}_1, \ldots, \mathcal{L}_{2^n}) + \delta$$

With the roles of $\mathcal{L}_i$ and $\mathcal{N}_i$, switched it also follows that

$$F(\mathcal{L}_1, \ldots, \mathcal{L}_{2^n}) \le F(\mathcal{N}_1, \ldots, \mathcal{N}_{2^n}) + \delta$$

that is

$$|F(\mathcal{N}_1, \ldots, \mathcal{N}_{2^n}) - F(\mathcal{L}_1, \ldots, \mathcal{L}_{2^n})| < \delta$$

which proves the continuity of $F(\mathcal{L}_1, \ldots, \mathcal{L}_{2^n})$. Obviously, $F(\mathcal{L}_1, \ldots, \mathcal{L}_{s^n}) > 0$ for all $\mathcal{L}_i$. As the set of all 2^n-tuples of subspaces in $\mathbb{C}^n$ is compact, there exists an $\epsilon > 0$ such that $F(\mathcal{L}_1, \ldots, \mathcal{L}_{2^n}) \ge \epsilon$ for all $\mathcal{L}_i$, $i = 1, \ldots, 2^n$. From the definition of $F(\mathcal{L}_1, \ldots, \mathcal{L}_{2^n})$ is it easily seen that ϵ does not depend on the choice of x and y (because any pair of orthonormal vectors in $\mathbb{C}^n$ can be mapped to any other pair of such vectors by a unitary transformation). Hence the theorem follows. $\square$

When the transformations A and B are both derogatory, or both non-derogatory, with different Jordan structures, the situation is more complicated. The following question arises naturally: if $\{B_m\}_{m=1}^\infty$ is a sequence of transformations converging to A and such that each B_m has Jordan structure different from that of A, does it follow that

$$\lim_{m \to \infty} \frac{\text{dist}(\text{Inv}(A), \text{Inv}(B_m))}{\|B - A\|} = \infty ? \tag{16.6.2}$$

The next example shows that the answer is, in general, negative.

EXAMPLE 16.6.1. For $m = 1, 2, \ldots$, let

$$A = \begin{bmatrix} 0 & 1 & 0 \\ 0 & 0 & 0 \\ 0 & 0 & 0 \end{bmatrix} : \mathbb{C}^3 \to \mathbb{C}^3 ; \qquad B_m = \begin{bmatrix} -m^{-1} & 1 & 0 \\ 0 & 0 & 0 \\ 0 & 0 & 0 \end{bmatrix} : \mathbb{C}^3 \to \mathbb{C}^3$$

Clearly, for all m, B_m and A have different derogatory Jordan structure (in particular, different Jordan structure).

One-dimensional A-invariant subspaces are $\text{Span}\{e_1 + \beta e_3\}$, $\beta \in \mathbb{C}$ and $\text{Span}\{e_3\}$. The orthogonal projector on $\text{Span}\{e_1 + \beta e_3\}$ is

$$P_\beta = (1 + |\beta|^2)^{-1} \begin{bmatrix} 1 & 0 & \bar{\beta} \\ 0 & 0 & 0 \\ \beta & 0 & |\beta|^2 \end{bmatrix}$$

One-dimensional B_m-invariant subspaces are $\text{Span}\{e_1\}$, $\text{Span}\{e_3\}$, and $\text{Span}\{e_1 + m^{-1}e_2 + \beta e_3\}$ where $\beta \in \mathbb{C}$. The orthogonal projector on $\text{Span}\{e_1 + m^{-1}e_2 + \beta e_3\}$ is

$$Q_{m,\beta} = (1 + m^{-2} + |\beta|^2)^{-1} \begin{bmatrix} 1 & m^{-1} & \bar{\beta} \\ m^{-1} & m^{-2} & \bar{\beta}m^{-1} \\ \beta & \beta m^{-1} & |\beta|^2 \end{bmatrix}$$

Now there exists a constant $L_1 > 0$ (independent of β and m) such that

$$\|P_\beta - Q_{m,\beta}\| \le L_1 m^{-1} \tag{16.6.3}$$

Two-dimensional A-invariant subspaces are $\mathrm{Span}\{e_1, e_2 + \beta e_3\}$ where $\beta \in \mathbb{C}$ and $\mathrm{Span}\{e_1, e_3\}$. Two-dimensional B_m-invariant subspaces are $\mathrm{Span}\{e_1 + m^{-1}e_2, e_3\}$, $\mathrm{Span}\{e_1, e_3\}$, and $\mathrm{Span}\{e_1, e_2 + \beta e_3\}$, where $\beta \in \mathbb{C}$. The orthogonal projector on $\mathrm{Span}\{e_1 + m^{-1}e_2, e_3\}$ is

$$R_m = (1 + m^2)^{-1} \begin{bmatrix} 1 & m^{-1} & 0 \\ m^{-1} & m^{-2} & 0 \\ 0 & 0 & 1 + m^2 \end{bmatrix}$$

There exists a constant $L_2 > 0$ (independent of m) such that

$$\left\| R_m - \begin{bmatrix} 1 & 0 & 0 \\ 0 & 0 & 0 \\ 0 & 0 & 1 \end{bmatrix} \right\| \le L_2 m^{-1} \tag{16.6.4}$$

Now the inequalities (16.6.3) and (16.6.4) ensure that for $m = 1, 2, \ldots,$

$$\mathrm{dist}(\mathrm{Inv}(B_m), \mathrm{Inv}(A)) \le m^{-1} \max(L_1, L_2). \quad \square$$

In the last example both A and B_m are derogatory. Taking

$$A = \begin{bmatrix} 0 & 1 \\ 0 & 0 \end{bmatrix} \qquad B_m = \begin{bmatrix} -m^{-1} & 1 \\ 0 & 0 \end{bmatrix}$$

we obtain an example contradicting (16.6.2) with both A and B_m non-derogatory.

16.7　CONJECTURES

In view of Example 16.6.1 the following question arises: Given a transformation $A: \mathbb{C}^n \to \mathbb{C}^n$ with a certain Jordan structure, it is true that for any other Jordan structure there exists a sequence of linear transformations $\{B_m\}_{m=1}^\infty$ that have this other Jordan structure, for which $B_m \to A$, and for which

$$\lim_{m \to \infty} \frac{\mathrm{dist}(\mathrm{Inv}(A), \mathrm{Inv}(B_m))}{\|B_m - A\|} = \infty? \tag{16.7.1}$$

A similar question arises for the case of derogatory Jordan structure, when (16.7.1) is replaced by

$$\lim_{m \to \infty} \frac{\text{dist}(\text{Inv}(A), \text{Inv}(B))}{\|B_m - A\|^{1/\alpha}} = \infty$$

and α is the height of A. Of course, certain conditions should be imposed on the Jordan structure (or on the derogatory Jordan structure) of $\{B_m\}_{m=1}^{\infty}$ to ensure the existence of a sequence $\{B_m\}_{m=1}^{\infty}$ converging to A. A complete set of such conditions is given in Theorem 15.10.2.

Let us describe the Jordan structure of transformations on $\mathbb{C}^n$ in terms of sequences as in (16.1.1), and let Φ be the set of all such sequences. As in Section 15.10, for

$$\Omega =$$

$$\{s; r_1, r_2, \ldots, r_s; m_{11}, \ldots, m_{1r_1}; m_{21}, \ldots, m_{2r_2}; \ldots; m_{s1}, \ldots, m_{sr_s}\} \in \Phi$$

$$(16.7.2)$$

and for every nonempty set $\Delta \subset \{1, \ldots, s\}$ define

$$k_j(\Omega; \Delta) = \sum_{p \in \Delta} m_{pj}, \qquad j = 1, 2, \ldots.$$

Further, for Ω given by (16.7.2) denote by $P(\Omega)$ the set of all sequences

$$\Omega' =$$

$$\{s'; r_1', r_2', \ldots, r_s'; m_{11}', \ldots, m_{1r_1'}'; m_{21}', \ldots, m_{2r_2'}'; \ldots; m_{s'1}', \ldots, m_{s'r_s}'\}$$

$$\in \Phi$$

for which there is a partition of $\{1, \ldots, s'\}$ into s disjoint nonempty sets $\Delta_1, \ldots, \Delta_s$ such that the following relations hold:

$$\sum_{j=1}^{t} k_j(\Omega; \{p\}) \leq \sum_{j=1}^{t} k_j(\Omega'; \Delta_p); \qquad t = 1, 2, \ldots; \qquad p = 1, \ldots, s$$

$$\sum_{j=1}^{\infty} k_j(\Omega; \{p\}) = \sum_{j=1}^{\infty} k_j(\Omega'; \Delta_p); \qquad p = 1, \ldots, s$$

Note that $\Omega \in P(\Omega)$ always (one takes $\Delta_p = \{p\}, p = 1, \ldots, s$). The set $P(\Omega)$ consists of Ω if and only if Ω represents the Jordan structure corresponding to n distinct eigenvalues, that is, $s = n$.

Note that by Theorem 15.10.2, $P(\Omega)$ represents exactly those Jordan structures for which there is a sequence of transformations converging to a given transformation with the Jordan structure Ω.

We propose the following conjecture.

Conjecture 16.7.1

Let $A: \mathfrak{C}^n \to \mathfrak{C}^n$ be a transformation with the Jordan structure $\Omega \in \Phi$. Then for any sequence Ω' that belongs to $P(\Omega)$ and is different from Ω, there exists a sequence of transformations $\{B_m\}_{m=1}^{\infty}$ that converges to A, for which each B_m has the Jordan structure Ω', and for which

$$\lim_{m \to \infty} \frac{\text{dist}(\text{Inv}(A), \text{Inv}(B_m))}{\|B_m - A\|} = \infty$$

It is not difficult to verify this conjecture when A is nonderogatory. Indeed, without loss of generality we can assume that A is the $n \times n$ Jordan block with eigenvalue zero. In view of Theorem 15.10.2, any sequence Ω' belonging to $P(\Omega)$ (here Ω is the Jordan structure of A) has the form

$$\Omega' = \{s; 1, 1, \ldots, 1; m_1; \ldots; m_s\}$$

where $s > 1$ and m_i are positive integers with $\Sigma_{i=1}^{s} m_i = n$. Given such Ω', consider the following $n \times n$ matrix (we denote by 0_m and I_m the $m \times m$ zero and identity matrices, respectively):

$$B = A + \text{diag}[0_s, \eta_1 I_{m_1-1}, \ldots, \eta_s I_{m_s-1}] + A_\epsilon, \qquad \epsilon > 0$$

where $\eta_1, \ldots, \eta_s$ are the sth roots of ϵ, and the $n \times n$ matrix A_ϵ has ϵ in the $(s, 1)$ entry and zeros elsewhere. It is easy to see [by considering, e.g., $\det(\lambda I - B_\epsilon)$] that, at least for ϵ close enough to zero, the matrix B_ϵ has the Jordan structure Ω'. Clearly, $\eta_1, \ldots, \eta_s$ are the eigenvalues of B_ϵ, and $\langle 1, \eta_i, \ldots, \eta_i^{s-1}, 0, \ldots, 0 \rangle$ is the only eigenvector of B (up to multiplication by a nonzero scalar) corresponding to η_i for $i = 1, \ldots, s$. It follows (cf. the remark following Theorem 16.5.1) that

$$\liminf_{\epsilon \to 0} \frac{\text{dist}(\text{Inv}(A), \text{Inv}(B))}{\|B_\epsilon - A\|^{1/s}} > 0$$

and Conjecture 16.7.1 is verified for the matrix A.

To formulate the corresponding conjecture for derogatory Jordan structure, we introduce one more notion. Let

$$\Omega = \{s; r_1, \ldots, r_s; m_{11}, \ldots, m_{1r_1}; \ldots; m_{s1}, \ldots, m_{sr_s}\}$$

and

$$\Omega' = \{t; r_1', \ldots, r_t'; m_{11}', \ldots, m_{1r_1'}'; \ldots; m_{t1}', \ldots, m_{tr_t'}'\}$$

be two sequences from Φ. We say that Ω and Ω' have the *same derogatory part* if the number (say, u) of indices j, $1 \le j \le s$ such that $r_j \ge 2$ coincides with the number of indices j, $1 \le j \le t$ such that $r_j' \ge 2$, and, moreover, $r_j = r_j'$, $j = 1, \ldots, u$; $m_{jq} = m_{jq}'$, $q = 1, \ldots, r_j$; $j = 1, \ldots, u$. If it does not happen that Ω and Ω' have the same derogatory part, we say that Ω and Ω' have *different derogatory parts*.

Conjecture 16.7.2

Let the transformation $A: \mathbb{C}^n \to \mathbb{C}^n$ have the Jordan structure $\Omega \in \Phi$. Then for every sequence Ω' that belongs to $P(\Omega)$ and such that Ω' and Ω have different derogatory parts there exists a sequence of transformations $\{B_m\}_{m=1}^{\infty}$ that converges to A, for which each B_m has the Jordan structure Ω', and for which

$$\lim_{m \to \infty} \frac{\text{dist}(\text{Inv}(A), \text{Inv}(B_m))}{\|B_m - A\|^{1/\alpha}} = \infty$$

where α is the height of A.

16.8   EXERCISES

16.1   Given an $n \times n$ upper triangular Toeplitz matrix A, find all possible Jordan structures of upper triangular Toeplitz $n \times n$ matrices that are arbitrarily close to A. Are there additional Jordan structures if the perturbed matrix is not necessarily upper triangular Toeplitz?

16.2   Solve Exercise 16.1 for the class of $n \times n$ companion matrices.

16.3   Solve Exercise 16.1 for the class of $n \times n$ circulant matrices.

16.4   Solve Exercise 16.1 for the class of $n \times n$ matrices A such that $A^2 = 0$.

16.5   Prove or disprove each one of the following statements (a), (b), and (c): for every transformation $A: \mathbb{C}^n \to \mathbb{C}^n$ there exists an $\epsilon > 0$ such that any transformation $B: \mathbb{C}^n \to \mathbb{C}^n$ with $\|B - A\| < \epsilon$ has the property that (a) the height of B is equal to the height of A; (b) the height of B is not greater than the height of A; (c) the height of B is not smaller than the height of A.

16.6   Prove Conjecture 16.7.1 for the case when $A = J_3(0)$.

16.7   Given a transformation $A: \mathbb{C}^n \to \mathbb{C}^n$ and a number $\alpha > 0$, an A-invariant subspace $\mathcal{M}$ is called α *stable* if there exist positive constants K and ϵ such that every transformation $B: \mathbb{C}^n \to \mathbb{C}^n$ with $\|B - A\| < \epsilon$ has an invariant subspace $\mathcal{N}$ satisfying

$$\theta(\mathcal{M}, \mathcal{N}) \leq K \|B - A\|^{1/\alpha}$$

Show that all invariant subspaces of the Jordan block $J_n(\lambda_0)$ are α stable if $\alpha \geq n$. (*Hint*: Use Lemma 16.5.2.)

16.8   (a) For every $\alpha > 1$, give an example of an α-stable A-invariant subspace that is not Lipschitz stable. (b) For every $\alpha \geq 1$, give an example of a stable A-invariant subspace that is not α stable.

16.9   Are there α-stable invariant subspaces with $0 < \alpha < 1$?

Chapter Seventeen

Applications

Chapters 13–16 provide us with tools for the study of stability of divisors for monic matrix polynomials and rational matrix functions. In this chapter we develop a complete description of stable divisors in terms of their corresponding invariant subspaces and supporting projectors. Special attention is paid to Lipschitz stable and isolated divisors. We consider also the stability and isolatedness properties of solutions of matrix quadratic equations as well as stability of linear fractional decompositions of rational matrix functions.

17.1 STABLE FACTORIZATIONS OF MATRIX POLYNOMIALS: PRELIMINARIES

Let $L(\lambda)$ be an $n \times n$ monic matrix polynomial, and let

$$L(\lambda) = L_1(\lambda)L_2(\lambda) \cdots L_r(\lambda) \qquad (17.1.1)$$

be a factorization of $L(\lambda)$ into a product of $n \times n$ monic polynomials $L_1(\lambda), \ldots, L_r(\lambda)$. We say that the factorization (17.1.1) is stable if, after sufficiently small changes in the coefficients of $L(\lambda)$, the new matrix polynomial again admits a factorization of type (17.1.1) with only small changes in the factors $L_j(\lambda)$. In the next section we study stability of the factorization of type (17.1.1) in terms of invariant subspaces for the linearization of the matrix polynomial $L(\lambda)$. In this section we establish the framework for this study and prove results on continuity of the correspondence between factorizations and invariant subspaces to be used in the next section.

Let C_L be the companion matrix for $L(\lambda)$:

$$C_L = \begin{bmatrix} 0 & I & 0 & \cdots & 0 \\ 0 & 0 & I & \cdots & 0 \\ \vdots & \vdots & \vdots & & \vdots \\ 0 & 0 & 0 & \cdots & I \\ -A_0 & -A_1 & -A_2 & \cdots & -A_{l-1} \end{bmatrix}$$

where $L(\lambda) = I\lambda^l + \Sigma_{j=0}^{l-1} A_j \lambda^j$. As we have seen in Chapter 5, the triple (X_0, C_L, Y_0), where

$$X_0 = [I \quad 0 \quad \cdots \quad 0], \qquad Y_0 = \begin{bmatrix} 0 \\ 0 \\ \vdots \\ 0 \\ I \end{bmatrix}$$

is a standard triple for $L(\lambda)$. Further, there is a one-to-one correspondence between the factorizations (17.1.1) of $L(\lambda)$ and chains of C_L-invariant subspaces

$$\{0\} \subset \mathcal{M}_r \subset \cdots \subset \mathcal{M}_2 \subset \mathbb{C}^{nl} \tag{17.1.2}$$

with the property that the transformations

$$\begin{bmatrix} X_0 \\ X_0 C_L \\ \vdots \\ X_0 C_L^{l_i-1} \end{bmatrix}_{|\mathcal{M}_i} : \mathcal{M}_i \to \mathbb{C}^{nl_i}, \qquad i = 2, \ldots, r \tag{17.1.3}$$

are invertible (see Section 5.6). Here, $l_r < \cdots < l_2 < l$ are some positive integers. The correspondence between factorizations (17.1.1) and chains of C_L-invariant subspaces is given by the formulas from Theorem 5.6.1. Namely, let $\mathcal{N}_j$ be a direct complement to $\mathcal{M}_{j+1}$ in $\mathcal{M}_j$ ($j = 1, \ldots, r-1$) (by definition, $\mathcal{M}_1 = \mathbb{C}^{nl}$), and let $P_{\mathcal{N}_j} : \mathcal{M}_j \to \mathcal{N}_j$ be the projector on $\mathcal{N}_j$ along $\mathcal{M}_{j+1}$. For $j = 1, \ldots, r-1$, let p_j be the difference $l_{j+1} - l_j$ where, by definition, $l_1 = l$. Here l is the degree of $L(\lambda)$. Then for $j = 1, \ldots, r-1$ we have

$$L_j(\lambda) = \lambda^{p_j} I - (W_{j1} + \lambda W_{j2} + \cdots + \lambda^{p_j-1} W_{j,p_j})(P_{\mathcal{N}_j} C_L)^{p_j} P_{\mathcal{N}_j} \tilde{Y}_j$$

where

$$\tilde{Y}_j = (\mathrm{col}[X_0 C_L^i]_{i=0|\mathcal{M}_j}^{l_j-1})^{-1} \mathrm{col}[\delta_{im_j} I]_{i=1}^{m_j}$$

and the transformations $W_{ji} : \mathcal{N}_j \to \mathbb{C}^n$, $i = 1, \ldots, m_j$ are determined by

$$\text{col}[W_{ji}]_{i=1}^{p_j} = [P_{\mathcal{N}_j}\tilde{Y}_j,\ P_{\mathcal{N}_j}C_{L|\mathcal{N}_j}P_{\mathcal{N}_j}\tilde{Y}_j,\ \ldots,\ (P_{\mathcal{N}_j}C_{L|\mathcal{N}_j}P_{\mathcal{N}_j})^{p_j-1}P_{\mathcal{N}j}\tilde{Y}_j]$$

(As usual, δ_{uv} denotes the Kronecker symbol: $\delta_{uv} = 1$ if $u = v$ and $\delta_{uv} = 0$ if $u \neq v$.) For the last factor $L_r(\lambda)$ we have

$$L_r(\lambda) = \lambda^{l_r}I - X_{0|\mathcal{M}_r}(C_{L|\mathcal{M}_r})^{l_r}(V_{r1} + V_{r2}\lambda + \cdots + V_{r,l_r}\lambda^{l_r-1})$$

where

$$[V_{r1} \quad V_{r2} \cdots V_{r,l_r}] = (\text{col}[X_0C_L^i]_{i=0|\mathcal{M}_r}^{l_r-1})^{-1}\colon \mathbb{C}^{nl_r} \to \mathcal{M}_r$$

Also, it is convenient to use the formulas for the products $L_1(\lambda)L_2(\lambda)\cdots L_{i-1}(\lambda)$ and $L_i(\lambda)L_{i+1}(\lambda)\cdots L_r(\lambda)$ (cf. the proof of Theorem 5.6.1). We have for $i = 2, \ldots, r$:

$$L_i(\lambda)\cdots L_r(\lambda) = I\lambda^{l_i} - X_0(C_{L|\mathcal{M}_i})^{l_i}(V_{i1} + V_{i2}\lambda + \cdots + V_{i,l_i}\lambda^{l_i-1})$$

$$(17.1.4)$$

where

$$[V_{i1} \quad V_{i2} \cdots V_{i,l_i}] = [\text{col}[X_0(C_{L|\mathcal{M}_i})^{j-1}]_{j=1}^{l_i}]^{-1}$$

(Observe that when $i = r$ formula (17.1.4) coincides with the preceding formula for L_r.) Also, for $i = 2, \ldots, r$:

$$L_1(\lambda)\cdots L_{i-1}(\lambda) = I\lambda^{l-l_i}$$
$$- (Z_{i1} + Z_{i2}\lambda + \cdots + Z_{i,l-l_i}\lambda^{l-l_i-1})\cdot(P_iC_{L|\mathcal{M}_i})^{l-l_i}Y_0$$

$$(17.1.5)$$

where $\mathcal{M}_i'$ is a direct complement to $\mathcal{M}_i$ in $\mathbb{C}^{nl}$, P_i is the projector on $\mathcal{M}_i'$ along $\mathcal{M}_i$, and

$$\begin{bmatrix} Z_{i1} \\ Z_{i2} \\ \vdots \\ Z_{i,l-l_i} \end{bmatrix} = [P_iY_0,\ P_iC_{L|\mathcal{M}_i}P_iY_0,\ \ldots,\ (P_iC_{L|\mathcal{M}_i})^{l-l_i-1}P_iY_0]^{-1}$$

Our next step is to show that this correspondence between factorizations of monic matrix polynomials $L(\lambda)$ and chains of certain C_L-invariant subspaces is continuous. To this end define a metric σ_k on the set $\mathscr{P}_k$ of all $n \times n$ monic matrix polynomials of degree k:

$$\sigma_k\left(I\lambda^k + \sum_{i=1}^{k-1} B_i\lambda^i,\ I\lambda^k + \sum_{i=0}^{k-1} B_i'\lambda^i\right) = \sum_{i=0}^{k-1} \|B_i - B_i'\|$$

Now fix a positive integer l. Consider the set $\mathcal{W}_r$ of all r-tuples $(\mathcal{M}_r, \mathcal{M}_{r-1}, \ldots, \mathcal{M}_2, L(\lambda))$, where $L(\lambda)$ is a monic matrix polynomial of degree l, and $\mathcal{M}_r \subset \mathcal{M}_{r-1} \subset \cdots \subset \mathcal{M}_2$ is a chain of C_L-invariant subspaces. The set $\mathcal{W}_r$ is a metric space with the metric

$$\theta_r((\mathcal{M}_r, \ldots, \mathcal{M}_2, L(\lambda)), (\mathcal{M}'_r, \ldots, \mathcal{M}'_2, L'(\lambda)))$$

$$= \sum_{i=2}^{r} \theta(\mathcal{M}_i, \mathcal{M}'_i) + \sigma_l(L(\lambda), L'(\lambda))$$

For every increasing sequence $\xi = \{l_r < l_{r-1} < \cdots < l_2\}$ of positive integers l_i with $l_2 < l$, define the subset $\mathcal{W}_{r,\xi}$ of $\mathcal{W}_r$ consisting of the elements $(\mathcal{M}_r, \ldots, \mathcal{M}_2, L(\lambda))$ from $\mathcal{W}_r$ with the additional property that the transformations (17.1.3) are invertible.

Theorem 17.1.1

For each ξ the set $\mathcal{W}_{r,\xi}$ is open in $\mathcal{W}_r$.

Proof. Define the subspace $\mathcal{G}_{l-p}$ of $\mathbb{C}^{nl}$ by the condition $x = \langle x_1, \ldots, x_l \rangle \in \mathcal{G}_{l-p}$ if and only if $x_1 = \cdots = x_p = 0$ (here $x_i \in \mathbb{C}^n$). As

$$\begin{bmatrix} X_0 \\ X_0 C_L \\ \vdots \\ X_0 C_L^{p-1} \end{bmatrix} = \begin{bmatrix} I_{np} & 0 \end{bmatrix}$$

for $p = 1, \ldots, l$, it follows that the transformation (17.1.3) is invertible if and only if $\mathcal{M}_i$ is a direct complement to $\mathcal{G}_{l-l_i}$ in $\mathbb{C}^{nl}$. From Theorem 13.1.3 it follows that, if $\mathcal{M}_i \dotplus \mathcal{G}_{l-l_i} = \mathbb{C}^{nl}$, then for $\epsilon > 0$ sufficiently small we also have $\mathcal{M}'_i \dotplus \mathcal{G}_{l-l_i} = \mathbb{C}^{nl}$ for every subspace $\mathcal{M}'_i$ in $\mathbb{C}^{nl}$ with $\theta(\mathcal{M}_i, \mathcal{M}'_i) < \epsilon$. Hence $\mathcal{W}_{r,\xi}$ is open in $\mathcal{W}_r$. $\square$

Now define a map

$$F_\xi : \mathcal{W}_{r,\xi} \to \mathcal{P}_{l-l_2} \times \mathcal{P}_{l_2-l_3} \times \cdots \times \mathcal{P}_{l_{r-1}-l_r} \times \mathcal{P}_{l_r}$$

where $\xi = \{l_r, \ldots, l_2\}$ is an increasing sequence of positive integers $l_r, l_{r-1}, \ldots, l_2$ with $l_2 < l$, as follows. Given $(\mathcal{M}_r, \mathcal{M}_{r-1}, \ldots, \mathcal{M}_2, L(\lambda)) \in \mathcal{W}_{r,\xi}$, the image of this element is $(L_1(\lambda), \ldots, L_r(\lambda))$, where the monic matrix polynomials $L_i(\lambda)$ are taken from the factorization

$$L(\lambda) = L_1(\lambda) L_2(\lambda) \cdots L_r(\lambda)$$

which corresponds to the chain $\mathcal{M}_r \subset \cdots \subset \mathcal{M}_2$ of C_L-invariant subspaces. It

is evident that F_ξ is one-to-one and surjective, so that the map F_ξ^{-1} exists.
Make the set $\mathcal{P}_\xi = \mathcal{P}_{l-l_2} \times \mathcal{P}_{l_2-l_3} \times \cdots \times \mathcal{P}_{l_{r-1}-l_r} \times \mathcal{P}_{l_r}$ into a metric space
by defining

$$\rho((\mathcal{L}_1, \ldots, L_r), (L_1', \ldots, L_r')) = \sigma_{l-l_2}(L_1, L_1')$$
$$+ \sigma_{l_2-l_3}(L_2, L_2') + \cdots + \sigma_{l_r}(L_r, L_r')$$

If X_1, X_2 are topological spaces with metrics ρ_1, ρ_2, defined on X_1 and X_2,
respectively, the map $G: X_1 \to X_2$ is said to be *locally Lipschitz continuous*
if, for every $x \in X_1$, there is a deleted neighbourhood U_x of x for which

$$\sup_{y \in U_x} \left(\frac{\rho_2(Gx, Gy)}{\rho_1(x, y)} \right) < \infty$$

Obviously, a locally Lipschitz continuous map is continuous. It is easy to see
that the composition of two locally Lipschitz continuous maps is again locally
Lipschitz continuous.

Theorem 17.1.2

The maps F_ξ and F_ξ^{-1} are locally Lipschitz continuous.

Proof. Given $(\mathcal{M}_r, \ldots, \mathcal{M}_2, L(\lambda)) \in \mathcal{W}_{r,\xi}$, write $M_i(\lambda) = L_1(\lambda) \cdots L_{i-1}(\lambda)$, $N_i(\lambda) = L_i(\lambda) \cdots L_r(\lambda)$, where the products $L_1 \cdots L_{i-1}$
and $L_i \cdots L_r$ are given by (17.1.5) and (17.1.4), respectively. Then

$$L(\lambda) = M_2(\lambda) N_2(\lambda) = \cdots = M_r(\lambda) N_r(\lambda)$$

We show first that the coefficients of $M_i(\lambda)$, $N_i(\lambda)$, $i = 1, \ldots, r-1$ are
locally Lipschitz continuous. Observe that in the representations (17.1.4)
and (17.1.5) the coefficients of M_i and N_i are uniformly bounded in some
neighbourhood of $(\mathcal{M}_r, \ldots, \mathcal{M}_2, L(\lambda))$. It is then easily seen that in order
to establish the local Lipschitz continuity of the coefficients of M_i and N_i it is
sufficient to verify the following assertion: for a fixed
$(\mathcal{M}_r, \ldots, \mathcal{M}_2, L(\lambda)) \in \mathcal{W}_{r,\xi}$ there exist positive constants δ and C such that,
for a set of subspaces $\mathcal{L}_{r-1}, \ldots, \mathcal{L}_1$ satisfying $\theta(\mathcal{L}_i, \mathcal{M}_i) < \delta$ for $i = 2, \ldots, r$,
it follows that

$$\mathcal{L}_i \dotplus \mathcal{G}_{l-l_i} = \mathbb{C}^{nl}$$

Here $\mathcal{G}_{l-l_i} = \{ \langle 0, \ldots, 0, \alpha_1, \ldots, \alpha_{n(l-l_i)} \rangle \in \mathbb{C}^{nl} \mid \alpha_j \in \mathbb{C} \}$ and $\| \tilde{P}_{\mathcal{L}_i} - \tilde{P}_{\mathcal{M}_i} \| \leq C\theta(\mathcal{L}_i, \mathcal{M}_i)$, where $\tilde{P}_{\mathcal{L}_i}$ (resp. $\tilde{P}_{\mathcal{M}_i}$) is the projector on $\mathcal{L}_i$ (resp. $\mathcal{M}_i$)
along $\mathcal{G}_{l-l_i}$. But this conclusion follows from Theorem 13.1.3. Hence the
coefficients of $M_i(\lambda)$ and $N_i(\lambda)$ are locally Lipschitz continuous functions of
an element in $\mathcal{W}_{r,\xi}$. In particular, $L_1 = M_2$ and $L_r = N_r$ are locally Lipschitz
continuous.

To prove this property for $L_2, \ldots, L_{r-1}$, note that

$$M_i(\lambda) L_i(\lambda) = M_{i+1}(\lambda), \qquad i = 2, \ldots, r - 1 \qquad (17.1.6)$$

Regard the equalities (17.1.6) as a system of linear equations

$$Ax = b \qquad (17.1.7)$$

where A and b are formed by the entries of coefficients of $M_i(\lambda)$ and $M_{i+1}(\lambda)$ for $i = 2, \ldots, r - 1$, and the unknown vector x is formed by the entries of the coefficients of $L_2, \ldots, L_{r-1}$. The system (17.1.7) has a unique solution x; hence the matrix A is left invertible. So $x = A'b$, where A' is a left inverse of A. Observe that every matrix B with $\|B - A\| < \frac{1}{2}\|A'\|^{-1}$ is also left invertible with a left inverse B' satisfying

$$\|B' - A'\| \leq 2\|A'\| \cdot \|B - A\|$$

(cf. the proof of Theorem 13.5.1). This inequality shows that x is a locally Lipschitz continuous function of $(\mathcal{M}_r, \ldots, \mathcal{M}_2, L(\lambda)) \in \mathcal{W}_{r,\xi}$, because A and b have this property.

To establish the local Lipschitz continuity of F_ξ^{-1}, we consider a fixed element $(L_1, \ldots, L_r) \in \mathcal{P}_\xi$. It is apparent that the polynomial $L = L_1 L_2 \cdots L_r$ will be a Lipschitz continuous function of $L_1, \ldots, L_r$ in a neighbourhood of this fixed element. Further, let $\mathcal{M}_r \subset \cdots \subset \mathcal{M}_2$ be the chain of C_L-invariant subspaces corresponding to the factorization $L = L_1 L_2 \cdots L_r$. Let $N_i = L_i L_{i+1} \cdots L_r$ for $i = 2, \ldots, r$, and let

$$\mathcal{G}_{l-m_i} = \{\langle 0, \ldots, 0, \alpha_1, \ldots, \alpha_{n(l-m_i)} \rangle \in \mathbb{C}^{nl} \mid \alpha_j \in \mathbb{C}\}$$

where l is the degree of L and m_i is the degree of N_i. The projector $P_{\mathcal{M}_i}$ on $\mathcal{M}_i$ along $\mathcal{G}_{l-m_i}$ is given by the formula

$$P_{\mathcal{M}_i} = \begin{bmatrix} I & 0 \\ F_i & 0 \end{bmatrix}, \qquad F_i = \begin{bmatrix} X_0(C_{N_i})^{m_i} \\ \vdots \\ X_0(C_{N_i})^{l-1} \end{bmatrix} \qquad (17.1.8)$$

where $X_0 = [I \ 0 \ \cdots \ 0]$ and C_{N_i} is the companion matrix of N_i. Indeed, obviously, $P_{\mathcal{M}_i}$ is a projector and $\operatorname{Ker} P_{\mathcal{M}_i} = \mathcal{G}_{l-m_i}$. Let us check that $\operatorname{Im} P_{\mathcal{M}_i} = \mathcal{M}_i$. Recall (see the proof of the converse statement of Theorem 5.3.2) that $\mathcal{M}_i$ is given by the formula

$$\mathcal{M}_i = \operatorname{Im}\{[\operatorname{col}[X_0 C_L^{j-1}]_{j=1}^l]^{-1} \operatorname{col}[X_0(C_{N_i})^{j-1}]_{j=1}^l\}$$

As

$$\operatorname{col}[X_0 C_L^{j-1}]_{j=1}^l = I$$

and

$$\text{col}[X_0(C_{N_i})^{j-1}]_{j=1}^{m_i} = I$$

we find that $\mathcal{M}_1 = \text{Im}\begin{bmatrix} I \\ F_i \end{bmatrix} = \text{Im } P_{\mathcal{M}_i}$. Formula (17.1.8) implies the local Lipschitz continuity of $P_{\mathcal{M}_i}$ (as a function of $(L_1, \ldots, L_r)$) and, therefore, also of $\mathcal{M}_i$ (cf. Theorem 13.1.1). $\square$

17.2 STABLE FACTORIZATIONS OF MATRIX POLYNOMIALS: MAIN RESULTS

We say that a factorization

$$L(\lambda) = L_1(\lambda)L_2(\lambda)\cdots L_r(\lambda) \tag{17.2.1}$$

of a monic matrix polynomial $L(\lambda)$, where $L_i(\lambda)$ are monic matrix polynomials as well, is *stable* if for any $\epsilon > 0$ there exists a $\delta > 0$ such that any monic matrix polynomial $\tilde{L}(\lambda)$ with $\sigma_i(\tilde{L}, L) < \delta$ admits a factorization $\tilde{L}(\lambda) = \tilde{L}_1(\lambda)\cdots\tilde{L}_r(\lambda)$, where $\tilde{L}_i(\lambda)$ are monic matrix polynomials satisfying

$$\max(\sigma_{l-l_2}(\tilde{L}_1, L_1), \sigma_{l_2-l_3}(\tilde{L}_2, L_2), \ldots, \sigma_{l_{r-1}-l_r}(\tilde{L}_{r-1}, L_{r-1}), \sigma_{l_r}(\tilde{L}_r, L_r)) < \epsilon$$

Here l is the degree of L and $\tilde{L}$, whereas for $i = 2, \ldots, r$, l_i is the degree of the products $L_{i+1}\cdots L_r$ and $\tilde{L}_{i+1}\cdots\tilde{L}_r$.

Recall the definition of a stable chain of invariant subspaces given in Section 15.6.

Theorem 17.2.1

Let equality (17.2.1) be a factorization of the monic matrix polynomial $L(\lambda)$. Let $(\mathcal{M}_r, \ldots, \mathcal{M}_2, L(\lambda)) = F_\xi^{-1}(L_1, \ldots, L_r)$ be the corresponding chain of C_L-invariant subspaces. Then the factorization (17.2.1) is stable if and only if the chain

$$\mathcal{M}_r \subset \cdots \subset \mathcal{M}_2 \tag{17.2.2}$$

is stable.

Proof. If the chain (17.2.2) is stable, then by Theorem 17.1.2 the factorization (17.2.1) is stable.

Now conversely, suppose that the factorization (17.2.1) is stable but the chain (17.2.2) is not. Then there exists an $\epsilon > 0$ and a sequence of matrices $\{C_m\}_{m=1}^\infty$, such that $\lim_{m\to\infty} C_m = C_L$ and for any chain

$$\mathscr{L}_r^{(m)} \subset \cdots \subset \mathscr{L}_2^{(m)}$$

of C_m-invariant subspaces the inequality

$$\max_{2 \le i \le r} \theta(\mathscr{M}_i, \mathscr{L}_i^{(m)}) \ge \epsilon$$

holds. Put $Q = \mathrm{col}[\delta_{i1}I]_{i=1}^l$ and

$$S_m = \mathrm{col}[QC_m^{i-1}]_{i=1}^l, \qquad m = 1, 2, \ldots$$

Then S_m converges to $\mathrm{col}[QC^{i-1}]_{i=1}^l$, which is equal to the unit $nl \times nl$ matrix. So without loss of generality we may assume that S_m is nonsingular for all m. Let $S_m^{-1} = [U_{m1}, U_{m2}, \ldots, U_{ml}]$, and note that

$$U_{mi} \to \mathrm{col}[\delta_{ji}I]_{j=1}^l, \qquad i = 1, \ldots, l \tag{17.2.3}$$

A straightforward calculation shows that $S_m C_m S_m^{-1}$ is the companion matrix associated with the monic matrix polynomial

$$M_m(\lambda) = \lambda^l I - \sum_{i=0}^{l-1} \lambda^i QC_m^l U_{m,i+1}$$

From (17.2.3) and the fact that $C_m \to C_L$ it follows that $\sigma_l(M_m, L) \to 0$. But then we may assume that for all m the polynomial M_m admits a factorization

$$M_m(\lambda) = L_{1,m}(\lambda) \cdots L_{r,m}(\lambda) \tag{17.2.4}$$

where $\sigma_{p_i}(L_{i,m}(\lambda), L_i(\lambda)) \to 0$ for $i = 1, \ldots, r$ (here p_i is the degree of L_i, which is also equal to the degree of L_{im} for $m = 1, 2, \ldots$).

Let $\mathscr{M}_{r,m} \subset \cdots \subset \mathscr{M}_{2,m}$ be the chain of C_{M_m}-invariant subspaces corresponding to the factorization (17.2.4), that is

$$F_\xi(\mathscr{M}_{r,m}, \ldots, \mathscr{M}_{2,m}, M_m(\lambda)) = (L_{1,m}(\lambda), \ldots, L_{r,m}(\lambda))$$

By Theorem 17.1.2 we have

$$\lim_{m \to \infty} \left(\sum_{i=2}^r \theta(\mathscr{M}_{i,m}, \mathscr{M}_i) \right) = 0$$

Put $\mathscr{V}_{i,m} = S_m^{-1} \mathscr{M}_{i,m}$ for $i = 2, \ldots, r$ and $m = 1, 2, \ldots$. Then $\mathscr{V}_{i,m}$ is an invariant subspace for C_m for each m. Moreover, it follows from $S_m \to I$ that, for $i = 2, \ldots, r$, $\theta(\mathscr{V}_{i,m}, \mathscr{M}_{i,m}) \to 0$ as $m \to \infty$. (Indeed, by Theorem 13.1.1

$$\theta(\mathcal{V}_{i,m}, \mathcal{M}_{i,m}) \leq \|S_m^{-1} P_{im} S_m - P_{im}\|$$

where P_{im} is the orthogonal projector on $\mathcal{M}_{i,m}$. Now

$$\|S_m^{-1} P_{im} S_m - P_{im}\| \leq \|(S_m^{-1} - I) P_{im} S_m\| + \|P_{im}(I - S_m)\|$$
$$\leq \max_m \|S_m\| \cdot \|S_m^{-1} - I\| + \|I - S_m\|$$

which tends to zero as m tends to infinity.) But then $\theta(\mathcal{V}_{im}, \mathcal{M}_i) \to 0$ as $m \to \infty$, for $i = 2, \ldots, r$. This contradicts the choice of C_m, and the proof of Theorem 17.2.1 is complete. $\square$

Comparing Theorem 17.2.1 with Corollary 14.6.2 and Theorem 14.2.1, we obtain the next result.

Corollary 17.2.2

A factorization

$$L(\lambda) = L_1(\lambda) L_2(\lambda), \ldots, L_r(\lambda)$$

with monic matrix polynomials $L(\lambda), L_1(\lambda), \ldots, L_r(\lambda)$ is stable if and only if the corresponding chain

$$\mathcal{M}_r \subset \cdots \subset \mathcal{M}_2$$

of C_L-invariant subspaces satisfies the condition that for every eigenvalue λ_0 of C_L with $\dim \mathrm{Ker}(C_L - \lambda_0 I) > 1$ and for every i ($2 \leq i \leq r$) either $\mathcal{M}_i \supset \mathcal{R}_{\lambda_0}(C_L)$ or $\mathcal{M}_i \cap \mathcal{R}_{\lambda_0}(C_L) = \{0\}$.

One can formulate a criterion for stability of factorizations of this kind in terms of eigenvalues of the polynomials $L_i(\lambda)$ rather than the companion matrix (as we have done in Corollary 17.2.2), as follows.

Theorem 17.2.3

A factorization (17.2.1) is stable if and only if, for any common eigenvalue λ_0 of a pair $L_i(\lambda)$, $L_j(\lambda)$ ($i \neq j$) we have $\dim \mathrm{Ker}\, L(\lambda_0) = 1$.

The proof of Theorem 17.2.3 is based on the following lemma.

Lemma 17.2.4

Let

$$A = \begin{bmatrix} A_1 & A_0 \\ 0 & A_2 \end{bmatrix}$$

be a transformation from $\mathbb{C}^m$ *into* $\mathbb{C}^m$, *written in matrix form with respect to the decomposition* $\mathbb{C}^m = \mathbb{C}^{m_1} \oplus \mathbb{C}^{m_2}$ $(m_1 + m_2 = m)$. *Then* $\mathbb{C}^m$ *is a stable invariant subspace for A if and only if for each common eigenvalue* λ_0 *of* A_1 *and* A_2 *the condition* $\dim \text{Ker}(\lambda_0 I - A) = 1$ *is satisfied.*

Proof. It is clear that $\mathbb{C}^{m_1}$ is an invariant subspace for A. We know from Theorem 15.2.1 that $\mathbb{C}^{m_1}$ is stable if and only if for each Riesz projector P of A corresponding to an eigenvalue λ_0 with $\dim \text{Ker}(\lambda_0 I - A) \geq 2$, we have $P\mathbb{C}^{m_1} = 0$ or $P\mathbb{C}^{m_1} = \text{Im } P$.

Let P be a Riesz projector of A corresponding to an arbitrary eigenvalue λ_0. Also for $j = 1, 2$, let P_j be the Riesz projector associated with A_j and λ_0:

$$P_j = \frac{1}{2\pi i} \int_{|\lambda - \lambda_0| = \epsilon} (I\lambda - A_j)^{-1} \, d\lambda$$

for $j = 1, 2$, where $\epsilon > 0$ is sufficiently small. Then

$$P = \begin{bmatrix} P_1 & \dfrac{1}{2\pi i} \displaystyle\int_{|\lambda - \lambda_0| = \epsilon} (I\lambda - A_1)^{-1} A_0 (I\lambda - A_2)^{-1} \, d\lambda \\ 0 & P_2 \end{bmatrix}$$

Observe that for $i = 1, 2$, the Laurent expansion of $(I\lambda - A_i)^{-1}$ at λ_0 has the form

$$(I\lambda - A_i)^{-1} = \sum_{j=-1}^{-q} (\lambda - \lambda_0)^j P_i Q_{ij} P_i + \cdots + \qquad (17.2.5)$$

where Q_{ij} are some transformations of $\text{Im } P_i$ into itself and the ellipsis on the right-hand side of (17.2.5) represents a series in nonnegative powers of $(\lambda - \lambda_0)$. From (17.2.5) one sees that P has the form

$$P = \begin{bmatrix} P_1 & P_1 Q_1 + Q_2 P_2 \\ 0 & P_2 \end{bmatrix}$$

where Q_1 and Q_2 are certain transformations acting from $\mathbb{C}^{m_2}$ into $\mathbb{C}^{m_1}$. It follows that $\{0\} \neq P\mathbb{C}^{m_1} \neq \text{Im } P$ if and only if $\lambda_0 \in \sigma(A_1) \cap \sigma(A_2)$. Now appeal to Theorem 15.2.1 (see first paragraph of the proof) to finish the proof. $\square$

Proof of Theorem 17.2.3. Let $(\mathcal{M}_r, \ldots, \mathcal{M}_2, L(\lambda)) = F_\xi^{-1}(L_1, \ldots, L_r)$ be the chain of C_L-invariant subspaces corresponding to the factorization (17.2.1). From Theorem 17.2.1 (taking into account Corollary 17.2.2) we know that this factorization is stable if and only if $\mathcal{M}_2, \ldots, \mathcal{M}_r$ are stable C_L-invariant subspaces. Let l be the degree of L, let r_i be the degree of $L_1 L_2 \cdots L_i$, and let

$$\mathcal{G}_i = \{\langle x_1, \ldots, x_l \rangle \in \mathbb{C}^{nl} \mid x_1 = \cdots = x_{l-r_i} = 0\}$$

Then $\mathbb{C}^{nl} = \mathcal{M}_i \dotplus \mathcal{G}_i$. With respect to this decomposition, write

$$C_L = \begin{bmatrix} C_{1i} & * \\ 0 & C_{2i} \end{bmatrix}$$

As we know (see Corollary 5.3.3), $\sigma(L_{i+1} \cdots L_r) = \sigma(C_{1i})$ and $\sigma(L_1 \cdots L_i) = \sigma(C_{1i})$. Also $\sigma(C_L) = \sigma(L)$; the desired result is now obtained by applying Lemma 17.2.4. $\square$

Another characterization of stable factorizations of monic matrix polynomials can be given in terms of isolatedness. Consider a factorization

$$L(\lambda) = L_1(\lambda)L_2(\lambda) \cdots L_r(\lambda) \tag{17.2.6}$$

of a monic matrix polynomial $L(\lambda)$ into the product of monic polynomials $L_1(\lambda), \ldots, L_r(\lambda)$, and let p_i be the degree of L_i for $i = 1, \ldots, r$. This factorization is called *isolated* if there exists an $\epsilon > 0$ such that any factorization

$$L(\lambda) = M_1(\lambda)M_2(\lambda) \cdots M_r(\lambda)$$

of $L(\lambda)$ with monic polynomials $M_i(\lambda)$ satisfying $\sigma_{p_i}(L_i(\lambda), M_i(\lambda)) < \epsilon$ (it is assumed that the degree of M_i is p_i) coincides with (17.2.6), that is, $M_i(\lambda) = L_i(\lambda)$ for $i = 1, \ldots, r$.

Theorem 17.2.5

A factorization (17.2.6) *is stable if and only if it is isolated.*

Proof. Let $(\mathcal{M}_r, \ldots, \mathcal{M}_2, L(\lambda)) = F_\xi^{-1}(L_1, L_2, \ldots, L_r)$ be the corresponding chain of C_L-invariant subspaces. By Theorems 17.1.2 and 17.2.1, the factorization (17.2.6) is isolated if and only if each $\mathcal{M}_i$ satisfies the condition that either $\mathcal{M}_i \supset \mathcal{R}_{\lambda_0}(C_L)$ or $\mathcal{M}_i \cap \mathcal{R}_{\lambda_0}(C_L) = \{0\}$ for every eigenvalue λ_0 of C_L with $\dim \operatorname{Ker}(C_L - \lambda_0 I) > 1$. Now it remains to appeal to Corollary 17.2.2. $\square$

We conclude this section with a statement concerning stability of the property that a given factorization of a monic matrix polynomial is stable.

Theorem 17.2.6

Assume that

$$L(\lambda) = L_1(\lambda)L_2(\lambda) \cdots L_r(\lambda)$$

is a stable factorization with monic matrix polynomials $L_1(\lambda)$, $L_2(\lambda), \ldots, L_r(\lambda)$. Then there exists an $\epsilon > 0$ such that every factorization

$$M(\lambda) = M_1(\lambda) M_2(\lambda) \cdots M_r(\lambda)$$

with monic matrix polynomials $M_1(\lambda), \ldots, M_r(\lambda)$ is stable provided

$$\sigma_{l-l_2}(M_1, L_1) + \sigma_{l_2-l_3}(M_2, L_2) + \cdots + \sigma_{l_{r-1}-l_r}(M_{r-1}, L_{r-1}) + \sigma_{l_r}(M_r, L_r) < \epsilon$$

where for $i = 2, \ldots, r$, l_i is the degree of the products $L_i \cdots L_r$ and $M_i \cdots M_r$.

The proof of Theorem 17.2.6 is obtained by combining Theorem 17.2.1 and Corollary 15.4.2.

17.3 LIPSCHITZ STABLE FACTORIZATIONS OF MONIC MATRIX POLYNOMIALS

A factorization

$$L(\lambda) = L_1(\lambda) L_2(\lambda) \cdots L_r(\lambda) \tag{17.3.1}$$

of the monic matrix polynomial $L(\lambda)$, where $L_1(\lambda), \ldots, L_r(\lambda)$ are monic matrix polynomials as well, is called *Lipschitz stable* if there exist positive constants ϵ and K such that any monic matrix polynomial $\tilde{L}(\lambda)$ with $\sigma_l(\tilde{L}, L) < \epsilon$ admits a factorization $\tilde{L}(\lambda) = \tilde{L}_1(\lambda) \cdots \tilde{L}_r(\lambda)$ with monic matrix polynomials $\tilde{L}_i(\lambda)$ satisfying

$$\max\{\sigma_{l-l_2}(\tilde{L}_1, L_1), \sigma_{l_2-l_3}(\tilde{L}_2, L_2), \ldots, \sigma_{l_r}(\tilde{L}_r, L_r)\} \le K \sigma_l(\tilde{L}, L)$$

Obviously, every Lipschitz stable factorization is stable. The converse is not true in general, as one can see from the results of this section.

We start with the correspondence between the factorization (17.3.1) and chains of C_L-invariant subspaces, where C_L is the companion matrix for $L(\lambda)$, described in Section 17.1.

Theorem 17.3.1

The factorization (17.3.1) is Lipschitz stable if and only if the corresponding chain of C_L-invariant subspaces

$$\mathcal{M}_r \subset \mathcal{M}_{r-1} \subset \cdots \subset \mathcal{M}_2 \tag{17.3.2}$$

is Lipschitz stable.

The Lipschitz stability of (17.3.2) is understood in the sense of Lipschitz stability of lattices of invariant subspaces (Section 15.6). In the particular case of chains, the chain (17.3.2) is, by definition, Lipschitz stable if there exist positive constants ϵ and K [that depend on C_L and the chain (17.3.2)] with the property that every $nl \times nl$ matrix A with $\|A - C_L\| < \epsilon$ has a chain

$$\mathcal{L}_r \subset \cdots \subset \mathcal{L}_2$$

of invariant subspaces such that

$$\max(\theta(\mathcal{M}_r, \mathcal{L}_r), \ldots, \theta(\mathcal{M}_2, \mathcal{L}_2)) \leq K\|A - C_L\|$$

Proof. If the chain (17.3.2) is Lipschitz stable, then by Theorem 17.1.2 the factorization (17.3.1) is Lipschitz stable. Conversely, assume that the factorization (17.3.1) is Lipschitz stable but the chain (17.3.2) is not. Then there exists a sequence $\{C_m\}_{m=1}^{\infty}$ of $nl \times nl$ matrices such that $\|C_m - C_L\| < (1/m)$ and for every chain $\mathcal{L}_r \subset \cdots \subset \mathcal{L}_2$ of C_m-invariant subspaces the inequality

$$\max(\theta(\mathcal{M}_r, \mathcal{L}_r), \ldots, \theta(\mathcal{M}_2, \mathcal{L}_2)) \geq m\|C_m - C_L\| \qquad (17.3.3)$$

holds. We continue now with an argument analogous to that used in the proof of Theorem 17.2.1. Putting $S_m = \mathrm{col}[QC_m^{i-1}]_{i=1}^l$, where $Q = \mathrm{col}[\delta_{i1}]_{i=1}^l$, we verify that S_m is nonsingular (at least for large m) and that $S_m C_m S_m^{-1}$ is the companion matrix associated with the matrix polynomial

$$M_m(\lambda) = \lambda^l I - \sum_{i=0}^{l-1} \lambda^i QC_m^l U_{m,i+1}$$

where $[U_{m1}, U_{m2}, \ldots, U_{ml}] = S_m^{-1}$. We assume that S_m is nonsingular for $m = 1, 2, \ldots$. Observe that $\mathrm{col}[QC_L^{i-1}]_{i=1}^l$ is the unit matrix I; so it is not difficult to check that for $m = 1, 2, \ldots$

$$\sigma_l(M_m, L) \leq K_1\|C_m - C_L\| \qquad (17.3.4)$$

Here and in the sequel we denote certain positive constants independent of m by $K_1, K_2, \ldots$. As the factorization (17.3.1) is Lipschitz stable, for m sufficiently large the polynomial $M_m(\lambda)$ admits a factorization

$$M_m(\lambda) = M_{1m}(\lambda) \cdots M_{rm}(\lambda) \qquad (17.3.5)$$

with monic matrix polynomials $M_{1m}(\lambda), \ldots, M_{rm}(\lambda)$ such that

$$\max(\sigma_{p_1}(M_{1m}, L_1), \ldots, \sigma_{p_r}(M_{rm}, L_r)) \leq K_2\sigma_l(M_m, L) \qquad (17.3.6)$$

Let $\mathcal{M}_{r,m} \subset \cdots \subset \mathcal{M}_{2,m}$ be the chain of C_{M_m}-invariant subspaces corresponding to the factorization (17.3.5). By Theorem 17.1.2 we have

$$\sum_{j=2}^{r} \theta(\mathcal{M}_{j,m}, \mathcal{L}_j) + \sigma_l(M_m, L) \le K_3 \left[\sum_{j=1}^{r} \sigma_{p_j}(M_{jm}, L_j) \right] \quad (17.3.7)$$

From (17.3.4), (17.3.6), and (17.3.7) one obtains

$$\sum_{j=2}^{r} \theta(\mathcal{M}_{j,m}, \mathcal{L}_j) \le r K_1 K_2 K_3 \| C_m - C_L \| \quad (17.3.8)$$

Put $\mathcal{V}_{i,m} = S_m^{-1} \mathcal{M}_{i,m}$ for $i = 2, \ldots, r$ and $m = 1, 2, \ldots$. Then $\mathcal{V}_{i,m}$ is C_m invariant for each m. Further, the formula for S_m shows that

$$\| I - S_m \| \le K_4 \| C_m - C_L \| \quad (17.3.9)$$

Indeed

$$I - S_m = \text{col}[Q(C_L^{i-1} - C_m^{i-1})]_{i=1}^{l}$$
$$= \text{col}[Q(C_L^{i-2}(C_L - C_m) + C_L^{i-3}(C_L - C_m)C_m$$
$$+ \cdots + C_L(C_L - C_m)C_m^{i-3} + (C_L - C_m)C_m^{i-2})]_{i=1}^{l}$$

and (17.3.9) follows. Now (cf. the proof of Theorem 17.2.1)

$$\theta(\mathcal{V}_{i,m}, \mathcal{M}_{i,m}) \le \max_{m} \| S_m \| \cdot \| S_m^{-1} - I \| + \| I - S_m \|$$
$$\le (\max_{m} \| S_m \| \max_{m} \| S_m^{-1} \| + 1) \| I - S_m \| \le K_5 \| C_m - C_L \|$$

Using this inequality and (17.3.8), we obtain

$$\sum_{j=2}^{r} \theta(\mathcal{V}_{i,m}, \mathcal{L}_j) \le \sum_{j=2}^{r} [\theta(\mathcal{V}_{i,m}, \mathcal{M}_{i,m}) + \theta(\mathcal{M}_{i,m}, \mathcal{L}_i)] \le K_6 \| C_m - C_L \|$$

a contradiction with (17.3.3). $\square$

Combining Theorem 17.3.1 with Theorems 15.6.2 and 15.5.1, we obtain the following corollary.

Corollary 17.3.2

For the factorization (17.3.1) and the corresponding chain of C_L-invariant subspaces (17.3.2), the following statements are equivalent: (a) the factorization (17.3.1) is Lipschitz stable; (b) all the C_L-invariant subspaces

$\mathcal{M}_2, \ldots, \mathcal{M}_r$ are spectral; (c) for every $\epsilon > 0$ sufficiently small there exists a $\delta > 0$ with the property every $nl \times nl$ matrix B with $\|B - C_L\| < \delta$ has a unique chain of invariant subspaces $\mathcal{N}_r \subset \mathcal{N}_{r-1} \subset \cdots \subset \mathcal{N}_2$ such that $\max(\theta(\mathcal{M}_r, \mathcal{N}_r), \ldots, \theta(\mathcal{M}_2, \mathcal{N}_2)) < \epsilon$.

Now we are ready to state and prove the main result of this section, namely, the description of Lipschitz stable factorizations. (Recall the definition of the metric σ_k on matrix polynomials given in Section 17.1.)

Theorem 17.3.3

The following statements are equivalent for a factorization

$$L(\lambda) = L_1(\lambda) \cdots L_r(\lambda) \tag{17.3.10}$$

of the monic $n \times n$ matrix polynomial $L(\lambda)$ of degree l, where $L_1(\lambda), \ldots, L_r(\lambda)$ are also monic matrix polynomials of degrees $p_1, \ldots, p_r$, respectively: (a) the factorization (17.3.10) is Lipschitz stable; (b) $\sigma(L_j) \cap \sigma(L_k) = \emptyset$ for $j \neq k$; (c) for every $\epsilon > 0$ sufficiently small there exists a $\delta > 0$ such that any monic matrix polynomial $\tilde{L}(\lambda)$ with $\sigma_l(L, \tilde{L}) < \delta$ has a unique factorization $\tilde{L}(\lambda) = \tilde{L}_1(\lambda) \cdots \tilde{L}_r(\lambda)$ with the property that $\max(\sigma_{p_1}(L_1, \tilde{L}_1), \ldots, \sigma_{p_r}(L_r, \tilde{L}_r)) < \epsilon$.

Proof. Observe that for $j = 2, \ldots, r$,

$$\sigma(C_L|_{\mathcal{M}_j}) = \sigma(L_j \cdots L_r)$$

where $\mathcal{M}_r \subset \cdots \subset \mathcal{M}_2$ is the chain of C_L-invariant subspaces corresponding to the factorization (17.3.10) (see formula (17.1.4)). Also, denoting by $\mathcal{M}_j'$ a direct complement to $\mathcal{M}_j$ in $\mathcal{M}_{j-1}$ for $j = 2, \ldots, r$, defining $\mathcal{M}_1 = \mathbb{C}^{nl}$, and letting $P_j : \mathcal{M}_{j-1} \to \mathcal{M}_j'$ be the projector on $\mathcal{N}_j'$ along $\mathcal{M}_j$, we have

$$\sigma(P_j C_L|_{\mathcal{M}_j'}) = \sigma(L_{j-1})$$

So, the subspaces $\mathcal{M}_j$ are spectral if and only if $\sigma(L_j) \cap \sigma(L_k) = \emptyset$ for $j \neq k$. Hence the equivalence (a) $\Leftrightarrow$ (b) in Theorem 17.3.3 follows from the equivalence (a) $\Leftrightarrow$ (b) in Corollary 17.3.2. Similarly, the equivalence (a) $\Leftrightarrow$ (c) in Theorem 17.3.3 follows from the corresponding equivalence in Corollary 17.3.2, taking account of Theorem 17.1.2. $\square$

17.4 STABLE MINIMAL FACTORIZATIONS OF RATIONAL MATRIX FUNCTIONS: THE MAIN RESULT

Throughout this section $W_0(\lambda)$, $W_{01}(\lambda)$, $W_{02}(\lambda), \ldots, W_{0k}(\lambda)$ are rational $n \times n$ matrix functions that take the value I at infinity. We assume that

$W_0(\lambda) = W_{01}(\lambda)W_{02}(\lambda)\cdots W_{0k}(\lambda)$ and that this factorization is minimal. The following notion of stability of this factorization is natural. Let

$$W_0(\lambda) = I_n + C_0(\lambda I_\delta - A_0)^{-1}B_0 \tag{17.4.1}$$

$$W_{0i}(\lambda) = I_n + C_{0i}(\lambda I_{\delta_i} - A_{0i})^{-1}B_{0i}, \qquad i = 1,\ldots,k \tag{17.4.2}$$

be the minimal realizations for W_0 and $W_{01}, W_{02}, \ldots, W_{0k}$ (so δ is the McMillan degree of W_0, and δ_i is the McMillan degree of W_i for $i = 1, \ldots, k$). The minimal factorization $W_0 = W_{01} \cdots W_{0k}$ is called *stable* if for each $\epsilon > 0$ there exists a $\omega > 0$ such that $\|A - A_0\| + \|B - B_0\| + \|C - C_0\| < \omega$ implies that the realization

$$W(\lambda) = I_n + C(\lambda I_\delta - A)^{-1}B$$

is minimal and W admits a minimal factorization $W = W_1 W_2, \ldots, W_k$, where for $i = 1, \ldots, k$, the rational matrix function $W_i(\lambda)$ has a minimal realization

$$W_i(\lambda) = I_n + C_i(\lambda I_{\delta_i} - A_i)^{-1}B_i$$

with the extra property that $\|A_i - A_{0i}\| + \|B_i - B_{0i}\| + \|C_i - C_{0i}\| < \epsilon$. Since all minimal realizations of a given rational matrix function are mutually similar (Theorems 7.1.4 and 7.1.5), this definition does not depend on the choice of the minimal realizations (17.4.1) and (17.4.2).

The next theorem characterizes stability of minimal factorizations in terms of spectral data.

Theorem 17.4.1

The minimal factorization $W_0(\lambda) = W_{01}(\lambda)W_{02}(\lambda)\cdots W_{0k}(\lambda)$ is stable if and only if each common pole (zero) of W_{0j} and W_{0p} ($j \neq p$) is a pole (zero) of W_0 of geometric multiplicity 1.

The *geometric multiplicity* of a pole (zero) λ_0 of a rational matrix function $W(\lambda)$ is the number of negative (positive) partial multiplicities of $W(\lambda)$ at λ_0 (see Section 7.2).

We need some preliminary discussion before starting the proof of Theorem 17.4.1. As we have seen in Theorem 7.5.1, the minimal factorizations

$$W_0(\lambda) = W_{01}(\lambda)\cdots W_{0k}(\lambda) \tag{17.4.3}$$

of $W_0(\lambda)$ are in one-to-one correspondence with those direct sum decompositions

$$\mathbb{C}^\delta = \mathcal{L}_1 \dotplus \cdots \dotplus \mathcal{L}_k \tag{17.4.4}$$

for which the subspaces $\mathscr{L}_1 \dotplus \cdots \dotplus \mathscr{L}_p$ $(p = 1, \ldots, k)$ are A_0-invariant and the subspaces $\mathscr{L}_k \dotplus \mathscr{L}_{k+1} \dotplus \cdots \dotplus \mathscr{L}_p$ $(p = k, \ldots, 1)$ are $A_0^\times$ invariant, where $A_0^\times = A_0 - B_0 C_0$. Moreover, the minimal factorization (17.4.3) corresponding to the direct sum decomposition (17.4.4) is given by

$$W_{0j}(\lambda) = I + C_0 \pi_j (\lambda I - A_0)^{-1} \pi_j B_0, \qquad j = 1, \ldots, k \quad (17.4.5)$$

where π_j is the projector on $\mathscr{L}_j$ along $\mathscr{L}_1 \dotplus \cdots \dotplus \mathscr{L}_{j-1} \dotplus \mathscr{L}_{j+1} \dotplus \cdots \dotplus \mathscr{L}_k$; note that the realizations (17.4.5) are necessarily minimal. In the formula (17.4.5) the transformations $C_0 \pi_j : \mathscr{L}_j \to \mathfrak{C}^n$, $\pi_j A_0 \pi_j : \mathscr{L}_j \to \mathscr{L}_j$, and $\pi_j B_0 : \mathfrak{C}^n \to \mathscr{L}_j$ are understood as matrices of sizes $n \times l_j$, $l_j \times l_j$, and $l_j \times n$, respectively, where $l_j = \dim \mathscr{L}_j$, with respect to some basis in $\mathscr{L}_j$.

Let (A, B, C) be a triple of matrices of sizes $\delta \times \delta$, $\delta \times n$, $n \times \delta$, respectively. Consider the ordered k-tuple $\Pi = (\pi_1, \ldots, \pi_k)$ of projectors in $\mathfrak{C}^\delta$. We say that Π is a *supporting k-tuple of projectors* with respect to the triple of matrices (A, B, C) if $\pi_j \pi_i = \pi_i \pi_j = 0$ for $i \neq j$, $\pi_1 + \cdots + \pi_k = I$, the subspaces $\mathrm{Im}(\pi_1 + \cdots + \pi_p)$ for $p = 1, 2, \ldots, k$ are A invariant, and the subspaces $\mathrm{Im}(\pi_p + \pi_{p+1} + \cdots + \pi_k)$, $p = 1, \ldots, k$, are $A^\times$ invariant, where $A^\times = A - BC$. Clearly, Π is a supporting k-tuple of projectors with respect to (A_0, B_0, C_0) if and only if the subspaces $\mathscr{L}_j = \mathrm{Im}\,\pi_j (j = 1, \ldots, k)$ form a direct sum decomposition of $\mathfrak{C}^\delta$ as in (17.4.4).

A supporting k-tuple of projectors $\Pi = (\pi_1, \ldots, \pi_k)$ with respect to (A, B, C) will be called *stable* if for every $\epsilon > 0$ there exists an $\omega > 0$ such that, for any triple of matrices (A', B', C') of sizes $\delta \times \delta$, $\delta \times n$, $n \times \delta$, respectively, with $\|A - A'\| + \|B - B'\| + \|C - C'\| < \omega$, there exists a supporting k-tuple of projectors $\Pi' = (\pi_1', \ldots, \pi_k')$ with respect to (A', B', C') such that

$$\sum_{i=1}^{k} \|\pi_i' - \pi_i\| < \epsilon$$

The first step in the proof of Theorem 17.4.1 is the following lemma.

Lemma 17.4.2

Let (17.4.1) *be a minimal realization for* $W_0(\lambda)$, *and let* $\Pi = (\pi_1, \ldots, \pi_k)$ *be a supporting k-tuple of projectors with respect to* (A_0, B_0, C_0), *with the corresponding minimal factorization*

$$W_0(\lambda) = W_{01}(\lambda) W_{02}(\lambda) \cdots W_{0k}(\lambda) \quad (17.4.6)$$

(so that, for $j = 1, \ldots, k$, $W_{0j}(\lambda) = I + C_0 \pi_j (\lambda I - A_0)^{-1} \pi_j B_0$ *with respect to some basis* $x_1^{(i)}, \ldots, x_{l_j}^{(j)}$ *in* $\mathrm{Im}\,\pi_j$). *Then* Π *is stable if and only if the factorization* (17.4.6) *is stable.*

The proof of Lemma 17.4.2 is rather long and technical and is given in the next section.

Next, we make the connection with stable invariant subspaces.

Lemma 17.4.3

Let $\Pi = (\pi_1, \ldots, \pi_k)$ *be a supporting k-tuple of projectors with respect to* (A_0, B_0, C_0). *Then* Π *is stable if and only if the* A_0-*invariant subspaces* $\text{Im}(\pi_1 + \cdots + \pi_j)$, $j = 1, \ldots, k$ *are stable and the* $A_0^\times$-*invariant subspaces* $\text{Im}(\pi_j + \pi_{j+1} + \cdots + \pi_k)$, $j = 1, \ldots, k$ *are stable as well (as before,* $A_0^\times = A_0 - B_0 C_0$).

Again, it will be convenient to relegate the proof of Lemma 17.4.3 to the next section.

Proof of Theorem 17.4.1. Let $\Pi = (\pi_1, \ldots, \pi_k)$ be the supporting k-tuple of projectors with respect to (A_0, B_0, C_0) that corresponds to the minimal factorization

$$W_0(\lambda) = W_{01}(\lambda)W_{02}(\lambda) \cdots W_{0k}(\lambda) \tag{17.4.7}$$

By Lemmas 17.4.2 and 17.4.3 the factorization (17.4.7) is stable if and only if the A_0-invariant subspaces $\mathcal{L}_j \stackrel{\text{def}}{=} \text{Im}(\pi_1 + \cdots + \pi_j)$, $j = 1, \ldots, k$ are stable and the $A_0^\times$-invariant subspaces $\mathcal{M}_j \stackrel{\text{def}}{=} \text{Im}(\pi_j + \pi_{j+1} + \cdots + \pi_k)$, $j = 1, \ldots, k$ are stable as well.

With respect to the decomposition $\mathbb{C}^\delta = \text{Im } \pi_1 \dotplus \text{Im } \pi_2 \dotplus \cdots \dotplus \text{Im } \pi_k$, write

$$A_0 = \begin{bmatrix} A_{11} & A_{12} & \cdots & A_{1k} \\ 0 & A_{22} & \cdots & A_{2k} \\ \vdots & \vdots & & \vdots \\ 0 & 0 & \cdots & A_{kk} \end{bmatrix}, \quad A_0^\times = \begin{bmatrix} A_{11}^\times & 0 & \cdots & 0 \\ A_{21}^\times & A_{22}^\times & \cdots & 0 \\ \vdots & \vdots & & \vdots \\ A_{k1}^\times & A_{k2}^\times & \cdots & A_{kk}^\times \end{bmatrix}$$

In view of Lemma 17.2.4, $\mathcal{L}_j$ is stable if and only if, for every common eigenvalue λ_0 of

$$\begin{bmatrix} A_{11} & A_{12} & \cdots & A_{1j} \\ 0 & A_{22} & \cdots & A_{2j} \\ \vdots & \vdots & & \vdots \\ 0 & 0 & \cdots & A_{jj} \end{bmatrix} \quad \text{and} \quad \begin{bmatrix} A_{j+1,j+1} & A_{j+1,j+2} & \cdots & A_{j+1,k} \\ 0 & A_{j+2,j+2} & \cdots & A_{j+2,k} \\ \vdots & \vdots & & \vdots \\ 0 & 0 & \cdots & A_{kk} \end{bmatrix}$$

we have dim $\text{Ker}(\lambda_0 I - A) = 1$. So all the subspaces $\mathcal{L}_1, \ldots, \mathcal{L}_k$ are stable if and only if every common eigenvalue of A_{jj} and A_{pp} ($j \neq p$) is an eigenvalue of A_0 of geometric multiplicity 1. Similarly, all the subspaces $\mathcal{M}_1, \ldots, \mathcal{M}_k$

are stable if and only if every common eigenvalue of $A_{jj}^{\times}$ and $A_{pp}^{\times}$ with $j \neq p$ is an eigenvalue of $A_0^{\times}$ of geometric multiplicity 1. It follows that the factorization (17.4.7) is stable if and only if every common eigenvalue of A_{jj} and A_{pp} (resp. of $A_{jj}^{\times}$ and $A_{pp}^{\times}$) with $j \neq p$ is an eigenvalue of A_0 (resp. of $A_0^{\times}$) of geometric multiplicity 1. To finish the proof, observe that the realizations (17.4.5) are minimal and hence, by Theorem 7.2.3, the poles (resp. zeros) of $W_{0j}(\lambda)$ coincide with the eigenvalues of $\pi_j A_0 \pi_j = A_{jj}$ (resp. eigenvalues of $\pi_j A_0^{\times} \pi_j = A_{jj}^{\times}$). Also, the partial multiplicities of a pole (resp. zero) λ_0 of W_{0j} are equal to the partial multiplicities of λ_0 as an eigenvalue of A_{jj} (resp. $A_{jj}^{\times}$). Analogous statements hold for the poles and zeros of $W_0(\lambda)$ and eigenvalues of A_0 and $A_0^{\times}$. □

17.5   PROOF OF THE AUXILIARY LEMMAS

We start with the proof of Lemma 17.4.2.

Assume that Π is stable. Given $\epsilon > 0$, let ϵ' be a positive number that we fix later. By Lemma 13.3.2 there exists an $\omega_1 > 0$ with the property that, for any projector π_j' such that $\| \pi_j' - \pi_j \| < \omega_1$, there exists an invertible transformation $S_j \colon \mathbb{C}^{\delta} \to \mathbb{C}^{\delta}$ with $S_j(\operatorname{Im} \pi_j) = \operatorname{Im} \pi_j'$ and $\| I - S_j \| < \epsilon'$. We also assume that $\omega_1 \leq \min(\epsilon', 1)$. Further, let ω_2 be the number corresponding to ω_1 as defined by the stability of Π.

As the realization (17.4.1) is minimal, in view of Theorem 7.1.5 the matrix $\operatorname{col}[C_0 A_0^j]_{j=0}^{p-1}$ is left invertible, where p is the degree of the minimal polynomial for A_0, and the matrix $[B_0, A_0 B_0, \ldots, A_0^{p-1} B_0]$ is right invertible. Since the left (right) invertibility of a matrix X is stable under small perturbations [indeed, if $\| Y - X \| < \| X^I \|^{-1}$, then Y is also left (right) invertible], there exists a $\tau > 0$ such that the realization

$$W(\lambda) = I_n + C(\lambda I_{\delta} - A)^{-1} N \qquad (17.5.1)$$

is minimal provided $\| A - A_0 \| + \| B - B_0 \| + \| C - C_0 \| < \tau$.

Now put $\omega = \min(\omega_1, \omega_2, \tau, \epsilon')$ and let (A, B, C) be such that

$$\| A - A_0 \| + \| B - B_0 \| + \| C - C_0 \| < \omega$$

Then the realization (17.5.1) is minimal. By the stability of Π, there exists a supporting k-tuple of projectors $\Pi' = (\pi_1', \ldots, \pi_k')$ with respect to (A, B, C) such that

$$\sum_{j=1}^{k} \| \pi_i' - \pi_i \| < \omega_1$$

For $j = 1, \ldots,$ let $S_j \colon \mathbb{C}^{\delta} \to \mathbb{C}^{\delta}$ be invertible transformations with $S_j(\operatorname{Im} \pi_j) = \operatorname{Im} \pi_j'$ and $\| I - S_j \| < \epsilon'$. Now put

$$\tilde{W}_j(\lambda) = I_n + C\pi'_j S_j (\lambda I - S_j^{-1} \pi'_j A \pi'_j S_j)^{-1} S_j^{-1} \pi'_j B \qquad (17.5.2)$$

for each j, where the transformation S_j is understood as $S_j : \text{Im } \pi_j \to \text{Im } \pi'_j$. Also, we regard the rational functions (17.5.2) as matrix functions with respect to the basis introduced for Im π_j. We have the minimal factorization

$$W(\lambda) = \tilde{W}_1(\lambda)\tilde{W}_2(\lambda) \cdots \tilde{W}_k(\lambda)$$

Moreover, writing $\rho = \max(\|A_0\|, \|B_0\|, \|C_0\|, \|\pi_1\|, \ldots, \|\pi_k\|)$, we obtain

$$\|C_0\pi_j - S\pi'_j S_j\| + \|\pi_j A_0 \pi_j - S_j^{-1}\pi'_j A\pi'_j S_j\| + \|\pi_j B_0 - S_j^{-1}\pi'_j B\|$$
$$\leq \|C_0\pi_j(I - S_j)\| + \|C_0(\pi_j - \pi'_j)S_j\| + \|(C - C_0)\pi'_j S_j\|$$
$$\quad + \|(I - S_j^{-1})\pi_j A_0 \pi_j\| + \|S_j^{-1}\pi_j A_0 \pi_j(I - S_j)\| + \|S_j^{-1}(\pi_j - \pi'_j)A_0\pi_j S_j\|$$
$$\quad + \|S_j^{-1}\pi'_j A_0(\pi_j - \pi'_j)S_j\| + \|S_j^{-1}\pi'_j(A_0 - A)\pi'_j S_j\| + \|(I - S_j^{-1})\pi_j B_0\|$$
$$\quad + \|S_j^{-1}(\pi_j - \pi'_j)B_0\| + \|S_j^{-1}\pi'_j(B_0 - B)\|$$
$$\leq \rho^2\epsilon' + \rho\omega_1(1 + \epsilon') + \omega(\omega_1 + \rho)(1 + \epsilon')$$
$$\quad + \|(I - S_j^{-1})\|\rho^3 + \|S_j^{-1}\|\rho^3\epsilon' + \|S_j^{-1}\|\omega_1\rho^2(1 + \epsilon')$$
$$\quad + \|S_j^{-1}\|(\omega_1 + \rho)\rho\omega_1(1 + \epsilon') + \|S_j^{-1}\|(\omega_1 + \rho)^2\omega(1 + \epsilon') + \|I - S_j^{-1}\|\rho^2$$
$$\quad + \|S_j^{-1}\|\omega_1\rho + \|S_j^{-1}\|(\omega_1 + \rho)\omega$$

Use the inequalities $\omega_1 \leq \epsilon'$, $\omega \leq \epsilon'$ and the inequalities $\|S_j^{-1}\| \leq (1 - \epsilon')^{-1}$, $\|I - S_j^{-1}\| \leq \epsilon'(1 - \epsilon')^{-1}$ (assuming $\epsilon' < 1$; cf. the proof of Theorem 16.2.1) to get

$$\|C_0\pi_j - C\pi'_j S_j\| + \|\pi_j A_0 \pi_j - S_j^{-1}\pi'_j A\pi'_j S_j\| + \|\pi_j B_0 - S_j^{-1}\pi'_j B\|$$
$$\leq \rho^2\epsilon' + \rho\epsilon'(1 + \epsilon') + \epsilon'(\epsilon' + \rho)(1 + \epsilon') + 2\epsilon'(1 - \epsilon')^{-1}\rho^3$$
$$\quad + (1 - \epsilon')^{-1}\epsilon'\rho^2(1 + \epsilon') + (1 - \epsilon')^{-1}(\epsilon' + \rho)\rho\epsilon'(1 + \epsilon')$$
$$\quad + (1 - \epsilon')^{-1}(\epsilon' + \rho)^2\epsilon'(1 + \epsilon') + \epsilon'(1 - \epsilon')^{-1}\rho^2 + (1 - \epsilon')^{-1}\epsilon'\rho$$
$$\quad + (1 - \epsilon')^{-1}(\epsilon' + \rho)\epsilon'$$

It remains to choose $\epsilon' < 1$ in such a way that this expression is less than ϵ, and the stability of factorization (17.4.6) is proved.

Conversely, let the factorization (17.4.6) be stable and assume that Π is not stable. Then there exist an $\epsilon > 0$ and sequences $\{A_m\}_{m=1}^{\infty}$, $\{B_m\}_{m=1}^{\infty}$, $\{C_m\}_{m=1}^{\infty}$ such that

$$\lim_{m \to \infty} \{\|A_m - A_0\| + \|B_m - B_0\| + \|C_m - C_0\|\} = 0 \qquad (17.5.3)$$

and

$$\sum_{i=1}^{k} \|\pi'_i - \pi_i\| \geq \epsilon$$

where $\Pi = (\pi'_1, \ldots, \pi'_k)$ is any supporting k-tuple of projectors with respect to at least one of the triples (A_m, B_m, C_m), $m = 1, 2, \ldots$. Since (17.5.1) is a minimal realization, we can assume (using Theorem 7.1.5 and the fact that the full-range and null kernel properties of a pair of transformations are preserved under sufficiently small perturbation of this pair) that

$$W_m(\lambda) \overset{\text{def}}{=} I_n + C_m(\lambda I_\delta - A_m)^{-1} B_m$$

is minimal for all m. In view of the stability of (17.4.6), we can also assume that each $W_m(\lambda)$ admits a minimal factorization

$$W_m(\lambda) = W_{m1}(\lambda) W_{m2}(\lambda) \cdots W_{mk}(\lambda) \qquad (17.5.4)$$

where for $j = 1, 2, \ldots k$, we obtain

$$W_{mj}(\lambda) = I + C_{mj}(\lambda I - A_{mj})^{-1} B_{mj} \qquad (17.5.5)$$

and

$$C_{mj} : \text{Im } \pi_j \to \mathbb{C}^n, \qquad A_{mj} : \text{Im } \pi_j \to \text{Im } \pi_j, \ B_{mj} : \mathbb{C}^n \to \text{Im } \pi_j$$

are transformations written as matrices with respect to the basis introduced for $\text{Im } \pi_j$ with the property that

$$\lim_{m \to \infty} \{ \|C_{mj} - C_0 \pi_j\| + \|A_{mj} - \pi_j A_0 \pi_j\| + \|B_{mj} - \pi_j B_0\| \} = 0 \qquad (17.5.6)$$

For fixed m, consider the minimal realization

$$W_m(\lambda) = I + \tilde{C}_m(\lambda I - \tilde{A}_m)^{-1} \tilde{B}_m$$

where

$$\tilde{C}_m = [C_{m1}, C_{m2}, \ldots, C_{mk}]$$

$$\tilde{A}_m = \begin{bmatrix} A_{m1} & B_{m1}C_{m2} & \cdots & B_{m1}C_{mk} \\ 0 & A_{m2} & \cdots & B_{m2}C_{mk} \\ \vdots & \vdots & & \vdots \\ 0 & 0 & \cdots & A_{mk} \end{bmatrix}, \qquad \tilde{B}_m = \begin{bmatrix} B_{m1} \\ B_{m2} \\ \vdots \\ B_{mk} \end{bmatrix}$$

obtained from the minimal factorization (17.5.3) [cf. formula (7.3.4)]. As any two minimal realizations of $W_m(\lambda)$ are similar, there exists an invertible transformation $S_m: \mathrm{Im}\,\pi_1 \,\dot{+}\, \cdots \,\dot{+}\, \mathrm{Im}\,\pi_k \to \mathbb{C}^\delta$ such that

$$C_m S_m = \tilde{C}_m, \qquad S_m^{-1} A_m S_m = \tilde{A}_m, \qquad S_m^{-1} B_m = \tilde{B}_m$$

Actually, such an S_m is unique, and from the explicit formula for S_m (Theorem 7.1.3) we find, using (17.5.3) and (17.5.6), that $S_m \to I$ as $m \to \infty$.

Now let $\Pi^{(m)} = (\pi_1^{(m)}, \ldots, \pi_k^{(m)})$ be the supporting k-tuple of projectors with respect to (A_m, B_m, C_m), which corresponds to the minimal factorization (17.5.4). Thus, for $j = 1, \ldots, k$ we have

$$W_{mj}(\lambda) = I + C_m \pi_j^{(m)} (\lambda I - \pi_j^{(m)} A_m \pi_j^{(m)})^{-1} \pi_j^{(m)} B_m \qquad (17.5.7)$$

and hence $\pi_j^{(m)} = S_m \pi_j S_m^{-1}$. We find that $\Sigma_{j=1}^k \|\pi_j^{(m)} - \pi_j\| \to 0$ as $m \to \infty$, a contradiction with the choice of (A_m, B_m, C_m). Lemma 17.4.2 is proved.

We pass on to the proof of Lemma 17.4.3. Assume that the subspaces $\mathrm{Im}(\pi_1 + \cdots + \pi_j)$, $j = 1, \ldots, k$ are stable A_0-invariant subspaces and that $\mathrm{Im}(\pi_j + \pi_{j+1} + \cdots + \pi_k)$ are stable $A_0^\times$-invariant subspaces. Arguing by contradiction, assume that Π is not stable. Then there exist an $\epsilon > 0$ and sequences $\{A_m\}_{m=1}^\infty$, $\{B_m\}_{m=1}^\infty$, and $\{C_m\}_{m=1}^\infty$ such that

$$\lim_{m \to \infty} \{\|A_m - A_0\| + \|B_m - B_0\| + \|C_m - C_0\|\} = 0$$

and

$$\sum_{j=1}^k \|\pi_j' - \pi_j\| \geq \epsilon \qquad (17.5.8)$$

for every supporting k-tuple of projectors $(\pi_1', \ldots, \pi_k')$ with respect to (A_m, B_m, C_m), $m = 1, 2, \ldots$. Then clearly $A_m \to A_0$ and $A_m^\times \stackrel{\mathrm{def}}{=} A_m - B_m C_m \to A_0^\times$ as $m \to \infty$. By assumption, and using Theorem 15.6.1, for each positive integer m there exists a sequence of chains of subspaces $\{0\} \subset \mathcal{L}_1^{(m)} \subset \cdots \subset \mathcal{L}_{k-1}^{(m)} \subset \mathcal{L}_k^{(m)} = \mathbb{C}^\delta$, such that $\mathcal{L}_1^{(m)}, \ldots, \mathcal{L}_k^{(m)}$ are A_m invariant and

$$\lim_{m \to \infty} \theta(\mathcal{L}_j^{(m)}, \mathrm{Im}(\pi_1 + \cdots + \pi_j)) = 0 \quad \text{for} \quad j = 1, \ldots, k \quad (17.5.9)$$

Similarly, there exists a sequence of chains of subspaces

$$\mathbb{C}^\delta = \mathcal{M}_1^{(m)} \supset \mathcal{M}_2^{(m)} \supset \cdots \supset \mathcal{M}_k^{(m)} \supset \{0\}, \qquad m = 1, 2, \ldots$$

such that $\mathcal{M}_j^{(m)}$, $j = 1, \ldots, k$ are $A_m^\times$ invariant and

$$\lim_{m \to \infty} \theta(\mathcal{M}_j^{(m)}, \mathrm{Im}(\pi_j + \pi_{j+1} + \cdots + \pi_k)) = 0, \, j = 1, \ldots, k$$

$$(17.5.10)$$

As $\operatorname{Im}(\pi_1 + \cdots + \pi_j) \dotplus \operatorname{Im}(\pi_{j+1} + \cdots + \pi_k) = \mathbb{C}^\delta$, for $j = 1, \ldots, k-1$ and sufficiently large m, we find, using Lemma 13.3.2, that

$$\mathscr{L}_j^{(m)} \dotplus \mathscr{M}_{j+1}^{(m)} = \mathbb{C}^\delta , \qquad j = 1, \ldots, k-1$$

Now let

$$\mathscr{N}_j^{(m)} = \mathscr{L}_j^{(m)} \cap \mathscr{M}_j^{(m)} , \qquad j = 1, \ldots, k$$

It is easy to see that

$$\mathscr{N}_j^{(m)} \cap \mathscr{L}_{j-1}^{(m)} = \{0\} , \qquad j = 2, \ldots, k \qquad (17.5.11)$$

Furthermore

$$\mathscr{N}_1^{(m)} + \cdots + \mathscr{N}_j^{(m)} = \mathscr{L}_j^{(m)} , \qquad j = 1, \ldots, k \qquad (17.5.12)$$

Indeed, (17.5.12) obviously holds for $j = 1$. Assuming that (17.5.12) is proved for $j = p - 1$, we have

$$\mathscr{N}_1^{(m)} + \cdots + \mathscr{N}_p^{(m)} = \mathscr{L}_{p-1}^{(m)} + (\mathscr{L}_p^{(m)} \cap \mathscr{M}_p^{(m)})$$

where is clearly contained in $\mathscr{L}_p^{(m)}$. Take $x \in \mathscr{L}_p^{(m)}$, and write $x = y + z$, where $y \in \mathscr{L}_{p-1}^{(m)}$ and $z \in \mathscr{M}_p^{(m)}$. Then $z = x - y \in \mathscr{L}_p^{(m)}$, and $x \in \mathscr{L}_{p-1}^{(m)} + (\mathscr{L}_p^{(m)} \cap \mathscr{M}_p^{(m)})$. So (17.5.12) is proved. Combining (17.5.11) and (17.5.12), we find that

$$\mathscr{N}_1^{(m)} \dotplus \cdots \dotplus \mathscr{N}_k^{(m)} = \mathbb{C}^\delta$$

Developing an analog of the proof of (17.5.12), one proves that

$$\mathscr{N}_j^{(m)} + \mathscr{N}_{j+1}^{(m)} + \cdots + \mathscr{N}_k^{(m)} = \mathscr{M}_j^{(m)} , \qquad j = 1, \ldots, k$$

For sufficiently large m, let $\pi_j^{(m)}$ be the projector on $\mathscr{N}_j^{(m)}$ along $\mathscr{N}_1^{(m)} \dotplus \cdots \dotplus \mathscr{N}_{j-1}^{(m)} \dotplus \mathscr{N}_{j+1}^{(m)} \dotplus \cdots \dotplus \mathscr{N}_k^{(m)}$. Then the k-tuple of projectors $(\pi_1^{(m)}, \pi_2^{(m)}, \ldots, \pi_k^{(m)})$ is supporting for (A_m, B_m, C_m). Denoting by $\tau_j^{(m)}$ the projector on $\mathscr{L}_j^{(m)}$ along $\mathscr{M}_{j+1}^{(m)}$ ($j = 1, \ldots, k-1$), we have $\tau_j^{(m)} = \pi_1^{(m)} + \cdots + \pi_j^{(m)}$. On the other hand, (17.5.9) and (17.5.10) imply, in view of Theorem 13.4.3, that for $j = 1, \ldots, k-1$.

$$\lim_{m \to \infty} \| \tau_j^{(m)} - (\pi_1 + \cdots + \pi_j) \| = 0$$

and so $\lim_{m \to \infty} \| \pi_j^{(m)} - \pi_j \| = 0$, a contradiction with (17.5.8).

Conversely, assume that Π is stable, but one of the A_0-invariant subspaces $\operatorname{Im} \pi_1, \ldots, \operatorname{Im}(\pi_1 + \cdots + \pi_k)$, say, $\operatorname{Im}(\pi_1 + \cdots + \pi_j)$, is not stable.

Then there exist an $\epsilon > 0$ and a sequence $\{A_m\}_{m=1}^{\infty}$ such that $\| A_m - A_0 \| \to 0$ as $m \to \infty$ and

$$\theta(\mathcal{M}, \mathrm{Im}(\pi_1 + \cdots + \pi_j)) \geq \epsilon \qquad (17.5.13)$$

for every A_m-invariant subspace $\mathcal{M}$ $(m = 1, 2, \ldots)$. As Π is stable, there exists a sequence of k-tuples of projectors $\Pi^{(m)} = (\pi_1^{(m)}, \ldots, \pi_k^{(m)})$, $m = 1, 2, \ldots$ such that $\Pi^{(m)}$ is supporting for (A_m, B_0, C_0) and

$$\lim_{m \to \infty} \left[\sum_{p=1}^{k} \| \pi_p^{(m)} - \pi_p \| \right] = 0$$

Hence for the A_m-invariant subspace $\mathrm{Im}(\pi_1^{(m)} + \cdots + \pi_j^{(m)})$ we have

$$\lim_{m \to \infty} \theta(\mathrm{Im}(\pi_1^{(m)} + \cdots + \pi_j^{(m)}), \mathrm{Im}(\pi_1 + \cdots + \pi_j)) = 0$$

a contradiction with (17.5.13). In a similar way, one arrives at a contradiction if Π is stable but one of the $A_0^{\times}$-invariant subspaces $\mathrm{Im}(\pi_j + \pi_{j+1} + \cdots + \pi_k)$, $j = 1, \ldots, k$, is not stable.

Lemma 17.4.3 is proved completely.

17.6 STABLE MINIMAL FACTORIZATIONS OF RATIONAL MATRIX FUNCTIONS: FURTHER DEDUCTIONS

In this section we use Theorem 17.4.1 and its proof to derive some useful information on stable minimal factorizations of rational matrix functions. First, let us make Theorem 17.4.1 more precise in the sense that if the minimal factorization

$$W_0(\lambda) = W_{01}(\lambda) \cdots W_{0k}(\lambda) \qquad (17.6.1)$$

is stable, then so is every minimal factorization sufficiently close to (17.6.1).

Theorem 17.6.1

Assume that (17.6.1) *is a stable minimal factorization, and let*

$$W_0(\lambda) = I_n + C_0(\lambda I_\delta - A_0)^{-1} B_0 \qquad (17.6.2)$$

and

$$W_{0j}(\lambda) = I_n + C_{0j}(\lambda I_{l_j} - A_{0j})^{-1} B_{0j}, \qquad j = 1, \ldots, k \qquad (17.6.3)$$

be minimal realizations of $W_0(\lambda)$ and $W_{0j}(\lambda)$. Then every minimal factorization

$$W(\lambda) = W_1(\lambda) \cdots W_k(\lambda)$$

with minimal realizations

$$W(\lambda) = I_n + C(\lambda I_\delta - A)^{-1}B$$

and

$$W_j(\lambda) = I_n + C_j(\lambda I_{l_j} - A_j)^{-1}B_j, \qquad j = 1, \ldots, k$$

is stable provided

$$\|A - A_0\| + \|B - B_0\| + \|C - C_0\|$$
$$+ \sum_{j=1}^{k} \{\|A_j - A_{0j}\| + \|B_j - B_{0j}\| + \|C_j - C_{0j}\|\}$$

is small enough.

The proof of this result is obtained by combining Corollary 15.4.2 with Lemmas 17.4.2 and 17.4.3.

Let us clarify the connection between isolatedness and stability for minimal factorizations. The minimal factorization (17.6.1) is called *isolated* if the following holds: given minimal realizations

$$W_{0j}(\lambda) = I_n + C_{0j}(\lambda I_{l_j} - A_{0j})^{-1}B_{0j}$$

for $j = 1, \ldots, k$, there exists $\epsilon > 0$ such that, if

$$W_0(\lambda) = \tilde{W}_{01}(\lambda) \cdots \tilde{W}_{0k}(\lambda)$$

is a minimal factorization with rational matrix functions $\tilde{W}_{01}(\lambda) \cdots \tilde{W}_{0k}(\lambda)$ that admit minimal realizations

$$\tilde{W}_{0j}(\lambda) = I_n + \tilde{C}_{0j}(\lambda I_{l_j} - \tilde{A}_{0j})^{-1}\tilde{B}_{0j} \qquad (17.6.4)$$

such that

$$\sum_{j=1}^{k} \{\|\tilde{A}_{0j} - A_{0j}\| + \|\tilde{B}_{0j} - B_{0j}\| + \|\tilde{C}_{0j} - C_{0j}\|\} \le \epsilon$$

then necessarily $\tilde{W}_{0j}(\lambda) = W_{0j}(\lambda)$ for each j. It is easily seen that this definition does not depend on the choice of the minimal realization (17.6.4).

From the proof of Theorem 17.4.1 and the fact that the stable invariant subspaces coincide with the isolated ones (Section 14.3), it is found that this property also holds for stable minimal factorizations:

Theorem 17.6.2

The minimal factorization (17.6.1) is stable if and only if it is isolated.

Consider again the minimal factorization (17.6.1) with given minimal realizations (17.6.2) and (17.6.3) for $W_0(\lambda)$ and $W_{01}(\lambda), \ldots, W_{0k}(\lambda)$. We say that (17.6.1) is *Lipschitz stable* if there exist positive constants ϵ and K with the following property: for every triple of matrices (A, B, C) with appropriate sizes and with $\|A - A_0\| + \|B - B_0\| + \|C - C_0\| < \epsilon$, the realization

$$W(\lambda) = I_n + C(\lambda I_\delta - A)^{-1}B$$

is minimal and $W(\lambda)$ admits a minimal factorization $W = W_1 W_2 \cdots W_k$ such that, for $j = 1, \ldots, k$, $W_j(\lambda)$ has a minimal realization

$$W_j(\lambda) = I_n + C_j(\lambda I_{l_j} - A_j)^{-1}B_j$$

where, for each j

$$\|A_j - A_{0j}\| + \|B_j - B_{0j}\| + \|C_j - C_{0j}\|$$
$$\leq K\{\|A - A_0\| + \|B - B_0\| + \|C - C_0\|\}$$

Again, the proof of Theorem 17.4.1, together with the description of Lipschitz stable invariant subspaces (Theorem 15.5.1), yields a characterization of Lipschitz stable minimal factorizations, as follows.

Theorem 17.6.3

For the minimal factorization (17.6.1), the following statements are equivalent: (a) equation (17.6.1) is Lipschitz stable; (b) for every pair of indices $j \neq p$, the rational functions $W_{0j}(\lambda)$ and $W_{0p}(\lambda)$ have no common zeros and no common poles; (c) given minimal realizations (17.6.2) and (17.6.3) of $W_0(\lambda)$ and $W_{01}(\lambda), \ldots, W_{0k}(\lambda)$, for every sufficiently small $\epsilon > 0$ there exists an $\omega > 0$ such that for any triple (A, B, C) with $\|A - A_0\| + \|B - B_0\| + \|C - C_0\| < \omega$ the realization

$$W(\lambda) = I_n + C(\lambda I_\delta - A)^{-1}B$$

is minimal and $W(\lambda)$ admits a unique minimal factorization $W(\lambda) = W_1(\lambda)W_2(\lambda) \cdots W_k(\lambda)$ with the property that for $j = 1, \ldots, k$ each $W_j(\lambda)$ has a minimal realization

$$W_j(\lambda) = I_n + C_j(\lambda I_{l_j} - A_j)^{-1}B_j$$

satisfying

$$\|A_j - A_{0j}\| + \|B_j - B_{0j}\| + \|C_j - C_{0j}\| < \epsilon$$

17.7 STABILITY OF LINEAR FRACTIONAL DECOMPOSITIONS OF RATIONAL MATRIX FUNCTIONS

Let $U(\lambda)$ be a rational $q \times s$ matrix function with finite value at infinity. In this section we study stability of minimal linear fractional decompositions

$$U(\lambda) = \mathcal{F}_W(V) \tag{17.7.1}$$

where $W(\lambda)$ and $V(\lambda)$ are rational matrix functions of suitable sizes that take finite values at infinity. (See Sections 7.6–7.8 for the definition and basic facts on linear fractional decompositions.)

In informal terms, the stability of (17.7.1) means that any rational matrix function $\tilde{U}(\lambda)$ sufficiently close to $U(\lambda)$ admits a minimal linear fractional decomposition $\tilde{U}(\lambda) = \mathcal{F}_{\tilde{W}}(\tilde{V})$, where the rational matrix functions $\tilde{W}(\lambda)$ and $\tilde{V}(\lambda)$ are as close as we wish to $W(\lambda)$ and $V(\lambda)$, respectively. To make this notion precise, we resort to minimal realizations for the matrix functions involved. Thus let

$$U(\lambda) = \delta + \gamma(\lambda I - \alpha)^{-1}\beta \tag{17.7.2}$$

be a minimal realization of $U(\lambda)$, where α, β, γ, and δ are matrices of sizes $l \times l$, $l \times s$, $q \times l$, and $q \times s$, respectively. Also, let

$$W(\lambda) = D + C(\lambda I - A)^{-1}B$$

and

$$V(\lambda) = d + c(\lambda I - a)^{-1}b$$

be minimal realizations of $W(\lambda)$ and $V(\lambda)$. We say that the minimal linear fractional decomposition (17.7.1) is *Lipschitz stable* if there exist positive constants ϵ and K such that any $q \times s$ rational matrix function $\tilde{U}(\lambda)$ that admits a realization

$$\tilde{U}(\lambda) = \tilde{\delta} + \tilde{\gamma}(\lambda I - \tilde{\alpha})^{-1}\tilde{\beta} \tag{17.7.3}$$

with

$$\max\{\|\tilde{\delta} - \delta\|, \|\tilde{\gamma} - \gamma\|, \|\tilde{\beta} - \beta\|, \|\tilde{\alpha} - \alpha\|\} < \epsilon \qquad (17.7.4)$$

has a minimal linear fractional decomposition

$$\tilde{U}(\lambda) = \mathscr{F}_{\tilde{W}}(\tilde{V})$$

where the rational matrix functions $\tilde{W}(\lambda)$ and $\tilde{V}(\lambda)$ admit realizations

$$\tilde{W}(\lambda) = \tilde{D} + \tilde{C}(\lambda I - \tilde{A})^{-1}\tilde{B}, \qquad \tilde{V}(\lambda) = \tilde{d} + \tilde{c}(\lambda I - \tilde{a})^{-1}\tilde{b}$$

with the property that

$$\max\{\|\tilde{D} - D\|, \|\tilde{C} - C\|, \|\tilde{B} - B\|, \|\tilde{A} - A\|, \|\tilde{d} - d\|, \|\tilde{c} - c\|,$$
$$\|\tilde{b} - b\|, \|\tilde{a} - a\|\} \le K \max\{\|\tilde{\delta} - \delta\|, \|\tilde{\gamma} - \gamma\|,$$
$$\|\tilde{\beta} - \beta\|, \|\tilde{\alpha} - \alpha\|\} \qquad (17.7.5)$$

It is assumed, of course, that the sizes of two matrices coincide each time their difference appears in the preceding inequalities.

Since any two minimal realizations of the same rational matrix function are similar (Theorems 7.1.4 and 7.1.5), it is easily seen that the definition of Lipschitz stability does not depend on the particular choice of minimal realizations for $U(\lambda)$, $W(\lambda)$, and $V(\lambda)$.

It is remarkable that a large class of minimal linear fractional decompositions is Lipschitz stable, as opposed to the factorization of monic matrix polynomials and the minimal factorization of rational matrix functions, where Lipschitz stability is exceptional in a certain sense (Sections 17.3 and 17.6).

Theorem 17.7.1

Let

$$U(\lambda) = \mathscr{F}_W(V) = W_{21}(\lambda) + W_{22}(\lambda)V(\lambda)(I - W_{12}(\lambda)V(\lambda))^{-1}W_{11}(\lambda) \qquad (17.7.6)$$

be a minimal linear fractional decomposition, where

$$W(\lambda) = \begin{bmatrix} W_{11}(\lambda) & W_{12}(\lambda) \\ W_{21}(\lambda) & W_{22}(\lambda) \end{bmatrix}$$

is a suitable partition of $W(\lambda)$. Assume that the rational matrix functions $W(\lambda)$ and $U(\lambda)$ take finite values at infinity, and assume, in addition, that the matrices $W_{11}(\infty)$ and $W_{22}(\infty)$ are invertible. Then (17.7.6) is Lipschitz stable.

Proof. We make use of Theorem 7.7.1, which describes minimal linear fractional decompositions in terms of reducing pairs of subspaces with respect to the minimal realization (17.7.2). Thus there exists an $[\alpha \ \beta]$-invariant subspace $\mathcal{M}_1 \subset \mathbb{C}^l$ and an $\begin{bmatrix} \alpha \\ \gamma \end{bmatrix}$-invariant subspace $\mathcal{M}_2 \subset \mathbb{C}^l$, which are direct complements to each other and such that for some transformations $F: \mathbb{C}^l \to \mathbb{C}^s$ and $G: \mathbb{C}^q \to \mathbb{C}^l$ with $(\alpha + \beta F)\mathcal{M}_1 \subset \mathcal{M}_1$ and $(\alpha + G\gamma)\mathcal{M}_2 \subset \mathcal{M}_2$ the formulas (7.7.5)–(7.7.10) hold.

Moreover, one can choose F and G in such a way that $\mathcal{M}_1$ is a spectral invariant subspaces (i.e., a sum of root subspaces) for $\alpha + \beta F$ and $\mathcal{M}_2$ is a spectral invariant subspace for $\alpha + G\gamma$. Indeed, Theorem 7.7.2 shows that the linear fractional decomposition (17.7.6) depends on $(\mathcal{M}_1, \mathcal{M}_2; F|_{\mathcal{M}_1}, Q_{\mathcal{M}_1}G)$ only, where $Q_{\mathcal{M}_1}$ is the projector on $\mathcal{M}_1$ along $\mathcal{M}_2$. [Of course, it is assumed that the minimal realization (17.7.2) of $U(\lambda)$ is fixed.] But the proof of Theorem 15.8.1 shows that there exists a transformation $F': \mathbb{C}^l \to \mathbb{C}^s$ such that $F'x = 0$ for all $x \in \mathcal{M}_1$ and the $(\alpha + \beta(F + F'))$-invariant subspace $\mathcal{M}_1$ is spectral. So we can replace F by $F + F'$. Similarly, one proves that G can be chosen with spectral $(\alpha + G\gamma)$-invariant subspace $\mathcal{M}_2$. In the rest of the proof we assume that F and G satisfy this additional property.

Now let $\tilde{U}(\lambda)$ be another rational $q \times s$ matrix function with finite value at infinity that admits a realization (17.7.3) with the property (17.7.4). Here the positive number $\epsilon > 0$ is sufficiently small and is chosen later.

First, observe that for $\epsilon > 0$ small enough the realization (17.7.3) is also minimal. Indeed, by Theorem 7.6.1 we have

$$\sum_{j=0}^{l-1} \text{Im}(\alpha^j \beta) = \mathbb{C}^l, \qquad \bigcap_{j=0}^{l-1} \text{Ker}(\gamma \alpha^j) = \{0\}$$

which means the right invertibility of $[\beta, \alpha\beta, \ldots, \alpha^{l-1}\beta]$ and the left invertibility of

$$\begin{bmatrix} \gamma \\ \gamma\alpha \\ \vdots \\ \gamma\alpha^{l-1} \end{bmatrix}$$

Since one-sided invertibility of a matrix is a property that is preserved under small perturbations of that matrix, our conclusion concerning minimality of (17.7.3) follows.

Recall (Theorem 15.8.1) that the spectral invariant subspaces $\mathcal{M}_1$ and $\mathcal{M}_2$ for $(\alpha + \beta F)$ and for $(\alpha + G\gamma)$, respectively, are Lipschitz stable. It follows that there exists a constant $K_1 > 0$ such that $\tilde{\alpha} + \tilde{\beta}F$ and $\tilde{\alpha} + G\tilde{\gamma}$ have invariant subspaces $\tilde{\mathcal{M}}_1$ and $\tilde{\mathcal{M}}_2$, respectively, with the property that

$$\theta(\mathcal{M}_1, \tilde{\mathcal{M}}_1) + \theta(\mathcal{M}_2, \tilde{\mathcal{M}}_2) \leq K_1 \max\{\|\tilde{\alpha} - \alpha\|, \|\tilde{\beta} - \beta\|, \|\tilde{\gamma} - \gamma\|, \|\tilde{\delta} - \delta\|\}$$

provided ϵ is small enough. By Lemma 13.3.2, by choosing sufficiently small ϵ we ensure that $\tilde{\mathcal{M}}_1$ and $\tilde{\mathcal{M}}_2$ are again direct complements to each other. In other words, $(\tilde{\mathcal{M}}_1, \tilde{\mathcal{M}}_2)$ is a reducing pair with respect to the realization (17.7.3). Let $\tilde{d} = d$, $\tilde{D}_{11} = D_{11}$, $\tilde{D}_{22} = D_{22}$, $\tilde{D}_{12} = D_{12}$, and

$$\tilde{D}_{21} = \tilde{\delta} - D_{22}d(I - D_{12}d)^{-1}D_{11}$$

Also, put $\tilde{F} = F$, $\tilde{G} = G$. By Theorem 7.7.1 we obtain a minimal linear fractional decomposition $\tilde{U}(\lambda) = \mathcal{F}_{\tilde{W}}(\tilde{V})$, where the functions $\tilde{W}(\lambda)$ and $\tilde{V}(\lambda)$ are given by formulas (7.7.5)–(7.7.10) except that each letter (with the exception of $\mathbb{C}^q$, $\mathbb{C}^s$, $\mathbb{C}^l$) has a tilde. These formulas show that for $\epsilon > 0$ small enough there is a positive constant K satisfying (17.7.5) provided F and G satisfy the following property: given a basis $f_1, \ldots, f_k$ in $\mathcal{M}_1$, there exists a positive constant K_2 (which depends on this basis only) such that

$$\|\tilde{F}_1 - F_1\| + \|\tilde{G}_1 - G_1\| \leq K_2\{\theta(\mathcal{M}_1, \tilde{\mathcal{M}}_1) + \theta(\mathcal{M}_2, \tilde{\mathcal{M}}_2)\} \quad (17.7.7)$$

Here $F_1 = F|_{\mathcal{M}_1}: \mathcal{M}_1 \to \mathbb{C}^s$ and $G_1 = Q_{\mathcal{M}_1}G: \mathbb{C}^q \to \mathcal{M}_1$, where $Q_{\mathcal{M}_1}$ stands for the projector on $\mathcal{M}_1$ along $\mathcal{M}_2$, are transformations written as matrices with respect to the basis $f_1, \ldots, f_k$ (and the standard orthonormal bases in $\mathbb{C}^s$ and $\mathbb{C}^q$), and $\tilde{F}_1 = F|_{\tilde{\mathcal{M}}_1}: \tilde{\mathcal{M}}_1 \to \mathbb{C}^s$, $\tilde{G}_1 = Q_{\tilde{\mathcal{M}}_1}G: \mathbb{C}^q \to \tilde{\mathcal{M}}_1$ are similarly defined matrices with respect to some basis $g_1, \ldots, g_k$ in $\tilde{\mathcal{M}}_1$, where $Q_{\tilde{\mathcal{M}}_1}$ is the projector on $\tilde{\mathcal{M}}_1$ along $\tilde{\mathcal{M}}_2$.

To prove the existence of a constant $K_2 > 0$ with the property (17.7.7), we appeal to Lemma 13.3.2. In view of this lemma, in case $\tilde{\mathcal{M}}_1$ and $\tilde{\mathcal{M}}_2$ are sufficiently close to $\mathcal{M}_1$ and $\mathcal{M}_2$, respectively, there exists a constant $K_3 > 0$ (depending on $\mathcal{M}_1$ and $\mathcal{M}_2$ only) such that

$$\max(\|I - S\|, \|I - S^{-1}\|) \leq K_3\{\theta(\mathcal{M}_1, \tilde{\mathcal{M}}_1) + \theta(\mathcal{M}_2, \tilde{\mathcal{M}}_2)\}$$

for some invertible transformation $S: \mathbb{C}^l \to \mathbb{C}^l$ such that $S\mathcal{M}_1 = \tilde{\mathcal{M}}_1$ and $S\mathcal{M}_2 = \tilde{\mathcal{M}}_2$. It remains to choose $g_1 = Sf_1, \ldots, g_k = Sf_k$. $\square$

It is instructive to compare Theorem 17.7.1 with Theorems 17.4.1 and 17.6.3. Thus any minimal factorization $U(\lambda) = U_1(\lambda)U_2(\lambda)$, where $U_1(\lambda)$ and $U_2(\lambda)$ are $n \times n$ rational matrix functions with value I at infinity, is Lipschitz stable in the class of minimal linear fractional decompositions. In contrast, this minimal factorization need not be Lipschitz stable (or even stable) in the class of minimal factorizations. The following example illustrates this point:

EXAMPLE 17.7.1. Let

$$U(\lambda) = \begin{bmatrix} 1 + \lambda^{-1} & 0 \\ 0 & 1 + \lambda^{-1} \end{bmatrix}$$

It is easily seen that $U(\lambda)$ admits a minimal factorization

$$U(\lambda) = \begin{bmatrix} 1 + \lambda^{-1} & 0 \\ 0 & 1 \end{bmatrix}\begin{bmatrix} 1 & 0 \\ 0 & 1 + \lambda^{-1} \end{bmatrix} \qquad (17.7.8)$$

This minimal factorization is not stable because the perturbed rational matrix function

$$U_\epsilon(\lambda) = \begin{bmatrix} 1 + \lambda^{-1} & \epsilon\lambda^{-2} \\ 0 & 1 + \lambda^{-1} \end{bmatrix}, \qquad \epsilon \neq 0$$

does not have nontrivial minimal factorizations at all. On the other hand, (17.7.8) can be represented as a minimal linear fractional decomposition $U(\lambda) = \mathcal{F}_W(V)$ with

$$W(\lambda) = \mathrm{diag}[1, 1, 1 + \lambda^{-1}, 1]; \qquad V(\lambda) = \begin{bmatrix} 1 & 0 \\ 0 & 1 + \lambda^{-1} \end{bmatrix}$$

Observe that $W(\lambda)$ has a minimal realization

$$W(\lambda) = I + \begin{bmatrix} 0 \\ 0 \\ 1 \\ 0 \end{bmatrix}(\lambda - 0)^{-1}[0 \quad 0 \quad 1 \quad 0]$$

Now $U_\epsilon(\lambda)$ also admits a minimal linear fractional decomposition $U_\epsilon(\lambda) = \mathcal{F}_{W_\epsilon}(V)$, where

$$W_\epsilon(\lambda) = \begin{bmatrix} 1 & 0 & 0 & 0 \\ 0 & 1 & 0 & 0 \\ 0 & -\epsilon\lambda^{-1} & 1 + \lambda^{-1} & \epsilon\lambda^{-1} \\ 0 & 0 & 0 & 1 \end{bmatrix}$$

Moreover, $W_\epsilon(\lambda)$ has a minimal realization

$$W_\epsilon(\lambda) = I + \begin{bmatrix} 0 \\ 0 \\ 1 \\ 0 \end{bmatrix}(\lambda - 0)^{-1}[0, -\epsilon, 1, \epsilon]$$

Hence, as predicted by Theorem 17.7.1, the minimal factorization (17.7.8) is Lipschitz stable when understood as a minimal linear fractional decomposition. □

17.8 ISOLATED SOLUTIONS OF MATRIX QUADRATIC EQUATIONS

Consider the matrix quadratic equation

$$XBX + XA - DX - C = 0 \qquad (17.8.1)$$

where A, B, C, D are known matrices of sizes $n \times n$, $n \times m$, $m \times n$, $m \times m$, respectively, and X is a matrix of size $m \times n$ to be found.

For any $m \times n$ matrix X, let

$$G(X) = \left\{ \begin{bmatrix} x \\ Xx \end{bmatrix} \,\middle|\, x \in \mathbb{C}^n \right\} \subset \mathbb{C}^n \oplus \mathbb{C}^m$$

be the *graph* of X. The following proposition connects the solutions of (17.8.1) with invariant subspaces of the $(m + n) \times (m + n)$ matrix

$$T = \begin{bmatrix} A & B \\ C & D \end{bmatrix}$$

Proposition 17.8.1

For an $m \times n$ matrix X, the subspace $G(X)$ is T invariant if and only if $X satisfies (17.8.1).

Proof. Assume that $G(X)$ is T invariant. So for every $x \in \mathbb{C}^n$ there exists a $y \in \mathbb{C}^n$ such that

$$T \begin{bmatrix} x \\ Xx \end{bmatrix} = \begin{bmatrix} y \\ Xy \end{bmatrix}$$

The correspondence $x \to y$ is clearly linear; so $y = Zx$ for some $n \times n$ matrix Z, and we have

$$T \begin{bmatrix} x \\ Xx \end{bmatrix} = \begin{bmatrix} Zx \\ XZx \end{bmatrix}$$

for all $x \in \mathbb{C}^n$, or

$$\begin{bmatrix} A & B \\ C & D \end{bmatrix} \begin{bmatrix} I \\ X \end{bmatrix} = \begin{bmatrix} Z \\ XZ \end{bmatrix} \qquad (17.8.2)$$

This implies $Z = A + BX$ and

$$C + DX = XZ = X(A + BX)$$

which means that (17.8.1) holds.

Conversely, if (17.8.1) holds and $Z \overset{\text{def}}{=} A + BX$, then (17.8.2) holds. This implies the T invariance of $G(X)$. $\square$

To take advantage of Proposition 17.8.1 in describing isolated solutions of (17.8.1), we need a preliminary result.

Lemma 17.8.2

Define a function $\hat{G}$ from the set $M_{m \times n}$ of all $m \times n$ matrices to the set of all subspaces in $\mathbb{C}^n \oplus \mathbb{C}^m$ by $\hat{G}(X) = G(X)$. Then $\hat{G}$ is a homeomorphism (i.e., a bijective map that is continuous together with its inverse) between $M_{m \times n}$ and the set of all subspaces $\mathcal{M} \subset \mathbb{C}^n \oplus \mathbb{C}^m$ with the property that $\theta(\mathcal{M}, \mathcal{H}) < 1$, where $\mathcal{H} = \mathbb{C}^n \oplus \{0\}$.

Here $\theta(\mathcal{M}, \mathcal{N})$ is the gap between $\mathcal{M}$ and $\mathcal{N}$ (see Chapter 13).

Proof. The continuity of $\hat{G}$ and $\hat{G}^{-1}$ follows from the easily verified fact that the orthogonal projector P on $G(X)$ is given by

$$P = \begin{bmatrix} L & LX^* \\ XL & XLX^* \end{bmatrix} \tag{17.8.3}$$

where $L = (I + X^*X)^{-1}$. Let us check that $\theta(G(X), \mathcal{H}) < 1$. By Theorem 13.1.1

$$\theta(G(X), \mathcal{H}) = \max\{ \sup_{\substack{x \in \mathcal{H} \\ \|x\| = 1}} \|(I - P)x\|, \sup_{\substack{x \in G(X) \\ \|x\| = 1}} \|(I - P_{\mathcal{H}})x\|\} \tag{17.8.4}$$

where $P_{\mathcal{H}}$ is the orthogonal projector on $\mathcal{H}$. The second supremum is

$$\sup\left\|(I - P_{\mathcal{H}})\begin{bmatrix} y \\ Xy \end{bmatrix}\right\| = \sup\|Xy\|$$

where $\left\|\begin{bmatrix} y \\ Xy \end{bmatrix}\right\| = 1$, that is, $\|Xy\|^2 = 1 - \|y\|^2$. As $\|Xy\| / \|y\| \leq \|X\|$ is uniformly bounded, it follows that $\|y\|$ is bounded away from zero. Hence the second supremum in (17.8.4) is less than 1.

To show that the first supremum in (17.8.4) is also less than 1, assume (arguing by contradiction) that

$$\sup_{\|x\| = 1} \left\|(I - P)\begin{bmatrix} x \\ 0 \end{bmatrix}\right\| = 1$$

As $\left\|(I - P)\begin{bmatrix} x \\ 0 \end{bmatrix}\right\|^2 + \left\|P\begin{bmatrix} x \\ 0 \end{bmatrix}\right\|^2 = \left\|\begin{bmatrix} x \\ 0 \end{bmatrix}\right\|^2 = 1$, it follows that

$$\inf_{\|x\| = 1} \left\|P\begin{bmatrix} x \\ 0 \end{bmatrix}\right\| = 0$$

and by formula (17.8.3)

$$\inf_{\|x\|=1} \left\| \begin{bmatrix} Lx \\ XLx \end{bmatrix} \right\| = 0 \qquad (17.8.5)$$

But L is invertible, so

$$\|x\| = \|L^{-1}Lx\| \le \|L^{-1}\| \cdot \|Lx\|$$

and (17.8.5) is impossible. Thus $\theta(G(X), \mathcal{H}) < 1$ as claimed.

Now we must show that every subspace $\mathcal{M} \subset \mathbb{C}^n \oplus \mathbb{C}^m$ with $\theta(\mathcal{M}, \mathcal{H}) = a < 1$ is a graph subspace, that is, $\mathcal{M} = G(X)$ for some X. First, Theorem 13.1.2 shows that $\dim \mathcal{M} = \dim \mathcal{H} = n$. Further, assume that $P_{\mathcal{H}} x = 0$ for some $x \in \mathcal{M}$. Denoting by P the orthogonal projector on $\mathcal{M}$, we have $\|x\| = \|(P_{\mathcal{M}} - P_{\mathcal{H}})x\|$, which, in view of the condition $\theta(\mathcal{M}, \mathcal{H}) = \|P_{\mathcal{M}} - P_{\mathcal{H}}\| < 1$, implies $x = 0$. Hence $Q \stackrel{\text{def}}{=} P_{\mathcal{H}|\mathcal{M}}: \mathcal{M} \to \mathcal{H}$ is an invertible linear transformation. Now $\mathcal{M} = G((I - P_{\mathcal{H}})Q^{-1})$. Indeed, if $x \in \mathcal{M}$, then

$$x = Qx + (I - P_{\mathcal{H}})Q^{-1} \cdot Qx \in G((I - P_{\mathcal{H}})Q^{-1})$$

On the other hand, if for some $u \in \mathcal{H}$

$$y = \begin{bmatrix} u \\ (I - P_{\mathcal{H}})Q^{-1}u \end{bmatrix}$$

then the vector $v = Q^{-1}u$ has the property that $v \in \mathcal{M}$, $P_{\mathcal{H}}y = u = P_{\mathcal{H}}v$ and $(I - P_{\mathcal{H}})y = (I - P_{\mathcal{H}})Q^{-1}u = (I - P_{\mathcal{H}})v$. So $y = v$; therefore, y belongs to $\mathcal{M}$. □

A solution X of (17.8.1) is called *isolated* if there exists a neighbourhood of X in the linear space $M_{m \times n}$ of all $m \times n$ matrices that does not contain other solutions of (17.6.1). A solution X is called *inaccessible* if the only continuous function $\varphi: [0, 1] \to M_{m \times n}$ such that $\varphi(0) = X$ and $\varphi(t)$ is a solution of (17.8.1) for every $t \in [0, 1]$, is the constant function $\varphi(t) \equiv X$. Clearly, every isolated solution is inaccessible.

We now have a characterization of isolated and inaccessible solutions of (17.8.1).

Theorem 17.8.3

The following statements are equivalent: (a) X_0 *is an isolated solution of* (17.8.1); (b) X_0 *is an inaccessible solution of* (17.8.1); (c) *for every eigenvalue* λ_0 *of the matrix*

$$T_0 = \begin{bmatrix} A + BX_0 & B \\ 0 & D - X_0 B \end{bmatrix}$$

with $\dim \mathrm{Ker}(T_0 - \lambda_0 I) > 1$, *either*

$$\mathcal{R}_{\lambda_0}(T_0) \cap \begin{bmatrix} \mathbb{C}^n \\ 0 \end{bmatrix} = \{0\}$$

or

$$\mathcal{R}_{\lambda_0}(T_0) \subset \begin{bmatrix} \mathbb{C}^n \\ 0 \end{bmatrix}$$

(d) *every common eigenvalue of $A + BX_0$ and $D - X_0 B$ has geometric multiplicity one as an eigenvalue of T_0.*

Proof. Making a change of variable $Y = X - X_0$, we see that X satisfies (17.8.1) if and only if Y satisfies the equation

$$YBY + Y(A + BX_0) - (D - X_0 B)Y = 0 \tag{17.8.6}$$

Hence X_0 is an isolated (or inaccessible) solution of (17.8.1) if and only if 0 is an isolated (or inaccessible) solution of (17.8.6). By Proposition 17.8.1 and Lemma 17.8.2, the correspondence

$$Y \to \left\{ \begin{bmatrix} x \\ Yx \end{bmatrix} \middle| x \in \mathbb{C}^n \right\}$$

is a homeomorphism between the set of all solutions Y of (17.8.5) and the set of T_0-invariant subspaces $\mathcal{M}$ such that $\theta(\mathcal{M}, \mathcal{H}) < 1$, where $\mathcal{H} = \begin{bmatrix} \mathbb{C}^n \\ 0 \end{bmatrix}$. Hence 0 is an isolated (resp. inaccessible) solution of (17.8.6) if and only if $\mathcal{H}$ is an isolated (resp. inaccessible) T_0-invariant subspace. An application of Theorem 14.3.1 and Proposition 14.3.3 shows that (a), (b), and (c) are equivalent.

Further, the characteristic polynomial of T_0 is the product of the characteristic polynomials of $A + BX_0$ and $D - X_0 B$. As the multiplicity of λ_0 as a zero of the characteristic polynomial of a matrix S is equal to the dimension of $\mathcal{R}_{\lambda_0}(S)$, it follows that λ_0 is a common eigenvalue of $A + BX_0$ and $D - X_0 B$ if and only if

$$\{0\} \neq \mathcal{R}_{\lambda_0}(T_0) \cap \mathcal{H} \neq \mathcal{R}_{\lambda_0}(T_0)$$

So (c) and (d) are equivalent. $\square$

An interesting particular case appears when $B = 0$. Then we have the equation

$$XA - DX = C \tag{17.8.7}$$

which is a system of linear equations in the entries of X. It is well known from the theory of linear equations that equation (17.8.7) either has no solutions, has a unique solution, or has infinitely many solutions. [In this case the homogeneous equation

$$YA - DY = 0 \qquad (17.8.8)$$

has nontrivial solutions, and the general form of solutions of (17.8.7) is $X_0 + Y$, where X_0 is a particular solutions of (17.8.7) and Y is the general solution of the homogeneous equation.] Clearly, a solution X of (17.8.7) is isolated if and only if (17.8.8) has only the trivial solution. Using the criterion of Theorem 17.8.3, we obtain the following well-known result.

Corollary 17.8.4

The equation $YA - DY = 0$ has only the trivial solution $Y = 0$ if and only if $\sigma(A) \cap \sigma(D) = \emptyset$.

Reconsidering the general case of equation (17.8.1), let us give some sufficient conditions for isolatedness of the solutions.

Corollary 17.8.5

If the matrix

$$T = \begin{bmatrix} A & B \\ C & D \end{bmatrix}$$

is nonderogatory [i.e., $\dim \operatorname{Ker}(T - \lambda_0 I) = 1$ for every eigenvalue λ_0 of T], then the number of solutions of (17.8.1) (if they exist) is finite and, consequently, every solution is isolated.

Proof. The matrix T has a finite number of invariant subspaces; namely, there are exactly $\Pi_{i=1}^r (\dim \mathcal{R}_{\lambda_i}(T) + 1)$ of them, where $\lambda_1, \ldots, \lambda_r$ are all the distinct eigenvalues of T. It remains to appeal to Proposition 17.8.1. $\square$

EXAMPLE 17.8.1. Consider the equation

$$\begin{bmatrix} x \\ y \end{bmatrix} \begin{bmatrix} 1 & 1 \end{bmatrix} \begin{bmatrix} x \\ y \end{bmatrix} + \begin{bmatrix} x \\ y \end{bmatrix} - \begin{bmatrix} 1 & 0 \\ 0 & 0 \end{bmatrix} \begin{bmatrix} x \\ y \end{bmatrix} = 0 \qquad (17.8.9)$$

We have $B = [1 \ \ 1]$, $A = 1$, $D = \begin{bmatrix} 1 & 0 \\ 0 & 0 \end{bmatrix}$, $C = \begin{bmatrix} 0 \\ 0 \end{bmatrix}$. So

$$T = \begin{bmatrix} A & B \\ C & D \end{bmatrix} = \begin{bmatrix} 1 & 1 & 1 \\ 0 & 1 & 0 \\ 0 & 0 & 0 \end{bmatrix}$$

The only one-dimensional T-invariant subspaces are $\mathcal{M}_1 = \operatorname{Span}\{e_1\}$ and $\mathcal{M}_2 = \operatorname{Span}\{e_1 - e_3\}$. Defining $\mathcal{H} = \operatorname{Span}\{e_1\}$, we have

$$\theta(\mathcal{M}_1, \mathcal{H}) = 0; \; \theta(\mathcal{M}_2, \mathcal{H}) = \left\| \begin{bmatrix} \frac{1}{2} & 0 & -\frac{1}{2} \\ 0 & 0 & 0 \\ -\frac{1}{2} & 0 & \frac{1}{2} \end{bmatrix} - \begin{bmatrix} 1 & 0 & 0 \\ 0 & 0 & 0 \\ 0 & 0 & 0 \end{bmatrix} \right\| = \frac{1}{\sqrt{2}} < 1$$

so by Proposition 17.8.1 and Lemma 17.8.2 there exist only two solutions $\begin{bmatrix} x_1 \\ y_1 \end{bmatrix}$ and $\begin{bmatrix} x_2 \\ y_2 \end{bmatrix}$ given by

$$\mathcal{M}_1 = \left\{ \begin{bmatrix} t \\ x_1 t \\ y_1 t \end{bmatrix} \middle| t \in \mathcal{C} \right\}; \qquad \mathcal{M}_2 = \left\{ \begin{bmatrix} t \\ x_2 t \\ y_2 t \end{bmatrix} \middle| t \in \mathcal{C} \right\}$$

Hence

$$\begin{bmatrix} x_1 \\ y_1 \end{bmatrix} = \begin{bmatrix} 0 \\ 0 \end{bmatrix}, \qquad \begin{bmatrix} x_2 \\ y_2 \end{bmatrix} = \begin{bmatrix} 0 \\ -1 \end{bmatrix}$$

As expected from Corollary 17.8.5, the number of solutions of (17.8.9) is finite. □

Another particular case of (17.8.1) is of interest. Consider the equation

$$X^2 + A_1 X + A_0 = 0 \qquad (17.8.10)$$

where A_1 and A_0 are given $n \times n$ matrices, and X is an $n \times n$ matrix to be found. Equation (17.8.10) is a particular case of (17.8.1) with $B = I$, $C = -A_0$, $D = -A_1$, and $A = 0$ and is sometimes described as "unilateral." The matrix T turns out to be just the companion matrix of the matrix polynomial $L(\lambda) \overset{\text{def}}{=} \lambda^2 I + \lambda A_1 + A_0$:

$$T = \begin{bmatrix} 0 & I \\ -A_0 & -A_1 \end{bmatrix}$$

Proposition 17.8.1 gives a one-to-one correspondence between the set of solutions X of (17.8.10) and the set of T-invariant subspaces of the form

$$\left\{ \begin{bmatrix} x \\ Xx \end{bmatrix} \middle| x \in \mathcal{C}^n \right\}.$$

We remark that a T-invariant subspace $\mathcal{M}$ has this form if and only if the transformation $[I \; 0]|_{\mathcal{M}} : \mathcal{M} \to \mathcal{C}^n$ is invertible. In this way we recover the description of right divisors of $L(\lambda)$ given in Section 5.3. Similarly, the equation

$$X^2 + X A_1 + A_0 = 0$$

considered as a particular case of (17.8.1) gives rise (by using Proposition 17.8.1) to a description of left divisors of the matrix polynomial $\lambda^2 I + \lambda A_1 + A_0$.

17.9 STABILITY OF SOLUTIONS OF MATRIX QUADRATIC EQUATIONS

Consider the equation

$$XBX + XA - DX - C = 0 \tag{17.9.1}$$

with the same assumptions on the matrices A, B, C, D as in the preceding section. We say that a solution X of (17.9.1) is *stable* if for any $\epsilon > 0$ there is $\delta > 0$ such that whenever A', B', C', D' are matrices of appropriate size with

$$\max\{\|A - A'\|, \|B - B'\|, \|C - C'\|, \|D - D'\|\} < \delta$$

the equation

$$YB'Y + YA' - D'Y - C' = 0$$

has a solution Y for which $\|Y - X\| < \epsilon$. It turns out that the situation with regard to stability and isolatedness is analogous to that for invariant subspaces.

Theorem 17.9.1

A solution X of equation (17.9.1) *is stable if and only if* X *is isolated.*

Proof. It is sufficient to prove the theorem for the case when $C = 0$ and the solution X is the zero matrix (see the proof of Theorem 17.8.3). In this case $G(X) = \mathbb{C}^n \oplus \{0\}$; so the homeomorphism described in Lemma 17.8.2 implies that $X = 0$ is a stable (resp. isolated) solution of

$$XBX + XA - DX = 0$$

if and only if $\mathbb{C}^n \oplus \{0\}$ is a stable (resp. isolated) $\begin{bmatrix} A & B \\ 0 & D \end{bmatrix}$-invariant subspace. Now use the fact that the isolated invariant subspaces for a linear transformation coincide with the stable ones (Theorems 15.2.1 and 14.3.1). $\square$

In view of Theorem 7.9.1, statements (c) and (d) in Theorem 17.8.3 describe the stable solutions of equation (17.9.1). In the particular case when $B = 0$ we find that the solution X of $XA - DX = C$ is stable if and only if $\sigma(A) \cap \sigma(D) = \emptyset$.

As a solution X of (17.9.1) is stable if and only if the subspace $\mathrm{Im}\begin{bmatrix} I \\ X \end{bmatrix}$ is stable as a T-invariant subspace, where

$$T = \begin{bmatrix} A & B \\ C & D \end{bmatrix}$$

we can deduce some properties of stable solutions of (17.9.1) from the corresponding properties of stable T-invariant subspaces. For instance, the set of stable solutions of (17.9.1) is always finite (it may also be empty), and the number of stable solutions of (17.9.1) does not exceed the number $\gamma(T)$ of n-dimensional stable T-invariant subspaces, which can be calculated as follows. Let $\lambda_1, \ldots, \lambda_p$ be all the distinct eigenvalues of T with algebraic multiplicities $m_1, \ldots, m_p$, respectively; then $\gamma(T)$ is the number of sequences of type $(q_1, \ldots, q_p)$, where q_j are nonnegative integers with the properties that $q_j \le m_j$, either $q_j = 0$ or $q_j = m_j$ for every j such that $\dim \mathrm{Ker}(\lambda_j I - T) > 1$, and $q_1 + \cdots + q_p = n$.

Using Corollary 15.4.2, we obtain the following property of stable solutions of (17.9.1).

Theorem 17.9.2

Let X be a stable solution of (17.9.1). *Then every solution Y of equation*

$$YB'Y + YA' - D'X - C' = 0$$

where A', B', C', and D' are matrices of appropriate sizes, is stable provided

$$\|Y - X\| + \|A - A'\| + \|B - B'\| + \|C - C'\| + \|D - D'\|$$

is small enough.

The notion of Lipschitz stability of solutions of (17.7.1) is introduced naturally: a solution X of (17.7.1) is called *Lipschitz stable* if there exist positive constants ϵ and K such that, for any matrices A', B', C', D' of appropriate sizes with

$$\max\{\|A - A'\|, \|B - B'\|, \|C - C'\|, \|D - D'\|\} < \epsilon$$

the equation

$$YB'Y + YA' - D'Y - C' = 0$$

has a solution Y satisfying

$$\|X - Y\| \le K(\|A - A'\| + \|B - B'\| + \|C - C'\| + \|D - D'\|)$$

Theorem 17.9.3

A solution of (17.9.1) *is Lipschitz stable if and only if* $\sigma(A + BX) \cap \sigma(D - XB) = \emptyset$.

Proof. Again, we can assume without loss of generality that $C = 0$ and $X = 0$. Formula (17.8.3) shows that the function $\hat{G}$ introduced in Lemma 17.8.2 is locally Lipschitz continuous; that is, for every $m \times n$ matrix Y there exists a neighbourhood $\mathcal{U}$ of Y and a positive constant K such that

$$\theta(\hat{G}(Z), \hat{G}(Y)) \le K\|Z - Y\|$$

for every $Z \in \mathcal{U}$. The inverse function $\hat{G}$ is locally Lipschitz continuous as well. So the zero matrix is a Lipschitz stable solution of (17.9.1) (where $C = 0$) if and only if the subspace $\mathcal{H} \overset{\text{def}}{=} \mathbb{C}^n \oplus \{0\}$ is Lipschitz stable as an invariant subspace for the matrix

$$T = \begin{bmatrix} A & B \\ 0 & D \end{bmatrix}$$

By Theorem 15.5.1, $\mathcal{H}$ is Lipschitz stable if and only if it is a spectral invariant subspace for T. This means that $\sigma(A) \cap \sigma(D) = \emptyset$. Indeed, if $\sigma(A) \cap \sigma(D) \ne \emptyset$, then there exists a T-invariant subspace $\mathcal{L}$ strictly bigger than $\mathcal{H}$ and such that $\sigma(T|_{\mathcal{L}}) = (T|_{\mathcal{H}})$ [e.g., $\mathcal{L} = \mathcal{H} + \text{Span}\{x_0\}$, where x_0 is an eigenvector of D corresponding to an eigenvalue $\lambda_0 \in \sigma(A) \cap \sigma(D)$]. So $\mathcal{H}$ is not spectral. Conversely, if $\sigma(A) \cap \sigma(D) = \emptyset$, then with the use of Lemma 4.1.3, it follows that $\mathcal{H}$ is spectral. $\square$

Similarly, one can obtain the following fact from Theorem 15.5.1: the solution X in (17.9.1) is Lipschitz stable if and only if for every sufficiently small $\epsilon > 0$ there exists a $\delta > 0$ such that

$$\|A - A'\| + \|B - B'\| + \|C - C'\| + \|D - D'\| < \delta$$

implies that the equation

$$YB'Y + YA' - D'Y - C' = 0$$

has a *unique* solution Y satisfying $\|Y - X\| < \epsilon$.

17.10   THE REAL CASE

In this section we quickly review some real analogs of the results obtained in this chapter.

Let $L(\lambda)$ be a monic matrix polynomial whose coefficients are real $n \times n$ matrices, and consider a factorization

$$L(\lambda) = L_1(\lambda)L_2(\lambda) \cdots L_r(\lambda) \qquad (17.10.1)$$

where $L_j(\lambda)$ are monic matrix polynomials with real coefficients. Using the results of Section 15.9 and the approach developed in the proof of Theorem 17.3.1, one obtains necessary and sufficient conditions for stability of the factorization (17.10.1) (the analog of Corollary 17.2.2). The definition of a stable factorization of real monic matrix polynomials is the same as in the complex case, except that now only real matrix polynomials are allowed as perturbations of $L(\lambda)$ and as factors in a factorization of the perturbed polynomial.

Theorem 17.10.1

Let C_L be the companion matrix of $L(\lambda)$, and let

$$\mathcal{M}_r \subset \mathcal{M}_{r-1} \subset \cdots \subset \mathcal{M}_2$$

be the chain of C_L-invariant subspaces in $\mathbb{R}^{nl}$ [where l is the degree of $L(\lambda)$] corresponding to the factorization (17.10.1). Then (17.10.1) is stable if and only if the following conditions are satisfied: (a) for every eigenvalue λ_0 of C_L with geometric multiplicity greater than 1 and for every i $(2 \le i \le r)$, either $\mathcal{M}_i \supset \mathcal{R}_{\lambda_0}(C_L)$ or $\mathcal{M}_i \cap \mathcal{R}_{\lambda_0}(C_L) = \{0\}$; (b) for every real eigenvalue λ_0 of C_L with geometric multiplicity of 1 and even algebraic multiplicity, the algebraic multiplicity of λ_0 as an eigenvalue of each restriction $C_{L|\mathcal{M}_i}$ (if λ_0 is an eigenvalue of $C_{L|\mathcal{M}_i}$ at all) is also even.

In contrast with the complex case (Theorem 17.2.5), not every isolated real factorization (17.10.1) is stable. Using the description of isolated invariant subspaces for real transformations (Section 15.9), one finds that (17.10.1) is isolated if and only if the condition (a) in Theorem 17.10.1 holds.

Now we pass to the stability of minimal factorizations

$$W_0(\lambda) = W_{01}(\lambda)W_{01}(\lambda) \cdots W_{0k}(\lambda) \qquad (17.10.2)$$

of a rational matrix function $W_0(\lambda)$ such that the entries of $W_0(\lambda)$ are real for real λ. (In short, such rational matrix functions are called real.) The functions $W_{0j}(\lambda)$ are also assumed to be real, and, in addition, we require that all rational matrix functions involved are $n \times n$ and take value I at infinity. Again, the stability of (17.10.2) is defined as in the complex case with only *real* rational matrix functions allowed. The main result on stability of (17.10.2) is the following analog of Theorem 17.4.1.

Theorem 17.10.2

The minimal factorization (17.10.2) *of the real rational matrix function* $W_0(\lambda)$ *with* $W_0(\infty) = I$, *where for* $j = 1, 2, \ldots, k$, $W_{0j}(\lambda)$ *is also a real rational matrix function with* $W_{0j}(\infty) = I$, *is stable if and only if the following conditions hold*: (a) *each common pole* (*zero*) *of* W_{0j} *and* W_{0p} ($j \neq p$) *is a pole* (*zero*) *of* W_0 *of geometric multiplicity* 1; (b) *each even order real pole* λ_0 *of* W_0 (*resp. of* W_0^{-1}) *is also a pole of each* W_{0j} (*resp. of each* W_{0j}^{-1}) *of even order* (*if* λ_0 *is a pole of* W_{0j} *or of* W_{0j}^{-1} *at all*).

Recall that the geometric multiplicity of a pole (zero) λ_0 of a rational matrix function $W(\lambda)$ is the number of negative (positive) partial multiplicities of $W(\lambda)$ at λ_0. In connection with condition (b), observe that the order of a pole λ_0 of $W_0(\lambda)$ is the least positive integer p such that $(\lambda - \lambda_0)^p W_0(\lambda)$ is analytic in a neighbourhood of λ_0. It coincides with the greatest absolute value of a negative partial multiplicity of $W_0(\lambda)$ at λ_0, as one can easily see using the local Smith form for $W_0(\lambda)$ at λ_0.

We omit the proof of Theorem 17.10.2. It can be obtained in a similar way to the proof of Theorem 17.4.1 by using the description of stable invariant subspaces for real transformations presented in Section 15.9.

As in the case of matrix polynomials, not every isolated minimal factorization of a real rational matrix function with real factors is stable (in the class of real factorizations). It is found that (17.10.2) is isolated if and only if condition (a) of Theorem 17.10.2 holds. Let us give an example of an isolated but not stable minimal factorization of real rational matrix functions.

EXAMPLE 17.10.1. Let

$$W_0(\lambda) = \begin{bmatrix} 1 & \lambda^{-1} + \lambda^{-2} \\ 0 & 1 + \lambda^{-1} \end{bmatrix}; \qquad W_{01}(\lambda) = \begin{bmatrix} 1 & \lambda^{-1} \\ 0 & 1 \end{bmatrix};$$

$$W_{02}(\lambda) = \begin{bmatrix} 1 & 0 \\ 0 & 1 + \lambda^{-1} \end{bmatrix}$$

One verifies easily that $W_0(\lambda) = W_{01}(\lambda)W_{02}(\lambda)$ and this factorization is minimal (indeed, the McMillan degree of $W_0(\lambda)$ is 2, whereas the McMillan degree of $W_{01}(\lambda)$ and $W_{02}(\lambda)$ is 1). Furthermore

$$W_{01}(\lambda)^{-1} = \begin{bmatrix} 1 & -\lambda^{-1} \\ 0 & 1 \end{bmatrix}, \qquad W_{02}(\lambda)^{-1} = \begin{bmatrix} 1 & 0 \\ 0 & \lambda(\lambda+1)^{-1} \end{bmatrix}$$

so $W_{01}(\lambda)$ and $W_{02}(\lambda)$ do not have common zeros. It is easily seen that $\lambda_0 = 0$ is a common pole of $W_0(\lambda)$, $W_{01}(\lambda)$, and $W_{02}(\lambda)$ and that the only negative partial multiplicities of $W_0(\lambda)$, $W_{01}(\lambda)$, and $W_{02}(\lambda)$ at λ_0 are -2, -1, and -1, respectively. Hence condition (a) of Theorem 17.10.2 is satisfied, but condition (b) is not. It follows that the factorization $W_0(\lambda) = W_{01}(\lambda)W_{02}(\lambda)$ is isolated but not stable in the class of minimal factorizations of real rational matrix functions. $\square$

Finally, consider the matrix quadratic equation

$$XBX + XA - DX - C = 0 \qquad (17.10.3)$$

where A, B, C, D are known real matrices of sizes $n \times n$, $n \times m$, $m \times n$, $m \times m$, respectively, and X is a real matrix of size $m \times n$ to be found. The solution of X of (17.10.3) is called *isolated* if there exists $\epsilon > 0$ such that the set of all real matrices Y satisfying $\|X - Y\| < \epsilon$ does not contain solutions of (17.10.3) other than X. The solution of (17.10.3) is called *stable* if for any $\epsilon > 0$ there is $\delta > 0$ such that whenever A', B', C', D' are real matrices of appropriate sizes with

$$\max\{\|A - A'\|, \|B - B'\|, \|C - C'\|, \|D - D'\|\} < \delta ,$$

the equation

$$YB'Y + YA' - D'Y - C' = 0$$

has a real solution Y for which $\|Y - X\| < \epsilon$. The isolated and stable solutions can be characterized as follows.

Theorem 17.10.3

The solution X_0 of (17.10.3) is isolated if and only if every common eigenvalue of $A + BX_0$ and $D - X_0B$ has geometric multiplicity 1 as an eigenvalue of the matrix

$$T = \begin{bmatrix} A & B \\ C & D \end{bmatrix}$$

The solution X_0 is stable if and only if it is isolated and, in addition, for every real eigenvalue λ_0 of T with even algebraic multiplicity the algebraic multiplicity of λ_0 as an eigenvalue of $A + BX_0$ (or of $D - X_0B$) is even (if λ_0 is an eigenvalue of $A + BX_0$, or of $D - X_0B$ at all).

In connection with the second statement in this theorem, observe that

$$\begin{bmatrix} I & 0 \\ -X_0 & I \end{bmatrix} T \begin{bmatrix} I & 0 \\ X_0 & I \end{bmatrix} = \begin{bmatrix} A + BX_0 & B \\ 0 & D - X_0B \end{bmatrix} \qquad (17.10.4)$$

and thus the algebraic multiplicity $m(T; \lambda_0)$ for the eigenvalue λ_0 of T is equal to the sum of the algebraic multiplicities $m(A + BX_0; \lambda_0)$ and $m(D - X_0B; \lambda_0)$. Consequently, if $m(T; \lambda_0)$ is even, then the evenness of one of the numbers $m(A + BX_0; \lambda_0)$ and $m(D - X_0B; \lambda_0)$ implies the evenness of the other.

Again, we omit the proof of Theorem 17.10.3. It can be obtained by using an argument similar to the proofs of Theorems 17.8.3 and 17.9.1, using the description of stable and isolated invariant subspaces for real transformations (Section 15.9) and taking into account equation (17.10.4).

17.11   EXERCISES

17.1   Find all stable factorizations (whose factors are linear matrix polynomials) of the monic matrix polynomial

$$L(\lambda) = \begin{bmatrix} \lambda^2 - \lambda & -\lambda + 1 \\ 0 & \lambda^2 - \lambda \end{bmatrix}$$

Does $L(\lambda)$ have a nonstable factorization?

17.2   Solve Exercise 17.1 for the matrix polynomial

$$L(\lambda) = \begin{bmatrix} \lambda^2 - 2\lambda & -\lambda + 1 \\ 0 & \lambda^2 - 2\lambda \end{bmatrix}$$

17.3   Let $L(\lambda)$ be a monic $n \times n$ matrix polynomial of degree l such that C_L has nl distinct eigenvalues. Show that any factorization of $L(\lambda)$ (whose factors are monic matrix polynomials as well) is stable.

17.4   Is any factorization of monic matrix polynomial $L(\lambda)$ stable if C_L is diagonable?

17.5   Show that the factorization $L = L_1 L_2 L_3$ of a monic matrix polynomial $L(\lambda)$ is stable if and only if each of the factorizations $L = L_2 M$, $M = L_2 L_3$ is stable, where $M = L_1^{-1} L$.

17.6   Is the property expressed in Exercise 17.5 true for Lipschitz stability?

17.7   Show that a factorization of 2×2 monic matrix polynomials $L = L_1 L_2$ is stable if and only if one of $L_1(\lambda_0)$ and $L_2(\lambda_0)$ is invertible for every $\lambda_0 \in \mathbb{C}$ such that $L(\lambda_0) = 0$.

17.8   Let $L(\lambda) = I\lambda^l + \Sigma_{j=0}^{l-1} A_j \lambda^j$ be an $n \times n$ matrix polynomial whose coefficients A_j are circulant matrices. Show that any factorization

$$L(\lambda) = L_1(\lambda)L_2(\lambda) \cdots L_r(\lambda)$$

where for $j = 1, \ldots, r$, $L_j(\lambda)$ is a monic matrix polynomial with circulant coefficients, is stable in the algebra of circulant matrices, in the following sense: for every $\epsilon > 0$ there exists a $\delta > 0$ such that every monic matrix polynomial $\tilde{L}(\lambda)$ of degree l with circulant coefficients that satisfies $\sigma_l(\tilde{L}, L) < \delta$ admits a factorization

$$\tilde{L}(\lambda) = \tilde{L}_1(\lambda)\tilde{L}_2(\lambda)\cdots\tilde{L}_r(\lambda)$$

where $\tilde{L}_1(\lambda), \ldots, \tilde{L}_r(\lambda)$ are monic matrix polynomials with circulant coefficients and such that

$$\sigma_{p_1}(\tilde{L}_1, L_1) + \cdots + \sigma_{p_r}(\tilde{L}_r, L_r) < \epsilon$$

(Here p_j is the degree of L_j and of $\tilde{L}_j$, for $j = 1, \ldots, r$.)

17.9 Give an example of a nonstable factorization of an $n \times n$ matrix polynomial with circulant coefficients.

17.10 Let $L(\lambda) = \text{diag}[M_1(\lambda), M_2(\lambda)]$, where $M_1(\lambda)$ and $M_2(\lambda)$ are monic matrix polynomials of sizes $n_1 \times n_1$ and $n_2 \times n_2$, respectively, and let

$$L(\lambda) = \text{diag}[M_{11}(\lambda), M_{21}(\lambda)]\cdots\text{diag}[M_{1r}(\lambda), M_{2r}(\lambda)] \quad (1)$$

be a factorization of $L(\lambda)$, where for $j = 1, \ldots, r$, $M_{1j}(\lambda)$ and $M_{2j}(\lambda)$ have sizes $n_1 \times n_1$ and $n_2 \times n_2$, respectively.

(a) Prove that if (1) is stable, then each factorization

$$M_1(\lambda) = M_{11}(\lambda)\cdots M_{1r}(\lambda); \qquad M_2(\lambda) = M_{21}(\lambda)\cdots M_{2r}(\lambda) \quad (2)$$

is stable as well.

(b) Show that the converse of statement (a) is generally false.

(c) Show that the factorization (1) is stable in the algebra of all matrices of type

$$\begin{bmatrix} A_1 & 0 \\ 0 & A_2 \end{bmatrix} \quad (3)$$

where A_1 (resp. A_2) is any $n_1 \times n_1$ (resp. $n_2 \times n_2$) matrix if and only if each factorization (2) is stable. (Stability in the algebra of all matrices of type (3) is understood in the same way as stability in the algebra of circulant matrices, as explained in Exercise 17.8.)

17.11 Let V be the algebra of all $n \times n$ matrices of type

$$\begin{bmatrix} \alpha_1 & 0 & \cdots & 0 & \beta_1 \\ 0 & \alpha_2 & \cdots & \beta_2 & 0 \\ \vdots & \vdots & & \vdots & \vdots \\ 0 & \beta_{n-1} & \cdots & \alpha_{n-1} & 0 \\ \beta_n & 0 & \cdots & 0 & \alpha_n \end{bmatrix}$$

where α_j and β_j are complex numbers, and let $L(\lambda)$ be a monic matrix polynomial with coefficients from the algebra V. Describe factorizations of $L(\lambda)$ that are stable in the algebra V. (*Hint:* Use Exercise 17.7.)

17.12 Find all stable minimal factorizations of the rational matrix function

$$\begin{bmatrix} 1 + \lambda^{-1} & (\lambda - 1)^{-1} + \lambda^{-2} \\ (\lambda - 1)^2 & 1 \end{bmatrix}$$

Is there a nonstable factorization of this function?

17.13 Prove that every minimal factorization of a scalar rational function with value I at infinity is stable. (It is assumed that the factors are scalar rational functions with value I at infinity as well.)

17.14 Let $W(\lambda)$ be a rational matrix function with value I at infinity. Assume that $W(\lambda)$ has δ distinct zeros and δ distinct poles, where δ is the McMillan degree of $W(\lambda)$. Show that every minimal factorization of $W(\lambda)$ is stable.

17.15 Let $W(\lambda)$ be an $n \times n$ rational matrix function with value I at infinity that is a circulant, that is, of type

$$\begin{bmatrix} w_1(\lambda) & w_2(\lambda) & \cdots & w_n(\lambda) \\ w_n(\lambda) & w_1(\lambda) & \cdots & w_{n-1}(\lambda) \\ \vdots & \vdots & & \vdots \\ w_2(\lambda) & w_3(\lambda) & \cdots & w_1(\lambda) \end{bmatrix}$$

where $w_1(\lambda), w_2(\lambda), \ldots, w_n(\lambda)$ are scalar rational functions. Show that every minimal factorization of $W(\lambda)$ is stable in the class of circulant rational matrix functions.

17.16 Give an example of nonstable minimal factorization of a circulant rational matrix function with value I at infinity whose factors are also from this class.

17.17 Let $W(\lambda)$ be a rational matrix function with $W(\infty) = I$, and let

$$W(\lambda) = W_1(\lambda) \cdots W_r(\lambda) \qquad (4)$$

be a factorization of $W(\lambda)$, where $W_j(\lambda)$ are also rational matrix functions with value I at infinity. Show that if

$$W(\lambda^2) = W_1(\lambda^2) \cdots W_r(\lambda^2)$$

is a minimal factorization, then (4) is also minimal. Is the converse true?

17.18 (a) Find all solutions of the matrix quadratic equation

$$X \begin{bmatrix} 0 & 0 \\ -1 & 0 \end{bmatrix} X + X \begin{bmatrix} 1 & 0 \\ 0 & 1 \end{bmatrix} - \begin{bmatrix} 0 & 0 \\ 1 & 0 \end{bmatrix} X = 0$$

(b) Find all stable solutions of this equation.
(c) Find all Lipschitz stable solutions of this equation.

17.19 (a)   Describe all circulant solutions of the equation

$$XBX + XA - DX - C = 0 \tag{5}$$

       with circulant matrices A, B, C, and D.

     (b)   Can one obtain all circulant solutions of (5), in the event that B is invertible, by the formula $\frac{1}{2}(D - A)B^{-1} + (\frac{1}{4}(D - A)^2 B^{-2} + 4BC)^{1/2}$?

17.20   Solve the quadratic equation

$$X^2 = \begin{bmatrix} 1 & 2 \\ 0 & -1 \end{bmatrix}$$

Notes to Part 3

Chapter 13. This chapter contains mainly well-known results. The main ideas and results concerning the metric space of subspaces appeared first in the infinite dimensional framework [see Krein, Krasnoselskii and Milman (1948); Gohberg and Markus (1959); and also Gohberg and Krein (1957)], and they are adapted here for the finite-dimensional case. The contents of Sections 13.1 and 13.4 are standard. The exposition presented here is based on that of Chapter S.4 in the authors' book (1982) [see also Kato (1976)]. Theorem 13.2.3 is from Gohberg and Markus (1959). The exposition in Section 13.3 follows Section 7.2 in Bart, Gohberg, and Kaashoek (1979). Theorem 13.6.3, along with other related results, was obtained in Gohberg and Leiterer (1972) as a consequence of general properties of cocycles in certain algebras of continuous matrix functions. Theorem 13.5.1 appears in the infinite dimensional framework in Gohberg and Krupnik (1979); here we follow the authors' book (1983b). The material on normed spaces presented in Section 13.8 is standard knowledge. For the first part of this section we made use of the exposition in Lancaster and Tismenetsky (1985).

Chapter 14. The description of connected components in the set of invariant subspaces (Sections 14.1 and 14.2) is found in Douglas and Pearcy (1968) [see also Shayman (1982)]. An identification of isolated invariant subspaces is given in Douglas and Pearcy (1968). Note that in the infinite-dimensional framework (Hilbert space and bounded linear operators) there exist inaccessible invariant subspaces that are not isolated [see Douglas and Pearcy (1968)]. Theorem 14.3.5 was originally proved in the infinite-dimensional case [Douglas and Pearcy (1968)]. The results on coinvariant and semiinvariant subspaces in Section 14.5 appear here for the first time.

Chapter 15. Theorem 15.2.1 appeared in Bart, Gohberg and Kaashoek (1978) and Campbell and Daughtry (1979). The proof presented here follows the exposition in Bart, Gohberg and Kaashoek (1979). Parts (a) ⇔ (b) of Theorem 15.5.1 was first proved in Kaashoek, van der Mee and Rodman (1982). The statement of Theorem 15.5.1 and the remaining proof is taken from Ran and Rodman (1983). Theorem 15.7.1 was proved in Conway and Halmos (1980). Theorem 15.8.1, although not stated in this way, was proved in Gohberg and Rubinstein (1985). The material of Section 15.9 is based on Bart, Gohberg and Kaashoek (1979). Theorem 15.10.1 was proved in den Boer and Thijsse (1980) and Markus and Parilis (1980). Theorem 15.10.2 is suggested by Theorem 2.4 in den Boer and Thijsse (1980).

The results of this chapter play an important role in explicit numerical computation of invariant subspaces. However, we do not touch the topic of numerical computation in this book, and refer the reader to the following sources: Bart, Gohberg, Kaashoek and van Dooren (1980); Golub and Wilkinson (1976); Ruhe (1970, 1970b); van Dooren (1981, 1983); and Golub and van Loan (1983).

Chapter 16. Most of the results and expositions of the material in this chapter is taken from Gohberg and Rodman (1986). Corollary 16.1.3 appeared in Brickman and Fillmore (1967). Lemma 16.5.1 is a particular case of a result due to Ostrowski [see pages 334–335 in Ostrowski (1973)].

Chapter 17. The main results of Section 17.2 (where the case of factorization into the product of two factors $L(\lambda) = L_1(\lambda)L_2(\lambda)$ was considered) are from Bart, Gohberg and Kaashoek (1978). The exposition of Sections 17.1 and 17.2 follows Gohberg, Lancaster, and Rodman (1982), where only the case of two factors was considered [see also the authors' paper (1979)]. The results of Section 17.3 are presented here probably for the first time. The main part of the contents of Section 17.4, as well as Theorems 17.6.1 and 17.6.2, is taken from Bart, Gohberg and Kaashoek (1979). Lemma 17.8.2 is taken from Campbell and Daughtry (1979). The main results of Section 17.7 are from Gohberg and Rubinstein (1985). Example 17.10.1 is taken from Chapter 9 in Bart, Gohberg and Kaashoek (1979).

Part Four

Analytic Properties of Invariant Subspaces

This part is devoted to the study of transformations that depend analytically on a parameter, and to the dependence of their invariant subspaces on the parameter. We begin with the simplest invariant subspaces, the kernel and image of the transformation, and this already requires the development of a theory of analytic families of invariant subspaces. Also, the solution of some basic problems is required, such as the existence of analytic bases and analytic complements for analytic families of subspaces. This material is all presented in Chapter 18 and is probably presented in a book on linear algebra for the first time. More generally, these results appeared first in the theory of analytic fibre bundles.

The study of more sophisticated objects and their dependence on the complex parameter z is the subject of Chapter 19. These include irreducible subspaces, the Jordan form, and Jordan bases. These results can be viewed as extensions of perturbation theory for analytic families of transformations.

The final chapter of Part 4 (and of the book) contains applications of the two preceding chapters to problems that have already appeared in earlier chapters, but now in the context of analytic dependence on a parameter. These applications include the factorization of matrix polynomials and rational matrix functions and the solution of quadratic matrix equations.

Chapter Eighteen

Analytic Families
of Subspaces

In this chapter we study analytic families of transformations and analytic families of their invariant subspaces. For this purpose, the basic notion of an analytic family of subspaces is introduced and studied. This notion is of a local character, and the analysis of its global properties is one of the main problems of this chapter. In the proofs of Lemmas 18.4.2 and 18.5.2 (only) we use some basic methods from the theory of infinite-dimensional spaces, and this leads us beyond the prerequisites in linear algebra required up to this point. It is shown that the kernel and image of an analytic family of transformations form two analytic families of subspaces (possibly after correction at a discrete set of points). Other classes of invariant subspaces whose behaviour is analytic (at least locally) are also studied. In Section 18.8 we analyze the case when the whole lattice of invariant subspaces behaves analytically. This occurs for analytic families of transformations with a fixed Jordan structure.

18.1 DEFINITION AND EXAMPLES

Let Ω be a domain (i.e., a connected open set) in the complex plane $\mathbb{C}$, and assume that for every $z \in \Omega$ a transformation $A(z)$: $\mathbb{C}^n \to \mathbb{C}^m$ is given. We say that $A(z)$ is an analytic family on Ω if in a neighbourhood U_{z_0} of each point $z_0 \in \Omega$ the transformation valued function $A(z)$ admits representation as a power series

$$A(z) = \sum_{j=0}^{\infty} A_j(z - z_0)^j, \qquad z \in U_{z_0}$$

where $A_0, A_1, \ldots,$ are transformations from $\mathbb{C}^n$ into $\mathbb{C}^m$. Equivalently, $A(z)$ is said to depend analytically on z in Ω if the entries in the matrix representing $A(z)$ in fixed bases in $\mathbb{C}^n$ and $\mathbb{C}^m$ are analytic functions of z on the domain Ω. Obviously, this definition does not depend on the choice of these bases.

Now let $\{\mathcal{M}(z)\}_{z \in \Omega}$ be a family of subspaces in $\mathbb{C}^n$. So for every z in Ω, $\mathcal{M}(z)$ is a subspace in $\mathbb{C}^n$. We say that the family $\{\mathcal{M}(z)\}_{z \in \Omega}$ is *analytic* on Ω if for every $z_0 \in \Omega$ there exists a neighbourhood $U_{z_0} \subset \Omega$ of z_0, a subspace $\mathcal{M} \subset \mathbb{C}^n$, and an invertible transformation $A(z): \mathbb{C}^n \to \mathbb{C}^n$ that depends analytically on z in U_{z_0} and

$$\mathcal{M}(z) = A(z)\mathcal{M}, \qquad z \in U_{z_0} \tag{18.1.1}$$

It is easily seen that for an analytic family of subspaces $\{\mathcal{M}(z)\}_{z \in \Omega}$ the dimension of $\mathcal{M}(z)$ is independent of z. Indeed, (18.1.1) shows that $\dim \mathcal{M}(z)$ is fixed for z belonging to the neighbourhood U_{z_0} of z_0. Since Ω is connected, for any two points $z', z'' \in \Omega$ there is a sequence $z_0 = z', z_1, \ldots, z_k = z''$ of points in Ω such that the intersections $U_{z_i} \cap U_{z_{i-1}}$, $i = 1, \ldots, k$ are not empty. Then obviously $\dim \mathcal{M}(z_i) = \dim \mathcal{M}(z_{i-1})$, $i = 1, \ldots, k$, and hence $\dim \mathcal{M}(z') = \dim \mathcal{M}(z'')$.

Let us give some examples of analytic families of subspaces.

Proposition 18.1.1

Let $x_1(z), \ldots, x_p(z)$ be analytic functions of z on the domain Ω whose values are n-dimensional vectors. If for every $z_0 \in \Omega$ the vectors $x_1(z_0), \ldots, x_p(z_0)$ are linearly independent, then

$$\mathrm{Span}\{x_1(z), \ldots, x_p(z)\}, \qquad z \in \Omega$$

is an analytic family of subspaces.

Proof. Take $z_0 \in \Omega$, and let $y_{p+1}, \ldots, y_n$ be vectors in $\mathbb{C}^n$ such that $x_1(z_0), \ldots, x_p(z_0), y_{p+1}, \ldots, y_n$ form a basis in $\mathbb{C}^n$. Then

$$\det[x_1(z_0) \cdots x_p(z_0) y_{p+1} \cdots y_n] \neq 0$$

As the determinant is a continuous function of its entries and $x_j(z)$, $j = 1, \ldots, p$ are analytic (and hence continuous) functions of z on Ω, it follows that

$$\det[x_1(z) \cdots x_p(z) y_{p+1} \cdots y_n] \neq 0$$

for all z belonging to some neighbourhood U of z_0. Hence

$$\mathrm{Span}\{x_1(z), \ldots, x_p(z)\} = [x_1(z) \cdots x_p(z) y_{p+1} \cdots y_n]\mathcal{M}, \qquad z \in U$$

where $\mathcal{M}$ is spanned by the first p coordinate unit vectors in $\mathbb{C}^n$, and $\mathrm{Span}\{x_1(z), \ldots, x_p(z)\}$ is, by definition, analytic on Ω. $\square$

We see later that the property described in Proposition 18.1.1 is characteristic in the sense that for every analytic family of subspaces, there exists a basis that consists of analytic vector functions.

Proposition 18.1.2

Let $A(z)$: $\mathbb{C}^n \to \mathbb{C}^m$ be an analytic family of transformations on Ω, and assume that $\dim \mathrm{Ker}\, A(z)$ is constant (i.e., independent of z for z in Ω). Then $\mathrm{Ker}\, A(z)$ is an analytic family of subspaces (of $\mathbb{C}^n$) on Ω, whereas $\mathrm{Im}\, A(z)$ is an analytic family of subspaces (of $\mathbb{C}^m$) on Ω.

Note that $\dim \mathrm{Ker}\, A(z)$ is constant on Ω if and only if the rank of $A(z)$ is constant, or, equivalently, the dimension of $\mathrm{Im}\, A(z)$ is constant.

Proof. Write $A(z)$ as an $m \times n$ matrix with respect to fixed bases in $\mathbb{C}^m$ and $\mathbb{C}^n$. Take $z_0 \in \Omega$. There exists a nonzero minor of size $p \times p$ of $A(z_0)$, where by assumption, $p = \mathrm{rank}\, A(z)$ is independent of z. For simplicity of notation assume that this minor is in the upper left corner of $A(z_0)$. As the entries of $A(z)$ depend analytically on z, this $p \times p$ minor is also nonzero for all z in a sufficiently small neighbourhood U_0 of z_0. So for any $z \in U_0$ [here we use the assumption that rank $A(z)$ is independent of z], we obtain

$$\mathrm{Im}\, A(z) = \mathrm{Span}\{a_1(z), \ldots, a_p(z)\}$$

where $a_i(z)$ is the ith column of $A(z)$. Let $b_{p+1}, \ldots, b_m$ be m-dimensional vectors such that $a_1(z_0), \ldots, a_p(z_0), b_{p+1}, \ldots, b_m$ form a basis in $\mathbb{C}^m$, that is

$$\det[a_1(z_0), \ldots, a_p(z_0), b_{p+1}, \ldots, b_m] \neq 0$$

Again, by the analyticity of $a_1(z), \ldots, a_p(z)$, there exists a neighbourhood $V_0 \subset U_0$ such that

$$\det[a_1(z), \ldots, a_p(z), b_{p+1}, \ldots, b_m] \neq 0$$

for all $z \in V_0$. Now for $z \in V_0$ we have

$$\mathrm{Im}\, A(z) = [a_1(z), \ldots, a_p(z), b_{p+1}, \ldots, b_m]\mathcal{M}$$

where $\mathcal{M} = \mathrm{Span}\{e_1, \ldots, e_p\} \subset \mathbb{C}^m$. So, by definition, $\mathrm{Im}\, A(z)$ is an analytic family of subspaces.

Now consider $\mathrm{Ker}\, A(z)$ and fix a z_0 in Ω. There exists a nonzero minor of size $p \times p$ of $A(z_0)$, which will be supposed to lie in the left upper corner of $A(z_0)$. Partition $A(z)$ accordingly:

$$A(z) = \begin{bmatrix} B(z) & C(z) \\ D(z) & E(z) \end{bmatrix}$$

where $B(z)$, $C(z)$, $D(z)$, and $E(z)$ are matrix functions of sizes $p \times p$, $p \times (n - p)$, $(m - p) \times m$, $(m - p) \times (n - p)$, respectively, and are analytic on Ω. For some neighbourhood U of z_0 we have $\det B(z) \neq 0$ for $z \in U$. If the vector $\begin{bmatrix} x \\ y \end{bmatrix}$, $x \in \mathbb{C}^p$, $y \in \mathbb{C}^{n-p}$ belongs to Ker $A(z)$ and $z \in U$, then

$$\begin{cases} B(z)x + C(z)y = 0 \\ D(z)x + E(z)y = 0 \end{cases}$$

or

$$\begin{cases} x = -B(z)^{-1}C(z)y \\ [-D(z)B(z)^{-1}C(z) + E(z)]y = 0 \end{cases}$$

It follows that dim Ker $A(z) = $ dim Ker$[-D(z)B(z)^{-1}C(z) + E(z)]$. But dim Ker $A(z)$ is independent of z and equal to $n - p$; consequently, $D(z)B(z)^{-1}C(z) + E(z) = 0$ for all $z \in U$. Now, obviously

$$\text{Ker } A(z) = \begin{bmatrix} I & -B(z)^{-1}C(z) \\ 0 & I \end{bmatrix} \mathcal{N}, \qquad z \in U$$

where $\mathcal{N} = \left\{ \begin{bmatrix} 0 \\ y \end{bmatrix} \, \middle| \, y \in \mathbb{C}^{n-p} \right\}$. Hence Ker $A(z)$ is an analytic family on Ω. $\square$

We see later that the examples of analytic families of subspaces given in Proposition 18.1.2 are basic. In fact, any analytic family of subspaces is the image (or the kernel) of an analytic transformation whose values are projectors.

More generally, without the extra assumption that the dimension of Ker $A(z)$ is independent of z, the families of subspaces Ker $A(z)$ and Im $A(z)$, where $A(z): \mathbb{C}^n \to \mathbb{C}^m$ is an analytic family on Ω, are not analytic on Ω. Let us give a simple example illustrating this fact.

EXAMPLE 18.1.1.   Let

$$A(z) = \begin{bmatrix} z & z^2 \\ z^2 & z^3 \end{bmatrix}, \qquad z \in \mathbb{C}$$

Obviously, $A(z): \mathbb{C}^2 \to \mathbb{C}^2$ is an analytic family on $\mathbb{C}$ (written as a matrix in the standard basis in $\mathbb{C}^2$). We have

$$\text{Im } A(z) = \begin{cases} \text{Span}\begin{bmatrix} 1 \\ z \end{bmatrix}, & z \neq 0 \\ \{0\}, & z = 0 \end{cases}$$

$$\text{Ker } A(z) = \begin{cases} \text{Span}\begin{bmatrix} -z \\ 1 \end{bmatrix}, & z \neq 0 \\ \math022^2, & z = 0 \end{cases}$$

As dim Im $A(z)$ is not constant, the family of subspaces Im $A(z)$ is not analytic on $\math022$. Similarly, Ker $A(z)$ is not analytic on $\math022$. Note, however, that by changing Im $A(z)$ at the single point $z = 0$ (replacing $\{0\}$ by $\text{Span}\begin{bmatrix} 1 \\ 0 \end{bmatrix}$, we obtain a family of one-dimensional subspaces $\text{Span}\begin{bmatrix} 1 \\ z \end{bmatrix}$ that is analytic on $\math022$) (indeed,

$$\text{Span}\begin{bmatrix} 1 \\ z \end{bmatrix} = \begin{bmatrix} 1 & 0 \\ z & 1 \end{bmatrix} \text{Span}\{e_1\})$$

Similarly, by changing Ker $A(z)$ at the single point $z = 0$ we obtain an analytic family of subspaces $\text{Span}\begin{bmatrix} -z \\ 1 \end{bmatrix}, z \in \math022$. $\square$

18.2 KERNEL AND IMAGE OF ANALYTIC FAMILIES OF TRANSFORMATIONS

We have observed in the preceding section that, if $A(z): \math022^n \to \math022^m$ is an analytic family of transformations, then, in general, Ker $A(z)$ and Im $A(z)$ are not analytic families of subspaces. However, Example 18.1.1 suggests that after a change at certain points Ker $A(z)$ and Im $A(z)$ become analytic families. It turns out that this is true in general. To make this statement more precise, it is convenient to introduce some terminology. Let $A(z): \math022^n \to \math022^m$ be an analytic family of transformations on Ω. The *singular set* $S(A)$ of $A(z)$ is the set of all $z_0 \in \Omega$ for which

$$\text{rank } A(z_0) < \max_{z \in \Omega} \text{rank } A(z)$$

Note that the singular set is *discrete*; that is, for every $z_0 \in S(A)$ there is a neighbourhood $U \subset \Omega$ of z_0 such that $U \cap S(A) = \{z_0\}$.

Theorem 18.2.1

Let $A(z): \math022^n \to \math022^m$ be an analytic family of transformations on Ω, and let $r = \max_{z \in \Omega} \text{rank } A(z)$. Then there exist m-dimensional vector-valued functions $y_1(z), \ldots, y_r(z)$ and n-dimensional vector-valued functions $x_1(z), \ldots, x_{n-r}(z)$ that are all analytic on Ω and have the following properties: (a) $y_1(z), \ldots, y_r(z)$ are linearly independent for every $z \in \Omega$; (b) $x_1(z), \ldots, x_{n-r}(z)$ are linearly independent for every $z \in \Omega$; (c) for every z not belonging to the singular set of $A(z)$

$$\text{Span}\{y_1(z), \ldots, y_r(z)\} = \text{Im } A(z) \qquad (18.2.1)$$

and

$$\text{Span}\{x_1(z), \ldots, x_{n-r}(z)\} = \text{Ker } A(z) \qquad (18.2.2)$$

For any z belonging to the singular set of $A(z)$ the inclusions

$$\text{Span}\{y_1(z), \ldots, y_r(z)\} \supset \text{Im } A(z) \qquad (18.2.3)$$

and

$$\text{Span}\{x_1(z), \ldots, x_{n-r}(z)\} \subset \text{Ker } A(z)$$

hold.

In particular (Proposition 18.1.1), $\text{Span}\{y_1(z), \ldots, y_r(z)\}$ is an analytic family of subspaces that coincides with Im $A(z)$ outside the singular set of $A(z)$. Similarly, $\text{Span}\{x_1(z), \ldots, x_{n-r}(z)\}$ is an analytic family of subspaces that coincides with Ker $A(z)$ outside $S(A)$.

The proof of Theorem 18.2.1 is based on the following lemma.

Lemma 18.2.2

Let $x_1(z), \ldots, x_r(z)$ be n-dimensional vector-valued functions that are analytic on a domain Ω in the complex plane. Assume that for some $z_0 \in \Omega$, the vectors $x_1(z_0), \ldots, x_r(z_0)$ are linearly independent. Then there exist n-dimensional vector functions $y_1(z), \ldots, y_r(z)$ with the following properties: (a) $y_1(z), \ldots, y_r(z)$ are analytic on Ω; (b) $y_1(z), \ldots, y_r(z)$ are linearly independent for every $z \in \Omega$; (c) $\text{Span}\{y_1(z), \ldots, y_r(z)\} = \text{Span}\{x_1(z), \ldots, x_r(z)\}$ $(\subset \mathbb{C}^n)$ for every $z \in \Omega \setminus \Omega_0$, where $\Omega_0 = \{z \in \Omega \mid x_1(z), \ldots, x_r(z)$ are linearly dependent$\}$. If, in addition, for some s $(\leq r)$ the vector functions $x_1(z), \ldots, x_s(z)$ are linearly independent for all $z \in \Omega$, then $y_i(z)$, $i = 1, \ldots, r$ can be chosen in such a way that (a)–(c) hold, and moreover, $y_1(z) = x_1(z), \ldots, y_s(z) = x_s(z)$ for all $z \in \Omega$.

In the proof of Lemma 18.2.2 we use two classical results (see Chapter 3 of Markushevich (1965), Vol. 3, for example) in the theory of analytic and meromorphic functions that are stated here for the reader's convenience.

Recall that a set $S \subset \Omega$ is called *discrete* if for every $z \in S$ there is a neighbourhood V of z such that $V \cap S = \{z\}$. (In particular, the empty set and the finite sets are discrete.) Note also that a discrete set is at most countable.

Lemma 18.2.3

(*Weierstrass's theorem*). *Let $S \subset \Omega$ be a discrete set, and for every $z_0 \in S$ let a positive integer $s(z_0)$ be given. Then there exists a (scalar) function $f(z)$ that is analytic on Ω and for which the set of zeros of $f(z)$ coincides with S, and for every $z_0 \in S$ the multiplicity of z_0 as a zero of $f(z)$ is exactly $s(z_0)$.*

Lemma 18.2.4

(*Mittag–Leffler theorem*). *Let $S \subset \Omega$ be a discrete set, and for every $z_0 \in S$ let a rational function of type*

$$q_{z_0}(z) = \sum_{j=1}^{k} \alpha_j (z - z_0)^{-j} \tag{18.2.4}$$

be given, where k is a positive integer (depending on z_0) and α_j are complex numbers (also depending on z_0). Then there exists a function $f(z)$ that is meromorphic on Ω, for which the set of poles of $f(z)$ coincides with S, and for every $z_0 \in S$, the singular part of $f(z)$ at z_0 coincides with $q_{z_0}(z)$; that is, $f(z) - q_{z_0}(z)$ is analytic at z_0.

Proof of Lemma 18.2.2. We proceed by induction on r. Consider first the case $r = 1$. Let $g(z)$ be an analytic scalar function on Ω with the property that every zero of $g(z)$ is also a zero of $x_1(z)$ having the same multiplicity, and vice versa. The existence of such a $g(z)$ is ensured by the Weierstrass theorem given above. Put $y_1(z) = (g(z))^{-1} x_1(z)$ to prove Lemma 18.2.2 in the case $r = 1$.

Now we can pass on to the general case. Using the induction assumption, we can suppose that $x_1(z), \ldots, x_{r-1}(z)$ are linearly independent for every $z \in \Omega$. Let $X_0(z)$ be an $r \times r$ submatrix of the $n \times r$ matrix $[x_1(z), \ldots, x_r(z)]$ such that $\det X_0(z_0) \neq 0$. It is well known in the theory of analytic functions that the set of zeros of the not identically zero analytic function $\det X_0(z)$ is discrete. Since $\det X_0(z_0) \neq 0$ implies that the vectors $x_1(z_0), \ldots, x_r(z_0)$ are linearly independent, it follows that the set

$$\Omega_0 = \{x \in \Omega \mid x_1(z), \ldots, x_r(z) \text{ are linearly dependent}\}$$

is also discrete. Disregarding the trivial case when Ω_0 is empty, we can write $\Omega_0 = \{\zeta_1, \zeta_2, \ldots\}$, where $\zeta_i \in \Omega$, $i = 1, 2, \ldots$, is a finite or countable sequence with no limit points inside Ω.

Let us show that for every $j = 1, 2, \ldots$, there exist a positive integer s_j

and scalar functions $a_{1j}(z), \ldots, a_{r-1,j}(z)$ that are analytic in a neighbourhood of ζ_j such that the system of n-dimensional analytic vector functions on Ω

$$x_1(z), \ldots, x_{r-1}(z), (z - \zeta_j)^{-s_j}\left[x_r(z) + \sum_{i=1}^{r-1} a_{ij}(z)x_i(z)\right] \quad (18.2.5)$$

has the following properties: for each $z \neq \zeta_j$ it is linearly equivalent to the system $x_1(z), \ldots, x_r(z)$ (i.e., both systems span the same subspace in $\mathbb{C}^n$); for $z = \zeta_j$ it is linearly independent. Indeed, consider the $n \times r$ matrix $B(z)$ whose columns are formed by $x_1(z), \ldots, x_r(z)$. By the induction hypothesis, there exists an $(r - 1) \times (r - 1)$ submatrix $B_0(z)$ in the first $r - 1$ columns of $B(z)$ such that $\det B_0(\zeta_j) \neq 0$. For simplicity of notation suppose that $B_0(z)$ is formed by the first $r - 1$ columns and rows in $B(z)$; so

$$B(z) = \begin{bmatrix} B_0(z) & B_1(z) \\ B_2(z) & B_3(z) \end{bmatrix}$$

where $B_1(z)$, $B_2(z)$, and $B_3(z)$ are of sizes $(r - 1) \times 1$, $(n - r + 1) \times (r - 1)$, and $(n - r + 1) \times 1$, respectively. Since $B_0(z)$ is invertible in a neighbourhood of ζ_j, we can write

$$B(z) = \begin{bmatrix} I & 0 \\ B_2(z)B_0^{-1}(z) & I \end{bmatrix}\begin{bmatrix} B_0(z) & 0 \\ 0 & W(z) \end{bmatrix}\begin{bmatrix} I & B_0^{-1}(z)B_1(z) \\ 0 & I \end{bmatrix}$$

$$(18.2.6)$$

where $W(z) = B_3(z) - B_2(z)B_0^{-1}(z)B_1(z)$ is an $(n - r + 1) \times 1$ matrix. Let s_j be the multiplicity of ζ_j as a zero of the vector function $W(z)$. Consider the matrix function

$$\tilde{B}(z) = \begin{bmatrix} I & 0 \\ B_2(z)B_0^{-1}(z) & I \end{bmatrix}\begin{bmatrix} B_0(z) & 0 \\ 0 & (z - \zeta_j)^{-s_j}W(z) \end{bmatrix}$$

Clearly, the columns $b_1(z), \ldots, b_r(z)$ of $\tilde{B}(z)$ are analytic and linearly independent vector functions in a neighbourhood $V(\zeta_j)$ of ζ_j. From formula (18.2.6) it is clear that $\mathrm{Span}\{x_1(z), \ldots, x_r(z)\} = \mathrm{Span}\{b_1(z), \ldots, b_r(z)\}$ for $z \in V(\zeta_j) \smallsetminus \zeta_j$. Further, from (18.2.6) we obtain

$$\tilde{B}(z) = \begin{bmatrix} B_0(z) & 0 \\ B_2(z) & (z - \zeta_j)^{-s_j}W(z) \end{bmatrix}$$

and

$$\begin{bmatrix} 0 \\ (z - \zeta_j)^{-s_j}W(z) \end{bmatrix} = (z - \zeta_j)^{-s_j}B(z)\begin{bmatrix} -B_0^{-1}(z)B_1(z) \\ I \end{bmatrix}$$

So the columns $b_1(z), \ldots, b_r(z)$ of $\tilde{B}(z)$ have the form (18.2.5), where $a_{ij}(z)$ are analytic scalar functions in a neighbourhood of ζ_j.

Now choose $y_1(z), \ldots, y_r(z)$ in the form

$$y_1(z) = x_1(z), \ldots, y_{r-1}(z) = x_{r-1}(z), \qquad y_r(z) = \sum_{i=1}^{r} g_i(z) x_i(z)$$

where the scalar functions $g_i(z)$ are constructed as follows: (a) $g_r(z)$ is analytic and different from zero in Ω except for the set of poles $\zeta_1, \zeta_2, \ldots$, with corresponding multiplicities $s_1, s_2, \ldots$; (b) the functions $g_i(z)$ (for $i = 1, \ldots, r-1$) are analytic in Ω except for the poles $\zeta_1, \zeta_2, \ldots$, and the singular part of $g_i(z)$ at ζ_j (for $j = 1, 2, \ldots$) is equal to the singular part of $a_{ij}(z) g_r(z)$ at ζ_j.

Let us check the existence of such functions $g_i(z)$. Let $g_r(z)$ be the inverse of an analytic function with zeros at $\zeta_1, \zeta_2, \ldots$, with corresponding multiplicities $s_1, s_2, \ldots$ (such an analytic function exists by Lemma 18.2.3). The functions $g_1(z), \ldots, g_{r-1}(z)$ are constructed by using the Mittag–Leffler theorem (Lemma 18.2.4).

Property (a) ensures that $y_1(z), \ldots, y_r(z)$ are linearly independent for every $z \in \Omega \smallsetminus \{\zeta_1, \zeta_2, \ldots\}$. In a neighbourhood of each ζ_j we have

$$y_r(z) = \sum_{i=1}^{r-1} (g_i(z) - a_{ij}(z) g_r(z)) x_i(z) + g_r(z) \left(x_r(z) + \sum_{i=1}^{r-1} a_{ij}(z) x_i(z) \right)$$

$$= (z - \zeta_j)^{-s_j} \left[x_r(z) + \sum_{i=1}^{r-1} a_{ij}(z) x_i(z) \right]$$

$$+ \{\text{linear combination of } x_1(\zeta_j), \ldots, x_{r-1}(\zeta_j)\} + \cdots +$$

$$(18.2.7)$$

where the final ellipsis denotes a vector function that is analytic in a neighbourhood of ζ_j and assumes the value zero at ζ_j. Formula (18.2.7) and the linear independence of vectors (18.2.5) for $z = \zeta_j$ ensures that $y_1(\zeta_j), \ldots, y_r(\zeta_j)$ are linearly independent. Finally, the last statement of Lemma 18.2.2 follows from the proof of the first part of this lemma. $\square$

Proof of Theorem 18.2.1. Let $A_0(z)$ be an $r \times r$ submatrix of $A(z)$ that is nonsingular for some $z \in \Omega$, that is, $\det A_0(z) \neq 0$. So the set Ω_0 of zeros of the analytic function $\det A_0(z)$ is either empty or consists of isolated points. In what follows we assume for simplicity that $A_0(z)$ is located in the top left corner of $A(z)$ of size $r \times r$.

Let $x_1(z), \ldots, x_r(z)$ be the first r columns of $A(z)$, and let $y_1(z), \ldots, y_r(z)$ be the vector functions constructed in Lemma 18.2.2. Then for each $z \in \Omega \smallsetminus \Omega_0$ we have

$$\text{Span}\{y_1(z), \ldots, y_r(z)\} = \text{Span}\{x_1(z), \ldots, x_r(z)\} = \text{Im } A(z)$$

$$(18.2.8)$$

[The last equality follows from the linear independence of $x_1(z), \ldots, x_r(z)$ for $z \in \Omega \smallsetminus \Omega_0$.] We now prove that

$$\text{Span}\{y_1(z), \ldots, y_r(z)\} \supset \text{Im } A(z), \qquad z \in \Omega \qquad (18.2.9)$$

Equality (18.2.8) means that for every $z \in \Omega \smallsetminus \Omega_0$ there exists an $r \times r$ matrix $B(z)$ such that

$$Y(z)B(z) = A(z), \qquad z \in \Omega \smallsetminus \Omega_0 \qquad (18.2.10)$$

where $Y(z) = [y_1(z), \ldots, y_r(z)]$. Note that $B(z)$ is necessarily unique. [Indeed, if $B'(z)$ also satisfies (18.2.10), we have $Y(z)(B(z) - B'(z)) = 0$, and, in view of the linear independence of the columns of $Y(z)$, $B(z) = B'(z)$.] Further, $B(z)$ is analytic in $\Omega \smallsetminus \Omega_0$. To check this, pick an arbitrary $z' \in \Omega \smallsetminus \Omega_0$, and let $Y_0(z)$ be an $r \times r$ submatrix of $Y(z)$ such that $\det(Y_0(z')) \neq 0$. [For simplicity of notation assume that $Y_0(z)$ occupies the top r rows of $Y(z)$.] Then $\det(Y_0(z)) \neq 0$ in some neighbourhood V of z', and $(Y_0(z))^{-1}$ is analytic on $z \in V$. Now $Y(z)^{-L} \overset{\text{def}}{=} [(Y_0(z))^{-1}, 0]$ is a left inverse of $Y(z)$; premultiplying (18.2.10) by $Y(z)^{-L}$, we obtain

$$B(z) = Y(z)^{-L}A(z), \qquad z \in V \qquad (18.2.11)$$

So $B(z)$ is analytic on $z \in V$; since $z' \in \Omega \smallsetminus \Omega_0$ was arbitrary, $B(z)$ is analytic on $\Omega \smallsetminus \Omega_0$.

Moreover, $B(z)$ admits analytic continuation to the whole of Ω, as follows. Let $z_0 \in \Omega_0$, and let $Y(z)^{-L}$ be a left inverse of $Y(z)$, which is analytic in a neighbourhood V_0 of z_0. [The existence of such $Y(z)$ is proved as above.] Define $B(z)$ as $Y(z)^{-L}A(z)$ for $z \in V_0$. Clearly, $B(z)$ is analytic on V_0, and for $z \in V_0 \smallsetminus \{z_0\}$, this definition coincides with (18.2.11) in view of the uniqueness of $B(z)$. So $B(z)$ is analytic on Ω.

Now it is clear that (18.2.10) holds also for $z \in \Omega_0$, which proves (18.2.9). Consideration of dimensions shows that in fact we have an equality in (18.2.9), unless rank $A(z) < r$. Thus (18.2.1) and (18.2.3) are proved.

We pass now to the proof of existence of $y_{r+1}(z), \ldots, y_n(z)$ such that (b), (18.2.2), and (18.2.4) hold. Let $a_1(z), \ldots, a_r(z)$ be the first r rows of $A(z)$. By assumption $a_1(\bar{z}), \ldots, a_r(\bar{z})$ are linearly independent for some $\bar{z} \in \Omega$. Apply Lemma 18.2.2 to construct n-dimensional analytic row functions $b_1(z), \ldots, b_r(z)$ such that for all $z \in \Omega$ the rows $b_1(z), \ldots, b_r(z)$ are linearly independent, and for $z \in \Omega \smallsetminus \Omega_0$,

$$\text{Span}\{b_1(z)^T, \ldots, b_r(z)^T\} = \text{Span}\{a_1(z)^T, \ldots, a_r(z)^T\} \qquad (18.2.12)$$

Fix $z_0 \in \Omega$, and let $b_{r+1}, \ldots, b_r$ be n-dimensional rows such that the vectors $b_1(z_0)^T, \ldots, b_r(z_0)^T, b_{r+1}^T, \ldots, b_n^T$ form a basis in $\mathbb{C}^n$. Applying Lemma 18.2.2 again [for $x_1(z) = b_1(z)^T, \ldots, x_r(z) = b_r(z)^T$, $x_{r+1}(z) = b_{r+1}^T, \ldots, x_n(z) = b_n^T$], we construct n-dimensional analytic row functions $b_{r+1}(z), \ldots, b_n(z)$ such that the $n \times n$ matrix

$$B(z) = \begin{bmatrix} b_1(z) \\ b_2(z) \\ \vdots \\ b_n(z) \end{bmatrix}$$

is nonsingular for all $z \in \Omega$. Then the inverse $B(z)^{-1}$ is analytic on Ω. Let $y_{r+1}(z), \ldots, y_n(z)$ be the last $(n - r)$ columns of $B(z)^{-1}$. We claim that (b), (18.2.2), and (18.2.4) are satisfied with this choice.

Indeed, (b) is evident. Take $z \in \Omega \smallsetminus \Omega_0$; from (18.2.12) and the construction of $y_{r+1}(z), \ldots, y_n(z)$ it follows that

$$\mathrm{Ker} \begin{bmatrix} a_1(z) \\ a_2(z) \\ \vdots \\ a_r(z) \end{bmatrix} \supset \mathrm{Span}\{y_{r+1}(z), \ldots, y_n(z)\}$$

But since $z \notin \Omega_0$, every row of $A(z)$ is a linear combination of the first r rows. So in fact

$$\mathrm{Ker}\, A(z) \supset \mathrm{Span}\{y_{r+1}(z), \ldots, y_n(r)\} \tag{18.2.13}$$

Now (18.2.13) implies that for $z \in \Omega \smallsetminus \Omega_0$

$$A(z)[y_{r+1}(z), \ldots, y_n(z)] = 0 \tag{18.2.14}$$

Passing to the limit when z approaches a point from Ω_0, we find that (18.2.14), as well as the inclusion (18.2.13), holds for every $z \in \Omega$. Consideration of dimensions shows that the equality holds in (18.2.13) if and only if rank $A(z) = r$. $\quad\square$

18.3 GLOBAL PROPERTIES OF ANALYTIC FAMILIES OF SUBSPACES

In the definition of an analytic family of subspaces the transformation $A(z)$ and the subspace $\mathcal{M}$ depend on z_0, so the definition of an analytic family of subspaces has a local character. However, it turns out that for a given analytic family of subspaces $\mathcal{M}(z)$ there exists an analytic family $A(z)$ and a subspace $\mathcal{M}$ independent of z_0 for which the equality $\mathcal{M}(z) = A(z)\mathcal{M}$ holds.

Theorem 18.3.1

Let $\{\mathcal{M}(z)\}_{z\in\Omega}$ be an analytic family of subspaces (of $\mathbb{C}^n$) on Ω. Then there exist invertible transformations $A(z)$: $\mathbb{C}^n \to \mathbb{C}^n$ that are analytic on Ω, and a subspace $\mathcal{M} \subset \mathbb{C}^n$ such that $\mathcal{M}(z) = A(z)\mathcal{M}$, for all $z \in \Omega$.

The lengthy proof of Theorem 18.3.1 is relegated to the next two sections. First, we wish to emphasize that this is a particularly important result concerning analytic families of subspaces and has many consequences, some of which we describe now.

Theorem 18.3.2

For an analytic family of subspaces $\mathcal{M}(z)$ (of $\mathbb{C}^n$) on Ω the following properties hold: (a) there exist n-dimensional vector functions $x_1(z), \ldots, x_p(z)$ that are analytic on Ω and such that, for each $z \in \Omega$, the vectors $x_1(z), \ldots, x_p(z)$ are linearly independent and

$$\mathcal{M}(z) = \mathrm{Span}\{x_1(z), \ldots, x_p(z)\}$$

(b) there is an analytic family of projectors $P(z)$ defined on Ω such that $\mathcal{M}(z) = \mathrm{Im}\, P(z)$ for all $z \in \Omega$; (c) for every $z \in \Omega$ there exists a direct complement $\mathcal{N}(z)$ to $\mathcal{M}(z)$ in $\mathbb{C}^n$ such that the family of subspaces $\mathcal{N}(z)$ is analytic.

Proof. Let $A(z)$ and $\mathcal{M}$ be as in Theorem 18.3.1, and let $x_1, \ldots, x_p$ be a basis in $\mathcal{M}$. Then $x_i(z) = A(z)x_i$, $i = 1, \ldots, p$ satisfy (a). To satisfy (b), put $P(z) = A(z)PA(z)^{-1}$, where P is a projector on $\mathcal{M}$. Finally, the family of subspaces $\mathcal{N}(z) = A(z)\mathcal{N}$, where $\mathcal{N}$ is a direct complement in $\mathcal{M}$ in $\mathbb{C}^n$, satisfies (c). $\square$

Note that property (b) [as well as property (a)] is characteristic for analytic families of subspaces. So, if $P(z)$ is an analytic family of projectors on Ω, then $\mathrm{Im}\, P(z)$ is an analytic family of subspaces. We leave the verification of this statement to the reader.

In connection with Theorem 18.3.2 (c), note that the orthogonal complement $\mathcal{M}(z)^{\perp}$ is usually not an analytic family, as the next example shows.

EXAMPLE 18.3.1.   For any $z \in \mathbb{C}$ let

$$\mathcal{M}(z) = \mathrm{Span}\begin{bmatrix} 1 \\ z \end{bmatrix} \subset \mathbb{C}^2$$

Then

$$\mathcal{M}(z)^{\perp} = \mathrm{Span}\begin{bmatrix} -\bar{z} \\ 1 \end{bmatrix}$$

which is not analytic. Indeed, if $\mathcal{M}(z)^\perp$ were analytic, then for z in a neighbourhood U of each point $z_0 \in \mathbb{C}$ we would have

$$\text{Span}\begin{bmatrix} -\bar{z} \\ 1 \end{bmatrix} = A(z)\mathcal{M}$$

where $A(z)$ is a 2×2 analytic family of invertible matrices and $\mathcal{M}$ is a fixed one-dimensional subspace that, without loss of generality, may be assumed equal to $\text{Span}\{e_1\}$. So

$$\text{Span}\begin{bmatrix} -\bar{z} \\ 1 \end{bmatrix} = \text{Span}\begin{bmatrix} a_1(z) \\ a_2(z) \end{bmatrix}$$

on U, where $\begin{bmatrix} a_1(z) \\ a_2(z) \end{bmatrix}$ is the first column of $A(z)$. Hence $a_2(z) \neq 0$ and for all $z \in U$

$$\bar{z} = -a_1(z)a_2(z)^{-1} \tag{18.3.1}$$

However, the function $\bar{z}$ is not analytic in U, so (18.3.1) cannot happen. $\square$

In the next section we will need the following generalization of Theorem 18.3.2.

Theorem 18.3.3

Let $\mathcal{M}(z)$ and $\mathcal{N}(z)$ be analytic families of subspaces (of $\mathbb{C}^n$) on Ω such that $\mathcal{M}(z) \subset \mathcal{N}(z)$ for all $z \in \Omega$. Then there exist n-dimensional vector functions $x_1(z), \ldots, x_p(z)$ [where $p = \dim \mathcal{N}(z) - \dim \mathcal{M}(z)$] that are analytic on Ω and such that, for each $z \in \Omega$, the vectors $x_1(z), \ldots, x_p(z)$ form a basis in $\mathcal{N}(z)$ modulo $\mathcal{M}(z)$.

Proof. By Theorem 18.3.2 there are bases $y_1(z), \ldots, y_s(z)$ in $\mathcal{M}(z)$ and $v_1(z), \ldots, v_t(z)$ in $\mathcal{N}(z)$ that are analytic on Ω. By Lemma 18.2.2 there exist analytic vector functions $y_{s+1}(z), \ldots, y_t(z)$ such that $y_1(z), \ldots, y_t(z)$ are linearly independent for each $z \in \Omega$ and

$$\text{Span}\{y_1(z), \ldots, y_t(z)\} = \text{Span}\{v_1(z), \ldots, v_t(z)\} = \mathcal{N}(z)$$

Obviously, $y_{s+1}(z), \ldots, y_t(z)$ is the desired analytic basis in $\mathcal{N}(z)$ modulo $\mathcal{M}(z)$. $\square$

We note one more consequence of Theorem 18.3.1.

Corollary 18.3.4

Let $\mathcal{M}_1(z), \ldots, \mathcal{M}_k(z)$ be analytic families of subspaces (of $\mathbb{C}^n$) on Ω, and assume that for each $z \in \Omega$, $\mathbb{C}^n$ is a direct sum of $\mathcal{M}_1(z), \ldots, \mathcal{M}_k(z)$. Then, *given $z_0 \in \Omega$, there exists a family of invertible transformations $S(z)$: $\mathbb{C}^n \to \mathbb{C}^n$ that is analytic on Ω and for which $S(z)\mathcal{M}_i(z_0) = \mathcal{M}_i(z)$ on Ω, and $S(z_0) = I$.*

Proof. It follows from Theorem 18.3.1 that there exist analytic families of invertible transformations $S_i(z)$: $\mathbb{C}^n \to \mathbb{C}^n$, $i = 1, \ldots, k$, such that $S_i(z_0) = I$ and $S_i(z)\mathcal{M}_i(z_0) = \mathcal{M}_i(z)$ for all $z \in \mathbb{C}$. Now the transformation $S(z)$: $\mathbb{C}^n \to \mathbb{C}^n$ defined by the property that $S(z)x = S_i(z)x$ for all $x \in \mathcal{M}_i(z_0)$ satisfies the requirements of Corollary 18.3.4.   □

18.4   PROOF OF THEOREM 18.3.1 (COMPACT SETS)

As a first step towards the proof of Theorem 18.3.1, a result is proved in this section that can be considered as a weaker version of that theorem. We say that a function $f(z)$ (whose values may be vectors, or transformations) is analytic on a compact set $K \subset \Omega$ if $f(z)$ is analytic on some open set containing K.

Theorem 18.4.1

Let $K \subset \Omega$ be a compact set, and let $\mathcal{M}(z) \subset \mathbb{C}^n$ be an analytic family of subspaces on Ω. Then there exist vector functions $f_1(z), \ldots, f_r(z) \in \mathbb{C}^n$ that are analytic on K and such that $f_1(z), \ldots, f_r(z)$ is a basis in $\mathcal{M}(z)$ for every $z \in K$.

In turn, we need some preliminaries for the proof of Theorem 18.4.1. First, we introduce the notion of an incomplete factorization. Let $A(z)$ be an $n \times n$ matrix function that is analytic on a neighbourhood of the unit circle and is nonsingular on the unit circle. An *incomplete factorization* of $A(z)$ is a representation of the form

$$A(z) = \,^-A(z)^+A(z) \qquad (18.4.1)$$

that holds whenever $|z| = 1$ and the family $^+A(z)$ is nonsingular and analytic on the disc $|z| \le 1$, and the family $^-A(z)$ is nonsingular and analytic on the annulus $1 \le |z| < \infty$.

Lemma 18.4.2

Every $n \times n$ matrix function $A(z)$ that is analytic and nonsingular on a neighbourhood of the unit circle admits an incomplete factorization.

Proof. Consider first the case when $A(z)$ is analytic on the disc $|z| \le 1$. Let z_0 be a zero of det $A(z)$ with $|z_0| < 1$. Then for some invertible matrix T_0 the first row of $T_0 A(z)$ is zero at the point z_0. Put

$$^+A_1(z) = \text{diag}[(z - z_0)^{-1}, 1, \ldots, 1] T_0 A(z)$$

$$^-A_1(z) = T_0^{-1}[\text{diag}(z - z_0), 1, \ldots, 1]$$

Then $A(z) = {}^-A_1(z) \, {}^+A_1(z)$; moreover, $^-A_1(z)$ is analytic and invertible for $1 \leq |z| < \infty$, $^+A_1(z)$ is analytic and invertible for $|z| \leq 1$, and the number of zeros of $\det {}^+A_1(z)$ inside the unit circle is strictly less than that of $\det A(z)$. If $\det {}^+A_1(z) \neq 0$ for $|z| \leq 1$, then $A(z) = {}^-A_1(z) \, {}^+A_1(z)$ is an incomplete factorization of $A(z)$. Otherwise, we apply the construction above to $^+A_1(z)$, and after a finite number of steps an incomplete factorization of $A(z)$ is obtained.

Now it is easy to prove Lemma 18.4.2 for the case that $A(z)$ is meromorphic in the disc $|z| \leq 1$ (more exactly, admits a meromorphic continuation into the disc). Indeed, let $z_1, \ldots, z_k$ be all the poles of $A(z)$ inside the unit disc with orders $\alpha_1, \ldots, \alpha_k$, respectively. Then the function $B(z) = \Pi_{i=1}^{k} (z - z_i)^{\alpha_i} A(z)$ is analytic for $|z| \leq 1$ and thus (according to the assertion proved in the preceding paragraph) admits an incomplete factorization: $B(z) = {}^-B(z) \, {}^+B(z)$. So (18.4.1) with $^-A(z) = \{ \Pi_{s=1}^{k} (z - z_i)^{-\alpha_i} \} \, {}^-B(z)$; $^+A(z) = {}^+B(z)$ is an incomplete factorization of $A(z)$.

Now consider the general case. Let $\epsilon > 0$ be such that $A(z)$ is analytic and invertible in the closed annulus $\bar{\Phi} = \{ z \in \mathbb{C} \,|\, 1 - \epsilon \leq |z| \leq 1 + \epsilon \}$. In the sequel we use some basic and elementary facts about the structure of the set C_ω of all $n \times n$ matrix functions $X(z)$ that are continuous in the closed annulus $\bar{\Phi}$ and analytic in the open annulus $\Phi = \{ z \in \mathbb{C} \,|\, 1 - \epsilon < |z| < 1 + \epsilon \}$.

The set C_ω is an algebra with pointwise addition and multiplication of matrices and multiplication by scalars, that is, for $z \in \bar{\Phi}$ and $X(z), Y(z) \in C_\omega$ we define

$$(XY)(z) = X(z)Y(z), \qquad (X + Y)(z) = X(z) + Y(z), \qquad (\alpha X)(z) = \alpha X(z)$$

Introduce the following norm in C_ω:

$$\|X\|_{C_\omega} = \max_{z \in \bar{\Phi}} \|X(z)\|$$

where $X(z) \in C_\omega$. It is easily seen that this is indeed a norm; that is, the axioms (a)–(c) of Section 13.8 are satisfied. Moreover

$$\|XY\|_{C_\omega} \leq \|X\|_{C_\omega} \|Y\|_{C_\omega}$$

for $X, Y \in C_\omega$. In fact, the normed algebra C_ω is a Banach algebra, which means that each Cauchy sequence converges in the norm $\| \cdot \|_{C_\omega}$ to some function in C_ω. This follows from the fact that the uniform limit of continuous functions on $\bar{\Phi}$ is itself a continuous function on $\bar{\Phi}$, and the limit of analytic function on Φ which is uniform on each compact set in Φ is itself analytic on Φ.

Let $\mathcal{M}_+$ be the set of all matrix functions from C_ω that admit an analytic continuation to the set $\{z \in \mathbb{C} \mid |z| \le 1 - \epsilon\}$ and let $\mathcal{M}_-$ be the set of all matrix functions from C_ω that admit an analytic continuation to the set $\{z \in \mathbb{C} \mid |z| \ge 1 + \epsilon\} \cup \{\infty\}$ and assume the zero value at infinity. It is easily seen (as for C_ω) that $\mathcal{M}_+$ and $\mathcal{M}_-$ are closed subspaces in the norm $\|\cdot\|_{C_\omega}$. Clearly, $\mathcal{M}_+ \cap \mathcal{M}_- = \{0\}$ (here 0 stands for the identically zero $n \times n$ matrix function on $\bar{\Phi}$). Furthermore, $\mathcal{M}_+ + \mathcal{M}_- = C_\omega$. Indeed, recall that every function $X(z) \in C_\omega$ can be developed into the Laurent series

$$X(z) = \sum_{j=-\infty}^{\infty} z^j X_j \quad (1 - \epsilon < |z| < 1 + \epsilon)$$

where the functions

$$X_+(z) \overset{\text{def}}{=} \sum_{j=0}^{\infty} z^j X_j \quad \text{and} \quad X_-(z) \overset{\text{def}}{=} \sum_{j=-\infty}^{-1} z^j X_j$$

belong to $\mathcal{M}_+$ and $\mathcal{M}_-$, respectively. Denoting $P_+(X(z)) = X_+(z)$, we obtain a projector $P_+ : C_\omega \to C_\omega$ with $\operatorname{Im} P_+ = \mathcal{M}_+$ and $\operatorname{Ker} P_+ = \mathcal{M}_-$. It turns out that P_+ is bounded, that is

$$\|P_+\| \overset{\text{def}}{=} \sup\{\|P_+(X)\|_{C_\omega} \mid X(z) \in C_\omega, \|X\|_{C_\omega} = 1\} < \infty$$

[See page 225 in Gohberg and Goldberg (1981), for example; the proof is based on Banach's theorem that every bounded linear operator that maps a Banach space onto itself, and is one-to-one, has a bounded inverse.]

Return to our original matrix function $A(z)$. Clearly $A(z)^{-1} \in C_\omega$, and the Laurent series $A(z)^{-1} = \sum_{j=-\infty}^{\infty} z^j A_j$ converges uniformly in the annulus $1 - \epsilon \le |z| \le 1 + \epsilon$. Therefore, for some N the matrix function $A_N(z) = \sum_{j=-N}^{N} z^j A_j$ has the following properties: $\det A_N(z) \ne 0$ for $1 - \epsilon \le |z| \le 1 + \epsilon$ and

$$A(z)^{-1} = A_N(z)(I - M(z))$$

where $M(z) \in C_\omega$ and

$$\|M\|_{C_\omega} < (4\|P_+\|)^{-1} \tag{18.4.2}$$

Let

$${}^+N = P_+ M + P_+((P_+ M)M) \cdots \in C_\omega$$

Because of (18.4.2), $\|{}^+N\|_{C_\omega} < \frac{1}{2}$, and hence $I + {}^+N$ is invertible in the algebra C_ω. (Here I represents the constant $n \times n$ identity matrix.) Denote ${}^+G = (I + {}^+N)^{-1}$. Then ${}^+G$ and $({}^+G)^{-1}$ belong to the image of P_+. In particular, ^+G and $(^+G)^{-1}$ are analytic in the disc $|z| \leq 1$. Furthermore, one checks easily that

$$P[(I + {^+}N)(I - M)] = I$$

so the function $^-G = (I + {^+}N)(I - M)$ is analytic for $1 \leq |z| < \infty$ and at infinity. As

$$\|{^-}G - I\|_{C_\omega} \leq \|{^+}N\|_{C_\omega} + \|M\|_{C_\omega} + \|{^+}NM\|_{C_\omega} < \tfrac{1}{2} + \tfrac{1}{4} + \tfrac{1}{8} < 1$$

^-G is invertible in C_ω. Since both ^-G and I belong to the (closed) subalgebra $C_\omega^- = \{\alpha I + \operatorname{Ker} P_+ \mid \alpha \in \mathbb{C}\}$ of C_ω, also $(^-G)^{-1} \in C_\omega^-$. Now write

$$A(z)^{-1} = A_N(z){^+}G(z){^-}G(z), \quad \text{or} \quad A(z) = ({^-}G(z))^{-1}({^+}G(z))^{-1}(A_N(z))^{-1}$$

and use the fact (proved in the preceding paragraph) that the function $({^+}G(z))^{-1}(A_N(z))^{-1}$, which is meromorphic on the unit disc, admits an incomplete factorization. $\quad\Box$

Lemma 18.4.3

Let $f_1, \ldots, f_r \in \mathbb{C}^n$ and $g_1, \ldots, g_r \in \mathbb{C}^n$ be two systems of analytic and linearly independent vectors on Ω such that

$$\operatorname{Span}\{f_1(z), \ldots, f_r(z)\} = \operatorname{Span}\{g_1(z), \ldots, g_r(z)\} \qquad (18.4.3)$$

for $z \subset \Omega_0$, where $\Omega_0 \subset \Omega$ is a set with at least one limit point inside Ω. Then

$$\operatorname{Span}\{f_1(z), \ldots, f_r(z)\} = \operatorname{Span}\{g_1(z), \ldots, g_r(z)\}$$

for every $z \in \Omega$ and

$$[f_1(z) \cdots f_r(z)] = [g_1(z) \cdots g_r(z)] \cdot A(z) \qquad (18.4.4)$$

where $A(z)$ is an $r \times r$ matrix function that is invertible and analytic on Ω.

Proof. Consider the system $\Pi = \{f_1, \ldots, f_r, g_1, \ldots, g_r\}$ of $2r$ n-dimensional vectors. Then rank $\Pi(z) = r$ for $z \in \Omega_0$. On the other hand, the set $\{z_0 \in \Omega \mid \operatorname{rank} \Pi(z_0) < \max_{z \in \Omega} \operatorname{rank} \Pi(z)\}$ is discrete. Thus $r = \max_{x \in \Omega} \operatorname{rank} \Pi(z)$, and (18.4.3) holds for every $z \in \Omega$ because both systems $f_1(z), \ldots, f_r(z)$ and $g_1(z), \ldots, g_r(z)$ are linearly independent. Consequently, there exists a unique matrix function $A(z)$ such that (18.4.4) holds. It remains to prove that $A(z)$ is analytic on Ω. Let $z_0 \in \Omega$, and suppose, for example, that the square matrix $X(z)$ formed by the upper r rows of

$[g_1(z), \ldots, g_r(z)]$ is invertible for $z = z_0$. Computing $A(z)$ in a neighbourhood of z_0 by Cramer's formulas, we see that $A(z)$ is analytic in a neighbourhood of z_0. Thus $A(z)$ is analytic on Ω. $\square$

Proof of Theorem 18.4.1. Without loss of generality, we can suppose that K is a connected set (otherwise consider a larger compact set). Fix $z_0 \in K$, and let $\mathcal{N}_0$ be some direct complement for $\mathcal{M}(z_0)$ in $\math6{C}^n$. Then

$$\math6{C}^n = \mathcal{M}(z) \dotplus \mathcal{N}_0 \tag{18.4.5}$$

is a direct sum decomposition for every $z \in K$ except maybe for a finite set of points $z_1, \ldots, z_k$. Indeed, by the definition of an analytic family of subspaces, for every $\eta \in K$ there exists a neighbourhood U_η of η and an analytic and invertible matrix function $B_\eta(z)$ defined on U_η such that $B_\eta(z)\mathcal{M} = \mathcal{M}(z)$ on U_η, where $\mathcal{M}$ is a fixed subspace in $\math6{C}^n$. We can assume [by changing $B_\eta(z)$ if necessary] that the subspace $\mathcal{M}$ is independent of η. [Here we use the fact that $\dim \mathcal{M}(z)$ is constant because of the connectedness of Ω.] Actually, we assume $\mathcal{M} = \mathcal{M}(z_0)$. Let $x_1, \ldots, x_r$ be some basis in $\mathcal{M}(z_0)$, and let $x_{r+1}, \ldots, x_n$ be a basis in $\mathcal{N}_0$. Then for $z \in U_\eta$ the subspaces $\mathcal{M}(z)$ and $\mathcal{N}_0$ are direct complements to each other if and only if

$$D_\eta(z) = \det[B_\eta(z)x_1, \ldots, B_\eta(z)x_r, x_{r+1}, \ldots, x_n] \neq 0$$

Two cases can occur: (a) $D_\eta(z) \equiv 0$ for $z \in U_\eta$; (b) $D_\eta(z) \neq 0$, and then we can suppose (taking U_η smaller if necessary) that $D_\eta(z) \neq 0$ only at a finite number of points of U_η. Let us call the points η for which (a) holds points of the first kind, and the points η for which (b) holds points of the second kind. Since K is connected, all $\eta \in K$ are of the same kind, and since z_0 is of the second kind, all $\eta \in K$ are of the second kind. Further, let $U_{\eta_1}, \ldots, U_{\eta_l}$ be a finite covering of the compact K. Since $D_{\eta_j}(z) \neq 0$ only at a finite number of points z in U_{η_j}, $j = 1, \ldots, l$, we find that (18.4.5) holds for every $z \in K$ except possibly for a finite number of points $z_1, \ldots, z_k \in K$.

By the definition of an analytic family of subspaces, there exist neighbourhoods $U(z_1), \ldots, U(z_k)$ of $z_1, \ldots, z_k$, respectively, and functions $B^{(1)}(z), \ldots, B^{(k)}(z)$ that are invertible and analytic on $U(z_1), \ldots, U(z_k)$, respectively, such that

$$B^{(j)}(z)\mathcal{M}(z_j) = \mathcal{M}(z) , \qquad z \in U(z_j) , \qquad j = 1, \ldots, k$$

Let $x_1^{(j)}, \ldots, x_r^{(j)}$ be some basis of the subspace $\mathcal{M}(z_j)$, and let $g_i^{(j)}(z) = B^{(j)}(z)x_i^{(j)}$, $(i = 1, \ldots, r; \ z \in U(z_j); \ j = 1, \ldots, k)$. Then for $\rho > 0$ small enough we have

$$\text{Span}\{g_1^{(j)}(z), \ldots, g_r^{(j)}(z)\} = \mathcal{M}(z)$$

as long as $|z - z_j| \leq \rho$ for $j = 1, \ldots, k$. Let

$$S_j = \{z \in \mathbb{C} \mid |z - z_j| < \rho\}; \qquad S = \bigcup_{j=1}^{k} S_j$$

For every $z \in K \smallsetminus S$ let $P(z)$ be the projector on $\mathcal{M}(z)$ along $\mathcal{N}_0$. Then we claim that $P(z)$ is an analytic function on $K \smallsetminus S$.

Indeed, we have to prove this assertion in a neighbourhood of every $\mu_0 \in K \smallsetminus S$. Let U_0 be a neighbourhood of μ_0 in the set $K \smallsetminus S$ such that, when $z \in U_0$, $\mathcal{M}(z) = B(z)\mathcal{M}(\mu_0)$ for some analytic and invertible matrix function $B(z)$ on U_0. The matrix function $\tilde{B}(z)$ defined on U_0 by the properties that $\tilde{B}(z)x = B(z)x$ for all $x \in \mathcal{M}(\mu_0)$ and $\tilde{B}(z)y = y$ for all $y \in \mathcal{N}_0$ is analytic and invertible. As $P(z) = B(z)P_0(B(z))^{-1}$, where P_0 is the projector on $\mathcal{M}(\mu_0)$ along $\mathcal{N}_0$, the analyticity of $P(z)$ on U_0 follows.

Let us now prove that there exist vector functions $f_1^{(0)}(z), \ldots, f_r^{(0)}(z)$ that are analytic on $K \smallsetminus S$ and for which

$$\mathrm{Span}\{f_1^{(0)}(z), \ldots, f_r^{(0)}(z)\} = \mathrm{Im}\, P(z) = \mathcal{M}(z)$$

where $z \in K \smallsetminus S$. Indeed, let $z_0 \in K \smallsetminus S$ be a fixed point. Then dim $\mathrm{Im}\, P(z_0) = r$; let $g_1^{(0)}(z), \ldots, g_r^{(0)}(z)$ be columns of $P^{(0)}(z)$ that are linearly independent for $z = z_0$. In view of Lemma 18.2.2, there exist analytic and linearly independent vector functions $f_1^{(0)}(z), \ldots, f_r^{(0)}(z)$ defined on $K \smallsetminus S$ such that

$$\mathrm{Span}\{f_1^{(0)}(z), \ldots, f_r^{(0)}(z)\} = \mathrm{Span}\{g_1^{(0)}(z), \ldots, g_r^{(0)}(z)\}$$

for every $z \in K \smallsetminus S$, except maybe for a finite set of points. (The set of exceptional points is at most finite because of the compactness of $K \smallsetminus S$.) But from the choice of $g_1^{(0)}, \ldots, g_r^{(0)}$ it follows that

$$\mathrm{Span}\{g_1^{(0)}(z), \ldots, g_r^{(0)}(z)\} = \mathrm{Im}\, P(z) = \mathcal{M}(z)$$

for every $z \in K \smallsetminus S$, except perhaps for a finite set of points [viz., those points z for which the vectors $g_1^{(0)}(z), \ldots, g_r^{(0)}(z)$ are not linearly independent]. Thus

$$\mathrm{Span}\{f_1^{(0)}(z), \ldots, f_r^{(0)}(z)\} = \mathcal{M}(z) \tag{18.4.6}$$

for every $z \in K \smallsetminus S$ except maybe for a finite number of points. As both sides of (18.4.6) are analytic families of subspaces on $K \smallsetminus S$ (Proposition 18.1.1), it is easily seen that, in fact, (18.4.6) holds for every $z \in K \smallsetminus S$.

Consider now the systems $\{f_1^{(0)}(z), \ldots, f_r^{(0)}(z)\}$ and $\{g_1^{(1)}(z), \ldots, g_r^{(1)}(z)\}$. These systems form two bases for $\mathcal{M}(z)$ that are analytic in a neighbourhood of the circle $|z - z_1| = \rho$. Therefore, by Lemma

18.4.3 there exists an $r \times r$ matrix function $A(z)$ analytic and invertible on a neighbourhood U of the set $\{z \in \mathbb{C} \mid |z - z_1| = \rho\}$ and such that, for all $z \in U$

$$[g_1^{(1)}(z), \ldots, g_r^{(1)}(z)] = [f_1^{(0)}(z), \ldots, f_r^{(0)}(z)]A(z) \qquad (18.4.7)$$

By Lemma 18.4.2, the function $A(z)$ admits an incomplete factorization relative to the circle $\{z \mid |z - z_1| = \rho\}$: $A(z) = {}^-A(z) \cdot {}^+A(z)$ ($|z - z_1| = \rho$). In view of (18.4.7), we find that, when $|z - z_1| = \rho$

$$[f_1^{(1)}(z) \cdots f_r^{(1)}(z)] \overset{\text{def}}{=} [g_1^{(1)}(z) \cdots g_r^{(1)}(z)]({}^+A(z))^{-1}$$
$$= [f_1^{(0)}(z) \cdots f_r^{(0)}(z)] \, {}^-A(z)$$

    Clearly, the functions $f_1^{(1)}(z), \ldots, f_r^{(1)}(z)$ can be continued analytically to the set $K \smallsetminus (S_2 \cup \cdots \cup S_k)$. Moreover, since ${}^+A(z)$ [resp. ${}^-A(z)$] is invertible for $|z - z_1| \le \rho$ (resp. $|z - z_1| \ge \rho$), the set $f_1^{(1)}(z), \ldots, f_r^{(1)}(z)$ is linearly independent for every $z \in K \smallsetminus (S_2 \cup \cdots \cup S_k)$. Furthermore, for any $z \in K \smallsetminus (S_2 \cup \cdots \cup S_k)$, we obtain

$$\text{Span}\{f_1^{(1)}(z), \ldots, f_r^{(1)}(z)\} = \mathcal{M}(z)$$

Now take the point z_2 and apply similar arguments, and so on. After k steps one obtains the conclusion of Theorem 18.4.1. $\quad\square$

18.5   *PROOF OF THEOREM 18.3.1 (GENERAL CASE)*

In this section we finish the proof of Theorem 18.3.1. The main idea is to pass from the case of compact sets (Theorem 18.4.1) to the case of a general domain Ω. To this end we need some approximation theorems.

    A set $M \subset \mathbb{C}$ is called *finitely connected* if M is connected and $\mathbb{C} \smallsetminus M$ consists of a finite number of connected components. A set $N \subset M$ is called *simply connected* relative to M if for every connected component Y of $\mathbb{C} \smallsetminus N$ the set $Y \cap (\mathbb{C} \smallsetminus M)$ is not empty. The first of the necessary approximation theorems is the following.

Lemma 18.5.1

Let $K \subset \Omega$ be a finitely connected compact set that is also simply connected relative to Ω. Let $Y_1, \ldots, Y_s$ be all the bounded components of $\mathbb{C} \smallsetminus K$, and, for $j = 1, \ldots, s$ let $z_j \in Y_j \smallsetminus \Omega$ be fixed points. Let $A(z)$ be an $m \times n$ matrix function that is analytic on K. Then for every $\epsilon > 0$ there exists a rational matrix function $B(z)$ of size $m \times n$ such that $B(z)$ is analytic on $\mathbb{C} \smallsetminus \{z_1, \ldots, z_s\}$ and, for any $z \in K$,

$$\|B(z) - A(z)\| < \epsilon \qquad (18.5.1)$$

Proof.   Without loss of generality we will suppose that $m = n = 1$, that is, the functions $A(z)$ and $B(z)$ are scalars. We prove that it is possible to choose a rational function of the form

$$R(z) = \sum_{\nu=0}^{k_0} z^\nu x_{0\nu} + \sum_{j=1}^{s} \sum_{\nu=1}^{k_i} (z - z_j)^{-\nu} x_{j\nu} \qquad (18.5.2)$$

where the $x_{j\nu} \in \mathbb{C}$ and such that $|A(z) - R(z)| < \epsilon$ for any $z \in K$. Let $U \subset \mathbb{C} \smallsetminus \{z_1, \ldots, z_s\}$ be a neighbourhood of K whose boundary ∂U consists of $s + 1$ closed simple rectifiable contours. Then for $z \in K$, we obtain

$$A(z) = \frac{1}{2\pi i} \int_{\partial U} \frac{A(\eta)}{\eta - z} \, d\eta$$

Since this integral can be uniformly approximated by Riemann sums, we have to prove only that the function $(\eta - z)^{-1}$ can be uniformly approximated by functions of the form $\sum_{\nu=0}^{k} (z - z_j)^{-\nu} x_\nu$, $x_\nu \in \mathbb{C}$, where $\eta \in \partial U \cap Y_j$ $(j = 1, \ldots, s)$, and $(\eta - z)^{-1}$ can be approximated uniformly by the polynomials $\sum_{\nu=0}^{k} z^\nu x_\nu$, $(x_\nu \in \mathbb{C})$ where $\eta \in \partial U \cap (\mathbb{C} \smallsetminus (K \cup Y_1 \cup \cdots \cup Y_s))$. But this assertion follows from Runge's theorem [Chapter 4 of Markushevich (1965), Vol. 1], which states that, given a simply connected domain Γ in $\mathbb{C} \cup \{\infty\}$ and a point ζ in the interior of $(\mathbb{C} \cup \{\infty\}) \smallsetminus \Gamma$, any analytic function $f(z)$ on Γ is the limit of a sequence of rational functions with their only pole at ζ, and the convergence of this sequence to $f(z)$ is uniform on every compact subset of Γ. Indeed, for $j = 1, \ldots, s$ the set bounded by the contour $\partial U \cap Y_j$ is simply connected, as well as the set $(\mathbb{C} \cup \{\infty\}) \smallsetminus (K \cup Y_1 \cup \cdots Y_s)$. $\square$

Lemma 18.5.2

Let K and $z_1, \ldots, z_s$ be as in Lemma 18.5.1. If $A(z)$ is an $n \times n$ matrix function that is analytic on K and invertible for every $z \in K$, then for every $\epsilon > 0$ there exists an analytic and invertible matrix function $B(z)$ defined on $\mathbb{C} \smallsetminus \{z_1, \ldots, z_s\}$ such that (18.5.1) holds for any $z \in K$.

Proof.   Denote by G the group of all $n \times n$ matrix functions $M(z)$ that are analytic on K and invertible for every $z \in K$, together with the topology induced by the norm $\|M\|_G = \max_{z \in K} \|M(z)\|$. Let G_I be the connected component of the topological space G that contains I, the constant $n \times n$ identity matrix. In fact

$$G_I = \left\{ X \in G \,\middle|\, \text{there exist an integer } v > 0 \text{ and } M_1, \ldots, M_v \in G, \|M_j\|_G < 1 \right.$$

$$\left. j = 1, \ldots, v \text{ such that } X = \prod_{j=1}^{v} (I - M_j) \right\} \qquad (18.5.3)$$

Indeed, denoting the right-hand side of (18.5.3) by G_0, let us prove first that G_0 is both a closed and an open set in G. Let $F \in G_0$ and $H \in G$ be such that $\|H - F\|_G < \|F^{-1}\|_G^{-1}$. Then $H = (I - M)F$, where $M = I - HF^{-1}$. We have

$$\|M\|_G = \|I - HF^{-1}\|_G = \|(H - F)F^{-1}\|_G \le \|H - F\|_G \|F^{-1}\|_G < 1$$

that is, $H \in G_0$. So G_0 is open. Suppose now that $F_j \in G_0$, $j = 1, 2, \ldots$ and $\|F_j - F\| \to 0$ for some $F \in G$. Let j_0 be large enough such that $\|F_{j_0} - F\| \le \|F_{j_0}^{-1}\|^{-1}$. Then $F = (I - M)F_{j_0}$ where $\|M\|_G = \|I - FF_{j_0}^{-1}\|_G < 1$, that is, $F \in G_0$. So G_0 is a closed set.

Now let us prove that G_0 is connected. Let

$$X = \prod_{j=1}^{v} (I - M_j) \in G_0 , \qquad \|M_j\|_G < 1$$

then

$$X(t) = \prod_{j=1}^{v} (I - tM_j) , \qquad t \in [0, 1]$$

is a continuous function that connects X and I in G_0. So G_0 is connected and thus is the connected component of G that contains I. So (18.5.3) is proved. As a side observation, note that G_0 is also a subgroup of G. Indeed, let $X, Y \in G_0$. Then the set $X \cdot G_0^{-1}$ is connected and contains I; therefore, $X \cdot G_0^{-1} \subset G_0$. In particular, $XY^{-1} \in G_0$, which means that G_0 is a subgroup of G.

Now let $A(z)$ be as in Lemma 18.5.2, and suppose first that $A \in G_I$. Then

$$A = (I - M_1) \cdots (I - M_v)$$

for some $M_1, \ldots, M_v \in G$ with $\|M_j\|_G < 1$ for $j = 1, \ldots, v$. Rewrite this representation in the form

$$A = \exp(\ln(I - M_1)) \cdots \exp(\ln(I - M_v))$$

where

$$\ln(I - M_j) = \sum_{k=1}^{\infty} \frac{1}{k} M_j^k$$

By Lemma 18.5.1, for each $j = 1, \ldots, v$ there exists a rational $n \times n$ matrix function D_j whose poles are contained in $\{z_1, \ldots, z_s, \infty\}$, with the property that D_j approximates the analytic function $\ln(I - M_j(z))$ well enough to ensure that the analytic matrix function $B(z) = \exp(D_1(z)) \cdots \exp(D_v(z))$

satisfies (18.5.1) for every $z \in K$. Clearly, $B(z)$ is invertible for every $z \in \mathbb{C} \smallsetminus \{z_1, \ldots, z_k\}$, so the lemma is proved in the case $A(z) \in G_I$.

We now pass to the general case. Let G_A be the connected component of G which contains $A(z)$. It suffices to show that there exists an $n \times n$ matrix function $D(z)$ that is analytic and invertible in $\mathbb{C} \smallsetminus \{z_1, \ldots, z_s\}$ and such that $D(z) \in G_A$. Indeed, then $A(z)D(z)^{-1} \in G_I$, and as we have seen already, there exists an analytic and invertible matrix function $\tilde{B}(z)$ in $\mathbb{C} \smallsetminus \{z_1, \ldots, z_s\}$ with the property that $\|\tilde{B} - AD^{-1}\|_G < \epsilon \|D\|_G^{-1}$. The matrix function $B(z) = \tilde{B}(z)D(z)$ is the desired one.

Thus let us prove the existence of $D(z)$. According to Lemma 18.5.1, for every $\delta > 0$ there exists a rational matrix function $D_0(z)$ that is analytic on $\mathbb{C} \smallsetminus \{z_1, \ldots, z_s\}$ and such that $\|D_0(z) - A(z)\| < \delta$ when $z \in K$. Choose $\delta > 0$ small enough to ensure that $D_0(z)$ is invertible for $z \in K$ and $D_0 \in G_A$. Since $D_0(z)$ is a rational function, $\det D_0(z) \neq 0$ for every $z \in \mathbb{C} \smallsetminus \{z_1, \ldots, z_s\}$ except perhaps for a finite set of points $\eta_1, \ldots, \eta_m \in \mathbb{C} \smallsetminus \{z_1, \ldots, z_s\}$, which do not belong to K.

Denote by $Y(\eta_1)$ the connected component of $\mathbb{C} \smallsetminus K$ that contains η_1, and let $z(\eta_1)$ be the point from $\{\infty, z_1, \ldots, z_s\}$ that belongs to $Y(\eta_1)$. Let $\rho > 0$ be such that the disc $\{z \in \mathbb{C} \mid |z - \eta_1| \leq \rho\}$ is contained in $Y(\eta_1) \smallsetminus \{z(\eta_1), \eta_2, \ldots, \eta_m\}$. By Lemma 18.4.2 there exists an incomplete factorization of $D_0(z)$ with respect to the circle $|z - \eta_1| = \rho$:

$$D_0(z) = {}^-D_0(z) \cdot {}^+D_0(z) \qquad (|z - \eta_1| = \rho) \qquad (18.5.4)$$

where ${}^+D_0(z)$ is analytic and invertible in the disc $\{z \in \mathbb{C} \mid |z - \eta_1| \leq \rho\}$ and ${}^-D_0(z)$ is analytic and invertible for $\rho \leq |z - \eta_1| < \infty$. The equality ${}^-D_0 = D_0({}^+D_0)^{-1}$ shows that ${}^-D_0$ admits analytic continuation to the whole of $\mathbb{C}$ and ${}^-D_0(z)$ is invertible for all $z \neq \eta_1$. Also, ${}^+D_0$ is analytic and invertible on $\mathbb{C} \smallsetminus \{z_1, \ldots, z_s, \eta_2, \ldots, \eta_m\} \supset K$.

Let $Y(t)$, $0 \leq t \leq 1$ be a continuous function with values in $Y(\eta_1)$ such that $y(0) = \eta_1$, $y(1) = z(\eta_1)$. Then the formula

$$F_t(z) = {}^-D_0(z + \eta_1 - y(t)), \qquad z \in K, \qquad 0 \leq t \leq 1$$

defines a continuous map $F: [0, 1] \to G$ with $F_0 = {}^-D_0$. Hence $D_1 \overset{\text{def}}{=} F_1 \, {}^+D_0 \in G_A$. As ${}^+D_0$ is invertible on $\mathbb{C} \smallsetminus \{z_1, \ldots, z_s, \eta_2, \ldots, \eta_m\}$ and $F_1(z) = {}^-D_0(z + \eta_1 - z(\eta_1))$ is invertible on $\mathbb{C} \smallsetminus \{z(\eta_1)\}$, it follows that $D_1(z)$ is analytic and invertible on $\mathbb{C} \smallsetminus \{z_1, \ldots, z_s, \eta_2, \ldots, \eta_m\}$. Repeating this argument $m - 1$ times with respect to the points $\eta_2, \ldots, \eta_m$, we obtain the desired function $D(z)$. $\square$

The following lemma is the main approximation result that will be used in the transition from compact sets in Ω to the domain Ω itself.

Lemma 18.5.3

Let $K \subset \Omega$ be a finitely connected compact set that is also simply connected relative to Ω. Let $\mathcal{M} \subset \mathbb{C}^n$ be a fixed subspace and $A(z)$ be an $n \times n$ matrix function that is analytic and invertible on K and such that $A(z)\mathcal{M} = \mathcal{M}$ for $z \in K$. Then for every $\epsilon > 0$ there exists a matrix function $B(z)$ that is analytic and invertible on Ω and such that

$$\|B(z) - A(z)\| < \epsilon$$

for all $z \in K$ and $B(z)\mathcal{M} = \mathcal{M}$ for all $z \in \Omega$.

Proof. Without loss of generality, we can assume that $\mathcal{M} = \mathrm{Span}\{e_1, \ldots, e_r\}$, for some r. Then in the 2×2 block matrix formed by representation with respect to the direct sum decomposition $\mathcal{M} \dotplus \mathcal{M}^{\perp} = \mathbb{C}^n$ we have

$$A(z) = \begin{bmatrix} A_1(z) & A_{12}(z) \\ 0 & A_2(z) \end{bmatrix}$$

Because $A(z)$ is invertible when $z \in K$, so are $A_1(z)$ and $A_2(z)$. Use Lemma 18.5.2 to find matrix functions $B_1(z)$ and $B_2(z)$ that are analytic and invertible on Ω and such that $\|B_i(z) - A_i(z)\| < \epsilon/3$ for $z \in K$; $i = 1, 2$. By Lemma 18.5.1 there exists an analytic matrix function $B_{12}(z)$ on Ω such that $\|B_{12}(z) - A_{12}(z)\| < \epsilon/3$ for $z \in K$. Then

$$B(z) = \begin{bmatrix} B_1(z) & B_{12}(z) \\ 0 & B_2(z) \end{bmatrix}$$

satisfies the requirements of Lemma 18.5.3. $\square$

The following result allows us to pass from the compact sets in Ω to Ω itself.

Lemma 18.5.4

Let $K_1 \subset K_2 \subset \cdots \subset \Omega$ be a sequence of finitely connected compact sets K_j, which are also simply connected relative to Ω. For $m = 1, 2, \ldots$, let $G_m(z)$ be an $n \times n$ matrix function that is analytic and invertible on K_m and satisfies $G_m(z)\mathcal{M} = \mathcal{M}$ for $z \in K_m$ and for some fixed subspace $\mathcal{M} \subset \mathbb{C}^n$. Then for $m = 1, 2, \ldots$, there exists an $n \times n$ matrix function $D_m(z)$ that is analytic and invertible on K_m and such that, whenever $z \in K_m$

$$D_m(z)\mathcal{M} = \mathcal{M} \qquad and \qquad G_m(z) = D_m(z)D_{m+1}^{-1}(z)$$

Proof. We need the following simple assertion. Let $X_1, X_2, \ldots$, be a sequence of $n \times n$ matrices such that

$$\alpha \overset{\text{def}}{=} \sum_{m=1}^{\infty} \|X_m\| < \infty \tag{18.5.5}$$

Then the infinite product $Y = \Pi_{m=1}^{\infty} (I + X_m)$ converges and $\|I - Y\| \le \alpha e^{\alpha}$. Indeed, for the matrices $Y_m = \Pi_{j=1}^{m} (I + X_j)$ we have the estimates:

$$\|Y_m\| \le \exp\left(\sum_{j=1}^{m} \|X_j\|\right) \le e^{\alpha} \qquad (m = 1, 2, \dots)$$

$$\|Y_m - Y_{m+1}\| = \|Y_m - Y_m(I + X_{m+1})\| = \|Y_m X_{m+1}\| \le \|Y_m\| \cdot \|X_{m+1}\|$$
$$\le \|X_{m+1}\| e^{\alpha}$$

Thus, in view of (18.5.5) the infinite product $Y = \Pi_{m=1}^{\infty} (I + X_m)$ converges. Moreover

$$\|I - Y\| \le \|I - Y_1\| + \sum_{m=1}^{\infty} \|Y_m - Y_{m+1}\| \le \|X_1\| + \sum_{m=1}^{\infty} \|X_{m+1}\| e^{\alpha} \le \alpha e^{\alpha}$$

We now prove Lemma 18.5.4 itself. Applying Lemma 18.5.3 repeatedly, we find for $m = 1, 2, \dots$ a matrix function $H_m(z)$ that is analytic and invertible on K_m, for which $H_1(z) \equiv I$, and for $z \in K_m$, $H_m(z)\mathcal{M} = \mathcal{M}$ and

$$\|H_m^{-1}(z) G_m(z) H_{m+1}(z)\| < 2^{-(m+1)}$$

The assertion proved in the preceding paragraph ensures that for every $m = 1, 2, \dots$, the infinite product

$$E_m = \prod_{j=0}^{\infty} (H_{m+j}^{-1} G_{m+j} H_{m+1+j})$$

converges uniformly on K_m, and $\|I - E_m(z)\| \le 2^{-m} \exp(2^{-m}) < 1$ for $z \in K_m$. Consequently, $E_m(z)$ is invertible for every $z \in K_m$. Further, $E_m(z)\mathcal{M} = \mathcal{M}(z \in K_m; m = 1, 2, \dots)$. Indeed, since $E_m(z)$ is invertible, it is sufficient to prove that $E_m(z)\mathcal{M} \subset \mathcal{M}$. But this follows from the equalities $H_m(z)\mathcal{M} = \mathcal{M}$, $G_m(z)\mathcal{M} = \mathcal{M}$ and the definition of E_m. Now we can put $D_m(z) = H_m(z) E_m(z)$, because $E_m = H_m^{-1} G_m H_{m+1} E_{m+1}$ and consequently $G_m = (H_m E_m)(H_{m+1} E_{m+1})^{-1}$. $\square$

We are now prepared to prove Theorem 18.3.1.

Proof of Theorem 18.3.1. Let us show first that there exists a sequence of compact sets $K_1 \subset K_2 \subset \cdots$ that are finitely connected, simply connected relative to Ω, and for which $\cup_{i=1}^{\infty} K_i = \Omega$. To this end choose a sequence of closed discs $S_m \subset \Omega$, $m = 1, 2, \dots$, such that $\cup_{m=1}^{\infty} S_m = \Omega$. It is sufficient to construct K_m in such a way that $K_m \supset S_m$, $m = 1, 2, \dots$. Put $K_1 = S_1$, suppose that $K_1, \ldots, K_m$ are already constructed, and $K_j \supset S_j$ for $j = 1, \ldots, m$. Let M be a connected compact set such that $M \supset K_m \cup S_{m+1}$, and let $V_1, \ldots, V_k \subset \Omega$ be a finite set of closed discs from $\{S_m\}_{m=1}^{\infty}$ such that $N \stackrel{\text{def}}{=} \cup_{j=1}^k V_j \supset M$. Clearly, N is a finitely connected compact set. If N is also simply connected relative to Ω, then put $K_{m+1} = N$. Otherwise, put $K_{m+1} = N \cup Y_1 \cup \cdots \cup Y_s$, where $Y_1, \ldots, Y_s$ are all the bounded connected components of the set $\mathbb{C} \smallsetminus N$, which are entirely in Ω.

Given the sequence $K_1 \subset K_2 \subset \cdots$ constructed in the preceding paragraph. Choose $z_0 \in K_1$ and put $\mathcal{M}_0 = \mathcal{M}(z_0)$ [here $\mathcal{M}(z)$ is the analytic family of subspaces (of $\mathbb{C}^n$) on Ω given in Theorem 18.3.1]. Without loss of generality we can assume that $\mathcal{M}_0 = \text{Span}\{e_1, \ldots, e_r\}$. By Theorem 18.4.1, there exist analytic vector functions $f_1^{(m)}(z), \ldots, f_r^{(m)}(z)$ in K_m that form a basis in $\mathcal{M}(z)$ for every $z \in K_m$. Using Lemma 18.2.2, we find analytic vector functions $f_{r+1}^{(m)}(z), \ldots, f_n^{(m)}(z)$ defined on K_m such that the vectors $f_1^{(m)}(z), \ldots, f_n^{(m)}(z)$ form a basis in $\mathbb{C}^n$ for every $z \in K_m$ (indeed, apply Lemma 18.2.2 with $x_1(z) = f_1^{(m)}(z), \ldots, x_r(z) = f_r^{(m)}(z)$, $x_{r+1}(z) \equiv g_1, \ldots, x_n(z) \equiv g_{n-r}$, where $g_1, \ldots, g_{n-r}$ is a basis in a fixed direct complement to $\mathcal{M}_0$). Then the matrix function $A_m(z) = [f_1^{(m)}(z), f_2^{(m)}(z), \ldots, f_n^{(m)}(z)]$ is analytic and invertible on K_m and satisfies

$$\mathcal{M}(z) = A_m(z)\mathcal{M}_0 \qquad (18.5.6)$$

where $z \in K_m$. Put $G_m(z) = A_m^{-1}(z)A_{m+1}(z)$ for $z \in K_m$. Then (18.5.6) ensures that $G_m(z)\mathcal{M}_0 = \mathcal{M}_0$ ($z \in K_m, m = 1, 2, \ldots$). By Lemma 18.5.4 (for $m = 1, 2, \ldots$) there exists an analytic and invertible matrix function $D_m(z)$ on K_m such that $G_m = D_m D_{m+1}^{-1}$ and, for $z \in K_m$

$$D_m(z)\mathcal{M}_0 = \mathcal{M}_0 \qquad (18.5.7)$$

Since $A_{m+1}(z)D_{m+1}(z) = A_m(z)D_m(z)$ ($z \in K_m; m = 1, 2, \ldots$) the relation $A(z) = A_m(z)D_m(z)$, which holds for all $z \in K_m$, defines an analytic and invertible matrix function $A(z)$ on Ω. Now the relation $A(z)\mathcal{M}_0 = \mathcal{M}(z)$ for $z \in \Omega$ follows from (18.5.6) and (18.5.7). $\square$

18.6 DIRECT COMPLEMENTS FOR ANALYTIC FAMILIES OF SUBSPACES

Let $\mathcal{M}(z)$ be an analytic family of subspaces of $\mathbb{C}^n$ defined on a domain Ω. If $\mathcal{N}$ is a direct complement to $\mathcal{M}(z_0)$ and $z_0 \in \Omega$, then the results of Chapter 13 (Theorem 13.1.3) imply that $\mathcal{N}$ is also a direct complement to $\mathcal{M}(z)$ as long as z is sufficiently close to z_0. This local property of direct complements raises the corresponding global question: does there exist a subspace $\mathcal{N}$ of $\mathbb{C}^n$ that is a direct complement to $\mathcal{M}(z)$ for all $z \in \Omega$? The simple example below shows that the answer is generally no.

EXAMPLE 18.6.1. Let

$$M(z) = \text{Im}\begin{bmatrix} z^2 - z + 1 \\ z^2 + z \end{bmatrix} \subset \mathbb{C}^2, \qquad z \in \mathbb{C}$$

As the polynomials $z^2 - z + 1$ and $z^2 + z$ do not have common zeros, it follows that $M(z)$ is an analytic family of subspaces. Indeed, if z_0 is such that $z_0^2 + z_0 \neq 0$, then in a neighbourhood of z_0 we have

$$M(z) = \begin{bmatrix} 1 & z^2 - z + 1 \\ 0 & z^2 + z \end{bmatrix}\left(\text{Span}\begin{bmatrix} 0 \\ 1 \end{bmatrix}\right)$$

and if z_0 is such that $z_0^2 - z_0 + 1 \neq 0$, then there is a neighbourhood of z_0 in which

$$M(z) = \begin{bmatrix} 0 & z^2 - z + 1 \\ 1 & z^2 + z \end{bmatrix}\left(\text{Span}\begin{bmatrix} 0 \\ 1 \end{bmatrix}\right)$$

However, there is no one-dimensional subspace $\text{Span}\begin{bmatrix} a \\ b \end{bmatrix}$ (with at least one of the complex numbers a, b nonzero) such that

$$\text{Span}\begin{bmatrix} a \\ b \end{bmatrix} \dotplus M(z) = \mathbb{C}^2 \qquad\qquad (18.6.1)$$

for all $z \in \mathbb{C}$. Indeed, (18.6.1) means

$$\det\begin{bmatrix} a & z^2 - z + 1 \\ b & z^2 + z \end{bmatrix} = (a - b)z^2 + (a + b)z - b \neq 0$$

for all $z \in \mathbb{C}$, which is impossible. $\quad\square$

It turns out that although one common direct complement for an analytic family of subspaces may not exist, only two subspaces are needed to serve as "alternate" direct complements for each member of the analytic family.

Theorem 18.6.1

For an analytic family of subspaces $\{M(z)\}_{z \in \Omega}$ of $\mathbb{C}^n$ there exist two subspaces $N_1, N_2 \subset \mathbb{C}^n$ such that for each $z \in \Omega$ either $M(z) \dotplus N_1 = \mathbb{C}^n$ or $M(z) \dotplus N_2 = \mathbb{C}^n$ holds.

Proof. To prove this we first need the following observation: for any k-dimensional subspace $L \subset \mathbb{C}$, the set $DC(L)$ of all direct complements to L in $\mathbb{C}^n$ is open and dense in the set of all $(n - k)$-dimensional subspaces. Indeed, the openness of $DC(L)$ follows immediately from Theorem 13.1.3. To prove denseness, let N be an $(n - k)$-dimensional subspace in $\mathbb{C}^n$ with basis $f_1, \ldots, f_{n-k}$, and let $\mathcal{N}_0$ be a direct complement to $\mathcal{L}$ with basis $g_1, \ldots, g_{n-k}$. For a complex number ϵ put $\mathcal{N}(\epsilon) = \mathrm{Span}\{f_1 + \epsilon g_1, \ldots, f_{n-k} + \epsilon g_{n-k}\}$. Clearly, the vectors $f_i + \epsilon g_i$, $i = 1, \ldots, n-k$ are linearly independent for ϵ close enough to 0, so dim $\mathcal{N}(\epsilon) = n - k$. Moreover, Theorem 13.4.2 shows that

$$\lim_{\epsilon \to 0} \theta(\mathcal{N}(\epsilon), \mathcal{N}) = 0$$

It remains to show that $\mathcal{N}(\epsilon)$ belongs to $DC(\mathcal{L})$. To this end pick a basis $h_1, \ldots, h_k$ in $\mathcal{L}$, and consider the $n \times n$ matrix

$$G(\epsilon) = [f_1 + \epsilon g_1, \ldots, f_{n-k} + \epsilon g_{n-k}, h_1, \ldots, h_k]$$

As

$$\det[g_1, \ldots, g_{n-k}, h_1, \ldots, h_k] \neq 0$$

(recall that $\mathcal{N}_0 \dotplus \mathcal{L} = \mathbb{C}^n$); also

$$\det\left[\frac{1}{\epsilon}f_1 + g_1, \ldots, \frac{1}{\epsilon}f_{n-k} + g_{n-k}, h_1, \ldots, h_k\right] \neq 0$$

for $|\epsilon|$ sufficiently large. Hence det $G(\epsilon) \neq 0$ for $|\epsilon|$ large enough. We find that det $G(\epsilon) \not\equiv 0$, and since det $G(\epsilon)$ is a polynomial in ϵ it follows that det $G(\epsilon) \neq 0$ for $\epsilon \neq 0$ and sufficiently close to zero. Obviously, $\mathcal{N}(\epsilon) \in DC(\mathcal{L})$ for such ϵ.

Now we start to prove Theorem 18.6.1 itself. Fix $z_0 \in \Omega$, and let $\mathcal{N}_1$ be a direct complement to $\mathcal{M}(z_0)$ in $\mathbb{C}^n$. By Theorem 18.3.2 it is possible to pick vector functions $x_1(z), \ldots, x_p(z) \in \mathbb{C}^n$ that are analytic on Ω and such that, for every $z \in \Omega$, the vectors $x_1(z), \ldots, x_p(z)$ form a basis in $\mathcal{M}(z)$. Letting $f_1, \ldots, f_{n-p}$ be a basis in $\mathcal{N}_1$, consider the $n \times n$ matrix function

$$G(z) = [f_1, \ldots, f_{n-p}, x_1(z), \ldots, x_p(z)]$$

which is analytic on Ω. As det $G(z_0) \neq 0$, the determinant of $G(z)$ is not identically zero, and thus the number of distinct zeros of det $G(z)$ is at most countable. Let $z_1, z_2, \ldots \in \Omega$ be all of these zeros. Then $\mathcal{N}_1$ is a direct complement to $\mathcal{M}(z)$ for $z \notin \{z_1, z_2, \ldots\}$. On the other hand, we have seen that, for $i = 1, 2, \ldots$, the sets $DC(\mathcal{M}(z_i))$ are open and dense in the set of all $(n - p)$-dimensional subspaces in $\mathbb{C}^n$. As the latter set is a complete metric space in the gap topology (Section 13.4), it follows that the intersection $\cap_{i=1}^{\infty} DC(\mathcal{M}(z_i))$ is again dense [the Baire category theorem; e.g., see Kelley (1955)]. In particular, this intersection is not empty, so there exists a subspace $\mathcal{N}_2 \subset \mathbb{C}^n$ that is simultaneously a direct complement to all of $\mathcal{M}(z_1), \mathcal{M}(z_2), \ldots$. $\quad \square$

The following result shows that for analytic families of subspaces that appear as the kernel or the image of a *linear* matrix function there exists a common direct complement. As Example 18.6.1 shows, the result is not necessarily valid for nonlinear matrix functions.

Theorem 18.6.2

Let T_1 and T_2 be $m \times n$ matrices such that the dimension of $\mathrm{Ker}(T_1 + zT_2)$ is constant, that is, it is independent of z on $\mathbb{C}$ [and the same is automatically true for $\dim \mathrm{Im}(T_1 + zT_2)$]. Then there exist subspaces $\mathcal{N}_1 \subset \mathbb{C}^n$, $\mathcal{N}_2 \subset \mathbb{C}^m$ such that

$$\mathcal{N}_1 \dotplus \mathrm{Ker}(T_1 + zT_2) = \mathbb{C}^n , \qquad \mathcal{N}_2 \dotplus \mathrm{Im}(T_1 + zT_2) = \mathbb{C}^m$$

for all $z \in \mathbb{C}$.

Note that in view of Proposition 18.1.1 and Theorem 18.2.1 the families of subspaces $\mathrm{Ker}(T_1 + zT_2)$ and $\mathrm{Im}(T_1 + zT_2)$ are analytic on $\mathbb{C}$.

Proof. For the proof of Theorem 18.6.2 we use the Kronecker canonical form for linear matrix polynomials under strict equivalence (which is developed in the appendix to this book).

As $\dim \mathrm{Ker}(T_1 + zT_2)$ is independent of $z \in \mathbb{C}$, the canonical form of $T_1 + zT_2$ does not have the term $zI + J$. So, in the notation of Theorem A.7.3, there exist invertible matrices Q_1 and Q_2 such that

$$Q_1(T_1 + zT_2)Q_2 = 0_{u \times v} \oplus L_{p_1} \oplus \cdots \oplus L_{p_k} \oplus L_{q_1}^T \oplus \cdots \oplus L_{q_l}^T$$
$$\oplus (I_{r_1} + \lambda J_{r_1}(0)) \oplus \cdots \oplus (I_{r_s} + \lambda J_{r_s}(0)) \qquad (18.6.2)$$

It is easily seen that

$$\mathrm{Ker}\, L_{q_i}^T = \{0\} , \qquad \mathrm{Ker}(I_{r_j} + \lambda J_{r_j}(0)) = \{0\}$$

for all $z \in \mathbb{C}$, and that

$$\mathrm{Ker}\, L_{p_j} = \mathrm{Span}\{e_1 - ze_2 + z^2 e_3 - \cdots \pm z^{p_j - 1} e_{p_j}\} , \qquad z \in \mathbb{C}$$

So there exists a direct complement $\mathcal{M}_1$ to $\mathrm{Ker}[Q_1(T_1 + zT_2)Q_2]$ for all $z \in \mathbb{C}$ given as follows:

$$\mathcal{M}_1 = \mathrm{Span}\,\{e_{v+2}, \ldots, e_{v+p_1}, e_{v+p_1+2}, \ldots, e_{v+p_1+p_2}, \ldots,$$

$$e_{v+p_1+\cdots+p_{k-1}+2}, \ldots, e_{v+p_1+\cdots+p_k}, e_j \text{ with } j > v + p_1 + \cdots + p_k\}$$

As

$$\mathrm{Ker}(T_1 + zT_2) = Q_2(\mathrm{Ker}(Q_1(T_1 + zT_2)Q_2))$$

it follows that

$$Q_2 \mathcal{M}_1 \dotplus \mathrm{Ker}(T_1 + zT_2) = \mathbb{C}^n , \qquad z \in \mathbb{C}$$

The part of Theorem 18.6.2 concerning $\mathrm{Im}(T_1 + zT_2)$ is proved similarly, taking into account the facts that

$$\mathrm{Im}\, L_{p_i} = \mathbb{C}^{p_i - 1} , \qquad \mathrm{Im}(I_{r_j} + \lambda J_{r_j}(0)) = \mathbb{C}^{r_j}$$

and for each $z \in \mathbb{C}$, $\mathrm{Im}\, L_{q_i}^T$ has a direct complement $\mathrm{Span}\{e_1\}$. $\quad\square$

18.7   ANALYTIC FAMILIES OF INVARIANT SUBSPACES

Let $A(z)$: $\mathbb{C}^n \to \mathbb{C}^n$ be an analytic family of transformations on Ω. Our next topic concerns the analytic properties (as functions on z) of certain invariant subspaces of $A(z)$.

We have already seen some first results in this direction in Section 18.1. Namely, if the rank of $A(z)$ is independent of z, then $\mathrm{Im}\, A(z)$ and $\mathrm{Ker}\, A(z)$ are analytic families of subspaces. In the general case, $\mathrm{Im}\, A(z)$ and $\mathrm{Ker}\, A(z)$ become analytic families of subspaces if corrected on the singular set of $A(z)$. The next theorem is mainly a reformulation of this statement.

For convenience, let us introduce another definition: an analytic family of subspaces $\{\mathcal{M}(z)\}_{z \in \Omega}$ is called $A(z)$ invariant on Ω if the subspace $\mathcal{M}(z)$ is $A(z)$ invariant for every $z \in \Omega$.

Theorem 18.7.1

There exist $A(z)$-invariant analytic families $\{\mathcal{M}(z)\}_{z \in \Omega}$ and $\{\mathcal{N}(z)\}_{z \in \Omega}$ such that $\mathcal{M}(z) = \mathrm{Im}\, A(z)$ and $\mathcal{N}(z) = \mathrm{Ker}\, A(z)$ for every z not belonging to the singular set of $A(z)$.

Proof.   In view of Theorem 18.2.1 we have only to prove that $\mathcal{M}(z_0)$ and $\mathcal{N}(z_0)$ are $A(z_0)$ invariant for every $z_0 \in S(A)$. But this follows from Theorem 15.1.1 because $\lim_{z \to z_0} A(z) = A(z_0)$ and

$$\lim_{z \to z_0} \theta(\mathcal{M}(z), \mathcal{M}(z_0)) = \lim_{z \to z_0} \theta(\mathcal{N}(z), \mathcal{N}(z_0)) = 0 \quad \square$$

Another class of $A(z)$-invariant subspaces whole behaviour is analytic (at least locally) includes spectral subspaces, as follows.

Theorem 18.7.2

Let Γ be a contour in the complex plane such that $\Gamma \cap \sigma(A(z_0)) = \emptyset$ for a fixed $z_0 \in \Omega$. Then the sum $\mathcal{M}_\Gamma(z)$ of the root subspaces of $A(z)$ correspond-

ing to the eigenvalues inside Γ, is an $A(z)$-invariant analytic family of subspaces in a neighbourhood U of z_0.

Proof. As $A(z)$ is a continuous function of z on Ω, the eigenvalues of $A(z)$ also depend continuously on z. Hence there is a neighbourhood U of z_0 such that $A(z)$ has no eigenvalues on Γ for any z in the closure of U. Now for $z \in U$ we have

$$\mathcal{M}_\Gamma(z) = \operatorname{Im}\left[\frac{1}{2\pi i} \int_\Gamma (\lambda I - A(z))^{-1}\, d\lambda \right] \tag{18.7.1}$$

We have seen in Section 2.4 that

$$P(z) = \frac{1}{2\pi i} \int_\Gamma (\lambda I - A(z))^{-1}\, d\lambda \tag{18.7.2}$$

is a projector for every $z \in U$. So, to prove that $\mathcal{M}_\Gamma(z)$ is an analytic family in U, it is sufficient to check that $P(z)$ is an analytic function on U. Indeed, $|\det(\lambda I - A(z))| \geq \delta > 0$ for every $\lambda \in \Gamma$ and $z \in U$, where δ is independent of λ and z. Hence $\|(\lambda I - A(z))^{-1}\|$ is bounded for $\lambda \in \Gamma$ and $z \in U$, and consequently the Riemann sums

$$\frac{1}{2\pi i} \sum_{j=0}^{m-1} (\lambda_{j+1} - \lambda_j)(\lambda_j I - A(z))^{-1}$$

where $\lambda_0, \ldots, \lambda_m$ are consecutive points in the positive direction on Γ with $\lambda_m = \lambda_0$, converge to the integral (18.7.2) uniformly on every compact set in U. As each Riemann sum is obviously analytic on U, so is the integral (18.7.2). $\square$

In view of Theorems 18.7.1 and 18.7.2, the following question arises naturally: does there exist an $A(z)$-invariant analytic family that is nontrivial (i.e., different from $\{0\}$ and $\mathbb{C}^n$)? Without restrictions on $A(z)$ the answer is no, as the following example shows.

EXAMPLE 18.7.1.   Define an analytic family on $\mathbb{C}$ by

$$A(z) = \begin{bmatrix} 0 & 1 \\ z & 0 \end{bmatrix} : \mathbb{C}^2 \to \mathbb{C}^2$$

Here the $A(z)$-invariant subspaces (for a fixed z) are easy to find: the only nontrivial invariant subspace of $A(0)$ is $\operatorname{Span}\{e_1\}$, and, when $z \neq 0$, the only nontrivial invariant subspaces of $A(z)$ are

$$\operatorname{Span}\begin{bmatrix} 1 \\ u_1 \end{bmatrix} \quad \text{and} \quad \operatorname{Span}\begin{bmatrix} 1 \\ u_2 \end{bmatrix}$$

where u_1 and u_2 are the square roots of z. It is easily seen that there is no nontrivial, $A(z)$-invariant, analytic family of subspaces on $\mathbb{C}$. $\square$

In the next section we study $A(z)$-invariant analytic families of subspaces under the extra condition that $A(z)$ have the same Jordan structure for all $z \in \Omega$. We see that, in this case, nontrivial $A(z)$-invariant analytic families of subspaces always exist. On the other hand, we have seen in Example 18.7.1 that there exists a nontrivial $A(z)$-invariant family of subspaces that is analytic in $\mathbb{C}$ except for the branch point at zero. Such phenomena occur more generally and are studied in detail in Chapter 19.

18.8   ANALYTIC DEPENDENCE OF THE SET OF INVARIANT SUBSPACES AND FIXED JORDAN STRUCTURE

Given a family of transformations $A(z)$: $\mathbb{C}^n \to \mathbb{C}^n$ that depends analytically on the parameter z in a domain $\Omega \in \mathbb{C}$, we say that the lattice $\text{Inv}(A(z))$ *depends analytically* on $z \in \Omega$ if there exists an invertible transformation $S(z)$: $\mathbb{C}^n \to \mathbb{C}^n$ that is analytic on Ω and such that $\text{Inv}(A(z)) = S(z)(\text{Inv}(A(z_0)))$ for all $z \in \Omega$ and some fixed point $z_0 \in \Omega$. This definition does not depend on the choice of z_0. Indeed, if

$$\text{Inv}(A(z)) = S(z)(\text{Inv}(A(z_0)))$$

then for every $z_0' \in \Omega$ we have

$$\text{Inv}(A(z)) = S(z)(S(z_0'))^{-1}(\text{Inv}(A(z_0')))$$

Also, replacing $S(z)$ by $S(z)S(z_0)^{-1}$, we can require in the definition of analytic dependence of $\text{Inv}(A(z))$ that $S(z_0) = I$.

Since $\text{Inv}(A)$, $\text{Inv}(B)$ are linearly isomorphic if and only if A and B have the same Jordan structure (Theorem 16.1.2), a necessary condition for analytic dependence of $\text{Inv}(A(z))$ on z is that $A(z)$ have *fixed Jordan structure*, that is, the number m of different eigenvalues of $A(z)$ is independent of z on Ω, and for every pair $z_1, z_2 \in \Omega$ the different eigenvalues $\lambda_1(z_1), \ldots, \lambda_m(z_1)$ and $\lambda_1(z_2), \ldots, \lambda_m(z_2)$ of $A(z_1)$ and $A(z_2)$, respectively, can be enumerated so that the partial multiplicities of $\lambda_j(z_1)$ [as an eigenvalue of $A(z_1)$] coincide with the partial multiplicities of $\lambda_j(z_2)$ [as an eigenvalue of $A(z_2)$], for $j = 1, \ldots, m$.

Using Theorem 16.1.2, we find that the family $A(z)$ has fixed Jordan structure if and only if, for every $z_1, z_2 \in \Omega$ the lattices $\text{Inv}(A(z_1))$ and $\text{Inv}(A(z_2))$ are isomorphic. Clearly, this property is necessary for the lattice $\text{Inv}(A(z))$ to depend analytically on $z \in \Omega$. The following result shows that this property is also sufficient as long as Ω is simply connected.

Theorem 18.8.1

Let Ω be a simply connected domain in $\mathbb{C}$, and let $A(z)$: $\mathbb{C}^n \to \mathbb{C}^n$ be an analytic family of transformations on Ω. Then $\text{Inv}(A(z))$ depends analytically on $z \in \Omega$ if and only if $A(z)$ have fixed Jordan structure.

In particular, the condition of a fixed Jordan structure ensures existence of at least as many $A(z)$-invariant analytic families of subspaces as there are $A(z_0)$-invariant subspaces.

Proof. We assume that $A(z)$ is represented as a matrix-valued function with respect to some basis in $\mathbb{C}^n$ that is independent of z on Ω. Fix a z_0 in Ω. Let $\lambda_1, \ldots, \lambda_p$ be all the distinct eigenvalues of $A(z_0)$, and let Γ_i be a circle around λ_i chosen so small that $\Gamma_i \cap \Gamma_j = \emptyset$ for $i \neq j$. As the proof of Theorem 16.3.1 shows, there exists an $\epsilon > 0$ with the property that if $B: \mathbb{C}^n \to \mathbb{C}^n$ is a transformation with the same Jordan structure as $A(z_0)$, and if $\| B - A(z_0) \| < \epsilon$, then there is a unique eigenvalue $\mu_i(B)$ of B in each circle Γ_i $(1 \leq i \leq p)$, and, moreover, the partial multiplicities of $\mu_i(B)$ (as an eigenvalue of B) coincide with the partial multiplicities of λ_i (as an eigenvalue of $A(z_0)$). Hence, for every z from some neighbourhood U_1 of z_0, there is a unique eigenvalue [denoted by $\mu_i(z)$] of $A(z)$ in the circle Γ_i $(1 \leq i \leq p)$, and the partial multiplicities of $\mu_i(z)$ coincide with those of λ_i. Obviously, $\mu_i(z_0) = \lambda_i$.

Let us prove that $\mu_i(z)$ is analytic on U_1. Indeed, denoting by m_i the algebraic multiplicity of λ_i [as an eigenvalue of $A(z_0)$], we have

$$\mu_i(z) = \frac{1}{m_i} \operatorname{tr}\left[\frac{1}{2\pi i} \int_{\Gamma_i} (\lambda I - A(z))^{-1} \, d\lambda \right]$$

which is an analytic function of z on U_1.

We have proved that in a neighbourhood of each point $z_0 \in \Omega$ the distinct eigenvalues of $A(z)$ are analytic functions of z. It follows that the eigenvalues of $A(z)$ admit analytic continuation along any curve in Ω. By the monodromy theorem [see, e.g. Rudin (1974); this is where the simple connectedness of Ω is used] the distinct eigenvalues $\mu_1(z), \ldots, \mu_p(z)$ of $A(z)$ are analytic functions on Ω.

Now fix $z_0 \in \Omega$ and define the family of transformations $B(z): \mathbb{C}^n \to \mathbb{C}^n$, $z \in \Omega$ by the requirement that $B(z)x = [\mu_i(z_0) - \mu_i(z)]x$ for any x belonging to the root subspace of $A(z)$ corresponding to the eigenvalue $\mu_i(z)$. It is easily seen that $B(z)$ is analytic on Ω. Indeed, for every $z_1 \in \Omega$ let $\Gamma_1', \ldots, \Gamma_p'$ be circles around $\mu_1(z_1), \ldots, \mu_p(z_1)$, respectively, so small that $\mu_j(z_1)$ is the only eigenvalue of $A(z_1)$ inside or on the circle Γ_j' for $j = 1, \ldots, p$. There is a neighbourhood V of z_1 such that any $A(z)$ with $z \in V$ has the unique eigenvalue $\mu_j(z)$ inside the circle Γ_j', $j = 1, \ldots, p$. Then

$$B(z) = \sum_{j=1}^{p} \frac{1}{2\pi i} \int_{\Gamma_j'} [\mu_j(z_0) - \mu_j(z)](\lambda I - A(z))^{-1} \, d\lambda, \qquad z \in V$$

which is analytic on V in view of the analyticity of $A(z)$ and $\mu_j(z)$ for $j = 1, \ldots, p$. Put $\tilde{A}(z) = A(z) + B(z)$. Obviously, the set of $A(z)$-invariant subspaces coincides with the set of $\tilde{A}(z)$-invariant subspaces for all $z \in \Omega$, so it is sufficient to prove Theorem 18.8.1 for $\tilde{A}(z)$ instead of $A(z)$. From the definition of $\tilde{A}(z)$ it is clear that the eigenvalues of $\tilde{A}(z)$ are $\mu_1(z_0), \ldots, \mu_p(z_0)$, that is, they do not depend on z, and, moreover, the partial multiplicities of $\mu_j(z_0)$ as eigenvalues of $\tilde{A}(z)$ do not depend on z, either. In other words, in Theorem 18.8.1 we may assume that $A(z)$ is similar to $A(z_0)$ for all $z \in \Omega$.

For $j = 1, \ldots, p$, let m_j be the maximal partial multiplicity of $\mu_j(z_0)$ as an eigenvalue of $A(z_0)$ [and hence as an eigenvalue of $A(z)$ for all z in Ω]. Note that since $A(z)$ is similar to $A(z_0)$ for all $z \in \Omega$, by Proposition 18.1.2 there is an analytic basis in $\mathrm{Ker}(A(z) - \mu_j(z_0)I)^m$ for $m = 0, 1, 2, \ldots$ (i.e., for each fixed j and m). By Theorem 18.3.3 there exists a basis $x_{11}^{(j)}(z), \ldots, x_{q_j1}^{(j)}(z)$ in $\mathrm{Ker}(A(z) - \mu_j(z_0)I)^{m_j}$ modulo $\mathrm{Ker}(A(z) - \mu_j(z_0))^{m_j-1}$ that is analytic on Ω. It is easily seen that the vectors

$$(A(z) - \mu_j(z_0)I)x_{r1}^{(j)}(z), \qquad r = 1, \ldots, q_j$$

are linearly independent for all $z \in \Omega$ and belong to $\mathrm{Ker}(A(z) - \mu_j(z_0)I)^{m_j-1}$. Hence by Theorem 18.3.3 again there is a basis $x_{12}^{(j)}(z), \ldots, x_{r_j2}^{(j)}(z)$ in $\mathrm{Ker}(A(z) - \mu_j(z_0)I)^{m_j-1}$ modulo

$$\mathrm{Span}\{(A(z) - \mu_j(z_0)I)x_{r1}^{(j)}(z), \qquad r = 1, \ldots, q_j\}$$

which is analytic on Ω. Next we find an analytic basis

$$x_{13}^{(j)}(z), \ldots, x_{s_j3}^{(j)}(z)$$

in $\mathrm{Ker}(A(z) - \mu_j(z_0)I)^{m_j-2}$ modulo

$$\mathrm{Span}\{(A(z) - \mu_j(z_0)I)^2 x_{r1}^{(j)}(z), \qquad r = 1, \ldots, q_j;$$
$$(A(z) - \mu_j(z_0)I)x_{s2}^{(j)}(z), \qquad s = 1, \ldots, r_j\}$$

and so on. Now define the $n \times n$ matrix $T(z)$ formed by the columns

$$(A(z) - \mu_j(z_0)I)^{m_j-1}x_{11}^{(j)}(z), \ldots, (A(z) - \mu_j(z_0)I)x_{11}^{(j)}(z), x_{11}^{(j)}(z),$$
$$(A(z) - \mu_j(z_0)I)^{m_j-1}x_{21}^{(j)}(z), \ldots, x_{21}^{(j)}(z), \ldots,$$
$$(A(z) - \mu_j(z_0)I)^{m_j-1}x_{q_j1}^{(j)}(z), \ldots, x_{q_j1}^{(j)}(z), (A(z) - \mu_j(z_0)I)^{m_j-2}$$
$$\times x_{12}^{(j)}(z), \ldots, x_{12}^{(j)}(z), \ldots, (A(z) - \mu_j(z_0)I)^{m_j-2}x_{r_j2}^{(j)}(z), \ldots, x_{r_j2}^{(j)}(z), \ldots$$

$$(18.8.1)$$

where $j = 1, \ldots, p$. As the proof of the Jordan form of a matrix shows (see Section 2.3), the columns of $T(z)$ form a Jordan basis of $A(z)$. In particular,

$T(z)$ is invertible for all $z \in \Omega$. Clearly, $T(z)$ is analytic on Ω. As $T(z)^{-1}A(z)T(z)$ is a constant matrix (i.e., independent of z) and is in the Jordan form, the assertion of Theorem 18.8.1 follows. $\square$

In the course of the proof of Theorem 18.8.1 we have also proved the following result on analytic families of similar transformations.

Corollary 18.8.2

Let $A(z): \mathbb{C}^n \to \mathbb{C}^n$ be an analytic family of transformations on Ω, where Ω is a simply connected domain. Assume that, for a fixed point $z_0 \in \Omega$, $A(z)$ is similar to $A(z_0)$ for all $z \in \Omega$. Then there exists an invertible transformation $T(z): \mathbb{C}^n \to \mathbb{C}^n$, which is analytic on Ω and such that $T(z_0) = I$ and $T(z)^{-1}A(z)T(z) = A(z_0)$ for all $z \in \Omega$.

The assumption that Ω is simply connected in Theorem 18.8.1 is necessary, as the next example shows.

EXAMPLE 18.8.1.   Let $\Omega = \mathbb{C} \smallsetminus \{0\}$, and let

$$A(z) = \begin{bmatrix} 0 & z \\ 1 & 0 \end{bmatrix}$$

Clearly, $A(z)$ has fixed Jordan structure on Ω (the eigenvalues being the two square roots of z). The nontrivial $A(z)$-invariant subspaces are

$$\mathrm{Span}\begin{bmatrix} \sqrt{z} \\ 1 \end{bmatrix} \quad \text{and} \quad \mathrm{Span}\begin{bmatrix} -\sqrt{z} \\ 1 \end{bmatrix}$$

Clearly, there is no (single-valued) invertible 2×2 matrix function $S(z)$ that is analytic on $\mathbb{C} \smallsetminus \{0\}$ and satisfies the conditions of Theorem 18.8.1. $\square$

Note that in the proof of Theorem 18.8.1 the existence of an analytic Jordan basis [Formula (18.8.1)] of $A(z)$ also follows from a general result on analytic perturbations of matrices (see Section 19.2).

18.9   ANALYTIC DEPENDENCE ON A REAL VARIABLE

The results presented in Sections 18.1–18.8 include the case when the families of transformations $\mathbb{C}^n \to \mathbb{C}^m$ and subspaces of $\mathbb{C}^n$ are analytic in a *real* variable on an open interval (a, b) of the real axis. The definition of analyticity is analogous to that in the complex case: representation as a power series (this time with *real* coefficients) in a real neighbourhood of each point $t_0 \in (a, b)$. As the radius of convergence of this power series is positive, it converges also in some complex neighbourhood of t_0. Consequently, a family of transformations from $\mathbb{C}^n$ into $\mathbb{C}^m$ (or of subspaces of $\mathbb{C}^n$) that is analytic on (a, b) can be extended to a family of linear transformations (or subspaces) that is analytic in some complex neighbourhood Ω of (a, b), and the results presented in Sections 18.1–18.8 do apply.

It is noteworthy that, in contrast to the complex variable case, the orthogonal complement preserves analyticity, as follows.

Theorem 18.9.1

Let $\mathcal{M}(t)$ be a family of subspaces (of $\mathbb{C}^n$) that is analytic in the real variable t on (a, b). Then the orthogonal complement $\mathcal{M}(t)^\perp$ is an analytic family of subspaces on (a, b) as well.

Proof. Let $t_0 \in (a, b)$. Then in some real neighbourhood U_1 of t_0 there exists an analytic family of invertible transformations $A(t)$: $\mathbb{C}^n \to \mathbb{C}^n$ such that $\mathcal{M}(t) = A(t)\mathcal{M}$, $t \in U_1$ for a fixed subspace $\mathcal{M} \subset \mathbb{C}^n$. Assume (without loss of generality) that $\mathcal{M} = \mathrm{Span}\{e_1, \ldots, e_p\}$ for some p, and write $A(t)$ as the $n \times n$ matrix with entries that are analytic on (a, b) with respect to the standard basis in $\mathbb{C}^n$. Then $\mathcal{M}(t) = \mathrm{Im}\, B(t)$ for $t \in U_1$, where $B(t)$ is formed by the first p columns of $A(t)$. As $A(t)$ is invertible, the columns of $B(t)$ are linearly independent. For notational simplicity, assume that the top p rows of $B(t_0)$ are linearly independent and hence form a nonsingular $p \times p$ matrix. Then there is a real neighbourhood $U_2 \subset U_1$ of t_0 such that the top rows of $B(t)$ form a nonsingular $p \times p$ matrix $C(t)$ as well. So for $t \in U_2$, we obtain

$$\mathcal{M}(t) = \mathrm{Im}\begin{bmatrix} C(t) \\ D(t) \end{bmatrix} = \mathrm{Im}\begin{bmatrix} I \\ D(t)C(t)^{-1} \end{bmatrix}$$

where $D(t)$ is the $(n - p) \times p$ matrix formed by the bottom $n - p$ rows of $B(t)$. Denoting $X(t) = D(t)C(t)^{-1}$, consider the $p \times p$ matrix function $S(t) = (I + X(t)^* X(t))^{-1}$ for $t \in U_2$. Note that $I + X(t)^* X(t)$ is positive definite and thus invertible. Clearly, $S(t)$ is positive definite and analytic on U_2. Let Γ be a contour that lies in the open right half plane, is symmetrical with respect to the real axis, and contains all the eigenvalues of $S(t_0)$ in its interior. Then all eigenvalues of $S(t)$, where t is taken from some neighbourhood $U_3 \subset U_2$ of t_0, will also be in the interior of Γ. For such a t the integral

$$Z(t) = \frac{1}{2\pi i}\int_\Gamma \lambda^{1/2}(\lambda I - S(t))^{-1}\, d\lambda$$

where $\lambda^{1/2}$ is the analytic branch of the square root that takes positive values for λ positive, is well defined and $Z(t)^2 = S(t)$ (see Section 2.10). Moreover, because of the symmetry of Γ, the matrix $Z(t)$ is positive definite for all $t \in U_3$. Also, $Z(t)$ is an analytic family of matrices on U_3. Now one sees easily that, for $t \in U_3$

$$P(t) \overset{\text{def}}{=} \begin{bmatrix} Z(t) & Z(t)X(t)^* \\ X(t)Z(t) & X(t)Z(t)X(t)^* \end{bmatrix}$$

is the orthogonal projector on $\mathcal{M}(t)$. Indeed, a straightforward computation verifies that $P(t)^2 = P(t) = P(t)^*$. So $P(t)$ is an orthogonal projector. Furthermore, it is clear that

$$\text{Im } P(t) \supset \text{Im} \begin{bmatrix} I \\ X(t) \end{bmatrix} (= \mathcal{M}(t)) \tag{18.9.1}$$

and since rank $P(t)$ is easily seen to be p, equality (rather than inclusion) holds in (18.9.1). Consequently, $\mathcal{M}(t)^\perp$ is the image of the analytic family of projectors $I - P(t)$, and thus $\mathcal{M}(t)^\perp$ is analytic on U_3. As $t_0 \in (a, b)$ was arbitrary, the analyticity of $\mathcal{M}(t)^\perp$ on (a, b) follows. $\square$

One can also consider families of real transformations from $\mathcal{R}^n$ into $\mathcal{R}^m$, as well as families of subspaces in the real vector space $\mathcal{R}^n$, which are analytic in a real variable t on (a, b). For such families of real linear transformations and subspaces the results of Sections 18.1–18.8 hold also. However, in Theorem 18.7.2 the contour Γ should be symmetrical with respect to the real axis; and in the definition of fixed Jordan structure one has to require, in addition, that the enumeration $\lambda_1(z_1), \ldots, \lambda_m(z_1)$ and $\lambda_1(z_2), \ldots, \lambda_m(z_2)$ of distinct eigenvalues of $A(z_1)$ and $A(z_2)$, respectively, is such that $\lambda_i(z_1) = \overline{\lambda_j(z_1)}$ holds if and only if $\lambda_i(z_2) = \overline{\lambda_j(z_2)}$.

18.10 EXERCISES

18.1 Let

$$A(z) = \begin{bmatrix} 1 & 1 \\ z & 2z \\ z^2 & 4z^2 \end{bmatrix} : \mathcal{C}^2 \to \mathcal{C}^3, z \in \mathcal{C}$$

be an analytic family of transformations written as a matrix in the standard orthonormal bases in $\mathcal{C}^2$ and $\mathcal{C}^3$.

(a) Are Im $A(z)$ and Ker $A(z)$ analytic families of subspaces?

(b) Find an analytic vector function $y(z)$ such that $y(z) \neq 0$ for all $z \in \mathcal{C}$ and Span$\{y(z)\} = $ Ker $A(z)$ for all $z \in \mathcal{C}$ with the exception of a discrete set.

(c) Find linearly independent and analytic (in $\mathcal{C}$) vector functions $y_1(z), y_2(z)$ such that Span$\{y_1(z), y_2(z)\} = $ Im $A(z)$ for all $z \in \mathcal{C}$ with the exception of a discrete set. [*Hint*: Use the Smith form for the matrix polynomial $A(z)$.]

18.2   Solve Exercise 18.1 for

$$A(z) = \begin{bmatrix} z+1 & z \\ z & z-1 \\ 1 & z \end{bmatrix}$$

18.3   Let $P(z)$ be an analytic family of projectors. Show that Im $P(z)$ is an analytic family of subspaces.

18.4   Let

$$A(z) = \text{diag}[A_1(z), A_2(z), \ldots, A_k(z)]$$

where for $j = 1, \ldots, k$, $A_j(z)$ is an analytic family of transformations on a domain Ω. Prove that the following statements are equivalent:

(a) Im $A(z)$ and Ker $A(z)$ are analytic families of subspaces.
(b) Im $A_j(z)$ is an analytic family of subspaces, for $j = 1, \ldots, k$.
(c) Ker $A_j(z)$ is an analytic family of subspaces, for $j = 1, \ldots, k$.

18.5   Let $A(z): \mathbb{C}^n \to \mathbb{C}^n$ be an analytic family of transformations on Ω such that $A(z)^2 = I$ for all $z \in \Omega$. Prove that the families of subspaces $\text{Im}(A(z) - I)$ and $\text{Im}(A(z) + I)$ are analytic on Ω.

18.6   Let $A(z)$ be an analytic family of transformations on Ω such that $p(A(z)) = 0$ for all $z \in \Omega$, where $p(\lambda)$ is a scalar polynomial of degree m with distinct zeros $\lambda_1, \ldots, \lambda_m$. Prove that the families of subspaces $\text{Ker}(\lambda_j I - A(z))$, $j = 1, \ldots, m$ are analytic on Ω.

18.7   Does the result of Exercise 18.6 hold if $p(\lambda)$ has less than m distinct zeros?

18.8   Given matrices A and B of sizes $n \times n$ and $n \times m$, respectively, show that $\text{Ker}[\lambda I + A, B]$ is an analytic family of subspaces if and only if (A, B) is a full-range pair.

18.9   Given matrices C and A of sizes $p \times n$ and $n \times n$, respectively, show that $\text{Im} \begin{bmatrix} \lambda I + A \\ C \end{bmatrix}$ is an analytic family of subspaces if and only if (C, A) is a null kernel pair.

18.10   Given an analytic $n \times n$ matrix function $A(z)$ on Ω that is upper triangular for all $z \in \Omega$, when is Ker $A(z)$ analytic on Ω?

18.11   For the following analytic vector functions $x_1(z)$, $x_2(z)$, where $z \in \mathbb{C}$, find analytic vector functions $y_1(z)$, $y_2(z)$ of $z \in \mathbb{C}$ such that $y_1(z)$ and $y_2(z)$ are linearly independent for every $z \in \mathbb{C}$ and

$$\text{Span}\{x_1(z), x_2(z)\} = \text{Span}\{y_1(z), y_2(z)\}$$

for every $z \in \mathbb{C}$ except for a discrete set:

(a) $x_1(z) = \langle z^2, 1 - z, 0 \rangle$, $x_2(z) = \langle z^3, 1 - z^2, z^2 - z \rangle$
(b) $x_1(z) = \langle 1, -z, z \rangle$, $x_2(z) = \langle 1, z^2, z^2 + z \rangle$

[*Hint*: Use the Smith form for the matrix polynomial $[x_1(z), x_2(z)]$.]

18.12 Let $x_1(z), \ldots, x_k(z)$ be n-dimensional vector polynomials such that, for at least one value $z_0 \in \mathcal{C}$, the vectors $x_1(z_0), \ldots, x_k(z_0)$ are linearly independent. Prove that one can construct n-dimensional vector polynomials $y_1(z), \ldots, y_k(z)$ such that $y_1(z), \ldots, y_k(z)$ are linearly independent for all $z \in \mathcal{C}$ and

$$\text{Span}\{y_1(z), \ldots, y_k(z)\} = \text{Span}\{x_1(z), \ldots, x_k(z)\}$$

for all $z \in \mathcal{C}$ with the possible exception of a finite set, as follows. Let

$$[x_1(z), \ldots, x_k(z)] = E(z)D(z)F(z)$$

be the Smith form of the $n \times k$ matrix $[x_1(z), \ldots, x_k(z)]$; then put

$$[y_1(z), \ldots, y_k(z)] = E(z)F(z)$$

18.13 Complete the following linearly independent analytic families of vectors in $\mathcal{C}^4$ (depending on the complex variable $z \in \mathcal{C}$) to analytic families of vectors that form a basis in $\mathcal{C}^4$ for every $z \in \mathcal{C}$:

(a) $x_1(z) = \langle 1, z, z^2, z^3 \rangle$; $\quad x_2(z) = \langle 1, 2z, 4z^2, 1 \rangle$
(b) $x_1(z) = \langle 1, z, 0, 1 \rangle$; $\quad x_2(z) = \langle -1, 0, z-1, 2 \rangle$;
$\qquad\qquad x_3(z) = \langle -1, 0, 0, z+1 \rangle$

18.14 For the following analytic families $\mathcal{M}(z)$ of subspaces in $\mathcal{C}^n$ that depend on $z \in \mathcal{C}$, find two subspaces $\mathcal{N}_1$ and $\mathcal{N}_2$ such that for every $z \in \mathcal{C}$ at least one of

$$\mathcal{M}(z) \dotplus \mathcal{N}_1 = \mathcal{C}^n \quad \text{or} \quad \mathcal{M}(z) \dotplus \mathcal{N}_2 = \mathcal{C}^n$$

holds:

(a) $\quad \mathcal{M}(z) = \text{Im} \begin{bmatrix} 1 & 1 \\ z & 2z \\ z^2 & 4z^2 \\ z^3 & 1 \end{bmatrix}$

(b) $\quad \mathcal{M}(z) = \text{Im} \begin{bmatrix} 1 & -1 & -1 \\ z & 0 & 0 \\ 0 & z-1 & 0 \\ 1 & 2 & z+1 \end{bmatrix}$

18.15 For each $n \geq 2$ give an example of an analytic family of transformations $A(z): \mathcal{C}^n \to \mathcal{C}^n$ defined on Ω that has no nontrivial $A(z)$-invariant analytic family of subspaces on Ω.

18.16 Let $A(z)$ be an analytic family of transformations defined on Ω such that $p(A(z)) = 0$ for all $z \in \Omega$, where $p(\lambda)$ is a scalar polynomial of degree m with m distinct zeros. Prove that there are at least 2^m $A(z)$-invariant analytic families of subspaces on Ω.

Chapter Nineteen

Jordan Form of
Analytic Matrix Functions

In this chapter we study the behaviour of eigenvalues and eigenvectors of a transformation that depends analytically on a parameter in both the local and global frameworks. It turns out that this behaviour is analytic except for isolated singularities that are described in detail. The results obtained allow us to solve (at least partially) the problem of analytic extendability of an invariant subspace. In turn, the solution of this problem is used in Chapter 20 for the solution of various problems concerning divisors of monic matrix polynomials, minimal factorization of rational matrix functions, and solutions of matrix quadratic equations, all of which involve analytic dependence on a parameter. Clearly, the material of this chapter relies on more advanced complex analysis than does that of the preceding chapters. However, this is not a prerequisite for understanding the main results.

19.1 LOCAL BEHAVIOUR OF EIGENVALUES AND EIGENVECTORS

Let $A(z): \mathbb{C}^n \to \mathbb{C}^n$ be a family of transformations that is analytic on a domain Ω. In this section we study the behaviour of eigenvalues and eigenvectors as functions of z in a neighbourhood of a fixed point $z_0 \in \Omega$. First let us state the main result in this direction.

Theorem 19.1.1

Let $\mu_1, \ldots, \mu_k$ be all the distinct eigenvalues of $A(z_0)$, that is, the distinct zeros of the equation $\det(\mu I - A(z_0)) = 0$, where $k \leq n$, and let $r_i (i = 1, \ldots, k)$ be the multiplicity of μ_i as a zero of $\det(\mu I - A(z_0)) = 0$ (so $r_1 + \cdots + r_k = n$). Then there is a neighbourhood $\mathcal{U}$ of z_0 in Ω with the following properties: (a) there exist positive integers $m_{11}, \ldots, m_{1s_1};$

$m_{21}, \ldots, m_{2,s_2}; \ldots; m_{k1}, \ldots, m_{ks_k}$ *such that the n eigenvalues (not necessarily distinct) of $A(z)$ for $z \in \mathcal{U} > \{z_0\}$, are given by the fractional power series*:

$$\mu_{ij\sigma}(z) = \mu_i + \sum_{\alpha=1}^{\infty} a_{\alpha ij}[(z - \mu_i)_\sigma^{1/m_{ij}}]^\alpha ; \qquad \sigma = 1, \ldots, m_{ij}; \, j = 1, \ldots, s_i$$

$$i = 1, \ldots, k \qquad\qquad (19.1.1)$$

where $a_{\alpha ij} \in \mathbb{C}$ and for $\sigma = 1, \ldots, m_{ij}$

$$(z - \mu_i)_\sigma^{1/m_{ij}} = |z - \mu_i|^{1/m_{ij}}\left(\cos\left(\frac{2\pi\sigma}{m_{ij}}\right) + i \sin\left(\frac{2\pi\sigma}{m_{ij}}\right)\right)$$

(b) the dimension γ_{ij} of $\mathrm{Ker}(A(\lambda) - \mu_{ij\sigma}(z)I)$, as well as the partial multiplicities $m_{ij}^{(1)} \geq \cdots \geq m_{ij}^{(\gamma_{ij})}$ (>0) of the eigenvalue $\mu_{ij\sigma}(z)$ of $A(\lambda)$, do not depend on z (for $z \in \mathcal{U} \smallsetminus \{z_0\}$) and do not depend on σ; (c) for each $i = 1, \ldots, k$ and $j = 1, \ldots, s_i$ there exist vector-valued fractional power series converging for $z \in \mathcal{U}$:

$$x_{ij\sigma}^{(\eta\delta)}(z) = \sum_{\alpha=0}^{\infty} x_{\alpha ij}^{(\gamma\delta)}[(z - \mu_i)_\sigma^{1/m_{ij}}]; \qquad \delta = 1, \ldots, m_{ij}^{(\gamma)}$$

$$\gamma = 1, \ldots, \gamma_{ij} ; \qquad \sigma = 1, \ldots, m_{ij} \qquad\qquad (19.1.2)$$

where $x_{\alpha ij}^{(\gamma\delta)} \in \mathbb{C}^n$, such that for each γ and each $z \in \mathcal{U} \smallsetminus \{z_0\}$ the vectors $x_{ij\sigma}^{(\gamma,1)}(z), \ldots, x_{ij\sigma}^{(\gamma,m_{ij}^{(\gamma)})}(z)$ form a Jordan chain of $A(z)$ corresponding to $\mu_{ij}(z)$:

$$(A(z) - \mu_{ij\sigma}(z)I)x_{ij\sigma}^{(\gamma p)}(z) = x_{ij\sigma}^{(\gamma, p-1)}(z), \, p = 1, \ldots, m_{ij}^{(\gamma)};$$

$$\gamma = 1, \ldots, \gamma_{ij} , \qquad \sigma = 1, \ldots, m_{ij} \qquad\qquad (19.1.3)$$

where by definition $x_{ij\sigma}^{(\gamma 0)}(z) = 0$, and $x_{ij\sigma}^{(\gamma 1)}(z) \neq 0$. Moreover, for every $z \in \mathcal{M} \smallsetminus \{z_0\}$ the vectors

$$x_{ij\sigma}^{(\gamma\delta)}(z) ; \qquad \delta = 1, \ldots, m_{ij}^{(\gamma)} ; \qquad \gamma = 1, \ldots, \gamma_{ij} ; \qquad \sigma = 1, \ldots, m_{ij}$$

$$j = 1, \ldots, s_i ; \qquad i = 1, \ldots, k$$

form a basis in $\mathbb{C}^n$.

The full proof of Theorem 19.1.1 is too long to be presented here. We refer the reader to the book of Baumgartel (1985), and especially Section IX.3 there, for a complete proof.

Let us make some remarks concerning this important theorem. First, in the expansion (19.1.3), if $m_{ij} > 1$, then the greatest common divisor of all positive integers α such that $a_{\alpha ij} \neq 0$ is 1 (so $\mu_{ij\sigma}(z)$, $\sigma = 1, \ldots, m_{ij}$ have a branch point at μ_i of branch multiplicity m_{ij} and not less). If $m_{ij} = 1$, then $\mu_{ij\sigma}(z)$ are analytic on a neighbourhood of μ_i; it may even happen that $\mu_{ij\sigma}(z)$ is the constant function μ_i (see Example 19.1.2). Second, the theorem does not say anything explicit about the partial multiplicities $p_{i1} \geq \cdots \geq p_{i,t_i}$ of the eigenvalue μ_i of $A(z_0)$ (we know only that $\Sigma_{j=1}^{t_i} p_{ij} = r_i$ for $i = 1, \ldots, k$). However, there is a connection between the partial multiplicities $m_{ij}^{(1)} \geq \cdots \geq m_{ij}^{(\gamma_{ij})}$ of the eigenvalues $\mu_{ij\sigma}(z)$ of $A(z)$ ($z \in \mathcal{U} \smallsetminus \{z_0\}$) and the partial multiplicities of the eigenvalue μ_i of $A(z_0)$. This connection is given by the following formula (see Theorem 15.10.2):

$$\sum_{q=1}^{l} p_{iq} \leq \sum_{q=1}^{l} \sum_{j=1}^{s_i} m_{ij} m_{ij}^{(q)}, \qquad l = 1, 2, \ldots,$$

where p_{iq} is interpreted as zero for $k > t_i$, and similarly for $m_{ij}^{(q)}$ when $q > \gamma_{ij}$. As the total sum of partial multiplicities of eigenvalues near μ_i does not change after small perturbation of the transformation, we also have the equality

$$\sum_{q=1}^{t_i} p_{iq} = \sum_{j=1}^{s_i} \sum_{q=1}^{\gamma_{ij}} m_{ij} m_{ij}^{(q)}$$

Let us illustrate Theorem 19.1.1 with an example.

EXAMPLE 19.1.1.   Let $z_0 = 0$ and

$$A(z) = \begin{bmatrix} 0 & 1 & 1 & 0 \\ z & 0 & 0 & 1 \\ 0 & 0 & 0 & 1 \\ 0 & 0 & z & 0 \end{bmatrix}$$

The only eigenvalue of $A(0)$ is zero, with partial multiplicities 3 and 1. [The easiest way to find the partial multiplicities of $A(0)$ is to observe that rank $A(0) = 2$ and $A(0)^2 \neq 0$.] To find the eigenvalues of $A(z)$, we have to solve the equation $\det(\mu I - A(z)) = 0$, which gives (in the notation of Theorem 19.1.1)

$$\mu_{ij\sigma}(z) = z^{1/2}, \qquad j = 1; \qquad i = 1; \qquad \sigma = 1, 2$$

(so we have $k = 1$, $s_1 = 1$, $m_{11} = 2$). It is not difficult to see that the only partial multiplicity of $\mu_{ij\sigma}(z)$ is $m_{ij}^{(1)} = 2$. The Jordan chain of $A(z)$ corresponding to $\mu_{ij\sigma}(z)$ is

$$x_{ij\sigma}^{(11)}(z) = \langle 1, z^{1/2}, 0, 0 \rangle ; \qquad x_{ij\sigma}^{(12)}(z) = \langle 0, 0, 1, z^{1/2} \rangle \quad \square$$

An important particular case of Theorem 19.1.1 appears when the eigenvalues of $A(z)$ are analytic in a neighbourhood of z_0, that is, all integers m_{ij} are equal to 1, as follows.

Corollary 19.1.2

Assume that all the eigenvalues of $A(z)$ are analytic in a neighbourhood of z_0. Then the distinct eigenvalues $\mu_1(z), \ldots, \mu_k(z)$ of $A(z)$, $z \neq z_0$ can be enumerated so that $\mu_i(z)$ is an analytic function in a neighbourhood $\mathcal{U}_1$ of z_0. Further, assuming that the enumeration of the distinct eigenvalues of $A(z)$ for $z \neq z_0$ is as above, there exist analytic n-dimensional vector functions

$$y_{ij}^{(\gamma)}(z) ; \qquad i = 1, \ldots, k ; \qquad j = 1, \ldots, s_i ; \qquad \gamma = 1, \ldots, \gamma_{ij}$$

$$(19.1.4)$$

in a neighbourhood $\mathcal{U}_2 \subset \mathcal{U}_1$ of z_0 with the following properties: (a) for every $z \in \mathcal{U}_2 \smallsetminus \{z_0\}$, and for $i = 1, \ldots, k$; $j = 1, \ldots, s_i$ the vectors $y_{ij}^{(1)}(z), \ldots, y_{ij}^{(\gamma_{ij})}(z)$ form a Jordan chain of $A(z)$ corresponding to the eigenvalue $\mu_i(z)$; (b) for every $z \in \mathcal{U}_2 \smallsetminus \{z_0\}$ the vectors (19.1.4) form a basis in $\mathbb{C}^n$.

The following example illustrates this corollary.

EXAMPLE 19.1.2. Let

$$A(z) = \begin{bmatrix} 0 & z \\ 0 & 0 \end{bmatrix}, \qquad z \in \mathbb{C}$$

Obviously, the eigenvalues of $A(z)$ are analytic (even constant). It is easy to find analytic vector functions $y_{ij}^{(\gamma)}(z)$ as in Corollary 19.1.2: we have $k = 1$, $s_1 = 1$, $\gamma_{11} = 2$, and

$$y_{11}^{(1)}(z) = \begin{bmatrix} z \\ 0 \end{bmatrix}, \qquad y_{11}^{(2)}(z) = \begin{bmatrix} 0 \\ 1 \end{bmatrix}$$

Note that $y_{11}^{(1)}(z)$, $y_{11}^{(2)}(z)$ do not form a basis in $\mathbb{C}^2$ for $z = 0$; also, $y_{11}^{(1)}(0)$, $y_{11}^{(2)}(0)$ do not form a Jordan chain of $A(0)$. This shows that in (a) and (b) in Corollary 19.1.2 one cannot, in general, replace $\mathcal{U}_2 \smallsetminus \{z_0\}$ by $\mathcal{U}_2$. $\square$

19.2 GLOBAL BEHAVIOUR OF EIGENVALUES AND EIGENVECTORS

The result of Theorem 19.1.1 allows us to derive some global properties of eigenvalues and eigenvectors of an analytic family of transformations $A(z): \mathbb{C}^n \to \mathbb{C}^n$ defined on Ω. As before, Ω is a domain in the complex plane.

For a transformation $X: \mathbb{C}^n \to \mathbb{C}^n$ we denote by $\nu(X)$ the number of distinct eigenvalues of X. Obviously, $1 \leq \nu(X) \leq n$.

Theorem 19.2.1

Let $A(z): \mathbb{C}^n \to \mathbb{C}^n$ be an analytic family of transformations on Ω. Then for all $z \in \Omega$ except a discrete set S_0 we have

$$\nu(A(z)) = \max_{z \in \Omega} \nu(A(z))$$

for $z_0 \in S_0$ we have

$$\nu(A(z_0)) < \max_{z \in \Omega} \nu(A(z))$$

Proof. Theorem 19.1.1 shows that for every $z_0 \in \Omega$ there is a neighbourhood $\mathcal{U}_{z_0}$ of z_0 such that $\nu(A(z))$ is constant (equal to ν_0, say) for $z \in \mathcal{U}_{z_0} \smallsetminus \{z_0\}$ and $\nu(A(z_0)) \leq \nu_0$. *A priori*, it appears that ν_0 may depend on z_0. Let us show that actually ν_0 is independent of z_0. For $\nu = 1, \ldots, n$, let $\mathcal{V}_\nu = U \mathcal{U}_{z_0}$, where the union is taken over all $z_0 \in \Omega$ such that $\nu(A(z)) = \nu$ in a deleted neighbourhood $\mathcal{U}_{z_0} \smallsetminus \{z_0\}$ of z_0. Obviously, $\mathcal{V}_1, \ldots, \mathcal{V}_n$ are open sets whose union is Ω, and it is easily seen that they are mutually disjoint. This can happen only if all $\mathcal{V}_j$ are empty except for $\mathcal{V}_{\nu_0}$; therefore, $\mathcal{V}_{\nu_0} = \Omega$. It is clear also that

$$\nu_0 = \max_{z \in \Omega} \nu(A(z))$$

Now if $\nu(A(z')) < \nu_0$ for some $z' \in \Omega$, then by Theorem 19.1.1 we have $\nu(A(z)) = \nu_0$ in a deleted neighbourhood of z'. This shows that the set S_0 of all $z \in \Omega$ for which $\nu(A(z)) < \nu_0$ is indeed discrete. $\square$

The points from S_0 will be called the *multiple points* of the analytic family of transformations $A(z)$, because at these points the eigenvalues of $A(z)$ attain higher multiplicity than "usual."

Another way to prove Theorem 19.2.1 is by examining a suitable resultant matrix. Let

$$\det(\mu I - A(z)) = \mu^n + \sum_{j=0}^{n-1} a_j(z) \mu^j$$

for some scalar functions $a_j(z)$ that are analytic on Ω, and consider the $(2n-1) \times (2n-1)$ matrix whose entries are analytic functions on Ω:

$$R(z) = \begin{bmatrix} a_0(z) & a_1(z) & \cdots & a_{n-1}(z) & 1 & 0 & \cdots & 0 \\ 0 & a_0(z) & a_1(z) & \cdots & a_{n-1}(z) & 1 & \cdots & 0 \\ \vdots & \vdots & \vdots & & & & & \vdots \\ 0 & 0 & 0 & \cdots & & a_0(z) & a_1(z) & \cdots & 1 \\ a_1(z) & 2a_2(z) & \cdots & (n-1)a_{n-1}(z) & n & 0 & & 0 \\ 0 & a_1(z) & 2a_2(z) & \cdots & & n & \cdots & 0 \\ \vdots & \vdots & \vdots & & & & & \vdots \\ 0 & 0 & 0 & \cdots & & a_1(z) & 2a_2(z) & \cdots & n \end{bmatrix}$$

This is the resultant matrix of two scalar polynomials on μ: $\det(\mu I - A(z))$ and $(\partial/\partial u)(\det(\mu I - A(z)))$. A well-known property of resultant matrices [see, e.g., Gohberg and Heinig (1975)] states that $2n - 1 - \mathrm{rank}\, R(z)$ is equal to the number of common zeros of these two polynomials in μ (counting multiplicities). In other words

$$2n - 1 - \mathrm{rank}\, R(z) = n - \nu(A(z))$$

or

$$\mathrm{rank}\, R(z) = n - 1 + \nu(A(z)) \qquad (19.2.1)$$

Now let k ($n \le k \le 2n - 1$) be the largest size of a square submatrix in $R(z)$ whose determinant is not identically zero. Denoting by $S_1(z), \ldots, S_l(z)$ all such submatrices in $R(z)$, we obviously have $\mathrm{rank}\, R(z) = k$ if at least one of $\det S_1(z), \ldots, \det S_l(z)$ is different from zero; and $\mathrm{rank}\, R(z) < k$ otherwise. Comparing with (19.2.1), we obtain:

$$\nu(A(z)) = k - n + 1$$

if not all numbers $\det S_1(z), \ldots, \det S_l(z)$ are zeros;

$$\nu(A(z)) < k - n + 1$$

otherwise. Since the set of common zeros of $\det S_1(z), \ldots, \det S_l(z)$ is discrete, Theorem 19.2.1 follows. $\square$

Theorem 19.1.1 shows that the distinct eigenvalues of $A(z)$, $\mu_1(z), \ldots, \mu_\nu(z)$ (where $\nu = \max_{z \in \Omega} \nu(A(z))$) are analytic on $\Omega \smallsetminus S_0$, where S_0 is taken from Theorem 19.2.1, and have at most algebraic branch points is S_0. [Some of the functions $\mu_1(z), \ldots, \mu_\nu(z)$ may also be analytic at certain points in S_0.] Denote by S_1 the subset of S_0 consisting of all the points z_0 such that at least one of the functions $\mu_j(z)$, $j = 1, \ldots, \nu$ is not analytic at z_0. As a subset of a discrete set, S_1 is itself discrete. The set S_1 will be called the *first exceptional set* of the analytic family of linear transformations $A(z)$, $z \in \Omega$.

It may happen that $S_1 \ne S_0$, as shown in the following example.

EXAMPLE 19.2.1.   Let $\Omega = \mathbb{C}$ and

$$A(z) = \begin{bmatrix} 0 & z \\ z & 0 \end{bmatrix}$$

The eigenvalues of $A(z)$ are $\pm z$, so in this case $S_0 = \{0\}$ but $S_1 = \emptyset$.   □

Example 19.1.2 shows that in general, when $z \in \Omega \smallsetminus S_1$, one cannot expect that there will be a Jordan basis of $A(z)$ that depends analytically on z. To achieve that we must exclude from consideration a second exceptional set, which is described now.

Theorem 19.2.2

Let $A(z)$: $\mathbb{C}^n \to \mathbb{C}^n$ be an analytic family of transformations on Ω with the set S_0 of multiple points and let $\mu_1(z), \ldots, \mu_\nu(z)$ be the distinct eivenvalues of $A(z)$ analytic on $\Omega \smallsetminus S_0$ and having at most branch points in S_0. Let $m_{j1}(z) \geq \cdots \geq m_{j\gamma}(z)$, $\gamma = \gamma(j, z)$ be the partial multiplicities of the eigenvalue $\mu_j(z)$ of $A(z)$ for $j = 1, \ldots, \nu$; $z \notin S_0$. Then there exists a discrete set $\tilde{S}_2$ in Ω such that $\tilde{S}_2 \subset \Omega \smallsetminus S_0$ and the number $\gamma(j, z)$ of partial multiplicities and the partial multiplicities $m_{jk}(z)$ themselves, $k = 1, \ldots, \gamma(j, z)$ do not depend on z in $\Omega \smallsetminus (S_0 \cup \tilde{S}_2)$, for $1, \ldots, \nu$.

Proof.   The proof follows the pattern of the proof of Theorem 19.2.1. In view of Theorem 19.1.1, for every $z_0 \in \Omega$, there is a neighbourhood $\mathcal{U}_{z_0}$ of z_0 such that the number of distinct eigenvalues $\nu = \nu(z_0)$, as well as the number $\gamma_j = \gamma_j(z_0)$ of partial multiplicities and the partial multiplicities themselves $m_{j1} \geq \cdots \geq m_{j\gamma_j}$, $m_{jk} = m_{jk}(z_0)$, corresponding to the jth eigenvalue, are constant for $z \in \mathcal{U}_{z_0} \smallsetminus \{z_0\}$. It is assumed that the distinct eigenvalues of $A(z)$ for $z \in \mathcal{U}_{z_0} \smallsetminus \{z_0\}$ are enumerated so that they are analytic and $\gamma_1 \geq \cdots \geq \gamma_\nu$. Denote by Δ the (finite) set of all sequences of type

$$\delta = \{\nu; \gamma_1, \ldots, \gamma_\nu; m_{11}, \ldots, m_{1\gamma_1}; \ldots; m_{\nu 1}, \ldots, m_{\nu\gamma_\nu}\} \quad (19.2.2)$$

where ν, γ_j, m_{jk} are positive integers with the properties that $\nu \leq n$; $\gamma_1 \geq \cdots \geq \gamma_\nu$; $m_i \geq \cdots \geq m_{i\gamma_i}$, $i = 1, \ldots, \nu$; $\Sigma_{i,j} m_{ij} = n$. For any sequence $\delta \in \Delta$ as in (19.2.2) let $\mathcal{V}_\delta = \cup \mathcal{U}_{z_0}$, where the union is taken over all $z_0 \in \Omega$ such that $\nu = \nu(z_0)$; $\gamma_j = \gamma_j(z_0)$, $j = 1, \ldots, \nu$; $m_{ij} = m_{ij}(z_0)$, $j = 1, \ldots, \gamma_i$; $i = 1, \ldots, \nu$. Obviously, $\mathcal{V}_\delta$ is open and $\cup_{\delta \in \Delta} \mathcal{V}_\delta = \Omega$. Also, the sets $\mathcal{V}_\delta$, $\delta \in \Delta$ are mutually disjoint. As Ω is connected, this means that all $\mathcal{V}_\delta$, except for one of them, are empty. So Theorem 19.2.2 follows.   □

The set $S_2 = \tilde{S}_2 \cup (S_0 \smallsetminus S_1)$, where $\tilde{S}_2$ is taken from Theorem 19.2.2 and S_0 and S_1 are the set of multiple points and the first exceptional set of $A(z)$, respectively, is called the *second exceptional set* of $A(z)$. Note that $S_2 \cap S_1 =$

$\emptyset$. The second exceptional set is characterized by the properties that the distinct analytic eigenvalues of $A(z)$ can be continued analytically into any point $z_0 \in S_2$, but for every $z_0 \in S_2$, either $\nu(A(z_0)) < \max_{z \in \Omega} \nu(A(z))$, or $\nu(A(z_0)) = \max_{z \in \Omega} \nu(A(z))$ and for at least one analytic eigenvalue $\mu_j(z)$ of $A(z)$ the partial multiplicities of $\mu_j(z_0)$ are different from the partial multiplicities of $\mu_j(z)$, $z \neq z_0$, in a neighbourhood of z_0.

EXAMPLE 19.2.2.   Let

$$A(z) = \begin{bmatrix} p_1(z) & q_1(z) & \cdots & & 0 \\ & p_2(z) & q_2(z) & & \vdots \\ \vdots & & \ddots & \ddots & \\ & & & & q_{n-1}(z) \\ 0 & & \cdots & & p_n(z) \end{bmatrix}, \quad z \in \mathbb{C}$$

where $p_j(z)$ and $q_j(z)$ are not identically zero polynomials such that

$$p_1(z) = \cdots = p_{k_1}(z); \; p_{k_1+1}(z) = \cdots = p_{k_2}(z); \ldots; \; p_{k_{q-1}+1}(z) = \cdots = p_{k_q}(z)$$

for all $z \in \mathbb{C}$, where $1 \leq k_1 < k_2 < \cdots < k_{q-1} < k_q = n$. We also assume that the polynomials $p_{k_1}(z), \ldots, p_{k_q}(z)$ are all different. We have the set of multiple points

$$S_0 = \{z \in \mathbb{C} \mid p_{k_i}(z) = p_{k_j}(z) \quad \text{for some} \quad i \neq j\}$$

the first exceptional set S_1 is empty, and the second exceptional set S_2 is the union of S_0 and the set $\{z \in \mathbb{C} \setminus S_0 \mid q_i(z) = 0 \text{ for some } k_p + 1 \leq i \leq k_{p+1} - 1$ and some $p\}$. $\square$

Now we state the result on existence of an analytic Jordan basis for an analytic family of transformations.

Theorem 19.2.3

Let $A(z) \colon \mathbb{C}^n \to \mathbb{C}^n$ be an analytic family of transformations in Ω with the first exceptional set S_1 and the second exceptional set S_2. Let $\mu_1(z), \ldots, \mu_\nu(z)$ be the distinct eigenvalues of $A(z)$ (apart from the multiple points), which are analytic on $\Omega \setminus S_1$ and have at most algebraic branch points in S_1. Then there exist n-dimensional vector functions

$$x_{11}^{(j)}(z), \ldots, x_{1,m_{j1}}^{(j)}(z), x_{21}^{(j)}(z), \ldots, x_{2,m_{j2}}^{(j)}(z), \ldots, x_{\gamma_j 1}^{(j)}(z), \ldots, x_{\gamma_j m_{j\gamma_j}}^{(j)}(z)$$

$$(19.2.3)$$

$j = 1, \ldots, \nu$, where $m_{j1} \geq \cdots \geq m_{j\gamma_j}$ are positive integers, with the following properties: (a) the functions (19.2.3) are analytic on $\Omega \setminus S_1$ and have at most

algebraic branch points in S_1; *(b) for every* $z \in \Omega \smallsetminus (S_1 \cup S_2)$ *the vectors* (19.2.3) *form a basis in* $\mathbb{C}^n$; *(c) for every* $z \in \Omega \smallsetminus (S_1 \cup S_2)$ *the vectors*

$$x_{k1}^{(j)}(z), \ldots, x_{k,m_{jk}}^{(j)}(z)$$

form a Jordan chain of the transformation $A(z)$ *corresponding to the eigenvalue* $\mu_j(z)$, *for* $k = 1, \ldots, \gamma_j; j = 1, \ldots, \nu$.

It is easily seen that if $\mu_j(z)$ has an algebraic branch point at $z_0 \in S_1$, then all eigenvectors

$$x_{11}^{(j)}(z), x_{21}^{(j)}(z), \ldots, x_{\gamma_j 1}^{(j)}(z)$$

of $A(z)$ corresponding to $\mu_j(z)$ also have an algebraic branch point at z_0. Indeed, let $y(z)$ be some (say, the kth) coordinate of $x_{11}^{(j)}(z)$ that is not identically zero. The equality $[A(z) - \mu_j(z)]x_{11}^{(j)}(z) = 0$ for z in $\Omega \smallsetminus S_2$ implies that

$$\mu_j(z) = \frac{a_k(z)x_{11}^{(j)}(z)}{y(z)} \qquad (19.2.4)$$

where $a_k(z)$ is the kth row of $A(z)$. If $x_{11}^{(j)}(z)$ were analytic at z_0, then (19.2.4) implies that $\mu_j(z)$ is also analytic at z_0, a contradiction.

The proof of Theorem 19.2.3 is given in the next section.

In the particular case when $A(z)$ is diagonable (i.e., similar to a diagonal matrix) for every $z \notin S_1 \cup S_2$, the conclusions of Theorem 19.2.3 can be strengthened, as follows.

Theorem 19.2.4

Let $A(z)$ *be as in Theorem* 19.2.3, *and assume that* $A(z)$ *is diagonable for all* $z \notin S_1 \cup S_2$. *Then there exist n-dimensional vector functions*

$$x_1^{(j)}(z), \ldots, x_{\gamma_j}^{(j)}(z), \qquad j = 1, \ldots, \nu \qquad (19.2.5)$$

with the following properties: (a) the functions (19.2.5) *are analytic on* $\Omega \smallsetminus S_1$ *and have at most algebraic branch points in* S_1; *(b) for every* $z \in \Omega$ *and every* $j = 1, \ldots, \nu$ *the vectors* $x_1^{(j)}(z), \ldots, x_{\gamma_j}^{(j)}(z)$ *are linearly independent; (c) for every* $z \in \Omega \smallsetminus (S_1 \cup S_2)$ *the vectors* $x_1^{(j)}(z), \ldots, x_{\gamma_j}^{(j)}(z)$ *form a basis in* $\mathrm{Ker}(\mu_j(z)I - A(z))$. *In particular, the vectors* (19.2.5) *form a basis in* $\mathbb{C}^n$ *for every* $z \in \Omega \smallsetminus (S_1 \cup S_2)$.

The strengthening of Theorem 19.2.3 arises in statement (b), where the linear independence is asserted for all $z \in \Omega$ and not only for $z \in \Omega \smallsetminus (S_1 \cup S_2)$ as asserted in Theorem 19.2.3. The proof of Theorem 19.2.4 is obtained in the course of the proof of Theorem 19.2.3.

We illustrate Theorem 19.2.4 with a simple example.

EXAMPLE 19.2.3. Let

$$A(z) = \begin{bmatrix} 0 & z \\ 0 & z^2 \end{bmatrix} : \mathbb{C}^2 \to \mathbb{C}^2$$

Here $S_1 = \emptyset$; $S_2 = \{0\}$. The eigenvectors $x_1(z) = \begin{bmatrix} 1 \\ 0 \end{bmatrix}$ and $x_2(z) = \begin{bmatrix} 1 \\ z \end{bmatrix}$ corresponding to the eigenvalues 0 and z^2 of $A(z)$, respectively, are analytic and nonzero for all $z \in \mathbb{C}$ (including the point $z = 0$), as ensured by Theorem 19.2.4. However, $x_1(z)$ and $x_2(z)$ are not linearly independent for $z = 0$. $\square$

19.3   PROOF OF THEOREM 19.2.3

We need some preparation for the proof of Theorem 19.2.3.

A family of transformations $B(z)$: $\mathbb{C}^n \to \mathbb{C}^n$ is called *branch analytic* on Ω if $B(z)$ is analytic on Ω except for a discrete set of algebraic (as opposed to logarithmic) branch points. The same definition applies to n-dimensional vector functions as well. The *singular set* of a family of transformations $B(z)$: $\mathbb{C}^n \to \mathbb{C}^n$, which is branch analytic on Ω, is, by definition, the set of all $z_0 \in \Omega$ such that

$$\dim \text{Im } B(z_0) < \max_{z \in \Omega} \dim \text{Im } B(z)$$

It is easily seen that the singular set is discrete and coincides with the set of all $z_0 \in \Omega$ with

$$\dim \text{Ker } B(z_0) > \min_{z \in \Omega} \dim \text{Ker } B(z)$$

We use the notation $S(B)$ to designate the singular set of $B(z)$.

Lemma 19.3.1

Let $B(z)$: $\mathbb{C}^n \to \mathbb{C}^m$ be a branch analytic family of transformations on Ω. Then there exist m-dimensional branch analytic vector-valued functions $y_1(z), \ldots, y_r(z)$ on Ω, and n-dimensional branch analytic vector-valued functions $x_1(z), \ldots, x_{n-r}(z)$ on Ω with the following properties: (a) each branch point for any function $y_j(z), j = 1, \ldots, r$ or $x_k(z), k = 1, \ldots, n - r$ is also a branch point of $B(z)$; (b) $y_1(z), \ldots, y_r(z)$ are linearly independent for every $z \in \Omega$; (c) $x_1(z), \ldots, x_{n-r}(z)$ are linearly independent for every $z \in \Omega$; (d) $\text{Span}\{y_1(z), \ldots, y_r(z)\} = \text{Im } B(z)$ and $\text{Span}\{x_1(z), \ldots, x_{n-r}(z)\} = \text{Ker } B(z)$ for every z not belonging to $S(B)$.

The proof of this lemma can be obtained by repeating the proofs of Lemma 18.2.2 and Theorem 18.2.1 with the following modification: in place of the Weierstrass and Mittag–Leffler theorems (Lemmas 18.2.3 and 18.2.4), one must use the branch analytic and branch meromorphic versions of these theorems. [In the context of Riemann surfaces, these versions can be found in Kra (1972).]

Lemma 19.3.2

Let $B_1(z): \mathbb{C}^n \to \mathbb{C}^m$ and $B_2(z): \mathbb{C}^n \to \mathbb{C}^m$ be branch analytic families of transformations on Ω, such that

$$\operatorname{Ker} B_1(z) \supset \operatorname{Ker} B_2(z)$$

for every $z \in \Omega$ that does not belong to the union of the singular sets of $B_1(z)$ and $B_2(z)$. Then there exist branch analytic n-dimensional vector functions $x_1(z), \ldots, x_s(z)$, $z \in \Omega$ with the following properties: (a) every branch point of any $x_j(z)$, $j = 1, \ldots, s$ is also a branch point of at least one of $B_1(z)$ and $B_2(z)$; (b) $x_1(z), \ldots, x_s(z)$ are linearly independent for every $z \in \Omega$; (c) for every $z \in \Omega$ that does not belong to $S(B_1) \cup S(B_2)$ the vectors $x_1(z), \ldots, x_s(z)$ form a basis in $\operatorname{Ker} B_1(z)$ modulo $\operatorname{Ker} B_2(z)$.

An analogous result also holds in case $\operatorname{Im} B_1(z) \supset \operatorname{Im} B_2(z)$, for all $z \in \Omega$ with the possible exception of singular points of $B_1(z)$ and $B_2(z)$.

Proof. We regard $B_1(z)$ and $B_2(z)$ as $m \times n$ matrix functions, with respect to fixed bases in $\mathbb{C}^n$ and $\mathbb{C}^m$. By Lemma 19.3.1, find linearly independent branch analytic vector-valued functions $y_1(z), \ldots, y_v(z)$ on Ω such that

$$\operatorname{Span}\{y_1(z), \ldots, y_v(z)\} = \operatorname{Ker} B_2(z) \tag{19.3.1}$$

for all $z \in \Omega$ not belonging to the singular set of $B_2(z)$. Fix $z_0 \in \Omega$, and choose $x_{v+1}, \ldots, x_n$ in such a way that $y_1(z_0), \ldots, y_v(z_0), x_{v+1}, \ldots, x_n$ form a basis in $\mathbb{C}^n$. Using the branch analytic version of Lemma 18.2.2 (cf. the paragraph following Lemma 19.3.1), find branch analytic vector functions $y_{v+1}(z), \ldots, y_n(z)$, $z \in \Omega$ such that $y_1(z), \ldots, y_v(z)$, $y_{v+1}(z), \ldots, y_n(z)$ form a basis in $\mathbb{C}^n$ for every $z \in \Omega$. If necessary, replace $B_i(z)$ by $B_i(z)S(z)$, $i = 1, 2$, where $S(z) = [y_1(z) \cdots y_n(z)]$ is an invertible $n \times n$ matrix function, and we can assume that

$$B_i(z) = [0_{m \times v} \; \tilde{B}_i(z)], \qquad i = 1, 2$$

where $\tilde{B}_i(z)$ are branch analytic $m \times (n - v)$ matrix functions, and $\operatorname{Ker} \tilde{B}_2(z) = 0$ for all $z \in \Omega$ with the possible exception of a discrete set of points. By Lemma 19.3.1 again, find branch analytic linearly independent

$\mathbb{C}^{n-v}$-valued functions $\tilde{x}_1(z), \ldots, \tilde{x}_s(z)$, $z \in \Omega$ such that $\tilde{x}_1(z), \ldots, \tilde{x}_s(z)$ is a basis in $\operatorname{Ker} \tilde{B}_1(z)$ for all $z \in \Omega$ except for the singular points of $\tilde{B}_1(z)$. Then the vector functions

$$x_j(z) = \begin{bmatrix} 0 \\ \tilde{x}_j(z) \end{bmatrix} \in \mathbb{C}^n, \qquad j = 1, \ldots, s$$

satisfy the requirements of Lemma 19.3.2. $\square$

Lemma 19.3.3

Let $B_1(z)$ and $B_2(z)$ be as in Lemma 19.3.2, and let $x_1(z), \ldots, x_t(z)$ be branch analytic n-dimensional vector functions with the following properties: (a) every branch point of any $x_j(z)$, $j = 1, \ldots, t$ is also a branch point of at least one of $B_1(z)$ and $B_2(z)$; (b) there exists a discrete set $T \supset S(B_1) \cup S(B_2)$ such that $x_1(z), \ldots, x_t(z)$ belong to $\operatorname{Ker} B_1(z)$ and are linearly independent modulo $\operatorname{Ker} B_2(z)$ for every $z \in \Omega \smallsetminus T$. Then there exist branch analytic n-dimensional vector functions $x_{t+1}(z), \ldots, x_s(z)$ such that every point of any $x_j(z)$, $j = t + 1, \ldots, s$ is a branch point of at least one of $B_1(z)$ and $B_2(z)$ and for every $z \in \Omega \smallsetminus T$ the set $x_1(z), \ldots, x_t(z), x_{t+1}(z), \ldots, x_s(z)$ forms a basis in $\operatorname{Ker} B_1(z)$ modulo $\operatorname{Ker} B_2(z)$.

The case $t = 0$ [when the set $x_1(z), \ldots, x_t(z)$ does not appear] is not excluded in Lemma 19.3.3.

Proof. Arguing as in the proof of Lemma 19.3.2, we can assume that $\operatorname{Ker} B_2(z) = 0$ for every $z \notin S(B_2)$. Replacing T by $T \cup S(B_2)$, we can assume that $S(B_2) = \emptyset$.

Further, by the branch analytic version of Lemma 18.2.2, there exist branch analytic and linearly independent vector functions $y_1(z), \ldots, y_t(z)$ with

$$\operatorname{Span}\{x_1(z), \ldots, x_t(z)\} = \operatorname{Span}\{y_1(z), \ldots, y_t(z)\}, \qquad z \in \Omega \smallsetminus T$$

There exist branch analytic vector functions $y_{t+1}(z), \ldots, y_n(z)$ such that $y_1(z), \ldots, y_n(z)$ form a basis in $\mathbb{C}^n$ for every $z \in \Omega$ (cf. the proof of Lemma 19.3.2). By replacing $B_1(z)$ by $B_1(z)[y_1(z) \cdots y_n(z)]$, we can assume that

$$B_1(z) = [0_{n \times t} \; \tilde{B}_1(z)]$$

and the proof is reduced to the case $t = 0$. But then Lemma 19.3.1 is applicable. $\square$

We are ready now to prove Theorem 19.2.3. The main idea is to mimic the proof of the Jordan form for a transformation (Section 2.3) using Lemma 19.3.2 when necessary.

Proof of Theorem 19.2.3. For a fixed j ($j = 1, \ldots, \nu$) let m_{j1} be the maximal positive integer p such that

$$\text{Ker}(\mu_j(z)I - A(z))^p \neq \text{Ker}(\mu_j(z)I - A(z))^{p-1}$$

for all $z \notin S_1 \cup S_2$. By Theorem 19.2.1 and the definition of S_2 the number m_{j1} is well defined. By Lemma 19.3.2, there exist branch analytic vector functions $x^{(j)}_{1,m_{j1}}(z), \ldots, x^{(j)}_{k,m_{j1}}(z)$ on Ω that are linearly independent for every $z \in \Omega$, can have branch points only in S_1, and such that

$$x^{(j)}_{1,m_{j1}}(z), \ldots, x^{(j)}_{k_1,m_{j1}}(z)$$

form a basis in $\text{Ker}(\mu_j(z)I - A(z))^{m_{j1}}$ modulo $\text{Ker}(\mu_j(z)I - A(z))^{m_{j1}-1}$, for every $z \in \Omega$ that does not belong to $S((\mu_j(z)I - A(z))^{m_{j1}}) \cup S((\mu_j(z)I - A(z))^{m_{j1}-1})$. As we have seen in the proof of the Jordan form, the vectors

$$x^{(j)}_{q,m_{j1}-1}(z) \overset{\text{def}}{=} (-\mu_j(z)I + A(z))x^{(j)}_{q,m_{j1}}(z), \qquad q = 1, \ldots, k_1$$

are linearly independent modulo $\text{Ker}(\mu_j(z)I - A(z))^{m_{j1}-2}$ for every $z \notin S_1 \cup S_2$ (we assume here that $m_{j1} \geq 2$). By Lemma 19.3.3, there exist branch analytic vector functions on Ω:

$$x^{(j)}_{k_1+1,m_{j1}-1}(z), \ldots, x^{(j)}_{k_2,m_{j1}-1}(z)$$

with branch points only in S_1 and such that for every $z \notin S_1 \cup S_2$ the vectors

$$x^{(j)}_{1,m_{j1}-1}(z), x^{(j)}_{2,m_{j1}-1}(z), \ldots, x^{(j)}_{k_1,m_{j1}-1}(z), x^{(j)}_{k_1+1,m_{j1}-1}(z), \ldots, x^{(j)}_{k_2,m_{j1}-1}(z)$$

form a basis in $\text{Ker}(\mu_j(z)I - A(z))^{m_{j1}-1}$ modulo $\text{Ker}(\mu_j(z)I - A(z))^{m_{j1}-2}$. Continuing this process as in the proof of the Jordan form, we obtain the vector function (19.2.3) with the desired properties. $\quad\square$

19.4   ANALYTIC EXTENDABILITY OF INVARIANT SUBSPACES

In this section we study the following problem: given an analytic family of transformations $A(z)$ on Ω and an invariant subspace $\mathcal{M}_0$ of $A(z_0)$, when is there a family of subspaces $\mathcal{M}(z)$ that is analytic in some domain $\Omega' \subset \Omega$ with $z_0 \in \Omega'$, and such that $\mathcal{M}(z_0) = \mathcal{M}_0$ and $\mathcal{M}(z)$ is $A(z)$ invariant for all $z \in \Omega'$? (As before, Ω is a domain in $\mathbb{C}$.) If this happens, we say that $\mathcal{M}_0$ is *extendable* to an analytic $A(z)$-invariant family of subspaces on Ω'. The main result in this direction is given in the following theorem.

Theorem 19.4.1

Let $A(z): \mathbb{C}^n \to \mathbb{C}^n$ be an analytic family of transformations on Ω with the first and second exceptional sets S_1 and S_2, respectively. Then, provided $z_0 \in \Omega \smallsetminus (S_2 \cup S_1)$, every $A(z_0)$-invariant subspace $\mathcal{M}_0$ is extendable to an analytic $A(z)$-invariant family of subspaces on $\Omega \smallsetminus S_1$.

Proof. For $j = 1, \ldots, \nu$, let

$$x_{11}^{(j)}(z), \ldots, x_{1,m_{j1}}^{(j)}(z), x_{21}^{(j)}(z), \ldots, x_{2,m_{j2}}^{(j)}(z), \ldots, x_{\gamma_j 1}^{(j)}(z), \ldots, x_{\gamma_j, m_{j\gamma_j}}^{(j)}(z)$$

$$(19.4.1)$$

be n-dimensional vector functions as in Theorem 19.2.3. We consider $A(z)$ and vectors (19.4.1) as an $n \times n$ matrix function and n-dimensional vector functions, respectively, written in the standard orthonormal basis in $\mathbb{C}^n$.

Let $z_0 \in \Omega \smallsetminus (S_2 \cup S_1)$, and let J_0 be the Jordan form of $A(z_0)$:

$$J = \operatorname{diag}[J_1, \ldots, J_\nu]$$

where

$$J_j = \operatorname{diag}[J_{m_{j1}}(\mu_j(z_0)), \ldots, J_{m_{j\gamma_j}}(\mu_j(z_0))]$$

and $J_k(\mu)$ is the $k \times k$ Jordan block with eigenvalue μ. For $z \in \Omega \smallsetminus S_1$ let $T(z)$ be the $n \times n$ matrix whose columns are the vectors (19.4.1) (in this order). Observe that $T(z)$ is analytic on $\Omega \smallsetminus S_1$ with algebraic branch points in S_1 and $T(z)$ is invertible for $z \in \Omega \smallsetminus (S_2 \cup S_1)$ [the function $T(z)$ is analytic but not necessarily invertible at points in S_2]. Then we have $A(z_0)T(z_0) = T(z_0)J$. Given an $A(z_0)$-invariant subspace $\mathcal{M}_0$, and any $z \in \Omega \smallsetminus (S_1 \cup S_2)$, define

$$\mathcal{M}(z) = T(z)T(z_0)^{-1}\mathcal{M}_0$$

Clearly, $\mathcal{M}(z)$ is analytic and $A(z)$ invariant for $z \in \Omega \smallsetminus (S_1 \cup S_2)$, and also $\mathcal{M}(z_0) = \mathcal{M}_0$. We show that $\mathcal{M}(z)$ admits analytic and $A(z)$-invariant continuation into the set S_2. Let $f_1, \ldots, f_k$ be a basis in $\mathcal{M}_0$; then the vectors

$$g_1(z) = T(z)T(z_0)^{-1}f_1, \ldots, g_k(z) = T(z)T(z_0)^{-1}f_k$$

form a basis in $\mathcal{M}(z)$ for every $z \in \Omega \smallsetminus (S_1 \cup S_2)$. Note that $g_1(z), \ldots, g_k(z)$ are analytic in $\Omega \smallsetminus S_1$. By Lemma 18.2.2 there exist n-dimensional vector functions $h_1(z), \ldots, h_k(z)$ that are analytic on $\Omega \smallsetminus S_1$, linearly independent for every $z \in \Omega \smallsetminus S_1$, and for which

$$\mathrm{Span}\{h_1(z), \ldots, h_k(z)\} = \mathrm{Span}\{g_1(z), \ldots, g_k(z)\}$$

whenever $z \notin S_1 \cup S_2$. Putting $\mathcal{M}(z) = \mathrm{Span}\{h_1(z), \ldots, h_k(z)\}$ for $z \in S_2$, we clearly obtain an analytic extension of the analytic family $\{\mathcal{M}(z)\}_{z \in \Omega \smallsetminus (S_1 \cup S_2)}$ to the points in S_2. As for a fixed $z_0 \in S_2$ we have

$$\lim_{z \to z_0} A(z) = A(z_0), \qquad \lim_{z \to z_0} \theta(\mathcal{M}(z), \mathcal{M}(z_0)) = 0$$

it follows in view of Theorem 13.4.2 that $\mathcal{M}(z_0)$ is $A(z_0)$ invariant. $\square$

The proof of Theorem 19.4.1 shows that the analytic $A(z)$-invariant family of subspaces $\mathcal{M}(z)$ on $\Omega \smallsetminus S_1$ with $\mathcal{M}(z_0) = \mathcal{M}_0$, has at most algebraic branch points in S_1, in the following sense. For every $z' \in S_1$, either $\mathcal{M}(z)$ can be analytically continued into z' (i.e., there exists a subspace $\mathcal{M}'$, which is necessarily $A(z')$ invariant, and for which the family of subspaces $\mathcal{N}(z)$, $z \in (\Omega \smallsetminus S_1) \cup \{z'\}$ defined by $\mathcal{N}(z) = \mathcal{M}(z)$ on $\Omega \smallsetminus S_1$, $\mathcal{N}(z') = \mathcal{M}'$ is analytic on $(\Omega \smallsetminus S_1) \cup \{z'\}$), or $\mathcal{M}(z) = S(z)\mathcal{M}_0$ in a neighbourhood of z', where $S(z)$ is an invertible family of transformations that is analytic on a deleted neighbourhood of z' and has an algebraic branch point at z'.

Looking ahead to the applications of the next chapter, we introduce the notion of analytic extendability of chains of invariant subspaces. Let $A(z)\colon \mathbb{C}^n \to \mathbb{C}^n$ be an analytic family of transformations on Ω, and let

$$\Lambda_0 = \{\mathcal{M}_{01} \subset \mathcal{M}_{02} \subset \cdots \subset \mathcal{M}_{0r}\}$$

be a chain of $A(z_0)$-invariant subspaces. We say that Λ_0 is *extendable* to an analytic chain of $A(z)$-invariant subspaces on a set $\Omega' \subset \Omega$ containing z_0 if there exist analytic families of subspaces $\mathcal{M}_{01}(z), \ldots, \mathcal{M}_{0r}(z)$ on Ω' such that $\mathcal{M}_{0j}(z_0) = \mathcal{M}_{0j}$ for $j = 1, \ldots, r$, $\mathcal{M}_{0j}(z) \subset \mathcal{M}_{0k}(z)$ for $j < k$ and $z \in \Omega'$, and $\mathcal{M}_{0j}(z)$ is $A(z)$ invariant for all $z \in \Omega'$. Clearly, this is a generalization of the notion of extendability of a single invariant subspace dealt with in Theorem 19.4.1. The arguments used in the proof of Theorem 19.4.1 also prove the following result on analytic extendability of chains of invariant subspaces.

Theorem 19.4.2

Let $A(z)$, S_1 and S_2 be as in Theorem 19.4.1. Then every chain of $A(z_0)$-invariant subspaces, where $z_0 \in \Omega \smallsetminus (S_2 \cup S_1)$, is extendable to an analytic chain of $A(z)$-invariant subspaces on $\Omega \smallsetminus S_1$. Moreover, the analytic families of subspaces that form this analytic chain have at most algebraic branch points at S_1 (in the sense explained after the proof of Theorem 19.4.1).

Chains consisting of spectral subspaces are important examples of chains of subspaces that are always analytically extendable. Recall that an A-invariant subspace $\mathcal{M}$ is called *spectral* if $\mathcal{M}$ is a sum of root subspaces of A.

Theorem 19.4.3

Let $A(z)$ and S_1 be as in Theorem 19.4.1. Then every chain $\Lambda_0 = \{\mathcal{M}_{01} \subset \cdots \subset \mathcal{M}_{0r}\}$ of spectral subspaces of $A(z_0)$, where $z_0 \in \Omega$, is extendable to an analytic chain of $A(z)$-invariant subspaces on $\Omega \smallsetminus (S_1 \smallsetminus \{z_0\})$ that has at most algebraic branch points at $S_1 \smallsetminus \{z_0\}$.

Proof. For $j = 1, \ldots, r$ write $\mathcal{M}_{0j} = \operatorname{Im} P_{\Gamma_j}(z_0)$, where

$$P_{\Gamma_j}(z_0) = \frac{1}{2\pi i} \int_{\Gamma_j} (\lambda I - A(z_0))^{-1} \, d\lambda$$

is the Riesz projector of $A(z_0)$ corresponding to a suitable simple rectifiable contour Γ_j. We can assume that Γ_j lies in the interior of Γ_k, for $j > k$. Let $\mathcal{U} \subset \Omega$ be a neighbourhood of z_0 that is so small that $A(z)$ has no eigenvalues on $\Gamma_1 \cup \cdots \cup \Gamma_r$, for $z \in \mathcal{U}$. Clearly, for $z \in \mathcal{U}$, we find that

$$\Lambda(z) = \{\mathcal{M}_1(z) \subset \cdots \subset \mathcal{M}_r(z)\}$$

where $\mathcal{M}_j(z) = \operatorname{Im} P_{\Gamma_j}(z)$ form an analytic chain of $A(z)$-invariant subspaces in $\mathcal{U}$. Fix $\tilde{z} \in \mathcal{U} \smallsetminus (S_1 \cup S_2)$, and let $\tilde{\mathcal{M}}_j(z)$ be the analytic $A(z)$-invariant family of subspaces (cf. the proof of Theorem 19.4.1) to which $\mathcal{M}_j(\tilde{z})$ is extendable. It is easily seen that $\tilde{\mathcal{M}}_j(z) = \mathcal{M}_j(z)$ for $z \in \mathcal{U} \smallsetminus (S_1 \cup S_2)$, so Λ_0 admits the desired extension. $\square$

To analyze the extendability of $A(z_0)$-invariant subspaces when $z \in S_2$, we need the following notion. An invariant subspace $\mathcal{M}_0$ of $A(z_0)$, $z_0 \in \Omega$ is called *sequentially isolated* (in Ω) if there is no sequence $z_m \neq z_0$, $m = 1, 2, \ldots$ of points in Ω tending to z_0 such that, for some $A(z_m)$-invariant subspace $\mathcal{M}_m$ $(m = 1, 2, \ldots)$, we have $\lim_{m \to \infty} \theta(\mathcal{M}_m, \mathcal{M}_0) = 0$. Theorem 19.4.1 shows, in particular, that every $A(z_0)$-invariant subspace with $z_0 \in \Omega \smallsetminus (S_1 \cup S_2)$ is sequentially nonisolated. However, certain $A(z_0)$-invariant subspaces with $z_0 \in S_2$ may be sequentially isolated, as follows.

EXAMPLE 19.4.1. Let

$$A(z) = \begin{bmatrix} 0 & z \\ 0 & 0 \end{bmatrix}, \qquad z \in \mathbb{C}$$

Here S_1 is empty, $S_2 = \{0\}$. Any $A(0)$-invariant subspace of the form $\operatorname{Span}\begin{bmatrix} y \\ 1 \end{bmatrix}$, where y is a complex number, is sequentially isolated. On the other hand, the $A(0)$-invariant subspace $\operatorname{Span}\begin{bmatrix} 1 \\ 0 \end{bmatrix}$ is sequentially nonisolated. $\square$

Clearly, a sequentially isolated $A(z_0)$-invariant subspace is not extendable to an analytic $A(z)$-invariant family of subspaces on a neighbourhood of z_0.

We conjecture that these are the only nonextendable invariant subspaces.

Conjecture 19.4.4

Let $A(z)$, S_1, and S_2 be as in Theorem 19.4.1. Then every sequentially nonisolated $A(z_0)$-invariant subspace $\mathcal{M}_0$, where $z_0 \in S_2$, is extendable to an analytic $A(z)$-invariant family of subspaces on $\Omega \smallsetminus S_1$ that has at most algebraic branch points in S_1 (in the same sense as the remark following the proof of Theorem 19.4.1).

Theorem 19.4.3 verifies this conjecture in case $\mathcal{M}_0$ is a spectral subspace.

19.5 ANALYTIC MATRIX FUNCTIONS OF A REAL VARIABLE

The results of Sections 19.1–19.4 hold also for $n \times n$ matrix functions $A(t)$ that are analytic in the real variable t on an open interval Ω on the real line. Of particular interest is the case when all eigenvalues of $A(t)$ are real, as follows.

Theorem 19.5.1

Let $A(t)$ be an $n \times n$ matrix function that is analytic in the real variable t on Ω. Assume that, for all $t \in \Omega$, all eigenvalues of $A(t)$ are real. Then the eigenvalues of $A(t)$ are also analytic functions of t on Ω.

Proof. Let $t_0 \in \Omega$. By Theorem 19.1.1, all eigenvalues of $A(t)$, for t in a neighbourhood of t_0, are given by fractional power series of the form

$$\lambda(t) = \lambda_0 + \sum_{j=1}^{\infty} c_j[(t-t_0)^{1/\alpha}]^j$$

where c_j are complex numbers. Let j_1 be the first index such that $c_{j_1} \neq 0$. [If all c_j are zeros, then $\lambda(t) \equiv \lambda_0$ is obviously analytic at t_0.] Then

$$c_{j_1} = \lim_{t \to t_0} \frac{\lambda(t) - \lambda_0}{[(t-t_0)^{1/\alpha}]^{j_1}} \tag{19.5.1}$$

Take $t > t_0$ and $(t-t_0)^{1/\alpha}$ positive. Since $\lambda(t)$ and λ_0 are real, we find that c_{j_1} must be real. In (19.5.1) we now take $t < t_0$ and $(t-t_0)^{1/\alpha} = |t - t_0|^{1/\alpha} \cdot (\cos(2\pi j/\alpha) + i \sin(2\pi j/\alpha))$. We obtain a contradiction with the fact that c_{j_1} is real unless j_1 is a multiple of α. If $j_2 > j_1$ is the minimal integer with $c_{j_2} \neq 0$, then

$$c_{j_2} = \lim_{t \to t_0} \frac{\lambda(t) - \lambda_0 - c_{j_1}(t - t_0)^{j_1/\alpha}}{[(t - t_0)^{1/\alpha}]^{j_2}}$$

and the preceding argument shows that c_{j_2} is real and j_2 is a multiple of α. Continue in this way to conclude that $\lambda(t)$ is analytic in a neighbourhood of t_0. As t_0 was arbitrary in Ω, the analyticity of $\lambda(t)$ on Ω follows. $\quad\square$

Combining this result with Theorems 19.2.3 and 19.4.1, we have the following corollary.

Corollary 19.5.2

Let $A(t)$ be an analytic $n \times n$ matrix function of a real variable t on Ω, and assume that all eigenvalues of $A(t)$ are real when $t \in \Omega$. Let S_2 be the discrete set of points in Ω defined by the property that either

$$\nu(t_0) < \max_{t \in \Omega} \nu(t), \qquad t_0 \in S_2$$

where $\nu(t)$ is the number of distinct eigenvalues of $A(t)$, or

$$\nu(t_0) = \max_{t \in \Omega} \nu(t), \qquad t_0 \in S_2$$

but at least for one analytic eigenvalue $\mu_j(t)$ of $A(t)$ the partial multiplicities of $\mu_j(t_0)$ are different from the partial multiplicities of $\mu_j(t)$, $t \neq t_0$ in a real neighbourhood of t_0. Then there exist analytic n-dimensional vector functions

$$x_{11}(t), \ldots, x_{1m_1}(t); \ldots; x_{r1}(t), \ldots, x_{rm_r}(t) \qquad (19.5.2)$$

on Ω such that for every $t \in \Omega \smallsetminus S_2$ the vectors (19.5.2) form a basis in $\mathbb{C}^n$ and, for $j = 1, \ldots, r$, $x_{j1}(t), \ldots, x_{jm_j}(t)$ is a Jordan chain of the transformation $A(t)$. Moreover, when $t_0 \in \Omega \smallsetminus S_2$, every $A(t_0)$-invariant subspace $\mathcal{M}_0$ is extendable to an analytic $A(t)$-invariant family of subspaces on Ω.

In particular, the conclusions of Corollary 19.5.2 hold for an analytic $n \times n$ matrix function $A(t)$ of the real variable $t \in \Omega$ that is diagonable and all eigenvalues of which are real for every $t \in \Omega$. These properties are satisfied, for example, if $A(t)$ is an analytic matrix function on Ω that is hermitian for all $t \in \Omega$.

19.6   EXERCISES

19.1   Find the first and second exceptional sets for the following analytic families of transformations:

(a)
$$A(z) = \begin{bmatrix} 1 & z \\ 1 & 0 \end{bmatrix}$$

(b)
$$A(z) = \begin{bmatrix} 0 & 1 & 0 \\ 0 & 0 & 1 \\ z^4 - 3z^2 & 3z & z - z^2 \end{bmatrix}$$

19.2   In Exercise 19.1 (a) find a basis in $\mathbb{C}^2$ that is analytic on $\mathbb{C}$ (with the possible exception of branch points) and consists of eigenvectors of $A(z)$ (with the possible exception of a discrete set of values of z).

19.3   Describe the first and second exceptional sets for the following types of analytic families of transformations $A(z)$: $\mathbb{C}^n \to \mathbb{C}^n$ on Ω:

(a) $A(z) = \text{diag}[a_1(z), \ldots, a_n(z)]$ is a diagonal matrix.
(b) $A(z)$ is a circulant matrix (with respect to a fixed basis in $\mathbb{C}^n$) for every $z \in \Omega$.
(c) $A(z)$ is an upper triangular Toeplitz matrix for every $z \in \Omega$.
(d)   For every $z \in \Omega$, all the entries in $A(z)$, with the possible exception of the entries (i, j) with $i = j$ or with $i + j = n + 1$, are zeros.

19.4   Show that the analytic matrix function of type $\alpha(z)I + \beta(z)A$, where $\alpha(z)$ and $\beta(z)$ are scalar analytic functions and A is a fixed $n \times n$ matrix, has all eigenvalues analytic.

19.5   Show that if $A(z) = \alpha(z)I + \beta(z)A$ is the function of Exercise 19.4 and $\beta(z)$ is a polynomial of degree l, then the second exceptional set of $A(z)$ contains not more than l points.

19.6   Prove that the number of exceptional points of a polynomial family of transformations $\sum_{j=0}^{k} z^j A_j$, $z \in \mathbb{C}$, is always finite. [*Hint*: Use the approach based on the resultant matrix (Section 19.2).]

19.7   Let $A(z)$ be an analytic $n \times n$ matrix function defined on Ω whose values are circulant matrices. When is every $A(z_0)$-invariant subspace analytically extendable for every $z_0 \in \Omega$?

19.8   Describe the analytically extendable $A(z_0)$-invariant subspaces, where $A(z)$ is an analytic $n \times n$ matrix function on Ω with upper traingular Toeplitz values, and $z_0 \in \Omega$.

19.9   Let $A(z)$: $\mathbb{C}^n \to \mathbb{C}^n$ be an analytic family of transformations defined on Ω, and assume that $A(z_0)$ is nonderogatory for some $z_0 \in \Omega$. Prove that every $A(z_0)$-invariant subspace is sequentially nonisolated. (*Hint*: Use Theorem 15.2.3.)

19.10 Let $A(z)$ be an analytic $n \times n$ matrix function of the real variable $z \in \Omega$, where Ω is an open interval on the real line, such that $A(z)$ is hermitian for every $z \in \Omega$. Prove that there exist analytic families $x_1(z), \ldots, x_n(z)$ of n-dimensional vectors on Ω such that for every $z_0 \in \Omega$ the vectors $x_1(z_0), \ldots, x_n(z_0)$ form an orthonormal basis of eigenvectors of $A(z_0)$. [*Hint*: Let $\lambda_0(z)$ be an eigenvalue of $A(z)$ that is analytic on Ω (one exists by Theorem 19.5.1). Choose an analytic vector function $x_1(z) \in \operatorname{Ker}(A(z) - \lambda_0(z)I)$ on Ω with $\|x(z)\| = 1$. Repeat this argument for the restriction of $A(z)$ to $\operatorname{Span}\{x_1(z)\}^{\perp}$—recall that $\operatorname{Span}\{x_1(z)\}^{\perp}$ is an analytic family of subspaces on Ω—and so on.]

19.11 Let A and B be hermitian $n \times n$ matrices, and assume that A has n distinct eigenvalues $\lambda_1, \lambda_2, \ldots, \lambda_n$. Show that in the power series

$$\lambda_k(z) = \lambda_k + \sum_{j=1}^{\infty} z^j \lambda_k^{(j)}$$

and

$$f_k(z) = f_k + \sum_{j=1}^{\infty} z^j f_k^{(j)}$$

representing the eigenvalue $\lambda_k(z)$ of $A + zB$ and the corresponding eigenvector $f_k(z)$ of $A + zB$, for z sufficiently close to zero, we have

$$\lambda_k^{(1)} = (Bf_k, f_k), \quad f_k^{(1)} = \sum_{i \neq k}^{n} \frac{(Bf_k, f_i)}{\lambda_k - \lambda_i} f_i + \alpha_k f_k$$

where α_k are pure imaginary numbers. It is assumed that $\|f_k(z)\| = 1$ for real z sufficiently close to zero. [*Hint*: By Exercise 19.10, the eigenvalue $\lambda_k(z)$ and the corresponding eigenvector $f_k(z)$ are analytic functions of z. Show that the equality

$$Af_k^{(1)} + Bf_k = \lambda_k f_k^{(1)} + \lambda_k^{(1)} f_k \tag{1}$$

holds. Find $\lambda_k^{(1)}$ by taking the scalar product of (1) with f_k. By taking the scalar product of (1) with f_i ($i \neq k$) it is found that

$$(f_k^{(1)}, f_i) = \frac{(Bf_k, f_i)}{\lambda_k - \lambda_i} \quad (i \neq k)$$

The condition $\|f_k(z)\| = 1$ gives $(f_k^{(1)}, f_k) + (f_k, f_k^{(1)}) = 0$.]

Chapter Twenty

Applications

This chapter contains applications of the results of the previous two chapters. These applications are concerned with problems of factorizations of monic matrix polynomials and rational matrix functions depending analytically on a parameter. The main problem is the analysis of analytic properties of divisors. Solutions of a matrix quadratic equation with coefficients depending analytically on a parameter are also analyzed.

20.1 FACTORIZATION OF MONIC MATRIX POLYNOMIALS

Consider a monic matrix polynomial $L(\lambda) = I\lambda^l + \Sigma_{j=0}^{l-1} A_j \lambda^j$, where $A_0, \ldots, A_{l-1}$ are $n \times n$ matrices that depend analytically on the parameter z for $z \in \Omega$, and Ω is a domain in the complex plane. We write $A_j = A_j(z)$ and $L(\lambda) = L(\lambda, z)$. In this section we study the behaviour of factorizations $L(\lambda, z) = L_1(\lambda, z) \cdots L_r(\lambda, z)$ of $L(\lambda, z)$ as functions of z. Our attention is focused on the problem of analytic extension of factorizations from a given $z_0 \in \Omega$.

Let

$$
C(z) = \begin{bmatrix}
0 & I & 0 & \cdots & 0 \\
0 & 0 & I & \cdots & 0 \\
\vdots & \vdots & \vdots & & \vdots \\
-A_0(z) & -A_1(z) & -A_2(z) & \cdots & -A_{l-1}(z)
\end{bmatrix}
$$

be the companion matrix of $L(\lambda, z)$. Obviously, $C(z)$ is an analytic $n \times n$ matrix function on Ω. The first (resp. second) exceptional set of $C(z)$ is called the first (resp. second) exceptional set of $L(\lambda, z)$. In other words (see Chapter 19), $z_0 \in \Omega$ belongs to the first exceptional set S_1 of $L(\lambda, z)$ if and only if not all solutions of $\det L(\lambda, z) = 0$ (as functions of z) are analytic at z_0. The point z_0 belongs to the second exceptional set S_2 of $L(\lambda, z)$ if and only if all solutions of det $L(\lambda, z) = 0$ are analytic in a neighbourhood of z_0 and, denoting by $\lambda_1(z), \ldots, \lambda_r(z)$ all the different analytic functions in a neighbourhood of z_0 satisfying det $L(\lambda_j(z), z) = 0$, $j = 1, \ldots, r$, we have either (a) $\lambda_j(z_0) = \lambda_k(z_0)$ for some $j \neq k$ or (b) all the numbers $\lambda_1(z_0), \ldots, \lambda_r(z_0)$ are different, but for at least one $\lambda_j(z)$ the partial multiplicities of $L(\lambda, z)$ at $\lambda_j(z)$ are not the same when $z = z_0$ and when $z \neq z_0$ (and z is sufficiently close to z_0).

Now we state the main result on analytic extendability of factorizations of $L(\lambda, z)$.

Theorem 20.1.1

Let $z_0 \in \Omega \smallsetminus (S_1 \cup S_2)$ and

$$L(\lambda, z_0) = L_1(\lambda) \cdots L_r(\lambda) \tag{20.1.1}$$

where $L_j(\lambda)$, $j = 1, \ldots, r$ are monic matrix polynomials and S_1 (resp. S_2) is the first (resp. second) exceptional set of $L(\lambda, z)$. Then there exist monic matrix polynomials $L_1(\lambda, z), \ldots, L_r(\lambda, z)$ whose coefficients are analytic functions on $\Omega \smallsetminus (S_1 \cup S)$ (where S is some discrete subset of $\Omega \smallsetminus \{z_0\}$), having at most poles in S, and at most algebraic branch points in S_1, and such that

$$L(\lambda, z) = L_1(\lambda, z) \cdots L_r(\lambda, z)$$

for $z \in \Omega \smallsetminus S$, and $L_j(\lambda, z_0) = L_j(\lambda)$ for $j = 1, \ldots, r$.

Note that the case when $S_1 \cap S \neq \emptyset$ is not excluded. This means that the coefficients $A_{jk}(z)$ of $L_j(\lambda, z)$ may have an algebraic branch point and a pole at the same point z' simultaneously, that is, there is a power series representation of type

$$A_{jk}(z) = \sum_{l=-p}^{\infty} B_l(z - z')^{l/q}$$

in a deleted neighbourhood of z', where p and q are positive integers.

Proof. We use the description of factorizations of monic matrix polynomials in terms of invariant subspaces developed in Chapter 5. Let

$$X = \begin{bmatrix} I & 0 & \cdots & 0 \end{bmatrix}, \qquad C(z), \qquad Y = \begin{bmatrix} 0 \\ \vdots \\ 0 \\ I \end{bmatrix} \tag{20.1.2}$$

be a standard triple for $L(\lambda)$, and let

$$\mathcal{M}_1 \subset \cdots \subset \mathcal{M}_{r-1} \tag{20.1.3}$$

be the chain of $C(z_0)$-invariant subspaces corresponding to the factorization (20.1.1) [with respect to the triple (20.1.2)]. In particular, for $j = 1, \ldots, r-1$, the transformations

$$\begin{bmatrix} X|_{\mathcal{M}_j} \\ X|_{\mathcal{M}_j} C(z_0')|_{\mathcal{M}_j} \\ \vdots \\ X|_{\mathcal{M}_j} [(C(z_0))|_{\mathcal{M}_j}]^{p_j-1} \end{bmatrix} : \mathcal{M}_j \to \mathbb{C}^{np_j}$$

are invertible, where p_j is the sum of degrees of the matrix polynomials $L_{r-j+1}(\lambda), \ldots, L_r(\lambda)$. By Theorem 19.4.2 the chain (20.1.3) is extendable to a chain $\mathcal{M}_1(z) \subset \cdots \subset \mathcal{M}_{r-1}(z)$ of $C(z)$-invariant subspaces that is analytic in $\Omega \smallsetminus S_1$ and has at most algebraic branch points in S_1. Let $S = S^{(1)} \cup \cdots \cup S^{(r-1)}$, where $S^{(j)}$ is the discrete set of all $z \in \Omega$ for which the transformation

$$\begin{bmatrix} X|_{\mathcal{M}_j(z)} \\ X|_{\mathcal{M}_j(z)} C(z)|_{\mathcal{M}_j(z)} \\ \vdots \\ X|_{\mathcal{M}_j(z)} [(C(z))|_{\mathcal{M}_j(z)}]^{p_j-1} \end{bmatrix} : \mathcal{M}_j(z) \to \mathbb{C}^{np_j}$$

is not invertible. For $z \in \Omega \smallsetminus S$, let

$$L(\lambda, z) = L_1(\lambda, z) \cdots L_r(\lambda, z)$$

be the factorization of $L(\lambda, z)$ that corresponds to the chain $\mathcal{M}_1(z) \subset \cdots \subset \mathcal{M}_{r-1}(z)$ of $C(z)$-invariant subspaces [with respect to the triple (20.1.2)]. Formulas (5.6.3) and (5.6.5) show that the coefficients of $L_j(\lambda, z)$ have all the desired properties. $\square$

In the same way (using Theorem 19.4.3 in place of Theorem 19.4.2) one proves the analytic extendability of spectral factorization, as follows.

Theorem 20.1.2

Let $z_0 \in \Omega$ and

$$L(\lambda, z_0) = L_1(\lambda) \cdots L_r(\lambda)$$

where $\sigma(L_j) \cap \sigma(L_k) = \emptyset$ for $j \neq k$. Then there exist monic matrix polynomials $L_1(\lambda, z), \ldots, L_r(\lambda, z)$ with the same properties as in Theorem 20.1.1, and whose coefficients are, in addition, analytic at z_0.

We say that a factorization

$$L(\lambda, z_0) = L_1(\lambda) \cdots L_r(\lambda), \qquad z_0 \in \Omega \qquad (20.1.4)$$

of monic matrix polynomials $L_j(\lambda) = I\lambda^{l_j} + \Sigma_{k=0}^{l_j-1} A_{jk}\lambda^k$, $j = 1, \ldots, r$ is *sequentially nonisolated* if there is a sequence of points $\{z_m\}_{m=1}^\infty$ in $\Omega \smallsetminus \{z_0\}$ such that $\lim_{m\to\infty} z_m = z_0$ and a sequence of factorizations

$$L(\lambda, z_m) = L_1^{(m)}(\lambda) \cdots L_r^{(m)}(\lambda), \qquad m = 1, 2, \ldots$$

where

$$L_j^{(m)}(\lambda) = I\lambda^{l_j} + \sum_{k=0}^{l_j-1} A_{jk}^{(m)}\lambda^k, \qquad j = 1, \ldots, r$$

with $\lim_{m\to\infty} A_{jk}^{(m)} = A_{jk}$ for $k = 0, \ldots, l-1$ and $j = 1, \ldots, r$. Theorem 20.1.1 shows, in particular, that every factorization (20.1.4) with $z_0 \notin S_1 \cup S_2$ is sequentially nonisolated. Simple examples show that sequentially isolated factorizations do exist, for instance:

EXAMPLE 20.1.1. Let $C(z)$ be any matrix depending analytically on z in a domain Ω with the property that for $z = z_0 \in \Omega$, $C(z_0)$ has a square root and for $z \neq z_0$, z in a neighbourhood of z_0, $C(z)$ has no square root. The prime example here is

$$z_0 = 0, \qquad C(z) = \begin{bmatrix} 0 & 0 \\ z & 0 \end{bmatrix}$$

Then define $L(\lambda, z) = I\lambda^2 - C(z)$. It is easily seen that if $L(\lambda, z)$ has a right divisor $I\lambda - A(z)$, then $L(\lambda, z) = I\lambda^2 - A^2(z)$ and hence that $L(\lambda, z)$ has a monic right divisor if and only if $C(z)$ has a square root. Thus, under the hypotheses stated, $L(\lambda, z)$ has an isolated divisor at z_0. $\square$

It is an open question whether every sequentially nonisolated factorization $L(\lambda, z_0) = L_1(\lambda) \cdots L_r(\lambda)$ of monic matrix polynomials with z_0 belonging to the second exceptional set S_2 of $L(\lambda, z)$ is analytically extendable in the sense of Theorem 20.1.1. (It is clear that the sequential nonisolatedness is a necessary condition for the analytic extendability.) A proof of Conjecture 19.4.4 will answer this question in the affirmative.

20.2 RATIONAL MATRIX FUNCTIONS DEPENDING ANALYTICALLY ON A PARAMETER

In this section we study the realizations and exceptional points of rational matrix functions that depend analytically on a parameter. This will serve as

a background for the study of analytic extendability of mimimal factorizations of such functions to be dealt with in the next section.

Let $W(\lambda, z) = [w_{ij}(\lambda, z)]_{i,j=1}^{n}$ be a rational $n \times n$ matrix function that
depends analytically on the parameter z for $z \in \Omega$, where Ω is a domain in
$\mathbb{C}$. That is, each entry $w_{ij}(\lambda, z)$ is a function of type $p_{ij}(\lambda, z)/q_{ij}(\lambda, z)$,
where $p_{ij}(\lambda, z)$ and $q_{ij}(\lambda, z)$ are (scalar) polynomials in λ whose coefficients
are analytic functions of z on Ω. We assume that:

(a)   For each i and j and for all $z \in \Omega$, the polynomial $q_{ij}(\lambda, z)$ in λ is not
      identically zero, so the rational matrix function $W(\lambda, z)$ is well
      defined for every $z \in \Omega$.

(b)   It is convenient to make the further assumption, namely, that for
      each pair of indices i, j $(1 \le i, j \le n)$ there exists a $z_0 \in \Omega$ such that
      the leading coefficient of $q_{ij}(\lambda, z)$ is nonzero at $z = z_0$ and the
      polynomials $p_{ij}(\lambda, z_0)$ and $q_{ij}(\lambda, z_0)$ are coprime, that is, have no
      common zeros. In particular, this assumption rules out the case
      when $p_{ij}(\lambda, z)$ and $q_{ij}(\lambda, z)$ have a nontrivial common divisor whose
      coefficients depend analytically on z for $z \in \Omega$.

(c)   Finally, we assume that for every $z \in \Omega$ the rational matrix function
      $W(\lambda, z)$ (as a function of λ) is analytic at infinity and $W(\infty, z) = I$.

Assumptions (a), (b), and (c) are maintained throughout this section.

It can happen that $W(\lambda, z)$ has zeros and poles tending to infinity when z
tends to a certain point $z_0 \in \Omega$. This is illustrated in the next example.

EXAMPLE 20.2.1.   Let

$$W(\lambda, z) = \frac{1 + \lambda z}{z + 1 + \lambda z}, \qquad z \in \mathbb{C}$$

Obviously, $W(\lambda, z)$ satisfies conditions (a), (b), and (c). Specifically,
$W(\lambda, z)$ depends analytically on z for $z \in \mathbb{C}$, $W(\infty, z) = 1$ for all $z \in \mathbb{C}$, and
the polynomials $1 + \lambda z$ and $z + 1 + \lambda z$ have no common zeros for $z = 1$.
However, $W(\lambda, z)$ has a zero at $\lambda = -z^{-1}$ and a pole at $\lambda = -(z + 1)z^{-1}$,
and both tend to infinity as $z \to 0$. $\square$

A convenient criterion for boundedness of zeros and poles in a neighbourhood of each point in Ω can be given in terms of the entries of $W(\lambda, z)$,
as follows.

Proposition 20.2.1

*The poles and zeros of $W(\lambda, z)$ are bounded in a neighbourhood of each
point in Ω if and only if, for each entry $p_{ij}(\lambda, z)/q_{ij}(\lambda, z)$ of $W(\lambda, z)$, the
leading coefficient of the polynomial $q_{ij}(\lambda, z)$ has no zeros in Ω (as an
analytic function of z on Ω).*

Proof. Assume that the leading coefficient of each $q_{ij}(\lambda, z)$ has no zeros in Ω. Fix $z_0 \in \Omega$. Write $q_{ij}(\lambda, z) = \sum_{k=0}^{s} q_{ijk}(z)\lambda^k$, and, in general, s depends on i and j. As $q_{ijs}(z_0) \neq 0$, the zeros of $q_{ij}(\lambda, z)$ are bounded in a neighbourhood of z_0. Indeed, writing $r_k(z) = q_{ijk}(z)/q_{ijs}(z)$ for $k = 0, \ldots, s-1$, the zeros of the polynomial

$$\lambda^s + \sum_{k=0}^{s-1} r_k(z)\lambda^k, \qquad z \in \mathcal{U}$$

are all in the disc $|\lambda| \leq 1 + \max_{z \in \mathcal{U}} (|r_0(z)|, \ldots, |r_{s-1}(z)|)$, where $\mathcal{U}$ is a suitably chosen neighbourhood of z_0. As the poles of $W(\lambda, z)$ must also be zeros of at least one of the polynomials $q_{ij}(\lambda, z)$, $i, j = 1, \ldots, n$, it follows that there exists an $M > 0$ such that the poles of $W(\lambda, z)$ are all in the disc $|\lambda| \leq M$ for every $z \in \mathcal{U}$. Arguing by contradiction, assume that the zeros of $W(\lambda, z)$ are not bounded in any neighbourhood of z_0. So there exist sequences $\{z_m\}_{m=1}^{\infty}$ and $\{\lambda_m\}_{m=1}^{\infty}$ such that $z_m \to z_0$, $z_m \in \mathcal{U} \smallsetminus \{z_0\}$, $\lambda_m \to \infty$, $|\lambda_m| > M$ and λ_m is a zero of $W(z_m)$. Then $W(\lambda_m, z_m)x_m = 0$ for some vector $x_m \in \mathbb{C}^n$ of norm 1. [Here we use the fact that λ_m is not a pole of $W(\lambda, z_m)$.] Passing to a subsequence, if necessary, we can assume that $x_m \to x_0$ for some $x_0 \in \mathbb{C}^n$, $\|x_0\| = 1$. Using also the fact that $W(\infty, z) = I$ for all $z \in \mathcal{U}$, it follows that $W(\lambda, z)$ is continuous on the set $(\lambda, z) \in (\{\lambda \in \mathbb{C} \,|\, |\lambda| > M\} \cup \{\infty\}) \times \mathcal{U}$. A quick way to verify this is by using a general result that says that if a function $f(z_1, \ldots, z_m)$ of complex variables $z_1, \ldots, z_m$ defined on $V_1 \times \cdots \times V_m$ where each V_j is a domain in $\mathbb{C}$, is analytic in each variable separately (when all other variables are fixed), then $f(z_1, \ldots, z_m)$ is analytic (in particular, continuous) on $V_1 \times \cdots \times V_m$. For the proof of this result, see, for example, Bochner and Martin (1948). Now the continuity of $W(\lambda, z)$ implies $W(\infty, z_0)x_0 = 0$, a contradiction with the fact that $W(\infty, z_0) = I$.

Conversely, let $z_0 \in \Omega$ be a zero of the leading coefficient of some $q_{ij}(\lambda, z)$. Then there is a zero $\lambda_0 = \lambda_0(z)$ of the polynomial $q_{ij}(\lambda, z)$ such that $\lambda_0(z)$ tends to infinity as z tends to z_0. As $\lambda_0(z)$ is a pole of $W(\lambda, z)$ provided $p_{ij}(\lambda_0(z), z) \neq 0$, we have only to show that $\lambda_0(z)$ is not a zero of $p_{ij}(\lambda, z)$ for $z \neq z_0$ sufficiently close to z_0.

To this end, use the existence of a point $z_1 \in \Omega$ such that $q_{ijs}(z_1) \neq 0$ and the polynomials $p_{ij}(\lambda, z_1)$ and $q_{ij}(\lambda, z_1)$ are coprime. The coprimeness of $p_{ij}(\lambda, z) = \sum_{j=0}^{t} p_{ij}(z)\lambda^j$ and $q_{ij}(\lambda, z)$ is equivalent to the invertibility of the $(s + t) \times (s + t)$ resultant matrix

$$R_{ij}(z) = \begin{bmatrix} q_{ij0}(z) & q_{ij1}(z) & \cdots & q_{ijs}(z) & 0 & 0 & \cdots & 0 \\ 0 & q_{ij0}(z) & \cdots & q_{ijs-1}(z) & q_{ijs}(z) & 0 & \cdots & 0 \\ \vdots & \vdots & & \ddots & & \ddots & & \vdots \\ 0 & 0 & \cdots & & q_{ij0}(z) & q_{ij1}(z) & \cdots & q_{ijs}(z) \\ p_{ij0}(z) & p_{ij1}(z) & \cdots & p_{ijt}(z) & 0 & 0 & \cdots & 0 \\ 0 & p_{ij0}(z) & \cdots & p_{ijt-1}(z) & p_{ijt}(z) & 0 & \cdots & 0 \\ \vdots & \vdots & & \ddots & & \ddots & & \vdots \\ 0 & 0 & \cdots & & p_{ij0}(z) & p_{ij1}(z) & \cdots & p_{ijt}(z) \end{bmatrix}$$

as long as $q_{ijs}(z) \neq 0$ [e.g., see Uspensky (1978)]. So $\det R(z_1) \neq 0$, and since $\det R(z)$ is an analytic function of z on Ω, it follows that $\det R(z) \neq 0$ for all $z \neq z_0$ sufficiently close to z_0. Hence indeed, $p_{ij}(\lambda_0(z), z) \neq 0$ for $z \neq z_0$ in some neighbourhood of z_0. $\square$

It turns out that the boundedness of the poles and zeros of $W(\lambda, z)$ is precisely the condition needed for existence of an analytic minimal realization in the following sense.

Theorem 20.2.2

Let $W(\lambda, z)$ be a rational $n \times n$ matrix function that depends analytically on the parameter $z \in \Omega$ and satisfies assumptions (a), (b), and (c). Let the zeros and poles of $W(\lambda, z)$ be bounded in a neighbourhood of every point in Ω. Then there exist analytic matrix functions on Ω, $A(z)$, $B(z)$, and $C(z)$ of sizes $m \times m$, $m \times n$, and $n \times m$, respectively, such that

$$W(\lambda, z) = I + C(z)(\lambda I - A(z))^{-1}B(z), \qquad z \in \Omega \qquad (20.2.1)$$

and for every $z \in \Omega$, with the possible exception of a discrete set S, the realization (20.2.1) is minimal.

Conversely, if (20.2.1) holds for some matrix functions $A(z)$, $B(z)$, and $C(z)$ of appropriate sizes that are analytic on Ω, then the zeros and poles of $W(\lambda, z)$ are bounded in a neighbourhood of every point in Ω.

Proof. By Theorem 7.1.2, for every $z \in \Omega$ there exists a realization

$$W(\lambda, z) = I + C_0(z)(\lambda I - A_0(z))^{-1}B_0(z)$$

for some matrices $C_0(z)$, $A_0(z)$, and $B_0(z)$. Further, by Proposition 20.2.1, the leading coefficients of the denominators of the entries in $W(\lambda, z)$ have no zeros in Ω. According to this fact, the proof of Theorem 7.1.2 shows that $A_0(z)$, $B_0(z)$, and $C_0(z)$ can be chosen to be analytic matrix functions of z on Ω. Let $p \times p$ be the size of $A_0(z)$. By Theorem 18.2.1 we can find families of subspaces of $\mathbb{C}^p$, $\mathcal{K}(z)$, and $\mathcal{J}(z)$, which are analytic on Ω and are such that, for every $z \in \Omega$ with the possible exception of a discrete set S_1, we have

$$\mathcal{K}(z) = \bigcap_{i=0}^{p-1} \mathrm{Ker}(C_0(z)(A_0(z))^i) = \mathrm{Ker} \begin{bmatrix} C_0(z) \\ C_0(z)A_0(z) \\ \vdots \\ C_0(z)(A_0(z))^{p-1} \end{bmatrix}$$

and

$$\mathcal{J}(z) = \sum_{i=0}^{p-1} \text{Im}((A_0(z))^i B_0(z))$$

$$= \text{Im}[B_0(z), A_0(z)B_0(z), \dots, (A_0(z))^{p-1}B_0(z)]$$

For $z \in S_1$ we have

$$\mathcal{K}(z) \subset \bigcap_{i=0}^{p-1} \text{Ker}(C_0(z)(A_0(z))^i)$$

and

$$\mathcal{J}(z) \supset \sum_{i=0}^{p-1} \text{Im}((A_0(z))^i B_0(z))$$

By Theorem 18.3.2, when $z \in \Omega$ we may write

$$\mathcal{K}(z) = \text{Im } P(z) = \text{Ker}(I - P(z))$$
$$\mathcal{J}(z) = \text{Ker}(I - Q(z))$$

where $P(z)$: $\mathbb{C}^p \to \mathbb{C}^p$ and $Q(z)$: $\mathbb{C}^p \to \mathbb{C}^p$ are analytic families of projectors on Ω. Using the same Theorem 18.2.1, we find an analytic family of subspaces $\mathcal{L}(z)$ on Ω such that

$$\mathcal{L}(z) = \mathcal{K}(z) \cap \mathcal{J}(z) = \text{Ker}\begin{bmatrix} I - P(z) \\ I - Q(z) \end{bmatrix}$$

for every $z \in \Omega$ except possibly for a discrete set $S_2 \subset \Omega$. For each $z \in S_2$ we have

$$\mathcal{L}(z) \subset \mathcal{K}(z) \cap \mathcal{J}(z)$$

In view of Theorem 18.3.2 there exists an analytic family of subspaces $\mathcal{N}(z)$ on Ω such that

$$\mathbb{C}^p = \mathcal{J}(z) \dotplus \mathcal{N}(z)$$

for all $z \in \Omega$. Also, Lemma 19.3.2, with

$$B_2(z) = \begin{bmatrix} I - P(z) \\ I - Q(z) \end{bmatrix} \quad \text{and} \quad B_1(z) = \begin{bmatrix} I - P(z) \\ I \end{bmatrix}$$

ensures the existence of an analytic family of subspaces $\mathcal{M}(z)$ on Ω such that

$$\mathcal{J}(z) = \mathcal{L}(z) \dotplus \mathcal{M}(z)$$

for all $z \in \Omega$. Let

$$A(z) = P_{\mathcal{M}(z)} A_0 |_{\mathcal{M}(z)} \colon \mathcal{M}(z) \to \mathcal{M}(z)$$

$$B(z) = P_{\mathcal{M}(z)} B_0(z) \colon \mathbb{C}^n \to \mathcal{M}(z)$$

$$C(z) = C_0(z) |_{\mathcal{M}(z)} \colon \mathcal{M}(z) \to \mathbb{C}^n$$

where $P_{\mathcal{M}(z)}$ is the projector on $\mathcal{M}(z)$ along $\mathcal{L}(z) \dotplus \mathcal{N}(z)$. We regard $A(z)$, $B(z)$, and $C(z)$ as matrices with respect to a fixed basis $x_1(z), \ldots, x_m(z)$ in $\mathcal{M}(z)$ such that $x_j(z)$ are analytic functions on Ω (such a basis exists in view of Theorem 18.3.2). It is easily seen that $A(z)$, $B(z)$, and $C(z)$ are analytic on Ω. The proof of Theorem 6.1.3, together with Theorem 6.1.5, shows that

$$W(\lambda, z) = I + C(z)(\lambda I - A(z))^{-1} B(z) \qquad (20.2.2)$$

for every $z \in \Omega \smallsetminus (S_1 \cup S_2)$, and that (20.2.2) is a minimal realization for $W(\lambda, z)$ when $z \in S_1 \cup S_2$. By continuity, equation (20.2.2) holds also for $z \in S_1 \cup S_2$, and the first part of Theorem 20.2.2 is proved.

Assume now that (20.2.1) holds for some analytic matrix functions $A(z)$, $B(z)$, and $C(z)$. It follows from Theorem 7.2.3 that every pole of $W(\lambda, z)$ is an eigenvalue of $A(z)$ and every zero of $W(\lambda, z)$ is an eigenvalue of $A(z) - B(z)C(z)$ (although the converse need not be true). As the eigenvalues of $A(z)$ and of $A(z) - B(z)C(z)$ depend continuously on z, they are bounded in a neighbourhood of each point in Ω, and the converse statement of Theorem 20.2.2 follows. $\square$

As the proof of Theorem 20.2.2 shows, the converse statement of this theorem remains true if the matrix functions $A(z)$, $B(z)$, and $C(z)$ satisfying (20.2.1) are merely assumed to be continuous on Ω.

The discrete set S from Theorem 20.2.2 consists of exactly those points where the McMillan degree of $W(\lambda, z)$ is less than m. This follows from Theorems 7.1.3 and 7.1.5. Note also that the McMillan degree of $W(\lambda)$ is equal to m for every $z \in \Omega \smallsetminus S$.

From now on it will be assumed (in addition to the assumptions made in the beginning of this section) that the zeros and poles of $W(\lambda, z)$ are bounded in a neighbourhood of each point in Ω. Let

$$W(\lambda, z) = I + C(z)(\lambda I - A(z))^{-1} B(z) \qquad (20.2.3)$$

be a minimal realization of $W(\lambda, z)$ for $z \in \Omega \smallsetminus S$, as in Theorem 20.2.2. Here S is the set of all $z \in \Omega$ such that the realization (20.2.3) is not minimal. Denote by S_1 and S_2 the first and second exceptional sets, respectively, of the analytic matrix function $A(z)$, as defined in Section 19.2. Similarly, let $S_1^{\times}$ and $S_2^{\times}$ be the first and second exceptional sets, respectively, of $A(z)^{\times} \overset{\text{def}}{=} A(z) - B(z)C(z)$, $z \in \Omega$. The set $S_1 \cup S_1^{\times}$ will be called the *first exceptional set* T_1 of $W(\lambda, z)$. As the poles (resp. zeros) of $W(\lambda, z)$, when $z \in \Omega \smallsetminus S$, are exactly the eigenvalues of $A(z)$ [resp. of

$A(z)^\times]$ (see Section 7.2), it follows that the point $z_0 \in \Omega$ belongs to the first exceptional set of $W(\lambda, z)$ if and only if there is a pole or a zero $\lambda_0(z)$ of $W(\lambda, z)$, where $z \in \mathcal{U} \smallsetminus \{z_0\}$, where $\mathcal{U}$ is a neighbourhood of z_0, such that z_0 is an algebraic branch point of $\lambda_0(z)$. Note that it can happen that (20.2.3) is not a minimal realization for some z belonging to the first exceptional set of $W(\lambda)$ (see Example 20.2.2). The set

$$(S_2 > S_1^\times) \cup (S_2^\times \smallsetminus S_1) \cup (S \smallsetminus (S_1^\times \cup S_1))$$

will be called the *second exceptional set* T_2 of $W(\lambda, z)$. Denoting by $\delta(z)$ the McMillan degree of $W(\lambda, z)$, we obtain the following description of the points in the second exceptional set: $z \in T_2$ if and only if all poles and zeros of $W(\lambda, z)$ can be continued analytically (as functions of z) to z_0, and either

$$\delta(z_0) < \max_{z \in \Omega} \delta(z) \qquad \text{or} \qquad \delta(z_0) = \max_{z \in \Omega} \delta(z)$$

and for at least one zero (or pole) $\lambda_0(z)$ that is analytic in a neighbourhood $\mathcal{U}$ of z_0, the zero (or pole) multiplicities of $W(\lambda, z)$ corresponding to $\lambda_0(z)$ ($z \in \mathcal{U} \smallsetminus \{z_0\}$) are different from the zero (or pole) multiplicities of $W(\lambda, z_0)$ at $\lambda_0(z_0)$. Again, it can happen that T_2 intersects with the set of points where the realization (20.2.3) is not minimal. Clearly, both T_1 and T_2 are discrete sets. Note also that the set $T_1 \cup T_2$ contains all the points z_0 for which $\delta(z_0) < \max_{z \in \Omega} \delta(z)$.

EXAMPLE 20.2.2.   Let

$$W(\lambda, z) = 1 + [0, z, 0, z] \left(\lambda I - \begin{bmatrix} 0 & 1 & 0 & 0 \\ z & 0 & 0 & 0 \\ 0 & 0 & 0 & 1 \\ 0 & 0 & -z+2 & 0 \end{bmatrix} \right)^{-1} \begin{bmatrix} 1 \\ 0 \\ 1 \\ 0 \end{bmatrix}$$

$$= 1 + \frac{z^2}{\lambda^2 - z} + \frac{z(z-2)}{\lambda^2 - z(z-2)}, \quad z \in \mathbb{C}$$

be a scalar rational function depending analytically on z for $z \in \mathbb{C}$. Clearly, $W(\infty, z) = 1$ and the zeros and poles of $W(\lambda, z)$ are bounded in a neighbourhood of each point $z \in \mathbb{C}$ (cf. Proposition 20.2.2). In the notation introduced above we have $S_1 = \{0, 2\}$; $S_2 = \{1\}$; $S = \{0\}$. Further

$$A(z)^\times = \begin{bmatrix} 0 & 1-z & 0 & -z \\ z & 0 & 0 & 0 \\ 0 & -z & 0 & 1-z \\ 0 & 0 & -z+2 & 0 \end{bmatrix}$$

and a calculation shows that

$$\det(\lambda I - A(z)^\times) = \lambda^4 + 2\lambda^2(z-1) + z(z-2)(2z-1)$$

So the eigenvalues of $A(z)^\times$ are given by the formula

$$\lambda = (-(z-1) \pm \sqrt{-2z^3 + 6z^2 - 4z + 1})^{1/2}$$

It is easily seen that $S_1^\times = \{0, 2, \frac{1}{2}, z_1, z_2, z_3\}$, where z_1, z_2, z_3 are the zeros of the polynomial $-2z^3 + 6z^2 - 4z + 1$, and $S_2^\times$ is empty. The first exceptional set of $W(\lambda, z)$ is $\{0, 2, \frac{1}{2}, z_1, z_2, z_3\}$, whereas the second exceptional set of $W(\lambda, z)$ consists of one point $\{1\}$. $\square$

20.3 MINIMAL FACTORIZATIONS OF RATIONAL MATRIX FUNCTIONS

Let $W(\lambda, z)$ be a rational $n \times n$ matrix function depending analytically on the parameter z for $z \in \Omega$, as in the preceding section. Let

$$W(\lambda, z_0) = W_{10}(\lambda) \cdots W_{r0}(\lambda) \qquad (20.3.1)$$

be a minimal factorization of $W(\lambda, z_0)$, for some $z_0 \in \Omega$. Here $W_{10}(\lambda), \ldots, W_{r0}(\lambda)$ are $n \times n$ rational matrix functions with value I at infinity. We study the problem of continuation of (20.3.1) to an analytic family of minimal factorizations. In case z_0 does not belong to the exceptional sets of $W(\lambda, z)$, such a continuation is always possible, as the following theorem shows.

Theorem 20.3.1

Let $W(\lambda, z)$ be a rational $n \times n$ matrix function that depends analytically on z for $z \in \Omega$ and such that $W(\infty, z) = I$ for $z \in \Omega$. Assume that the denominator and numerator of each entry in $W(\lambda, z)$ are coprime for some $z_0 \in \Omega$ that is not a zero of the leading coefficient of the denominator. Assume, in addition, that the zeros and poles of $W(\lambda, z)$ are bounded in a neighbourhood of each point in Ω. Let

$$S = \{z_0 \in \Omega \mid \delta(z_0) < \max_{z \in \Omega} \delta(z)\}$$

where $\delta(z)$ is the McMillan degree of $W(\lambda, z)$, and let T_1 and T_2 be the first and second exceptional sets of $W(\lambda, z)$, respectively. Consider a minimal factorization (20.3.1) with $z_0 \in \Omega \smallsetminus (T_1 \cup T_2)$. Then there exist rational matrix functions $W_1(\lambda, z), \ldots, W_r(\lambda, z)$, the entries of which depend analytically on z in Ω (with the possible exception of algebraic branch points in T_1 and of a discrete set $D \subset \Omega$ of poles), and having the following properties: (a) $W_j(\infty, z) = I$ for $j = 1, \ldots, r$ and every $z \in \Omega \smallsetminus D$; (b) the point z_0 does not *belong to D and $W_j(\lambda, z_0) = W_{j0}(\lambda)$ for $j = 1, \ldots, r$; (c) $W(\lambda, z) = W_1(\lambda, z) \cdots W_r(\lambda, z)$ for every $z \in \Omega \smallsetminus D$. Moreover, this factorization is minimal for every $z \in \Omega \smallsetminus (D \cup S)$.*

The set D of poles of $W_j(\lambda, z)$ in Theorem 20.3.1 generally depends on the factorization (20.3.1), and not only on the original function $W(\lambda, z)$. This is in contrast with the sets T_1 and T_2 that depend on $W(\lambda, z)$ only.

Proof. Let $A(z)$, $B(z)$, and $C(z)$ be as in Theorem 20.2.2, so the realization (20.2.1) is minimal for all $z \in \Omega \smallsetminus S$. Using Theorem 7.5.1, let

$$\mathcal{C}^m = \mathcal{L}_{10} \dotplus \cdots \dotplus \mathcal{L}_{r0}$$

be the direct sum decomposition corresponding to the minimal factorization (20.3.1), with respect to the minimal realization

$$W(\lambda, z_0) = I + C(z_0)(\lambda I - A(z_0))^{-1} B(z_0)$$

Thus, for $j = 1, \ldots, r - 1$ the subspaces

$$\mathcal{M}_{j0} = \mathcal{L}_{10} \dotplus \cdots \dotplus \mathcal{L}_{j0}$$

are $A(z_0)$ invariant, whereas the subspaces

$$\mathcal{N}_{20} = \mathcal{L}_{20} \dotplus \cdots \dotplus \mathcal{L}_{r0}, \ldots, \quad \mathcal{N}_{r-1,0} = \mathcal{L}_{r-1,0} \dotplus \mathcal{L}_{r0}, \quad \mathcal{N}_{r0} = \mathcal{L}_{r0}$$

are $A(z_0)^\times$ invariant. [Here, as usual, $A(z)^\times = A(z) - B(z)C(z)$.] Now by Theorem 19.4.2, there exist families of subspaces $\mathcal{M}_j(z)$ for $j = 1, \ldots, r - 1$, and $\mathcal{N}_j(z)$ for $j = 2, \ldots, r$, which are analytic on Ω, except possibly for algebraic branch points in T_1, and have the following properties: (a) $\mathcal{M}_1(z) \subset \cdots \subset \mathcal{M}_{r-1}(z)$ for all $z \in \Omega$; (b) $\mathcal{N}_2(z) \supset \cdots \supset \mathcal{N}_r(z)$ for all $z \in \Omega$; (c) $\mathcal{M}_j(z)$ are $A(z)$ invariant, and $\mathcal{N}_j(z)$ are $A(z)^\times$ invariant; (d) $\mathcal{M}_j(z_0) = \mathcal{M}_{j0}$, $j = 1, \ldots, r - 1$; $\mathcal{N}_j(z_0) = \mathcal{N}_{j0}$, $j = 2, \ldots, r$.

Let m_j be the dimension of $\mathcal{L}_{j0}$, $j = 1, \ldots, r$ (so $m_1 + \cdots + m_r = m$). It follows from the proof of Theorem 19.4.2 that

$$\mathcal{M}_j(z) = \text{Span}\{x_1^{(j)}(z), \ldots, x_{p_j}^{(j)}(z)\}, \qquad z \in \Omega, \qquad j = 1, \ldots, r - 1$$

$$(20.3.2)$$

$$\mathcal{N}_j(z) = \text{Span}\{y_1^{(j)}(z), \ldots, y_{q_j}^{(j)}(z)\}, \qquad z \in \Omega, \qquad j = 2, \ldots, r \quad (20.3.3)$$

where for each j, the vector functions $x_1^{(j)}(z), \ldots, x_{p_j}^{(j)}(z)$, as well as $y_1^{(j)}(z), \ldots, y_{q_j}^{(j)}(z)$, are linearly independent for every $z \in \Omega$ and analytic on Ω except possibly for algebraic branch points in T_1. Here $p_j = m_1 +$

$\cdots + m_j$ is the dimension of $\mathcal{M}_j(z)$ and $q_j = m_j + m_{j+1} + \cdots + m_r$ is the dimension of $\mathcal{N}_j(z)$.

Our next observation is that

$$\mathcal{M}_j(z) \dotplus \mathcal{N}_{j+1}(z) = \mathbb{C}^m , \qquad z \in \Omega \smallsetminus D_j , \qquad j = 1, \ldots, r-1 \qquad (20.3.4)$$

where D_j is a discrete set in Ω. [Note that the sum in (20.3.4) is direct.] Indeed, by (20.3.2) and (20.3.3) we have

$$\dim(\mathcal{M}_j(z) + \mathcal{N}_j(z)) = \operatorname{rank} F_j(z) , \qquad z \in \Omega$$

where

$$F_j(z) = [x_1^{(j)}(z) \cdots x_{p_i}^{(j)}(z) \quad y_1^{(j+1)}(z) \cdots y_{q_{j+1}}^{(j+1)}(z)] \qquad (20.3.5)$$

is a matrix function of size $m \times (p_j + q_{j+1}) = m \times m$. It remains to observe that $\det F_j(z)$ is not identically zero [because $\det F_j(z_0) \neq 0$] and (20.3.4) holds with D_j being the set of zeros of $\det F_j(z)$.

Let $D = D_1 \cup \cdots \cup D_{r-1}$. In particular, we have

$$\mathcal{M}_j(z) \dotplus \mathcal{N}_j(z) = \mathbb{C}^m , \qquad z \in \Omega \smallsetminus D , \qquad j = 2, \ldots, r-1$$
$$(20.3.6)$$

Note also that z_0 does not belong to D.

Consider the subspaces $\mathcal{L}_j(z) = \mathcal{M}_j(z) \cap \mathcal{N}_j(z)$ for $j = 2, \ldots, r-1$. First, it is clear that

$$\mathcal{L}_j(z_0) = \mathcal{L}_{j0} , \qquad j = 2, \ldots, r-1$$

Second

$$\mathcal{L}_1(z) \dotplus \cdots \dotplus \mathcal{L}_r(z) = \mathbb{C}^m , \qquad z \in \Omega \smallsetminus D \qquad (20.3.7)$$

where we put $\mathcal{L}_1(z) = \mathcal{M}_1(z)$ and $\mathcal{L}_r(z) = \mathcal{N}_r(z)$. Indeed, it is sufficient to verify that

$$\mathcal{M}_{j-1}(z) \dotplus \mathcal{L}_j(z) = \mathcal{M}_j(z) , \qquad z \in \Omega \smallsetminus D , \qquad j = 2, \ldots, r$$
$$(20.3.8)$$

[By definition, $\mathcal{M}_r(z) = \mathbb{C}^m$.] The inclusion $\subset$ in (20.3.8) is evident from the definition of $\mathcal{L}_j(z)$. Further, for $z \in \Omega \smallsetminus D$, we have

$$\mathcal{M}_{j-1}(z) \cap \mathcal{L}_j(z) \subset \mathcal{M}_{j-1}(z) \cap \mathcal{N}_j(z) \cap \mathcal{M}_j(z) = \{0\}$$

in view of (20.3.4). Now, using (20.3.6), we have for $z \in \Omega \smallsetminus D$

$$\dim \mathcal{L}_j(z) = \dim \mathcal{M}_j(z) + \dim \mathcal{N}_j(z) - m = m_j$$

so

$$\dim \mathcal{M}_{j-1}(z) + \dim \mathcal{L}_j(z) = p_{j-1} + m_j = p_j = \dim \mathcal{M}_j(z)$$

and (20.3.8) follows.

By Theorem 7.5.1, for $z \in \Omega \smallsetminus (D \cup S)$, there exists a minimal factorization

$$W(\lambda, z) = W_1(\lambda, z) \cdots W_r(\lambda, z) \qquad (20.3.9)$$

which corresponds to the direct sum decomposition (20.3.7), with respect to the minimal realization

$$W(\lambda, z) = I + C(z)(\lambda I - A(z))^{-1} B(z)$$

If we show that each projector $\pi_j(z)$ on $\mathcal{L}_j(z)$ along $\mathcal{L}_1(z) \dotplus \cdots \dotplus \mathcal{L}_{j-1}(z) \dotplus \mathcal{L}_{j+1}(z) \dotplus \cdots \dotplus \mathcal{L}_r(z)$ is analytic in Ω, except possibly for algebraic branch points in T_1 and poles in D, then formula (7.5.5) shows that $W_j(\lambda, z)$ have all the properties required in Theorem 20.3.1. [Note that, by continuity, factorization (20.3.9) holds also for $z \in S \smallsetminus D$, but it is not minimal at these points.]

To verify these properties of $\pi_j(z)$, introduce for $z \in \Omega \smallsetminus D$ the projector $Q_j(z)$ on $\mathcal{M}_j(z)$ along $\mathcal{N}_{j+1}(z)$ for $j = 1, \ldots, r-1$. Define also $Q_0(z) = 0$ and $Q_r(z) = I$. One checks easily that for $j = 1, \ldots, r$

$$Q_j(z) Q_{j-1}(z) = Q_{j-1}(z) Q_j(z), \qquad z \in \Omega \smallsetminus D \qquad (20.3.10)$$

[Indeed, both sides of (20.3.10) take the value 0 on vectors from $\mathcal{N}_{j+1}(z)$ and from $\mathcal{L}_j(z)$, and take the value x on each vector x from $\mathcal{M}_{j-1}(z)$.] Therefore, $(I - Q_{j-1}(z)) Q_j(z)$ is a projector that coincides with $\pi_j(z)$ for $j = 1, \ldots, r$. But

$$Q_j(z) = F_j(z) \begin{bmatrix} I_{p_j} & 0 \\ 0 & 0 \end{bmatrix} (F_j(z))^{-1}, \qquad j = 1, \ldots, r-1$$

where $F_j(z)$ is given by (20.3.5); so $Q_j(z)$ is analytic on Ω except possibly for algebraic branch points in T_1 and poles in D. Hence $\pi_j(z)$ also enjoys these properties. $\square$

Consider now an important case of analytic continuation of minimal factorizations that can also be achieved when $z_0 \in T_1 \cup T_2$.

Theorem 20.3.2

Let $W(\lambda, z)$ be as in Theorem 20.3.1, and let

$$W(\lambda, z_0) = W_{10}(\lambda) \cdots W_{r0}(\lambda)$$

be a minimal factorization of $W(\lambda, z_0)$ where $z_0 \in \Omega$. [As usual, $W_{j0}(\lambda)$ are rational matrix functions with value I at infinity.] Assume that $W_{j0}(\lambda)$ and $W_{k0}(\lambda)$ have no common zeros and no common poles when $j \neq k$. Then there exist rational matrix functions $W_j(\lambda, z)$, $j = 1, \ldots, r$ with the properties described in Theorem 20.3.1 and that, in addition, are analytic on a neighbourhood of z_0.

The proof is obtained in the same way as the proof of Theorem 20.3.1, by using Theorem 19.4.3 in place of Theorem 19.4.2.

To conclude this section we discuss minimal factorizations (20.3.1) that cannot be continued analytically (as in Theorem 20.3.1). We say that the minimal factorization (20.3.1) is *sequentially nonisolated*, if there is a sequence of points $\{z_m\}_{m=1}^{\infty}$ in $\Omega \smallsetminus \{z_0\}$ such that $z_m \to z_0$, and sequences of rational matrix functions $\{W_{jm}(\lambda)\}_{m=1}^{\infty}$, $j = 1, \ldots, r$ with value I at infinity such that

$$W(\lambda, z_m) = W_{1m}(\lambda) \cdots W_{rm}(\lambda)$$

is a minimal factorization of $W(\lambda, z_m)$, $m = 1, 2, \ldots$, and for $j = 1, \ldots, r$

$$\lim_{m \to \infty} W_{jm}(\lambda) = W_{j0}(\lambda) \tag{20.3.11}$$

Equation (20.3.11) is understood in the sense that for each pair of indices k, l $(1 \le k, l \le n)$ the (k, l) entry of $W_{jm}(\lambda)$ has the form

$$\frac{\sum_{p=0}^{u} \alpha_{pm} \lambda^p}{\sum_{q=0}^{v} \beta_{qm} \lambda^q}$$

where α_{pm} and β_{pm} are complex numbers (depending, of course, on j, k, and l) such that $\lim_{m \to \infty} \alpha_{pm} = \alpha_p$, $\lim_{m \to \infty} \beta_{qm} = \beta_q$ $(p = 1, \ldots, u;\ q = 1, \ldots, v)$, and the (k, l)-entry in $W_{j0}(\lambda)$ is

$$\frac{\sum_{p=0}^{u} \alpha_p \lambda^p}{\sum_{q=0}^{v} \beta_q \lambda^q}$$

Clearly, if an analytic continuation (as in Theorem 20.3.1) of the minimal factorization (20.3.1) exists, then this factorization is sequentially nonisolated. In particular, Theorem 20.3.1 shows that every minimal factorization (20.3.1) with $z_0 \in \Omega \smallsetminus (T_1 \cup T_2)$ is sequentially nonisolated. Also, Theorem 20.3.2 shows that a minimal factorization (20.3.1) is sequentially nonisolated provided $W_{j0}(\lambda)$ and $W_{k0}(\lambda)$ have no common zeros and no common poles when $j \neq k$.

It turns out that not every minimal factorization of $W(\lambda, z_0)(z_0 \in \Omega)$ can be continued analytically; indeed, we exhibit next a sequentially isolated minimal factorization.

EXAMPLE 20.3.1. Let

$$W(\lambda, z) = \begin{bmatrix} 1 + (\lambda - z)^{-1} & 0 \\ 0 & 1 + \lambda^{-1} \end{bmatrix}, \quad z \in \mathbb{C}$$

and consider the minimal factorization of $W(\lambda, 0)$:

$$W(\lambda, 0) = \begin{bmatrix} 1 + \lambda^{-1} & 0 \\ 0 & 1 + \lambda^{-1} \end{bmatrix} = \begin{bmatrix} 1 & \lambda^{-1} \\ 0 & 1 + \lambda^{-1} \end{bmatrix} \begin{bmatrix} 1 + \lambda^{-1} & -\lambda^{-1} \\ 0 & 1 \end{bmatrix}$$

$$(20.3.12)$$

We verify that this factorization is sequentially isolated. To this end we find all minimal factorizations of $W(\lambda, z)$, where $z \neq 0$. A minimal realization of $W(\lambda, z)$ is easily found:

$$W(\lambda, z) = I + I\left(\lambda I - \begin{bmatrix} z & 0 \\ 0 & 0 \end{bmatrix}\right)^{-1} I$$

In the notation of Theorem 20.2.1, we have

$$A(z) = \begin{bmatrix} z & 0 \\ 0 & 0 \end{bmatrix}, \qquad B(z) = C(z) = I, \qquad A(z)^x = \begin{bmatrix} z - 1 & 0 \\ 0 & -1 \end{bmatrix}$$

Theorem 7.5.1 shows that all nontrivial minimal factorizations of $W(\lambda, z)(z \neq 0)$ are given by the formulas

$$W(\lambda, z) = \begin{bmatrix} 1 + (\lambda - z)^{-1} & 0 \\ 0 & 1 \end{bmatrix} \begin{bmatrix} 1 & 0 \\ 0 & 1 + \lambda^{-1} \end{bmatrix}$$

$$W(\lambda, z) = \begin{bmatrix} 1 & 0 \\ 0 & 1 + \lambda^{-1} \end{bmatrix} \begin{bmatrix} 1 + (\lambda - z)^{-1} & 0 \\ 0 & 1 \end{bmatrix}$$

So the minimal factorization (20.3.12) is indeed sequentially isolated. $\quad\square$

20.4 MATRIX QUADRATIC EQUATIONS

Consider the matrix quadratic equation

$$XBX + XA - DX - C = 0$$

where A, B, C, D are known matrices of sizes $n \times n$, $n \times m$, $m \times n$, $m \times m$, respectively, and X is an $m \times n$ matrix to be found. We assume that

$A = A(z)$, $B = B(z)$, $C = C(z)$, and $D = D(z)$ are analytic functions of z on Ω, where Ω is a domain in the complex plane. The analytic properties of the solutions X as functions of z are studied.

Let

$$T(z) = \begin{bmatrix} A(z) & B(z) \\ C(z) & D(z) \end{bmatrix}, \qquad z \in \Omega$$

be the $(m + n) \times (m + n)$ analytic matrix function, and let S_1 and S_2 be the first and second exceptional sets of $T(z)$ as defined in Section 20.2. We have the following main result.

Theorem 20.4.1

For every $z_0 \in \Omega \smallsetminus (S_1 \cup S_2)$ and every solution X_0 of

$$XB(z_0)X + XA(z_0) - D(z_0)X - C(z_0) = 0 \qquad (20.4.1)$$

there exists an $m \times n$ matrix function $X(z)$ that is analytic on Ω, except possibly for algebraic branch points in S_1 and a discrete set of poles in Ω, and such that $X(z_0) = X_0$ and

$$X(z)B(z)X(z) + X(z)A(z) - D(z)X(z) - C(z) = 0 \qquad (20.4.2)$$

for every $z \in \Omega$ that is not a pole of $X(z)$. [The case when a point $z_0 \in S_1$ is also a pole of $X(z)$ is not excluded.]

Proof. By Proposition 17.8.1, the subspace

$$\mathcal{M}_0 = \mathrm{Im}\begin{bmatrix} I \\ X_0 \end{bmatrix} \subset \mathbb{C}^{n+m}$$

is $T(z_0)$ invariant. By Theorem 19.4.1, there is a family of subspaces $\mathcal{M}(z)$ that is analytic on Ω except possibly for algebraic branch points in S_1, for which $\mathcal{M}(z_0) = \mathcal{M}_0$, and for which $\mathcal{M}(z)$ is $T(z)$ invariant for all $z \in \Omega$. By Theorem 18.3.2 there exists an $(m + n) \times n$ analytic matrix function $S(z)$ on Ω with linearly independent columns such that, for all $z \in \Omega$, $\mathcal{M}(z) = \mathrm{Im}\, S(z)$.

Write

$$S(z) = \begin{bmatrix} S_1(z) \\ S_2(z) \end{bmatrix}$$

where $S_1(z)$ is of size $n \times n$ and $S_2(z)$ is of size $m \times n$, and observe that $\det S_1(z) \not\equiv 0$ [because $S_1(z_0) = I$]. Now by the same Proposition 17.8.1

$$X(z) = S_2(z)S_1(z)^{-1}$$

is the desired solution of (20.4.2).   □

Consider an example.

EXAMPLE 20.4.1.   Let $C(z)$ be an $n \times n$ analytic matrix function on Ω with det $C(z) \neq 0$, and assume that the eigenvalues of $C(z)$ are analytic functions. [This will be the case if, for instance, $C(z)$ has an upper triangular form.] Assume in addition that $C(z)$ has n distinct eigenvalues for every $z \in \Omega$. Consider the equation

$$X^2 = C(z) \tag{20.4.3}$$

Here

$$T(z) = \begin{bmatrix} 0 & I \\ C(z) & 0 \end{bmatrix}$$

and it is easily seen that $\det(\lambda I - T(z)) = \det(\lambda^2 I - C(z))$. So λ_0 is an eigenvalue of $T(z_0)$ if and only if λ_0^2 is an eigenvalue of $C(z_0)$. It follows that the first exceptional set of $T(z_0)$ is contained in the set $S \overset{\text{def}}{=} \{z \in \Omega \mid \det C(z) = 0\}$. As for every $z \in \Omega \smallsetminus S$ the matrix $T(z)$ has $2n$ distinct eigenvalues, it follows that the second exceptional set of $T(z)$ is also contained in S. By Theorem 20.4.1 every solution X_0 of (20.4.3) with $z = z_0 \in \Omega \smallsetminus S$ can be extended to a family of solutions $X(z)$ of (20.4.3) that is meromorphic on Ω except possibly for algebraic branch points in S.   □

In addition, let us indicate a case when an analytic extension of a solution of (20.4.1) is always possible.

Theorem 20.4.2

Let X_0 be a solution of (20.4.1) and $z_0 \in \Omega$. Furthermore, assume that the $T(z_0)$-invariant subspace $\operatorname{Im}\begin{bmatrix} I \\ X_0 \end{bmatrix}$ is spectral. Then there exists an $m \times n$ matrix function $X(z)$ with the properties described in Theorem 20.4.1 and, in addition, $X(z)$ is analytic in a neighbourhood of z_0.

The proof of Theorem 20.4.2 is obtained in the same way as the proof of Theorem 20.4.1, but using Theorem 19.4.3 in place of Theorem 19.4.1.

In connection with Theorem 20.4.2, note the following fact. Assume that $m = n$. If X_1 and X_2 are solutions of (20.4.1) such that

$$\sigma(A(z_0) + B(z_0)X_1) \cap \sigma(A(z_0) + B(z_0)X_2) = \emptyset$$

then both $T(z_0)$-invariant subspaces

$$\mathcal{M}_j = \operatorname{Im}\begin{bmatrix} I \\ X_j \end{bmatrix}, \qquad j = 1, 2$$

are spectral. Indeed

$$T(z_0)\begin{bmatrix} I \\ X_j \end{bmatrix} = \begin{bmatrix} I \\ X_j \end{bmatrix}(A(z_0) + B(z_0)X_j), \qquad j = 1, 2$$

so $\sigma(T(z_0)|_{\mathcal{M}_1}) \cap \sigma(T(z_0)|_{\mathcal{M}_2}) = \emptyset$. In particular, $\mathcal{M}_1 \cap \mathcal{M}_2 = \{0\}$. As dim $\mathcal{M}_1 = \dim \mathcal{M}_2 = n$, it follows that $\mathcal{M}_1 \dotplus \mathcal{M}_2 = \mathbb{C}^{2n}$. (Here we use the assumption that $m = n$.) Hence both $\mathcal{M}_1$ and $\mathcal{M}_2$ are spectral.

The following example shows that not every solution of the equation

$$XB(z_0)X + XA(z_0) - D(z_0)X - C(z_0) = 0, \qquad z_0 \in \Omega$$

can be continued analytically as in Theorem 20.4.1. (Of course, it is then necessary that $z_0 \in S_1 \cup S_2$.)

EXAMPLE 20.4.2.   Consider the scalar equation

$$zx^2 = 0 \tag{20.4.4}$$

The solution $x = 1$ of (20.4.4) with $z_0 = 0$ cannot be continued analytically. $\square$

20.5   *EXERCISES*

20.1   Let

$$L(\lambda, z) = \begin{bmatrix} \lambda^2 & 0 \\ -\lambda z & \lambda^2 - 1 \end{bmatrix}$$

Find the analytic continuation (as in Theorem 20.1.1) of the factorization

$$L(\lambda, 1) = \left(I\lambda - \begin{bmatrix} 0 & 1 \\ 1 & 0 \end{bmatrix}\right)\left(I\lambda - \begin{bmatrix} 0 & 1 \\ 0 & 0 \end{bmatrix}\right)$$

What are the poles of this analytic continuation?

20.2   Let $L(\lambda, z)$ be a monic $n \times n$ matrix polynomial of degree l whose coefficients are analytic on Ω, and assume that for every $z \in \Omega$ det $L(\lambda, z)$ has nl distinct zeros. Prove that for every factorization $L(\lambda, z_0) = L_1(\lambda) \cdots L_r(\lambda)$, where $z_0 \in \Omega$ and $L_j(\lambda)$ are monic matrix polynomials, there exist monic matrix polynomials $L_1(\lambda, z), \ldots, L_r(\lambda, z)$ whose coefficients are analytic on Ω and such that $L_j(\lambda, z_0) = L_j(\lambda)$ for $j = 1, \ldots, r$.

20.3 Show that if by Theorem 20.1.1 the polynomial $L(\lambda, z)$ is scalar, then the analytic continuations of $L_j(\lambda)$ do not have poles in Ω (i.e., $S = \emptyset$ in the notation of Theorem 20.1.1).

20.4 Let $L(\lambda, z)$ be a monic matrix polynomial whose coefficients are circulant matrices analytic on Ω. Prove that the analytic continuation of every factorization $L(\lambda, z_0) = L_1(\lambda) \cdots L_r(\lambda)$ where $z_0 \in \Omega$ (as in Theorem 20.1.1) has no poles in Ω.

20.5 Prove that every factorization of a monic scalar polynomial $L(\lambda, z)$ with coefficients depending analytically on $z \in \Omega$ is sequentially nonisolated. (*Hint*: Use Exercise 19.9.)

20.6 Find the first and second exceptional sets for the following rational matrix functions depending analytically on a parameter $z \in \mathbb{C}$:

(a)
$$W(\lambda, z) = 1 + \frac{z-1}{\lambda - z} + \frac{z+1}{\lambda + z}$$

(b)
$$W(\lambda, z) = \begin{bmatrix} 1 + \dfrac{z^2}{\lambda^2 - z^2} & \dfrac{\lambda + 1}{\lambda^2} \\ 0 & 1 + \dfrac{z^2}{\lambda^2 + z^2} \end{bmatrix}$$

20.7 Let $W(\lambda, z)$ be as in Exercise 20.6 (a). Find the analytic continuations (as in Theorem 20.3.1) of all minimal factorizations of the rational matrix function $W(\lambda, z)$.

20.8 Let $W(\lambda, z)$ be a rational matrix function that satisfies the hypotheses of Theorem 20.3.1. Assume that for some $z_0 \in \Omega$, $W(\lambda, z_0)$ has δ distinct zeros and δ distinct poles, where δ is the maximum of the McMillan degrees of $W(\lambda, z)$ for $z \in \Omega$. Prove that every minimal factorization

$$W(\lambda, z_0) = W_1(\lambda) \cdots W_r(\lambda)$$

admits an analytic continuation into a neighbourhood of z_0, that is, there exist rational matrix functions $W_1(\lambda, z), \ldots, W_r(\lambda, z)$ that are analytic in z on a neighbourhood $\mathcal{U}$ of z_0 such that

$$W(\lambda, z) = W_1(\lambda, z) \cdots W_r(\lambda, z)$$

is a minimal factorization for every $z \in \mathcal{U}$, and $W_j(\lambda, z_0) = W_j(\lambda)$ for $j = 1, \ldots, r$.

20.9 Let

$$XB(z)X + XA(z) - D(z)X - C(z) = 0 \tag{1}$$

be a matrix equation, where $A(z)$, $B(z)$, $C(z)$, and $D(z)$ are analytic matrix functions (of appropriate sizes) on a domain $\Omega \subset \mathbb{C}$. Assume that all eigenvalues of the matrix

$$\begin{bmatrix} A(z) & B(z) \\ C(z) & D(z) \end{bmatrix}$$

are distinct, for every $z \in \Omega$. Prove that given a solution X_0 of (1) with $z = z_0 \in \Omega$, there exists an analytic matrix function $X(z)$ on Ω such that $X(z)$ is a solution of (1) for every $z \in \Omega$ and $X(z_0) = X_0$.

20.10 We say that a solution X_0 of (1) with $z = z_0 \in \Omega$ is sequentially nonisolated if there exist a sequence $\{z_m\}_{m=1}^{\infty}$ such that $z_m \to z_0$ as $m \to \infty$ and $z_m \neq z_0$ for $m = 1, 2, \ldots$, and a sequence $\{X_m\}_{m=1}^{\infty}$ such that

$$X_m B(z_m) X_m - X_m A(z_m) - D(Z_m) X_m - C(z_m) = 0$$

for $m = 1, 2, \ldots$, which satisfies

$$\lim_{m \to \infty} X_m = X_0$$

Prove that if the matrix

$$\begin{bmatrix} A(z_0) & B(z_0) \\ C(z_0) & D(z_0) \end{bmatrix}$$

is nonderogatory, then every solution of (1) with $z = z_0$ is sequentially nonisolated.

20.11 Give an example of a solution of (1) that is sequentially isolated.

Notes to
Part 4

Chapter 18. This chapter is an introduction to the basic facts on analytic families of subspaces. The main result is Theorem 18.3.1, which connects the local and global properties of an analytic family of subspaces. This result (in a more general framework) appeared first in the theory of analytic fibre bundles [Grauert (1958), Allan (1967), Shubin (1979)]. Here, we follow Gohberg and Leiterer (1972, 1973) in the proof of this theorem.

The result of Theorem 18.2.1 goes back to Shmuljan (1957) [see also Gohberg and Rodman (1981)]. The proof of Theorem 18.2.1 presented here is from the authors' book (1982). The results of Section 18.6 seem to be new. In case of a function $\lambda I - A$, where A is a bounded linear operator acting in infinite dimensional Banach space, the result of Theorem 18.6.2 was proved in Saphar (1965).

Chapter 19. The starting point for the material in this chapter (Theorem 19.1.1) is taken from the book by Baumgartel (1985). Theorem 19.5.1 was proved in Porsching (1968). The analytic extendability problem for invariant subspaces is probably treated here for the first time.

Chapter 20. We consider in this chapter some of the applications dealt with in Chapters 5, 7, and 17, but in the new circumstances when the matrices involved depend analytically on a complex parameter. All the results (except those in Section 20.1) seem to be new. In Section 20.1 we adapt and generalize the results developed in Chapter 5 of Gohberg, Lancaster, and Rodman (1982). Example 20.1.1 is Example 20.5.4 of the authors' book (1982).

Appendix

Equivalence of Matrix Polynomials

To make this work more self-contained, we present in this appendix the basic facts about equivalence of matrix polynomials that are used in the main body of the book. Two concepts of equivalence are discussed. For the first of these, two matrix polynomials $A(\lambda)$ and $B(\lambda)$ are said to be equivalent if one is obtained from the other by premultiplication and postmultiplication with square matrix polynomials having constant nonzero determinant. Elementary divisors (or, alternatively, invariant polynomials) form the full set of invariants for this concept of equivalence, and the Smith form (which is diagonal) is the canonical form. This equivalence is studied in detail in Sections A.1–A.4.

The second concept of equivalence is the strict equivalence of linear matrix polynomials $A + \lambda B$ and $A_1 + \lambda B_1$. This means that $P(A + \lambda B)Q = A_1 + \lambda B_1$ for some invertible matrices P and Q. For strict equivalence the full set of invariants comprises minimal column indices, minimal row indices, elementary divisors, and elementary divisors at infinity. The Kronecker form (which is block diagonal) is the canonical form. A thorough treatment of strict equivalence is presented in Sections A.5–A.7. The canonical form for equivalence of matrix polynomials is a natural prerequisite for this presentation.

A.1   THE SMITH FORM: EXISTENCE

In this and subsequent sections we consider matrix polynomials $A(\lambda) = \sum_{j=0}^{s} A_j \lambda^j$, where A_j are $m \times n$ matrices whose entries are complex numbers (so that we admit the case of rectangular matrices A_j). Of course, the sizes of all A_j must be the same. Two $m \times n$ matrix polynomials $A(\lambda)$ and $B(\lambda)$ are said to be *equivalent* if

$$A(\lambda) = E(\lambda)B(\lambda)F(\lambda) \quad \text{for all} \quad \lambda \in \mathbb{C} \tag{A.1.1}$$

and some matrix polynomials $E(\lambda)$ and $F(\lambda)$ of sizes $m \times n$ and $n \times n$, respectively, with constant nonzero determinants (i.e., independent of λ). We use the symbol $\sim$: $A(\lambda) \sim B(\lambda)$ to mean that $A(\lambda)$ and $B(\lambda)$ are equivalent.

It is easy to see that $\sim$ is an equivalence relation, that is: (a) $A(\lambda) \sim A(\lambda)$ for every matrix polynomial $A(\lambda)$; (b) $A(\lambda) \sim B(\lambda)$ implies $B(\lambda) \sim A(\lambda)$; (c) $A(\lambda) \sim B(\lambda)$ and $B(\lambda) \sim C(\lambda)$ implies $A(\lambda) \sim C(\lambda)$. Indeed, if $B(\lambda) = A(\lambda)$, then (A.1.1) holds with $E(\lambda) \equiv I_m$, $F(\lambda) \equiv I_n$. Further, assume that (A.1.1) holds for matrix polynomials $A(\lambda)$ and $B(\lambda)$. As $\det E(\lambda) \equiv \text{const} \neq 0$, the formula for the inverse matrix in terms of cofactors implies that $E(\lambda)^{-1}$ is a matrix polynomial as well and since $\det E(\lambda) \det E(\lambda)^{-1} = 1$, it follows that $\det E(\lambda)^{-1}$ is also a nonzero constant. Similarly, $F(\lambda)^{-1}$ is a matrix polynomial for which $\det F(\lambda)^{-1}$ is a nonzero constant. Now we have

$$B(\lambda) = E(\lambda)^{-1}A(\lambda)F(\lambda)^{-1}$$

which means that $B(\lambda) \sim A(\lambda)$. Finally, let us check (c). We have

$$A(\lambda) = E_1(\lambda)B(\lambda)F_1(\lambda), \qquad B(\lambda) = E_2(\lambda)C(\lambda)F_2(\lambda)$$

where $E_1(\lambda)$, $E_2(\lambda)$, $F_1(\lambda)$, $F_2(\lambda)$ have constant nonzero determinants. Then $A(\lambda) = E(\lambda)C(\lambda)F(\lambda)$ with $E(\lambda) = E_1(\lambda)E_2(\lambda)$ and $F(\lambda) = F_2(\lambda)F_1(\lambda)$. So $A(\lambda) \sim C(\lambda)$.

The central result on equivalence of matrix polynomials is the Smith form, which describes the simplest matrix polynomial in each equivalence class, as follows.

Theorem A.1.1

An $m \times n$ matrix polynomial $A(\lambda)$ is equivalent to a unique $m \times n$ matrix polynomial $D(\lambda)$ where

$$D(\lambda) = \begin{bmatrix} d_1(\lambda) & & & & & & 0 \\ & \ddots & & & & & \\ & & d_r(\lambda) & & & & \vdots \\ & & & \ddots & & & \\ & & & & 0 & & \\ & & & & & \ddots & \\ 0 & & & \cdots & & & 0 \end{bmatrix} \tag{A.1.2}$$

is a diagonal polynomial matrix with monic scalar polynomials $d_i(\lambda)$ such that $d_i(\lambda)$ is divisible by $d_{i-1}(\lambda)$ for $i = 2, \ldots, r$.

In other words, for every matrix polynomial $A(\lambda)$ there exist matrix polynomials $E(\lambda)$ and $F(\lambda)$ with constant nonzero determinants such that

$$E(\lambda)A(\lambda)F(\lambda) = D(\lambda) \tag{A.1.3}$$

has the form (A.1.2), and this form is uniquely determined by $A(\lambda)$. The matrix polynomial $D(\lambda)$ of (A.1.2) is called the *Smith form* of $A(\lambda)$ and plays an important role in the analysis of matrix polynomials. Note that $E(\lambda)$ and $F(\lambda)$ from (A.1.3) are not unique in general. Note also that the zeros on the main diagonal in $D(\lambda)$ are absent in case $A(\lambda_0)$ has full rank for some $\lambda_0 \in \mathbb{C}$. [In particular, this happens if $A(\lambda)$ is an $n \times n$ matrix polynomial with leading coefficient I.]

Proof of Theorem A.1.1 (First Part). Here we prove the existence of a $D(\lambda)$ of the form (A.1.2) that is equivalent to a given $A(\lambda)$. We use the following elementary transformations of a matrix polynomial $A(\lambda)$ of size $m \times n$: (a) interchange two rows, (b) add to some row another row multiplied by a scalar polynomial, and (c) multiply a row by a nonzero complex number, together with the three corresponding operations on columns.

Note that each of these transformations is equivalent to the multiplication of $A(\lambda)$ by an invertible matrix as follows. Interchange of rows (columns) i and j in $A(\lambda)$ is equivalent to multiplication on the left (right) by

$$
\begin{array}{c}
\\ \\ i \rightarrow \\ \\ \\ j \rightarrow \\ \\ \\
\end{array}
\begin{bmatrix}
1 & & & & & \cdots & 0 \\
\vdots & \ddots & & & & & \\
& & 0 & \cdots & 1 & \cdots & \\
\vdots & & \vdots & & \vdots & & \vdots \\
& \cdots & 1 & \cdots & 0 & & \\
\vdots & & & & & \ddots & \\
0 & \cdots & & & & & 1
\end{bmatrix}
\tag{A.1.4}
$$

Adding to the ith row of $A(\lambda)$ the jth row multiplied by the polynomial $f(\lambda)$ is equivalent to multiplication on the left by

$$
\begin{array}{c}
\\ \\ \\ \\ i \rightarrow \\ \\ \\ \\ \\
\end{array}
\begin{bmatrix}
1 & & & & & & & \\
& \ddots & & & & & & \\
& & 1 & & & & \vdots & \\
& & & \ddots & & & & \\
& \cdots & & & 1 & \cdots & f(\lambda) & \\
& & & & & \ddots & & \\
& & & & & & 1 & \vdots \\
& & & & & & & \ddots \\
& & & & & & & 1
\end{bmatrix}
\begin{array}{c} j\downarrow \end{array}
\tag{A.1.5}
$$

the same operation for columns is equivalent to multiplication on the right
by the matrix

$$
\begin{bmatrix}
1 & & & & & & & & \\
& 1 & & \vdots & & & & & \\
& & \ddots & & & & & & \\
i \rightarrow & \cdots & & 1 & & & & & \\
& & & \vdots & \ddots & & & & \\
& & & & & 1 & & & \\
& & & & & & \ddots & & \\
j \rightarrow & \cdots & & f(\lambda) & \cdots & & & 1 & \\
& & & & & & & & \ddots \\
& & & & & & & & & 1
\end{bmatrix}
\tag{A.1.6}
$$

Finally, multiplication of the ith row (column) in $A(\lambda)$ by a number $a \neq 0$ is
equivalent to the multiplication on the left (right) by

$$
\begin{bmatrix}
1 & & & & & & & \\
& \ddots & & & & & & \\
& & 1 & & & & & \\
& & & \ddots & & & & \\
i \rightarrow & \cdots & \cdots & & a & & & \\
& & & & & \ddots & & \\
& & & & & & 1 & \\
& & & & & & & \ddots \\
& & & & & & & & 1
\end{bmatrix}
\tag{A.1.7}
$$

[Empty spaces in (A.1.4)–(A.1.7) are assumed to be zeros.] Matrices of the
form (A.1.4)–(A.1.7) are be called *elementary*. It is apparent that the
determinant of any elementary matrix is a nonzero constant. Consequently,
it is sufficient to prove that, by applying a sequence of elementary transformations, every matrix polynomial $A(\lambda)$ can be reduced to a diagonal
form: $\mathrm{diag}[d_1(\lambda), \ldots, d_r(\lambda), 0, \ldots, 0]$, where $d_1(\lambda), \ldots, d_r(\lambda)$ are scalar
polynomials such that the quotients $d_i(\lambda)/d_{i-1}(\lambda)$, $i = 1, 2, \ldots, r-1$, are
also scalar polynomials. We prove this statement by induction on m and n.
For $m = n = 1$ it is evident.

Consider now the case $m = 1$, $n > 1$; that is

$$
A(\lambda) = [a_1(\lambda) a_2(\lambda) \cdots a_n(\lambda)]
$$

If all $a_j(\lambda)$ are zeros, there is nothing to prove. Suppose that not all the
$z_j(\lambda)$ are zeros, and let $a_{j_0}(\lambda)$ be a polynomial of minimal degree among the
nonzero entries of $A(\lambda)$. We can suppose that $j_0 = 1$. [Otherwise, interchange columns in $A(\lambda)$.] By elementary transformations it is possible to

replace all the other entries in $A(\lambda)$ by zero. Indeed, let $a_j(\lambda) \neq 0$. Divide $a_j(\lambda)$ by $a_1(\lambda)$: $a_j(\lambda) = b_j(\lambda)a_1(\lambda) + r_j(\lambda)$, where $r_j(\lambda)$ is the remainder and its degree is less than the degree of $a_1(\lambda)$, or $r_j(\lambda) \equiv 0$. Add to the jth column the first column multiplied by $-b_j(\lambda)$. Then $r_j(\lambda)$ will appear in the jth position of the new matrix. If $r_j(\lambda) \neq 0$, then put $r_j(\lambda)$ in the first position, and if there is still a nonzero entry [different from $r_j(\lambda)$], apply the same argument again. Namely, divide this (say, the kth) entry by $r_j(\lambda)$ and add to the kth column the first multiplied by minus the quotient of the division, and so on. Since the degrees of the remainders decrease, after a finite number of steps [not more than the degree of $a_1(\lambda)$] we find that all the entries in our matrix, except the first, are zeros. This proves Theorem A.1.1 in the case $m = 1$, $n > 1$. The case $m > 1$, $n = 1$ is treated in a similar way.

Assume now that $m, n > 1$, and assume that the theorem is proved for matrices with $m - 1$ rows and $n - 1$ columns. We can suppose that the $(1, 1)$ entry of $A(\lambda)$ is nonzero and has the minimal degree among the nonzero entries of $A(\lambda)$. [Indeed, if $A(\lambda) \neq 0$, we can reach this condition by interchanging rows and/or columns in $A(\lambda)$. If $A(\lambda) \equiv 0$, Theorem A.1.1 is trivial.] With the help of the procedure described in the previous paragraph [applied for the first row and the first column of $A(\lambda)$], by a finite number of elementary transformations we reduce $A(\lambda)$ to the form

$$A_1(\lambda) = \begin{bmatrix} a_{11}^{(1)}(\lambda) & 0 & \cdots & 0 \\ 0 & a_{22}^{(1)}(\lambda) & \cdots & a_{2n}^{(1)}(\lambda) \\ \vdots & \vdots & & \vdots \\ 0 & a_{m2}^{(1)}(\lambda) & \cdots & a_{mn}^{(1)}(\lambda) \end{bmatrix}$$

Suppose that for some $i, j > 1$, $a_{ij}^{(1)}(\lambda) \neq 0$ and is not divisible by $a_{11}^{(1)}(\lambda)$ (without remainder). Then add to the first row the ith row and apply the above arguments again. We obtain a matrix polynomial of the form

$$A_2(\lambda) = \begin{bmatrix} a_{11}^{(2)}(\lambda) & 0 & \cdots & 0 \\ 0 & a_{22}^{(2)}(\lambda) & \cdots & a_{2n}^{(2)}(\lambda) \\ \vdots & \vdots & & \vdots \\ 0 & a_{m2}^{(2)}(\lambda) & \cdots & a_{mn}^{(2)}(\lambda) \end{bmatrix}$$

where the degree of $a_{11}^{(2)}(\lambda)$ is less than the degree of $a_{11}^{(1)}(\lambda)$. If there still exists some entry $a_{ij}^{(2)}(\lambda)$ that is not divisible by $a_{11}^{(2)}(\lambda)$, repeat the same procedure once more, and so on. After a finite number of steps we obtain the matrix

$$A_3(\lambda) = \begin{bmatrix} a_{11}^{(3)}(\lambda) & 0 & \cdots & 0 \\ 0 & a_{22}^{(3)}(\lambda) & \cdots & a_{2n}^{(3)}(\lambda) \\ \vdots & \vdots & & \vdots \\ 0 & a_{m2}^{(3)}(\lambda) & \cdots & a_{mn}^{(3)}(\lambda) \end{bmatrix}$$

where every $a_{ij}^{(3)}(\lambda)$ is divisible by $a_{11}^{(3)}(\lambda)$. Multiply the first row (or column) by a nonzero constant to make the leading coefficient of the polynomial $a_{11}^{(3)}(\lambda)$ equal to 1. Now define the $(m-1) \times (n-1)$ matrix polynomial

$$A_4(\lambda) = \frac{1}{a_{11}^{(3)}(\lambda)} \begin{bmatrix} a_{22}^{(3)}(\lambda) & \cdots & a_{2n}^{(3)}(\lambda) \\ \vdots & & \vdots \\ a_{m_2}^{(3)}(\lambda) & \cdots & a_{mn}^{(3)}(\lambda) \end{bmatrix}$$

and apply the induction hypothesis for $A_4(\lambda)$ to complete the proof of existence of a Smith form $D(\lambda)$. $\square$

A.2 THE SMITH FORM: UNIQUENESS

We need some preparations to prove the uniqueness of the Smith form $D(\lambda)$ in Theorem A.1.1. Let $A = [a_{ij}]_{i,j=1}^{m,n}$ be an $m \times n$ matrix with complex entries. Choose k rows, $1 \le i_1 < \cdots < i_k \le m$, and k columns, $1 \le j_1 < \cdots < j_k \le n$, in A, and consider the determinant $\det[a_{i_l, j_q}]_{l,q=1}^k$ of the $k \times k$ submatrix of A formed by these rows and columns. This determinant is called a *minor* of A. Loosely speaking, we can say that this minor is of order k and is composed of the rows $i_1, \ldots, i_k$ and columns $j_1, \ldots, j_k$ of A. It is denoted by $A\begin{pmatrix} i_1 & \cdots & i_k \\ j_1 & \cdots & j_k \end{pmatrix}$. We establish the important Binet–Cauchy formula, which expresses the minors of a product of two matrices in terms of the minors of each factor, as follows.

Theorem A.2.1

Let $A = BC$, where B is a $m \times p$ matrix, and C is a $p \times n$ matrix. Then for every k, $1 \le k \le \min(m, n)$ and every minor $A\begin{pmatrix} i_1 & \cdots & i_k \\ j_1 & \cdots & j_k \end{pmatrix}$ of order k we have

$$A\begin{pmatrix} i_1 & \cdots & i_k \\ j_1 & \cdots & j_k \end{pmatrix} = \sum B\begin{pmatrix} i_1 & i_2 & \cdots & i_k \\ \alpha_1 & \alpha_2 & \cdots & \alpha_k \end{pmatrix} C\begin{pmatrix} \alpha_1 & \alpha_2 & \cdots & \alpha_k \\ j_1 & j_2 & \cdots & j_k \end{pmatrix}$$

$$(A.2.1)$$

where the sum is taken over all sequences $\{\alpha_q\}_{q=1}^k$ of integers satisfying $1 \le \alpha_1 < \alpha_2 < \cdots < \alpha_k \le p$. In particular, if $k > p$, then the sum on the right-hand side of (A.2.1) is empty and the equation is interpreted as $A\begin{pmatrix} i_1 & \cdots & i_k \\ j_1 & \cdots & j_k \end{pmatrix} = 0$.

Note that for $k = 1$ formula (A.2.1) is just the rule of multiplication of two matrices. On the other hand, if $m = p = n$ and $k = n$, then (A.2.1) gives the familiar multiplication formula for determinants: $\det(BC) = \det B \cdot \det C$.

Proof. As the rank of A does not exceed p, we have $A\begin{pmatrix} i_1 & \cdots & i_k \\ j_1 & \cdots & j_k \end{pmatrix} = 0$ as long as $k > p$. So we can assume $k \le p$. For simplicity of notation assume also $i_q = j_q = q$, $q = 1, \ldots, k$. Letting $A = [a_{ij}]_{i,j=1}^{m,n}$, $B = [b_{ij}]_{i,j=1}^{m,p}$, $C = [c_{ij}]_{i,j=1}^{p,n}$, we may write $A\begin{pmatrix} 1 & 2 & \cdots & k \\ 1 & 2 & \cdots & k \end{pmatrix}$ in the form

$$\det \begin{bmatrix} \sum_{\alpha_1=1}^{p} b_{1\alpha_1} c_{\alpha_1 1} & \sum_{\alpha_2=1}^{p} b_{1\alpha_2} c_{\alpha_2 2} & \cdots & \sum_{\alpha_k=1}^{p} b_{1\alpha_k} c_{\alpha_k k} \\ \sum_{\alpha_1=1}^{p} b_{2\alpha_1} c_{\alpha_1 1} & \sum_{\alpha_2=1}^{p} b_{2\alpha_2} c_{\alpha_2 2} & \cdots & \sum_{\alpha_k=1}^{p} b_{2\alpha_k} c_{\alpha_k k} \\ \vdots & \vdots & & \vdots \\ \sum_{\alpha_1=1}^{p} b_{k\alpha_1} c_{\alpha_1 1} & \sum_{\alpha_2=1}^{p} b_{k\alpha_2} c_{\alpha_2 2} & \cdots & \sum_{\alpha_k=1}^{p} b_{k\alpha_k} c_{\alpha_k k} \end{bmatrix}$$

and using the linearity of the determinant as a function of each column, this expression is easily seen to be equal to

$$\sum \det \begin{bmatrix} b_{1\alpha_1} c_{\alpha_1 1} & b_{1\alpha_2} c_{\alpha_2 2} & \cdots & b_{1\alpha_k} c_{\alpha_k k} \\ b_{2\alpha_1} c_{\alpha_1 1} & b_{2\alpha_2} c_{\alpha_2 2} & \cdots & b_{2\alpha_k} c_{\alpha_k k} \\ \vdots & \vdots & & \vdots \\ b_{k\alpha_1} c_{\alpha_1 1} & b_{k\alpha_2} c_{\alpha_2 2} & \cdots & b_{k\alpha_k} c_{\alpha_k k} \end{bmatrix}$$

$$= \sum B\begin{pmatrix} 1 & 2 & \cdots & k \\ \alpha_1 & \alpha_2 & \cdots & \alpha_k \end{pmatrix} c_{\alpha_1 1} c_{\alpha_2 2} \cdots c_{\alpha_k k} \qquad (A.2.2)$$

where the sum is taken over all k-tuples of integers $(\alpha_1, \ldots, \alpha_k)$ such that $1 \le \alpha_k \le p$. (Here we use the notation $B\begin{pmatrix} 1 & 2 & \cdots & k \\ \alpha_1 & \alpha_2 & \cdots & \alpha_k \end{pmatrix}$ to denote $\det[b_{l\alpha_q}]_{l,q=1}^{k}$ even when the the sequence $\{\alpha_q\}_{q=1}^{k}$ is not increasing, or when it contains repetitions of numbers.) If not all $\alpha_1, \alpha_2, \ldots, \alpha_k$ are different, then clearly $B\begin{pmatrix} 1 & 2 & \cdots & k \\ \alpha_1 & \alpha_2 & \cdots & \alpha_k \end{pmatrix} = 0$. Ignoring these summands in (A.2.2), split the remaining terms into groups of $k!$ terms each in such a way that the summands in the same group differ only in the order of indices $\alpha_1, \alpha_2, \ldots, \alpha_k$. We obtain:

$$A\begin{pmatrix} 1 & 2 & \cdots & k \\ 1 & 2 & \cdots & k \end{pmatrix}$$

$$= \sum_{1 \le \alpha_1 < \cdots < \alpha_k \le p} \sum_{\pi} B\begin{pmatrix} 1 & 2 & \cdots & k \\ \alpha_{\pi(1)} & \alpha_{\pi(2)} & \cdots & \alpha_{\pi(k)} \end{pmatrix} c_{\alpha_{\pi(1)}1} \cdots c_{\alpha_{\pi(k)}k}$$

$$\text{(A.2.3)}$$

where the internal summation is over all permutations π of $\{1, 2, \ldots, k\}$. Denoting by $\epsilon(\pi)$ the sign of π ($\epsilon(\pi)$ is 1 if π is even and -1 if π is odd), we find that the right-hand side of (A.2.3) is

$$\sum_{1 \le \alpha_1 < \cdots < \alpha_k \le p} \sum_{\pi} \epsilon(\pi) B\begin{pmatrix} 1 & 2 & \cdots & k \\ \alpha_1 & \alpha_2 & \cdots & \alpha_k \end{pmatrix} c_{\alpha_{\pi(1)}1} \cdots c_{\alpha_{\pi(k)}k}$$

$$= \sum_{1 \le \alpha_1 < \cdots < \alpha_k \le p} B\begin{pmatrix} 1 & 2 & \cdots & k \\ \alpha_1 & \alpha_2 & \cdots & \alpha_k \end{pmatrix} C\begin{pmatrix} \alpha_1 & \cdots & \alpha_k \\ 1 & \cdots & k \end{pmatrix}$$

and the theorem is proved. □

Returning to matrix polynomials observe that the minors of a matrix polynomial $A(\lambda)$ are (scalar) polynomials, so we can speak about their greatest common divisors.

Theorem A.2.2

Let $A(\lambda)$ be an $m \times n$ matrix polynomial. Let $p_k(\lambda)$ be the greatest common divisor (with leading coefficient 1) of the minors of $A(\lambda)$ of order k, if not all of them are zeros, and let $p_k(\lambda) \equiv 0$ if all the minors of order k of $A(\lambda)$ are zeros. Let $p_0(\lambda) = 1$ and $D(\lambda) = \text{diag}[d_1(\lambda), \ldots, d_r(\lambda), 0, \ldots, 0]$ be a Smith form of $A(\lambda)$ (which exists by the part of Theorem A.1.1 already proved). Then r is the maximal integer such that $p_r(\lambda) \not\equiv 0$, and

$$d_i(\lambda) = \frac{p_i(\lambda)}{p_{i-1}(\lambda)}, \qquad i = 1, \ldots, r \qquad \text{(A.2.4)}$$

Proof. Let us show that if $A_1(\lambda)$ and $A_2(\lambda)$ are equivalent matrix polynomials, then the greatest common divisors $p_{k,1}(\lambda)$ and $p_{k,2}(\lambda)$ of the minors of order k of $A_1(\lambda)$ and $A_2(\lambda)$, respectively, are equal. Indeed, we have

$$A_1(\lambda) = E(\lambda) A_2(\lambda) F(\lambda)$$

for some matrix polynomials $E(\lambda)$ and $F(\lambda)$ with constant nonzero determinants. Apply Theorem A.2.1 twice to express a minor of $A_1(\lambda)$ of order k as a linear combination of minors of $A_2(\lambda)$ of the same order. Therefore, it follows that $p_{k,2}(\lambda)$ is a divisor of $p_{k,1}(\lambda)$. But the equation

$$A_2(\lambda) = E^{-1}(\lambda)A(\lambda)F^{-1}(\lambda)$$

implies that $p_{k,1}(\lambda)$ is a divisor of $p_{k,2}(\lambda)$. So $p_{k,1}(\lambda) = p_{k,2}(\lambda)$. In the same way one shows that the maximal integer r_1 such that $p_{r,1}(\lambda) \not\equiv 0$ coincides with the maximal integer r_2 such that $p_{r_2,2}(\lambda) \not\equiv 0$.

Now apply this observation for the matrix polynomials $A(\lambda)$ and $D(\lambda)$. It follows that we have to prove Theorem A.2.2 only in the case that $A(\lambda)$ itself is in the diagonal form $A(\lambda) = D(\lambda)$. From the structure of $D(\lambda)$ it is clear that

$$d_1(\lambda)d_2(\lambda)\cdots d_s(\lambda), \qquad s = 1, \ldots, r$$

is the greatest common divisor of the minors of $D(\lambda)$ of order s. So $p_s(\lambda) = d_1(\lambda)\cdots d_s(\lambda)$, $s = 1, \ldots, r$, and (A.2.4) follows. $\square$

Theorem A.2.2 immediately implies the uniqueness of the Smith form (A.1.2). Indeed, Theorem A.2.2 shows that the number r of not identically zero entries in the Smith form of $A(\lambda)$, as well as the entries $d_1(\lambda), \ldots, d_r(\lambda)$ themselves, can be expressed explicitly in terms of $A(\lambda)$, that is, r and $d_1(\lambda), \ldots, d_r(\lambda)$ are uniquely determined by $A(\lambda)$.

A.3   INVARIANT POLYNOMIALS, ELEMENTARY DIVISORS, AND PARTIAL MULTIPLICITIES

In this section we study various invariants appearing in the Smith form for the matrix polynomials. Let $A(\lambda)$ be an $m \times n$ matrix polynomial with the Smith form $D(\lambda)$. The diagonal elements $d_1(\lambda), \ldots, d_r(\lambda)$ in $D(\lambda)$ are called the *invariant polynomials* of $A(\lambda)$. The number r of invariant polynomials can be defined as

$$r = \max_{\lambda \in \mathcal{C}}\{\operatorname{rank} A(\lambda)\} \tag{A.3.1}$$

Indeed, since $E(\lambda)$ and $F(\lambda)$ from (A.1.3) are invertible matrices for every λ, we have $\operatorname{rank} A(\lambda) = \operatorname{rank} D(\lambda)$ for every $\lambda \in \mathcal{C}$. On the other hand, it is clear that $\operatorname{rank} D(\lambda) = r$ if λ is not a zero of one of the invariant polynomials, and $\operatorname{rank} D(\lambda) < r$ otherwise. So (A.3.1) follows.

The set of invariant polynomials forms a complete invariant for equivalence of matrix polynomials of the same size.

Theorem A.3.1

Matrix polynomials $A(\lambda)$ and $B(\lambda)$ of the same size are equivalent if and only if the invariant polynomials of $A(\lambda)$ and $B(\lambda)$ are the same.

Proof. Suppose the invariant polynomials of $A(\lambda)$ and $B(\lambda)$ are the same. Then their Smith forms are equal:

$$A(\lambda) = E_1(\lambda)D(\lambda)F_1(\lambda), \qquad B(\lambda) = E_2(\lambda)D(\lambda)F_2(\lambda)$$

where $\det E_i(\lambda) \equiv \text{const} \neq 0$, $\det F_i(\lambda) \equiv \text{const} \neq 0$, $i = 1, 2$. Consequently

$$(E_1(\lambda))^{-1}A(\lambda)(F_1(\lambda))^{-1} = (E_2(\lambda))^{-1}B(\lambda)(F_2(\lambda))^{-1}(=D(\lambda))$$

and

$$A(\lambda) = E(\lambda)B(\lambda)F(\lambda)$$

where $E(\lambda) = E_1(\lambda)(E_2(\lambda))^{-1}$, $F(\lambda) = F_1(\lambda)(F_2(\lambda))^{-1}$. Since $E_2(\lambda)$ and $F_2(\lambda)$ are matrix polynomials with constant nonzero determinants, the same is true for $E^{-1}(\lambda)$ and $F^{-1}(\lambda)$, and, consequently, for $E(\lambda)$ and $F(\lambda)$. So $A(\lambda) \sim B(\lambda)$.

Conversely, suppose $A(\lambda) = E(\lambda)B(\lambda)F(\lambda)$, where $\det E(\lambda) \equiv \text{const} \neq 0$, $\det F(\lambda) \equiv \text{const} \neq 0$. Let $D(\lambda)$ be the Smith form for $B(\lambda)$:

$$B(\lambda) = E_1(\lambda)D(\lambda)F_1(\lambda)$$

Then $D(\lambda)$ is also the Smith form for $A(\lambda)$:

$$A(\lambda) = E(\lambda)E_1(\lambda)D(\lambda)F_1(\lambda)F(\lambda)$$

By the uniqueness of the Smith form for $A(\lambda)$ [more exactly, by the uniqueness of the invariant polynomials of $A(\lambda)$], it follows that the invariant polynomials of $A(\lambda)$ are the same as those of $B(\lambda)$. $\square$

We now take advantage of the fact that the polynomial entries of $A(\lambda)$ and its Smith form $D(\lambda)$ are over $\mathbb{C}$ to represent each invariant polynomial $d_i(\lambda)$ as a product of linear factors:

$$d_i(\lambda) = (\lambda - \lambda_{i1})^{\alpha_{i1}} \cdots (\lambda - \lambda_{i,k_i})^{\alpha_{ik_i}}, \qquad i = 1, \ldots, r$$

where $\lambda_{i1}, \ldots, \lambda_{i,k_i}$ are different complex numbers and $\alpha_{i1}, \ldots, \alpha_{ik_i}$ are positive integers. The factors $(\lambda - \lambda_{ij})^{\alpha_{ij}}$, $j = 1, \ldots, k_i$, $i = 1, \ldots, r$ are called the *elementary divisors* of $A(\lambda)$.

Some different elementary divisors may contain the same polynomial $(\lambda - \lambda_0)^{\alpha}$ (this happens, for example, in case $d_i(\lambda) = d_{i+1}(\lambda)$ for some i); the total number of elementary divisors of $A(\lambda)$ is thus $\Sigma_{i=1}^{r} k_i$.

The degrees α_{ij} of the elementary divisors form an important characteristic of the matrix polynomial $A(\lambda)$. Here we mention only the following simple property of the elementary divisors, whose verification is left to the reader.

Proposition A.3.2

Let $A(\lambda)$ be an $n \times n$ matrix polynomial such that $\det A(\lambda) \neq 0$. Then the sum $\sum_{i=1}^{r} \sum_{j=1}^{k_i} \alpha_{ij}$ of degrees of its elementary divisors $(\lambda - \lambda_{ij})^{\alpha_{ij}}$ coincides with the degree of $\det A(\lambda)$.

Note that the knowledge of the elementary divisors of $A(\lambda)$ and the number r of its invariant polynomials $d_1(\lambda), \ldots, d_r(\lambda)$ is sufficient to construct $d_1(\lambda), \ldots, d_r(\lambda)$. In this construction we use the fact that $d_i(\lambda)$ is divisible by $d_{i-1}(\lambda)$. Let $\lambda_1, \ldots, \lambda_p$ be all the different complex numbers that appear in the elementary divisors, and let $(\lambda - \lambda_i)^{\alpha_{i1}}, \ldots, (\lambda - \lambda_i)^{\alpha_{i,k_i}}$ $(i = 1, \ldots, p)$ be the elementary divisors containing the number λ_i, and ordered in the descending order of the degrees $\alpha_{i1} \geq \cdots \geq \alpha_{i,k_i} > 0$. Clearly, the number r of invariant polynomials must be greater than or equal to $\max\{k_1, \ldots, k_p\}$. Under this condition, the invariant polynomials $d_1(\lambda), \ldots, d_r(\lambda)$ are given by the formulas

$$d_j(\lambda) = \prod_{i=1}^{p} (\lambda - \lambda_i)^{\alpha_{i,r+1-j}}, \qquad j = 1, \ldots, r$$

where we put $(\lambda - \lambda_i)^{\alpha_{ij}} = 1$ for $j > k_i$.

The following property of the elementary divisors is used subsequently.

Proposition A.3.3

Let $A(\lambda)$ and $B(\lambda)$ be matrix polynomials, and let $C(\lambda) = \text{diag}[A(\lambda), B(\lambda)]$, a block-diagonal matrix polynomial. Then the set of elementary divisors of $C(\lambda)$ is the union of the elementary divisors of $A(\lambda)$ and $B(\lambda)$.

Proof. Let $D_1(\lambda)$ and $D_2(\lambda)$ be the Smith forms of $A(\lambda)$ and $B(\lambda)$, respectively. Then clearly

$$C(\lambda) = E(\lambda) \begin{bmatrix} D_1(\lambda) & 0 \\ 0 & D_2(\lambda) \end{bmatrix} F(\lambda)$$

for some matrix polynomials $E(\lambda)$ and $F(\lambda)$ with constant nonzero determinant. Let $(\lambda - \lambda_0)^{\alpha_1}, \ldots, (\lambda - \lambda_0)^{\alpha_p}$, and $(\lambda - \lambda_0)^{\beta_1}, \ldots, (\lambda - \lambda_0)^{\beta_q}$ be the elementary divisors of $D_1(\lambda)$ and $D_2(\lambda)$, respectively, corresponding to the same complex number λ_0. Arrange the set of exponents $\alpha_1, \ldots, \alpha_p, \beta_1, \ldots, \beta_q$, in a nonincreasing order:

$$\{\alpha_1, \ldots, \alpha_p, \beta_1, \ldots, \beta_q\} = \{\gamma_1, \ldots, \gamma_{p+q}\}$$

where $0 < \gamma_1 \leq \cdots \leq \gamma_{p+q}$. Using Theorem A.2.2 it is clear that in the Smith form $D = \text{diag}[d_1(\lambda), \ldots, d_r(\lambda), 0, \ldots, 0]$ of $\text{diag}[D_1(\lambda), D_2(\lambda)]$, the invariant polynomial $d_r(\lambda)$ is divisible by $(\lambda - \lambda_0)^{\gamma_{p+q}}$ but not by $(\lambda -$

$\lambda_0)^{\gamma_{p+q}+1}$, $d_{r-1}(\lambda)$ is divisible by $(\lambda - \lambda_0)^{\gamma_{p+q}-1}$ but not by $(\lambda - \lambda_0)^{\gamma_{p+q-1}+1}$, and so on. It follows that the elementary divisors of

$$\begin{bmatrix} D_1(\lambda) & 0 \\ 0 & D_2(\lambda) \end{bmatrix}$$

[and thus also those of $C(\lambda)$] corresponding to λ_0, are just $(\lambda - \lambda_0)^{\gamma_1}, \ldots, (\lambda - \lambda_0)^{\gamma_{p+q}}$, and Proposition A.3.3 is proved. $\square$

In the rest of this section we assume that (as in Proposition A.3.2) the matrix polynomial $A(\lambda)$ is square and that the determinant of $A(\lambda)$ is not identically zero. In this case, complex numbers λ_0 such that $\det A(\lambda_0) = 0$ are called the *eigenvalues* of $A(\lambda)$. Clearly, the set of eigenvalues is finite [it contains not more than degree $(\det A(\lambda))$ points], and λ_0 is an eigenvalue of $A(\lambda)$ if and only if there is an elementary divisor of $A(\lambda)$ of type $(\lambda - \lambda_0)^\alpha$.

Let λ_0 be an eigenvalue of $A(\lambda)$, and let $(\lambda - \lambda_0)^{\alpha_1}, \ldots, (\lambda - \lambda_0)^{\alpha_p}$ be all the elementary divisors of $A(\lambda)$ that are divisible by $\lambda - \lambda_0$. The exponents $\alpha_1, \ldots, \alpha_p$ are called the *partial multiplicities* of $A(\lambda)$ corresponding to λ_0. Recall that some of the numbers $\alpha_1, \ldots, \alpha_p$ may be equal; the number α_j appears in the list of partial multiplicities as many times as there are elementary divisors $(\lambda - \lambda_0)^{\alpha_i}$ of $A(\lambda)$. The partial multiplicities play an important role in the following representation of matrix polynomials.

Theorem A.3.4

Let $A(\lambda)$ be an $n \times n$ matrix polynomial with $\det A(\lambda) \not\equiv 0$. Then for every $\lambda_0 \in \mathbb{C}$, $A(\lambda)$ admits the representation

$$A(\lambda) = E_{\lambda_0}(\lambda) \begin{bmatrix} (\lambda - \lambda_0)^{\kappa_1} & & 0 \\ & \ddots & \\ 0 & & (\lambda - \lambda_0)^{\kappa_n} \end{bmatrix} F_{\lambda_0}(\lambda) \qquad (A3.2)$$

where $E_{\lambda_0}(\lambda)$ and $F_{\lambda_0}(\lambda)$ are matrix polynomials invertible at λ_0, and $\kappa_1 \leq \cdots \leq \kappa_n$ are nonnegative integers, which coincide (after striking off zeros) with the partial multiplicities of $A(\lambda)$ corresponding to λ_0.

Proof. The existence of representation (A.3.2) follows easily from the Smith form. Namely, let $D(\lambda) = \operatorname{diag}[d_1(\lambda), \ldots, d_n(\lambda)]$ be the Smith form of $A(\lambda)$, and let

$$A(\lambda) = E(\lambda)D(\lambda)F(\lambda) \qquad (A.3.3)$$

where $\det E(\lambda) \equiv \operatorname{const} \neq 0$, $\det F(\lambda) \equiv \operatorname{const} \neq 0$. Represent each $d_i(\lambda)$ in the form

$$d_i(\lambda) = (\lambda - \lambda_0)^{\kappa_i}\bar{d}_i(\lambda), \qquad i = 1, \ldots, n$$

where $\bar{d}_i(\lambda_0) \neq 0$ and $\kappa_i \geq 0$. Since $d_i(\lambda)$ is divisible by $d_{i-1}(\lambda)$, we have $\kappa_i \geq \kappa_{i-1}$. Now (A.3.2) follows from (A.3.3), where

$$E_{\lambda_0}(\lambda) = E(\lambda) \operatorname{diag}[\bar{d}_1(\lambda), \ldots, \bar{d}_n(\lambda)], \qquad F_{\lambda_0}(\lambda) = F(\lambda)$$

It remains to show that the κ_i coincide (after striking off zeros) with the degrees of elementary divisors of $A(\lambda)$ corresponding to λ_0. To this end we show that any factorization of $A(\lambda)$ of type (A.3.2) with $\kappa_1 \leq \cdots \leq \kappa_n$ implies that κ_j is the multiplicity of λ_0 as a zero of $d_j(\lambda)$, $j = 1, \ldots, n$, where $D(\lambda) = \operatorname{diag}[d_1(\lambda), \ldots, d_n(\lambda)]$ is the Smith form of $A(\lambda)$. Indeed, let

$$A(\lambda) = E(\lambda)D(\lambda)F(\lambda)$$

where $E(\lambda)$ and $F(\lambda)$ are matrix polynomials with constant nonzero determinants. Comparing with (A.3.2), write

$$\operatorname{diag}[d_1(\lambda), \ldots, d_n(\lambda)] = \bar{E}_{\lambda_0}(\lambda) \operatorname{diag}[(\lambda - \lambda_0)^{\kappa_1}, \ldots, (\lambda - \lambda_0)^{\kappa_n}]\bar{F}_{\lambda_0}(\lambda)$$

$$\text{(A.3.4)}$$

where $\bar{E}_{\lambda_0}(\lambda) = (E(\lambda))^{-1}E_{\lambda_0}(\lambda)$, $\bar{F}_{\lambda_0}(\lambda)(F(\lambda))^{-1}$ are matrix polynomials invertible at λ_0. Applying Theorem A.2.1, we obtain

$$d_1(\lambda)d_2(\lambda)\cdots d_{i_0}(\lambda) = \sum_{i,j,k} m_{i,\bar{E}}(\lambda) \cdot m_{j,D_{\lambda_0}}(\lambda) \cdot m_{k,\bar{F}}(\lambda), \qquad i = 1, 2, \ldots, n$$

$$\text{(A.3.5)}$$

where $m_{i,\bar{E}}(\lambda)$ [resp. $m_{j,D_{\lambda_0}}(\lambda)$, $m_{k,\bar{F}}(\lambda)$] is a minor of order i_0 of $\bar{E}(\lambda)$ [resp. $\operatorname{diag}[(\lambda - \lambda_0)^{\kappa_1}, \ldots, (\lambda - \lambda_0)^{\kappa_n}]$, $\bar{F}(\lambda)$], and the sum in (A.3.5) is taken over a certain set of triples (i, j, k). It follows from (A.3.5) and the condition $\kappa_1 \leq \cdots \leq \kappa_n$, that λ_0 is a zero of the product $d_1(\lambda)d_2(\lambda)\cdots d_{i_0}(\lambda)$ of multiplicity at least $\kappa_1 + \kappa_2 + \cdots + \kappa_{i_0}$. Rewrite (A.3.4) in the form

$$(\bar{E}_{\lambda_0}(\lambda))^{-1} \operatorname{diag}[d_1(\lambda), \ldots, d_n(\lambda)](\bar{F}_{\lambda_0}(\lambda))^{-1}$$
$$= \operatorname{diag}[(\lambda - \lambda_0)^{\kappa_1}, \ldots, (\lambda - \lambda_0)^{\kappa_n}]$$

and apply Theorem A.2.1 again. Using the fact that $(\bar{E}_{\lambda_0}(\lambda))^{-1}$ and $(\bar{F}_{\lambda_0}(\lambda))^{-1}$ are rational matrix functions that are defined and invertible at $\lambda = \lambda_0$, and that $d_i(\lambda)$ is a divisor of $d_{i+1}(\lambda)$, we deduce that

$$(\lambda - \lambda_0)^{\kappa_1 + \cdots + \kappa_{i_0}} = d_1(\lambda)d_2(\lambda)\cdots d_{i_0}(\lambda)\Phi_{i_0}(\lambda)$$

where $\Phi_{i_0}(\lambda)$ is a rational function defined at $\lambda = \lambda_0$ (i.e., λ_0 is not a pole of

$\Phi_{i_0}(\lambda)$). It follows that λ_0 is a zero of $d_1(\lambda)d_2(\lambda)\cdots d_{i_0}(\lambda)$ of multiplicity exactly $\kappa_1 + \kappa_2 + \cdots + \kappa_{i_0}$, $i = 1, \ldots, n$. Hence κ_i is exactly the multiplicity of λ_0 as a zero of $d_i(\lambda)$ for $i = 1, \ldots, n$; that is, the nonzero numbers (if any) among $\kappa_1, \ldots, \kappa_n$ are the partial multiplicities of $A(\lambda)$ corresponding to λ_0. $\square$

As a consequence of Theorem A.3.4, note that there are nonzero κ_i in the representation (A.3.2) if and only if λ_0 is an eigenvalue of $A(\lambda)$.

A.4 EQUIVALENCE OF LINEAR MATRIX POLYNOMIALS

We study here equivalence and the Smith form for matrix polynomials of type $I\lambda - A$, where A is an $n \times n$ matrix. It turns out that for such matrix polynomials the notion of equivalence is closely related to similarity.

Theorem A.4.1

$I\lambda - A \sim I\lambda - B$ if and only if A and B are similar.

To prove this theorem, we have to introduce division of matrix polynomials.

We restrict ourselves to the case when the dividend is a general matrix polynomial $A(\lambda) = \sum_{j=0}^{l} A_j \lambda^j$, and the divisor is a matrix polynomial of type $I\lambda + X$, where X is a constant $n \times n$ matrix. In this case the following representation holds:

$$A(\lambda) = Q_r(\lambda)(I\lambda + X) + R_r \qquad (A.4.1)$$

where $Q_r(\lambda)$ is a matrix polynomial, which is called the *right quotient*, and R_r is a constant matrix, which is called the *right remainder*, on division of $A(\lambda)$ by $I\lambda + X$. Also

$$A(\lambda) = (I\lambda + X)Q_l(\lambda) + R_l \qquad (A.4.2)$$

where $Q_l(\lambda)$ is the left quotient, and the constant matrix R_l is the left remainder.

Let us check the existence of representation (A.4.1); (A.4.2) can be checked in a similar way. If $l = 0$ [i.e., $A(\lambda)$ is constant], put $Q_r(\lambda) \equiv 0$ and $R_r = A(\lambda)$. So we can suppose $l \geq 1$. Write $Q_r(\lambda) = \sum_{j=0}^{l-1} Q_j^{(r)} \lambda^j$. Comparing the coefficients of powers of λ on the right- and left-hand sides of (A.4.1), we can rewrite this relation as follows:

$$A_l = Q_{l-1}^{(r)}, \qquad A_{l-1} = Q_{l-2}^{(r)} + Q_{l-1}^{(r)}X, \ldots, A_1 = Q_0^{(r)} + Q_1^{(r)}X,$$

$$A_0 = Q_0^{(r)}X + R_r$$

Clearly, these relations define $Q_{l-1}^{(r)}, \ldots, Q_1^{(r)}, Q_0^{(r)}$, and R_r, sequentially.

It follows from this argument that the left and right quotient and remainder are uniquely defined.

Proof of Theorem A.4.1. In one direction this result is immediate: if $A = SBS^{-1}$ for some nonsingular S, then the equality $I\lambda - A = S(I\lambda - B)S^{-1}$ proves the equivalence of $I\lambda - A$ and $I\lambda - B$. Conversely, suppose $I\lambda - A \sim I\lambda - B$. Then for some matrix polynomials $E(\lambda)$ and $F(\lambda)$ with constant nonzero determinant we have

$$E(\lambda)(I\lambda - A)F(\lambda) = I\lambda - B$$

Suppose that division of $(E(\lambda))^{-1}$ on the left by $I\lambda - A$ and of $F(\lambda)$ on the right by $I\lambda - B$ yield

$$(E(\lambda))^{-1} = (I\lambda - A)S(\lambda) + E_0$$

$$F(\lambda) = T(\lambda)(I\lambda - B) + F_0 \tag{A.4.3}$$

Substituting in the equation

$$(E(\lambda))^{-1}(I\lambda - B) = (I\lambda - A)F(\lambda)$$

we obtain

$$\{(I\lambda - A)S(\lambda) + E_0\}(I\lambda - B) = (I\lambda - A)\{T(\lambda)(I\lambda - B) + F_0\}$$

whence

$$(I\lambda - A)(S(\lambda) - T(\lambda))(I\lambda - B) = (I\lambda - A)F_0 - E_0(I\lambda - B)$$

Since the degree of the matrix polynomial on the right-hand side here is 1, it follows that $S(\lambda) = T(\lambda)$; otherwise, the degree of the matrix polynomial on the left is at least 2. Hence

$$(I\lambda - A)F_0 = E_0(\lambda I - B)$$

so that

$$F_0 = E_0, \qquad AF_0 = E_0B, \qquad AE_0 = E_0B_0$$

It remains only to prove that E_0 is nonsingular. To this end divide $E(\lambda)$ on the left by $I\lambda - B$:

$$E(\lambda) = (I\lambda - B)U(\lambda) + R_0 \tag{A.4.4}$$

Then, using (A.4.3) and (A.4.4), we have

$$I = (E(\lambda))^{-1}E(\lambda) = \{(I\lambda - A)S(\lambda) + E_0\}\{(I\lambda - B)U(\lambda) + R_0\}$$

$$= (I\lambda - A)\{S(\lambda)(I\lambda - B)U(\lambda)\} + (I\lambda - A)F_0U(\lambda)$$

$$\quad + (I\lambda - A)S(\lambda)R_0 + E_0R_0$$

$$= (I\lambda - A)[S(\lambda)(I\lambda - B)U(\lambda) + F_0U(\lambda) + S(\lambda)R_0] + E_0R_0$$

Hence the matrix polynomial in the square brackets is zero, and $E_0R_0 = I$. It follows that E_0 is nonsingular. $\square$

The definitions of eigenvalues and partial multiplicities made in the preceding section can be applied to an $n \times n$ matrix polynomial of the form $I\lambda - A$. On the other hand, as an $n \times n$ matrix (or as a transformation represented by this matrix in the standard basis $e_1, \ldots, e_n$), A has eigenvalues and partial multiplicities as defined in Sections 1.2 and 2.2. It is an important fact that these notions for $I\lambda - A$ and for A coincide.

Theorem A.4.2

A complex number λ_0 is an eigenvalue of $I\lambda - A$ if and only if it is an eigenvalue of A. Moreover, the partial multiplicities of $I\lambda - A$ corresponding to its eigenvalue λ_0 coincide with the partial multiplicities of A corresponding to λ_0.

Proof. The first statement follows from the definitions: λ_0 is an eigenvalue of $I\lambda - A$ if and only if $\det(I\lambda - A) = 0$, which is exactly the definition of an eigenvalue of A. For the proof of the second statement, we can assume that A is in the Jordan form. Further, using Proposition A.3.3, we reduce the proof to the case when A is a single Jordan block of size $n \times n$:

$$A = \begin{bmatrix} \lambda_0 & 1 & 0 & \cdots & 0 \\ 0 & \lambda_0 & 1 & \cdots & 0 \\ \vdots & \vdots & & \ddots & \vdots \\ & & & & 1 \\ 0 & 0 & & \cdots & \lambda_0 \end{bmatrix}$$

The partial multiplicity of A is clearly n, corresponding to the eigenvalue λ_0. To find the partial multiplicities of $I\lambda - A$, observe that

$$I\lambda - A = \begin{bmatrix} \lambda - \lambda_0 & -1 & 0 & \cdots & 0 \\ 0 & \lambda - \lambda_0 & -1 & \cdots & 0 \\ \vdots & \vdots & & \ddots & \vdots \\ & & & & -1 \\ 0 & 0 & & \cdots & \lambda - \lambda_0 \end{bmatrix}$$

has a nonzero minor of order $n - 1$ that is independent of λ (namely, the

minor formed by crossing out the first column and the last row in $I\lambda - A$).
As $\det(I\lambda - A) = (\lambda - \lambda_0)^n$, Theorem A.2.2 implies that the Smith form of
$I\lambda - A$ is $\text{diag}[1, 1, \ldots, 1, (\lambda - \lambda_0)^n]$. So the only partial multiplicity of
$I\lambda - A$ is n, which corresponds to λ_0. □

We also need the following connection between the partial multiplicities
of a matrix A and submatrices of $I\lambda - A$.

Theorem A.4.3

Let A be an $n \times n$ matrix. Let $\alpha_1 \geq \cdots \geq \alpha_m$ be the partial multiplicities of an
eigenvalue λ_0 of A, and put $\alpha_i = 0$ for $i = m + 1, \ldots, n$. Then $\alpha_n + \alpha_{n-1} +$
$\cdots + \alpha_{n-p+1}$ is the minimal multiplicity of λ_0 as a zero of the determinant
(considered as a polynomial in λ) of any $p \times p$ submatrix in $I\lambda - A$.

Proof. By Theorems A.4.2 and A.3.4 we have the following representation:

$$I\lambda - A = E_{\lambda_0}(\lambda) \, \text{diag}[(\lambda - \lambda_0)^{\alpha_n}, (\lambda - \lambda_0)^{\alpha_{n-1}}, \ldots, (\lambda - \lambda_0)^{\alpha_1}] F_{\lambda_0}(\lambda)$$

$$(A.4.5)$$

where $E_{\lambda_0}(\lambda)$ and $F_{\lambda_0}(\lambda)$ are matrix polynomials invertible for $\lambda = \lambda_0$. Now
the Binet–Cauchy formula (Theorem A.2.1) implies that the multiplicity of
λ_0 as a zero of the determinant of any $p \times p$ submatrix in $I\lambda - A$ is at least
$\alpha_n + \alpha_{n-1} + \cdots + \alpha_{n-p+1}$. Rewriting (A.4.5) in the form

$$E_{\lambda_0}(\lambda)^{-1}(I\lambda - A)F_{\lambda_0}(\lambda)^{-1} = \text{diag}[(\lambda - \lambda_0)^{\alpha_n}, (\lambda - \lambda_0)^{\alpha_{n-1}}, \ldots, (\lambda - \lambda_0)^{\alpha_1}]$$

and using the Binet–Cauchy formula again, we find that

$$(\lambda - \lambda_0)^{\alpha_n + \alpha_{n-1} + \cdots + \alpha_{n-p+1}} = \sum_{i=1}^{s} \varphi_i(\lambda) \det(A_i(\lambda)) \qquad (A.4.6)$$

where $A_1(\lambda), \ldots, A_s(\lambda)$ are certain $p \times p$ submatrices in $I\lambda - A$, and
$\varphi_i(\lambda)$, $i = 1, \ldots, s$ are rational functions defined at λ_0 [so λ_0 is not a pole of
any $\varphi_i(\lambda)$]. It follows from equation (A.4.6) that at least one of the minors
$\det(A_i(\lambda))$ has a zero at λ_0 with multiplicity exactly equal to $\alpha_n + \alpha_{n-1} +$
$\cdots + \alpha_{n-p+1}$. □

A.5   STRICT EQUIVALENCE OF LINEAR MATRIX POLYNOMIALS: REGULAR CASE

Let $A + \lambda B$ and $A_1 + \lambda B_1$ be two linear matrix polynomials of the same size
$m \times n$. We say that $A + \lambda B$ and $A_1 + \lambda B_1$ are *strictly equivalent* if there exist

invertible matrices P and Q of sizes $m \times m$ and $n \times n$, respectively, independent of λ, such that, for all $\lambda \in \mathbb{C}$, we obtain

$$P(A + \lambda B)Q = A_1 + \lambda B_1$$

We denote strict equivalence by $A + \lambda B \overset{s}{\sim} A_1 + \lambda B_1$. It is easily seen that strict equivalence is indeed an equivalence relation, that is, that the three following properties hold: $A + \lambda B \overset{s}{\sim} A + \lambda B$ for every polynomial $A + \lambda B$. If $A + \lambda B \overset{s}{\sim} A_1 + \lambda B_1$, then also $A_1 + \lambda B_1 \overset{s}{\sim} A + \lambda B$. If $A + \lambda B \overset{s}{\sim} A_1 + \lambda B_1$ and $A_1 + \lambda B_1 \overset{s}{\sim} A_2 + \lambda B_2$, then $A + \lambda B \overset{s}{\sim} A_2 + \lambda B_2$.

Obviously, strict equivalence of linear matrix polynomials implies their equivalence. The converse is not true in general, as we see later in this section.

In this and subsequent sections we find the invariants of strict equivalence, as well as the simplest representative (the canonical form) in each class of strictly equivalent linear matrix polynomials. This section is devoted to the *regular* case. That is, when A and B are square matrices and $\det(A + \lambda B)$ does not vanish identically. In particular, the polynomials $A + \lambda B$ with squares matrices A and B and $\det B \neq 0$ are regular. This hypothesis is used in our first result.

Proposition A.5.1

Two regular polynomials $A + \lambda B$ and $A_1 + \lambda B_1$ with $\det B \neq 0$, $\det B_1 \neq 0$ are strictly equivalent if and only if they have the same invariant polynomials (or, equivalently, the same elementary divisors).

The proof is easily obtained by combining Theorems A.3.1 and A.4.1. However, the result of Proposition A.5.1 is false, in general, if we omit the conditions $\det B \neq 0$, $\det B_1 \neq 0$ and require only that the polynomials are regular.

EXAMPLE A.5.1. Let

$$A = A_1 = I_2, \qquad B = \begin{bmatrix} 0 & 1 \\ 0 & 0 \end{bmatrix}, \qquad B_1 = 0$$

The polynomials

$$A + \lambda B = \begin{bmatrix} 1 & \lambda \\ 0 & 1 \end{bmatrix}, \qquad \text{and} \qquad A_1 + \lambda B_1 = \begin{bmatrix} 1 & 0 \\ 0 & 1 \end{bmatrix}$$

are obviously regular, and both have the Smith form $\begin{bmatrix} 1 & 0 \\ 0 & 1 \end{bmatrix}$, that is, the same invariant polynomials. However, they cannot be strictly equivalent because B and B_1 have different ranks. (If $A + \lambda B$ and $A_1 + \lambda B_1$ were strictly equivalent, we would have $B = PB_1Q$ for some invertible P and Q, and this would imply the equality of the ranks of B and B_1.)   □

To extend the result of Proposition A.5.1 to the class of all regular polynomials $A + \lambda B$, we must introduce the elementary divisors at infinity. We say that λ^p is an elementary divisor at infinity of a regular polynomial $A + \lambda B$ if λ^p is an elementary divisor of $\lambda A + B$. Clearly, there exist elementary divisors at infinity of $A + \lambda B$ if and only if $\det B = 0$.

Theorem A.5.2

Two regular polynomials $A + \lambda B$ and $A_1 + \lambda B_1$ are strictly equivalent if and only if the elementary divisors of $A + \lambda B$ and $A_1 + \lambda B_1$ are the same and their elementary divisors at infinity are the same.

Proof.   Assume that $A + \lambda B$ and $A_1 + \lambda B_1$ are strictly equivalent. Then obviously $A + \lambda B$ and $A_1 + \lambda B_1$ are equivalent, so by Theorem A.3.1 they have the same elementary divisors. Moreover, $\lambda A + B$ and $\lambda A_1 + B_1$ are equivalent as well, so, by the same Theorem A.3.1, $A + \lambda B$ and $A_1 + \lambda B_1$ have the same elementary divisors at infinity.

To prove the second part of the theorem, we introduce homogeneous linear matrix polynomials. Thus we consider the polynomial $\mu A + \lambda B$ where $\mu, \lambda \in \mathbb{C}$. Note that every minor $m(\lambda, \mu)$ of order r of $\mu A + \lambda B$ is a polynomial of two complex variables μ and λ that is homogeneous of order r in the sense that

$$m(\alpha\lambda, \alpha\mu) = \alpha^r m(\lambda, \mu)$$

for every $\alpha, \lambda, \mu \in \mathbb{C}$. For a fixed $r, 1 \leq r \leq n$, let $p_r(\lambda, \mu)$ be the greatest common divisor in the set of homogeneous polynomials of all the nonzero minors $m_1(\lambda, \mu), \ldots, m_s(\lambda, \mu)$ of order r of $\mu A + \lambda B$. In other words, $p_r(\lambda, \mu)$ is a homogeneous polynomial that divides each $m_i(\lambda, \mu)$, and if $q(\lambda, \mu)$ is another homogeneous polynomial with this property, then $q(\lambda, \mu)$ divides $p_r(\lambda, \mu)$. Clearly, $p_{i-1}(\lambda, \mu)$ divides $p_i(\lambda, \mu)$. The polynomials $p_1(\lambda, \mu), \ldots, p_n(\lambda, \mu)$ are called the *invariant polynomials* of $\mu A + \lambda B$. As each minor $m(\lambda, \mu)$ of $\mu A + \lambda B$ is a homogeneous polynomial in λ and μ, it admits factorizations of the form

$$m(\lambda, \mu) = \mu^\beta \prod_{j=1}^{q} (\lambda + \alpha_j\mu)^{\gamma_j} = \lambda^{\beta'} \prod_{j=1}^{q'} (\mu + \alpha_j'\lambda)^{\gamma_j'}$$

for some complex numbers α_j and α_j'. (In fact, the nonzero α_j' values are the reciprocals of the nonzero α_j values.) Using factorizations of this kind, it is easily seen that $p_1(\lambda, 1), \ldots, p_n(\lambda, 1)$ are the invariant polynomials of $A + \lambda B$, whereas $p_1(1, \mu), \ldots, p_n(1, \mu)$ are the invariant polynomials of $\mu A + B$.

Returning to the proof of Theorem A.5.2, assume that the elementary divisors of $A + \lambda B$ and $A_1 + \lambda B_1$, including those at infinity, are the same. This means that the invariant polynomials of $A + \lambda B$ and $A_1 + \lambda B_1$ are the same, and so are the invariant polynomials of $\mu A + B$ and $\mu A_1 + B_1$. Since a homogeneous polynomial $p(\lambda, \mu)$ of λ and μ is uniquely defined by $p(\lambda, 1)$ and $p(1, \mu)$, it follows from the discussion in the preceding paragraph that the invariant polynomials of $\mu A + \lambda B$ and of $\mu A_1 + \lambda B_1$ are the same. Now we make a change of variables: $\lambda = x_1 \tilde{\lambda} + x_2 \tilde{\mu}$, $\mu = y_1 \tilde{\lambda} + y_2 \tilde{\mu}$, where $x_1 y_2 - x_2 y_1 \neq 0$. Then the invariant polynomials of $\tilde{\mu} \tilde{A} + \tilde{\lambda} \tilde{B}$ and of $\tilde{\mu} \tilde{A}_1 + \tilde{\lambda} \tilde{B}_1$ are again the same, where $\tilde{A} = y_2 A + x_2 B$, $\tilde{B} = y_1 A + x_1 B$, $\tilde{A}_1 = y_2 A_1 + x_2 B_1$, $\tilde{B}_1 = y_1 A_1 + x_1 B_1$. As the polynomials $A + \lambda B$ and $A_1 + \lambda B_1$ are regular, we can choose x_1 and y_1 in such a way that $\det \tilde{B} \neq 0$ and $\det \tilde{B}_1 \neq 0$. Apply Proposition A.5.1 to deduce that $\tilde{A} + \tilde{\lambda} \tilde{B}$ and $\tilde{A}_1 + \tilde{\lambda} \tilde{B}_1$ are strictly equivalent: $P\tilde{A}Q = \tilde{A}_1$, $P\tilde{B}Q = \tilde{B}_1$ for some invertible matrices P and Q. Since

$$A = \frac{1}{\Delta}(x_1 \tilde{A} - x_2 \tilde{B}), \qquad B = \frac{1}{\Delta}(-y_1 \tilde{A} + y_2 \tilde{B})$$

where $\Delta = x_1 y_2 - x_2 y_1$, and similarly for A_1 and B_1, we obtain $PAQ = A_1$, $PBQ = B_1$, and the strict equivalence of $A + \lambda B$ and $A_1 + \lambda B_1$ follows. $\square$

Theorem A.5.2 allows us to obtain the canonical form for strict equivalence of regular linear matrix polynomials, as follows.

Theorem A.5.3

Every regular, linear, matrix polynomial $A + \lambda B$ is strictly equivalent to a linear polynomial of the form

$$(I_{k_1} + \lambda J_{k_1}(0)) \oplus \cdots \oplus (I_{k_p} + \lambda J_{k_p}(0)) \oplus (\lambda I_{l_1} + J_{l_1}(\lambda_1)) \oplus \cdots \oplus (\lambda I_{l_q} + J_{l_q}(\lambda_q))$$

$$(A.5.1)$$

where $J_k(\lambda)$ is the $k \times k$ Jordan block with eigenvalue λ. The linear polynomial (A.5.1) is uniquely determined by $A + \lambda B$. In fact, $\lambda^{k_1}, \ldots, \lambda^{k_p}$ are the elementary divisors at infinity of $A + \lambda B$, whereas $(\lambda + \lambda_i)^{l_i}$, $i = 1, \ldots, q$ are the elementary divisors of $A + \lambda B$.

Proof. Let $A_1 + \lambda B_1$ be the polynomial (A.5.1). Using Proposition A.3.3, we see immediately that $(\lambda + \lambda_i)^{l_i}$, $i = 1, \ldots, q$ are the elementary divisors of $A_1 + \lambda B_1$ and λ^{k_i}, $i = 1, \ldots, p$ are its elementary divisors at infinity. If the strict equivalence claimed by the theorem holds, it follows from Theorem A.5.2 that (A.5.1) is uniquely determined by $A + \lambda B$, and that $A + \lambda B$ must have the specified elementary divisors.

It remains to prove that there is a strict equivalence of the required form. Let $c \in \mathbb{C}$ be such that $\det(A + cB) \neq 0$. Write $A + \lambda B = (A + cB) + (\lambda - c)B$, multiply on the left by $(A + cB)^{-1}$, and apply a similarity transformation reducing $(A + cB)^{-1}B$ to the Jordan form. We obtain

$$A + \lambda B \overset{s}{\sim} (I + (\lambda - c)J_0) \oplus (I + (\lambda - c)J_1) \qquad (A.5.2)$$

where J_0 is a nilpotent Jordan matrix (i.e., $J_0^l = 0$ for some l) and J_1 is an invertible Jordan matrix.

Multiply the first diagonal block on the right-hand side of (A.5.2) by $(I - cJ_0)^{-1}$. It is easily verified that

$$\{I + (\lambda - c)J_0\}(I - cJ_0)^{-1} = I + \lambda J_0(I - cJ_0)^{-1}$$

and since $J_0(I - cJ_0)^{-1}$ is also nilpotent, $I + \lambda J_0(I - cJ_0)^{-1}$ is similar to a matrix of the form

$$(I_{k_1} + \lambda J_{k_1}(0)) \oplus \cdots \oplus (I_{k_p} \oplus \lambda J_{k_p}(0))$$

Multiply the second diagonal block on the right-hand side of (A.5.2) by J_1^{-1} and reduce J_1^{-1} to its Jordan form by similarity. We find that

$$I + (\lambda - c)J_1 \overset{s}{\sim} (\lambda I_{l_1} + J_{l_1}(\lambda_1)) \oplus \cdots \oplus (\lambda I_{l_q} + J_{l_q}(\lambda_q))$$

for some complex numbers $\lambda_1, \ldots, \lambda_q$ and some positive integers $l_1, \ldots, l_q$. $\square$

A.6 THE REDUCTION THEOREM FOR SINGULAR POLYNOMIALS

Consider now the singular polynomial $A + \lambda B$, where A and B are $m \times n$ matrices. Singularity means that either $m \neq n$ or $m = n$ but $\det(A + \lambda B)$ is identically zero. Let r be the *rank* of $A + \lambda B$, that is, the size of the largest minors in $A + \lambda B$ that do not vanish identically. Then either $r < m$ or $r < n$ holds (or both).

Assume $r < n$. Then the columns of the matrix polynomial $A + \lambda B$ are linearly dependent, that is, the equation

$$(A + \lambda B)x = 0, \qquad \lambda \in \mathbb{C} \qquad (A.6.1)$$

where x is an unknown vector, has a nonzero solution.

Let us check first that there is a vector polynomial $x = x(\lambda) \not\equiv 0$ for which (A.6.1) is satisfied. For this purpose we can use the Smith form $D(\lambda)$ of $A + \lambda B$ in place of $A + \lambda B$ itself (see Theorem A.1.1). But because of the assumption $r < n$, the last column of $D(\lambda)$ is zero. Hence $D(\lambda)x = 0$ is satisfied with $x = \langle 0, \ldots, 0, 1 \rangle$.

The following example is important in the sequel.

EXAMPLE A.6.1.   Let

$$
L_\epsilon(\lambda) = \begin{bmatrix} \lambda & 1 & 0 & \cdots & 0 & 0 \\ 0 & \lambda & 1 & \cdots & 0 & 0 \\ \vdots & \vdots & \vdots & & \vdots & \vdots \\ 0 & 0 & 0 & \cdots & \lambda & 1 \end{bmatrix}
$$

be an $\epsilon \times (\epsilon + 1)$ linear matrix polynomial ($\epsilon = 1, 2, \ldots$). We claim that the minimal degree of a nonzero vector polynomial solution $x(\lambda)$ of the equation

$$
L_\epsilon(\lambda)x(\lambda) = 0
$$

is ϵ. Indeed, rewrite this equation in the form $\lambda x_1(\lambda) + x_2(\lambda) = 0$, $\lambda x_2(\lambda) + x_3(\lambda) = 0, \ldots, \lambda x_\epsilon(\lambda) + x_{\epsilon+1}(\lambda) = 0$, where $x_j(\lambda)$ is the jth coordinate of $x(\lambda)$. So

$$
x_k(\lambda) = (-1)^{k-1}\lambda^{k-1}x_1(\lambda), \qquad k = 1, 2, \ldots, \epsilon + 1
$$

and the minimal degree for $x(\lambda)$ (which is equal to ϵ) is obtained by taking $x_1(\lambda)$ to be a nonzero constant. $\square$

Among all not identically zero polynomial solutions $x(\lambda)$ of (A.6.1), we choose one of least degree ϵ and write

$$
x(\lambda) = x_0 - \lambda x_1 + \lambda^2 x_2 - \cdots + (-1)^\epsilon \lambda^\epsilon x_\epsilon, \qquad x_\epsilon \neq 0 \qquad \text{(A.6.2)}
$$

The following reduction theorem holds.

Theorem A.6.1

If ϵ is the minimal degree of a nonzero polynomial solution of (A.6.1), and if $\epsilon > 0$, then $A + \lambda B$ is strictly equivalent to a linear matrix polynomial of the form

$$
\begin{bmatrix} L_\epsilon & 0 \\ 0 & \hat{A} + \lambda \hat{B} \end{bmatrix} \qquad \text{(A.6.3)}
$$

where

$$L_\epsilon = \begin{bmatrix} \lambda & 1 & 0 & \cdots & 0 & 0 \\ 0 & \lambda & 1 & \cdots & 0 & 0 \\ \vdots & \vdots & \vdots & \ddots & \vdots & \vdots \\ 0 & 0 & & \cdots & \lambda & 1 \end{bmatrix} \qquad (A.6.4)$$

is an $\epsilon \times (\epsilon + 1)$ matrix, and the equation

$$(\hat{A} + \lambda \hat{B})x = 0 \qquad (A.6.5)$$

has no nonzero polynomial solutions of degree less than ϵ.

It is convenient to state and prove a lemma to be used in the proof of Theorem A.6.1. For an $m \times n$ matrix polynomial $U + \lambda V$, let

$$M_i[U + \lambda V] = \begin{bmatrix} U & 0 & \cdots & 0 \\ V & U & \cdots & 0 \\ 0 & V & & \vdots \\ \vdots & \vdots & & U \\ 0 & 0 & \cdots & V \end{bmatrix}$$

be a matrix of size $m(i + 2) \times n(i + 1)$ for $i = 0, 1, 2, \ldots$.

Lemma A.6.2

Assume that the rank of $U + \lambda V$ is less than n. Then ϵ is the minimal degree of nonzero polynomial solutions $y(\lambda)$ of

$$(U + \lambda V)y(\lambda) = 0, \qquad \lambda \in \mathbb{C} \qquad (A.6.6)$$

if and only if

$$\text{rank } M_i[U + \lambda V] = (i + 1)n; \qquad i = 0, \ldots, \epsilon - 1$$

and

$$\text{rank } M_\epsilon[U + \lambda V] < (\epsilon + 1)n \qquad (A.6.7)$$

Proof. Let $y(\lambda) = \Sigma_{j=0}^{\epsilon} \lambda^j y_j$ be a nonzero polynomial solution of (A.6.6) of the least degree. Then

$$Uy_0 = 0; \qquad Vy_0 + Uy_1 = 0, \ldots, Vy_{\epsilon-1} + Uy_\epsilon = 0; \qquad Vy_\epsilon = 0$$

or, equivalently

$$M_\epsilon [U + \lambda V] \begin{bmatrix} y_0 \\ y_1 \\ \vdots \\ y_\epsilon \end{bmatrix} = 0$$

Not all the vectors y_j are zero, and so (A.6.7) follows. Conversely, if (A.6.7) holds, we may reverse the argument and obtain a nonzero polynomial solution of (A.6.6) of degree ϵ. $\square$

Proof of Theorem A.6.1.   The proof is given in three steps. In the first step we show that

$$A + \lambda B \stackrel{s}{\sim} \begin{bmatrix} L_\epsilon & D + \lambda F \\ 0 & \hat{A} + \lambda \hat{B} \end{bmatrix}$$

for suitable matrices $\hat{A}$, $\hat{B}$, D, and F, then we show that $\hat{A} + \lambda \hat{B}$ satisfies the conclusions of Theorem A.6.1, and finally we prove that

$$\begin{bmatrix} L_\epsilon & D + \lambda F \\ 0 & \hat{A} + \lambda \hat{B} \end{bmatrix} \stackrel{s}{\sim} \begin{bmatrix} L_\epsilon & 0 \\ 0 & \hat{A} + \lambda \hat{B} \end{bmatrix}$$

(a) Let (A.6.2) be a vector polynomial satisfying (A.6.1):

$$(A + \lambda B)(x_0 - \lambda x_1 + \lambda^2 x_2 - \cdots + (-1)^\epsilon \lambda^\epsilon x_\epsilon) = 0, \qquad \lambda \in \mathbb{C}$$

where $x_\epsilon \neq 0$. This is equivalent to

$$Ax_0 = 0, \; Ax_1 = Bx_0, \ldots, \; Ax_\epsilon = Bx_{\epsilon-1}, \; Bx_\epsilon = 0 \qquad \text{(A.6.8)}$$

We claim that the vectors

$$Ax_1, Ax_2, \ldots, Ax_\epsilon \qquad \text{(A.6.9)}$$

are linearly independent. Assume the contrary, and let Ax_h ($h \geq 1$) be the first vector in (A.6.9) that is linearly dependent on the preceding ones:

$$Ax_h = \alpha_1 Ax_{h-1} + \alpha_2 Ax_{h-2} + \cdots + \alpha_{h-1} Ax_1$$

By (A.6.8) this equation can be rewritten as follows:

$$Bx_{h-1} = \alpha_1 Bx_{h-2} + \alpha_2 Bx_{h-3} + \cdots + \alpha_{h-1} Bx_0$$

that is, $Bx_{h-1}^* = 0$, where

$$x^*_{h-1} = x_{h-1} - \alpha_1 x_{h-2} - \alpha_2 x_{h-3} - \cdots - \alpha_{h-1} x_0$$

Furthermore, again by (A.6.8), we have

$$Ax^*_{h-1} = B(x_{h-2} - \alpha_1 x_{h-3} - \cdots - \alpha_{h-2} x_0) = Bx^*_{h-2}$$

where

$$x^*_{h-2} = x_{h-2} - \alpha_1 x_{h-3} - \cdots - \alpha_{h-2} x_0$$

Continuing the process and introducing the vectors

$$x^*_{h-3} = x_{h-3} - \alpha_1 x_{h-4} - \cdots - \alpha_{h-3} x_0, \quad \ldots, \quad x^*_1 = x_1 - \alpha_1 x_0, \quad x^*_0 = x_0$$

we obtain the equations

$$Bx^*_{h-1} = 0, \quad Ax^*_{h-1} = Bx^*_{h-2}, \quad \ldots, \quad Ax^*_1 = Bx^*_0, \quad Ax^*_0 = 0 \quad \text{(A.6.10)}$$

From (A.6.10) it follows that

$$x^*(\lambda) = x^*_0 - \lambda x^*_1 + \cdots + (-1)^{h-1} x^*_{h-1}$$

is a nonzero solution of (A.6.1) with degree not exceeding $h - 1 < \epsilon$, which is impossible. [The fact that this solution is not identically zero follows because $x^*_0 = x_0 \neq 0$; for if x_0 were zero, then $\lambda^{-1} x(\lambda)$ would be a polynomial solution of (A.6.1) of degree less than ϵ.] Thus the vectors (A.6.9) are linearly independent.

But then the vectors $x_0, \ldots, x_\epsilon$ are linearly independent as well. Indeed, if $\Sigma_{i=0}^{\epsilon} \alpha_i x_i = 0$, then $\Sigma_{i=1}^{\epsilon} \alpha_i Ax_i = 0$, and by the linear independence of (A.6.9) $\alpha_1 = \cdots = \alpha_\epsilon = 0$. So $\alpha_0 x_0 = 0$ and since $x_0 \neq 0$ we find that also $\alpha_0 = 0$.

Now write $A + \lambda B$ in a basis in $\mathbb{C}^n$ whose first $\epsilon + 1$ vectors are $x_0, x_1, \ldots, x_\epsilon$ and in a basis in $\mathbb{C}^m$ whose first ϵ vectors are $Ax_1, \ldots, Ax_\epsilon$. In view of equations (A.6.8), the polynomial $A + \lambda B$ in the new bases has the form

$$\begin{bmatrix} L_\epsilon & D + \lambda F \\ 0 & \hat{A} + \lambda \hat{B} \end{bmatrix}$$

for some $D, F, \hat{A},$ and $\hat{B}$.

In the second step we show that the equation $(\hat{A} + \lambda \hat{B})\hat{x} = 0$ has no nonzero polynomial solutions of degree less than ϵ. Note that

$$M_{\epsilon-1} \begin{bmatrix} L_\epsilon & D + \lambda F \\ 0 & \hat{A} + \lambda \hat{B} \end{bmatrix} \quad \text{(A.6.11)}$$

is obtained from

$$\begin{bmatrix} M_{\epsilon-1}[L_\epsilon] & M_{\epsilon-1}[D + \lambda F] \\ 0 & M_{\epsilon-1}[\hat{A} + \lambda\hat{B}] \end{bmatrix} \tag{A.6.12}$$

by a suitable permutation of rows and columns. By Lemma A.6.2 the rank of (A.6.11) is equal to ϵn; that is, the columns of (A.6.11) are linearly independent. By the same lemma, taking into account Example A.6.1, rank $M_{\epsilon-1}[L_\epsilon] = \epsilon(\epsilon + 1)$; that is, the square $\epsilon(\epsilon + 1) \times \epsilon(\epsilon + 1)$ matrix $M_{\epsilon-1}[L_\epsilon]$ is invertible. As the columns of (A.6.12) are linearly independent as well, we find that the columns of $M_{\epsilon-1}[\hat{A} + \lambda\hat{B}]$ are linearly independent, that is, rank $M_{\epsilon-1}[\hat{A} + \lambda\hat{B}] = \epsilon(n - (\epsilon + 1))$. Using Lemma A.6.2 again, we find that $(\hat{A} + \lambda\hat{B})\hat{x} = 0$ has no solutions of degree less than ϵ.

In the third step, replacing

$$\begin{bmatrix} L_\epsilon & D + \lambda F \\ 0 & \hat{A} + \lambda\hat{B} \end{bmatrix}$$

by

$$\begin{bmatrix} I & Y \\ 0 & I \end{bmatrix}\begin{bmatrix} L_\epsilon & D + \lambda F \\ 0 & \hat{A} + \lambda\hat{B} \end{bmatrix}\begin{bmatrix} I & -X \\ 0 & I \end{bmatrix} = \begin{bmatrix} L_\epsilon & D + \lambda F + Y(\hat{A} + \lambda\hat{B}) - L_\epsilon X \\ 0 & \hat{A} + \lambda\hat{B} \end{bmatrix}$$

for suitable matrices X and Y, we see that Theorem A.6.1 will be completely proved if we can show that the matrices X and Y can be chosen so that the matrix equation

$$L_\epsilon X = D + \lambda F + Y(\hat{A} + \lambda\hat{B}) \tag{A.6.13}$$

holds.

We introduce a notation for the elements of D, F, X and also for the rows of Y and the columns of $\hat{A}$ and $\hat{B}$:

$$D = [d_{ik}]_{i,k=1}^{\epsilon, n-\epsilon-1}, \qquad F = [f_{ik}]_{i,k=1}^{\epsilon, n-\epsilon-1}, \qquad X = [x_{jk}]_{j,k=1}^{\epsilon+1, n-\epsilon-1}$$

$$Y = \begin{bmatrix} y_1 \\ y_2 \\ \vdots \\ y_\epsilon \end{bmatrix}, \qquad \hat{A} = [a_1, a_2, \ldots, a_{n-\epsilon-1}], \qquad \hat{B} = [b_1, b_2, \ldots, b_{n-\epsilon-1}]$$

Then the matrix equation (A.6.13) can be replaced by a system of scalar equations that expresses the equality of the elements of the kth column on the right- and left-hand sides of (A.6.13). For $k = 1, 2, \ldots, n - \epsilon - 1$, we obtain

$$x_{2k} + \lambda x_{1k} = d_{1k} + \lambda f_{1k} + y_1 a_k + \lambda y_1 b_k$$

$$x_{3k} + \lambda x_{2k} = d_{2k} + \lambda f_{2k} + y_2 a_k + \lambda y_2 b_k$$

$$x_{4k} + \lambda x_{3k} = d_{3k} + \lambda f_{3k} + y_3 a_k + \lambda y_3 b_k \qquad (A.6.14)$$

$$\cdots\cdots\cdots\cdots\cdots\cdots\cdots\cdots\cdots$$

$$x_{\epsilon+1,k} + \lambda x_{\epsilon k} = d_{\epsilon k} + \lambda f_{\epsilon k} + y_\epsilon a_k + \lambda y_\epsilon b_k$$

The left-hand sides of these equations are linear polynomials in λ. The free term of each of the first $\epsilon - 1$ of these polynomials is equal to the coefficient of λ in the next polynomial. But then the right-hand sides must also satisfy this condition. Therefore, for $k = 1, 2, \ldots, n - \epsilon - 1$, we obtain

$$y_1 a_k - y_2 b_k = f_{2k} - d_{1k}$$

$$y_2 a_k - y_3 b_k = f_{3k} - d_{2k}$$

$$\cdots\cdots\cdots\cdots\cdots\cdots\cdots \qquad (A.6.15)$$

$$y_{\epsilon-1} a_k - y_\epsilon b_k = f_{\epsilon k} - d_{\epsilon-1,k}$$

If (A.6.15) holds, then the required elements of X can obviously be determined from (A.6.14).

It now remains to show that the system of equations (A.6.15) for the elements of Y always has a solution for arbitrary d_{ik} and f_{ik} ($i = 1, 2, \ldots, \epsilon$; $k = 1, 2, \ldots, n - \epsilon - 1$). Rewrite (A.6.15) in the form

$$[y_1, -y_2, y_3, \ldots, (-1)^\epsilon y_\epsilon] M_{\epsilon-2}[\hat{A} + \lambda \hat{B}] = [H_1 \cdots H_{\epsilon-1}]$$

where

$$H_j = [f_{j+1,1} - d_{j1}, \ldots, f_{j+1,n-\epsilon-1} - d_{j,n-\epsilon-1}], \qquad j = 1, \ldots, \epsilon - 1$$

and use the left invertibility of $M_{\epsilon-2}[\hat{A} + \lambda \hat{B}]$ (ensured by Lemma A.6.2) to verify that (A.6.15) has a solution $[y_1, -y_2, \ldots, (-1)^\epsilon y_\epsilon] = [H_1 \cdots H_{\epsilon-1}]\{M_{\epsilon-2}[\hat{A} + \lambda \hat{B}]\}^{-L}$, where the subscript "$L$" denotes a left inverse. Theorem A.6.1 is now proved completely. $\square$

A.7 MINIMAL INDICES AND STRICT EQUIVALENCE OF LINEAR MATRIX POLYNOMIALS (GENERAL CASE)

We introduce the important notion of minimal indices for linear matrix polynomials. Let $A + \lambda B$ be an arbitrary linear matrix polynomial of size $m \times n$. Then the k polynomial columns $x_1(\lambda), x_2(\lambda), \ldots, x_k(\lambda)$ that are solutions of the equation

$$(A + \lambda B)x = 0 \qquad (A.7.1)$$

are called *linearly dependent* if the rank of the polynomial matrix formed from these columns $X(\lambda) = [x_1(\lambda), x_2(\lambda), \ldots, x_k(\lambda)]$ is less than k! In that case there exist k polynomials $p_1(\lambda), p_2(\lambda), \ldots, p_k(\lambda)$, not all identically zero, such that

$$p_1(\lambda)x_1(\lambda) + p_2(\lambda)x_2(\lambda) + \cdots + p_k(\lambda)x_k(\lambda) \equiv 0 \qquad (A.7.2)$$

Indeed, let

$$X(\lambda) = E(\lambda)D(\lambda)F(\lambda)$$

be the Smith form of $X(\lambda)$, where $E(\lambda)$ [resp. $F(\lambda)$] is an $n \times n$ (resp. $k \times k$) matrix polynomial with constant nonzero determinant, and

$$D(\lambda) = \begin{bmatrix} \text{diag}[d_1(\lambda), \ldots, d_r(\lambda)] & 0 \\ 0 & 0 \end{bmatrix}$$

with nonzero polynomials $d_1(\lambda), \ldots, d_r(\lambda)$. As the rank r of $X(\lambda)$ is less than k, the last column of $D(\lambda)$ is zero. One verifies that (A.7.2) is satisfied with $\langle p_1(\lambda), \ldots, p_k(\lambda) \rangle = F(\lambda)^{-1}\langle 0, 0, \ldots, 0, 1 \rangle$. If polynomials $p_i(\lambda)$ (not all zero) with the property (A.7.2) do not exist, then the rank of X is k and we say that the solutions $x_1(\lambda), \ldots, x_k(\lambda)$ are *linearly independent*.

Among all the polynomial solutions of (A.7.1) we choose a nonzero solution $x_1(\lambda)$ of least degree ϵ_1. Among all polynomial solutions $x(\lambda)$ of the same equation for which $x_1(\lambda)$ and $x(\lambda)$ are linearly independent, we take a solution $x_2(\lambda)$ of least degree ϵ_2. Obviously, $\epsilon_1 \le \epsilon_2$. We continue the process, choosing from the polynomial solutions $x(\lambda)$ for which $x_1(\lambda)$, $x_2(\lambda)$, and $x(\lambda)$ are linearly independent a solution $x_3(\lambda)$ of minimal degree ϵ_3, and so on. Since the number of linearly independent solutions of (A.7.1) is always at most n, the process must come to an end. We obtain a *fundamental series of solutions* of (A.7.1)

$$x_1(\lambda), x_2(\lambda), \ldots, x_p(\lambda) \qquad (A.7.3)$$

having the degrees

$$\epsilon_1 \le \epsilon_2 \le \cdots \le \epsilon_p \qquad (A.7.4)$$

Note that it may happen that some degrees $\epsilon_1, \ldots, \epsilon_j$ are zeros. [This is the case when (A.7.1) admits constant nonzero solutions.] In general, a fundamental series of solutions is not uniquely determined (to within scalar factors) by the pencil $A + \lambda B$. However, note the following.

Proposition A.7.1

Two distinct fundamental series of solutions always have the same series of degrees $\epsilon_1, \ldots, \epsilon_p$.

Proof. In addition to (A.7.3), consider another fundamental series of solutions $\bar{x}_1(\lambda), \bar{x}_2(\lambda), \ldots$ with the degrees $\bar{\epsilon}_1, \bar{\epsilon}_2, \ldots$. Suppose that in (A.7.4)

$$\epsilon_1 = \cdots = \epsilon_{n_1} < \epsilon_{n_1+1} = \cdots = \epsilon_{n_2} < \cdots$$

and similarly, in the series $\bar{\epsilon}_1, \bar{\epsilon}_2, \ldots$

$$\bar{\epsilon}_1 = \cdots \bar{\epsilon}_{m_1} < \bar{\epsilon}_{m_1+1} = \cdots = \bar{\epsilon}_{m_2} < \cdots$$

Obviously, $\epsilon_1 = \bar{\epsilon}_1$. For every vector $\bar{x}_i(\lambda)$ $(i = 1, \ldots, m_1)$ there exists a polynomial $\bar{q}_i(\lambda) \neq 0$ such that

$$\bar{q}_i(\lambda)\bar{x}_i(\lambda) = \sum_{j=1}^{n_1} \bar{p}_{ij}(\lambda)x_j(\lambda), \qquad i = 1, \ldots, m_1 \tag{A.7.5}$$

for some polynomials $\bar{p}_{ij}(\lambda)$. (Otherwise, $\bar{x}_i, x_1, \ldots, x_{n_1}$ would be linearly independent and one could replace x_{n_1+1} by $\bar{x}_i$, which is of smaller degree, contrary to the definition of x_{n_1+1}.) Rewrite (A.7.5) in the form

$$[\bar{x}_1(\lambda) \cdots \bar{x}_{m_1}(\lambda)]\bar{Q}(\lambda) = [x_1(\lambda) \cdots x_{n_1}(\lambda)]\bar{P}(\lambda) \tag{A.7.6}$$

where $\bar{Q}(\lambda) = \mathrm{diag}[\bar{q}_1(\lambda), \ldots, \bar{q}_{m_1}(\lambda)]$ and $\bar{P}(\lambda) = [\bar{p}_{ij}(\lambda)]_{i,j=1}^{n_1,m_1}$ is an $n_1 \times m_1$ matrix polynomial. As $\bar{x}_1(\lambda), \ldots, \bar{x}_{m_1}(\lambda)$ are linearly independent, there is a nonzero minor $f(\lambda)$ of order m_1 of $[\bar{x}_1(\lambda) \cdots \bar{x}_{m_1}(\lambda)]$. So for every $\lambda \in \mathbb{C}$ that is not a zero of one of the polynomials $f(\lambda), \bar{q}_1(\lambda), \ldots, \bar{q}_{m_1}(\lambda)$ the rank of the matrix on the left-hand side of (A.7.6) is m_1. Hence (A.7.6) implies $m_1 \leq n_1$. Interchanging the roles of $\bar{x}_i(\lambda)$ and $x_i(\lambda)$, we find the opposite inequality $m_1 \geq n_1$. As $m_1 = n_1$, we have $\epsilon_{n_1+1} = \bar{\epsilon}_{m_1+1}$, and we can repeat the above argument with n_2 and m_2 in place of n_1 and m_1, respectively, and so on. $\square$

The degrees $\epsilon_1 \leq \cdots \leq \epsilon_p$ of polynomials in any fundamental series of polynomial solutions of (A.7.1) are called the *minimal column indices* of $A + \lambda B$. As Proposition A.7.1 shows, the number p of the minimal column indices and the indices themselves do not depend on the choice of the fundamental series. If there are no nonzero solutions of (A.7.1) (i.e., the rank of $A + \lambda B$ is equal to n), we say that the number of minimal column indices is zero, in this case no such indices are defined.

We define the *minimal row indices* of $A + \lambda B$ as the minimal column indices of $A^* + \lambda B^*$.

EXAMPLE A.7.1.　Let L_ϵ be as in Example A.6.1. The polynomial L_ϵ has the single minimal column index ϵ, whereas the minimal row indices are absent. Indeed, as in Example A.6.1, observe that every nonzero polynomial solution $x(\lambda) = \langle x_1(\lambda), \dots, x_{\epsilon+1}(\lambda)\rangle$ of

$$L_\epsilon x(\lambda) = 0 \qquad\qquad (A.7.7)$$

has the form

$$x_k(\lambda) = (-1)^{k-1}\lambda^{k-1}x_1(\lambda), \qquad k = 1, \dots, \epsilon + 1 \qquad (A.7.8)$$

and a solution $\hat{x}(\lambda)$ of minimal degree ϵ is obtained by taking $x_1(\lambda) \equiv 1$. Hence the first minimal column index of L_ϵ is ϵ. As (A.7.8) shows, every other solution $x(\lambda)$ of (A.7.7) has the form $x(\lambda) = x_1(\lambda)\hat{x}(\lambda)$, where $x_1(\lambda)$ is the first coordinate of $x(\lambda)$. So $x(\lambda)$ and $x_1(\lambda)$ are linearly dependent, which means that there are no more minimal column indices.

As the rows of L_ϵ are linearly independent for every λ, the minimal row indices are absent.

Similarly, we conclude that the transposed polynomial L_ϵ^T has the single minimal row index ϵ and no minimal column indices.　□

The importance of minimal indices stems from their invariance under strict equivalence, as follows.

Proposition A.7.2

If $A + \lambda B \overset{s}{\sim} A_1 + \lambda B_1$, then the minimal column indices of the polynomials $A + \lambda B$ and $A_1 + \lambda B_1$ are the same, and the minimal row indices of these polynomials are also the same.

The proof is immediate: if $P(A + \lambda B)Q = A_1 + \lambda B_1$ for invertible matrices P and Q, then the solutions of $(A + \lambda B)x(\lambda) = 0$ are obtained from the solutions of $(A_1 + \lambda B_1)y(\lambda) = 0$ by multiplication by Q: $x(\lambda) = Qy(\lambda)$, which preserves linear dependence and independence and also implies that $x(\lambda)$ and $y(\lambda)$ have the same degree.

We are now in a position to state and prove the main result concerning strict equivalence of linear matrix polynomials in general. We denote by L_ϵ the $\epsilon \times (\epsilon + 1)$ linear polynomial

$$\begin{bmatrix} \lambda & 1 & 0 & \cdots & 0 & 0 \\ 0 & \lambda & 1 & \cdots & 0 & 0 \\ \vdots & \vdots & & \ddots & \vdots & \vdots \\ 0 & 0 & & \cdots & \lambda & 1 \end{bmatrix} \qquad\qquad (A.7.9)$$

and L_ϵ^T is its transpose [which is an $(\epsilon + 1) \times \epsilon$ linear polynomial]. Then $0_{u \times v}$

will denote the zero $u \times v$ matrix. As before, $J_k(\lambda_0)$ represents the $k \times k$ Jordan block with eigenvalue λ_0.

Theorem A.7.3

Every $m \times n$ linear matrix polynomial $A + \lambda B$ is strictly equivalent to a unique linear matrix polynomial of type

$$0_{u \times v} \oplus L_{\epsilon_1} \oplus \cdots \oplus L_{\epsilon_p} \oplus L_{\eta_1}^T \oplus \cdots \oplus L_{\eta_q}^T \oplus (I_{k_1} + \lambda J_{k_1}(0)) \oplus \cdots$$
$$\oplus (I_{k_r} + \lambda J_{k_r}(0)) \oplus (\lambda I_{l_1} + J_{l_1}(\lambda_1)) \oplus \cdots \oplus (\lambda I_{l_s} + J_{l_s}(\lambda_s))$$

$$(A.7.10)$$

Here $\epsilon_1 \leq \cdots \leq \epsilon_p$ and $\eta_1 \leq \cdots \leq \eta_q$ are positive integers; $k_1, \ldots, k_r$ and $l_1, \ldots, l_s$ are positive integers; $\lambda_1, \ldots, \lambda_s$ are complex numbers.

The uniqueness of the linear matrix polynomial of type (A.7.10) to which $A + \lambda B$ is strictly equivalent means that the parameters u, v, p, q, r, s, $\{\epsilon_j\}_{j=1}^p$, $\{\eta_j\}_{j=1}^q$, $\{k_j\}_{j=1}^r$, and $\{\lambda_j\}_{j=1}^s$ are uniquely determined by the polynomial $A + \lambda B$. It may happen that some of the numbers u, v, p, q, r, and s are zeros. This means that the corresponding part is missing from formula (A.7.10).

Proof of Theorem A.7.3. Let $x_1, \ldots, x_v \in \math<0xC7>^n$ be a basis in the linear space of all constant solutions of the equation

$$(A + \lambda B)x = 0, \qquad \lambda \in \math<0xC7> \tag{A.7.11}$$

that is, all solutions that are independent of λ. Note that (A.7.11) is equivalent to the simultaneous equations

$$Ax = 0, \qquad Bx = 0$$

Likewise, let $y_1, \ldots, y_u \in \math<0xC7>^m$ be a basis in the linear space of all constant solutions of

$$(A^* + \lambda B^*)y = 0, \qquad \lambda \in \math<0xC7>$$

or, what is the same, the simultaneous equations

$$A^*y = 0, \qquad B^*y = 0$$

Write $A + \lambda B$ (understood for each $\lambda \in \math<0xC7>$ as a transformation written in the standard orthonormal bases in $\math<0xC7>^n$ and $\math<0xC7>^m$) as a matrix with respect to the basis in $\math<0xC7>^n$ whose first v vectors are $x_1, \ldots, x_v$ and the basis in $\math<0xC7>^m$ whose first u vectors are $y_1, \ldots, y_u$ and the others are orthogonal to $\mathrm{Span}\{y_1, \ldots, y_u\}$. Because $\mathrm{Im}\, A = (\mathrm{Ker}\, A^*)^\perp \subset (\mathrm{Span}\{y_1, \ldots, y_u\})^\perp$ and also $\mathrm{Im}\, B \subset (\mathrm{Span}\{y_1, \ldots, y_u\})^\perp$, it follows that, with respect to the indicated bases, $A + \lambda B$ has the form $0_{u \times v} \oplus (A_1 + \lambda B_1)$. Here $A_1 + \lambda B_1$ has the property that neither $(A_1 + \lambda B_1)x = 0$ nor $(A_1^* + \lambda B_1^*)y = 0$ has constant nonzero solutions.

If the rank of $A_1 + \lambda B_1$ is less than the number of columns in $A_1 + \lambda B_1$, apply the reduction theorem (Theorem A.6.1) several times to show that

$$A_1 + \lambda B_1 \overset{s}{\sim} L_{\epsilon_1} \oplus \cdots \oplus L_{\epsilon_p} \oplus (A_2 + \lambda B_2)$$

where $A_2 + \lambda B_2$ is such that the equation $(A_2 + \lambda B_2)x = 0$ has no nonzero polynomial solutions $x = x(\lambda)$. From the property of $\hat{A} + \lambda \hat{B}$ in Theorem A.6.1 it is clear that $\epsilon_1 \leq \cdots \leq \epsilon_p$. It is also clear that the process of consecutive applications of Theorem A.6.1 must terminate for the simple reason that the size of $A_1 + \lambda B_1$ is finite. The Smith form of $A_2 + \lambda B_2$ (Theorem A.1.1) shows that the number of columns of the polynomial $A_2 + \lambda B_2$ coincides with its rank.

If it happens that the rank of $A_2 + \lambda B_2$ is less than the number of its rows, apply the above procedure to $A_2^* + \lambda B_2^*$. After taking adjoints, we find that

$$A_2 + \lambda B_2 \overset{s}{\sim} L_{\eta_1}^T \oplus \cdots \oplus L_{\eta_q}^T \oplus (A_3 + \lambda B_3)$$

where $0 < \eta_1 \leq \cdots \leq \eta_q$ and the rank of $A_3 + \lambda B_3$ coincides with the number of columns and the number of rows of $(A_3 + \lambda B_3)$. In other words, $A_3 + \lambda B_3$ is regular. It remains to apply Theorem A.5.3 in order to show that the original polynomial $A + \lambda B$ is strictly equivalent to a polynomial of type (A.7.10).

It remains to show that such a polynomial (A.7.10) is unique. Proposition A.7.2 and Example A.7.1 show that the minimal column indices of $A + \lambda B$ are $0, \ldots, 0, \epsilon_1, \ldots, \epsilon_p$ (where 0 appears u times) and the minimal row indices of $A + \lambda B$ are $0, \ldots, 0, \eta_1, \ldots, \eta_q$ (where 0 appears v times). Hence the parameters u, v, p, q, $\{\epsilon_i\}_{i=1}^p$, and $\{\eta_j\}_{j=1}^q$ are uniquely determined by $A + \lambda B$. Further, observe that L_ϵ and L_ϵ^T have no elementary divisors; that is, their Smith forms are $[I_\epsilon \; 0]$ and $\begin{bmatrix} I_\epsilon \\ 0 \end{bmatrix}$, respectively. (This follows from Theorem A.2.2 since both L_ϵ and L_ϵ^T have an $\epsilon \times \epsilon$ minor that is equal to 1.) Using Proposition A.3.3, we see that the elementary divisors of (A.7.10) are $(\lambda + \lambda_1)^{l_1}, \ldots, (\lambda + \lambda_s)^{l_s}$, which must coincide with the elementary divisors of $A + \lambda B$ because of the strict equivalence of $A + \lambda B$ and (A.7.10) (Theorem A.3.1). Hence the parameters s, $\{l_i\}_{i=1}^s$, and $\{\lambda_i\}_{i=1}^s$ are also uniquely determined by $A + \lambda B$. Applying this argument for $\lambda A + B$ in place of $A + \lambda B$, we see that r and $\{k_i\}_{i=1}^r$ are uniquely determined by $A + \lambda B$ as well. $\quad\square$

The matrix polynomial (A.7.10) is called the *Kronecker canonical form* of $A + \lambda B$. Here $0, \ldots, 0, \epsilon_1, \ldots, \epsilon_p$ (u times 0) are the minimal column indices of $A + \lambda B$; $0, \ldots, 0, \eta_1, \ldots, \eta_q$ (v times 0) are the minimal row indices of $A + \lambda B$; $\lambda^{k_1}, \ldots, \lambda^{k_r}$ are the elementary divisors of $A + \lambda B$ at infinity; and $(\lambda + \lambda_1)^{l_1}, \ldots, (\lambda + \lambda_s)^{l_s}$ are the (finite) elementary divisors of $A + \lambda B$. We obtain the following corollary from Theorem A.7.3.

Corollary A.7.4

We have $A + \lambda B \overset{s}{\sim} A_1 + \lambda B_1$ if and only if the polynomials $A + \lambda B$ and $A_1 + \lambda B_1$ have the same minimal column indices, minimal row indices, elementary divisors, and elementary divisors at infinity.

Thus Corollary A.7.4 describes the full set of invariants for strict equivalence of linear matrix polynomials.

A.8   NOTES TO THE APPENDIX

This appendix contains well-known results on matrix polynomials. Essentially the entire material can be found in Chapters 6 and 12 of Gantmacher (1959), for example. In our exposition of Sections A.5–A.7 we follow this book. In the exposition of Sections A.1–A.4 we follow Gohberg, Lancaster, and Rodman (1982).

List of Notations
and Conventions

$X \subset Y$	inclusion between sets X and Y (equality not excluded)
$\mathbb{R}$	the field of real numbers
$\mathbb{R}^n$	the space of all n-dimensional real column vectors
$\mathbb{C}$	the field of complex numbers
$\mathbb{C}^n$	the space of all n-dimensional complex column vectors
$\bar{x}$	complex conjugate of complex number x
$\mathcal{Re}\ x = \frac{1}{2}(x + \bar{x})$	the real part of x
$\mathcal{Im}\ x = \dfrac{1}{2i}(x - \bar{x})$	the imaginary part of x
$\langle a_1, \ldots, a_n \rangle$	the n-dimensional column vector $\begin{bmatrix} a_1 \\ \vdots \\ a_n \end{bmatrix} \in \mathbb{C}^n$
$(\cdot, \cdot)$	the standard scalar product in $\mathbb{C}^n$: $(\langle a_1, \ldots, a_n \rangle, \langle b_1, \ldots, b_n \rangle)$ $= \sum_{i=1}^{n} a_i \bar{b}_i$
$\|x\| = (x, x)^{1/2} = \left(\sum^n |x_i|^2 \right)^2$	the norm of a vector $x = \langle x_1, \ldots, x_n \rangle \in \mathbb{C}^n$

$e_i = \langle 0, \ldots, 0, 1, 0, \ldots, 0 \rangle$	(with 1 in the ith place) the ith unit coordinate vector in $\mathbb{C}^n$; its size n will be clear from the context
"Linear transformation"	often abbreviated to "transformation"—when convenient, a linear transformation from $\mathbb{C}^m$ into $\mathbb{C}^n$ is assumed to be given by an $n \times m$ matrix with respect to the bases $e_1, \ldots, e_n$ in $\mathbb{C}^n$ and $e_1, \ldots, e_m$ in $\mathbb{C}^m$, consequently, when convenient, an $n \times m$ matrix will be considered as a linear transformation written in the standard bases $e_1, \ldots, e_n$ and $e_1, \ldots, e_m$
$[a_{ij}]_{i,j=1}^{m,n}$	$m \times n$ matrix whose entry in the (i, j) place is a_{ij}
I	unit matrix; identity linear transformation (the size of I is understood from the context)
I_k	the $k \times k$ unit matrix
A^T	the transpose of a matrix A
A^*	the adjoint of a transformation A; the conjugate transpose of a matrix A
$\bar{A}$	complex conjugate in every entry of a matrix A
A^{-L}	left inverse of a matrix (or transformation) A
A^{-R}	right inverse of a matrix (or transformation) A
A^I	one-sided inverse (left or right) of A; generalized inverse of A
tr A	the trace of a matrix (or transformation) A
$\|A\| = \max\limits_{x \neq 0} \dfrac{\|Ax\|}{\|x\|}$	the norm of a transformation A
$A_{\vert \mathcal{M}}$	the restriction of a transformation A to its invariant subspace $\mathcal{M}$
Im $A = \{Ay \mid y \in \mathbb{C}^m\}$	the image of a transformation $A: \mathbb{C}^m \to \mathbb{C}^n$
Ker $A = \{y \mid Ay = 0\}$	the kernel of a transformation A
$\sigma(A) = \{\lambda \in \mathbb{C} \mid \text{Ker}(A - \lambda I) \neq \{0\}\}$	the spectrum of a matrix (or transformation) A
$\mathcal{R}_\lambda(A)$	the root subspace of A corresponding to its eigenvalue λ

$J_k(\lambda)$	the Jordan block of size $k \times k$ with eigenvalue λ
$\text{diag}[A_1, \dots, A_p] = A_1 \oplus \cdots \oplus A_p$	the block diagonal matrix with the matrices $A_1, \dots, A_p$ along the main diagonal; or, the direct sum of the linear transformations $A_1, \dots, A_p$
$\text{col}[Z_i]_{i=1}^n = \begin{bmatrix} Z_1 \\ Z_2 \\ \vdots \\ Z_n \end{bmatrix},$	a block column matrix
$\text{Inv}(A)$	the set of all A-invariant subspaces
$\text{Inv}_p(A)$	the set of all p-dimensional A-invariant subspaces
$\text{Cinv}(A)$	the set of all coinvariant subspaces for A
$\text{Sinv}(A)$	the set of all semiivariant subspaces for A
$\text{Rinv}(A)$	the set of all reducing invariant subspaces for A
$\text{Rinv}_p(A)$	the set of all p-dimensional reducing invariant subspaces for A
$\text{Hinv}(A)$	the set of all hyperinvariant subspaces for A
$\text{Inv}^R(A)$	the set of all real invariant subspaces for a real transformation A
$\mathscr{C}(A)$	the set of all transformations (or matrices) that commute with a transformation (or matrix) A
$\{0\}$	the zero subspace
$\mathscr{M}^\perp$	the orthogonal complement to a subspace $\mathscr{M}$
$\mathscr{M} \dotplus \mathscr{N}$	direct sum of subspaces $\mathscr{M}$ and $\mathscr{N}$
$\mathscr{M} \oplus \mathscr{N}$	orthogonal sum of subspaces $\mathscr{M}$ and $\mathscr{N}$
$S_\mathscr{M}$	the unit sphere in a subspace $\mathscr{M}$
$d(x, Z) = \inf_{y \in Z} \|x - y\|$	the distance between a point $x \in \mathbb{C}^n$ and a set $Z \subset \mathbb{C}^n$
$d(X, Y)$	the distance between sets X and Y
$\theta(\mathscr{L}, \mathscr{M})$	the gap between $\mathscr{L}$ and $\mathscr{M}$
$\eta(\mathscr{L}, \mathscr{M})$	the minimal opening between $\mathscr{L}$ and $\mathscr{M}$
$\bar{\theta}(\mathscr{L}, \mathscr{M})$	the spherical gap between $\mathscr{L}$ and $\mathscr{M}$
$\varphi_{\min}(\mathscr{L}, \mathscr{M})$	the minimal angle between subspaces $\mathscr{L}$ and $\mathscr{M}$

$\mathbb{C}(\mathbb{C}^n)$	the metric space of all subspaces in $\mathbb{C}^n$
$\mathbb{G}_m(\mathbb{C}^n)$	the set of all m-dimensional subspaces of $\mathbb{C}^n$
$\text{Span}\{x_1, \ldots, x_k\}$	the subspace spanned by vectors $x_1, \ldots, x_k$
$M_{n,n}$	the algebra of all $n \times n$ matrices
$M(\mathcal{L})$	the algebra of all transformations on a linear space $\mathcal{L}$
T_j	the algebra of all upper triangular Toeplitz matrices of size $j \times j$
$\text{Inv}(V)$	the lattice of all invariant subspaces for an algebra V
$\text{Alg}(\Lambda)$	the algebra of all transformations for which every subspace from a lattice Λ is invariant
$U(\mathbb{C}^n)$	the set of all $n \times n$ unitary matrices
$U_+(\mathbb{R}^n)$	the set of all $n \times n$ real orthogonal matrices with determinant 1
$GL_r(n)$	the set of all real invertible $n \times n$ matrices
$\delta(W)$	the McMillan degree of a rational matrix function $W(\lambda)$
$S(A)$	the singular set of an analytic family of transformations $A(z)$
σ_{ij}	Kronecker index: $\delta_{ij} = 0$ if $i \neq j$; $\delta_{ij} = 1$ if $i = j$
$\left(\dfrac{u}{v}\right) = \dfrac{u!}{(u-v)!v'}$	($u \geq v$ are positive integers; $0! = 1$)
$K^{\#}$	the number of distinct elements in a finite set K
$\square$	end of a proof or an example

References

Allen, G. R., "Holomorphic vector-valued functions on a domain of holomorphy," *J. London Math. Soc.* **42**, 509–513 (1967).

Bart, H., I. Gohberg, and M. A. Kaashoek, "Stable factorization of monic matrix polynomials and stable invariant subspaces," *Integral Equations and Operator Theory* **1**, 496–517 (1978).

Bart, H., I. Gohberg, and M. A. Kaashoek, *Minimal Factorization of Matrix and Operator Functions* (Operator Theory: Advances and Applications, Vol. 1) Birkhaüser, Basel, 1979.

Bart, H., I. Gohberg, M. A. Kaashoek, and P. Van Dooren, "Factorizations of transfer functions," *SIAM J. Control Optim.* **18**(6), 675–696 (1980).

Baumgartel, H. *Analytic Perturbation Theory for Matrices and Operators* (Operator Theory: Advances and Applications, Vol. 15) Birkhauser, Basel–Boston–Stuttgart, 1985.

den Boer, H., and G. Ph. A. Thijsse, "Semistability of sums of partial multiplicities under additive perturbations," *Integral Equations and Operator Theory* **3**, 23–42 (1980).

Bochner, S., and W. T. Martin, *Several Complex Variables*, Princeton University Press, Princeton, NJ, 1948.

Brickman, L., and P. A. Fillmore, "The invariant subspace lattice of a linear transformaton," *Canad. J. Math.* **19**, 810–822 (1967).

Brockett, R., *Finite Dimensional Linear Systems*, John Wiley & Sons, New York, 1970.

Brunovsky, P., "A classification of linear controllable systems," *Kybernetika* (Praha) **3**, 173–187 (1970).

Campbell, S., and J. Daughtry, "The stable solutions of quadratic matrix equations," *Proc. AMS* **74**, 19–23 (1979).

Choi, M.-D., C. Laurie, and H. Radjavi, "On commutators and invariant subspaces," *Linear and Multilinear Algebra* **9**, 329–340 (1981).

Coddington, E. A., and N. Levinson, *Theory of Ordinary Differential Equations*, McGraw-Hill, New York, 1955.

Conway, J. B., and P. R. Halmos, "Finite-dimensional points of continuity of Lat," *Linear Algebra Appl.* **31**, 93–102 (1980).

Djaferis, T. E., and S. K. Mitter, "Some generic invariant factor assignment results using dynamic output feedback," *Linear Algebra Appl.* **50**, 103–131 (1983).

Donnellan, T., *Lattice Theory*, Pergamon Press, Oxford, 1968.

Douglas, R. G., and C. Pearcy, "On a topology for invariant subspaces," *J. Functional Analy.* **2**, 323–341 (1968).

Fillmore, P. A., D. A. Herrero, and W. E. Longstaff, "The hyperinvariant subspaces lattice of a linear transformation," *Linear Algebra Appl.* **17**, 125–132 (1977).

Gantmacher, F. R., *The Theory of Matrices*, Vols. I and II, Chelsea, New York, 1959.

Gochberg, I. Z., and J. Leiterer, "Über Algebren stetiger Operatorfunctionen," *Studia Mathematica*, Vol. LVII, 1–26, 1976.

Gohberg, I., and S. Goldberg, *Basic Operator Theory*, Birkhäuser, Basel, 1981.

Gohberg, I., and G. Heinig, "The resultant matrix and its generalizations, I. The resultant operator for matrix polynomials," *Acta Sci. Math.* (Szeged) **37**, 41–61 (Russian) (1975).

Gohberg, I., and M. A. Kaashoek, "Unsolved problems in matrix and operator theory, II. Partial multiplicities of a product," *Integral Equations and Operator Theory* **2**, 116–120 (1979).

Gohberg, I., M. A. Kaashoek, and F. van Schagen, "Similarity of operator blocks and canonical forms. I. General results, feedback equivalence and Kronecker indices," *Integral Equations and Operator Theory* **3**, 350–396 (1980).

Gohberg, I., M. A. Kaashoek, and F. van Schagen, "Similarity of operator blocks and canonical forms. II. Infinite dimensional case and Wiener–Hopf factorization," in *Topics in Modern Operator Theory. Operator Theory: Advances and Applications*, Vol. 2, Birkhäuser-Verlag, 1981, pp. 121–170.

Gohberg, I., M. A. Kaashoek, and F. van Schagen, "Rational matrix and operator functions with prescribed singularities," *Integral Equations and Operator Theory* **5**, 673–717 (1982).

Gohberg, I. C., and M. G. Krein, "The basic propositions on defect numbers, root numbers and indices of linear operators," *Uspehi Mat. Nauk* **12**, 43–118 (1957); translation, *Russian Math. Surveys* **13**, 185–264 (1960).

Gohberg, I., and N. Krupnik, *Einführung in die Theorie der eindimensionalen singulären Integraloperatoren*, Birkhäuser, Basel, 1979.

Gohberg, I., P. Lancaster, and L. Rodman, "Perturbation theory for divisors of operator polynomials," *SIAM J. Math. Anal.* **10**, 1161–1183 (1979).

Gohberg, I., P. Lancaster, and L. Rodman, *Matrix Polynomials*, Academic Press, New York, 1982.

Gohberg, I., P. Lancaster, and L. Rodman, "A sign characteristic for self-adjoint meromorphic matrix functions," *Applicable Analysis* **16**, 165–185 (1983a).

Gohberg, I., P. Lancaster, and L. Rodman, *Matrices and Indefinite Scalar Products* (Operator Theory: Advances and Applications, Vol. 8) Birkhäuser-Verlag, Basel, 1983b.

Gohberg, I., and Ju. Leiterer, "On holomorphic vector-functions of one variable, I. Functions on a compact set," *Matem. Issled.* **7**, 60–84 (Russian) (1972).

Gohberg, I., and Ju. Leiterer, "On holomorphic vector-functions of one variable, II. Functions on domains," *Matem. Issled.* **8**, 37–58 (Russian) (1973).

Gohberg, I. C. and A. S. Markus, "Two theorems on the gap between subspaces of a Banach space," *Uspehi Mat. Nauk* **14**, 135–140 (Russian) (1959).

Gohberg, I., and L. Rodman, "Analytic matrix functions with prescribed local data," *J. d'Analyse Math.* **40**, 90–128 (1981).

Gohberg, I., and L. Rodman, "On distance between lattices of invariant subspaces of matrices," *Linear Algebra Appl.* **76**, 85–120 (1986).

Gohberg, I., and S. Rubinstein, "Stability of minimal fractional decompositions of rational matrix functions," in *Operator Theory: Advances and Applications*, Vol. 18, Birkhäuser, Basel, 1986, pp. 249–270.

Golub, G. H., and C. F. van Loan, *Matrix Computations*, The Johns Hopkins University Press, Baltimore, 1983.

Golub, G. H., and J. H. Wilkinson, "Ill-conditioned eigensystems and the computation of the Jordan canonical form," *SIAM Review* **18**, 578–619 (1976).

Grauert, H., "Analytische Faserungen über holomorph vollständigen Räumen," *Math. Ann.* **135**, 263–273 (1958).

Guralnick, R. M., "A note on pairs of matrices with rank one commutator," *Linear and Multilinear Algebra* **8**, 97–99 (1979).

Halmos, P. R., "Reflexive lattices of subspaces," *J. London Math. Soc.* **4**, 257–263 (1971).

Halperin, I., and P. Rosenthal, "Burnside's theorem on algebras of matrices," *Am. Math. Monthly* **87**, 810 (1980).

Harrison, K. J., "Certain distributive lattices of subspaces are reflexive," *J. London Math. Soc.* **8**, 51–56 (1974).

Hautus, M. L. J., "Controllability and observability conditions of linear autonomous systems," *Ned. Akad. Wet. Proc.*, Ser. A, **12**, 443–448 (1969).

Helton, J. W., and J. A. Ball, "The cascade decompositions of a given system vs the linear fractional decompositions of its transfer function," *Integral Equations and Operator Theory* **5**, 341–385 (1982).

Hoffman, K., and R. Kunze, *Linear Algebra*, Prentice-Hall of India, New Delhi, 1967.

Jacobson, N., *Lectures in Abstract Algebra II: Linear Algebra*, Van Nostrand, Princeton, NJ, 1953.

Johnson, R. E., "Distinguished rings of linear transformations," *Trans. Am. Math. Soc.* **111**, 400–412 (1964).

Kaashoek, M. A., C. V. M. van der Mee, and L. Rodman, "Analytic operator functions with compact spectrum, II. Spectral pairs and factorization," *Integral Equations and Operator Theory* **5**, 791–827 (1982).

Kailath, T., *Linear Systems*, Prentice-Hall, Englewood Cliffs, NJ, 1980.

Kalman, R. E., "Mathematical description of linear dynamical systems," *SIAM J. Control* **1**, 152–192 (1963).

Kalman, R. E., "Kronecker invariants and feedback," *Proceedings of Conference on Ordinary Differential Equations*, Math. Research Center, Naval Research Laboratory, Washington, DC, 1971.

Kalman, R. E., P. L. Falb, and M. A. Arbib, *Topics in Mathematical System Theory*, McGraw-Hill, New York, 1969.

Kato, T., *Perturbation Theory for Linear Operators*, 2nd ed., Springer-Verlag, Berlin, 1976.

Kelley, J. L., *General Topology*, van Nostrand, New York, 1955.

Kra, I., *Automorphic Forms and Kleinian Groups*, Benjamin, Reading, MA, 1972.

Krein, M. G., "Introduction to the geometry of indefinite *J*-spaces and to the theory of operators in these spaces," *Am. Math. Soc. Translations* (2) **93**, 103–176 (1970).

Krein, M. G., M. A. Krasnoselskii, and D. P. Milman, "On the defect numbers of linear operators in Banach space and on some geometric problems," *Sbornik Trud. Inst. Mat. Akad. Nauk Ukr. SSR* **11**, 97–112 (Russian) (1948).

Kurosh, A. G., *Lectures in General Algebra*, Pergamon Press, Oxford, 1965.

Laffey, T. J., "Simultaneous triangularization of matrices—low rank cases and the non-derogatory case," *Linear and Multilinear Algebra* **6**, 269–305 (1978).

Lancaster, P., *Theory of Matrices*, Academic Press, New York, 1969.

Lancaster, P., and M. Tismenetsky, *The Theory of Matrices with Applications*, Academic Press, New York, 1985.

Lidskii, V. B., "Inequalities for eigenvalues and singular values," appendix in F. R. Gantmacher, *The Theory of Matrices*, Moscow, Nauka, 1966, pp. 535–559 (Russian).

Markus, A. S., and E. E. Parilis, "Change in the Jordan structure of a matrix under small perturbations," *Matem. Issled.* **54**, 98–109 (Russian) (1980).

Markushevich, A. I., *Theory of Analytic Functions*, Vols. I–III, Prentice-Hall, Englewood Cliffs, NJ, 1965.

Marsden, J. E., *Basic Complex Analysis*, Freeman, San Francisco, 1973.

Ostrowski, A. M., *Solution of Equations in Euclidean and Banach Space*, Academic Press, New York, 1973.

Porsching, T. A., "Analytic eigenvalues and eigenvectors," *Duke Math. J.* **35**, 363–367 (1968).

Radjavi, H., and P. Rosenthal, *Invariant Subspaces*, Springer-Verlag, Berlin, 1973.

Ran, A. C. M., and L. Rodman, "Stability of neutral invariant subspaces in indefinite inner products and stable symmetric factorizations," *Integral Equations and Operator Theory* **6**, 536–571 (1983).

Rodman, L., and M. Schaps, "On the partial multiplicities of a product of two matrix polynomials," *Integral Equations and Operator Theory* **2**, 565–599 (1979).

Rosenbrock, H. H., *State Space and Multivariate Theory*, Nelson, London, 1970.

Rosenbrock, H. H., and C. E. Hayton, "The general problem of pole assignment," *Intern. J. Control* **27**, 837–852 (1978).

Rosenthal, E., "A remark on Burnside's theorem on matrix algebras," *Linear Algebra Appl.* **63**, 175–177 (1984).

Rudin, W., *Real and Complex Analysis*, 2nd ed., Tata McGraw-Hill, New Delhi.

Ruhe, A., "Perturbation bounds for means of eigenvalues and invariant subspaces," *Nordisk Tidskrift fur Informations Behandlung* (BIT) **10**, 343–354 (1970a).

Ruhe, A., "An algorithm for numerical determination of the structure of a general matrix," *Nordisk Tidskrift fur Informations Behandlung* (BIT) **10**, 196–216 (1970b).

Saphar, P., "Sur les applications linéaires dans un espace de Banach. II," *Ann. Sci. École Norm. Sup.* **82**, 205–240 (1965).

Sarason, D., "On spectral sets having connected complement," *Acta Sci. Math.* (Szeged) **26**, 289–299 (1965).

Shayman, M. A., "On the variety of invariant subspaces of a finite-dimensional linear operator," *Trans. AMS* **274**, 721–747 (1982).

Shmuljan, Yu. L., "Finite dimensional operators depending analytically on a parameter," *Ukrainian Math. J.* **9**(2), 195–204 (Russian) (1957).

Shubin, M. A., "On holomorphic families of subspaces of a Banach space," *Integral Equations and Operator Theory* **2**, 407–420 (translation from Russian) (1979).

Sigal, E. I., "Partial multiplicities of a product of operator functions," *Matem. Issled.* **8**(3), 65–79 (Russian) (1973).

Soltan, V. P., "The Jordan form of matrices and its connection with lattice theory," *Matem. Issled.* **8**(27), 152–170 (Russian) (1973a).

Soltan, V. P., "On finite dimensional linear operators with the same invariant subspaces," *Matem. Issled.* **8**(30), 80–100 (Russian) (1973b).

Soltan, V. P., "On finite dimensional linear operators in real space with the same invariant subspaces," *Matem. Issled.* **9**, 153–189 (Russian) (1974).

Soltan, V. P., "The structure of hyperinvariant subspaces of a finite dimensional operator," in *Nonselfadjoint Operators*, Kishinev, Stiinca, 1976, pp. 192–203 (Russian).

Soltan, V. P., "The lattice of hyperinvariant subspaces for a real finite dimensional operator," *Matem. Issled.* **61**, 148–154, Stiinca, Kishinev (Russian) (1981).

Thijsse, G. Ph. A., "Rules for the partial multiplicities of the product of holomorphic matrix functions," *Integral Equations and Operator Theory* **3**, 515–528 (1980).

Thijsse, G. Ph. A., *Partial Multiplicities of Products of Holomorphic Matrix Functions*, Habilitationschrift, Dortmund, 1984.

Thompson, R. C., "Author vs. referee: A case history for middle level mathematicians," *Am. Math. Monthly*, **90**(10), 661–668 (1983).

Thompson, R. C., "Some invariants of a product of integral matrices," in *Proceedings of the 1984 Joint Summer Research Conference on Linear Algebra and its Role in Systems Theory*, 1985.

Uspensky, J. V., *Theory of Equations*, McGraw-Hill, New York, 1978.

Van Dooren, P., "The generalized eigenstructure problem in linear system theory," *IEEE Trans. Aut. Contr.* **AC-26**, 111–129 (1981).

Van Dooren, P., "Reducing subspaces: Definitions, properties and algorithms," in A. Ruhe and B. Kagstrom, Eds., *Matrix Pencils, Lecture Notes in Mathematics*, Vol. 973, Springer, New York, 1983, pp. 58–73.

Wells, R. O., *Differential Analysis on Complex Manifolds*, Springer-Verlag, New York, 1980.

Wonham, W. M., *Linear Multivariable Control: A Geometric Approach*, Springer-Verlag, Berlin, 1979.

Author Index

Subject Index